Angiogenesis in Health and Disease

ANGIOGENESIS IN HEALTH AND DISEASE

Basic Mechanisms and Clinical Applications

edited by
Gabor M. Rubanyi
Berlex Biosciences
Richmond, California

MARCEL DEKKER, INC. NEW YORK • BASEL

ISBN: 0-8247-8102-3

This book is printed on acid-free paper.

Headquarters
Marcel Dekker, Inc.
270 Madison Avenue, New York, NY 10016
tel: 212-696-9000; fax: 212-685-4540

Eastern Hemisphere Distribution
Marcel Dekker AG
Hutgasse 4, Postfach 812, CH-4001 Basel, Switzerland
tel: 41-61-261-8482; fax: 41-61-261-8896

World Wide Web
http://www.dekker.com

The publisher offers discounts on this book when ordered in bulk quantities. For more information, write to Special Sales/Professional Marketing at the headquarters address above.

Current printing (last digit):
10 9 8 7 6 5 4 3 2 1

PRINTED IN THE UNITED STATES OF AMERICA

Preface

Significant progress has been made in recent years in elucidating the cellular and molecular mechanisms of how new blood vessels are formed in health and disease. Vasculogenesis (development of the circulatory system early in the embryo's development) and normal angiogenesis (formation of new blood vessels from existing capillaries in late embryonic development and in the postnatal period) are essential physiological phenomena. Examples of insufficient or abnormal angiogenesis are (1) lack of sufficient angiogenesis (new capillary development) and collateral formation (as a consequence of the remodeling of existing blood vessels) in adult tissues during hypoxia and ischemia and (2) aberrant (excessive) angiogenesis in solid tumors, diabetic retinopathy, and chronic immuno/inflammatory disorders, such as rheumatoid arthritis, psoriasis, and atherosclerosis. The therapeutic strategies are opposite in the following conditions: stimulation/facilitation of angiogenesis and collateral formation in ischemic conditions (''therapeutic angiogenesis''), while inhibition of aberrant angiogenesis is the therapeutic goal in cancer, diabetes, and chronic inflammation.

The past few years have witnessed unprecedented progress in our understanding of most aspects of the angiogenic process—of the various factors and their receptors regulating and modulating normal and aberrant angiogenesis. These developments in basic sciences have allowed the establishment of novel therapeutic strategies. Indeed, angiogenesis research has shifted recently from basic research to practical clinical applications.

This book provides a comprehensive and in-depth overview of recent progress made in basic sciences (Parts I–V) and also the exciting new therapeutic strategies and clinical applications (Parts VI–IX) of angiogenesis research. The main topics covered by this comprehensive overview include (1) blood vessels and development; (2) angiogenic growth factors and their receptors; (3) novel angiogenic genes; (4) modulators of angiogenesis; (5) animal models of angiogenesis; (6) inhibition of angiogenesis in experimental models; (7) clinical applications of antiangiogenic therapy in cancer; (8) inflammation, wound healing, and rheumatoid arthritis; and (9) therapeutic angiogenesis in ischemia.

Each chapter is written by leading experts in their field, so that the reader will not only receive a brief review of each topic but also be presented with the latest information about cutting-edge basic and clinical research. Some of the exciting new basic discoveries (e.g., novel angiogenic regulators such as angiopoietins, TIE-receptors, and novel angiogenic genes such as thymosine β-4, EDG-1, and ARNT proteins) are presented, along with the observations made by means of the most recent technologies (transgenic and knockout mice, receptor protein mutations, and others). The book also describes the latest technologies used in the clinic, including gene therapy.

Based on a wide scope of basic scientific topics, technologies, and therapeutic applications, this unique book is highly recommended for those actively involved, as well as interested professionals, in a wide variety of disciplines involved in basic research or clinical applications from both academia and the biotech and pharamaceutic industries.

This one-of-a-kind book, which gives in-depth background knowledge and information about most of the latest breakthroughs of basic and clinical sciences of angiogenesis research and its clinical applications, will provide essential and fascinating reading material for all interested readers for many years to come.

I thank the authors for their excellent contributions and the publisher, Marcel Dekker, Inc., for its highly professional work and timely publication of this book. I would like to express my special gratitude to Graham Garratt, the late Executive Vice President and Publisher of Marcel Dekker, Inc., whose enthusiastic support, encouragement, and active help from the very beginning of this project made the original idea of this book become a reality.

Gabor M. Rubanyi

Contents

Contributors

Maria Luiza C. Albuquerque, M.D. Assistant Professor, Department of Pediatrics, Northwestern University Medical School, Chicago, Illinois

Nicolas Ancellin Department of Physiology, Center for Vascular Biology, University of Connecticut School of Medicine, Farmington, Connecticut

Takayuki Asahara, M.D., Ph.D. Assistant Professor of Medicine, Cardiovascular Medicine and Research, St. Elizabeth's Medical Center, Tufts University School of Medicine, Boston, Massachusetts

Anthony W. Ashton, Ph.D. Postdoctoral Fellow, Division of Cardiology, Department of Medicine, Albert Einstein College of Medicine, Bronx, New York

Hellmut G. Augustin, D.V.M., Ph.D. Cell Biology Laboratory, Department of Gynecology and Obstetrics, University of Göttingen Medical School, Göttingen, Germany

Andrew Baird, Ph.D. Vice President, Research and Development, Ciblex Corporation, San Diego, California

Pankaj Bhargava, M.D. Instructor of Medicine, Lombardi Cancer Center, Georgetown University Medical Center, Washington, D.C.

Guido Bold, Ph.D. Department of Oncology Research, Novartis Pharma AG, Basel, Switzerland

Laurence M. Boon, M.D. Plastic Surgeon, Laboratory of Human Molecular Genetics, Christian de Duve Institute of Cellular Pathology, and Center for Vascular Anomalies, Division of Plastic Surgery, Cliniques Universitaires St. Luc, Université Catholique de Louvain, Brussels, Belgium

Margaret D. Brown, Ph.D. School of Sport and Exercise Sciences, University of Birmingham, Birmingham, England

Elisabeth Deindl, Ph.D. Department of Experimental Cardiology, Max Planck Institute for Physiological and Clinical Research, Bad Nauheim, Germany

Russell F. DeVore III, M.D. Associate Professor, Department of Medicine, Division of Hematology/Oncology, Vanderbilt University Medical Center, Nashville, Tennessee

Luisa A. DiPietro, D.D.S., Ph.D. Associate Professor, Department of Surgery, Loyola University Medical Center, Maywood, Illinois

Yuval Dor Department of Molecular Biology, Hadassah Medical School, The Hebrew University, Jerusalem, Israel

Margaret S. Dordal, M.D., Ph.D. Associate Medical Director, Oncology Department, TAP Holdings Inc., Deerfield, Illinois

Saijun Fan, M.D., Ph.D. Assistant Professor, Department of Radiation Oncology, Long Island Jewish Medical Center, New Hyde Park, New York

Borja Fernández, Ph.D. Department of Experimental Cardiology, Max Planck Institute for Physiological and Clinical Research, Bad Nauheim, Germany

Napoleone Ferrara, M.D. Staff Scientist, Department of Cardiovascular Research, Genentech, Inc., South San Francisco, California

William D. Figg, Pharm.D. Senior Investigator, Medicine Branch, National Cancer Institute, National Institutes of Health, Bethesda, Maryland

Seth P. Finklestein, M.D. Associate Neurologist, Massachusetts General Hospital, and Associate Professor, Department of Neurology, Harvard Medical School, Boston, Massachusetts

Ingo Flamme, M.D. Center for Molecular Medicine, University of Cologne, Cologne, Germany

William E. Fogler, Ph.D. Associate Director, Experimental Therapeutics, EntreMed, Inc., Rockville, Maryland

Yong Song Gho National Institute of Dental and Craniofacial Research, National Institutes of Health, Bethesda, Maryland

Itzhak D. Goldberg, M.D. Chairman, Department of Radiation Oncology, Long Island Jewish Medical Center, New Hyde Park, New York

Derrick S. Grant, Ph.D. Assistant Professor of Medicine, Cardeza Foundation for Hematologic Research, Department of Medicine, Jefferson Medical College, Thomas Jefferson University, Philadelphia, Pennsylvania

Michael J. Hawkins, M.D. Associate Director, Washington Cancer Institute, Washington Hospital Center, Washington, D.C.

Timothy Hla, Ph.D. Assistant Professor and Director, Department of Physiology, Center for Vascular Biology, University of Connecticut School of Medicine, Farmington, Connecticut

Imo E. Höfer Department of Experimental Cardiology, Max Planck Institute for Physiological and Clinical Research, Bad Nauheim, Germany

Susan C. Hubchak, M.D. Department of Pediatrics, Northwestern University Medical School, Chicago, Illinois

Olga Hudlicka, M.D., Ph.D., D.Sc. Professor Emeritus, Department of Physiology, University of Birmingham, Birmingham, England

M. Luisa Iruela-Arispe, Ph.D. Assistant Professor, Department of Molecular, Cell and Developmental Biology, and Molecular Biology Institute, University of California, Los Angeles, Los Angeles, California

Jeffrey M. Isner, M.D. Professor of Medicine, St. Elizabeth's Medical Center, Tufts University School of Medicine, Boston, Massachusetts

Jeffrey R. Jackson, Ph.D. Senior Investigator, Department of Oncology Research, SmithKline Beecham Pharmaceuticals, King of Prussia, Pennsylvania

Michael Jaffe Pre-Doctoral Fellow, Institute for Advanced Studies, The Angiogenesis Foundation, Cambridge, Massachusetts

Anitha S. John, Ph.D. Department of Pathology, MCP-Hahnemann University, Philadelphia, Pennsylvania

Anthony Johns, Ph.D. Senior Scientist, Department of Cardiovascular Research, Berlex Biosciences, Richmond, California

Leena Karttunen, Ph.D. Department of Cell Biology, Harvard Medical School, Boston, Massachusetts

Eli Keshet, Ph.D. Professor, Department of Molecular Biology, Hadassah Medical School, The Hebrew University, Jerusalem, Israel

Seunghee Kim-Schulze, M.D. Department of Medicine, Mount Sinai School of Medicine, New York, New York

Hynda K. Kleinman, Ph.D. Research Chemist, Cell Biology Section, National Institute of Dental and Craniofacial Research, National Institutes of Health, Bethesda, Maryland

Michael Kluk Department of Physiology, Center for Vascular Biology, University of Connecticut School of Medicine, Farmington, Connecticut

Alisa E. Koch, M.D. Gallagher Research Professor of Medicine, Department of Medicine/Rheumatology, Northwestern University Medical School, and Department of Veterans' Affairs, Chicago Healthcare System, Chicago, Illinois

Pieter Koolwijk, Ph.D. Senior Investigator, Department of Vascular and Connective Tissue Research, Gaubius Laboratory TNO-PG, Leiden, The Netherlands

Erwin A. Kruger National Cancer Institute, National Institutes of Health, Bethesda, Maryland

Roger J. Laham, M.D. Assistant Professor of Medicine, Department of Medicine/Cardiology, Beth Israel Deaconess Medical Center and Harvard Medical School, Boston, Massachusetts

Katrin Lamszus, M.D. Department of Radiation Oncology, Long Island Jewish Medical Center, New Hyde Park, New York

John Laterra, M.D., Ph.D. Associate Professor, Department of Neurology, The Johns Hopkins School of Medicine and Kennedy-Krieger Research Institute, Baltimore, Maryland

Menq-Jer Lee Department of Physiology, Center for Vascular Biology, University of Connecticut School of Medicine, Farmington, Connecticut

Peter I. Lelkes, Ph.D. Director, Laboratory of Cell Biology, Department of Medicine, Sinai Samaritan Medical Center, University of Wisconsin Medical School, Milwaukee, Wisconsin

Vincent W. Li, M.D. Scientific Director, Institute for Advanced Studies, The Angiogenesis Foundation, Cambridge, Massachusetts

William W. Li, M.D. Medical Director, Institute for Advanced Studies, The Angiogenesis Foundation, Cambridge, Massachusetts

Catherine H. Liu Department of Physiology, Center for Vascular Biology, University of Connecticut School of Medicine, Farmington, Connecticut

Katherine M. Malinda, Ph.D. Staff Scientist, Wound Healing Laboratory, National Institute of Nursing Research, National Institutes of Health, Bethesda, Maryland

Emin Maltepe, Ph.D. Nathan and Frances Goldblatt Society Fellow, Committee on Cancer Biology, University of Chicago, Chicago, Illinois

Dieter Marmé, Ph.D. Tumor Biology Center, Institute of Molecular Medicine, Freiburg, Germany

Georg Martiny-Baron, Ph.D. Tumor Biology Center, Institute of Molecular Medicine, Freiburg, Germany

Andreas Menrad, Ph.D. Research Laboratories of Schering AG, Berlin, Germany

John McGuire, M.D. Instructor, Department of Pediatrics, Washington University Medical School, St. Louis, Missouri

Mai Nguyen, M.D., F.A.C.S. Assistant Professor, Department of Surgery, University of California, Los Angeles, Los Angeles, California

Nicholas N. Nissen, M.D. Fellow, Liver Transplant Center, University of California, Los Angeles, Los Angeles, California

Bjorn R. Olsen, M.D., Ph.D. Hersey Professor, Department of Cell Biology, and Harvard-Forsyth Professor, Department of Oral Biology, Harvard Medical School and Harvard School of Dental Medicine, Boston, Massachusetts

Maria Asunción Ortega Department of Molecular, Cell and Developmental Biology, Molecular Biology Institute, University of California, Los Angeles, Los Angeles, California

Stacy M. Plum Research Associate, Experimental Therapeutics, EntreMed, Inc., Rockville, Maryland

Peter J. Polverini, D.D.S., Ph.D. Department of Oral Pathology, University of Michigan School of Dentistry, Ann Arbor, Michigan

M. Lourdes Ponce, Ph.D. Craniofacial Developmental Biology and Regeneration Branch, National Institute of Dental and Craniofacial Research, National Institutes of Health, Bethesda, Maryland

Eddie Reed, M.D. Chief, Medical Ovarian Cancer Section, Medicine Branch, National Cancer Institute, National Institutes of Health, Bethesda, Maryland

R. Wesley Rose, M.Sc. Assistant Professor of Medicine, Cardeza Foundation for Hematologic Research, Department of Medicine, Jefferson Medical College, Thomas Jefferson University, Philadelphia, Pennsylvania

Johannes Rösel, Ph.D. Department of Oncology Research, Novartis Pharma AG, Basel, Switzerland

Eliot M. Rosen, M.D., Ph.D. Professor, Radiation Oncology, Long Island Jewish Medical Center, New Hyde Park, and Albert Einstein College of Medicine, Bronx, New York

Constance Runyan, M.D. Department of Pediatrics, Northwestern University Medical School, Chicago, Illinois

Wolfgang Schaper, M.D., Ph.D. Department of Experimental Cardiology, Max Planck Institute for Physiological and Clinical Research, Bad Nauheim, Germany

Michael Schirner, M.D. Research Laboratories of Schering AG, Berlin, Germany

H. William Schnaper, M.D. Associate Professor, Department of Pediatrics, Northwestern University Medical School, Chicago, Illinois

Martin R. Schneider, Ph.D. Research Laboratories of Schering AG, Berlin, Germany

Dimitri Scholz, Ph.D. Department of Experimental Cardiology, Max Planck Institute for Physiological and Clinical Research, Bad Nauheim, Germany

Alan Z. Segal, M.D. Graduate Assistant in Neurology, Massachusetts General Hospital, and Fellow, Harvard Medical School, Boston, Massachusetts

Gerhard Siemeister, Ph.D. Research Laboratories of Schering AG, Berlin, and Tumor Biology Center, Institute of Molecular Medicine, Freiburg, Germany

M. Celeste Simon, Ph.D. Associate Professor/Assistant Investigator, Department of Medicine, Howard Hughes Medical Institute, University of Chicago, Chicago, Illinois

Michael Simons, M.D. Associate Professor of Medicine, Beth Israel Deaconess Medical Center and Harvard Medical School, Boston, Massachusetts

Zoltan Szekanecz, M.D., Ph.D. Assistant Professor of Medicine, Rheumatology, and Clinical Immunology, Third Department of Medicine, University Medical School of Debrecen, Debrecen, Hungary

Shobha Thangada Department of Physiology, Center for Vascular Biology, University of Connecticut School of Medicine, Farmington, Connecticut

Karl-Heinz Thierauch, Ph.D. Research Laboratories of Schering AG, Berlin, Germany

Brian Thompson Department of Physiology, Center for Vascular Biology, University of Connecticut School of Medicine, Farmington, Connecticut

James S. Tomlinson, M.D. Department of Surgical Oncology, University of California, Los Angeles, Los Angeles, California

Dmitris Tsakayannis, M.D., F.A.C.S. Institute for Advanced Studies, The Angiogenesis Foundation, and Department of Surgery, University of Athens School of Medicine, Athens, Greece

George P. Tuszynski, Ph.D. Professor, Departments of Surgery and Pathology, MCP-Hahnemann University, Philadelphia, Pennsylvania

Victor W. M. van Hinsbergh, Ph.D. Deputy Head, Department of Vascular and Connective Tissue Research, Gaubius Laboratory TNO-PG, Leiden, and Professor in Cellular Pathophysiology of the Vessel Wall, Department of Physiology, Institute for Cardiovascular Research, Vrije Universiteit, Amsterdam, The Netherlands

Niels van Royen, M.D. Department of Experimental Cardiology, Max Planck Institute for Physiological and Clinical Research, Bad Nauheim, Germany

Francisca Vázquez, Ph.D. Research Associate, Department of Adult Oncology, Dana-Farber Cancer Institute, Boston, Massachusetts

Miikka Vikkula, M.D., Ph.D. Associate Member, Laboratory of Human Molecular Genetics, Christian de Duve Institute of Cellular Pathology and Université Catholique de Louvain, Brussels, Belgium

Jing-Liang Wang, Ph.D. Post-Graduate Researcher, Division of Surgical Oncology, Department of Surgery, University of California, Los Angeles, Los Angeles, California

J. Anthony Ware, M.D. Chief, Cardiovascular Division, and Sidney L. and Miriam K. Olson Professor of Cardiology, Departments of Medicine and Molecular Pharmacology, Albert Einstein College of Medicine and Montefiore Medical Center, Bronx, New York

Christopher R. Waters, B.Sc. Research Assistant, Laboratory of Cell Biology, Department of Medicine, Sinai Samaritan Medical Center, University of Wisconsin Medical School, Milwaukee, Wisconsin

James D. Winkler, Ph.D. Associate Director, Department of Oncology Research, SmithKline Beecham Pharmaceuticals, King of Prussia, Pennsylvania

Jeanette M. Wood, Ph.D. Department of Oncology Research, Novartis Pharma AG, Basel, Switzerland

Stefan Zöllner, Ph.D. Department of Cardiovascular Research, Berlex Biosciences, Richmond, California

Angiogenesis in Health and Disease

1

Molecular Biology of Vasculogenesis and Early Angiogenesis

Ingo Flamme
Center for Molecular Medicine, University of Cologne, Cologne, Germany

I. INTRODUCTION

The emergence of a primordial vascular system during early stages of embryogenesis and its further growth and differentiation are obviously distinct processes. Vascular development is initiated when the precursors of endothelial cells differentiate from the mesoderm and coalesce into a primitive capillary plexus, including the primordia of the principal vessels of the body. Upon onset of the heart beat the primordial vasculature becomes perfused and the architecture of the plexus is refined and remodelled (1). In 1971 Gonzalez-Crussi and Haar and Ackerman used the term *vasculogenesis* to describe the formation of the primordia of blood vessels in the avian and mouse embryos (2,3). Today this term is widely used in the literature to discriminate the initial steps of vascular development from *angiogenesis*, which denotes growth of new blood vessels from preexisting ones (4–6). Endothelial proliferation, sprouting, and intussusception are the features of angiogenesis that occur very impressively when the primordial vascular plexus is expanded to fulfil its destiny, which is to supply the growing embryo with nutrients and oxygen. While angiogenesis depends on metabolic demands and is modulated by hemodynamic factors such as shear stress (7,8), the primordial vasculature is laid down when the embryo can still be nourished by diffusion. Therefore, distinct molecular mechanisms have been implicated in vasculogenesis. The present article is intended to review the recent findings of experimental and molecular embryology in this field.

II. ANGIOBLASTS AND HEMANGIOBLASTS

The vascular endothelium is a continuous layer of a highly specialized, simple squamous epithelium that lines the inner surfaces of all blood vessels and the cavities of the heart, where it is called *endocardium* (9,10). It is something of a dogma that in the adult vascular endothelial cells can originate only from vascular endothelial cells and that this cell lineage is founded very early during ontogeny, when the mesoderm is formed and cells with certain characteristics of vascular endothelial cells differentiate within the mesoderm. These endothelial precursors, which have the capacity to form a vascular lumen, have been named *angioblasts* (11).

The mesoderm originates from the epiblast: in embryos of higher vertebrates epiblast cells move to the midline of the embryo and enter the space between epiblast and hypoblast (endoderm) by transmigration through the primitive streak (12). In analogy to amphibia, this process of invagination is often called *gastrulation*. During gastrulation, the epiblast cells undergo an epitheliomesenchymal transformation (Fig. 1). By means of electron microscopy, angioblasts cannot be distinguished from the other migratory mesenchymal cells of the early mesoderm (2,3). Gastrulation is the temporal hallmark of the onset of angioblast differentiation, but, in contrast to axial structures, is not needed for their commitment (13–16).

In the extraembryonic mesoderm of the yolk sac angioblasts originate from blood islands. These are clusters of epitheloid cells, which, in avian embryos of the one-somite stage and in the mouse at E8, can be distinguished from the surrounding mesenchyme (2,3). The outer cells of those clusters differentiate into angioblasts, which gradually flatten and form a continuous coat of vascular endothelial cells around the inner cells which differentiate into primitive hematopoietic cells. These are initially epitheloid and then become separated from each other by interstitial fluid, the primordial blood plasma. The origin of hematopoietic and vascular endothelial cells from blood islands has led to the hypothesis that blood and endothelial cells are derived from a common embryonic population of precursor cells, the hemangioblastic cells (11,17,18). Although this hypothesis is still elusive, the term *hemangioblastic* is widely used in the literature to denote the embryonic endothelial and hematopoietic cell lineages. This hypothesis was supported by the fact that endothelial and at least certain subsets of hematopoietic cells have a variety of molecular markers in common: QH1 and MB1 in quail (4,19); thrombomucin in chicken (20); and CD34, CD31/PECAM, VEGF receptors, tie-1 and tie-2, and erythropoietin-receptor in mammals (21–26).

The emergence of angioblasts and their assembly into the primordial vascular plexus can be visualized in the quail embryo by use of a monoclonal antibody directed against the QH-1 epitope (4,6). While the extraembryonic vascular plexus is laid down by coalescence of angioblasts from neighboring blood islands, angioblasts within the embryo appear along the anterior intestinal portal, giving rise to the endocardial tubes, and along the lateral edges of the somites; only little later, they coalesce to the dorsal aorta and the large vessel primordia of the body (27). The primordial vasculature consists of sinusoidal capillaries, and capillaries are arranged in a polygonal pattern that resembles a honeycomb. In the avian embryo, at the two-somite stage, intraembryonic blood-vessel primordia are already interconnected with the extraembryonic vasculature. In contrast to the extraembryonic angioblasts, their intraembryonic counterparts originate from the mesoderm without concomitant hematopoietic cells (28). The reason for this is not clear. There is well-proven evidence that embryonic angioblasts originate as an autochthon population and are not derived from extraembryonic hemangioblasts, as suggested earlier (29,30). At least in the avian embryo, there is a distinct borderline between hemangioblastic and pure angioblastic populations, which correlates with the borderline between the endoderm of the embryo and the yolk sac. The yolk sac endoderm is a simple cuboidal epithelium with abundant microvilli on its apical surface and large phagocytotic vacuoles in the cytoplasm. The origin of blood islands is restricted to the mesodermal area overlying the yolk sac endo-

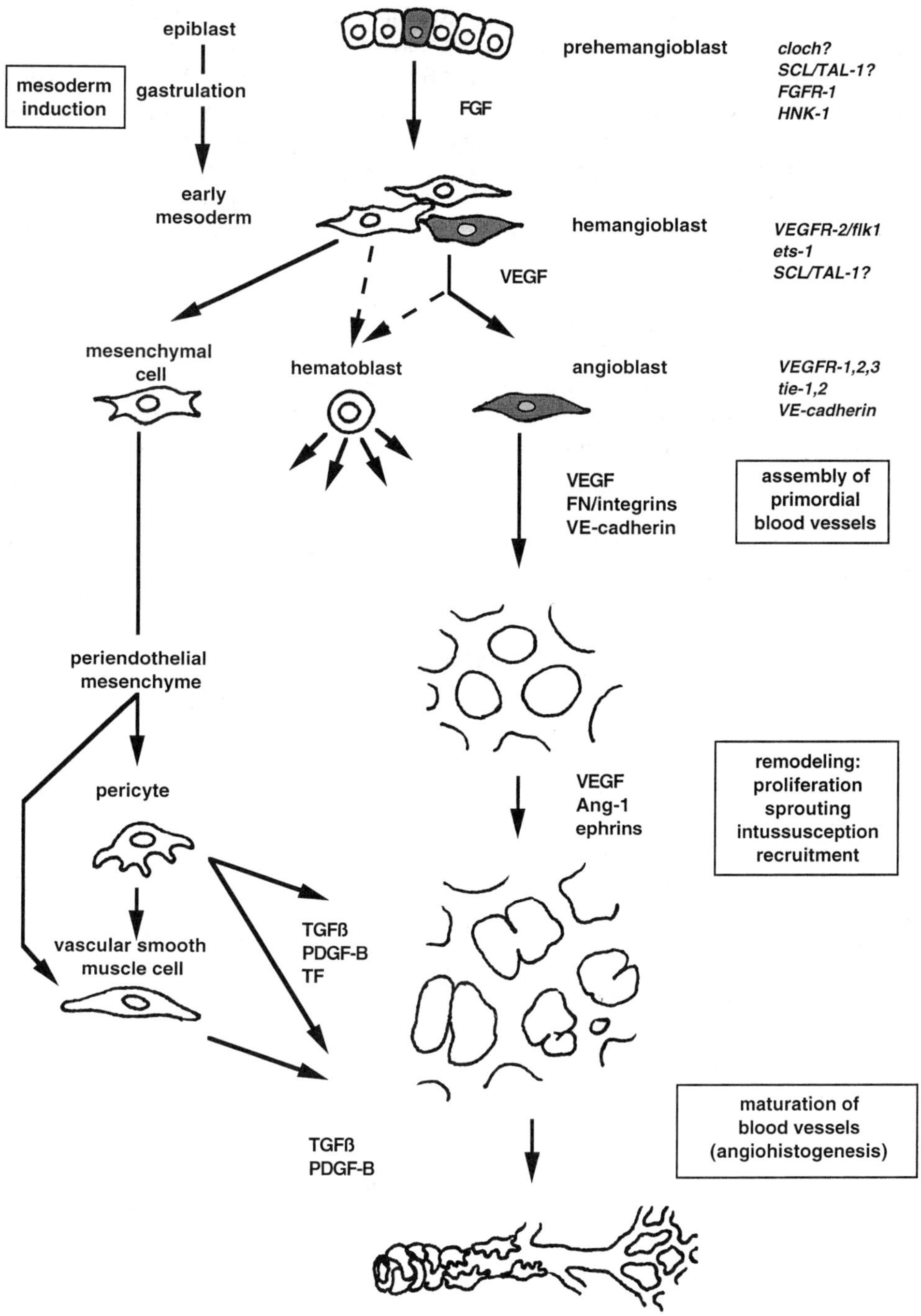

Figure 1 Schematic outline of the steps of vasculogenesis and early angiogenesis. The prehemangioblastic and hemangioblastic cells are still hypothetical. Factors that are relevant for the individual steps of vascular development are indicated next to the arrows. Molecules characteristic of prehemangioblastic, hemangioblastic, and angioblastic cells are indicated in italics.

derm (31,32). Initially the entire primordial vascular plexus develops in direct contact with the endoderm (33) but then becomes separated from the endoderm by a thin layer of mesodermal cells (34). Eventually the vascular plexus is embedded by mesenchymal cells, from which pericytes and vascular smooth muscle cells differentiate (35). Shortly after the primordial vasculature is established, the coelomic cavity forms that splits the lateral mesoderm into a somatopleura faced toward the ectoderm and a splanchnopleura faced towards the endoderm (32). By this process of mesoderm splitting, the primordial vascular plexus is allocated to the splanchnopleura. The role of endoderm during vasculogenesis is discussed in more detail later.

Later during development, hematopoietic cells also originate within the embryo proper. In birds, mice, and humans, these cells bud from the floor of the dorsal aorta at early limb bud stage, and it is believed that they originate from the ventral mesenchyme close to the dorsal aorta, where gonads and mesonephros are forming and which therefore was termed the AGM-region (36–39). The definitive origin of these hematopoietic cells is not clear and the possibility that they are directly derived from primordial germ cells that settle at that time to the AGM-region cannot be ruled out (40). An hemangioblastic origin of these cells was also assumed, as experiments in quail-chick chimeras have shown that the splanchnopleural mesenchyme contains precursors of both blood and endothelial cells, which contribute to the floor of the dorsal aorta, whereas the dorsal, paraxial mesoderm gives rise to endothelium only (41). Recently the splanchnic mesoderm of the allantoic bud of the avian embryo has also been shown to harbor hemangioblastic cells (42). Hematopoietic cells derived from the AGM-region are the definitive blood stem cells of the higher vertebrate which colonize the transient and definitive hematopoietic organs, liver, spleen, and bone marrow. This information comes from experiments in avian embryos using the yolk sac chimera technique and from long terminal reconstitution experiments in irradiated mice (39,43). However, the question whether primitive, yolk-sac-borne, and definitive AGM-derived, hematopoietic stem cells are derived from common or distinct populations of precursors is still controversial (44,45).

Angioblasts are widely distributed within the embryo and only excluded from certain regions and organs such as cartilage, epithelia (including the early neural epithelium), and the zone of primitive streak regression (sinus rhomboidalis) (4,46). Whether angioblastic cells are exclusively committed during and shortly after gastrulation stages or also later during development or even in the adult, is still controversial. Those organs that develop in close contact with the endoderm or are of endodermal origin contain an autochthon contingent of angioblasts, while those that develop from somatopleural mesenchyme, such as the extremities, are colonized by angioblasts from the trunk during their development (47). The reason for this may be that the somatopleural mesenchyme harbors only few angioblasts. There are several lines of evidence that the somites also contain angioblastic cells, which may contribute to the blood vessels around and within the paraxial mesoderm (48,49). However, it is not yet clear whether, in a given organ, the angioblasts are derived from still pluripotent mesenchymal cells or have been committed in the early mesoderm before the organ was formed. In general, angioblasts have the shape of mesenchymal cells and are migratory highly active and highly invasive "prevascular" endothelial cells (46). Angioblasts are involved in many angiogenetic processes in the embryo, such as the vascularization of the head mesenchyme, which is of neuroectodermal origin and initially avascular (46), and the vascularization of the embryonic central nervous system, which is vascularized by sprouting and immigration of individual angioblasts detaching from the perineural vascular plexus (50). In the extremities, individual angioblasts migrate in the avascular subectodermal zone (51). The coronary plexus of the heart derives from angioblasts that migrate together with presumptive epicardial cells from the sinus venosus over the surface of the myocardium. By a process very similar to that of early vasculogenesis, the coronary plexus is laid down and then becomes interconnected with the aortic outflow tract, thus establishing the primordia of the main coronary arteries (52,53). Angioblasts from early embryos are pluripotent as they are able to contribute to different vascular beds (54). Thus, angioblasts derived from the lateral mesoderm can contribute to endocardium as well as to extraembryonic vasculature (27).

The potency to form lumen is thought to be an intrinsic property of angioblasts. By forming a vessel lumen or by being integrated into a vessel, angioblasts acquire an epithelial phenotype and transform into the more sessile endothelial state. During angiogenesis, this process is reversed. Endothelial cells lose their epithelial phenotype and resume the property of angioblasts when they leave the vessel wall to form an angiogenetic sprout. They may even detach from the vessel and migrate as individual angioblasts into the tissue. It is not known whether and to what extent such angioblasts—for which the term *angiogenetic angioblasts* may be appropriate—are different from vasculogenetic angioblasts. That endothelial cells exceptionally can even dedifferentiate into mesenchymal cells has been demonstrated for endocardial cells of the chamber boundaries, which contribute to the mesenchyme of the endocardial cushions, the primordia of the cardiac valves (55). However, during maturation of blood vessels—which is accompanied by tissue- and organ-specific differentiation of endothelial cells (e.g., in kidney, muscle, and brain)—the potential of endothelial cells to dedifferentiate is thought to decrease (10).

In the adult, CD34 positive angioblast-like cells have been isolated from the blood. It was shown that such ''angioblasts'' floating in the blood are able to settle (e.g., to ischemic lesions) and to proliferate in their new location (56). However, it is not clear whether these angioblasts represent a distinct population of endothelial stem cells circulating in the blood or mature endothelial cells that have detached from the endothelial lining of blood vessels.

III. THE FIRST STEP OF VASCULOGENESIS: INDUCTION OF HEMANGIOBLASTS AND ANGIOBLAST—THE ROLE OF FGFs AND FGF RECEPTORS

As outlined above, the determination of hemangioblastic cells coincides with the formation of the embryonic mesoderm during gastrulation (Fig. 1). Therefore, the molecular mechanisms that are relevant for mesoderm formation may also pertain to the differentiation of hemangioblastic cells. These mechanisms have been examined in the most detail in the lower vertebrate *Xenopus laevis* using the animal cap assay. The animal cap comprising the pluripotent epiblast is separated from the vegetal pole, and unless reconstituted with the vegetal pole mesoderm formation does not occur. Among the factors that could substitute for the inductive influence of the vegetal pole, fibroblast growth factors (FGFs) have been identified as potent inducers of the so-called ventral mesoderm, which includes the blood islands (57). There is mounting evidence that the induction and fate determination of the mesoderm depends on balanced gradients of ventralizing and dorsalizing signals (58). Overexpression of dominant negative FGF receptor-1 mutants in *Xenopus* embryos (59) and FGF and FGF-receptor-gene targeting experiments in mice (60–62) have shown that FGF signaling is indispensable for mesoderm formation. The critical role of FGF signal transduction was further substantiated by the finding that hemangioblastic cells can be induced by FGF from quail embryos before gastrulation (63). In cultures of dissociated unincubated quail embryos, blood islands indistinguishable from those having differentiated in vivo are induced by basic and acidic FGF (FGF-1 and 2); in suspension culture, aggregates are formed that closely resemble embryoid bodies derived from mouse embryonic stem (ES) cells with a primitive capillary plexus in their walls (64). The endogenous inducer has not been determined. FGF-2 (basic FGF) has been detected in the avian embryo at the time of gastrulation (65,66), but this member of the FGF family lacks a signal peptide and is not secreted upon overexpression in the early *Xenopus* embryo and therefore is not likely to be involved in the induction of hemangioblastic mesoderm (67). The secreted family members XeFGF and FGF3 in *Xenopus* (68,69) and FGF-3, FGF-4, and FGF5 in the mouse (70–73) are expressed at the onset of gastrulation in the critical area of mesoderm formation and are therefore candidates for the inducers.

Since, in the quail vasculogenesis induction model, not all the epiblast cells but rather a small subset are biased toward hemangioblastic cells by exposure to FGF, it is feasible that in the epiblast a distinct population of prehemangioblastic cells exists that have the capacity to respond to stimulation by FGF with the formation of blood islands. The molecular outfit of these cells—apart from the fact that they have to possess receptors for FGF—is unknown. It was shown in avian embryos that those cells within the epiblast that immigrate via gastrulation express the HNK-1 epitope (74). The functional significance of this is not clear. Prehemangioblasts may be suspected among the HNK-1 expressing cells. Before gastrulation, HNK-1–expressing cells are distributed randomly over the avian epiblast; this is consistent with the finding that, from all areas of pregastrulation-epiblast blood islands can be induced in equal amounts (our own unpublished observations).

FGF receptor-1 is expressed in the early lateral mesoderm after gastrulation (75). Therefore hemangioblastic cells derived from their epiblast precursors may still express this receptor. However the role of FGF receptors for further differentiation and formation of the primordial vascular plexus is not understood. FGF receptors have not been detected on endothelial cells of embryos of higher vertebrates in vivo with the exception of the endothelial cells lining the lumen of large vessels (76–79). Therefore FGFs, although being crucial for the differentiation of angioblasts, are not likely to be involved in early embryonic angiogenesis. This is consistent with the finding that retroviral overexpression of a secreted form of bFGF in the avian limb bud causes fundamental changes in limb morphogenesis but does not interfere with vascularization of the target tissue (80).

IV. THE SECOND STEP OF VASCULOGENESIS: DIFFERENTIATION OF HEMANGIOBLASTS AND ANGIOBLAST AND THEIR ASSEMBLY INTO PRIMORDIAL BLOOD VESSELS—THE ROLE OF VEGF AND VEGF RECEPTORS

The induction of hemangioblastic cells leads to the differential expression of a whole series of genes, among which hematopoietic and endothelial marker genes are the candidates of hemangioblast differentiation. The earliest molecules specifically expressed by hemangioblastic cells are the high-affinity receptors for the vascular endothelial growth factor

(VEGF) (81–84). VEGF was discovered as an endothelial cell-specific mitogen and its role as an angiogenesis factor is well established (85,86). Different isoforms of VEGF generated by alternative splicing of mRNA have been described and found to be equally angiogenic (87,88). The specificity of the VEGF mitogen for vascular endothelial cells is due to the fact that VEGF receptors are almost solely expressed by vascular endothelial cells (89–91). The high expression levels of VEGF receptors in the hemangioblastic lateral mesoderm of early embryos of higher and lower vertebrates just after gastrulation (81–84,92,93) are indicative of a role of VEGF signaling in hemangioblast differentiation and are consistent with the observation that, in the quail blood island induction model, VEGFR-2 is induced soon after the addition of bFGF (83). Moreover, single VEGFR-2–expressing cells, sorted from mouse ES cell–derived embryoid bodies before the beginning of vasculogenesis, gave rise to both blood and endothelial cells in vitro (94,95). This was the first experimental evidence for the existence of cells with the potency of hemangioblasts.

The crucial importance of the VEGF signal transduction system for vasculogenesis was clearly demonstrated by targeted mutation of the genes of VEGF and VEGF receptors in mice (Fig. 1). Unexpectedly, already the loss of one allele of the VEGF gene was embryonically lethal (96,97). While differentiation of angioblasts obviously was not affected, the formation of intraembryonic blood vessels and remodeling of the yolk sac vasculature were severely impaired. The yolk sacs revealed the honeycomb-like pattern of the primordial plexus and lacked large vitelline vessels connecting the yolk sac vasculature with intraembryonic blood vessels. In homozygous mutants generated by aggregation of VEGF-/- ES cells with tetraploid ES cells (the latter ones only give rise to the yolk sac endoderm, whereas the knockout cells establish the embryo proper), the formation of intraembryonic blood vessels failed much more severely than in the heterozygous embryos (96). These data are indicative of the tight dose-dependent regulation of vasculogenesis by VEGF. Whether the importance of VEGF for vasculogenesis is dependent on its function of a survival factor or of a mitogen specific for endothelial cells is not yet clear.

The critical role of the VEGF dosage for vasculogenesis was also shown when in *Xenopus* embryos after overexpression of VEGF vascular hyperproliferation and fusion of vessels of the primordial vascular plexus was induced (92). Similar results were obtained by application of VEGF protein to early avian embryos (98,99). These data show that endothelial VEGF receptors are not saturated by their ligand in the early embryo. In contrast, concomitant upregulation of VEGFR-2 mRNA was observed after retroviral overexpression of VEGF in E4 chicken embryos when hypervascularization was induced (100). This upregulation of VEGFR-2 in response to the ligand also occurs in capillaries growing into VEGF-producing malignant tumors (101), and can be induced experimentally on brain capillaries (102). The mechanisms by which VEGF regulates the expression level of its own receptor have not yet been elucidated.

Two closely related receptor tyrosine kinases have been identified as high-affinity VEGF receptors: VEGFR-2 (flk-1 in mouse, KDR in human) and VEGFR-1/flt-1 (89,103–105). A third homologue, named VEGFR-3 or flt-4, is a receptor for the newly detected VEGF family members VEGF-C and VEGF-D (106,107). At early stages of embryonic development, VEGFR-3 is expressed by vascular endothelial cells (82,108), but later on its expression becomes restricted to the endothelium of lymphatic vessels (109,110). Consistently, VEGF-C was characterized as a lymphangiogenesis factor (111). The role of the VEGF signal transduction system during the two steps of vasculogenesis was dissected by targeted deletion of either VEGF receptor genes. Whereas disruption of the VEGFR-

2/flk-1 completely abrogated the differentiation of angioblasts and hemangioblasts and therefore homozygous mutants were completely avascular (112), disruption of the VEGFR-1/flt-1 led to augmentation of endothelial cells and disorganization and fusion of vessel lumina in the primordial vascular plexus; it was therefore also embryonically lethal but had no influence on angioblast differentiation (113). Mice lacking the VEGFR-3 failed to reorganize their yolk sac vasculature and showed pericardial bleeding (K. Alitalo, personal communication). The data clearly demonstrate that there is no redundancy in function of the VEGF receptors during vasculogenesis. Interestingly, mice with targeted mutation of only the tyrosine kinase domain of VEGFR-1 were viable and did not reveal any vascular defect (114). Therefore, it was concluded that the extracellular portion of VEGFR-1 but not signaling by VEGFR-1 is important for vascular integrity. Since the VEGFR-1 null phenotype is very similar to the results obtained by application of VEGF onto the early vasculature (98), it may be speculated that VEGFR-1 is necessary for balancing VEGF/VEGFR-2 signaling during vasculogenesis.

The discrepancy between the phenotypes of receptor and ligand knockout may be explained by the possibility that other, recently discovered members of the VEGF family can partially substitute for VEGF during differentiation of hemangioblastic cells. The lymphangiogenesis factor VEGF-C and VEGF-D, which are ligands for the VEGFR-3, also bind to VEGFR-2 (106,107). VEGFC was shown to bias VEGFR-2–expressing quail cells from the early mesoderm toward endothelial differentiation in the same manner as VEGF does (115). The role of VEGF-B (116) and the VEGF homologue's placenta growth factor (PIGF), which binds to VEGFR-1 but not to VEGFR-2 (117) and is expressed almost exclusively in throphoblast cells (118), is not yet clear.

Later during embryogenesis the expression pattern of VEGF and its receptors is indicative of a role in angiogenesis. For example, high levels of VEGF are expressed in the subependymal layers of the central nervous system, actually when the neuroepithelium becomes vascularized by sprouting and immigration of angioblasts from the leptomeningeal vascular plexus (119). The angiogenesis potential of VEGF was experimentally substantiated by retroviral and transgenic overexpression of the factor in mice and avian embryos (100,120–122). In addition, VEGF may serve as a vascular differentiation and survival factor in the embryo and the adult, which is possibly involved in induction and maintenance of the fenestrated endothelial phenotype (123–125). This is consistent with experimental data and correlates with the high expression levels of VEGF in the podocytes of kidney glomeruli and the epithelium of the choroid plexus of the brain ventricles (119).

In the early embryo, the endoderm is probably the main source of VEGF (119,126). This may account for the fact that a vascular plexus is established in the splanchnopleural mesoderm but not in the somatopleura, which is located beneath the ectoderm. The local concentration of endoderm-derived VEGF may be critical for the differentiation of endothelial cells at the periphery of blood islands, because in vitro VEGF was found to promote differentiation of VEGFR-2–positive cells from early mesoderm toward endothelium but did not support their differentiation into hematopoietic cells (127). There are several lines of evidence that endoderm-derived factors are necessary for both blood vessel and blood cell development (128–133). However, the endoderm is not sufficient to induce differentiation of hemangioblastic cells from their precursors (64). On the other hand, putative inhibitory factors derived from the ectoderm were hypothesized to explain the paucity of endothelial cells in the somatopleura, and hyaluronate secreted by the ectodermal epithelium is one possible candidate for those inhibitory molecules (134).

V. THE THIRD STEP OF VASCULOGENESIS: THE TRANSITION FROM VASCULOGENESIS TO ANGIOGENESIS—ROLE OF VASCULAR ENDOTHELIAL RECEPTORS AND THEIR LIGANDS

While VEGF is sufficient to induce angiogenesis in the embryo (100) other signaling systems have been found to be necessary for early embryonic angiogenesis as well (Figure 1). The receptor tyrosine kinases tie-1 and tie-2 are expressed by angioblasts at the stage of formation of the primordial vascular plexus only a little later than VEGF receptors (135–139) and are present also on hematopoietic cells (25,126,140). Expression of both gene products seems to be dispensable for the initial steps of vasculogenesis, as can be concluded from the results of their targeted mutation (141–143). The tie-2/tek RTK, however, exerts crucial functions in vascular sprouting and remodeling during early embryonic angiogenesis, and the phenotype of null mutations is complementary to that of null mutations of angiopoietin-1 (Ang-1), which has recently been identified to be the ligand for tie-2 (144,145). In the embryo and the yolk sac, formation of large vessels and sprouting of capillaries—e.g., in the central nervous system—is missing or disturbed and a homogeneous honeycomb-like vascular pattern is observed. In the heart, trabeculation is severely reduced, which is consistent with high expression levels of Ang-1 in the embryonic myocardium. Unlike the VEGF/VEGFR system, the activated Ang-1/tie-2 signal transduction system does not induce proliferation of endothelial cells. However, in an in vitro angiogenesis assay that employs fibrin gels (146), VEGF and Ang-1 synergistically induce endothelial cell sprouting (147). This synergism is likely to play an important role during early angiogenesis, but direct evidence is still missing. Interestingly the function of Ang-1 may be antagonized and balanced by a homologue, Ang-2, which is expressed during vascular regression (e.g., in the corpus luteum) and can inhibit embryonic angiogenesis when overexpressed as a transgene (148). In the embryo, Ang-2 is predominantly found in mesenchymal cells surrounding the aorta and may have a paracrine function during the maturation of large blood vessels.

Recently a novel but large class of receptor tyrosine kinases and their ligands, which are membrane-bound—the Eph-receptors and ephrins, respectively—have been discovered to play a fundamental role during vascular remodeling (149,150). The role of Eph/ephrin signaling was well established for axon guidance during maturation of the central nervous system (151). Targeted mutation of both ephrin-B2 and its receptor Eph-B4 revealed lethal vascular defects very similar to those obtained by knockout of the ang-1/tie-2 signal transduction system (149). It may be speculated that the two signaling systems cooperate during vascular remodeling. Intriguingly, Eph-B4 is expressed by venous endothelium, whereas the ligand is expressed by arterial endothelium. The Eph-receptor/ephrin interaction is likely to result in a bidirectional signaling (152). Thus, morphogenetic crosstalk at the arteriovenous border was postulated (150). However, the expression pattern does not sufficiently explain the general defects that result from the targeted mutations of the genes. In view of recent data from experiments in endothelial cell cultures (153,154), it may be speculated that additional related ligands and receptors may be involved in vascular remodeling and embryonic angiogenesis.

From the large family of transforming growth factor (TGF) proteins the TGF-beta 1 has been implicated in the differentiation of the primordial vascular plexus. TGF-beta 1 was found in the blood islands of the yolk sac and in the early mesoderm (155). Mice deficient in TGF-beta 1 revealed an impaired vasculogenesis and hematopoiesis and died at E10.5. Endothelial cells differentiated and assembled into a vascular plexus, but the endothelial lining of capillaries appeared to be dilated and endothelial adhesiveness to be disturbed (156). [The penetrance of this vasculogenesis phenotype was only about 50% and depended on the genetic background of the conceptus and the maternal mouse strain used (157)]. A complementary result indistinguishable from the TGF-beta 1 knockout phenotype was seen after targeted mutation of the TGF-beta receptor II (158). It is likely that for TGF-beta 1 signaling a heterodimer of TGF-beta receptor II with the activin receptor-like kinase (ALK-1) is used. Alk-1 is expressed at the sites of vasculogenesis in mouse embryos and by endothelial cells during embryonic angiogenesis (159).

TGF-beta 1 signaling from periendothelial cells during vasculogenesis may be important for maturation and stabilization of early blood vessels, which is completed by other mediators such as PDGFs necessary for recruitment of pericytes and establishment of periendothelial smooth muscle cells (160–162). Such cells are recruited from the perivascular mesenchyme quite early during embryonic development and give rise to smooth muscle cells around the dorsal aorta and other arteries from early limb bud stages onward (35,163,164). Disturbance of the cross-talk between endothelial and periendothelial cells may cause the phenotype observed in the knockout embryos. The role of TGF-beta for maturation and stability of blood vessels is also supported by the finding that mutations in the genes of the low affinity TGF-beta receptor endoglin and ALK-1 were found to be associated with the autosomal dominant inherited vascular malformations hereditary hemorrhaging telangiectasia type I and II, respectively (165,166).

The significance of early expression of erythropoietin receptor (EPOR) on hemangioblastic cells in the yolk sac is not clear (26). EPOR is expressed by endothelial cells in vitro and in vivo and EPO exerts a proliferative effect on endothelium (167,168). Loss of function mutations of both EPOR and EPO do not produce a vascular phenotype, but mice die at day 10.5 in consequence of severe anemia (169). Therefore a possible function during later development of the vasculature could not be analyzed.

Mice lacking tissue factor (TF), the high-affinity receptor, and cofactor for plasma factor VII/VIIa, and the primary cellular initiator of blood coagulation reveal a disturbed remodeling of the yolk sac vasculature characterized by the absence of vitelline vessels and die in utero between E9.5 and 10.5 (170). The differentiation of vascular smooth muscle cells is disturbed. Conversely, the knockout of factor VII did not interfere with vasculogenesis (171). Thus, other ligands than factor VII may be involved in TF function during vasculogenesis. Recent findings suggest that the expression of tissue factor supports the expression of VEGF (172). Consistently, TF is expressed in the yolk sac endoderm that produces VEGF (170,173). Thus, it may be speculated that the observed TF-/- phenotype is due to disturbed VEGF signaling in the extraembryonic membranes. Other cytokine receptors such as the seven-transmembrane-spanning G-protein-coupled CXCR4 receptor and its ligand PBSF/SDF1 may play a distinct role during vascularization of the gastrointestinal tract (174). But hematopoiesis, cardiogenesis, and cerebellar development are also impaired after deletion of these genes (175). CXC chemokines have been implicated in balancing angiogenesis—for example, IL-8 as an angiogenesis factor and PF-4 as angiostatic factor—since they respectively induce and inhibit vascularization in the cornea assay (176). Their role in vasculogenesis has not been defined. Other recently introduced and naturally occurring inhibitors of angiogenesis, such as angiostatin and endostatin, have not been examined during vasculogenesis and embryonic angiogenesis (177,178).

VI. ROLE OF TRANSCRIPTION FACTORS: THE MASTER REGULATOR OF VASCULOGENESIS NOT YET FOUND

Recently in zebrafish, a recessive, cell-autonomous mutant was identified and named *cloch*, which is characterized by the complete lack of endothelial, endocardial, and hematopoietic cells (179). Hematopoietic transcription factors are not expressed, and expression of VEGFR-2 is restricted to a small number of ill-defined cells (180). However, overexpression of the hematopoiesis transcription factor SCL (TAL-1) in cloch embryos almost completely rescued endothelial differentiation and hematopoiesis. Since the cloch mutation was localized outside the SCL gene, the as yet unidentified cloch gene product may act upstream of SCL (181). Overexpression of SCL in wild-type zebrafish embryos led to a respecification of paraxial and renal mesoderm toward hemangioblastic mesoderm (182). Initially the basic-helix-loop-helix transcription factor SCL was identified as a pathogenic agent in certain lymphatic leukemias, in which the gene product was constitutively overexpressed as the result of chromosomal translocation (183). Expression studies showed that SCL mRNA is expressed in both hematopoietic and endothelial cells in mice at early somite stages (184); in avian embryos of comparable stages, SCL protein could be localized in the nuclei of endothelial cells (185). Gene-targeting experiments revealed that SCL is essential for all hematopoietic lineages to differentiate (186–188). In order to gain a more detailed analysis of a potential vascular phenotype which in the SCL null mice may be masked by the early lethal defect in hematopoiesis, hematopoiesis was rescued by reintroduction of the SCL gene under the control of the regulatory sequence of the hematopoiesis-specific transcription factor GATA-1 (189). SCL-/- mice reconstituted in this way revealed normal hematopoiesis but died at E9.5 by failure of yolk sac angiogenesis. The honeycomb-like structure of the primary vascular plexus was not remodeled and vitelline vessels normally connecting the yolk sac vasculature with intraembryonic vessels failed to develop. Interestingly, in the intron enhancer of the VEGFR-2 gene of the mouse, there are potential SCL binding sites which may be important for the endothelial expression of VEGFR-2 (190). However, expression of VEGFR-2 in early hemangioblastic cells does not depend on the SCL transcription factor, because SCL knockout ES cells express VEGFR-2 upon differentiation as well (191). The knockout data are questionable as to whether SCL is also a master regulator of vasculogenesis in higher vertebrates, as is suggested from the results in zebrafish embryos. It may be speculated that, in higher vertebrates, SCL is either dispensable during vasculogenesis or replaceable by functionally related transcription factors.

Other candidates of transcriptional regulators of vasculogenesis are the ets family of transcription factors. In avian embryos, ets-1 was shown to be expressed in the early mesoderm just after gastrulation and later on to be restricted to endothelial cells mainly in the splanchnopleural mesoderm in avian and *Xenopus* embryos (192,193). Thus, the expression pattern of ets-1 coincides with the expression of VEGFR-2. Putative and functional ets binding sites have been detected in regulatory sequences of the genes of the VEGF receptors, in the core enhancer of the tie-2 gene, which is sufficient to drive endothelial specific gene expression, in the promoters of metalloproteinases which are expressed by migrating endothelial cells, and in many other endothelial genes (190,194–198). Thus, ets transcription factors may not only be involved in differentiation of angioblasts from their precursors but also be important for their migration in the early mesoderm, which invades the space between ectoderm and endoderm when moving outward from the primitive streak during gastrulation. Likewise, during physiological and tumor angiogenesis ets-1 is reexpressed by endothelium of sprouting capillaries in which VEGF receptors and metalloproteinases are highly up-regulated (199). Disruption of the ets-1 and ets-2 genes, however, did not result in defects of vasculogenesis but disturbed differentiation and survival of lymphocytes (when examined in ets-1-/-:RAG2-/- chimeras) and disturbed placentation, respectively (200–202). In contrast, the targeted mutation of the ets family member TEL resulted in disturbed yolk sac angiogenesis with lack of large vitelline vessels and consecutive intraembryonic apoptosis (203). However, differentiation of the primordial vascular plexus and hematopoiesis were not affected. Therefore, the function of ets family transcription factors during vasculogenesis is still not clear and functional redundancy of the individual family members during vasculogenesis cannot be excluded.

Oxygen tension and intracellular glucose concentrations, reflecting metabolical demands, are thought to be the key regulators of angiogenesis in tissues and tumors. VEGF is upregulated in response to hypoxia and low glucose concentrations by a dual mechanism, transcriptional activation and stabilization of its mRNA (204). Recently a heterodimeric transcription factor named HIF-1 (hypoxia inducible factor 1), consisting of the bHLH proteins HIF-1α and HIF-1β/ARNT (aryl hydrocarbon receptor nuclear translocator), was identified to mediate transcriptional activation of hypoxia-inducible genes such as EPO, VEGF, and enzymes of glycolysis (205–207). Although HIF-1 is a candidate for the regulation of VEGF gene expression also in vivo (208), a role of this factor during vasculogenesis is not established. Targeted mutation of HIF-1α in mice led to a series of malformations of the embryo, including abnormal neural folding and the complete lack of cephalic blood vessels (209,210). These defects may be at least in part due to a reduced expression of VEGF in locations where angiogenesis is needed to support tissue growth. Conversely, results of the two published knockouts of HIF-1β/ARNT were inconsistent, revealing either early vascular defects in the yolk sac and loss of cellular responsiveness to hypoxia and hypoglycemia or defects in placental differentiation (211,212). More recently a close homologue of HIF-1α was cloned by several groups and denominated EPAS-1/HRF/HLF/MOP-2 and HIF-2α, respectively (207,213–216). Unlike HIF-1α, which is expressed ubiquitously, HIF-2α is expressed most prominently in brain capillary endothelium and other blood vessels. High expression levels of HIF-2 alpha in kidney glomeruli, choroid plexus and lung—places where VEGF is highly expressed—suggest a role of this factor in VEGF gene activation (214,215,217). Since HIF-2α activates the transcription of VEGF, tie-2, and VEGFR-2, a role of this transcription factor as a master regulator of vascular development is likely (190,213,215). However, the role of HIF-2α for the transcriptional activation in vivo of these vascular target genes and its role in vasculogenesis remain to be established. Interestingly, vertebrate HIFs are homologous to the *Drosophila* trachealess (*trh*) gene product, which is a master regulator of tracheocyte differentiation (218,219). The tracheal system of insects that serves as a pipeline for gas consists of a unique cell type, the tracheocyte, grows by sprouting, and its growth is enhanced in response to hypoxia and metabolic demands, thus resembling the vascular system of vertebrates in many aspects (220). Moreover, human HIF-1α is functional in the *trh*-complex after ectopic expression in the *Drosophila* embryo, which is indicative of a highly conserved function of this factors during formation of vascular systems (221).

Other transcription factors that have been implicated in vasculogenesis and embryonic angiogenesis are the homeobox gene HEX/prh (222,223), the bHLH protein eHAND

(224), and the myocyte-specific enhancer factor MEF2a (225); these are not restricted to the endothelium. Whereas overexpression of HEX/prh in *Xenopus* embryos lead to changes in the primordial vascular plexus very similar to those obtained after overexpression of VEGF (226), a function of the other factors during embryonic vascular development remains to be determined.

VII. FROM ANGIOBLAST TO VESSEL LUMEN: CELL-TO-MATRIX CONTACT, CELL-TO-CELL CONTACT

Initially during vasculogenesis, angioblasts are either organized into the endothelial lining of the blood islands forming a vesicle around the central blood cell progenitors or they are found as migrating angioblasts in the mesoderm inside and outside the embryo. It is evident that angioblasts need the contact of a substratum when either migrating or settling during formation of blood vessels. In order to form a vascular lumen, angioblasts also need to contact each other and to establish stable cell-to-cell contacts. The molecular cues that trigger the formation of a vessel lumen in vivo have not been identified. Two principal morphogenetic mechanism for the formation of a vascular lumen have been encountered. The intracellular lumen formation is observed in endothelial cells that undergo so-called angiogenesis in vitro (227). In endothelial cells that have arranged themselves into cords, a lumen arises from an intracellular vacuole that eventually anastomoses with intracellular vacuoles in adjacent endothelial cells. In cross sections, those capillaries appear to be seamless. In vivo seamless capillaries have been observed in brain, kidney, and duodenal villi, and it is likely that this mechanism contributes to the formation of vascular lumen during angiogenesis in these organs (228). During vasculogenesis, only the second and probably more common type of vascular lumen formation has been described. The lumen emerges between several angioblasts that form a cord and finally a channel. This type of extracellular lumen formation has been described in detail for the formation of the dorsal aorta (229) but is also assumed for the other primordial blood vessels (230,231). Finally, upon lumen formation, the angioblast polarizes and becomes an endothelial cell, which, like other epithelial cells upon polarization, produces a basal lamina and needs stable interaction with this basal lamina (232,233). This interaction is executed by cell surface receptors that recognize compounds of the extracellular matrix. The major class of such receptors is represented by the integrins. A variety of integrins are expressed by endothelial cells. Insights into the role which integrins play during formation of blood vessels were gained from gene-targeting experiments and experiments employing inhibitory antibodies directed against distinct family members (234,235).

Intraembryonic application of antibodies against β_1 and $\alpha_v\beta_3$ integrins disturbed lumen formation and differentiation of the primordial vascular plexus, respectively (236,237). This may result from interference with the role these integrins play during the interaction of endothelial cells with the basement membrane components fibronectin, laminin, collagen IV, and vitronectin. Fibronectin is abundant around capillaries and blood islands of the primordial vasculature (238). As demonstrated by targeted mutation, fibronectin is required for vasculogenesis inside and outside the embryo but also for cardiac development and formation of axial structures such as somites and notochord (239). Conversely, embryos lacking the main fibronectin receptor after deletion of the α_5 integrin develop significantly further than fibronectin null embryos (240,241). Thus, the involvement of other fibronectin receptors than $\alpha_5\beta_1$ integrin in early vascular development is likely. Later, upon blood vessel maturation, the content of laminin is increased in the basal laminae of blood vessels, which is consistent with the role of laminin for the maintenance of epithelial polarity (238). Distinct functions of certain splice variants of fibronectin and isoforms of laminin for vascular development are likely but not yet clear (233,242).

Whereas the β_1 integrin knockout is lethal shortly after implantation (243)—as was to be expected from the central role of this molecule in cell adhesion as dimerization partner in a multitude of integrins—and β_1-/- ES cells do not give rise to functional angioblasts (244), loss of function mutations of β_3 and β_5 have no influence on vasculogenesis and not even on embryonic angiogenesis. Loss of function mutation of α_v integrin leads to embryonic and perinatal lethality by hemorrhage (234). This is in conflict with the findings from experiments using antibodies and peptidomimetics that show that $\alpha_v\beta_3$ and $\alpha_v\beta_5$ integrins are crucial for angiogenesis and survival of endothelial cells (235,245). Overlapping functions with other endothelial integrins during embryogenesis may be responsible for this discrepancy.

Formation and maintenance of the vessel lumen depends also on molecules that enables the endothelial cells to contact each other. The vascular endothelial specific cadherin VE-cadherin (cadherin-5) is responsible for establishment of the adherence junctions between endothelial cells by homotypic interaction (246,247) and inhibition of this interaction by monoclonal antibodies disrupts vascular integrity in vitro and in vivo leading to haemorrhage (248). VE-cadherin is expressed very early by angioblasts (249) and ES cells with a targeted mutation of the VE-cadherin gene cannot give rise to functional angioblasts (250). Whether the VE-cadherin/catenin complex in angioblasts is a target of VEGF signaling—as has been described for endothelial cells in vitro—is not known. In vascular endothelial cells, phosphorylation of the complex at tyrosine residues may account for loosening of cell-to-cell contacts during the initiation of angiogenesis and concomitant increase of vascular permeability in response to VEGF (251,252). A function of other endothelial cell adhesion molecules expressed early during vasculogenesis, such as CD34 and PECAM/CD31, has not been defined (21,23,253).

VIII. VASCULAR PATTERN FORMATION AND REMODELING

Transplantation experiments in early avian embryos have demonstrated that the pattern of blood vessels is determined by environmental factors and not an intrinsic program of the angioblasts (6,27,46). In many organs, the vascular architecture is adapted to the histological architecture of the organ and even the mode of vascular growth is influenced by this architecture. In skeletal muscle, for example, capillary blood vessels are aligned with the muscle fibers and grow by longitudinal splitting rather than by sprouting (254). Intersomitic arteries grow between the somites, thus conforming to the segmental pattern of the body wall (255). However, the pattern of primordia of the large body vessels is not necessarily the result of environmental factors, and it appears to be useful to distinguish between vascular macropattern and micropattern. It is tempting to believe that the vascular macropattern is determined by genetic factors, while the more variable micropattern depends on the environment. However, recent observations in zebrafish mutants have shown that the development of the macropattern is most likely an epigenetic phenomenon. Formation of the dorsal aorta at the midline most likely depends on signals from the notochord, because in zebrafish mutants that fail to form a notochord, no dorsal aorta is developed and formation of the aortic rudiment can be rescued in chimeras bearing wild-type notochord (93,256). The notochord-derived signals have not been identified, but VEGF is a candidate factor (257). The pathways for angioblasts destined to form the principal vessels are determined early during development and may be used in opposite directions. Unilateral inhibition of the migration of angioblasts along the anterior intestinal portal in avian embryos is compensated by angioblasts from the other side, which give rise to endocardium and connection with the dorsal aorta on the affected side (27). These findings are indicative of the high plasticity of early blood vessel primordia. Conversely, in the trunk, for reasons still unknown, angioblasts do not cross the sagittal body plane (49).

The structure of the vessel wall and the decision as to whether a blood vessel belongs to the arterial or venous part of the circulation depends on the location of the vessel and its distance to the heart. In the embryo in some locations, the same vessel that initially is used as an artery may later be used as a vein when the blood flow reverses after vascular remodeling (46). As nonperfused blood vessels regress and well-perfused vessels increase in size, hemodynamic forces have been implicated in the formation and reorganization of the vascular macropattern and experimental evidence was provided by ligation of vitelline vessels (258,259). Hemodynamic forces may upregulate the transcription of certain endothelial relevant genes such as PDGF-B that have shear-stress response elements in their promoters (260) and thus contribute to the differentiation and the recruitment of perivascular smooth muscle cells. This is in line with the observation that the aortas and the umbilical arteries, which have to withstand the highest hemodynamic forces in the embryo, are the first vessels that are coated with smooth muscle cells (163,164). Also, local factors are important for assembly of the vessel wall. Ablation of the neural crest leads to malformation of the outflow tract of the heart and the branchial arch arteries because neural crest cells contribute to the tunica media of these vessels (261,262). More recent findings showing differential expression of Eph-receptor B4 and its ligand on early embryonic venous and arterial endothelium, respectively, have suggested the possibility of a genetic influence on venous versus arterial determination of endothelial cells (149).

The primordial vascular micropattern is initially uniform in the embryo and the yolk sac. The polygonal distribution of blood vessels resembles a honeycomb. Upon onset of the heartbeat this pattern is subjected to a process now often termed *remodeling* in the literature (Figure 1). The term *remodeling* simplifies and subsumes a series of events. The small intervascular spaces are enlarged by tissue growth and primordial blood vessels begin to subdivide these spaces by sending out sprouts, or new intervascular spaces are formed by nonsprouting angiogenesis, called *intussusception*. Intravascular formation of endothelial pillars or posts that span the vessel lumen and in turn acquire a core of mesenchymal cells, and enlargement of this mesenchymal core to a new intervascular space, are the formal mechanism of intussusceptive vascular growth. While in lung and probably also liver intussusception is the predominant mode of vascular growth (263), sprouting and intussusception occur side by side in the yolk sac and other organs (264). The mechanisms that determine the spatial distance between sprouts and regulate the size of intervascular spaces are largely unknown. Also, regression of blood vessels and apoptosis of vascular endothelial cells may contribute to the remodeling of the primordial plexus. Focal regression of blood vessels precedes imminent cartilage differentiation in the limb mesenchyme (265–267) and is observed in the hyaloid vasculature (268). The corresponding feature of regression may be pruning of blood vessels, as is observed in the fetal retina (269). The mechanisms of embryonic vascular regression are not understood, but macrophages seem to be important for regression of the hyaloid vasculature (270). Another important component of remodeling is the formation of collecting blood vessels, which is most impressive in the yolk sac when the vitelline vessels are formed (271–273). The formation of larger collecting vessels involves the recruitment of periendothelial cells, which marks the beginning of angiohistogenesis. PDGF-B secreted by the endothelium may be the key signal for perivascular mesenchymal cells to differentiate into pericytes, which may contribute to the mechanical stability of the capillary wall (162). In the retina, recruitment of pericytes marks the transition from an immature to a mature capillary and makes capillaries independent on VEGF, which is a survival factor for immature capillaries of the retina (274).

Surprisingly, as described above in detail, many of the gene knockouts that are lethal due to vascular defects are characterized by disturbed remodeling of the yolk sac vasculature. The phenotypes of the yolk sacs of the different knockout mice closely resemble each other (VEGF, VEGFR-3, tissue factor, tie-2 and ang-1, Eph-B4, ephrin-B2, TEL, SCL/TAL-1; ARNT). Sprouting and formation of collecting vessels are impaired. In contrast, different phenotypes are observed in the corresponding embryos. The reason for this is not clear and may depend on the particular nature of extraembryonic angioblasts, which may be more sensitive to certain gene defects than their intraembryonic counterparts. This would suggest that the genes targeted act in concert during remodeling of extraembryonic blood vessels and are equally essential for sprouting and formation of vitelline vessels. Another explanation would be that remodeling of the yolk sac vasculature is dependent on hemodynamic forces generated by the embryonic heart, which, at least in VEGF and Tie-2 knockout mice, are missing or likely to be reduced, respectively. This is consistent with the old observation that after inhibiting the development of the chick embryo, a yolk sac vasculature is formed but no remodeling of this vasculature occurs (275). Conversely, in other knockouts in which regularly beating hearts have been observed, the disturbance of yolk sac vasculature appears to be the limiting factor for embryonic development.

IX. CONCLUSION

According to its early destination to nourish the growing embryo, the cardiovascular system is the first functional organ system of the vertebrate organism. The integrity of vascular development is a requirement for normal embryogenesis and targeted mutations of molecules, which are essential for formation, differentiation, and growth of blood vessels are embryonically lethal at an early stage. Gene-targeting experiments have substantiated the concept that there is a fundamental difference between the establishment of the primordial vascular plexus from angioblasts and the formation of new blood vessels from preexisting vessels. The induction and differentiation of angioblasts during mesoderm formation, their coalescence into primordial blood vessels, and the remodeling and growth of these vessels are the distinct steps of early vascular development. Endothelial growth factors and their receptors have been identified as the key regulators of vascular development. While signaling by fibroblast growth factors is crucial for the induction of endothelial precursors, thus initiating vascular development, further steps of vascular differentiation depend on signaling by vascular endothelial growth factor, angiopoietins, and ephrins. However, interaction, cooperation, and transcriptional regulation of the different signals during vascular development are still far from being understood and will be future issues of the developmental biology of blood vessels.

ACKNOWLEDGMENTS

I thank Werner Risau, Andreas Kappel, and Simon Bamforth (Max-Planck-Institut Bad Nauheim) for helpful discussions and critical reading of the manuscript. This work was supported by the Max-Planck-Gesellschaft, the Zentrum für Molekularbiologische Medizin der Universität zu Köln, and in part by a grant from the Deutsche Forschungsgemeinschaft (Fl 223/2-1).

REFERENCES

1. HM Evans. On the development of the aorta, cardinal and umbilical veins and the other blood vessels of vertebrate embryos from capillaries. Anat Rec 3:498–519, 1909.
2. F Gonzalez-Crussi. Vasculogenesis in the chick embryo: An ultrastructural study. Am J Anat 130:441–460, 1971.
3. JL Haar, G Ackerman. A phase and electron microscopic study of vasculogenesis and erythropoiesis in the yolk sac of the mouse. Anat Rec 170:199–224, 1971.
4. L Pardanaud, C Altmann, P Kitos, F Dieterlen-Lievre, CA Buck. Vasculogenesis in the early quail blastodisc as studied with a monoclonal antibody recognizing endothelial cells. Development 100:339–349, 1987.
5. W Risau, I Flamme. Vasculogenesis. Annu Rev Cell Dev Biol 11:73–91, 1995.
6. TJ Poole, JD Coffin. Vasculogenesis and angiogenesis: two distinct morphogenetic mechanisms establish embryonic vascular pattern. J Exp Zool 251:224–231, 1989.
7. J Folkman, Y Shing. Angiogenesis. J Biol Chem 267:10931–10934, 1992.
8. W Risau. Mechanisms of angiogenesis. Nature 386:671–674, 1997.
9. RC Wagner. Endothelial cell embryology and growth. Adv Microcirc 9:45–75, 1980.
10. W Risau. Differentiation of endothelium. FASEB J 9:926–933, 1995.
11. W His. Lecithoblast und Angioblast der Wirbelthiere. Abhandl K S Ges Wiss Math-Phys 22:171–328, 1900.
12. R Bellairs. The primitive streak. Anat Embryol 174:1–14, 1986.
13. Z Yablonka-Reuveni. The emergence of the endothelial cell lineage in the chick embryo can be detected by uptake of acetylated low density lipoprotein and the presence of a von Willebrand-like factor. Dev Biol 132:230–240, 1989.
14. N Zagris. Erythroid cell differentiation in unincubated chick blastoderm in culture. J Embryol Exp Morphol 58:209–216, 1980.
15. Y Azar, H Eyal-Giladi. Marginal zone cells–the primitive streak-inducing component of the primary hypoblast in the chick. J Embryol Exp Morphol 52:79–88, 1979.
16. B Christ, M Grim, J Wilting, K von Kirschhofer, F Wachtler. Differentiation of endothelial cells in avian embryos does not depend on gastrulation. Acta Histochem 91:193–199, 1991.
17. PDF Murray. The development in vitro of the blood of the early chick embryo. R Soc Lond Ser B III:1932.
18. FR Sabin. Origin and development of the primitive vessels of the chick and of the pig. Carnergie Contrib Embryol 6:61–124, 1917.
19. MC Labastie, TJ Poole, BM Peault, NM LeDouarin. MB1, a quail leukocyte-endothelium antigen: partial characterization of the cell surface and secreted forms in cultured endothelial cells. Proc Natl Acad Sci USA 83:9016–9020, 1986.
20. KM McNagny, I Pettersson, F Rossi, I Flamme, A Shevchenko, M Mann, T Graf. Thrombomucin, a novel cell surface protein that defines thrombozytes and hematopoietic progenitors. J Cell Biol 138:1395–1407, 1997.

21. S Baumhueter, N Dybdal, C Kyle, L Lasky. Global vascular expression of murine CD34, a sialomucine-like endothelial ligand for L-selectin. Blood 84:2554–2565, 1994.
22. H Wood, G May, L Healy, T Enver, G Morris-Kay. CD34 expression patterns during early mouse development are related to modes of blood vessel formation and reveal additional sites of hematopoiesis. Blood 90:2300–2311, 1997.
23. HM DeLisser, PJ Newman, SM Albelda. Molecular and functional aspects of PECAM-1/CD31. Immunology Today 15:490–495, 1994.
24. N Kabrun, H Buhring, K Choi, A Ullrich, W Risau, G Keller. Flk-1 expression defines a population of early embryonic hematopoietic precursors. Development 124:2039–2048, 1997.
25. M Yano, A Iwama, H Nishio, J Suda, G Takada, T Suda. Expression and function of murine receptor tyrosine kinases, TIE and TEK, in hematopoietic stem cells. Blood 89:4317–4326, 1997.
26. J McGann, L Silver, J Liesveld, J Palis. Erythropoietin-receptor expression and function during the initiation of murine yolk sac erythropoiesis. Exp Hematol 25:1149–1157, 1997.
27. JD Coffin, TJ Poole. Endothelial cell origin and migration in embryonic heart and cranial blood vessel development. Anat Rec 231:383–395, 1991.
28. F Dieterlen-Lievre. Hemopoietic cell progenitors in the avian embryo: origin and migrations. Ann NY Acad Sci 511:77–87, 1987.
29. P Wilms, B Christ, J Wilting, F Wachtler. Distribution and migration of angiogenic cells from grafted avascular intraembryonic mesoderm. Anat Embryol 183:371–377, 1991.
30. FP Reagen. Vascularization phenomena in fragments of embryonic bodies completely isolated from yolk-sac blastoderm. Anat Rec 9:329–341, 1915.
31. IG Mobbs, DB McMillan. Structure of the endodermal epithelium of the chick yolk sac during early stages of development. Am J Anat 155:287–309, 1979.
32. J Kessel, B Fabian. Graded morphogenetic patterns during the development of the extraembryonic blood system and coelom of the chick blastoderm: a scanning electron microscope and light microscope study. Am J Anat 173:99–112, 1985.
33. M Mato, E Aikawa, K Kishi. Some observations on interstice between mesoderm and endoderm in the area vasculosa of chick blastoderm. Exp Cell Res 35:426–428, 1964.
34. KZ Reiss, DM Noden. SEM characterization of a cellular layer separating blood vessels from endoderm in the quail embryo. Anat Rec 225:165–175, 1989.
35. ME Murphy, EC Carlson. An ultrastructural study of developing extracellular matrix in vitelline blood vessels of the early chick embryo. Am J Anat 151:345–375, 1978.
36. A Medvinsky, E Dzierzak. Definitive hematopoiesis is autonomously initiated by the AGM region. Cell 86:897–906, 1996.
37. I Olah, J Medgyes, B Glick. Origin of aortic cell clusters in the chicken embryo. Anat Rec 222:60–68, 1988.
38. M Tavian, L Coulombel, D Luton, H San Clemente, F Dieterlen-Lievre, B Peault. Aorta-associated CD34+ hematopoietic cells in the early human embryo. Blood 87:67–72, 1996.
39. LF Dieterlen, L Pardanaud, F Yassine, F Cormier. Early haemopoietic stem cells in the avian embryo. J Cell Sci Suppl 10:29–44, 1988.
40. I Rich. Primordial germ cells are capable of producing cells of the hematopoietic system in vitro. Blood 463–472, 1995.
41. L Pardanaud, D Luton, M Prigent, L Bourcheix, M Catala, F Dieterlen-Lièvre. Two distinct endothelial lineages in ontogeny, one of them related to hemopoiesis. Development 122: 1363–1371, 1996.
42. A Caprioli, T Jaffredo, R Gautier, C Dubourg, F Dieterlen-Lièvre. Blood-borne seeding by hematopoietic and endothelial precursors from the allantois. Proc Natl Acad Sci USA 95: 1641–1646, 1998.

43. AM Müller, A Medvinsky, S J., F Grosveld, E Dzierzak. Development of hematopoietic stem cell activity in the mouse embryo. Immunity 1:291–301, 1994.
44. M Kennedy, M Firpo, K Choi, C Wall, S Robertson, N Kabrun, G Keller. A common precursor for primitive erythropoiesis and definitive haematopoiesis. Nature 386:488–493, 1997.
45. T Nakano, H Kodama, T Honjo. In vitro development of primitive and definitive erythrocytes from different precursors. Science 272:722–724, 1996.
46. DM Noden. Development of craniofacial blood vessels. In: RN Feinberg, GK Sherer, R Auerbach, eds. The Development of the Vascular System. Basel: Karger, 1991, pp 1–24.
47. L Pardanaud, F Yassine, F Dieterlen-Lievre. Relationship between vasculogenesis, angiogenesis and haemopoiesis during avian ontogeny. Development 105:473–485, 1989.
48. L Pardanaud, F Dieterlen-Lievre. Does the paraxial mesoderm of the avian embryo have hemangioblastic capacity? Anat Embryol 192:301–308, 1995.
49. J Wilting, B Brand-Saberi, R Huang, Q Zhi, G Kontges, C Ordahl, B Christ. Angiogenic potential of the avian somite. Dev Dyn 202:165–171, 1995.
50. H Kurz, T Gartner, P Eggli, B Christ. First blood vessels in the avian neural tube are formed by a combination of dorsal angioblast immigration and ventral sprouting of endothelial cells. Dev Biol 133–147, 1996.
51. RN Feinberg, DM Noden. Experimental analysis of blood vessel development in the avian wing bud. Anat Rec 231:136–144, 1991.
52. AJJC Bogers, AC Gittenberger-de Groot, RE Poelmann, BM Péault, HA Huysmans. Development of the origin of the coronary arteries, a matter of ingrowth or outgrowth? Anat Embryol 180:437–441, 1989.
53. T Mikawa, DA Fischman. Retroviral analysis of cardiac morphogenesis: discontinuous formation of coronary vessels. Proc Natl Acad Sci USA 89:9504–9508, 1992.
54. DM Noden. Embryonic origins and assembly of blood vessels. Am Rev Respir Dis 140: 1097–1103, 1989.
55. M DeRuiter, R Poelmann, J VanMunstren, V Mironov, R Markwald, A Gittenberger-de Groot. Embryonic endothelial cells transdifferentiate into mesenchymal cells expressing smooth muscle actins in vivo and in vitro. Circ Res 80:444–451, 1997.
56. T Asahara, T Murohara, A Sullivan, M Silver, R van der Zee, T Li, B Witzenbichler, G Schatteman, J Isner. Isolation of putative progenitor endothelial cells for angiogenesis. Science 275:964–967, 1997.
57. JM Slack, BG Darlington, JK Heath, SF Godsave. Mesoderm induction in early Xenopus embryos by heparin-binding growth factors. Nature 326:197–200, 1987.
58. JC Smith. Mesoderm-inducing factors and mesodermal patterning. Curr Opin Cell Biol 7: 856–861, 1995.
59. E Amaya, TJ Musci, MW Kirschner. Expression of a dominant negative mutant of the FGF receptor disrupts mesoderm formation in Xenopus embryos. Cell 66:257–270, 1991.
60. B Feldmann, W Poueymirou, VE Papaioannou, TM DeChiara, M Goldfarb. Requirement of FGF-4 for postimplantation mouse development. Science 267:246–249, 1995.
61. B Ciruna, L Schwartz, K Harpal, T Yamaguchi, J Rossant. Chimeric analysis of fibroblast growth factor receptor-1 (FGFR1) function: a role for FGFR1 in morphogenetic movement through the primitive streak. Development 124:2829–2841, 1997.
62. T Yamaguchi, K Harpal, M Henkemeyer, J Rossant. FGFR-1 is required for embryonic growth and mesodermal patterning during mouse gastrulation. Genes Dev 8:3032–3044, 1994.
63. I Flamme, W Risau. Induction of vasculogenesis and hematopoiesis in vitro. Development 116:435–439, 1992.
64. K Krah, V Mironow, W Risau, I Flamme. Induction of vasculogenesis in quail blastodisc derived embryoid bodies. Dev Biol 164:123–132, 1994.
65. E Mitrani, Y Gruenbaum, H Shohat, T Ziv. Fibroblast growth factor during mesoderm induction in the early chick embryo. Development 109:387–393, 1990.

66. R Dono, R Zeller. Cell-type-specific nuclear translocation of fibroblast growth factor-2 isoforms during chicken kidney and limb morphogenesis. Dev Biol 163:316–330, 1994.
67. J Thompson, JMW Slack. Over-expression of fibroblast growth factors in Xenopus embryos. Mech Dev 38:175–182, 1992.
68. HV Isaacs, D Tannahill, JM Slack. Expression of a novel FGF in the Xenopus embryo: A new candidate inducing factor for mesoderm formation and anteroposterior specification. Development 114:711–720, 1992.
69. D Tannahill, HV Isaacs, MJ Close, G Peters, JM Slack. Developmental expression of the Xenopus int-2 (FGF-3) gene: activation by mesodermal and neural induction. Development 115:695–702, 1992.
70. DG Wilkinson, G Peters, C Dickson, AP McMahon. Expression of the FGF-related protooncogene int-2 during gastrulation and neurulation in the mouse. EMBO J 7:691–695, 1988.
71. L Niswander, GR Martin. Fgf-4 expression during gastrulation, myogenesis, limb and tooth development in the mouse. Development 114:755–768, 1992.
72. JM Hebert, M Boyle, GR Martin. mRNA localization studies suggest that murine FGF-5 plays a role in gastrulation. Development 112:407–415, 1991.
73. O Haub, M Goldfarb. Expression of the fibroblast growth factor-5 gene in the mouse embryo. Development 112:397–406, 1991.
74. CD Stern, DR Canning. Origin of cells giving rise to mesoderm and endoderm in chick embryo. Nature 343:273–275, 1990.
75. Tp Yamaguchi, R Conlon, J Rossant. Expression of the fibroblast growth factor receptor FGFR-1/flg during gastrulation and segmentation in the mouse embryo. Development 152: 75–88, 1992.
76. KG Peters, S Werner, G Chen, LT Williams. Two FGF receptor genes are differentially expressed in epithelial and mesenchymal tissues during limb formation and organogenesis in the mouse. Development 114:233–243, 1992.
77. JG Heuer, CS von Bartheld, Y Kinoshita, PC Evers, M Bothwell. Alternating phases of FGF receptor and NGF receptor expression in the developing chicken nervous system. Neuron 5: 283–296, 1990.
78. A Wanaka, J Milbrandt, EM Johnson, Jr. Expression of FGF receptor gene in rat development. Development 111:455–468, 1991.
79. G Patstone, EB Pasquale, P Maher. Different members of the fibroblast growth factor receptor family are specific to distinct cell types in the developing chicken embryo. Dev Biol 155: 107–123, 1993.
80. BB Riley, MP Savage, BK Simandl, BB Olwin, JF Fallon. Retroviral expression of FGF-2 (bFGF) affects patterning in chick limb bud. Development 118:95–104, 1993.
81. TP Yamaguchi, DJ Dumont, RA Conlon, ML Breitman, J Rossant. flk-1, an flt-related receptor tyrosine kinase is an early marker for endothelial cell precursors. Development 118:489–498, 1993.
82. A Eichmann, C Marcelle, C Breant, NM LeDouarin. Two molecules related to the VEGF receptor are expressed in early endothelial cells during avian embryonic development. Mech Dev 42:33–48, 1993.
83. I Flamme, G Breier, W Risau. Vascular endothelial growth factor (VEGF) and VEGF receptor 2 (flk-1) are expressed during vasculogenesis and vascular differentiation in the quail embryo. Dev Biol 169:699–712, 1995.
84. GH Fong, J Klingensmith, CR Wood, J Rossant, ML Breitman. Regulation of Flt-1 expression during mouse embryogenesis suggests a role in the establishment of vascular endothelium. Dev Dyn 207:1–10, 1996.
85. N Ferrara, KA Houck, LB Jakeman, J Winer, DW Leung. The vascular endothelial growth factor family of polypeptides. J Cell Biochem 47:211–218, 1991.

86. G Breier, W Risau. The role of vascular endothelial growth factor in blood vessel formation. Trends Cell Biol 6:454–456, 1996.
87. S Takeshita, Y Tsurumi, T Couffinahl, T Asahara, C Bauters, J Symes, N Ferrara, JM Isner. Gene transfer of naked DNA encoding for three isoforms of vascular endothelial growth factor stimulates collateral development in vivo. Lab Invest 75:487–501, 1996.
88. M Schmidt, I Flamme. The in vivo activity of vascular endothelial growth factor isoforms in the avian embryo. Growth Factors 15:183–197, 1998.
89. B Millauer, S Wizigmann Voos, H Schnurch, R Martinez, NP Moller, W Risau, A Ullrich. High affinity VEGF binding and developmental expression suggest Flk-1 as a major regulator of vasculogenesis and angiogenesis. Cell 72:835–846, 1993.
90. KG Peters, VC De, LT Williams. Vascular endothelial growth factor receptor expression during embryogenesis and tissue repair suggests a role in endothelial differentiation and blood vessel growth. Proc Natl Acad Sci USA 90:8915–8919, 1993.
91. G Breier, M Clauss, W Risau. Coordinate expression of vascular endothelial growth factor receptor-1 (Flt-1) and its ligand suggests a paracrine regulation of murine vascular development. Dev Dyn 204:228–239, 1995.
92. O Cleaver, K Tonissen, M Saha, P Krieg. Neovascularization of the Xenopus embryo. Dev Dyn 210:66–77, 1997.
93. B Fouquet, B Weinstein, F Serluca, M Fishman. Vessel patterning in the embryo of the zebrafish: guidance by notochord. Dev Biol 183:37–48, 1997.
94. K Choi, M Kennedy, A Kazarov, J Papadimitriou, G Keller. A common precursor for hematopoietic and endothelial cells. Development 125:725–732, 1998.
95. S Nishikawa, S Nishikawa, M Hirashima, N Matsuyoshi, H Kodama. Progressive lineage analysis by cell sorting and culture identifies FLK1+VE-cadherin+ cells at a diverging point of endothelial and hemopoietic lineages. Development 125:1747–1757, 1998.
96. P Carmeliet, V Ferreira, G Breier, S Pollefeyt, L Kieckens, M Gertsenstein, M Fahrig, A Vandenhoeck, K Harpal, C Eberhardt, C Declercq, J Pawling, L Moons, D Collen, W Risau, A Nagy. Abnormal blood vessel development and lethality in embryos lacking a single VEGF allele. Nature 380:435–439, 1996.
97. N Ferrara, K Carvermoore, H Chen, M Dowd, L Lu, KS Oshea, L Powellbraxton, KJ Hillan, MW Moore. Heterozygous embryonic lethality induced by targeted inactivation of the Vegf gene. Nature 380:439–442, 1996.
98. CJ Drake, CD Little. Exogenous vascular endothelial growth factor induces malformed and hyperfused vessels during embryonic neovascularization. Proc Natl Acad Sci USA 92:7657–7661, 1995.
99. R Tomanek, K Lotun, E Clark, P Suvarna, N Hu. VEGF and bFGF stimulate myocardial vascularization in embryonic chick. Am J Physiol 274:H1620–H1626, 1998.
100. I Flamme, M vonReutern, HCA Drexler, S Syedali, W Risau. Overexpression of vascular endothelial growth factor in the avian embryo induces hypervascularization and increased vascular permeability without alterations of embryonic pattern formation. Dev Biol 171:399–414, 1995.
101. KH Plate, G Breier, HA Weich, HD Mennel, W Risau. Vascular endothelial growth factor and glioma angiogenesis—coordinate induction of Vegf receptors, distribution of Vegf protein and possible in vivo regulatory mechanisms. Int J Cancer 59:520–529, 1994.
102. C Kremer, G Breier, W Risau, K Plate. Up-regulation of flk-1/vascular endothelial growth factor receptor 2 by its ligand in a cerebral slice culture system. Cancer Res 57:3852–3859, 1997.
103. BI Terman, M Dougher Vermazen, ME Carrion, D Dimitrov, DC Armellino, D Gospodarowicz, P Bohlen. Identification of the KDR tyrosine kinase as a receptor for vascular endothelial cell growth factor. Biochem Biophys Res Commun 187:1579–1586, 1992.
104. TP Quinn, KG Peters, VC De, N Ferrara, LT Williams. Fetal liver kinase 1 is a receptor for

vascular endothelial growth factor and is selectively expressed in vascular endothelium. Proc Natl Acad Sci USA 90:7533–7537, 1993.
105. C De Vries, JA Escobedo, H Ueno, K Houck, N Ferrara, LT Williams. The fms-like tyrosine kinase, a receptor for vasculare endothelial growth factor, Science 255:989–991, 1992.
106. V Joukov, K Pajusola, A Kaipainen, D Chilov, I Lahtinen, E Kukk, O Saksela, N Kalkkinen, K Alitalo. A novel vascular endothelial growth factor, VEGF-C, is a ligand for the Flt4 (VEGFR-3) and KDR (VEGFR-2) receptor tyrosine kinases. EMBO J 15:290–298, 1996.
107. M Achen, M Jeltsch, E Kukk, T Makinen, A Vitali, A Wilks, K Alitalo, S Stacker. Vascular endothelial growth factor D (VEGF-D) is a ligand for the tyrosine kinases VEGF receptor 2 (Flk1) and VEGF receptor 3 (Flt4). Proc Natl Acad Sci USA 95:548–553, 1998.
108. A Kaipainen, J Korhonen, K Pajusola, O Aprelikova, MG Persico, BI Terman, K Alitalo. The related FLT4, FLT1, and KDR receptor tyrosine kinases show distinct expression patterns in human fetal endothelial cells. J Exp Med 178:2077–2088, 1993.
109. E Kukk, A Lymboussaki, S Taira, A Kaipainen, M Jeltsch, V Joukov, K Alitalo. Vegf-C receptor binding and pattern of expression with Vegfr-3 suggests a role in lymphatic vascular development. Development 122:3829–3837, 1996.
110. A Kaipainen, J Korhonen, T Mustonen, V van Hinsbergh, G Fang, D Dumont, M Breitman, K Alitalo. Expression of the fms-like tyrosine kinase 4 gene becomes restricted to lymphatic endothelium during development. Proc Natl Acad Sci USA 92:3566–3570, 1995.
111. M Jeltsch, A Kaipainen, V Joukov, X Meng, M Lakso, H Rauvala, M Swartz, D Fukumura, R Jain, K Alitalo. Hyperplasia of lymphatic vessels in VEGF-C transgenic mice. Science 276:1423–1425, 1997.
112. F Shalaby, J Rossant, TP Yamaguchi, M Gertsenstein, XF Wu, ML Breitman, AC Schuh. Failure of blood-island formation and vasculogenesis ln Flk-1-deficient mice. Nature 376: 62–66, 1995.
113. GH Fong, J Rossant, M Gertsenstein, ML Breitman. Role Of the Flt-1 Receptor tyrosine kinase in regulating the assembly of vascular endothelium. Nature 376:66–70, 1995.
114. S Hiratsuka, O Minowa, J Kuno, T Noda, M Shibuya. Flt-1 lacking the tyrosine kinase domain is sufficient for normal development and angiogenesis in mice. Proc Natl Acad Sci USA 95:9349–9354, 1998.
115. A Eichmann, C Corbel, T Jaffredo, C Breant, V Joukov, V Kumar, K Alitalo, NM LeDouarin. Avian VEGF-C: cloning, embryonic expression pattern and stimulation of the differentiation of VEGFR2-expressing endothelial cell precursors. Development 125:743–752, 1998.
116. B Olofsson, K Pajusola, A Kaipainen, G von Euler, V Joukov, O Saksela, A Orpana, RF Pettersson, K Alitalo, U Eriksson. Vascular endothelial growth factor B, a novel growth factor for endothelial cells. Proc Natl Acad Sci USA 93:2576–2581, 1996.
117. BI Terman, L Khandke, M Doughervermazan, D Maglione, NJ Lassam, D Gospodarowicz, MG Persico, P Bohlen, M Eisinger. VEGF receptor subtypes KDR and Flt1 show different sensitivities to heparin and placenta growth factor. Growth Factors 11:187–195, 1994.
118. M Achen, J Gad, S Stacker, A Wilks. Placenta growth factor and vascular endothelial growth factor are co-expressed during early embryonic development. Growth Factors 15:69–80, 1997.
119. G Breier, U Albrecht, S Sterrer, W Risau. Expression of vascular endothelial growth factor during embryonic angiogenesis and endothelial cell differentiation. Development 114:521–532, 1992.
120. F Larcher, R Murillas, M Bolontrade, C Conti, J Jorcano. VEGF/VPF overexpression in skin of transgenic mice induces angiogenesis, vascular hyperpermeability and accelerated tumor development. Oncogene 17:303–311, 1998.
121. M Detmar, L Brown, M Schon, B Elicker, P Velasco, L Richard, D Fukumura, W Monsky, K Claffey, R Jain. Increased microvascular density and enhanced leukocyte rolling and adhesion in the skin of VEGF transgenic mice. J Invest Dermatol 111:1–6, 1998.

122. X Zeng, S Wert, R Federici, K Peters, J Whitsett. VEGF enhances pulmonary vasculogenesis and disrupts lung morphogenesis in vivo. Dev Dyn 211:215–227, 1998.
123. WG Roberts, GE Palade. Increased microvascular permeability and endothelial fenestration induced by vascular endothelial growth factor. J Cell Sci 108:2369–2379, 1995.
124. WG Roberts, GE Palade. Neovasculature induced by vascular endothelial growth factor is fenestrated. Cancer Res 57:765–772, 1997.
125. S Esser, K Wolburg, H Wolburg, G Breier, T Kurzchalia, W Risau. Vascular endothelial growth factor induces endothelial fenestrations in vitro. J Cell Biol 140:947–959, 1998.
126. DJ Dumont, GH Fong, MC Puri, G Gradwohl, K Alitalo, ML Breitman. Vascularization of the mouse embryo—a Study Of Flk-1, Tek, Tie, and vascular endothelial growth factor expression during development. Dev Dyn 203:80–92, 1995.
127. A Eichmann, C Corbel, V Nataf, P Vaigot, C Breant, N LeDouarin. Ligand-dependent development of the endothelial and hemopoietic lineages from embryonic mesodermal cells expressing vascular endothelial growth factor receptor 2. Proc Natl Acad Sci USA 94:5141–5146, 1997.
128. FH Wilt. Erythropoesis in the chick embryo: the role of endoderm. Science 147:1588–1590, 1964.
129. M Bielinska, N Narita, M Heikinheimo, SB Porter, DB Wilson. Erythropoiesis and vasculogenesis in embryoid bodies lacking visceral yolk sac endoderm. Blood 88:3720–3730, 1996.
130. Y Miura, FH Wilt. Tissue interaction and the formation of the first erythroblasts of the chick embryo. Dev Biol 19:201–211, 1969.
131. J Augustine. Influence of the entoderm on mesodermal expansion in the area vasculosa of the chick. J Embryol Exp Morphol 65:89–103, 1981.
132. J Kessel, BC Fabian. The pluripotency of the extraembryonic mesodermal cells of the early chick blastoderm: effects of the AP and AOV environments. Dev Biol 116:319–327, 1986.
133. I Flamme. Is extraembryonic angiogenesis in the chick embryo controlled by the endoderm? A morphology study. Anat Embryol 180:259–272, 1989.
134. RN Feinberg, MA Repo, JJ Saunders. Ectodermal control of the avascular zone of the peripheral mesoderm in the chick embryo. J Exp Zool 226:391–398, 1983.
135. J Korhonen, A Polvi, J Partanen, K Alitalo. The mouse tie receptor tyrosine kinase gene: expression during embryonic angiogenesis. Oncogene 9:395–403, 1994.
136. T Sato, Y Qin, C Kozak, K Audus. Tie-1 and tie-2 define another class of putative receptor tyrosine kinase genes expressed in early embryonic vascular system. Proc Natl Acad Sci USA 15:9355–9358, 1993.
137. H Schnürch, W Risau. Expression of tie-2, a member of a novel family of receptor tyrosine kinases, in the endothelial cell lineage. Development 119:957–968, 1993.
138. DJ Dumont, TP Yamaguchi, RA Conlon, J Rossant, ML Breitman. tek, a novel tyrosine kinase gene located on mouse chromosome 4, is expressed in endothelial cells and their presumptive precursors. Oncogene 7:1471–1480, 1992.
139. M Lyons, B Bell, D Stainier, K Peters. Isolation of the zebrafish homologues for the tie-1 and tie-2 endothelium-specific receptor tyrosine kinases. Dev Dyn 212:133–140, 1998.
140. A Sato, A Iwama, N Takakura, H Nishio, G Yancopoulos, T Suda. Characterization of TEK receptor tyrosine kinase and its ligands, angiopoietins, in human hematopoietic progenitor cells. Int Immunol 10:1217–1227, 1998.
141. TN Sato, Y Tozawa, U Deutsch, K Wolburgbuchholz, Y Fujiwara, M Gendronmaguire, T Gridley, H Wolburg, W Risau, Y Qin. Distinct roles of the receptor tyrosine kinases Tie-1 and Tie-2 in blood vessel formation. Nature 376:70–74, 1995.
142. D Dumont, G Gradwohl, G Fong, M Puri, M Gertsenstein A Auerbach, M Breitman. Dominant-negative and targeted null mutations in the endothelial receptor tyrosine kinase, tek, reveal a critical role in vasculogenesis of the embryo. Genes Dev 8:1897–1909, 1994.
143. M Puri, J Rossant, K Alitalo, A Bernstein, J Partanen. The receptor tyrosine kinase TIE is

required for integrity and survival of vascular endothelial cells. EMBO J 14:5884–5891, 1995.
144. S Davis, T Aldrich, P Jones, A Acheson, D Compton, V Jain, T Ryan, J Bruno, C Radziejewski, P Maisonpierre, G Yancopoulos. Isolation of angiopoietin-1, a ligand for the TIE2 receptor, by secretion-trap expression cloning. Cell 87:1161–1169, 1996.
145. C Suri, PF Jones, S Patan, S Bartunkova, PC Maisonpierre, S Davis, TN Sato, GD Yancopoulos. Requisite role of angiopoietin-1, a ligand for the tie2 receptor, during early embryonic angiogenesis. Cell 87:1171–1180, 1996.
146. V Nehls, D Drenckhahn. A novel, microcarrier-based in vitro assay for rapid and reliable quantification of three-dimensional cell migration and angiogenesis. Microvasc Res 50:311–322, 1995.
147. T Koblizek, C Weiss, G Yancopoulos, U Deutsch, W Risau. Angiopoietin-1 induces sprouting angiogenesis in vitro. Curr Biol 8:529–532, 1998.
148. P Maisonpierre, C Suri, P Jones, S Bartunkova, S Wiegand, C Radziejewski, D Compton, J McClain, T Aldrich, N Papadopoulos, T Daly, S Davis, T Sato, G Yancopoulos. Angiopoietin-2, a natural antagonist for Tie2 that disrupts in vivo angiogenesis. Science 277:55–60, 1997.
149. H Wang, Z Chen, D Anderson. Molecular distinction and angiogenic interaction between embryonic arteries and veins revealed by ephrin-B2 and its receptor Eph-B4. Cell 93:741–753, 1998.
150. G Yancopoulos, M Klagsbrun, J Folkman. Vasculogenesis, angiogenesis, and growth factors: ephrins enter the fray at the border. Cell 29:661–664, 1998.
151. J Flanagan, P Vanderhaeghen. The ephrins and Eph receptors in neural development. Annu Rev Neurosci 21:309–345, 1998.
152. S Holland, N Gale, G Mbamalu, G Yancopoulos, M Henkemeyer, T Pawson. Bidirectional signalling through the EPH-family receptor Nuk and its transmembrane ligands. Nature 383: 722–725, 1996.
153. E Stein, A Lane, D Cerretti, H Schoecklmann, A Schroff, R Van Etten, T Daniel. Eph receptors discriminate specific ligand oligomers to determine alternative signaling complexes, attachment, and assembly responses. Genes Dev 12:667–678, 1998.
154. A Pandey, H Shao, R Marks, P Polverini, V Dixit. Role of B61, the ligand for the Eck receptor tyrosine kinase, in TNF-alpha-induced angiogenesis. Science 268:567–569, 1995.
155. P Schmid, D Cox, G Bilbe, R Maier, G McMaster. Differential expression of TGF beta 1, beta 2 and beta 3 genes during mouse embryogenesis. Development 111:117–130, 1991.
156. M Dickson, J Martin, F Cousins, A Kulkarni, S Karlsson, R Akhurst. Defective haematopoiesis and vasculogenesis in transforming growth factor-beta 1 knock out mice. Development 121:1845–1854, 1995.
157. M Bonyadi, S Rusholme, F Cousins, H Su, C Biron, M Farrall, R Akhurst. Mapping of a major genetic modifier of embryonic lethality in TGF beta 1 knockout mice. Nature Genet 15:207–211, 1997.
158. M Oshima, H Oshima, M Taketo. TGF-beta receptor type II deficiency results in defects of yolk sac hematopoiesis and vasculogenesis. Dev Biol 179:297–302, 1996.
159. B Roelen, M van Rooijen, C Mummery. Expression of ALK-1, a type 1 serine/threonine kinase receptor, coincides with sites of vasculogenesis and angiogenesis in early mouse development. Dev Dyn 209:418–430, 1997.
160. J Folkman, P D'Amore. Blood vessel formation: what is its molecular basis? Cell 87:1153–1155, 1996.
161. K Hirschi, S Rohovsky, P D'Amore. PDGF, TGF-beta, and heterotypic cell-cell interactions mediate endothelial cell-induced recruitment of 10T1/2 cells and their differentiation to a smooth muscle fate. J Cell Biol 141:805–814, 1998.
162. P Lindahl, B Johansson, P Leveen, C Betsholtz. Pericyte loss and microaneurysm formation in PDGF-B-deficient mice. Science 277:242–245, 1997.

163. M de Ruiter, R Poelmann, L van Iperen, A Gittenberger-de Groot. The early development of the tunica media in the vascular system of rat embryos. Anat Embryol 181:341–349, 1990.
164. Y Takahashi, T Imanaka, T T. Spatial and temporal pattern of smooth muscle cell differentiation during development of the vascular system in the mouse embryo. Anat Embryol 194: 515–526, 1996.
165. D Johnson, J Berg, M Baldwin, C Gallione, I Marondel, S Yoon, T Stenzel, M Speer, M Pericak-Vance, A Diamond, A Guttmacher, C Jackson, L Attisano, R Kucherlapati, M Porteous, D Marchuk. Mutations in the activin receptor-like kinase 1 gene in hereditary haemorrhagic telangiectasia type 2. Nature Genet 13:189–195, 1996.
166. K McAllister, K Grogg, D Johnson, C Gallione, M Baldwin, C Jackson, E Helmbold, D Markel, W McKinnon, Jea Murrell. Endoglin, a TGF-beta binding protein of endothelial cells, is the gene for hereditary haemorrhagic telangiectasia type 1. Nature Genet 8:345–351, 1994.
167. A Anagnostou, E Lee, N Kessimian, R Levinson, M Steiner. Erythropoietin has a mitogenic and positive chemotactic effect on endothelial cells. Proc Natl Acad Sci USA 87:5978–5982, 1990.
168. A Anagnostou, Z Liu, M Steiner, K Chin, E Lee, N Kessimian, C Noguchi. Erythropoietin receptor mRNA expression in human endothelial cells. Proc Natl Acad Sci USA 91:3974–3978, 1994.
169. H Wu, X Liu, R Jaenisch, H Lodish. Generation of committed erythroid BFU-E and CFU-E progenitors does not require erythropoietin or the erythropoietin receptor. Cell 83:59–67, 1995.
170. P Carmeliet, N Mackman, L Moons, T Luther, P Gressens, I van Vlaenderen, H Demunck, M Kasper, G Breier, P Evrard, M Muller, W Risau, T Edgington, D Collen. Role of tissue factor in embryonic blood vessel development. Nature 383:73–75, 1996.
171. E Rosen, J Chan, E Idusogie, F Clotman, G Vlasuk, T Luther, L Jalbert, S Albrecht, L Zhong, A Lissens, L Schoonjans, L Moons, D Collen, F Castellino, P Carmeliet. Mice lacking factor VII develop normally but suffer fatal perinatal bleeding. Nature 390:290–294, 1997.
172. K Nakagawa, Y Zhang, H Tsuji, M Yoshizumi, T Kasahara, H Nishimura, P Nawroth, M Nakagawa. The angiogenic effect of tissue factor on tumors and wounds. Semin Thromb Hemost 24:207–210, 1998.
173. T Luther, C Flossel, N Mackman, A Bierhaus, M Kasper, S Albrecht, E Sage, L Iruela-Arispe, H Grossmann, A Strohlein, Y Zhang, P Nawroth, P Carmeliet, D Loskutoff, M Muller. Tissue factor expression during human and mouse development. Am J Pathol 149:101–113, 1996.
174. K Tachibana, S Hirota, H Iizasa, H Yoshida, K Kawabata, Y Kataoka, Y Kitamura, K Matsushima, N Yoshida, S Nishikawa, T Kishimoto, T Nagasawa. The chemokine receptor CXCR4 is essential for vascularization of the gastrointestinal tract. Nature 393:591–594, 1998.
175. Y Zou, A Kottmann, M Kuroda, I Taniuchi, D Littman. Function of the chemokine receptor CXCR4 in haematopoiesis and in cerebellar development. Nature 393:595–599, 1998.
176. B Moore, M Keane, C Addison, D Arenberg, R Strieter. CXC chemokine modulation of angiogenesis: the importance of balance between angiogenic and angiostatic members of the family. J Invest Med 46:113–120, 1998.
177. M O'Reilly, L Holmgren, Y Shing, C Chen, R Rosenthal, M Moses, W Lane, Y Cao, E Sage, J Folkman. Angiostatin: a novel angiogenesis inhibitor that mediates the suppression of metastases by a Lewis lung carcinoma. Cell 79:315–328, 1994.
178. M O'Reilly, T Boehm, Y Shing, N Fukai, G Vasios, W Lane, E Flynn, J Birkhead, B Olsen, J Folkman. Endostatin: an endogenous inhibitor of angiogenesis and tumor growth. Cell 88: 277–285, 1997.
179. D Stainier, B Weinstein, H Detrich, L Zon, M Fishman. Cloche, an early acting zebrafish gene, is required by both the endothelial and hematopoietic lineages. Development 121: 3141–3150, 1995.
180. W Liao, B Bisgrove, H Sawyer, B Hug, B Bell, K Peters, D Grunwald, D Stainier. The

zebrafish gene cloche acts upstream of a flk-1 homologue to regulate endothelial cell differentiation. Development 124:381–389, 1997.

181. E Liao, B Paw, A Oates, S Pratt, J Postlethwait, L Zon. SCL/Tal-1 transcription factor acts downstream of cloche to specify hematopoietic and vascular progenitors in zebrafish. Genes Dev 12:621–626, 1998.
182. M Gering, AR Rodaway, B Gottgens, RK Patient, AR Green. The SCL gene specifies haemangioblast development from early mesoderm. EMBO J 17:4029–4045, 1998.
183. CG Begley, PD Aplan, MP Davey, K Nakahara, K Tchorz, J Kurtzberg, MS Hershfield, BF Haynes, DI Cohen, TA Waldmann, et al. Chromosomal translocation in a human leukemic stem-cell line disrupts the T-cell antigen receptor delta-chain diversity region and results in a previously unreported fusion transcript. Proc Natl Acad Sci USA 86:2031–2035, 1989.
184. A Kallianpur, J Jordan, S Brandt. The SCL/TAL-1 gene is expressed in progenitors of both the hematopoietic and vascular systems during embryogenesis. Blood 1200–1208, 1994.
185. C Drake, S Brandt, T Trusk, C Little. TAL1/SCL is expressed in endothelial progenitor cells/angioblasts and defines a dorsal-to-ventral gradient of vasculogenesis. Dev Biol 192: 17–30, 1997.
186. C Porcher, W Swat, K Rockwell, Y Fujiwara, F Alt, S Orkin. The T cell leukemia oncoprotein SCL/tal-1 is essential for development of all hematopoietic lineages. Cell 86:47–57, 1996.
187. L Robb, N Elwood, A Elefanty, F Kontgen, R Li, L Barnett, C Begley. The scl gene product is required for the generation of all hematopoietic lineages in the adult mouse. EMBO J 15: 4123–4129, 1996.
188. R Shivdasani, E Mayer, S Orkin. Absence of blood formation in mice lacking the T-cell leukaemia oncoprotein tal-1/SCL. Nature 373:432–434, 1995.
189. J Visvader, Y Fujiwara, S Orkin. Unsuspected role for the T-cell leukemia protein SCL/tal-1 in vascular development. Genes Dev 12:473–479, 1998.
190. A Kappel, V Rönicke, A Damert, I Flamme, W Risau, G Breier. Identification of VEGF receptor-2 (Flk-1) promoter/enhancer sequences driving endothelium-specific transcription in transgenic mice. Blood 93(12), 1999. In press.
191. A Elefanty, L Robb, R Birner, C Begley. Hematopoietic-specific genes are not induced during in vitro differentiation of scl-null embryonic stem cells. Blood 90:1435–1447, 1997.
192. L Pardanaud, F Dieterlen-Lievre. Expression of C-ETS1 in early chick embryo mesoderm: relationship to the hemangioblastic lineage. Cell Adhesion Com 1:151–160, 1993.
193. D Meyer, M Durliat, F Senan, M Wolff, M Andre, J Hourdry, P Remy. Ets-1 and Ets-2 proto-oncogenes exhibit differential and restricted expression patterns during Xenopus laevis oogenesis and embryogenesis. Int J Dev Biol 41:607–620, 1997.
194. C Iwasaka, K Tanaka, M Abe, Y Sato. Ets-1 regulates angiogenesis by inducing the expression of urokinase-type plasminogen activator and matrix metalloproteinase-1 and the migration of vascular endothelial cells. J Cell Physiol 169:522–531, 1996.
195. T Watabe, K Yoshida, M Shindoh, M Kaya, K Fujikawa, H Sato, M Seiki, S Ishii, K Fujinaga. The Ets-1 and Ets-2 transcription factors activate the promoters for invasion-associated urokinase and collagenase genes in response to epidermal growth factor. Int J Cancer 77:128–137, 1998.
196. J Schwachtgen, N Janel, L Barek, M Duterque-Coquillaud, J Ghysdael, D Meyer, D Kerbiriou-Nabias. Ets transcription factors bind and transactivate the core promoter of the von Willebrand factor gene. Oncogene 15:3091–3102, 1997.
197. K Wakiya, A Begue, D Stehelin, M Shibuya. A cAMP response element and an Ets motif are involved in the transcriptional regulation of flt-1 tyrosine kinase (vascular endothelial growth factor receptor 1) gene. J Biol Chem 271:30823–30828, 1996.
198. T Schlaeger, S Bartunkova, J Lawitts, G Teichmann, W Risau, U Deutsch, T Sato. Uniform vascular-endothelial-cell-specific gene expression in both embryonic and adult transgenic mice. Proc Natl Acad Sci USA 94:3058–3063, 1997.
199. N Wernert, MB Raes, P Lassalle, MP Dehouck, B Gosselin, B Vandenbunder, D Stehelin.

c-ets1 proto-oncogene is a transcription factor expressed in endothelial cells during tumor vascularization and other forms of angiogenesis in humans. Am J Pathol 140:119–127, 1992.
200. N Muthusamy, K Barton, JM Leiden. Defective activation and survival of T cells lacking the Ets-1 transcription factor. Nature 377:639–342, 1995.
201. JC Bories, DM Willerford, D Grevin, L Davidson, A Camus, P Martin, D Stehelin, FW Alt. Increased T-cell apoptosis and terminal B-cell differentiation induced by inactivation of the Ets-1 proto-oncogene. Nature 377:635–638, 1995.
202. H Yamamoto, M Flannery, S Kupriyanov, J Pearce, S McKercher, G Henkel, R Maki, Z Werb, R Oshima. Defective trophoblast function in mice with a targeted mutation of Ets2. Genes Dev 12:1315–1326, 1998.
203. LC Wang, F Kuo, Y Fujiwara, DG Gilliland, TR Golub, SH Orkin. Yolk sac angiogenic defect and intra-embryonic apoptosis in mice lacking the Ets-related factor TEL. EMBO J 16:4374–4383, 1997.
204. E Ikeda, MG Achen, G Brier, W Risau. Hypoxia-induced transcriptional activation and increased mRNA stability of vascular endothelial growth factor in C6 glioma cells. J Biol Chem 270:19761–19766, 1995.
205. GL Wang, BH Jiang, EA Rue, GL Semenza. Hypoxia inducible factor 1 is a basic helix loop helix PAS heterodimer regulated by cellular O2 tension. Proc Natl Acad Sci USA 92:5510–5514, 1995.
206. GL Semenza. Transcriptional Regulation by hypoxia-inducible factor 1—molecular mechanisms of oxygen homeostasis. Trends Cardiovasc Med 6:151–157, 1996.
207. RH Wenger, M Gassmann. Oxygen(es) and the hypoxia-inducible factor-1. Biol Chem 378: 609–616, 1997.
208. JA Forsythe, B Jiang, NV Iyer, F Agani, SW Leung, RD Koos, GL Semenza. Activation of vascular endothelial growth factor gene transcription by hypoxia-inducible factor 1. Mol Cell Biol 16:4604–4613, 1996.
209. HE Ryan, J Lo, RS Johnson. HIF-1 alpha is required for solid tumor formation and embryonic vascularization. EMBO J 17:3005–3015, 1998.
210. NV Iyer, LE Kotch, F Agani, SW Leung, E Laughner, RH Wenger, M Gassmann, JD Gearhart, AM Lawler, AY Yu, GL Semenza. Cellular and developmental control of O2 homeostasis by hypoxia-inducible factor 1 alpha. Genes Dev 12:149–162, 1998.
211. KR Kozak, B Abbott, O Hankinson. ARNT-deficient mice and placental differentiation. Dev Biol 191:297–305, 1997.
212. E Maltepe, JV Schmidt, D Baunoch, CA Bradfield, MC Simon. Abnormal angiogenesis and responses to glucose and oxygen deprivation in mice lacking the protein ARNT. Nature 386: 403–407, 1997.
213. H Tian, SL McKnight, DW Russel. Endothelial PAS domain protein 1 (EPAS1), a transcription factor selectively expressed in endothelial cells. Genes Dev 11:72–82, 1997.
214. I Flamme, T Fröhlich, M von Reutern, A Kappel, A Damert, W Risau. HRF, a putative basic helix-loop-helix-PAS-domain transcription factor is closely related to Hypoxia-inducible factor-1a and developmentally expressed in blood vessels. Mech Dev 63:51–60, 1997.
215. M Ema, S Taya, N Yokotani, K Sogawa, Y Matsuda, Y Fujii-Kuriyama. A novel bHLH-PAS factor with close sequence similarity to hypoxia-inducible factor 1a regulates the VEGF expression and is potentially involved in lung and vascular development. Proc Natl Acad Sci USA 94:4273–4278, 1997.
216. JB Hogenesch, WK Chan, VH Jackiw, RC Brown, YZ Gu, M Pray-Grant, GH Perdew, CA Bradfield. Characterization of a subset of the basic-helix-loop-helix-PAS superfamily that interacts with components of the dioxin signaling pathway. J Biol Chem 272:8581–8593, 1997.
217. S Jain, E Maltepe, MM Lu, C Simon, CA Bradfield. Expression of ARNT, ARNT2, HIF1 alpha, HIF2 alpha and Ah receptor mRNAs in the developing mouse. Mech Dev 73:117–123, 1998.

218. DD Isaac, DJ Andrew. Tubulogenesis in drosophila—a requirement for the trachealess gene product. Genes Dev 10:103–117, 1996.
219. R Wilk, I Weizman, BZ Shilo. Trachealess encodes a bHLH-PAS protein that is an inducer of tracheal cell fates in drosophila. Genes Dev 10:93–102, 1996.
220. G Manning, M Krasnow. Development of the Drosophila tracheal system. In: The Development of Drosophila Melanogaster. Cold Spring Harbor, NY: Cold Spring Harbour Laboratory Press, 1993.
221. E Zelzer, P Wappner, BZ Shilo. The PAS domain confers target gene specificity of Drosophila bHLH/PAS proteins. Genes Dev 11:2079–2089, 1997.
222. VW Keng, KE Fujimori, Z Myint, N Tamamaki, Y Nojyo, T Noguchi. Expression of Hex mRNA in early murine postimplantation embryo development. FEBS Lett 426:183–186, 1998.
223. PQ Thomas, A Brown, RS Beddington. Hex: a homeobox gene revealing peri-implantation asymmetry in the mouse embryo and an early transient marker of endothelial cell precursors. Development 125:85–94, 1998.
224. DB Sparrow, S Kotecha, N Towers, TJ Mohun. Xenopus eHAND: a marker for the developing cardiovascular system of the embryo that is regulated by bone morphogenetic proteins. Mech Dev 71:151–163, 1998.
225. SV Subramanian, B Nadal-Ginard. Early expression of the different isoforms of the myocyte enhancer factor-2 (MEF2) protein in myogenic as well as non-myogenic cell lineages during mouse embryogenesis. Mech Dev 57:103–112, 1996.
226. CS Newman, F Chia, PA Krieg. The XHex homeobox gene is expressed during development of the vascular endothelium: overexpression leads to an increase in vascular endothelial cell number. Mech Dev 66:83–93, 1997.
227. J Folkman, C Haudenschild. Angiogenesis in vitro. Nature 288:551–556, 1980.
228. T Bär, F-H Güldner, JR Wolff. ''Seamless'' endothelial cells of blood capillaries. Cell Tissue Res 235:99–106, 1984.
229. R Hirakow, T Hiruma. TEM-studies on development and canalization of the dorsal aorta in the chick embryo. Anat Embryol 166:307–315, 1983.
230. R Hirakow, T Hiruma. Scanning electron microscopic study on the development of primitive blood vessels in chick embryos at the early somite-stage. Anat Embryol 163:299–306, 1981.
231. R Lanot. Formation of the early vascular network in chick embryo: microscopical aspects. Arch Biol Liege 91:423–438, 1980.
232. S Eaton, K Simons. Apical, basal, and lateral cues for epithelial polarization. Cell 82:5–8, 1995.
233. D Grant, H Kleinman. Regulation of capillary formation by laminin and other components of the extracellular matrix. Experientia 79:317–333, 1997.
234. R Hynes, B Bader. Targeted mutations in integrins and their ligands: their implications for vascular biology. Thromb Haemost 78:83–87, 1997.
235. S Strömblad, DA Cheresh. Cell adhesion and angiogenesis. Trends in cell biology 6:462–467, 1996.
236. CJ Drake, LA Davis, CD Little. Antibodies to beta 1-integrins cause alterations of aortic vasculogenesis, in vivo. Dev Dyn 193:83–91, 1992.
237. C Drake, D Cheresh, C Little. An antagonist of integrin alpha v beta 3 prevents maturation of blood vessels during embryonic neovascularization. J Cell Sci 108:2655–2661, 1995.
238. W Rilsau, V Lemmon. Changes in the vascular extracellular matrix during embryonic vasculogenesis and angiogenesis. Dev Biol 125:441–450, 1988.
239. EL George, EN Georges Labouesse, RS Patel King, H Rayburn, RO Hynes. Defects in mesoderm, neural tube and vascular development in mouse embryos lacking fibronectin. Development 119:1079–1091, 1993.
240. KL Goh, JT Yang, RO Hynes. Mesodermal defects and cranial neural crest apoptosis in alpha5 integrin-null embryos. Development 124:4309–4319, 1997.

241. JT Yang, RO Hynes. Fibronectin receptor functions in embryonic cells deficient in alpha 5 beta 1 integrin can be replaced by alpha V integrins. Mol Biol Cell 7:1737–1748, 1996.
242. L Sorokin, F Pausch, M Durbeej, P Ekblom. Differential expression of five laminin alpha (1-5) chains in developing and adult mouse kidney. Dev Dyn 210:446–462, 1997.
243. R Fässler, M Meyer. Consequences of lack of beta 1 integrin gene expression in mice. Genes Dev 9:1896–1908, 1995.
244. W Bloch, E Forsberg, S Lentini, C Brakebusch, K Martin, H Krell, U Weidle, K Addicks, R Fassler. Beta 1 integrin is essential for teratoma growth and angiogenesis. J Cell Biol 139: 265–278, 1997.
245. PC Brooks, RA Clark, DA Cheresh. Requirement of vascular integrin alpha v beta 3 for angiogenesis. Science 264:569–571, 1994.
246. T Bach, C Barsigian, D Chalupowicz, D Busler, C Yaen, D Grant, J Martinez. VE-Cadherin mediates endothelial cell capillary tube formation in fibrin and collagen gels. Exp Cell Res 238:324–334, 1998.
247. M Lampugnani, M Corada, L Caveda, F Breviario, O Ayalon, B Geiger, E Dejana. The molecular organization of endothelial cell to cell junctions: differential association of plakoglobin, beta-catenin, and alpha-catenin with vascular endothelial cadherin (VE-cadherin). J Cell Biol 129:203–217, 1995.
248. N Matsuyoshi, K Toda, Y Horiguchi, T Tanaka, S Nakagawa, M Takeichi, S Imamura. In vivo evidence of the critical role of cadherin-5 in murine vascular integrity. Proc Assoc Am Physicians 109:362–371, 1997.
249. G Breier, F Breviario, L Caveda, R Berthier, H Schnürch, U Gotsch, D Vestweber, W Risau, E Dejana. Molecular cloning and expression of murine vascular endothelial-cadherin in early stage development of cardiovascular system. Blood 87:630–641, 1996.
250. D Vittet, T Buchou, A Schweitzer, E Dejana, P Huber. Targeted null-mutation in the vascular endothelial-cadherin gene impairs the organization of vascular-like structures in embryoid bodies. Proc Natl Acad Sci USA 94:6273–6278, 1997.
251. C Kevil, D Payne, E Mire, J Alexander. Vascular permeability factor/vascular endothelial cell growth factor-mediated permeability occurs through disorganization of endothelial junctional proteins. J Biol Chem 273:15099–15103, 1998.
252. S Esser, M Lampugnani, M Corada, E Dejana, W Risau. Vascular endothelial growth factor induces VE-cadherin tyrosine phosphorylation in endothelial cells. J Cell Sci 111:1853–1865, 1998.
253. J Cheng, S Baumhueter, G Cacalano, K Carver-Moore, H Thibodeaux, R Thomas, H Broxmeyer, S Cooper, N Hague, M Moore, L Lasky. Hematopoietic defects in mice lacking the sialomucin CD34. Blood 87:479–490, 1996.
254. FM Hansen-Smith, O Hudlicka, S Egginton. In vivo angiogenesis in adult rat skeletal muscle: early changes in capillary network architecture and ultrastructure. Cell Tissue Res 286:123–136, 1996.
255. JD Coffin, TJ Poole. Embryonic vascular development: immunohistochemical identification of the origin and subsequent morphogenesis of the major vessel primordia in quail embryos. Development 102:735–748, 1988.
256. L Sumoy, J Keasey, T Dittman, D Kimelman. A role for notochord in axial vascular development revealed by analysis of phenotype and the expression of VEGR-2 in zebrafish flh and ntl mutant embryos. Mech Dev 63:15–27, 1997.
257. O Cleaver, P Krieg. VEGF mediates angioblast migration during development of the dorsal aorta in Xenopus. Development 125:3905–3914, 1998.
258. B Hogers, M DeRuiter, A Gittenberger-de Groot, R Poelmann. Unilateral vitelline vein ligation alters intracardiac blood flow patterns and morphogenesis in the chick embryo. Circ Res 80:473–481, 1997.
259. G Levenson, O Nelsen. Experimentally induced variations in vitelline artery development and circulatory pattern in the early chick embryo. J Morphol 123:313–327, 1967.

260. N Resnick, M Gimbrone. Hemodynamic forces are complex regulators of endothelial gene expression. FASEB J 9:874–882, 1995.
261. DE Bockman, ME Redmond, ML Kirby. Alteration of early vascular development after ablation of cranial neural crest. Anat Rec 225:209–217, 1989.
262. M Bergwerff, M Verberne, M DeRuiter, R Poelmann, A Gittenberger-de Groot. Neural crest cell contribution to the developing circulatory system: implications for vascular morphology? Circ Res 82:221–231, 1998.
263. PH Burri, MR Tarek. A novel mechanism of capillary growth in the rat pulmonary microcirculation. Anat Rec 228:35–45, 1990.
264. I Flamme, M Messerli, W Risau, M Jacob, HJ Jacob. Vascular growth in the extraembryonic mesoderm of avian embryos. In: Formation and Differentiation of Early Embryonic Mesoderm. New York: Plenum Press, 1992, pp. 323–335.
265. R Hallmann, RN Feinberg, CH Latker, J Sasse, W Risau. Regression of blood vessels precedes cartilage differentiation during chick limb development. Differentiation 34:98–105, 1987.
266. CH Latker, RN Feinberg, DC Beebe. Localized vascular regression during limb morphogenesis in the chicken embryo: II. Morphological changes in the vasculature. Anat Rec 214:410–417, 1986.
267. RN Feinberg, CH Latker, DC Beebe. Localized vascular regression during limb morphogenesis in the chicken embryo: I. Spatial and temporal changes in the vascular pattern. Anat Rec 214:405–409, 1986.
268. R Jack. Regression of the hyaloid vascular system: an ultrastructural analysis, Am J Ophthalmol 74:261–272, 1972.
269. S Connolly, T Hores, L Smith, P D'Amore. Characterization of vascular development in the mouse retina. Microvasc Res 36:275–290, 1988.
270. R Lang, J Bishop. Macrophages are required for cell death and tissue remodeling in the developing mouse eye. Cell 74:453–462, 1993.
271. A Hughes. Studies on the area vasculosa of the embryo chick: I. The first differentiation of the vitellina arteria. J Anat 70:76–122, 1935.
272. Z Grodzinski. Die Entwicklung der Venen in der Keimscheibe des Hühnchens. Bull Int Acad Pol Sci Lett Serie Bll:305–322, 1935.
273. D Popoff. Die Dottersackgefäße des Huhnes. Kreidel's Verlag, Wiesbaden, 1894.
274. L Benjamin, I Hemo, E Keshet. A plasticity window for blood vessel remodelling is defined by pericyte coverage of the preformed endothelial network and is regulated by PDGF-B and VEGF. Development 125:1591–1598, 1998.
275. W Chapman. The effect of the heart-beat upon the development of the vascular system in the chick. Am J Anat 23:175–203, 1918.

2

Arteriogenesis, Collateral Blood Vessels, and Their Development

Elisabeth Deindl, Borja Fernández, Imo E. Höfer, Niels van Royen, Dimitri Scholz, and Wolfgang Schaper
Max Planck Institute for Physiological and Clinical Research, Bad Nauheim, Germany

I. INTRODUCTION

Vascular occlusive diseases are the major cause of mortality worldwide–despite the tremendous progress in medical and surgical treatment. The treatment of cardiovascular diseases by procedures such as bypass surgery, stent implantation, and heart transplantation is expensive and has significantly contributed to the financial crises in health care in industrialized countries; moreover, such procedures carry some risk for the patient. It is therefore important to develop prophylactic therapeutics for patients at risk for cardiovascular disease.

The morphological reasons for these types of diseases are stenosing processes that result, in the worst case, in total occlusion of the vessel. Fatal sequelae such as cardiac infarction are well known. However, in some rare cases, patients have survived occlusions even of the main coronary artery with only minimal symptoms because of the timely development of collateral arteries from small preexisting arteriolar connections. This process, also called *arteriogenesis* (Schaper and Ito, 1996), is time-consuming (Fulton, 1965; Schaper et al., 1971; Spalteholz, 1907) and normally proceeds more slowly than thrombus formation. However, collateral artery growth—artifically induced in animals via experimental stenosis—provides a good way of investigating the mechanisms of arteriogenesis. An understanding of the mechanisms of arteriogenesis is the basic requirement for the development of efficacious drugs for the treatment of patients at risk.

II. DEVELOPMENT AND GROWTH OF COLLATERAL ARTERIES

The development and growth of vascular structures in embryonic and adult tissues is based on four different mechanisms: angiogenesis; vasculogenesis; differentiation of arteries and veins; and arteriogenesis.

Angiogenesis is the formation of new capillary vessels from a preexisting capillary bed. Single endothelial cells within a capillary tube acquire a migratory phenotype (sprouting), invade the surrounding tissue and proliferate, forming new tubular structures that expand the preexisting capillary bed. The process of angiogenesis is responsible for the normal growth of the capillary plexuses in most embryonic organs and in adult tissues under certain physiological conditions, such as hypoxia and tumor invasion.

Vasculogenesis is restricted to the embryonic life and consists in the formation of vascular structures from precursor cells, called angioblasts. Angioblasts are multipotent progenitor cells that differentiate in situ, forming a network of capillaries called the primary plexus. This primary plexus expands by angiogenesis in specific organs, giving rise to the primitive vasculature of embryonic tissues.

The formation of arteries and veins takes place during late embryonic stages and continues during the postnatal period in most organs. Endothelial cells within a capillary tube become invested with mesenchymal cells from the surrounding tissue, forming an immature media, the maturation of which gives rise to a fully developed artery or vein. The cellular and molecular mechanisms implicated in the differentiation of arteries and veins require a tightly regulated interaction between endothelial and smooth muscle cells; this is currently still poorly understood.

During the last years, a new term referring to a particular process of vascular development and growth in adult tissues has been coined: *arteriogenesis*. (Ito et al., 1997a; Takeshita et al., 1995) Arteriogenesis is defined as the in situ growth of muscular collateral arteries (arterioles that interconnect two feeding arteries) in response to a physiological stimulus caused by the occlusion of an artery. Collateral arteries that respond to the stimulus are able to expand their diameter by a factor of 20 and increase their flow capacity by a factor of 10, thus leading to a compensation up to 30% of the normal maximum flow in experimental animals (Schaper; 1979, Schaper et al., 1976b; Schaper and Ito, 1996).

In the adult organism, a deficit in the blood flow gives rise to two different mechanisms of vessel growth: angiogenesis and arteriogenesis. As already explained, both mechanisms are basically different and take place in two distinct types of vessels: capillaries and collateral arteries, respectively. The capacity to form new vessels by sprouting is restricted to capillaries and small venules (Folkman, 1995); however, only collateral arteries derived from preexisting arteriolar connections by arteriogenesis are able to provide enough blood flow to the ischemic area.

The process of arteriogenesis or collateral growth has been studied mainly in dog hearts subjected to progressive coronary artery stenosis and in rats and rabbit hindlimbs after acute femoral artery occlusion (Arras et al., 1998; Ito et al., 1997a; Paskins-Hurlburt and Hollenberg, 1992; Schaper et al., 1992). The studies carried out in the dog model suggest that arteriogenesis takes place by growth of preexisting arterioles rather than capillary sprouting, followed by neoformation of muscular arteries. This assumption is supported by experimental evidence on the speed of the process of arteriogenesis obtained from the hindlimb models, where the conductance of muscular collateral arteries rises by a factor of 6 during the first week after femoral artery occlusion (Ito et al., 1997a). However, recent studies in rats suggest that perfusion of the ischemic hindlimb may also be improved by enlargement of terminal arterioles through a mechanism probably similar to that underlying neonatal arteriolar growth (Price and Skalak, 1998).

The first or *initial* phase of collateral growth begins some minutes after occlusion of the femoral artery in the rabbit and consists of the activation of endothelial cells of responding arterioles. This activation is probably induced by the increase in shear stress forces caused by pressure differentials that occur after arterial occlusion. Activated endothelial cells acquire a rough surface and express the calcium-independent cell adhesion molecules intercellular cell adhesion molecule (ICAM) and vascular cell adhesion molecule (VCAM), which facilitate the attachment and migration of monocytes to the arterial wall. (Scholz et al., in press). Once into the arterial wall, the activated macrophages express growth factors and cytokines; these are probably the signals that trigger the whole process of arterial growth (Arras et al., 1998).

The second stage in the process of arteriogenesis is the *proliferative* phase (1 to 3 days after occlusion), which is characterized by a maximal mitotic activity of endothelial, smooth muscle cells, and fibroblast (Figs. 1 and 2). Proliferation of endothelial and smooth muscle cells in adult tissues is an uncommon event. The turnover of these cell types in normal tissues is extremely low [one mitosis every 6 months (Schaper, 1971)] and is only increased under certain pathological conditions, such as atherosclerosis and restenosis.

However, proliferation is not the only requisite for collateral growth. To enlarge their final diameter, collateral arteries must undergo a deep remodeling process. This is defined as the early disorganization (*synthetic* phase) and later reorganization (*maturation* phase) of the vessel wall.

During the *synthetic* phase (3 to 14 days), the lamina elastica interna, the main barrier to smooth muscle cell migration, is digested and fragmented by elastolytic enzymes facilitating positive remodeling (Fig. 3A). Smooth muscle cells migrate from the media to populate the subendothelial space, creating a thick intima. These smooth muscle cells show an undifferentiated or secretory phenotype with abundant organelles, loss of most of the actin filaments, loss of basement membrane, and proteolytic activity (Figs. 4 and

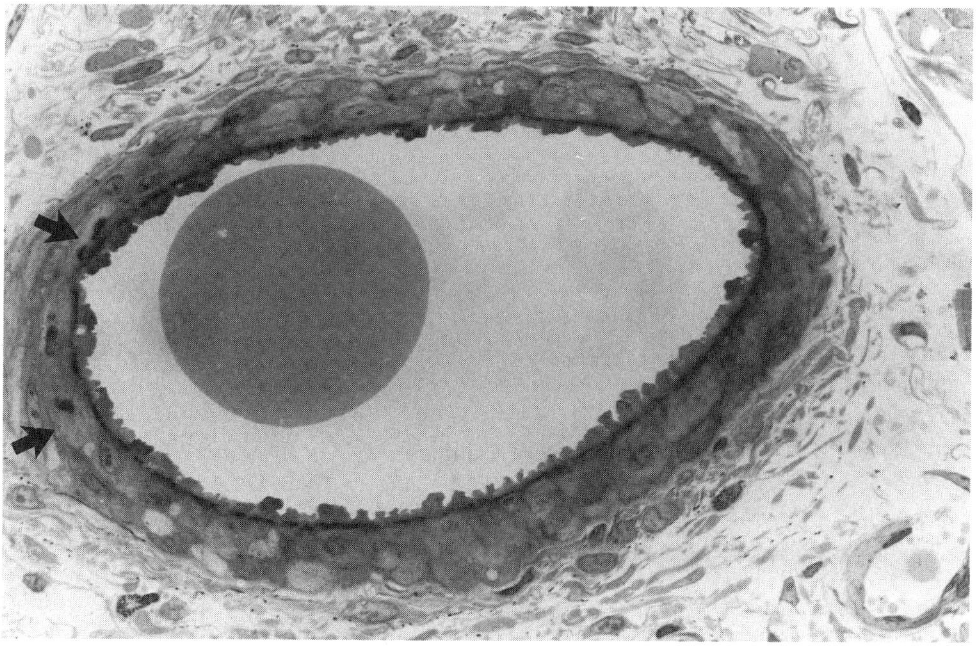

Figure 1 Mitotic activity (arrows) in a growing collateral artery 3 days after occlusion of the femoral artery. Semithin cross-section of one collateral artery. (Magn. ×472.5.)

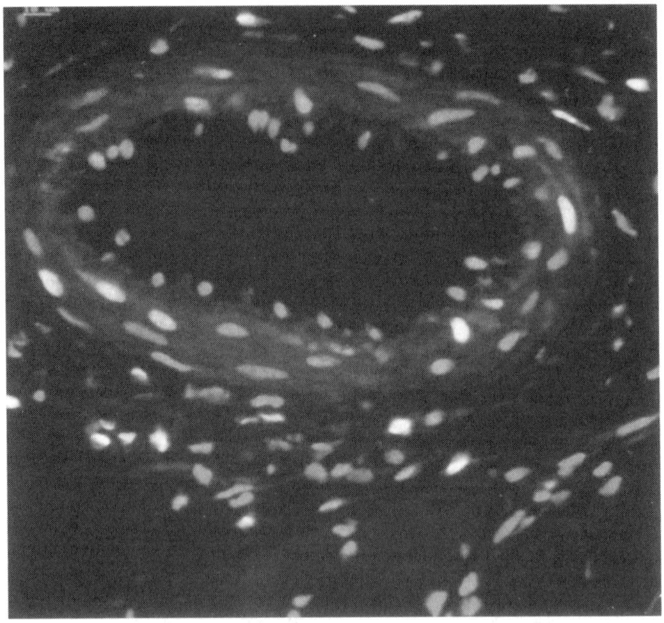

Figure 2 Cell cycle marker Ki-67 expression in the wall of the growing collateral artery 3 days after occlusion of the femoral artery. (Magn. ×340.)

5). At this phase, monocyte recruitment had ceased and adhesion molecules are no longer expressed. In some of the growing collateral arteries, the thickness of the neointimal layer increases in such a way that it virtually occludes the lumen, leading to regression of the vessel. The factors that regulate the regression or persistence of the growing collateral vessels are so far unknown, but shear forces have been postulated to play an important role (Schaper and Ito, 1996).

The last stage of collateral growth is the *maturation* phase, which begins between 14 and 21 days after occlusion. It is characterized by low levels of proliferation, migration, and proteolitic activity; restoration of the contractile phenotype of most smooth muscle cells; and neoformation of an internal elastic lamina beneath the endothelium (Fig. 3B).

During the *synthetic* and *maturation* phases of collateral growth, numerous figures of cell death and mitosis are found in the preexisting media. Proliferation is required to increase the number of vascular cells necessary to augment vessel size, while cell death in the media probably serves to eliminate superfluous smooth muscle cells, thus preventing an excessive increase in the vessel wall. Cell death of myocytes surrounding the growing collateral artery is another common event, probably important for opening the space needed for the new artery to develop.

Smooth muscle cells in the neointima of collateral arteries show a reduced amount of α-smooth muscle actin, vinculin, caldesmon, and calponin. The intermediate filament of these cells changes from desmin to vimentin during the period of migration and growth. Also the extracellular matrix of growing collaterals show a deep rearrangement, with increased collagen VI and fibronectin secretion. All these characteristics, together with the acquisition of an undifferentiated phenotype by smooth muscle cells, indicate an extensive vascular remodeling process typical of the embryonic development and differentiation of arteries. The embryonic development of arteries is a complex process, the basis of which

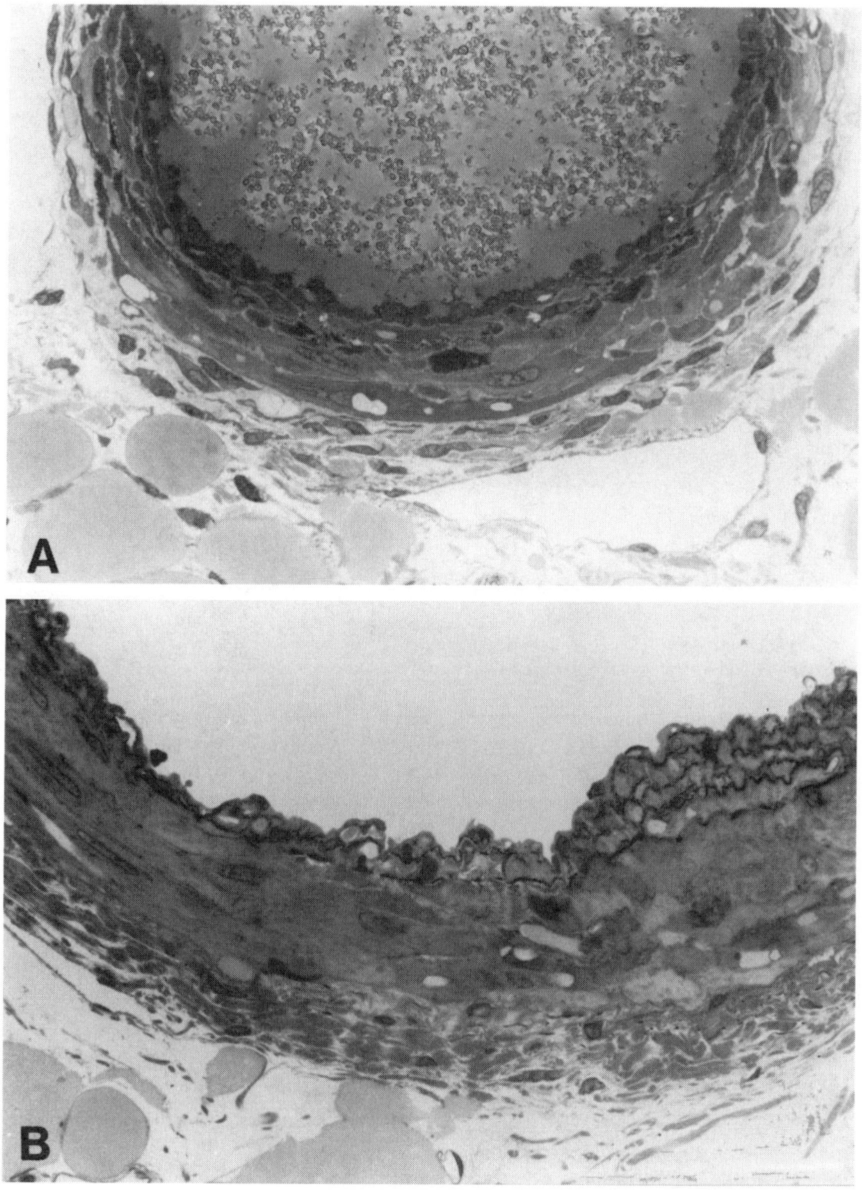

Figure 3 Neointima formation 7 days (A) and 42 days (B) after occlusion of the femoral artery. Semithin cross sections stained with toluidine blue. (Magn. ×448.)

is currently under an extensive research by vascular embryologists. Numerous homeobox genes and transcription factors implicated in the differentiation of arteries have been identified during the last years, but their particular function remains unknown. In the future, intensive research on the role of these genes in arterial differentiation, together with the establishment of a precise correlation between the processes of collateral growth and embryonic development of arteries, will improve our knowledge on collateral development and support the development of new therapies.

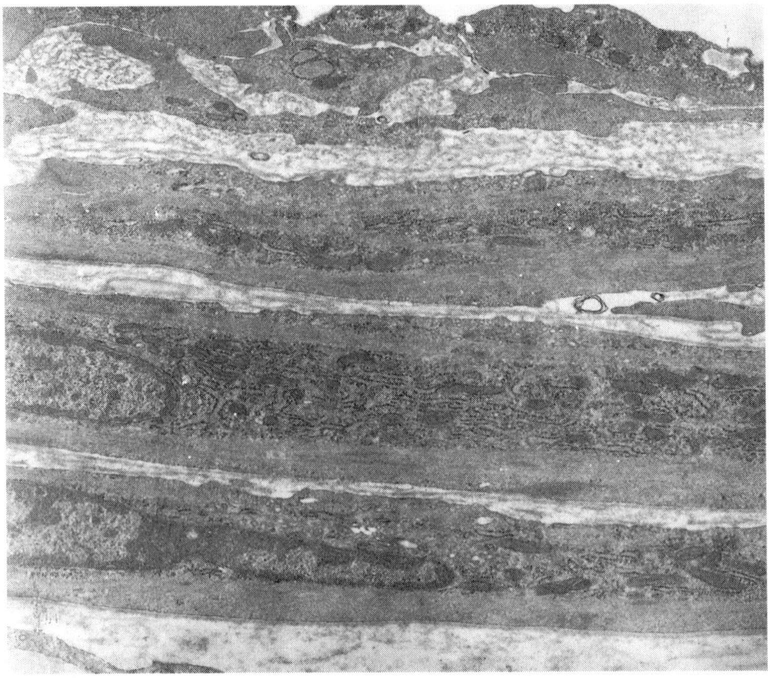

Figure 4 Ultrastructure of the collateral vessel wall 5 days after femoral artery occlusion. Smooth muscle cells contain much endoplasmic reticulum, ''synthetic'' phenotype. The lamina elastica interna is not distinguishable; the extracellular matrix is full of nonstructured debris. (Magn. ×6720.)

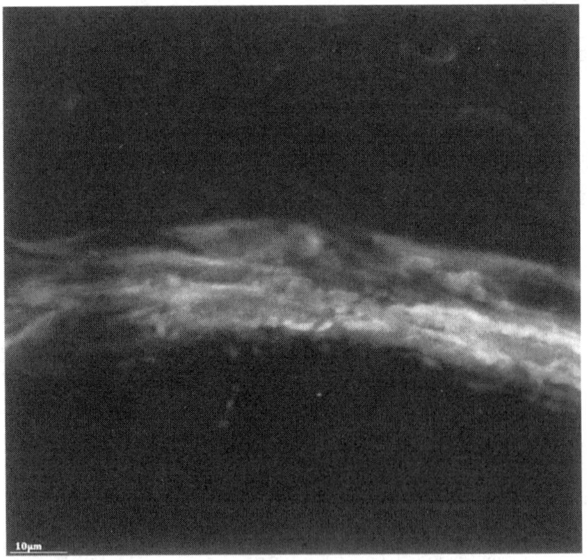

Figure 5 Matrix-metalloproteinase 3 (MMP-3) expression in the media of growing collateral artery 7 days after femoral artery occlusion. (Magn. ×1000.)

III. MOLECULAR MECHANISMS

Angiogenesis (capillary sprouting) and arteriogenesis are the two types of vessel growth existing in the postnatal organism. During the past two decades scientists have focused on the mechanisms of angiogenesis and found that this process involves the activation, migration, and proliferation of endothelial cells (D'Amore, 1987) and is regulated by certain growth factors (Folkman and Shing, 1992). The sprouting of new vessels is a tightly controlled process that occurs in the adult organism under normal physiological conditions only in the ovary and uterus throughout the menstrual cycle or placenta formation (Reynolds et al., 1992). However, neovascularization in the adult also occurs in response to injury or pathological processes such as tumor growth or diabetic retinopathy (Clark, 1993; Folkman, 1995; Kirsner and Eaglstein, 1993). Hypoxia represents an adequate stimulus for the upregulation of angiogenic factors. Vascular endothelial growth factor (VEGF) (Thomas, 1996) was shown to be upregulated under hypoxic and ischemic conditions in vitro (Shweiki et al., 1992) and in vivo (Tudor et al., 1995), both on the transcriptional and translational level (Ikeda et al., 1995); it represents an endothelial-specific mitogen that is involved in both adaptive and tumor angiogenesis. Arteriogenesis, however, occurs independent of ischemia or hypoxia (Ito et al., 1997a) and is in most models temporally and spatially dissociated from a reduced oxygen supply (Schaper and Schaper, 1993). Occlusion of an artery results in an increased flow in the adjacent normal arteriolar connections (Schaper et al., 1973), which, in turn, become wider (via remodeling or collateral artery growth) (Schaper et al., 1986). Pressure gradients developing along the arterioles increase shear stress, which may act as modeling force in the developement of arteries. (Schaper and Schaper, 1986). Collateral artery growth requires the proliferation of endothelial cells and smooth muscle cells.

We have developed an animal model (rabbit, see below) that permits the investigation of the mechanisms of arteriogenesis independent of ischemia and angiogenesis. It should therefore be possible to define the factors responsible for the growth of arterioles into true collateral arteries.

Growth factors are known to be important angiogenic factors; however, their role in arteriogenesis was unclear. We decided to start with the investigation of the role of VEGF as the most prominent angiogenic factor. The expression of growth factors and their receptors is changed very quickly in response to altered conditions. We therefore analyzed the expression of VEGF and its signaling receptors flt-1 and flk-1 during the first hours and days of collateral artery growth. However, our results did not show an increased expression either of VEGF or of its receptors (Deindl et al., 1998). We also analyzed the expression of members of the angiopoietin family, comprising the angiopoietins and the Tie receptors. Whereas the ligands for the Tie-1 receptor are unknown, for Tie-2 they are angiopoietin-1 and angiopoietin-2. Angiopoietin-1 is not a mitogen for endothelial cells but exerts a chemotactic effect. So VEGF, with its mitogenic activity, and angiopoietin-1, being involved in the elaboration of the vascular tree, act synergistically to induce sprout formation. The role of angiopoietin-2 is not completely elucidated as yet. However, one of its functions is to antagonize angiopoietin-1 by interfering with the kinase activity of Tie-2. The results we obtained for the VEGF family proved also to be true for the members of the angiopoietin family. We found no changed expression pattern for any of the corresponding mRNAs. The receptors for the angiopoietins are, like the receptors for VEGF, expressed on endothelial cells. This fact, the function of the two factors—VEGF as endothelial-specific mitogen and angiopoietin-1 with its role in the elaboration of the vascular tree—and our results on gene expression make it unlikely that one of these cytokines or both in combination are capable of inducing collateral artery growth directly.

Fibroblast growth factors (FGFs) are also involved in vessel growth. Their receptors are expressed on endothelial and smooth muscle cells. The high affinity receptors for the FGFs are a family of structurally related transmembrane tyrosine kinases (Burgess and Maciag, 1989). So far, four members of this family have been identified (Fernig and Gallagher, 1994, Partanen et al., 1992). For FGFR-1 and FGFR-2, a multitude of either cell-bound or secreted isoforms are characterized. Ligand-induced dimerization is a key event in transmembrane signaling. It results in autophosphorylation and the induction of the diverse biological responses (Schlessinger et al., 1995; Schlessinger and Ulrich, 1992). The FGFR-1 mRNA is the predominant form of the four known FGFRs expressed by proliferating smooth muscle cells (Xin et al., 1994). This strongly suggests that this is the type of the FGFRs that mediates signal transduction in proliferating smooth muscle cells. Our expression analysis showed that the mRNA of the FGFR-1 is significantly upregulated during the early phase of arteriogenesis (3 to 24 hr) (Deindl et al., 1997).

The ligands of the FGFRs are a family consisting of at least 15 heparin-binding polypeptide growth factors that mediate a wide range of biological activities (Baird and Klagsbrunn, 1991; Fernig and Gallagher, 1994; Givol and Yayon, 1992). Acidic FGF (aFGF) and basic FGF (bFGF), preferentially bound by the FGFR-1, are potent mitogens and play an important role in cell proliferation and differentiation (Basilico and Moscatelli, 1992; Burgess and Maciag; 1989, Jin et al., 1994; Johnson and Williams, 1993; Wang et al., 1994). Our results on aFGF and bFGF did not show a changed expression of these two ligands, pointing toward a receptor-bound regulation of the FGF system during collateral artery growth (Deindl et al., 1997).

To get a more comprehensive view of the molecular mechanisms of arteriogenesis, we are currently investigating the role of different genes in arteriogenesis. Via differential display reverse transcription polymerase chain reaction (DDRT-PCR), we observed over 200 differential expressed genes, including some of unknown function (Kampmann et al., 1999). At the moment we are further characterizing the role of the individual genes. An understanding of the molecular mechanisms that result in collateral formation should give us the tools to enhance and induce collateral artery growth in humans. In terms of therapies, it is important to recognize that capillary sprouting and collateral artery growth obey distinct and different mechanisms and to be aware that the application of angiogenic factors (especially when they show a specificity for endothelial cells) in patients must not result in capillary sprouting *and* collateral artery growth at once.

IV. STIMULATION OF ARTERIOGENESIS IN THE RABBIT HINDLIMB MODEL BY MONOCYTE ATTRACTION

Arguments in favor of the role of shear stress in the process of arteriogenesis are mounting, in contrast to angiogenesis which is stimulated by ischemia (Ito et al., 1996; Ito et al., 1997a). In order to study arteriogenesis, we developed a rabbit hindlimb model of femoral artery occlusion. In this model, regions of ischemia are restricted to the lower leg; the region in which arteriogenesis occurs, the upper leg, is not affected by ischemia.

In the upper leg region, shear stress is increased in preexisting arteriolar connections because of the large pressure differences created by the occlusion of the femoral artery.

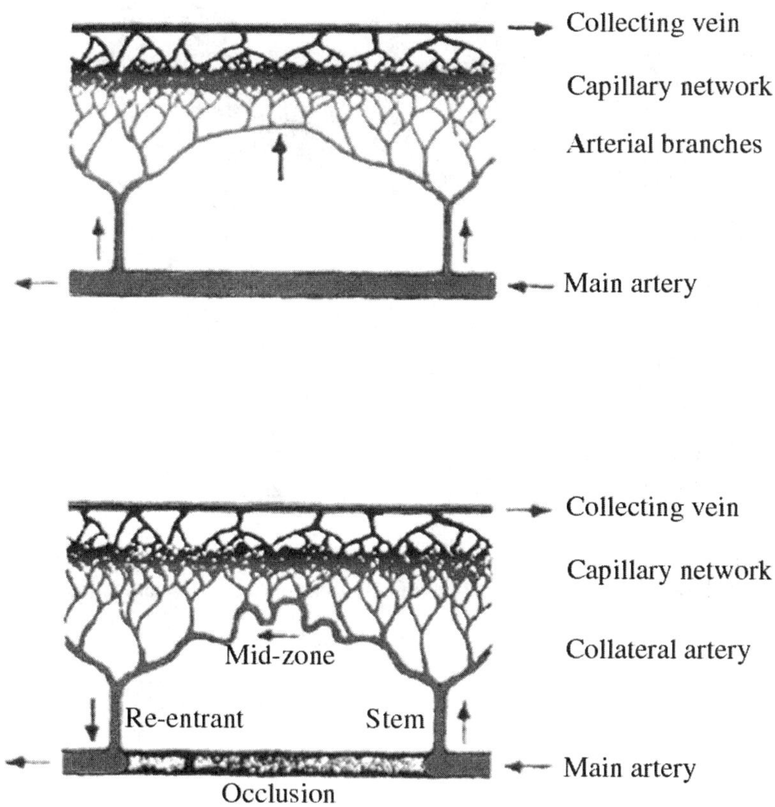

Figure 6 Collateral artery development from a preexisting arteriolar connection. The arrows indicate the direction of blood flow. (Modified from Longland, 1953.)

As described by Longland in 1953, the definition of a collateral vessel is its obligatory macroscopic appearance, with stem, midzone and reentrant areas (Fig. 6) (Longland, 1953). These preexisting arterioles originate from the arteriae femoralis circumflexa and profunda and reenter into the arteriae genualis and saphena parva. Because of the increased shear stress, endothelium is activated (Fig. 7) and adhesion molecules-like the selections and vascular cell adhesion molecules VCAM-1 and ICAM-1—are highly expressed (Davies, 1995; Gimbrone et al., 1997; Nagel et al., 1994; Resnick and Gimbrone, 1995). This leads to increased adhesion of monocytes, which, in turn, migrate through the vessel wall and differentiate to macrophages (Clinton and Libby, 1992; Schaper et al., 1976a). The macrophages produce large amounts of growth factors and chemokines [i.e., monocyte chemoattractant protein-1 (MCP-1), tumor necrosis factor-α (TNF-α), β-FGF, and granulocyte monocyte-colony stimulating factor (GM-CSF)] (Arras, Ito, 1998) and initiate the proliferation of the preexisting arteriolar connections leading to true collateral arteries, with several layers of vascular smooth muscle cells (Fig. 8) (Schaper et al., 1971; Schaper and Schaper, 1993).

Since monocytes/macrophages play such a dominant role in arteriogenesis, they are a potential therapeutic target through which this process might be influenced. MCP-1 is one of the most potent chemoattractants for circulating monocytes (Randolph and Furie,

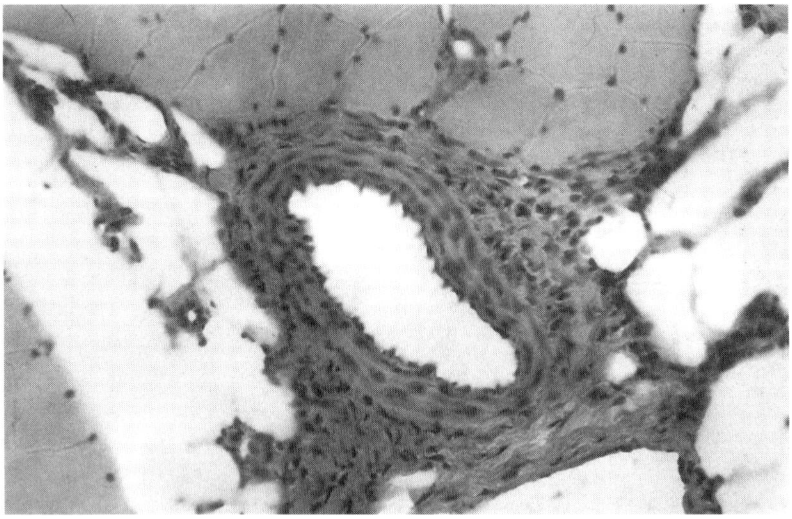

Figure 7 Hematoxylin and eosin (HE) staining of a proliferating collateral artery 7 days after femoral artery occlusion showing several layers of vascular smooth muscle cells and activation of the endothelium.

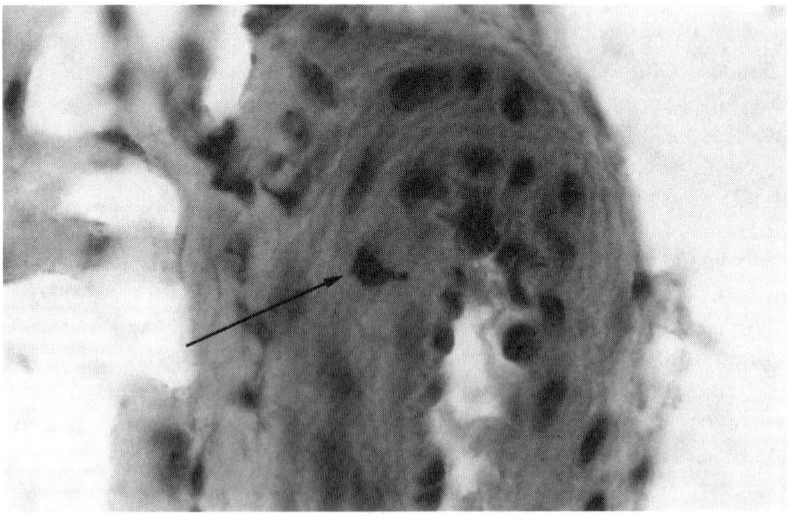

Figure 8 HE-staining of a proliferating collateral artery 3 days after femoral artery occlusion. The arrow indicates a vascular smooth muscle cell in mitosis.

1995) and is produced by shear stress–activated endothelium. After transmigration through the vessel wall of proliferating collateral arteries, macrophages themselves produce MCP-1, which, in turn, stimulates further homing of circulating monocytes to the growing vessels and their activation (MCP-1 is also known as monocyte chemoattractant and activating factor, or MCAF).

In order to study the natural course of arteriogenesis and the additional effects of intraarterial infusion of MCP-1 into the collateral circulation, angiograms were performed to detect changes in the vascular tree. In addition, we developed a model of pump-controlled in vivo perfusion to obtain quantitative functional data.

A. Radiographic Findings

Postmortem angiograms showed a typical corkscrew formation of collateral arteries spanning the occlusion site mainly in the musculi adductores longus and magnus and the vastus intermedius. Angiograms taken of animals infused with phosphate-buffered saline (PBS) for 7 days showed a significantly increased number of visible collateral arteries in comparison to animals without femoral artery ligation (Figs. 9 and 10) (no occlusion vs. PBS: 6.7 ± 1.2 vs. 16.2 ± 1.4; $p < 0.01$).

The application of MCP-1 led to an increased development of collateral arteries after femoral artery ligation (Fig. 11). In comparison to the PBS-group, the number of collateral arteries was significantly increased 7 days after ligation (PBS vs. MCP-1: 16.2 ± 1.4 vs. 30.2 ± 2.0; $p < 0.01$) (see also Fig. 12).

B. Collateral Conductance

Seven days after ligation, collateral conductance was significantly increased in comparison to the acute occlusion of the femoral artery (acute occlusion vs. PBS in ml/min per 100 mmHg: $0{,}94 \pm 0{,}19$ vs. 4.10 ± 0.48). In accordance with the angiographic data, the continuous infusion of MCP-1 significantly increased collateral conductance (PBS vs. MCP-1 4.10 ± 0.48 vs. 33.96 ± 1.76). The conductance in the hindlimb without femoral

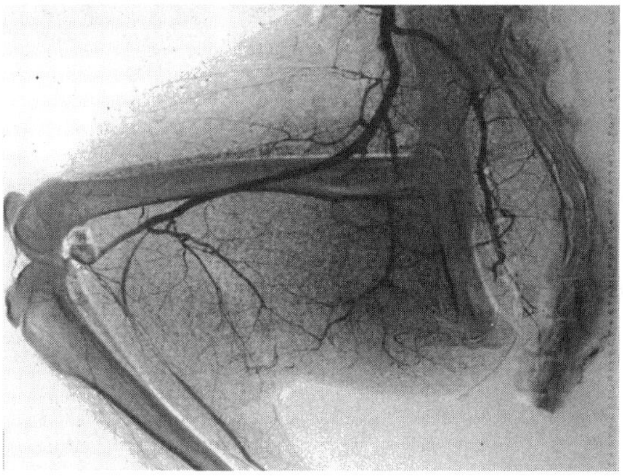

Figure 9 Angiogram of a rabbit hindlimb without occlusion of the femoral artery. Only few collateral vessels are visible.

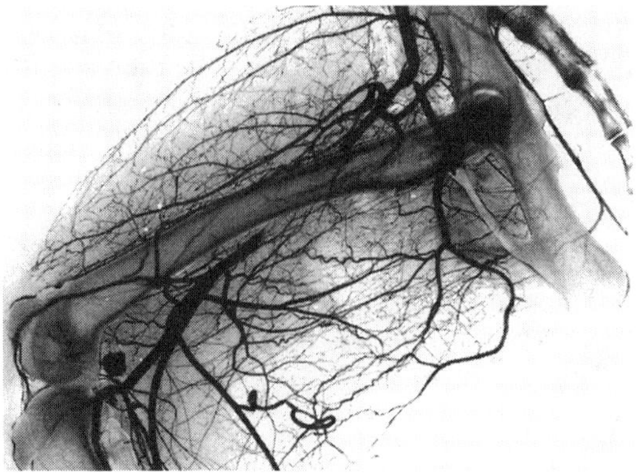

Figure 10 Angiogram of a rabbit hindlimb after 7 days of infusion of PBS. The typically corkscrew formation of the collateral arteries is clearly recognized, and the stem, midzone, and reentry areas are easily distinguishable.

artery ligation was 168.56 ± 15.99. All differences were statistically significant $p < 0.01$ (see also Fig. 13).

Occlusion of the femoral artery in the rabbit hindlimb leads to the proliferation of preexisting arteriolar connections building functional collateral arteries. Seven days after occlusion, the number of visible vessels had increased twofold in comparison to acute ligation. When rabbits are treated with MCP-1, the number of visible collateral vessels is more than doubled. The functional data of the in vivo perfusion measurements correlate well with the angiographic data. The natural process of flow restoration can be enhanced by MCP-1. In our model, MCP-1 leads to a more than 30-fold increase of collateral conductance in comparison to the acute occlusion.

Our data show that infusion of MCP-1, leading to an increased number of attracted monocytes to the activated endothelium of proliferating vessels, can at least accelerate

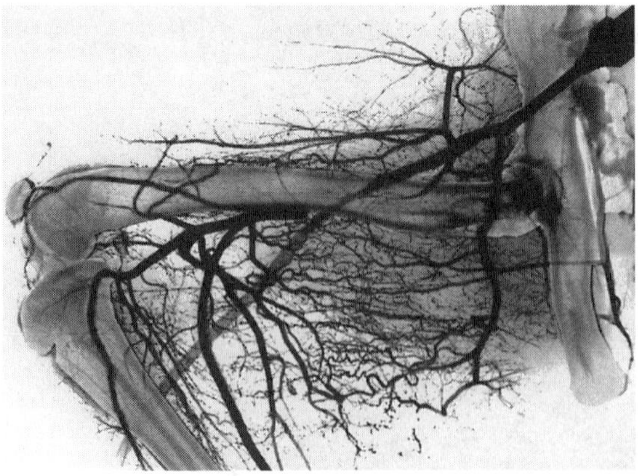

Figure 11 Angiogram of a rabbit hindlimb after 7 days of infusion of MCP-1. It is evident that the treatment with MCP-1 leads to an increased number of visible collateral arteries.

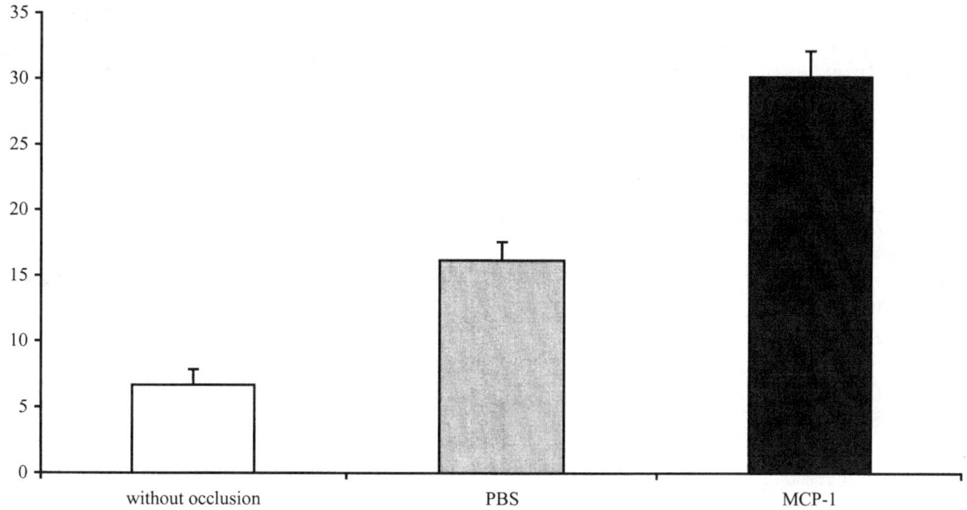

Figure 12 Diagram showing the number of visible collateral arteries. All observed differences are statistically significant ($p < 0.01$).

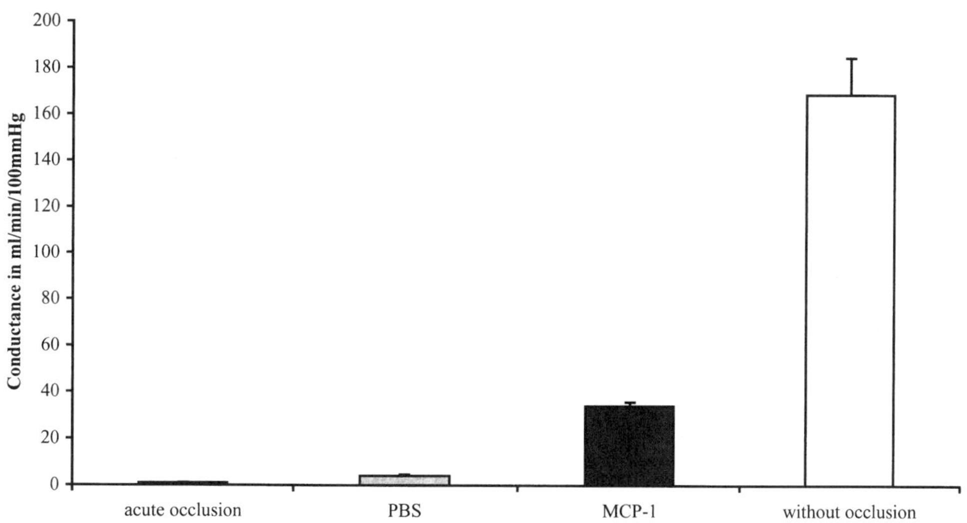

Figure 13 Diagram showing the conductance values in ml/min per 100 mmHg. All observed differences are statistically significant ($p < 0.01$).

the natural process of collateral artery growth. This offers new opportunities in the prevention and treatment of occlusive vascular diseases (e.g., coronary and peripheral artery disease). Its effects in the later course of proliferation and remodeling of collateral arteries is subject of further investigations.

ACKNOWLEDGMENT

We thank Circulation Research for giving us the permission to use Figs. 10 and 11 (Ito et al., 1997b).

REFERENCES

Arras M, Ito WD, Scholz D, Winkler B, Schaper J, Schaper W. Monocyte activation in angiogenesis and collateral growth in the rabbit hindlimb. J. Clin Invest 1998; 101(1):41–50.
Baird A, Klagsbrunn M. The fibroblast growth factor family. Cancer Cells 1991; 3(6):239–243.
Basilico C, Moscatelli D. The FGF family of growth factors and oncogenes. Adv Cancer Res 1992; 59:115–165.
Burgess WH, Maciag T. The heparin-binding (fibroblast) growth factor family of proteins. Annu Rev Biochem 1989; 58:575–606.
Clark RA. Biology of dermal wound healing. Clin Dermatol 1993; 11:647–666.
Clinton SK, Libby P. Cytokines and growth factors in atherogenesis. Arch Pathol Lab Med 1992; 116(12):1292–300.
Davies PF. Flow-mediated endothelial mechanotransduction. Physiol Rev 1995; 75(3):519–560.
Deindl E, Ito W, Zimmermann R, Schaper W. Increased FGFR1 expression is a prerequisite for angiogenesis and collateral growth (abstr). J Mol Med 1997; 75(5):13.
Deindl E, Ito W, Zimmermann RJ, Schaper W. VEGF, and important angiogenic factor, is not involved in arteriogenesis (abstr). J Mol Med 1998; 76(6):p-85.
D'Amore PA. Mechanisms of angiogenesis. Annu Rev Physiol 1987; 49:453–464.
Fernig DG, Gallagher JT. Fibroblast growth factors and their receptors: an information network controlling tissue growth, morphogenesis and repair. Prog Growth Factor Res 1994; 5:353–377.
Folkman J. Angiogenesis in cancer, vascular, rheumatoid and other diseases. Nature Med 1995; 1: 27–31.
Folkman J, Shing Y. Angiogenesis. J Biol Chem 1992; 267:10931–10934.
Fulton WFM. The Coronary Arteries. Springfield, IL: Charles C Thomas, 1965.
Gimbrone MAJ, Nagel T, Topper JN. Biomechanical activation: an emerging paradigm in endothelial adhesion biology. J Clin Invest 1997; 99:1809–1813.
Givol D, Yayon A. Complexity of FGF receptors: genetic basis for structural diversity and functional specificity. FASEB J 1992; 6:3362–3369.
Ikeda E, Achen MG, Breier G, Risau W. Hypoxia-induced transcriptional activation and increased mRNA stability of vascular endothelial growth factor in C6 glioma cells. J Biol Chem 1995; 270:19761–19766.
Ito W, Arras M, Winkler B, Htun P, Schaper W. Vessels grow by angiogenesis and vasculogenesis after femoral artery occlusion (abstr). Circulation 1996; 94(suppl):3560.
Ito WD, Arras M. Scholz D, Winkler B, Htun P, Schaper W. Angiogenesis but not collateral growth is associated with ischemia after femoral artery occlusion. Am J Physiol 1997a; 273:H1255–H1265.
Ito WD, Arras M, Winkler B, Scholz D, Schaper J. Schaper W. Monocyte chemotactic protein-1

increases collateral and peripheral conductance after femoral artery occlusion. Circ Res 1997b; 80:829–837.

Jin Y, Pasumarthi BS, Bock ME, Lytras A, Kardami E, Cattini PA. Cloning and expression of fibroblast growth factor receptor-1 isoforms in the mouse heart: evidence for isoform switching during heart development. J Mol Cell Cardiol 1994; 26:1449–1459.

Johnson DE, Williams LT. Structural and functional diversity in the FGF receptor multigene family growth factors and oncogenes. Adv Cancer Res 1993; 60:1–60.

Kampmann a, Hoefer I, Buschmann I, Schaper W. Genexpression bei adaptiver Arteriogenese (abstr). Z Kardiol 1999; 88(1):113.

Kirsner RS. Eaglstein WH. The wound healing process. Dermatol Clin 1993; 11:629–640.

Longland CJ. The collateral circulation of the limb. Ann R Coll Surg Engl 1953; 13:161–164.

Nagel T, Resnick N, Atkinson WJ, Dewey DF Jr, Gimbrone MA Jr. Shear stress selectively upregulates intercellular adhesion molecule-1 expression in cultured human vascular endothelial cells. J Clin Invest 1994; 94(2):885–891.

Partanen J, Vainikka S, Korhonen J, Armstrong E, Alitalo K. Diverse receptors for fibroblast growth factors. Prog Growth Factors Res 1992; 4(1):69–83.

Paskins-Hurlburt A, Hollenberg NK. ''Tissue need'' and limb collateral arterial growth: skeletal contractile power and perfusion during collateral development in the rat. Circ Res 1992; 70: 546–553.

Price RJ, Skalak TC. Arteriolar remodeling in skeletal muscle of rats exposed to chronic hypoxia. J Vasc Res 1998; 35:238–244.

Randolph GJ, Furie MB. A soluble gradient of endogenous monocyte chemoattractant protein-1 promotes the transendothelia migration of monocytes in vitro. J Immunol 1995; 155(7):3610–3618.

Resnick N, Gimbrone MA Jr. Hemodynamic forces are complex regulators of endothelial gene expression. FASEB J 1995; 9(10):874–882.

Reynolds RL, Killilea SD, Redme DA. Angiogenesis in the female reproductive system. FASEB J 1992; 6:886–892.

Schaper J, Koenig R, Franz D, Schaper W. The endiothelial surface of growing coronary collateral arteries. Intimal margination and diapedesis of monocytes: a combined SEM and TEM study. Virchows Arch A (Pathol Anat) 1976a; 370:193–205.

Schaper J, Schaper W, Winkler B. The collateral circulation of the heart (abstr). J Mol Cell Cardiol 1986; 18(suppl I):60.

Schaper W, ed. The Collateral Circulation of the Heart. Amsterdam London: Elsevier North Holland, 1971.

Schaper W. Collateral circulation. V. Mechanism of collateral enlargement. In: Schaper W, ed. The Pathophysiology of Myocardial Perfusion. Amsterdam, New York, Oxford: Elsevier/North-Holland, 1979:458–470.

Schaper W, Bernotat-Danielowski S, Nienaber C, Schaper J. Collateral circulation. In: Fozzard H, Haber E, Jennings R, Katz A, Morgan H, eds. The Heart and Cardiovascular System. Vol 2. New York: Raven Press, 1992: 1427–1464.

Schaper W, DeBrabander M, Lewi P. DNA-synthesis and mitoses in coronary collateral vessels of the dog. Circ Res 1971; 28:671–679.

Schaper W, Flameng W, Winkler B, Wuesten B, Tuerschmann W, Neugebauer G, Carl M, Pasyk S. Quantification of collateral resistance in acute and chronic experimental coronary occlusion in the dog. Circ Res 1976b; 39:371–377.

Schaper W, Flameng W, Wüsten B, Palmowski J. The distribution of coronary and of coronary collateral flow in normal hearts and after chronic coronary occlusion. In: Bloor CM, Olsson RA, eds. Current Topics in Coronary Research. New York and London: Plenum, 1973:151–160.

Schaper W, Ito W. Molecular mechanisms of coronary collateral vessel growth. Circ Res 1996; 79: 911–919.

Schaper W, Schaper J, eds. Collateral Circulation—Heart, Brain, Kidney, Limbs. Boston, Dordrecht and London: Kluwer, 1993.

Schlessinger J, Lax I, Lemmon M. Regualtion of growth factor activation by proteoglycans: what is the role of the low affinity receptors? Cell 1995; 83:357–360.

Schlessinger J, Ullrich A. Growth factor signaling by receptor tyrosine kinases. Neuron 1992; 9: 383–391.

Scholz D, Ito W, Fleming I, Deindl E, Sauer A, Babiak A, Buehler A, Wiesnet M, Busse R, Schaper J, Schaper W. Ultrastructure and molecular histology of rabbit hindlimb collateral artery growth (arteriogenesis). Virchows Arch. In press.

Shweiki D, Itin A, Sofer D, Keshet E. Vascular endothelial growth factor induced by hypoxia may mediate hypoxia-initiated angiogenesis. Nature 1992; 359:843–845.

Spalteholz W. Die Koronararterien des Herzens. Verhandl Anat Ges 1907; 21:141.

Takeshita S, Rossow ST, Kearney M, Zheng LP, Bauters C, Bunting S, Ferrara N, Symes JF, Isner JM. Time course of increased cellular proliferation in collateral arteries after administration of vascular endothelial growth factor in a rabbit model of lower limb vascular insufficiency. Am J Pathol 1995; 147: 1649–1660.

Thomas KA. Vascular endothelial growth factor, a potent and selective angiogenic agent. J Biol Chem 1996; 271:603–606.

Tudor RM, Flook BE, Volkel NF. Increased gene expression of VEGF and the VEGF receptors KDR/Flk and Flt in lungs exposed to acute or to chronic ischemia. J Clin Invest 1995; 95: 1789–1807.

Wang J-K, Gao G, Goldfarb M. Fibroblast growth factor receptors have different signaling and mitogenic potentials. Mol Cell Biol 1994; 14(1):181–188.

Xin X, Johnson AD, Scott-Burden T, Engler D, Casscells W. The predominant form of fibroblast growth factor receptor expressed by proliferating human arterial smooth muscle cells in culture is type I. Biochem Biophys Res Commun 1994; 204(2):557–564.

3
The Role of Vascular Endothelial Growth Factor in Angiogenesis

Napoleone Ferrara
Genentech, Inc., South San Francisco, California

I. INTRODUCTION

The development of a vascular supply is a fundamental requirement for organ development and differentiation during embryogenesis as well as for wound healing and reproductive functions in the adult (1). Angiogenesis is also implicated in the pathogenesis of a variety of disorders: proliferative retinopathies, age-related macular degeneration, tumors, rheumatoid arthritis, and psoriasis (1,2)

The search for positive regulators of angiogenesis has yielded several candidates, including aFGF, bFGF, TGF-α, TGF-β, HGF, TNF-α, angiogenin, IL-8 (3,4), and, more recently, the angiopoietins, the ligands of the Tie-2 receptor (5,6). The negative regulators so far identified include thrombospondin (7,8), the 16-kDa N-terminal fragment of prolactin (9), angiostatin (10), and endostatin (11).

This chapter discusses the role of vascular endothelial growth factor (VEGF) in angiogenesis. Over the last few years, several additional members of the VEGF gene family have been identified, including placenta growth factor (12), VEGF-B (13), VEGF-C (14,15), and VEGF-D (16). For a detailed review, see Ferrara and Davis-Smyth (17). The focus of this article is on the original VEGF, referred to also as *VEGF-A*. The finding that the loss of even a single VEGF allele results in embryonic lethality points to the irreplaceable role played by this factor in the development and differentiation of the vascular system (18,19). Work done by several laboratories over the last few years has elucidated the pivotal role of VEGF in pathological angiogenesis, such as that associated with tumors or ischemic retinal diseases (17). Furthermore, VEGF-induced angiogenesis has been shown to result in a therapeutic effect in animal models of coronary or limb ischemia (17). Presently, both therapeutic angiogenesis using recombinant VEGF or VEGF gene transfer and inhibition of VEGF-mediated pathological angiogenesis are being pursued clinically.

II. BIOLOGICAL ACTIVITIES OF VEGF

VEGF is a mitogen for vascular endothelial cells derived from arteries, veins, and lymphatics but is devoid of consistent and appreciable mitogenic activity for other cell types (17).

VEGF also elicits a pronounced angiogenic response in a variety of in vivo models, including the chick chorioallantoic membrane (20,21). Also, VEGF acts as a survival factor for cultured endothelial cells in serum-depleted conditions (22). Consistent with a prosurvival activity, VEGF induces expression of the antiapoptotic proteins Bc1-2 and A1 in human endothelial cells (23).

VEGF induces expression of the serine proteases urokinase-type and tissue-type plasminogen activators (PA) and also PA inhibitor 1 (PAI-1) in cultured bovine microvascular endothelial cells (24). Moreover, VEGF increases expression of the metalloproteinase interstitial collagenase in human umbilical vein endothelial cells but not in dermal fibroblasts (25).

VEGF is also known as vascular permeability factor (VPF) based on its ability to induce vascular leakage in guinea pig skin (26). Dvorak and colleagues proposed that an increase in microvascular permeability is a crucial step in angiogenesis associated with tumors and wounds (27). According to this hypothesis, a major function of VPF/VEGF in the angiogenic process is the induction of plasma protein leakage. This effect would result in the formation of an extravascular fibrin gel, a substrate for endothelial and tumor cell growth (28). Recent studies have also suggested that VEGF may also induce fenestrations in endothelial cells, both in vivo (29,30) and in a defined in vitro model, using adrenal cortex–derived capillary endothelial cells (31).

VEGF promotes expression of VCAM-1 and ICAM-1 in endothelial cells (32). This induction results in the adhesion of activated natural killer (NK) cells to endothelial cells, mediated by specific interaction of endothelial VCAM-1 and ICAM-1 with CD18 and VLA-4 on the surface of NK cells.

VEGF has been reported to have regulatory effects on blood cells. Clauss et al. reported that VEGF may promote monocyte chemotaxis (33). Broxmeyer et al. have shown that VEGF induces colony formation by mature subsets of granulocyte-macrophage progenitor cells (34). Furthermore, Gabrilovich et al. have reported that VEGF may have an inhibitory effect on the maturation of host professional antigen-presenting cells such as dendritic cells (35). VEGF was found to inhibit immature dendritic cells without having a significant effect on the function of mature cells. These findings led to suggestion that VEGF may facilitate tumor growth also by allowing the tumor to avoid the induction of an immune response (35).

VEGF induces vasodilatation in vitro in a dose-dependent fashion (36,37) and produces transient tachycardia, hypotension, and a decrease in cardiac output when injected intravenously in conscious, instrumented rats (37). Such effects appear to be caused by a decrease in venous return, mediated primarily by endothelial cell–derived nitric oxide (NO) (37). Accordingly, VEGF has no direct effect on contractility or rate in isolated rat heart in vitro (37). These hemodynamic effects, however, are not unique to VEGF: other angiogenic factors such as aFGF and bFGF also have the ability to induce NO-mediated vasodilatation and hypotension (38,39).

III. ORGANIZATION OF THE VEGF GENE AND CHARACTERISTICS OF THE VEGF PROTEINS

The human VEGF gene is organized in eight exons separated by seven introns. The coding region spans approximately 14 kilobases (kb) (40,41). The human VEGF gene maps to chromosome 6p21.3 (42). It is now well established that alternative exon splicing of a single VEGF gene results in the generation of four different molecular species, having respectively 121, 165, 189, and 206 amino acids following signal sequence cleavage ($VEGF_{121}$, $VEGF_{165}$, $VEGF_{189}$, $VEGF_{206}$). $VEGF_{165}$ lacks the residues encoded by exon 6, while $VEGF_{121}$ lacks the residues encoded by exons 6 and 7. Compared to $VEGF_{165}$, $VEGF_{121}$ lacks 44 amino acids; $VEGF_{189}$ has an insertion of 24 amino acids highly enriched in basic residues and $VEGF_{206}$ has an additional insertion of 17 amino acids (40). $VEGF_{165}$ is the predominant molecular species produced by a variety of normal and transformed cells. Transcripts encoding $VEGF_{121}$ and $VEGF_{189}$ are detected in the majority of cells and tissues expressing the VEGF gene (40). In contrast, $VEGF_{206}$ is a rare form (40).

Native VEGF is a basic, heparin-binding, homodimeric glycoprotein of 45,000 Da (43). These properties correspond to those of $VEGF_{165}$, the major isoform (44). $VEGF_{121}$ is a weakly acidic polypeptide that fails to bind to heparin (44). $VEGF_{189}$ and $VEGF_{206}$ are more basic and bind to heparin with greater affinity than $VEGF_{165}$ (44). $VEGF_{121}$ is a freely diffusible protein; $VEGF_{165}$ is also secreted although a significant fraction remains bound to the cell surface and the extracellular matrix (ECM). In contrast, $VEGF_{189}$ and $VEGF_{206}$ are almost completely sequestered in the ECM (45). However, these isoforms may be released in a soluble form by heparin or heparinase, suggesting that their binding site is represented by proteoglycans containing heparin-like moieties. The long forms may be released in an active form also by plasmin, following cleavage at the COOH terminus. This action generates a bioactive proteolytic fragment having molecular weight of ~34,000 Da (44). Plasminogen activation and generation of plasmin have been shown to play an important role in the angiogenesis cascade. Thus, proteolysis of VEGF is likely to occur also in vivo. Keyt et al. have shown that the bioactive product of plasmin action is comprised of the first 110 NH_2-terminal amino acids of VEGF (46). These findings suggest that the VEGF proteins may become available to endothelial cells by at least two different mechanisms: as freely diffusible proteins ($VEGF_{121}$, $VEGF_{165}$) or following protease activation and cleavage of the longer isoforms. However, loss of heparin binding, whether it is due to alternative splicing of RNA or plasmin cleavage, results in a substantial loss of mitogenic activity for vascular endothelial cells: compared to $VEGF_{165}$, $VEGF_{121}$ or $VEGF_{110}$ demonstrate 50- to 100-fold reduced potency when tested in endothelial cell growth assay (46). It has been suggested that the stability of VEGF-heparan sulfate-receptor complexes contributes to effective signal transduction and stimulation of endothelial cell proliferation (46). Ploüet et al. have recently proposed a role for urokinase in the generation of biactive $VEGF_{189}$ (47). Recombinant $VEGF_{189}$ from insect cells infected with a recombinant baculovirus was purified as a nonmitogenic 50-kDa precursor that binds to the receptor Flt-1 but not to Flk-1/KDR. However, it could be matured by urokinase into a 38-kDa fragment able to bind Flk-1/KDR and promote endothelial-cell proliferation (47). Recently, Poltorak et al. have provided evidence for the existence of an additional alternatively spliced isoform containing exons 1–6 and 8 of the VEGF gene, $VEGF_{145}$ (48). $VEGF_{145}$ is able to promote endothelial cell growth, albeit with a lower potency than $VEGF_{165}$.

Muller et al. have solved the crystal structure of VEGF at a resolution of 2.5 A (49). Overall, the VEGF monomer resembles that of PDGF, but its N-terminal segment is helical rather than extended. The dimerization mode of VEGF is similar to that of PDGF and unlike that of TGF-β (49).

IV. REGULATION OF VEGF GENE EXPRESSION

A. Oxygen Tension

Among the mechanisms that have been proposed to participate in the regulation of VEGF gene expression, oxygen tension plays a major role, both in vitro and in vivo. VEGF mRNA expression is rapidly and reversibly induced by exposure to low P_{O_2} in a variety of normal and transformed cultured cell types (50,51). Also, ischemia caused by occlusion of the left anterior descending coronary artery results in a dramatic increase in VEGF RNA levels in the pig and rat myocardium, suggesting the possibility that VEGF may mediate the spontaneous revascularization that follows myocardial ischemia (52, 53).

Similarities exist between the mechanisms leading to hypoxic regulation of VEGF and erythropoietin (Epo) (54). Hypoxia-inducibility is conferred on both genes by homologous sequences. By deletion and mutation analysis, a 28-base sequence has been identified in the 5′ promoter of the rat and human VEGF gene that mediated hypoxia-induced transcription (55,56). Such sequence reveals a high degree of homology and similar protein-binding characteristics as the hypoxia-inducible factor 1 (HIF-1) binding site within the Epo gene (57). HIF-1 has been identified as a mediator of transcriptional responses to hypoxia and is a basic, heterodimeric, helix-loop-helix protein consisting of two subunits, HIF-1α and aryl hydrocarbon receptor nuclear translocator (ARNT), known also as HIF-1β (58). Evidence for a critical role of HIF-1α has also been provided by gene knockout studies. Inactivation of HIF-1α in mice resulted in developmental arrest and lethality by E11 (59,60). Embryos manifested neural tube defects, cardiovascular malformations, and marked cell death within the cephalic mesenchyme. These results suggest that HIF-1α is a key regulator of cellular and developmental O_2 homeostasis (59). Also, the consequence of HIF-1α inactivation in ES cell on in vitro proliferation and in vivo tumorigenesis has been examined by two groups. Surprisingly, however, virtually opposite conclusions were reached. According to Ryan et al., HIF-1α inactivation leads to a dramatically decreased tumorigenesis and increased apoptosis, coincident with reduced hypoxia-induced expression of VEGF (60). In contrast, Carmeliet et al. reported that the growth of ES cell-derived tumors is paradoxically enhanced following loss of HIF-1α, possibly due to protection from hypoxia-induced apoptosis (61). Whether such discrepancies reflect differences in the genotype or clone of ES cells or other experimental variable, remains to be established.

It should be noted, however, that transcriptional activation is not the only mechanism leading to VEGF upregulation in response to hypoxia. Increased mRNA stability has been identified as a significant posttranscriptional component (62,63).

B. Hormones and Cytokines

Various cytokines or growth factors may upregulate VEGF mRNA expression. EGF, TGF-β, or KGF result in a marked induction of VEGF mRNA expression (64); EGF also stimulates VEGF release by cultured glioblastoma cells (65). In addition, treatment of quiescent cultures of epithelial and fibroblastic cell lines with TGF-β resulted in induction of VEGF mRNA and release of VEGF protein in the medium (66). Based on these findings, it has been proposed that VEGF may function as a paracrine mediator for indirect-acting angiogenic agents such as TGF-β (66). Both IL-1-α and PGE_2 have been shown to induce expression of VEGF in cultured synovial fibroblasts, suggesting the participation of such inductive mechanisms in inflammatory angiogenesis (67). IL-6 has been also shown to significantly induce VEGF expression in several cell lines (68). IGF-1 has also been shown to induce VEGF mRNA and protein in cultured colorectal carcinoma cells (69). Thyroid-stimulating hormone has been shown to induce VEGF expression in several thyroid carcinoma cell lines (70). Shifren et al. have examined the VEGF expression in the midgestation (16–20 weeks) human fetal adrenal cortex (71). Strong cytoplasmic immunostaining for VEGF was detected in clusters of fetal zone cells. In contrast, cells in the outer, less well vascularized, definitive zone of the cortex stained only weakly for VEGF.

C. Differentiation and Transformation

Cell differentiation has been shown to play an important role in the regulation of VEGF gene expression (72). The VEGF mRNA is upregulated during the conversion of 3T3 preadipocytes into adipocytes or during the myogenic differentiation of C2C12 cells.

Specific transforming events also result in induction of VEGF gene expression. A mutated form of the murine p53 tumor suppressor gene has been shown to result in induction of VEGF mRNA expression in NIH 3T3 cells in transient transfection assays (73). Likewise, oncogenic mutations or amplification of ras lead to VEGF upregulation (74–76). Moreover, the von Hippel-Lindau (VHL) tumor suppressor gene has been recently implicated in the regulation of VEGF gene expression (77–79). The VHL tumor suppressor gene is inactivated in patients with von Hippel-Lindau disease and in most sporadic clear-cell renal carcinomas. Although the function of the VHL protein remains to be fully elucidated, it is known that such protein interacts with the elongin BC subunits in vivo and regulates RNA polymerase II elongation activity in vitro by inhibiting formation of the elongin ABC complex. Human renal cell carcinoma cells either lacking endogenous wild-type VHL gene or expressing an inactive mutant demonstrated altered regulation of VEGF gene expression, which was corrected by introduction of wild-type VHL gene. Most of the endothelial cells' mitogenic activity released by tumor cells expressing mutant VHL gene was neutralized by anti-VEGF antibodies (77). These findings suggest that VEGF is a key mediator of the abnormal vascular proliferations and solid tumors characteristic of VHL syndrome. Iliopulos et al. have shown that a function of the VHL protein is to provide a negative regulation of a series of hypoxia-inducible genes, including the VEGF, platelet-derived growth factor B chain, and the glucose transporter GLUT1 genes (78). In the presence of a mutant VHL, mRNAs for such genes were produced under both normoxic and hypoxic conditions. Reintroduction of wild-type VHL resulted in inhibition of mRNA production under normoxic conditions and restored the characteristic hypoxia inducibility of those genes (78). Gnarra et al. suggested that the VHL protein regulates VEGF expression at a posttranscriptional level and that VHL inactivation in target cells causes a loss of VEGF suppression (79). However, Mukhopadyay et al. have recently provided evidence for an additional mechanism by which the VHL gene product suppresses VEGF (80). According to these authors, VHL interacts with Sp1 to repress vascular endothelial growth factor promoter activity as assessed by VEGF promoter-luciferase reporter studies. Deletion analysis defined a 144-bp region of the VEGF promoter necessary for VHL repression. This VHL-responsive element is GC rich and specifically binds the transcription factor Sp1 in crude nuclear extracts.

V. THE VEGF RECEPTORS

Ligand autoradiography studies on fetal and adult rat tissue sections demonstrated that high affinity VEGF binding sites are localized to the vascular endothelium of large or small vessels in situ (81,82), providing direct evidence for the hypothesis that VEGF is an endothelial cell-specific factor. VEGF binding was apparent not only on proliferating but also on quiescent endothelial cells (81,82). These studies also showed that the earliest developmental identification of high-affinity VEGF binding is in the hemangioblasts in the blood islands in the yolk sac (82).

VI. THE Flt-1 (VEGFR-1) AND Flk-1/KDR (VEGFR-2) TYROSINE KINASES

A. Binding Characteristics and Structure/Function Studies

Two VEGF receptor tyrosine kinases (RTKs) have been identified. The Flt-1 (fms-like-tyrosine kinase) (83) and KDR (kinase domain region) (84) receptors bind VEGF with high affinity. The murine homologue of KDR, Flk-1 (fetal liver kinase-1), shares 85% sequence identity with human KDR (85). Both Flt-1 and KDR/Flk-1 have seven immunoglobulin (Ig)-like domains in the extracellular domain (ECD), a single transmembrane region and a consensus tyrosine kinase sequence which is interrupted by a kinase-insert domain (85–87). Flt-1 has the highest affinity for rhVEGF$_{165}$, with a K_d of approximately 10–20 pM (83). KDR has a somewhat lower affinity for VEGF: the K_d has been estimated to be approximately 75–125 pM (84).

A cDNA coding an alternatively spliced soluble form of Flt-1 (sFlt-1), lacking the seventh Ig-like domain, transmembrane sequence and the cytoplasmic domain, has been identified in human umbilical vein endothelial cells (88,89). This sFlt-1 receptor binds VEGF with high affinity (K_d 10–20 pM) and is able to inhibit VEGF-induced mitogenesis and may be a physiological negative regulator of VEGF action (89).

An additional member of the family of RTKs with seven Ig-like domains in the ECD is Flt-4 (90–92) which, however, is not a receptor for VEGF but rather binds a newly identified ligand called VEGF-C or VEGF-related peptide (VRP) (14,15). This molecule has recently been shown to be involved in the regulation of lymphatic angiogenesis (93).

Recent studies have mapped the binding site for VEGF to the second immunoglobulin-like domain of Flt-1 and KDR (94,95). Deletion of the second domain of Flt-1 completely abolished the binding of VEGF. Introduction of the second domain of KDR into an Flt-1 mutant lacking the homologous domain restored VEGF binding (94). However, the ligand specificity was characteristic of the KDR receptor. Similar conclusions were reached in deletion experiments in KDR (96). These studies indicate that of the seven Ig-like domains in the extracellular domain, only domains 2–3 are needed for tight binding of VEGF to the KDR receptor (96). Recently, Wiesman et al. have solved the crystal structure of a VEGF-Flt-1 domain 2 complex (97). These studies have shown domain 2 in a predominantly hydrophobic interaction with the poles of the VEGF dimer. However, alanine-scanning mutagenesis analysis of the first three domains of Flt-1 has shown that several charged residues, especially Asp187, are important in maintaining the structural integrity of domain 2, although not directly involved in the ligand-receptor interface (98).

B. Signal Transduction

VEGF has been shown to induce the phosphorylation of at least 11 proteins in bovine aortic endothelial cells (99). PLC-γ, and two proteins that associate with PLC-γ were phosphorylated in response to VEGF. Furthermore, immunoblot analysis for mediators of signal transduction that contain SH2 domains demonstrated that VEGF induces phosphorylation of phosphaidylinositol 3-kinase, ras GTPase activating protein (GAP), and several others. These findings suggest that VEGF promotes the formation of multimeric aggregates of VEGF receptors with proteins that contain SH2 domains. These studies, however, did not identify which VEGF receptor(s) are involved in these events. Recently, it has been suggested that NO mediates, at least in part, the mitogenic effect of VEGF on cultured microvascular endothelium isolated from coronary venules (100).

Several studies have indicated that Flt-1 and KDR have different signal transduction properties (101,102). Porcine aortic endothelial cells lacking endogenous VEGF receptors display chemotaxis and mitogenesis in response to VEGF when transfected with a plasmid coding for KDR (101). In contrast, transfected cells expressing Flt-1 lack such responses (101,102). Flk-1/KDR undergoes strong ligand-dependent tyrosine phosphorylation in intact cells, while Flt-1 reveals a weak or undetectable response (101,102). Also, VEGF stimulation results in weak tyrosine phosphorylation that does not generate any mitogenic signal in transfected NIH 3T3 cells expressing Flt-1 (102). These findings agree with other studies showing that placental growth factor (PlGF), which binds with high affinity to Flt-1 but not to Flk-1/KDR, lacks direct mitogenic or permeability-enhancing properties or the ability to effectively stimulate tyrosine phosphorylation in endothelial cells (103). Interestingly, however, high concentrations of PlGF expected to saturate the Flt-1 sites are able to potentiate the activity of VEGF, both in vivo and in vitro (103). These findings led to the suggestion that Flt-1 may be not primarily a signaling receptor but rather a ''decoy'' receptor, able to regulate in a negative fashion the activity of VEGF on the vascular endothelium by sequestering and rendering this ligand less available to Flk-1/KDR (103). In apparent conflict with this hypothesis, subsequent studies indicated that Flt-1 is indeed able to interact with various signal transducing proteins, including the p85 subunit of the PI3 kinase and the MAP kinase, and generate under some circumstances a mitogenic signal, at least in transfected cell lines (101,104,105). Furthermore, constitutively active artificial chimeras of the human leukemia oncoprotein BCR-ABL with the ICD of Flt-1, with mutations within the Flt-1 sequence, transform Rat-1 fibroblasts, abrogate cytokine dependence in Ba/F3 cells, and induce neurite-like structures in PC12 cells, suggesting that the Flt-1 kinase has the potential to be activated and mediate differentiation/transformation signal(s), at least in such artificial systems. (106). However, very recent studies have provided evidence in support of the earlier hypothesis. Hirakutsa et al. have demonstrated that a targeted mutation resulting in a Flt-1 receptor lacking the tyrosine kinase (TK) domain but able to bind VEGF does not result in impaired embryonic development and angiogenesis in mice (107), while complete inactivation of the receptor results in embryonic lethality (see below). Furthermore, endothelial cells isolated from Flt-1($TK^{-/-}$) animals exhibited a normal mitogenic response to VEGF. These results suggest that Flt-1 plays its main role in angiogenesis as a ligand-binding molecule rather than as a signal transducing receptor, at least as defined by conventional criteria. Taken together, these studies strongly support the hypothesis that interaction with Flk-1/KDR is a critical requirement to induce the full spectrum of VEGF biological responses. In further support of this conclusion, VEGF mutants that bind selectively to Flk-1/KDR, are fully active endothelial cell mitogens (108) and anti-idiotypic antibodies that activate Flk-1/KDR promote tumor angiogenesis (109). Furthermore, Flk-1/KDR activation has been shown to be required for the anti-apoptotic effects of VEGF for HUVE cells (22). Such a survival effect of VEGF has been shown to be mediated by the PI3 kinase/Akt pathway (22). Nevertheless, at least one biological response, the migration of monocytes in response to VEGF (or PlGF), has been shown to be mediated by Flt-1 (110). VEGF-induced macrophage migration was suppressed in Flt-1 (TK$^{-/-}$ mice (107). Moreover, the ability of VEGF to inhibit maturation of dendritic cells has been associated with activation of Flt-1 (111). It is tempting to speculate that Flt-1 is able to mediate a specific regulatory signal in cell types other than endothelial cells.

Recently, Soker et al. have demonstrated the existence of an additional VEGF receptor (112) that binds $VEGF_{165}$ but not $VEGF_{121}$. This isoform-specific VEGF binding site is identical to human neuropilin-1, a receptor for the collapsin/semaphorin family that mediates neuronal cell guidance. When coexpressed in cells with KDR, neuropilin-1 enhanced the binding of $VEGF_{165}$ to KDR and $VEGF_{165}$-mediated chemotaxis. Conversely, inhibition of $VEGF_{165}$ binding to neuropilin-1 inhibits its binding to KDR and its mitogenic activity for endothelial cells. These findings suggest that neuropilin-1 may present VEGF to the KDR receptor in a manner that enhances the effectiveness of KDR-mediated signal transduction.

C. Regulation

The expression of Flt-1 and Flk-1/KDR genes is largely restricted to the vascular endothelium. The promoter region of Flt-1 has been cloned and characterized and a 1-kb fragment of the 5′ flanking region essential for endothelial-specific expression was identified (113). A 4-kb 5′ flanking sequence has been identified in the promoter of KDR that confers endothelial cell specific activation (114).

Similarly to VEGF, hypoxia has been proposed to play an important role in the regulation of VEGF receptor gene expression. Exposure to acute or chronic hypoxia led to pronounced up-regulation of both Flt-1 and Flk-1/KDR genes in the lung vasculature in a rat model (115). Also, Flk-1/KDR and Flt-1 mRNAs were substantially up-regulated throughout the heart following myocardial infarction in the rat (116). However, in vitro studies have yielded paradoxical results. Hypoxia increases VEGF receptor number by 50% in cultured bovine retinal capillary endothelial cells but the expression of KDR is not induced but paradoxically shows an initial downregulation (117). Brogi et al. have proposed that the hypoxic upregulation of KDR observed in vivo is not direct but requires the release of an unidentified paracrine mediator from ischemic tissues (118). Recent studies have provided evidence for a differential transcriptional regulation of the Flt-1 and KDR genes by hypoxia (119). These studies indicate that, unlike the KDR/Flk-1 gene, the Flt-1 receptor gene is directly upregulated by hypoxia via a hypoxia inducible enhancer element located in the promoter (119).

TGF-β has been reported to regulate in a negative fashion the expression of KDR mRNA in endothelial cells (120). According to Patterson et al., TNF-α potently reduced mRNA transcripts of both Flt-1 and Flk-1/KDR in venous and arterial endothelial cells in a dose- and time-dependent fashion (121). These authors concluded that TNF-α downregulates expression of both VEGF receptors in human endothelial cells and that this effect is transcriptional, at least for KDR. However, Giraudo et al. reported that TNF-α upregulates in a dose- and time-dependent manner the expression and the function of KDR as well as the expression of its coreceptor neuropilin-1 in human endothelium (122).

VII. ROLE OF VEGF AND ITS RECEPTORS IN PHYSIOLOGICAL ANGIOGENESIS

A. Normal Development

Gene targeting studies have demonstrated that both Flt-1 and Flk-1/KDR are essential for development of the embryonic vasculature in mice (123,124). However, their respective roles in endothelial cell proliferation and differentiation appear to be distinct. Mouse embryos homozygous for a targeted mutation in the Flt-1 locus died in utero between days 8.5 and 9.5 (123). Endothelial cells developed in both embryonic and extraembryonic sites but failed to organize in normal vascular channels. As previously noted, more recent studies uncovered the remarkable finding that normal blood vessel development and survival may occur in the absence of the tyrosine kinase domain of Flt-1 (107).

Mice in which the Flk-1 gene had been inactivated lacked vasculogenesis and also failed to develop blood islands. Hematopoietic precursors were severely disrupted and organized blood vessels failed to develop throughout the embryo or the yolk sac, resulting in death in utero between days 8.5 and 9.5 (124).

However, these findings do not necessarily imply VEGF as being equally essential, since other ligands might potentially activate the Flt-1 and Flk-1 receptors and thus substitute VEGF action. Two independent studies (18,19) have generated direct evidence for the role played by VEGF in embryonic vasculogenesis and angiogenesis. Inactivation of the VEGF gene in mice resulted in embryonic lethality in heterozygous embryos between days 11 and 12. The $VEGF^{\pm}$ embryos were growth-retarded and also exhibited a number of developmental anomalies. The forebrain region appeared significantly underdeveloped. In the heart, the outflow region was grossly malformed; the dorsal aortas were rudimentary, and the thickness of the ventricular wall was markedly decreased. The yolk sac revealed a markedly reduced number of nucleated red blood cells within the blood islands. Significant defects in the vasculature of other tissues and organs including placenta and nervous system were observed. In situ hybridization confirmed expression of VEGF mRNA in heterozygous embryos. Thus, the $VEGF^{+/-}$ phenotype appears to be due to gene dosage and not to maternal imprinting. While several heterozygous phenotypes have been described (125), this may be the first example of embryonic lethality following the loss of a single allele of a gene that is not maternally imprinted. Interestingly, inactivation of the PlGF gene does not result in embryonic lethality, even in the homozygous state (126). $PlGF^{-/-}$ mice are viable and fertile, although they may have some impairment of wound healing. These findings suggest that other members of the VEGF gene family are not equally critical for vascular development and survival.

To determine the role of VEGF in postnatal development, two independent strategies were employed: inducible, Cre-*loxP*–mediated gene targeting or administration of a soluble VEGF receptor chimeric protein [mFlt (1-3)-IgG] (127). Partial inhibition of VEGF achieved by inducible gene targeting resulted in increased mortality, stunted body growth and impaired organ development, most notably of the liver. Administration of mFlt

(1-3)-IgG, which achieves a higher degree of VEGF inhibition, resulted in growth arrest and mortality within 4 to 6 days when the treatment was initiated at day 1 or 8 postnatally. Decreased levels of proliferation of various cell types were detected in all organs studied and increased apoptosis was also observed in some tissues. Histological and biochemical changes consistent with renal failure were observed in treated animals. This was the major cause of mortality. These findings demonstrate that VEGF is essential not only for growth but also for homeostasis in a variety of organs in neonates. Furthermore, these findings are consistent with a critical dependence of glomerular development on VEGF (128–130). However, anti-VEGF treatment resulted in less significant alterations as the animal matured and the dependence on VEGF was eventually lost sometime after the fourth postnatal week. In the fully adult animal, VEGF appears to be required only for active angiogenic processes, such as corpus luteum angiogenesis.

B. Female Reproductive Tract Angiogenesis

The development and endocrine function of the ovarian corpus luteum (CL) are dependent on the growth of new capillary vessels (131). Earlier studies suggested the release of angiogenic factors from developing CL (132). The VEGF mRNA is temporally and spatially related to the proliferation of blood vessels in the rat, mouse, and primate ovary and in the rat uterus, suggesting that VEGF is a mediator of the cyclical growth of blood vessels that occurs in the female reproductive tract (133–136). Very recently, this hypothesis has been tested directly due to the availability of effective inhibitors of rodent VEGF, such as truncated soluble Flt-1 receptors (94,137). Treatment with Flt (1-3)-IgG resulted in virtually complete suppression of CL angiogenesis in a rat model of hormonally induced ovulation (137). This effect was associated with inhibition of CL development and progesterone release. Also, failure of maturation of the endometrium was observed, probably reflecting suppression of ovarian steroid production plus a direct inhibition of locally produced VEGF. Areas of ischemic necrosis were demonstrated in the CL of treated animals.

VIII. ROLE OF VEGF IN PATHOLOGICAL ANGIOGENESIS

A. Tumors

In 1945 Algire et al., based on the observation that the growth of tumor xenografts in transparent chambers in mice is always preceded by an increase in vascular density, concluded that ''the rapid growth of tumor explants is dependent upon the development of a rich vascular supply and that an outstanding characteristic of the tumor cell is its capacity to elicit continuously the growth of new capillary endothelium from the host'' (138). In 1971 Folkman proposed antiangiogenesis as a novel strategy for the treatment of solid tumors (139). Since then, extensive research has been dedicated to the identification and characterization of tumor angiogenesis factors (140). Although several molecules were initially implicated, there was considerable uncertainty as to their role as endogenous mediators of tumor growth and angiogenesis. For example, antibodies against bFGF, the molecule most consistently associated with tumor angiogenesis, did not result in inhibition of growth of several murine tumor cell lines implanted in syngeneic mice (141).

Many tumor cell lines secrete VEGF in vitro (142). In situ hybridization studies have demonstrated that the VEGF mRNA is markedly upregulated in the vast majority of human tumors so far examined, including carcinomas of the lung (143,144) breast (145,146), gastrointestinal tract (147–150), kidney (151,152), bladder (151), ovary (153), and endometrium (154) and several intracranial tumors including glioblastoma multiforme (155–157) and sporadic as well as VHL syndrome–associated capillary hemangioblastoma (158,159). In glioblastoma multiforme and other tumors with significant necrosis, the expression of VEGF mRNA is highest in hypoxic tumor cells adjacent to necrotic areas (155–157).

A correlation has been noted between VEGF expression and microvessel density in primary breast cancer sections (160). Postoperative survey indicated that the relapse-free survival rate of patients with VEGF-rich tumors was significantly worse than that of VEGF-poor tumors, suggesting that expression of VEGF is associated with stimulation of angiogenesis and with early relapse in primary breast cancer (161). A similar correlation has been described in gastric carcinoma patients (162). Furthermore, several studies have documented elevations in the VEGF plasma levels in tumor patients compared to tumor-free individuals, and it has been also reported that high VEGF levels prior to chemotherapy may be associated with poor outcome in non-Hodgkin's lymphoma as well as small cell lung carcinoma (163–166).

The availability of specific monoclonal antibodies capable of inhibiting VEGF-induced angiogenesis in vivo and in vitro (167) made it possible to generate direct evidence for a role of VEGF in tumorigenesis. Such antibodies were found to exert a potent inhibitory effect on the growth of three human tumor cell lines injected subcutaneously in nude mice, the SK-LMS-1 leiomyosarcoma, the G55 glioblastoma multiforme, and the A673 rhabdomyosarcoma (168). The growth inhibition ranged between 70 and >95%. Subsequently, many other tumor cell lines were found to be inhibited in vivo by this treatment (169–172).

Intravital videomicroscopy techniques have allowed a more direct verification of the hypothesis that anti-VEGF antibodies indeed block tumor angiogenesis (173). Noninvasive imaging of the vasculature revealed a nearly complete suppression of tumor angiogenesis in anti-VEGF treated animals as compared with controls at all time points examined (173). Intravital fluorescence microscopy and video imaging analysis have been also applied to address the important issue of the effects of VEGF on permeability and other properties of tumor vessels (174). Treatment with anti-VEGF monoclonal antibodies was initiated when tumor xenografts were already established and vascularized and resulted in time-dependent reductions in vascular permeability (174). These effects were accompanied by marked changes in the morphology of vessels, with reduction in diameter and tortuosity. A regression of blood vessels was observed after repeated administrations of anti-VEGF antibody.

An independent verification of the hypothesis that the VEGF action is required for tumor angiogenesis has been provided by the finding that retrovirus-mediated expression of a dominant negative Flk-1 mutant, which inhibits signal transduction through wild-type Flk-1 receptor, suppresses the growth of glioblastoma multiforme as well as other tumor cell lines in vivo (175,176). Furthermore, two recent studies have demonstrated that high local expression of sFlt-1, achieved by adenoviral-mediated gene transfer or by stable transfection of tumor cells, is able to significantly inhibit tumor growth, metastasis, and mortality rate (177,178).

A humanized version of a high-affinity anti-VEGF monoclonal antibody, which retains the same affinity and efficacy as the original murine antibody, has been generated by site-directed metagenesis of a human framework (179) and is being tested in humans as a treatment for solid tumors, alone or in combination with chemotherapy. Also, small molecules that inhibit Flk-1/KDR signal transduction have been identified and are undergoing clinical trials (180).

B. Angiogenesis Associated with Other Pathological Conditions

Diabetes mellitus, occlusion of central retinal vein, or prematurity with subsequent exposure to oxygen can all be associated with intraocular neovascularization (2). The new blood vessels may lead to vitreous hemorrhage, retinal detachment, neovascular glaucoma, and eventual blindness (2). Diabetic retinopathy is the leading cause of blindness in the working population (181). All of these conditions are known to be associated with retinal ischemia (182). In 1948, Michaelson proposed that a key event in the pathogenesis of these conditions is the release by the ischemic retina into the vitreous diffusible angiogenic factor(s) (''factor X'') responsible for retinal and iris neovascularization (183). Elevations of VEGF levels in the aqueous and vitreous of eyes with proliferative retinopathy have been described (184–186). In a large series, a strong correlation was found between levels of immunoreactive VEGF in the aqueous and vitreous humors and active proliferative retinopathy associated with diabetes, occlusion of central retinal vein, or prematurity (184). Furthermore, Okamoto et al. recently developed a transgenic mouse model in which the bovine rhodopsin promoter is coupled to the gene for human VEGF (187). This study demonstrates that overexpression of VEGF in the retina is sufficient to cause intraretinal and subretinal neovascularization and provides a new animal model to investigate this process.

More direct evidence for a role of VEGF as a mediator of intraocular neovascularization has been generated in a primate model of iris neovascularization and in a murine model of retinopathy of prematurity (188,189). In the former, intraocular administration of anti-VEGF antibodies dramatically inhibits the neovascularization that follows occlusion of central retinal veins (190). Likewise, soluble Flt-1 or Flk-1 fused to an IgG suppresses retinal angiogenesis in the mouse model (191).

Neovascularization is a major cause of visual loss also in age-related macular degeneration (AMD), the overall leading cause of blindness (2). Recent studies have demonstrated the immunohistochemical localization of VEGF in surgically resected choroidal neovascular membranes from AMD patients (192,193). These findings suggests a role for VEGF in the progression of AMD-related choroidal neovascularization, raising the possibility that a pharmacological treatment with monoclonal antibodies or other VEGF inhibitors may constitute a therapy for this condition.

Two independent studies have suggested that VEGF is involved in the pathogenesis of rheumatoid arthritis (RA), an inflammatory disease where angiogenesis plays a significant role (194,195). Levels of immunoreactive VEGF were found to be high in the synovial fluid of RA patients, while they were very low or undetectable in the synovial fluid of patients affected by other forms of arthritis or by degenerative joint disease (194, 195).

It has been shown that VEGF expression is increased in psoriatic skin (196). Increased vascularity and permeability are characteristic of psoriasis. Also, VEGF mRNA expression has been shown in three bullous disorders with subepidermal blister formation, bullous pemphigoid, erythema multiforme, and dermatitis herpetiformis (197).

IX. THERAPEUTIC APPLICATIONS OF VEGF

It has been shown that intraarterial or intramuscular administration of rhVEGF$_{165}$ may significantly augment perfusion and development of collateral vessels in a rabbit model, where chronic hindlimb ischemia was created by surgical removal of the femoral artery (198). These studies provided angiographic evidence of neovascularization in the ischemic limbs. Arterial gene transfer with cDNA encoding VEGF also led to revascularization in the same rabbit model to an extent comparable to that achieved with the recombinant protein (199,200). In addition, the hypothesis that the angiogenesis initiated by the administration of VEGF improved muscle function in ischemic limbs was tested by Walder et al. (201). A single intraarterial injection of rhVEGF$_{165}$ augmented muscle function in this rabbit model of peripheral limb ischemia. Similarly, Bauters et al. (202) have shown that both maximal flow velocity and maximal blood flow are significantly increased in ischemic limbs following VEGF administration. Other studies have shown that VEGF administration also leads to a recovery of normal endothelial reactivity in dysfunctional endothelium (203). Mack et al. have also shown that an adenovirus vector expressing VEGF$_{165}$ cDNA is capable of stimulating an angiogenic response that protects against acute vascular occlusion in the setting of preexisting ischemia in a rat model of hindlimb ischemia, suggesting that VEGF therapy might be useful in the prophylaxis of advancing arterial occlusive disease (204).

Banai et al. have shown that VEGF administration results in increased coronary blood flow in a dog model of coronary insufficiency (205). Following occlusion of the left circumflex coronary artery, daily intraluminal injections of rhVEGF distal to the occlusion resulted in a significant enhancement in collateral blood flow over a 4-week period. In addition, Harada et al. have demonstrated that extraluminal administration of as little as 2 μg of rhVEGF by an osmotic pump results in a significant increase in coronary blood flow in a pig model of chronic myocardial ischemia created by occlusion of the left proximal circumflex artery (206). Also, magnetic resonance imaging provided a noninvasive assessment of the benefits secondary to VEGF administration in the porcine model (207). Adenoviral-mediated gene transfer of VEGF$_{121}$ has been also shown to result in collateral vessel growth and functional improvement in a porcine model (208).

The hypothesis that VEGF may result in therapeutically significant angiogenesis in humans was initially tested by Isner et al. in a gene therapy trial in patients with severe limb ischemia (209). A case report of an interim analysis of this trial has been published (210). Arterial gene transfer of 2000 μg of naked plasmid DNA encoding VEGF$_{165}$, applied to the hydrogel polymer coating of an angioplasty balloon, resulted in angiographic and histological evidence of angiogenesis in the knee as well as the midtibial and ankle levels 4 weeks after the transfer. Such effects persisted at a 12-week view (210). In a subsequent trial, the VEGF$_{165}$ cDNA was injected intramuscularly (211). Gene transfer was performed in 10 limbs of 9 patients with nonhealing ischemic ulcers and/or rest pain due to peripheral arterial disease. The ankle-brachial index improved; newly visible collateral blood vessels were directly documented by contrast angiography in 7 limbs; and magnetic resonance angiography showed qualitative evidence of improved distal flow in 8 limbs.

Clinical trials using the recombinant protein (VEGF$_{165}$) are also ongoing. A phase I study in patients with coronary ischemia where VEGF was administered by intracoronary infusion has been reported, with encouraging initial results (212). The molecule was safely tolerated at all doses tested. However, patients experienced a dose-dependent transient decrease in systemic blood pressure, up to a maximum of 28 ± 8% at the dose of 0.167

μg/kg per minute. There was overall improvement in nuclear perfusion in 7/15 subjects and improved collateralization in 5/7 who underwent follow-up coronary angiography. Phase II studies are in progress.

X. PERSPECTIVES

The findings that heterozygous mutations inactivating the VEGF gene result in profound deficits in vasculogenesis and blood island formation, leading to early intrauterine death, emphasize the pivotal role played by this molecule in the development of the vascular system. However, numerous basic questions of the biology of VEGF are still incompletely answered, including the mechanisms of signal transduction of the VEGF receptors and the significance of the Flt-1 receptor and its unconventional mode of action.

The paradoxical finding that, in spite of the redundancy of angiogenic mediators, inactivation of VEGF alone substantially suppresses angiogenesis in a wide variety of physiological and pathological circumstances could be, at least in part, explained if one assumed the VEGF is a mediator of other angiogenic pathways. This hypothesis has recently received direct support by studies on FGF4 (213). The addition of conditioned medium of transfected cells expressing FGF4 to HUVEC resulted in a dose-dependent stimulation of VEGF, accompanied by a proliferative response. The induction of the angiogenic phenotype in HUVECs were completely suppressed by adding a neutralizing anti-VEGF antibody to FGF4-containing conditioned medium. These results suggest an indirect angiogenic activity of FGF-4 through the autocrine induction of VEGF secretion. Similarly, Seghezzi et al. have shown that bFGF induces VEGF expression in vascular endothelial cells through autocrine and paracrine mechanisms (214). Addition of recombinant bFGF to cultured endothelial cells results in increased VEGF expression. Neutralizing monoclonal antibody to VEGF inhibits bFGF-induced endothelial-cell proliferation. Remarkably, systemic administration of neutralizing anti-VEGF antibodies also reduces considerably bFGF-induced corneal angiogenesis (214). Whether this paradigm applies to growth factors other than bFGF or FGF4 remains to be established.

There is considerable experimental evidence that recombinant VEGF or gene therapy with VEGF gene may be used to promote endothelial cell growth and collateral vessel formation. This would represent a novel therapeutic modality for conditions that frequently are refractory to conservative measures and unresponsive to pharmacological therapy. As previously mentioned, clinical trials using $rhVEGF_{165}$ or gene therapy with VEGF cDNA are presently ongoing. It is unknown at the present time whether gene therapy or administration of recombinant VEGF would be the preferable approach. It is conceivable that gene therapy would minimize systemic effects of VEGF, such as hypotension. On the other hand, it is possible that the recombinant protein would be in other respects safer, minimizing or eliminating the risk of excessive or abnormal vascular proliferations that might result from high local expression of VEGF.

The high expression of VEGF mRNA in human tumors, the presence of the VEGF protein in ocular fluids of individuals with proliferative retinopathies and in the synovial fluid of RA patients, as well as the localization of VEGF in AMD lesions strongly supports the hypothesis that VEGF is a key mediator of angiogenesis associated with various disorders. Therefore, anti-VEGF antibodies or other inhibitors of VEGF, used alone or in combination with other agents, may be of therapeutic value for a variety of malignancies as well as for other disorders. The results of a phase I study using a humanized anti-VEGF monoclonal antibody in cancer patients have been recently reported (215). The treatment was found to be safe and tolerated and also showed some promising signs of clinical activity. Phase II studies in several tumor types are ongoing.

REFERENCES

1. Folkman J. Angiogenesis in cancer, vascular, rheumatoid and other disease. Nature Med 1: 27–31, 1995.
2. Garner A. Vascular Diseases, 2d ed. New York: Marcel Dekker, 1994.
3. Folkman J, Shing Y. Angiogenesis. J Biol Chem 267:10931–10934, 1992.
4. Risau W. Mechanisms of angiogenesis, Nature 386:671–674, 1997.
5. Suri C, Jones PF, Patan S, Bartunkova S, Maisonpierre PC, Davis S, Sato TN, Yancopoulos GD. Requisite role of angiopoietin-1, a ligand for the TIE2 receptor, during embryonic angiogenesis. Cell 87:1171–1180, 1996.
6. Maisonpierre PC, Suri C, Jones PF, Bartunkova S, Wiegend SJ, Radziejewski C, Compton D, McClain J, Aldrich TH, Papadopulos N, Daly TJ, Davis S, Sato TN, Yancopoulos GD. Angiopoietin-2, a natural antagonist for Tie-2 that disrupts in vivo angiogenesis. Science 277:55–60, 1997.
7. Good D, Polverini P, Rastinejad F, Beau M, Lemons R, Frazier W, Bouck N. A tumor suppressor-dependent inhibitor of angiogenesis is immunologically and functionally indistinguishable from a fragment of thrombospondin. Proc Natl Acad Sci USA 37:6624–6628, 1990.
8. DiPietro LA. Thrombospondin as a regulator of angiogenesis. In: Rosen E, Goldberg ID, eds. Regulation of Angiogenesis. Springer Verlag, 1997, pp. 295–314.
9. Ferrara N, Clapp C, Weiner R. The 16K fragment of prolactin specifically inhibits basal or fibroblast growth factor stimulated growth of capillary endothelial cells. Endocrinology 129: 896–900, 1991.
10. O'Reilly MS, Holmgren L, Shing Y, Chen C, Rosenthal RA, Moses M, Lane WS, Cao Y, Sage EH, Folkman J. Angiostatin: a novel angiogenesis inhibitor that mediates the suppression of metastases by a Lewis lung carcinoma. Cell 79:315–328, 1994.
11. O'Reilly MS, Boehm T, Shing Y, Fukai N, Vasios G, Lane WS, Flynn E, Birkhead JR, Olsen ER, Folkman J. Endostatin: an endogenous inhibitor of angiogenesis and tumor growth. Cell 88:277–285, 1997.
12. Maglione D, Guerriero V, Viglietto G, Delli-Bovi P, Persico MG. Isolation of a human placenta cDNA coding for a protein related to the vascular permeability factor. Proc Natl Acad Sci USA 88:9267–9271, 1991.
13. Olofsson B, Pajusola K, Kaipainen A, von Euler G, Joukov V, Saksela O, Orpana A, Pettersson RF, Alitalo K, Eriksson U. Vascular endothelial growth factor B, a novel growth factor for endothelial cells. Proc Natl Acad Sci USA 93:2576–2581, 1996.
14. Joukov V, Pajusola K, Kaipainen A, Chilov D, Lahtinen I, Kukk E, Saksela O, Kalkkinen N, Alitalo K. A novel vascular endothelial growth factor, VEGF-C, is a ligand for the Flt4 (VEGFR-3) and KDR (VEGFR-2) receptor tyrosine kinases. EMBO J 15:1751, 1996.
15. Lee J, Gray A, Yuan J, Luoh SM, Avraham H, Wood WI. Vascular endothelial growth factor-related protein: a ligand and specific activator of the tyrosine kinase receptor Flt4. Proc Natl Acad Sci USA 93:1988–1992, 1996.
16. Achen MG, Jeltsch M, Kukk E, Makinen T, Vitali A, Wilks AF, Alitalo K, Stacker SA. Vascular endothelial growth factor D (VEGF-D) is a ligand for the tyrosine kinases VEGF receptor 2 (Flk1) and VEGF receptor 3 (Flt4). Proc Natl Acad Sci USA 95:548–553, 1998.
17. Ferrara N, Davis-Smyth T. The biology of vascular endothelial growth factor. Endocr Rev 18:4–25, 1997.
18. Ferrara N, Carver Moore K, Chen H, Dowd M, Lu L, O'Shea KS, Powell Braxton L, Hillan

KJ, Moore MW. Heterozygous embryonic lethality induced by targeted inactivation of the VEGF gene. Nature 380:439–442, 1996.
19. Carmeliet P, Ferreira V, Breier G, Pollefeyt S, Kieckens L, Gertsenstein M, Fahrig M, Vandenhoeck A, Harpal K, Eberhardt C, Declercq C, Pawling J, Moons L, Collen D, Risau W, Nagy A. Abnormal blood vessel development and lethality in embryos lacking a single VEGF allele. Nature 380:435–439, 1996.
20. Leung DW, Cachianes G, Kuang WJ, Goeddel DV, Ferrara N. Vascular endothelial growth factor is a secreted angiogenic mitogen. Science 246:1306–1309, 1989.
21. Plouet J, Schilling J, Gospodarowicz D. Isolation and characterization of a newly identified endothelial cell mitogen produced by AtT20 cells. EMBO J 8:3801–3808, 1989.
22. Gerber HP, Kowalski J, McMurtrey A, Yan M, Keyt B, Dixit V, Ferrara N. VEGF regulates endothelial cell survival by the PI3-kinase/Akt signal transduction pathway: requirement for Flk-1/KDR activation. J Biol Chem 273:30336–30343, 1998.
23. Gerber HP, Dixit V, Ferrara N. Vascular endothelial growth factor induces expression of the antiapoptotic proteins Bcl-2 and A1 in vascular endothelial cells. J Biol Chem 273: 13313–13316, 1998.
24. Pepper MS, Ferrara N, Orci L, Montesano R. Vascular endothelial growth factor (VEGF) induces plasminogen activators and plasminogen activator inhibitor-1 in microvascular endothelial cells. Biochem Biophys Res Commun 181:902–906, 1991.
25. Unemori EN, Ferrara N, Bauer EA, Amento EP. Vascular endothelial growth factor induces interstitial collagenase expression in human endothelial cells. J Cell Physiol 153:557–562, 1992.
26. Dvorak HF, Brown LF, Detmar M, Dvorak AM. Vascular permeability factor/vascular endothelial growth factor, microvascular hyperpermeability, and angiogenesis. Am J Pathol 146: 1029–1039, 1995.
27. Dvorak HF. Tumors: wounds that do not heal. Similarities between tumor stroma generation and wound healing. N Engl J Med 315:1650–1659, 1986.
28. Dvorak HF, Harvey VS, Estrella P, Brown LF, McDonagh J, Dvorak AM. Fibrin containing gels induce angiogenesis. Implications for tumor stroma generation and wound healing. Lab Invest 57:673–686, 1987.
29. Roberts WG, Palade GE. Increased microvascular permeability and endothelial fenestration induced by vascular endothelial growth factor. J Cell Sci 108:2369–2379, 1995.
30. Roberts WG, Palade GE. Neovasculature induced by vascular endothelial growth factor is fenestrated. Cancer Res 57:765–772, 1997.
31. Esser S, Wolburg K, Wolburg H, Breier G, Kurzchalia T, Risau W. Vascular endothelial growth factor induces endothelial fenestrations in vitro. J Cell Biol 140:947–959, 1998.
32. Melder, RJ, Koenig GC, Witwer BP, Safabakhsh N, Munn LL, Jain RK. During angiogenesis, vascular endothelial growth factor and basic fibroblast growth factor regulate natural killer cell adhesion to tumor endothelium. Nature Med 2:992–997, 1996.
33. Clauss M, Gerlach M, Gerlach H, Brett J, Wang F, Familletti PC, Pan YC, Olander JV, Connolly DT, Stern D. Vascular permeability factor: a tumor-derived polypeptide that induces endothelial cell and monocyte procoagulant activity, and promotes monocyte migration. J Exp Med 172:1535–1545, 1990.
34. Broxmeyer HE, Cooper S, Li ZH, Lu L, Song HY, Kwon BS, Warren RE, Donner DB. Myeloid progenitor cell regulatory effects of vascular endothelial cell growth factor. Int J Hematol 62:203–215, 1995.
35. Gabrilovich DI, Chen HL, Girgis KR, Cunningham HT, Meny GM, Nadaf S, Kavanaugh D, Carbone DP. Production of vascular endothelial growth factor by human tumors inhibits the functional maturation of dendritic cells, Nature Med 2:1096–1103, 1996.
36. Ku DD, Zaleski JK, Liu S, Brock TA. Vascular endothelial growth factor induces EDRF-dependent relaxation in coronary arteries. Am J Physiol 265:H586–H592, 1993.
37. Yang R, Thomas GR, Bunting S, Ko A, Ferrara N, Keyt B, Ross J, Jin H. Effects of vascular

endothelial growth factor on hemodynamics and cardiac performance. J Cardiovasc Pharmacol 27:838–844, 1996.
38. Cuevas P, Carceller F, Ortega S, Zazo M, Nieto I, Gimenez-Gallego G. Hypotensive activity of fibroblast growth factor. Science 254:1208–1210, 1991.
39. Cuevas P, Garcia-Calvo M, Carceller F, Reimers D, Zazo M, Cuevas B, Munoz-Willery I, Martinez-Coso V, Lamas S, Gimenez-Gallego G. Correction of hypertension by normalization of endothelial levels of fibroblast growth factor and nitric oxide synthase in spontaneously hypertensive rats. Proc Natl Acad Sci USA 93:11996–12001, 1996.
40. Houck KA, Ferrara N, Winer J, Cachianes G, Li B, Leung DW. The vascular endothelial growth factor family: identification of a fourth molecular species and characterization of alternative splicing of RNA. Mol Endocrinol 5:1806–1814, 1991.
41. Tischer E, Mitchell R, Hartman T, Silva M, Gospodarowicz D, Fiddes JC, Abraham JA. The human gene for vascular endothelial growth factor: multiple protein forms are encoded through alternative exon splicing. J Biol Chem 266:11947–11954, 1991.
42. Vincenti V, Cassano C, Rocchi M, Persico G. Assignment of the vascular endothelial growth factor gene to human chromosome 6p21.3. Circulation 93:1493–1495, 1996.
43. Ferrara N, Henzel WJ. Pituitary follicular cells secrete a novel heparin-binding growth factor specific for vascular endothelial cells. Biochem Biophys Res Commun 161:851–858, 1989.
44. Houck KA, Leung DW, Rowland AM, Winer J, Ferrara N. Dual regulation of vascular endothelial growth factor bioavailability by genetic and proteolytic mechanisms. J Biol Chem 267:26031–26037, 1992.
45. Park JE, Keller H-A, Ferrara N. The vascular endothelial growth factor isoforms (VEGF): Differential deposition into the subepithelial extracellular matrix and bioactivity of extracellular matrix-bound VEGF. Mol Biol Cell 4:1317–1326, 1993.
46. Keyt BA, Berleau LT, Nguyen HV, Chen H, Heinsohn H, Vandlen R, Ferrara N. The carboxyl-terminal domain (111-165) of vascular endothelial growth factor is critical for its mitogenic potency. J Biol Chem 271:7788–7795, 1996.
47. Plouet J, Moro F, Bertagnolli S, Coldeboeuf N, Mazarguil H, Clamens S, Bayard F. Extracellular cleavage of the vascular endothelial growth factor 189-amino acid form by urokinase is required for its mitogenic effect. J Biol Chem 272:13390–13396, 1997.
48. Poltorak Z, Cohen T, Sivan R, Kandelis Y, Spira G, Vlodavsky I, Keshet E, Neufeld G. VEGF145, a secreted vascular endothelial growth factor isoform that binds to extracellular matrix. J Biol Chem 272:7151–7158, 1997.
49. Muller YA, Li B, Christinger HW, Wells JA, Cunningham BC, de Vos AM. Vascular endothelial growth factor: crystal structure and functional mapping of the kinase domain receptor binding site. Proc Natl Acad Sci USA 94:7192–7197, 1997.
50. Minchenko A, Bauer T, Salceda S, Caro J. Hypoxic stimulation of vascular endothelial growth factor expression in vivo and in vitro. Lab Invest 71:374–379, 1994.
51. Shima DT, Adamis AP, Ferrara N, Yeo KT, Yeo TK, Allende R, Folkman J, D'Amore PA. Hypoxic induction of endothelial cell growth factors in retinal cells: identification and characterization of vascular endothelial growth factor (VEGF) as the mitogen. Mol Med 1: 182–193, 1995.
52. Banai S, Shweiki D, Pinson A, Chandra M, Lazarovich G, Keshet E. Upregulation of vascular endothelial growth factor expression induced by myocardial ischemia: implications for coronary angiogenesis. Cardiovasc Res 28:1176–1179, 1994.
53. Hashimoto E, Ogita T, Nakaoka T, Matsuoka R, Takao A, Kira Y. Rapid induction of vascular endothelial growth factor expression by transient ischemia in rat heart. Am J Physiol 267: H1948–H1954, 1994.
54. Goldberg MA, Schneider TJ. Similarities between the oxygen-sensing mechanisms regulating the expression of vascular endothelial growth factor and erythropoietin. J Biol Chem 269:4355–4361, 1994.

55. Levy AP, Levy NS, Wegner S, Goldberg MA. Transcriptional regulation of the rat vascular endothelial growth factor gene by hypoxia. J Biol Chem 270:13333–13340, 1995.
56. Liu Y, Cox SR, Morita T, Kourembanas S. Hypoxia regulates vascular endothelial growth factor gene expression in endothelial cells: identification of a 5′ enhancer. Circ Res 77:638–643, 1995.
57. Madan A, Curtin PT. A 24-base-pair sequence 3′ to the human erythropoietin gene contains a hypoxia-responsive transcriptional enhancer. Proc Natl Acad Sci USA 90:3928–3932, 1993.
58. Wang GL, Semenza GL. Purification and characterization of hypoxia-inducible factor 1. J Biol Chem 270:1230–1237, 1995.
59. Iyer NV, Kotch LE, Agani F, Leung SW, Laughner E, Wenger RH, Gassmann M, Gearhart JD, Lawler AM, Yu AY, Semenza GL. Cellular and developmental control of O2 homeostasis by hypoxia-inducible factor 1 alpha. Genes Dev 12:149–162, 1998.
60. Ryan HE, Lo J, Johnson RS. HIF-1 alpha is required for solid tumor formation and embryonic vascularization. EMBO J 17:3005–3015, 1998.
61. Carmeliet P, Dor Y, Herbert JM, Fukumura D, Brusselmans K, Neeman M, Bono F, Abramovitch R, Maxwell P, Koch CJ, Ratcliffe P, Moons L, Jain RK, Collen D, Keshet E. Role of HIF-1-alpha in hypoxia-mediated apoptosis, cell proliferation and tumour angiogenesis. Nature 394:485–490, 1998.
62. Ikeda E, Achen MG, Breier G, Risau W. Hypoxia-induced transcriptional activation and increased mRNA stability of vascular endothelial growth factor in C6 glioma cells. J Biol Chem 270:19761–19766, 1995.
63. Levy AP, Levy NS, Goldberg MA. Post-transcriptional regulation of vascular endothelial growth factor by hypoxia. J Biol Chem 271:2746–2753, 1996.
64. Frank S, Hubner G, Breier G, Longaker MT, Greenhalg DG, Werner S. Regulation of VEGF expression in cultured keratinocytes: implications for normal and impaired wound healing. J Biol Chem 270:12607–12613, 1995.
65. Goldman C, Kim J, Wonf W-L, King V, Brock T, Gillespie Y. Epidermal growth factor stimulates vascular endothelial growth factor production by malignant glioma cells: a model of glioblastoma multiforme pathophysiology. Mol Biol Cell 4:121–133, 1993.
66. Pertovaara L, Kaipainen A, Mustonen T, Orpana A, Ferrara N, Saksela O, Alitalo K. Vascular endothelial growth factor is induced in response to transforming growth factor-beta in fibroblastic and epithelial cells. J Biol Chem 269:6271–6274, 1994.
67. Ben-Av P, Crofford LJ, Wilder RL, Hla T. Induction of vascular endothelial growth factor expression in synovial fibroblasts. FEBS Lett 372:83–87, 1995.
68. Cohen T, Nahari D, Cerem LW, Neufeld G, Levi BZ. Interleukin 6 induces the expression of vascular endothelial growth factor. J Biol Chem 271:736–741, 1996.
69. Warren RS, Yuan H, Matli MR, Ferrara N, Donner DB. Induction of vascular endothelial growth factor by insulin-like growth factor 1 in colorectal carcinoma. J Biol Chem 271: 29483–29488, 1996.
70. Soh EY, Sobhi SA, Wong MG, Meng YG, Siperstein AE, Clark OH, Duh QY. Thyroid-stimulating hormone promotes the secretion of vascular endothelial growth factor in thyroid cancer cell lines. Surgery 120:944–947, 1996.
71. Shifren JL, Mesiano S, Taylor RN, Ferrara N, Jaffe RB. Corticotropin regulates vascular endothelial growth factor expression in human fetal adrenal cortical cells. J Clin Endocrinol Metabolism 83:1342–1347, 1998.
72. Claffey KP, Wilkison WO, Spiegelman BM. Vascular endothelial growth factor: regulation by cell differentiation and activated second messenger pathways. J Biol Chem 267:16317–16322, 1992.
73. Kieser A, Weich H, Brandner G, Marme D, Kolch W. Mutant p53 potentiates protein kinase C induction of vascular endothelial growth factor expression. Oncogene 9:963–969, 1994.
74. Rak J, Mitsuhashi Y, Bayko L, Filmus J, Shirasawa S, Sasazuki T, Kerbel RS. Mutant ras

oncogenes upregulate VEGF/VPF expression: implications for induction and inhibition of tumor angiogenesis. Cancer Res 55:4575–4580, 1995.
75. Grugel S, Finkenzeller G, Weindel K, Barleon B, Marme D. Both v-Ha-Ras and v-Raf stimulate expression of the vascular endothelial growth factor in NIH 3T3 cells. J Biol Chem 270: 25915–25919, 1995.
76. Okada F, Rak JW, Croix BS, Lieubeau B, Kaya M, Roncari L, Shirasawa S, Sasazuki T, Kerbel RS. Impact of oncogenes in tumor angiogenesis: mutant K-ras up-regulation of vascular endothelial growth factor/vascular permeability factor is necessary, but not sufficient for tumorigenicity of human colorectal carcinoma cells. Proc Natl Acad Sci USA 95:3609–3614, 1998.
77. Siemeister G, Weindel K, Mohrs K, Barleon B, Martiny Baron G, Marme D. Reversion of deregulated expression of vascular endothelial growth factor in human renal carcinoma cells by von Hippel-Lindau tumor suppressor protein. Cancer Res 56:2299–2301, 1996.
78. Iliopoulos O, Levy AP, Jiang C, Kaelin WG, Jr, Goldberg MA. Negative regulation of hypoxia-inducible genes by the von Hippel-Lindau protein. Proc Natl Acad Sci USA 93: 10595–10599, 1996.
79. Gnarra JR, Zhou S, Merrill MJ, Wagner JR, Krumm A, Papavassiliou E, Oldfield EH, Klausner RD, Linehan WM. Post-transcriptional regulation of vascular endothelial growth factor mRNA by the product of the VHL tumor suppressor gene. Proc Natl Acad Sci USA 93:10589–10594, 1996.
80. Mukhopadhyay D, Knebelmann B, Cohen HT, Ananth S, Sukhatme VP. The von Hippel-Lindau tumor suppressor gene product interacts with Sp1 to repress vascular endothelial growth factor promoter activity. Mol Cell Biol 17:5629–5639, 1997.
81. Jakeman LB, Winer J, Bennett GL, Altar CA, Ferrara N. Binding sites for vascular endothelial growth factor are localized on endothelial cells in adult rat tissues. J Clin Invest 89:244–253, 1992.
82. Jakeman LB, Armanini M, Philips HS, Ferrara N. Developmental expression of binding sites and mRNA for vascular endothelial growth factor suggests a role for this protein in vasculogenesis and angiogenesis. Endocrinology 133:848–859, 1993.
83. de Vries C, Escobedo JA, Ueno H, Houck K, Ferrara N, Williams LT. The fms-like tyrosine kinase, a receptor for vascular endothelial growth factor. Science 255:989–991, 1992.
84. Terman BI, Dougher Vermazen M, Carrion ME, Dimitrov D, Armellino DC, Gospodarowicz D, Bohlen P. Identification of the KDR tyrosine kinase as a receptor for vascular endothelial cell growth factor. Biochem Biophys Res Commun 187:1579–1586, 1992.
85. Matthews W, Jordan CT, Gavin M, Jenkins NA, Copeland NG, Lemischka IR. A receptor tyrosine kinase cDNA isolated from a population of enriched primitive hematopoietic cells and exhibiting close genetic linkage to c-kit. Proc Natl Acad Sci USA 88:9026–9030, 1991.
86. Shibuya M, Yamaguchi S, Yamane A, Ikeda T, Tojo A, Matsushime H, Sato M. Nucleotide sequence and expression of a novel human receptor-type tyrosine kinase (flt) closely related to the fms family. Oncogene 8:519–527, 1990.
87. Terman BI, Carrion ME, Kovacs E, Rasmussen BA, Eddy RL, Shows TB. Identification of a new endothelial cell growth factor receptor tyrosine kinase. Oncogene 6:1677–1683, 1991.
88. Kendall RL, Thomas KA. Inhibition of vascular endothelial cell growth factor activity by an endogenously encoded soluble receptor. Proc Natl Acad Sci USA 90:10705–10709, 1993.
89. Kendall RL, Wang G, Thomas KA. Identification of a natural soluble form of the vascular endothelial growth factor receptor, FLT-1, and its heterodimerization with KDR. Biochem Biophys Res Commun 226:324–328, 1996.
90. Pajusola K, Aprelikova O, Korhonen J, Kaipainen A, Pertovaara L, Alitalo R, Alitalo K. FLT4 receptor tyrosine kinase contains seven immunoglobulin-like loops and is expressed in multiple human tissues and cell lines. Cancer Res 52:5738–5743, 1992.
91. Galland F, Karamysheva A, Mattei MG, Rosnet O, Marchetto S, Birnbaum D. Chromosomal

localization of FLT4, a novel receptor-type tyrosine kinase gene. Genomics 13:475–478, 1992.

92. Finnerty H, Kelleher K, Morris GE, Bean K, Merberg DM, Kriz R, Morris JC, Sookdeo H, Turner KJ, Wood CR. Molecular cloning of murine FLT and FLT4. Oncogene 8:2293–2298, 1993.
93. Jeltsch M, Kaipainen A, Joukov V, Meng X, Lakso M, Rauvala H, Awartz M, Fukumura D, Jain RK, Alitalo K. Hyperplasia of lymphatic vessels in VEGF-C transgenic mice. Science 276:1423–1425, 1997.
94. Davis-Smyth T, Chen H, Park J, Presta LG, Ferrara N. The second immunoglobulin-like domain of the VEGF tyrosine kinase receptor Flt-1 determines ligand binding and may initiate a signal transduction cascade. EMBO J 15:4919–4927, 1996.
95. Barleon B, Totzke F, Herzog C, Blanke S, Kremmer E, Siemeister G, Marme D, Martiny-Baron G. Mapping of the sites for ligand binding and receptor dimerization at the extracellular domain of the vascular endothelial growth factor receptor FLT-1. J Biol Chem 272:10382–10388, 1997.
96. Fuh G, Li B, Crowley C, Cunningham B, Wells JA. Requirements for binding and signaling of the kinase domain receptor for vascular endothelial growth factor. J Biol Chem 273: 11197–11204, 1998.
97. Wiesmann C, Fuh G, Christinger HW, Eigenbrot C, Wells JA, de Vos AM. Crystal structure at 1.7 A resolution of VEGF in complex with domain 2 of the Flt-1 receptor. Cell 91:695–704, 1997.
98. Davis-Smyth T, Presta LG, Ferrara N. Mapping the charged residues in the second immunoglobulin-like domain of the vascular endothelial growth factor/placenta growth factor receptor Flt-1 required for binding and structural stability. J Biol Chem 273:3216–3222, 1998.
99. Guo D, Jia Q, Song HY, Warren RS, Donner DB. Vascular endothelial cell growth factor promotes tyrosine phosphorylation of mediators of signal transduction that contain SH2 domains: association with endothelial cell proliferation. J Biol Chem 270:6729–6733, 1995.
100. Morbidelli L, Chang CH, Douglas JG, Granger HJ, Ledda F, Ziche M. Nitric oxide mediates mitogenic effect of VEGF on coronary venular endothelium. Am J Physiol 270:H411–415, 1996.
101. Waltenberger J, Claesson Welsh L, Siegbahn A, Shibuya M, Heldin CH. Different signal transduction properties of KDR and Flt1, two receptors for vascular endothelial growth factor. J Biol Chem 269:26988–26995, 1994.
102. Seetharam L, Gotoh N, Maru Y, Neufeld G, Yamaguchi S, Shibuya M. A unique signal transduction from FLT tyrosine kinase, a receptor for vascular endothelial growth factor VEGF. Oncogene 10:135–147, 1995.
103. Park JE, Chen HH, Winer J, Houck KA, Ferrara N. Placenta growth factor. Potentiation of vascular endothelial growth factor bioactivity, in vitro and in vivo, and high affinity binding to Flt-1 but not to Flk-1/KDR. J Biol Chem 269:25646–25654, 1994.
104. Cunningham SA, Waxham MN, Arrate PM, Brock TA. Interaction of the Flt-1 tyrosine kinase receptor with the p85 subunit of phosphatidylinositol 3-kinase: mapping of a novel site involved in binding. J Biol Chem 270:20254–20257, 1995.
105. Landgren E, Schiller P, Cao Y, Claesson-Welsh L. Placenta growth factor stimulates MAP kinase and mitogenicity but not phospholipase C-gamma and migration of endothelial cells expressing Flt 1. Oncogene 16:359–367, 1998.
106. Maru Y, Yamaguchi S, Shibuya M. Flt-1, a receptor for vascular endothelial growth factor, has transforming and morphogenic potentials. Oncogene 16:2585–2595, 1998.
107. Hiratsuka S, Minowa O, Kuno J, Noda T, Shibuya M. Flt-1 lacking the tyrosine kinase domain is sufficient for normal development and angiogenesis in mice. Proc Natl Acad Sci USA 4:9349–9354, 1998.
108. Keyt BA, Nguyen HV, Berleau LT, Duarte CM, Park J, Chen H, Ferrara N. Identification of vascular endothelial growth factor determinants for binding KDR and FLT-1 receptors:

generation of receptor-selective VEGF variants by site-directed mutagenesis. J Biol Chem 271:5638–5646, 1996.
109. Ortega N, Jonca F, Vincent S, Favard C, Ruchoux MM, Plouet J. Systemic activation of the vascular endothelial growth factor receptor KDR/flk-1 selectively triggers endothelial cells with an angiogenic phenotype. Am J Pathol 151:1215–1224, 1997.
110. Barleon B, Sozzani S, Zhou D, Weich HA, Mantovani A, Marme D. Migration of human monocytes in response to vascular endothelial growth factor (VEGF) is mediated via the VEGF receptor flt-1. Blood 87:3336–3343, 1996.
111. Oyama T, Ran S, Ishida T, Nadaf S, Kerr L, Carbone DP, Gabrilovich DI. Vascular endothelial growth factor affects dendritic cell maturation through the inhibition of nuclear factor-kappa B activation in hemopoietic progenitor cells. J Immunol 160:1224–1232, 1998.
112. Soker S, Takashima S, Miao HQ, Neufeld G, Klagsbrun M. Neuropilin-1 is expressed by endothelial and tumor cells as an isoform-specific receptor for vascular endothelial growth factor. Cell 92:735–745, 1998.
113. Morishita K, Johnson DE, Williams LT. A novel promoter for vascular endothelial growth factor receptor (flt-1) that confers endothelial-specific gene expression. J Biol Chem 270: 27948–27953, 1995.
114. Patterson C, Perrella MA, Hsieh CM, Yoshizumi M, Lee ME, Haber E. Cloning and functional analysis of the promoter for KDR/flk-1, a receptor for vascular endothelial growth factor. J Biol Chem 270:23111–23118, 1995.
115. Tuder RM, Flook BE, Voelkel NF. Increased gene expression for VEGF and the VEGF receptors KDR/Flk and Flt in lungs exposed to acute or to chronic hypoxia: modulation of gene expression by nitric oxide. J Clin Invest 95:1798–1807, 1995.
116. Li J, Brown LF, Hibberd MG, Grossman JD, Morgan JP, Simons M. VEGF, flk-1, and flt-1 expression in a rat myocardial infarction model of angiogenesis. Am J Physiol 270: H1803–H1811, 1996.
117. Takagi H, King GL, Ferrara N, Aiello LP. Hypoxia regulates vascular endothelial growth factor receptor KDR/Flk gene expression through adenosine A2 receptors in retinal capillary endothelial cells. Invest Ophthalmol Vis Sci 37:1311–1321, 1996.
118. Brogi E, Schatteman G, Wu T, Kim EA, Varticovski L, Keyt B, Isner JM. Hypoxia-induced paracrine regulation of vascular endothelial growth factor receptor expression. J Clin Invest 97:469–475, 1996.
119. Gerber HP, Condorelli F, Park J, Ferrara N. Differential transcriptional regulation of the two VEGF receptor genes: Flt-1, but not Flk-1/KDR, is up-regulated by hypoxia. J Biol Chem 272:23659–23667, 1997.
120. Mandriota SJ, Menoud PA, Pepper MS. Transforming growth factor beta 1 down-regulates vascular endothelial growth factor receptor 2/flk-1 expression in vascular endothelial cells. J Biol Chem 271:11500–11505, 1996.
121. Patterson C, Perrella MA, Endege WO, Yoshizumi M, Lee ME, Haber E. Downregulation of vascular endothelial growth factor receptors by tumor necrosis factor-alpha in cultured human vascular endothelial cells. J Clin Invest 98:490–496, 1996.
122. Giraudo E, Primo L, Audero E, Gerber HP, Koolwijk P, Soker S, Klagsbrun M, Ferrara N, Bussolino F. Tumor necrosis factor-alpha regulates expression of vascular endothelial growth factor receptor-2 and of its co-receptor neuropilin-1 in human vascular endothelial cells. J Biol Chem 273:22128–22135, 1998.
123. Fong GH, Rossant J, Gertsenstein M, Breitman ML. Role of the Flt-1 receptor tyrosine kinase in regulating the assembly of vascular endothelium. Nature 376:66–70, 1995.
124. Shalaby F, Rossant J, Yamaguchi TP, Gertsenstein M, Wu XF, Breitman ML, Schuh AC. Failure of blood-island formation and vasculogenesis in Flk-1-deficient mice. Nature 376: 62–66, 1995.
125. Brandon EP, Idzerda RL, McKnight GS. Targeting the mouse genome: a compendium of knockouts (part III). Curr Biol 5:873–881, 1995.

126. Carmeliet P, Collen D. Genetic analysis of blood vessel formation:role of endothelial versus smooth muscle cells. Trends Cardiovasc Med 8:271–281, 1997.
127. Gerber HP, Hillan KJ, Ryan AM, Kowalski J, Chen H, Wright BD, Radtke F, Aguet M, Ferrara N. VEGF is essential for somatic growth and survival in neonatal mice. Development. In press.
128. Simon M, Grone HJ, Johren O, Kullmer J, Plate KH, Risau W, Fuchs E. Expression of vascular endothelial growth factor and its receptors in human renal ontogenesis and in adult kidney. Am J Physiol 268:F240–F250, 1995.
129. Kitamoto Y, Tokunaga H, Tomita K. Vascular endothelial growth factor is an essential molecule for mouse kidney development: glomerulogenesis and nephrogenesis. J Clin Invest 99: 2351–2357, 1997.
130. Tufro-McReddie A, Norwood VF, Aylor KW, Botkin SJ, Carey RM, Gomez RA. Oxygen regulates vascular endothelial growth factor–mediated vasculogenesis and tubulogenesis. Dev Biol 183:139–149, 1997.
131. Bassett DL. The changes in the vascular pattern of the ovary of the albino rat during the estrous cycle. Am J Anat 73:251–278, 1943.
132. Gospodarowicz D, Takral TK. Production of a corpus luteum angiogenic factor responsible for proliferation of capillaries and neovascularization of the corpus luteum. Proc Natl Acad Sci USA 75:847–851, 1978.
133. Phillips HS, Hains J, Leung DW, Ferrara N. Vascular endothelial growth factor is expressed in rat corpus luteum. Endocrinology 127:965–967, 1990.
134. Ravindranath N, Little-Ihrig L, Phillips HS, Ferrara N, Zeleznik AJ. Vascular endothelial growth factor messenger ribonucleic acid expression in the primate ovary. Endocrinology 131:254–260, 1992.
135. Shweiki D, Itin A, Neufeld G, Gitay-Goren H, Keshet E. Patterns of expression of vascular endothelial growth factor (VEGF) and VEGF receptors in mice suggest a role in hormonally-mediated angiogenesis. J Clin Invest 91:2235–2243, 1993.
136. Cullinan-Bove K, Koos RD. Vascular endothelial growth factor/vascular permeability factor expression in the rat uterus: rapid stimulation by estrogen correlates with estrogen-induced increases in uterine capillary permeability and growth. Endocrinology 133:829–837, 1993.
137. Ferrara N, Chen H, Davis-Smyth T, Gerber H-P, Nguyen T-N, Peers D, Chisholm V, Hillan KJ, Schwall RH. Vascular endothelial growth factor is essential for corpus luteum angiogenesis. Nature Med 4:336–340, 1998.
138. Algire GH, Chalkley HW, Legallais FY, Park HD. Vascular reactions of normal and malignant tissues in vivo: I. Vascular reactions of mice to wounds and to normal and neoplastic transplants. J Natl Cancer Inst 6:73–85, 1945.
139. Folkman J. Tumor angiogenesis: therapeutic implications. N Engl J Med 285:1182–1186, 1971.
140. Folkman J, Klagsbrun M. Angiogenic factors. Science 235:442–447, 1987.
141. Dennis PA, Rifkin DB. Studies on the role of basic fibroblast growth factor in vivo: inability of neutralizing antibodies to block tumor growth. J Cell Physiol 144:84–98, 1990.
142. Ferrara N, Houck K, Jakeman L, Leung DW. Molecular and biological properties of the vascular endothelial growth facmily of proteins. Endocrinol Rev 13:18–32, 1992.
143. Volm M, Koomagi R, Mattern J, Stammler G. Angiogenic growth factors and their receptors in non-small cell lung carcinomas and their relationships to drug response in vitro. Anticancer Res 17:99–103, 1997.
144. Volm M, Koomagi R, Mattern J. Prognostic value of vascular endothelial growth factor and its receptor Flt-1 in squamous cell lung cancer. Int J Cancer 74:64–68, 1997.
145. Brown LF, Berse B, Jackman RW, Tognazzi K, Guidi AJ, Dvorak HF, Senger DR, Connolly JL, Schnitt SJ. Expression of vascular permeability factor (vascular endothelial growth factor) and its receptors in breast cancer. Hum Pathol 26:86–91, 1995.
146. Yoshiji H, Gomez DE, Shibuya M, Thorgeirsson UP. Expression of vascular endothelial

growth factor, its receptor, and other angiogenic factors in human breast cancer. Cancer Res 56:2013–2016, 1996.
147. Brown LF, Berse B, Jackman RW, Tognazzi K, Manseau EJ, Senger DR, Dvorak HF. Expression of vascular permeability factor (vascular endothelial growth factor) and its receptors in adenocarcinomas of the gastrointestinal tract. Cancer Res 53:4727–4735, 1993.
148. Suzuki K, Hayashi N, Miyamoto Y, Yamamoto M, Ohkawa K, Ito Y, Sasaki Y, Yamaguchi Y, Nakase H, Noda K, Enomoto N, Arai K, Yamada Y, Yoshihara H, Tujimura T, Kawano K, Yoshikawa K, Kamada T. Expression of vascular permeability factor/vascular endothelial growth factor in human hepatocellular carcinoma. Cancer Res 56:3004–3009, 1996.
149. Ellis LM, Takahashi Y, Fenoglio CJ, Cleary KR, Bucana CD, Evans DB. Vessel counts and vascular endothelial growth factor expression in pancreatic adenocarcinoma. Eur J Cancer 34:337–340, 1998.
150. Uchida S, Shimada Y, Watanabe G, Tanaka H, Shibagaki I, Miyahara T, Ishigami S, Imamura M. In oesophageal squamous cell carcinoma vascular endothelial growth factor is associated with p53 mutation, advanced stage and poor prognosis. Br J Cancer 77:1704–1709, 1998.
151. Brown LF, Berse B, Jackman RW, Tognazzi K, Manseau EJ, Dvorak HF, Senger DR. Increased expression of vascular permeability factor (vascular endothelial growth factor) and its receptors in kidney and bladder carcinomas. Am J Pathol 143:1255–1262, 1993.
152. Nicol D, Hii SI, Walsh M, Teh B, Thompson L, Kennett C, Gotley D. Vascular endothelial growth factor expression is increased in renal cell carcinoma. J Urol 157:1482–1486, 1997.
153. Olson TA, Mohanraj D, Carson LF, Ramakrishnan S. Vascular permeability factor gene expression in normal and neoplastic human ovaries. Cancer Res 54:276–280, 1994.
154. Guidi AJ, Abu Jawdeh G, Tognazzi K, Dvorak HF, Brown LF. Expression of vascular permeability factor (vascular endothelial growth factor) and its receptors in endometrial carcinoma. Cancer 78:454–460, 1996.
155. Shweiki D, Itin A, Soffer D, Keshet E. Vascular endothelial growth factor induced by hypoxia may mediate hypoxia-initiated angiogenesis. Nature 359:843–845, 1992.
156. Plate KH, Breier G, Weich HA, Risau W. Vascular endothelial growth factor is a potential tumour angiogenesis factor in human gliomas in vivo. Nature 359:845–848, 1992.
157. Phillips HS, Armanini M, Stavrou D, Ferrara N, Westphal M. Intense focal expression of vascular endothelial growth factor mRNA in human intracranial neoplasms: Association with regions of necrosis. Int J Oncol 2:913–919, 1993.
158. Berkman RA, Merrill MJ, Reinhold WC, Monacci WT, Saxena A, Clark WC, Robertson JT, Ali IU, Oldfield EH. Expression of the vascular permeability factor/vascular endothelial growth factor gene in central nervous system neoplasms. J Clin Invest 91:153–159, 1993.
159. Wizigmann Voos S, Breier G, Risau W, Plate KH. Up-regulation of vascular endothelial growth factor and its receptors in von Hippel-Lindau disease–associated and sporadic hemangioblastomas. Cancer Res 55:1358–1364, 1995.
160. Toi M, Kondo S, Suzuki H, Yamamoto Y, Inada K, Imazawa T, Taniguchi T, Tominaga T. Quantitative analysis of vascular endothelial growth factor in primary breast cancer. Cancer 77:1101–1106, 1996.
161. Gasparini G, Toi M, Gion M, Verderio P, Dittadi R, Hanatani M, Matsubara I, Vinante O, Bonoldi E, Boracchi P, Gatti C, Suzuki H, Tominaga T. Prognostic significance of vascular endothelial growth factor protein in node-negative breast carcinoma. J Natl Cancer Inst 89: 139–147, 1997.
162. Maeda K, Chung YS, Ogawa Y, Takatsuka S, Kang SM, Ogawa M, Sawada T, Sowa M. Prognostic value of vascular endothelial growth factor expression in gastric carcinoma. Cancer 77:858–863, 1996.
163. Kondo S, Asano M, Matsuo K, Ohmori I, Suzuki H. Vascular endothelial growth factor/vascular permeability factor is detectable in the sera of tumor-bearing mice and cancer patients. Biochim. Biophys Acta 1221:211–214, 1994.
164. Salven P, Teerenhovi L, Joensuu H. A high pretreatment serum vascular endothelial growth

factor concentration is associated with poor outcome in non-Hodgkin's lymphoma. Blood 90:3167–3172, 1997.

165. Salven P, Ruotsalainen T, Mattson K, Joensuu H. High pre-treatment serum level of vascular endothelial growth factor (VEGF) is associated with poor outcome in small-cell lung cancer. Int J Cancer 79:144–146, 1998.
166. Takigawa N, Segawa Y, Fujimoto N, Hotta K, Eguchi K. Elevated vascular endothelial growth factor levels in sera of patients with lung cancer. Anticancer Res 18:1251–1254, 1998.
167. Kim KJ, Li B, Houck K, Winer J, Ferrara N. The vascular endothelial growth factor proteins: identification of biologically relevant regions by neutralizing monoclonal antibodies. Growth Factors 7:53–64, 1992.
168. Kim KJ, Li B, Winer J, Armanini M, Gillett N, Phillips HS, Ferrara N. Inhibition of vascular endothelial growth factor-induced angiogenesis suppresses tumor growth in vivo. Nature 362: 841–844, 1993.
169. Warren RS, Yuan H, Matli MR, Gillett NA, Ferrara N. Regulation by vascular endothelial growth factor of human colon cancer tumorigenesis in a mouse model of experimental liver metastasis. J Clin Invest 95:1789–1797, 1995.
170. Melnyk O, Shuman MA, Kim KJ. Vascular endothelial growth factor promotes tumor dissemination by a mechanism distinct from its effect on primary tumor growth. Cancer Res 56:921–924, 1996.
171. Asano M, Yukita A, Matsumoto T, Kondo S, Suzuki H. Inhibition of tumor growth and metastasis by an immunoneutralizing monoclonal antibody to human vascular endothelial growth factor/vascular permeability factor 121. Cancer Res 55:5296–5301, 1995.
172. Borgstrom P, Bourdon MA, Hillan KJ, Sriramarao P, Ferrara N. Neutralizing anti-vascular endothelial growth factor antibody completely inhibits angiogenesis and growth of human prostate carcinoma micro tumors in vivo. Prostate 35:1–10, 1998.
173. Borgstrom P, Hillan KJ, Sriramarao P, Ferrara N. Complete inhibition of angiogenesis and growth of microtumors by anti-vascular endothelial growth factor neutralizing antibody: novel concepts of angiostatic therapy from intravital videomicroscopy. Cancer Res 56:4032–4039, 1996.
174. Yuan F, Chen Y, Dellian M, Safabakhsh N, Ferrara N, Jain RK. Time-dependent vascular regression and permeability changes in established human tumor xenografts induced by an anti-vascular endothelial growth factor/vascular permeability factor antibody. Proc Natl Acad Sci USA 93:14765–14770, 1996.
175. Millauer B, Shawver LK, Plate KH, Risau W, Ullrich A. Glioblastoma growth inhibited in vivo by a dominant-negative Flk-1 mutant. Nature 367:576–579, 1994.
176. Millauer B, Longhi MP, Plate KH, Shawver LK, Risau W, Ullrich A, Strawn LM. Dominant-negative inhibition of Flk-1 suppresses the growth of many tumor types in vivo. Cancer Res 56:1615–1620, 1996.
177. Kong HL, Hecht D, Song W, Kovesdi I, Hackett NR, Yayon A, Crystal RG. Regional suppression of tumor growth by in vivo transfer of a cDNA encoding a secreted form of the extracellular domain of the flt-1 vascular endothelial growth factor receptor. Hum Gene Ther 9:823–833, 1998.
178. Goldman CK, Kendall RL, Cabrera G, Soroceanu L, Heike Y, Gillespie GY, Siegal GP, Mao X, Bett AJ, Huckle WR, Thomas KA, Curiel DT. Paracrine expression of a native soluble vascular endothelial growth factor receptor inhibits tumor growth, metastasis, and mortality rate. Proc Natl Acad Sci USA 95:8795–8800, 1998.
179. Presta LG, Chen H, O'Connor SJ, Chisholm V, Meng YG, Krummen L, Winkler M, Ferrara N. Humanization of an anti-VEGF monoclonal antibody for the therapy of solid tumors and other disorders. Cancer Res 57:4593–4599, 1997.
180. Strawn LM, McMahon G, App H, Schreck R, Kuchler WR, Longhi MP, Hui TH, Tang C, Levitzki A, Gazit A, Chen I, Keri G, Orfi L, Risau W, Flamme I, Ullrich A, Hirth KP,

Shawver LK. Flk-1 as a target for tumor growth inhibition. Cancer Res 56:3540–3545, 1996.

181. Olk RJ, Lee CM. Diabetic Retinopathy: Practical Management Philadelphia: Lippincott, 1993.
182. Patz A. Studies on retinal neovascularization. Invest Ophthalmol Vis Sci 19:1133–1138, 1980.
183. Michaelson IC. The mode of development of the vascular system of the retina with some observations on its significance for certain retinal disorders. Trans Ophthalmol Soc UK 68: 137–180, 1948.
184. Aiello LP, Avery RL, Arrigg PG, Keyt BA, Jampel HD, Shah ST, Pasquale LR, Thieme H, Iwamoto MA, Park JE, Nguyen H, Aiello LM, Ferrara N, King GL. Vascular endothelial growth factor in ocular fluid of patients with diabetic retinopathy and other retinal disorders. N Engl J Med 331:1480–1487, 1994.
185. Adamis AP, Miller JW, Bernal MT, D'Amico DJ, Folkman J, Yeo TK, Yeo KT. Increased vascular endothelial growth factor levels in the vitreous of eyes with proliferative diabetic retinopathy. Am J Ophthalmol 118:445–450, 1994.
186. Malecaze F, Clemens S, Simorer-Pinotel V, Mathis A, Chollet P, Favard P, Bayard F, Plouet J. Detection of vascular endothelial growth factor mRNA and vascular endothelial growth factor-like activity in proliferative diabetic retinopathy, Arch Ophthalmol 112:1476–1482, 1994.
187. Okamoto N, Tobe T, Hackett SF, Ozaki H, Vinores MA, LaRochelle W, Zack DJ, Campochiaro PA. Transgenic mice with increased expression of vascular endothelial growth factor in the retina: a new model of intraretinal and subretinal neovascularization. Am J Pathol 151: 281–291, 1997.
188. Miller JW, Adamis AP, Shima DT, D'Amore PA, Moulton RS, O'Reilly MS, Folkman J, Dvorak HF, Brown LF, Berse B, et al. Vascular endothelial growth factor/vascular permeability factor is temporally and spatially correlated with ocular angiogenesis in a primate model. Am J Pathol 145:574–584, 1994.
189. Pierce EA, Avery RL, Foley ED, Aiello LP, Smith LE. Vascular endothelial growth factor/vascular permeability factor expression in a mouse model of retinal neovascularization. Proc Natl Acad Sci USA 92:905–909, 1995.
190. Adamis AP, Shima DT, Tolentino MJ, Gragoudas ES, Ferrara N, Folkman J, D'Amore PA, Miller JW. Inhibition of vascular endothelial growth factor prevents retinal ischemia-associated iris neovascularization in a nonhuman primate. Arch Ophthalmol 114:66–71, 1996.
191. Aiello LP, Pierce EA, Foley ED, Takagi H, Chen H, Riddle L, Ferrara N, King GL, Smith LE. Suppression of retinal neovascularization in vivo by inhibition of vascular endothelial growth factor (VEGF) using soluble VEGF-receptor chimeric proteins. Proc Natl Acad Sci USA 92:10457–10461, 1995.
192. Lopez PF, Sippy BD, Lambert HM, Thach AB, Hinton DR. Transdifferentiated retinal pigment epithelial cells are immunoreactive for vascular endothelial growth factor in surgically excised age-related macular degeneration–related choroidal neovascular membranes. Invest Ophthalmol Vis Sci 37:855–868, 1996.
193. Kvanta A, Algvere PV, Berglin L, Seregard S. Subfoveal fibrovascular membranes in age-related macular degeneration express vascular endothelial growth factor. Invest Ophthalmol Vis Sci 37:1929–1934, 1996.
194. Koch AE, Harlow L, Haines GK, Amento EP, Unemori EN, Wong W-L, Pope RM, Ferrara N. Vascular endothelial growth factor: a cytokine modulating endothelial function in rheumatoid arthritis. J Immunol 152:4149–4156, 1994.
195. Fava RA, Olsen NJ, Spencer-Green G, Yeo KT, Yeo TK, Berse B, Jackman RW, Senger DR, Dvorak HF, Brown LF. Vascular permeability factor/endothelial growth factor (VPF/VEGF): accumulation and expression in human synovial fluids and rheumatoid synovial tissue. J Exp Med 180:341–346, 1994.

196. Detmar M, Brown LF, Claffey KP, Yeo KT, Kocher O, Jackman RW, Berse B, Dvorak HF. Overexpression of vascular permeability factor/vascular endothelial growth factor and its receptors in psoriasis. J Exp Med 180:1141–1146, 1994.
197. Brown LF, Harrist TJ, Yeo KT, Stahle-Backdahl M, Jackman RW, Berse B, Tognazzi K, Dvorak HF, Detmar M. Increased expression of vascular permeability factor (vascular endothelial growth factor) in bullous pemphigoid, dermatitis herpetiformis, and erythema multiforme. J Invest Dermatol 104:744–749, 1995.
198. Takeshita S, Zhung L, Brogi E, Kearney M, Pu L-Q, Bunting S, Ferrara N, Symes JF, Isner JM. Therapeutic angiogenesis: a single intra-arterial bolus of vascular endothelial growth factor augments collateral vessel formation in a rabbit ischemic hind-limb model. J Clin Invest 93:622–670, 1994.
199. Takeshita S, Tsurumi Y, Couffinahl T, Asahara T, Bauters C, Symes J, Ferrara N, Isner JM. Gene transfer of naked DNA encoding for three isoforms of vascular endothelial growth factor stimulates collateral development in vivo. Lab Invest 75:487–501, 1996.
200. Takeshita S, Weir L, Chen D, Zheng LP, Riessen R, Bauters C, Symes JF, Ferrara N, Isner JM. Therapeutic angiogenesis following arterial gene transfer of vascular endothelial growth factor in a rabbit model of hindlimb ischemia. Biochem Biophys Res Commun 227:628–635, 1996.
201. Walder CE, Errett CJ, Bunting S, Lindquist P, Ogez JR, Heinsohn HG, Ferrara N, Thomas GR. Vascular endothelial growth factor augments muscle blood flow and function in a rabbit model of chronic hindlimb ischemia. J Cardiovasc Pharmacol 27:91–98, 1996.
202. Bauters C, Asahara T, Zheng LP, Takeshita S, Bunting S, Ferrara N, Symes JF, Isner JM. Physiological assessment of augmented vascularity induced by VEGF in ischemic rabbit hindlimb. Am J Physiol 267:H1263–H1271, 1994.
203. Bauters C, Asahara T, Zheng LP, Takeshita S, Bunting S, Ferrara N, Symes JF, Isner JM. Recovery of disturbed endothelium-dependent flow in the collateral-perfused rabbit ischemic hindlimb after administration of vascular endothelial growth factor. Circulation 91:2802–2809, 1995.
204. Mack CA, Magovern CJ, Budenbender KT, Patel SR, Schwarz EA, Zanzonico P, Ferris B, Sanborn T, Isom P, Ferris B, Sanborn T, Isom OW, Crystal RG, Rosengart TK. Salvage angiogenesis induced by adenovirus-mediated gene transfer of vascular endothelial growth factor protects against ischemic vascular occlusion. J Vasc Surg 27:699–709, 1998.
205. Banai S, Jaktlish MT, Shou M, Lazarous D, Scheinowitz M, Biro S, Epstein S, Unger E. Angiogenic-induced enhancement of collateral blood flow to ischemic myocardium by vascular endothelial growth factor. Circulation 2183–2189, 1994.
206. Harada K, Friedman M, Lopez JJ, Wang SY, Li J, Prasad PV, Pearlman JD, Edelman ER, Sellke FW, Simons M. Vascular endothelial growth factor administration in chronic myocardial ischemia. Am J Physiol 270:H1791–1802, 1996.
207. Pearlman JD, Hibberd MG, Chuang ML, Harada K, Lopez JJ, Gladstone SR, Friedman M, Sellke FW, Simons M. Magnetic resonance mapping demonstrates benefits of VEGF-induced myocardial angiogenesis. Nature Med 1:1085–1089, 1995.
208. Mack CA, Patel SR, Schwarz EA, Zanzonico P, Hahn RT, Ilercil A, Devereux RB, Goldsmith SJ, Christian TF, Sanborn TA, Kovesdi I, Hackett N, Isom OW, Crystal RG, Rosengart TK. Biologic bypass with the use of adenovirus-mediated gene transfer of the complementary deoxyribonucleic acid for vascular endothelial growth factor 121 improves myocardial perfusion and function in the ischemic porcine heart. J Thorac Cardiovasc Surg 115:168–176, 1998.
209. Isner JM, Walsh K, Symes J, Pieczek A, Takeshita S, Lowry J, Rosenfield K, Weir L, Brogi E, Jurayj D. Arterial gene transfer for therapeutic angiogenesis in patients with peripheral artery disease. Hum Gene Ther 7:959–988, 1996.
210. Isner JM, Pieczek A, Schainfeld R, Blair R, Haley L, Asahara T, Rosenfield K, Razvi S, Walsh K, Symes JF. Clinical evidence of angiogenesis after arterial gene transfer of phVEGF165 in patient with ischaemic limb. Lancet 348:370–374, 1996.

211. Baumgartner I, Pieczek A, Manor O, Blair R, Kearney M, Walsh K, Isner JM. Constitutive expression of phVEGF165 after intramuscular gene transfer promotes collateral vessel development in patients with critical limb ischemia. Circulation 97:1114–1123, 1998.
212. Henry TD, Rocha-Singh K, Isner JM, Kereiakes DJ, Giordano FJ, Simons M, Losordo DW, Hendel RC, Bonow RO, Rothman JM, Barbas ER, McCluskey ER. Results of intracoronary recombinant human vascular endothelial growth factor (rhVEGF) administration trial. J Am Coll Cardiol 31(2 suppl A):65A, 1998.
213. Deroanne CF, Hajitou A, Calberg-Bacq CM, Nusgens BV, Lapiere CM. Angiogenesis by fibroblast growth factor 4 is mediated through an autocrine up-regulation of vascular endothelial growth factor expression. Cancer Res 57:5590–5597, 1997.
214. Seghezzi G, Patel S, Ren CJ, Gualandris A, Pintucci G, Robbins ES, Shapiro RL, Galloway AC, Rifkin DB, Mignatti P. Fibroblast growth factor-2 (FGF-2) induces vascular endothelial growth factor (VEGF) expression in the endothelial cells of forming capillaries: an autocrine mechanism contributing to angiogenesis. J Cell Biol 141:1659–1673, 1998.
215. Gordon MS, Talpaz M, Margolin K, Holmgren E, Sledge GW, Benjamin R, Stalter S, Shak S, Adelman D. Phase I trial of recombinant humanized monoclonal anti-vascular endothelial growth factor in patients with metastatic cancer (abstr). 34th Annual Meeting of the American Society of Clinical Oncologists, Los Angeles, CA, May 16–19, 1998.

4
Fibroblast Growth Factors and Their Receptors

Andrew Baird
Ciblex Corporation, San Diego, California

I. OVERVIEW

Fibroblast growth factors (FGFs) represent a large family of polypeptides that are potent regulators of cell growth and differentiation (1–3). They play a major role in normal embryonic development and in tissue repair and regeneration and have also been implicated in numerous pathologies, including the development and progression of several malignancies. FGFs act on cells that are primarily of mesodermal origin but have broader effects on cells derived from the ectoderm and endoderm. Depending on the target cell, the conditions of cell culture, and the presence or absence of other trophic agents, FGFs alter migration, morphology, differentiation, and proliferation (4). While first discovered for their action on fibroblasts, hence their name, their activities and physiological roles are considerably more extensive than implied by their name. In fact, some members of the FGF family are not even mitogens for fibroblasts at all and would be better represented if thought of as epithelial growth factors.

To date the family of FGFs consists of 18 genes encoding proteins that have 30–80% structural homology at the amino acid level (Fig. 1). More may still be identified as the human genome is fully characterized. Interleukins 1α and β, which have 20% homology with FGF1 and FGF2, are not usually considered members of the FGF family even though they have a number of features in common with the FGFs, including a similar tertiary structure. While all of the members of the FGF family are designated FGF1 through FGF17, the names of acidic FGF (FGF1), basic FGF (FGF2), and KGF (keratinocyte growth factor, or FGF7) are also commonly used in the literature. Other names for the FGFs—like int2 (FGF3), hst1/ksFGF/kFGF (FGF4), hst2 (FGF6), androgen-induced growth factor (FGF8), and glial activating factor (FGF8)—are in rare use as the nomenclature describing the FGF family becomes standardized (2,5).

There are four distinct genes that encode for high-affinity receptors for FGFs. Called FGFR1, FGFR2, FGFR3, and FGFR4, respectively, these genes encode a multitude of structural variants that result from alternative splicing of their mRNAs (1). As a result, the specificity of ligand binding can be drastically modified by the generation of one or another alternatively spliced form. The best example is FGF7, also known as KGF, which

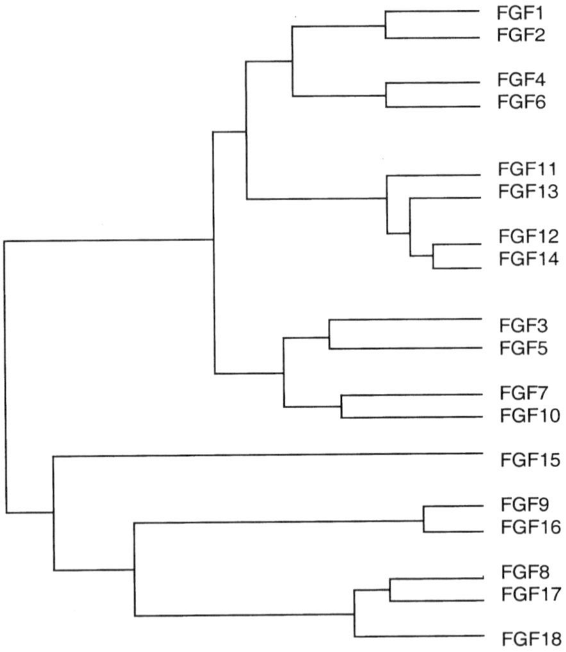

Figure 1 Structural homologies of the FGF family. Based on the sequences of rat FGF1, 2, 5, 7, 9, 10, 16, 17, and 18, and mouse FGF3, 4, 6, 8, 11, 12, 13, 14, and 15.

does not cross-react with any FGFR except a variant of FGFR2. Accordingly, it is very specific (6,7).

The basic structure of the FGFRs is all the same; they belong to the immunoglobulin (Ig)-like family of tyrosine kinases. All possess three Ig-like domains that can be processed to isoforms that contain 2 Ig-like domains connected to an intracellular tyrosine kinase domain that is activated by ligand binding. In some instances, genetic mutations in the transmembrane domain result in constitutive activation and in humans, various forms of dwarfism (8,9).

The FGF ligand–FGF receptor interaction is further complicated by the role played by proteoglycans in FGF action. Specifically, FGFs have an unusually high affinity for heparan sulfates, which translates into an ability to interact with glycosaminoglycans like heparin. This binding causes conformational changes in the FGFs that result in altered receptor binding and protects the ligand from proteolysis and acid denaturation (10,11). Most importantly, this interaction appears required for significant ligand-receptor binding. Because extracellular FGF binds to heparan sulfate proteoglycans in the matrix and at the cell surface, this binding is thought to protect the FGFs and serve as a reservoir of FGFs available after trauma, injury, or any pathophysiological event that might require a rapid mobilization of FGF activity (like angiogenesis). In this model, the release of matrix-degrading enzymes during embryonic development and tissue injury and after cell transformation to the cancer phenotype can serve to generate free, biologically active FGFs from these sites of sequestration (12,13). As some of the most powerful mitogens known, the FGFs appear to play key roles in significant human diseases like cancer.

II. THE FGF FAMILY OF GROWTH FACTORS

The number of ligands in the FGF family is increasing and, as sequencing of the human genome has accelerated, the discovery of new, structurally related proteins is likely to continue. They are thought to derive from a common ancestral gene (Fig. 1) and they all share a common structure of three exons separated by two short introns. They are located on different chromosomes (Table 1) and appear highly regulated, presumably to control malignancy. In normal, adult tissues, the predominant FGF gene products found are FGF1 and FGF2. While the functional roles played all FGFs are currently under active investigation, significantly less is known about the more recently characterized molecules.

A. FGF1 and FGF2

Acidic and basic FGFs (now called FGF1 and FGF2 respectively) are the prototypical members of the FGF family. First named for their different isoelectric points when first detected in, and purified from, tissue extracts, they have similar molecular weights and have over 50% structural homology. Both have been isolated from numerous tissues and have been the subject of extensive reviews (1–4).

Table 1 Chromosome Location of FGFs and Their Receptors

Protein	Alternate names	Chromosome (human)	Locus	Reference
FGF1	Acidic FGF	5	q31-33	61
FGF2	Basic FGF	4	q25-27	62
FGF3	Int-2	11	q13	63
FGF4	ksFGF/K-FGF/hst 1	11	q13	63
FGF5	hst-2	4	q21	64
FGF6	———	12	p13	65
FGF7	KGF	15	q15-21	66
FGF8	AIGF	10	q24-26	35
FGF9	GDGF	13	q11-12	66
FGF10	———	5	p13-12	39
FGF11	FHF-3	17	q21	67
FGF12	FHF-1	3	q28	67
FGF13	FHF-2	X	q26	68
FGF14	FHF-4	13	Q34	43
FGF15	———	———	———	———
FGF16	———	———	———	———
FGF17	———	———	———	———
FGF18	———	———	———	———
FGFR1	flt2, flg	8	p11.2-11.1	69
FGFR2	KGFR, bek	10	q26	70
FGFR3	———	4	p16.3	71
FGFR4	———	5	q35.1-qter	72

The gene for FGF1 is located on chromosome 5 and results in several mRNAs that differ in their 5′ untranslated region and not in their coding sequence (14). It is therefore likely that differential elements control mRNA translation in different tissues. In contrast, the mRNA encoding human FGF2 derives from a gene on chromosome 4 and encodes four proteins that all contain an identical core of the FGF2 protein sequence (15,16). The three amino terminally extended forms of FGF2 extend the 18-kDa molecule of 155 amino acids to proteins of 21, 22, and 25 kDa. These isoforms translocate to the nucleus, where they are thought to play a role in controlling transcription. The lowest-molecular-weight FGF2 is cytoplasmic and, while it can be released from cells, it appears to follow an unconventional release pathway (17,18). Unlike many of the other FGFs and all classically secreted proteins, both FGF1 and FGF2 lack a classical leader sequence and do not exit the cell through the endoplasmic reticulum Golgi apparatus complex.

Both FGF1 and FGF2 stimulate the proliferation mesodermal cells and many cells of neuroectodermal, ectodermal, and endodermal origin. They are also characterized by the wide array of activities that they can stimulate on different cell types (19). While they are perhaps best known as powerful angiogenic factors (because of their effects on endothelial cells), they also modulate differentiated function, delay senescence, inhibit apopotosis, and are chemotactic, depending on the in vitro cell system used, the conditions of cell culture, and the target cell evaluated. In vivo, these FGFs are active in angiogenesis, nerve regeneration, cartilage repair, and wound healing and are thought to play normal roles in tissue repair and homeostasis. The observation that an inappropriate expression of FGFs can result in tumor growth and development (see below) also suggests that they participate in the production of a large variety of pathogenic conditions that result in cell proliferation or are angiogenesis-dependent.

B. FGF3

With the structural characterization of FGF1 and FGF2, various databases could be evaluated for the presence of homologous proteins. The first of these FGFs so identified was int2 (20). Int-2 (now FGF3) was first recognized as a cellular gene that was activated after integration of the mouse mammary tumor virus into the mouse genome. The gene is located on chromosome 11 (Table 1) and encodes a protein of 240 amino acids that has about 40% homology to FGFs 1 and 2. An extension at the amino terminus includes a classical leader sequence that allows it to be secreted through the ER Golgi, but, interestingly, alternative forms of its mRNA encode an FGF3 that has no leader sequence. These isoforms remain inside the cell where they are translocated to the nucleus to play a role in regulating gene transcription. The elements that control the production of one over another form of FGF3 have not been defined. The secreted form of FGF3 derives from a classical AUG codon, while the cellular form is derived when this AUG is overridden by the use of an alternative start site at a CUG codon.

The regulation of FGF3 transcription appears to be coordinated by three distinct promoters and two alternative polyadenylation sites. The gene generates six different RNA species, all with the same coding capacity, but FGF3 has two alternative upstream initiation codons in which a CUG is the major start site for FGF3. A downstream AUG codon initiates translation of the shorter 30.5-kDa protein. A number of replacement and deletion mutations have shown that the amino-terminal extension is crucial for nuclear import, although the nuclear targeting signals are located elsewhere in the protein. The decision to enter the secretory pathway or nucleus appears to depend on a balance of competing signals involving the amino terminus—a signal peptide and the nuclear localization sequence. The relative position of the signaling motifs is also an important factor in establishing the proportion of FGF3 destined for the different intracellular compartments. Unfortunately, the physiological activities of FGF3 are ill defined, but it is thought to play a role in embryonic development (21,22).

C. FGF4

When the genomic fragments of the human tumor–derived oncogene *hst1* were sequenced and the oncogene derived from a Kaposi's sarcoma cell line were characterized, they were found to encode the same 206–amino acid protein that had 40% homology to FGF1, FGF2, and FGF3. First called K-FGF and hst1, it is now commonly known as FGF4 (20,23) and locates on human chromosome 11, in close proximity to FGF3 (Table 1). The amino terminus of FGF4 is extended when compared to FGF1 and FGF2 and the extension includes a leader sequence for secretion through the ER-Golgi. The mature protein is glycosylated, secreted, and transforming. It occurs with high frequency in some tumors, such as those of the stomach, and is found during normal embryonic development. Like FGF1 and FGF2, it is a highly angiogenic molecule, but unlike the prototypic FGFs, it is rare in adult tissues but thought to play a role in limb development (24) and somite patterning (25).

D. FGF5

The fifth member of the FGF family was identified from a cDNA library prepared from bladder cancer that was screened for transforming oncogenes (26). The protein encoded by the DNA isolated was found to have 40% homology to FGF1 and FGF2, hence making it FGF5. The 267–amino acid protein has a leader sequence and as such is secreted from cells through the ER-Golgi like FGF3 and 4. For this reason, it is a transforming molecule that is found in many human tumor-cell lines. It is a highly angiogenic protein that is normally found in embryonic development in brain, muscle, and the heart. The amount of this protein found in adult tissues, however, is low when compared to FGF1 and FGF2 but detectable.

The human FGF-5 gene has alternative polyadenylation sites that transcribe two main RNA species of 1.6 and 1.4 kb (26). The regulatory elements of transcription have yet to be identified, but there is evidence for translational control of FGF5 expression. The FGF5 mRNA contains a short out-of-frame open reading frame (ORF) upstream of the ORF coding for the growth factor. Deletion of the upstream ORF enhances FGF5 translation and transforming ability, supporting the observation that FGF5 is mitogenic and overexpressed in some human cancers.

E. FGF6

The sixth FGF was first detected as hst2 and as a cDNA that hybridized with probes encoding FGF4 (27,28). As a result, an FGF was sequenced that has 70% homology with FGF4 and about 30% with FGFs 1 and 2. While relatively little is known regarding the activities of FGF6 in vivo, it is angiogenic in vivo, transforming when expressed in cells, and normally found during embryonic development where it is presumably mediating many of the activities first ascribed to FGF1 and FGF2. But while FGF6 exhibits a restricted expression profile that is dominantly in the myogenic lineage (29), important functions have been suggested in wound healing and tissue regeneration (30). Like the FGFs with oncogenic potential, FGF6 has a leader sequence at its amino terminus that enables ER-Golgi dependent secretion.

F. FGF7

One of the most studied FGFs, FGF7, is unlike the other FGF mitogens in that it shows significant cell specificity for epithelial cells (6). First identified as keratinocyte growth factor (KGF), it has no effect on many target cells considered the usual targets for FGFs (i.e., endothelial cells). Because of its cell- and receptor-(FGFR2) specificity, KGF/FGF7 has no angiogenic activity. Its gene encodes an FGF7 with an amino terminus that has a leader sequence for secretion and a 22-kDa protein that has 35% homology to the other FGFs. Its mRNA is found in a number of stromal fibroblast cell lines derived from embryonic and adult tissues but is absent from epithelial cells. Of all the FGFs, FGF7 is the one that plays the clearest paracrine function, as it is always produced in stromal cells but acts only on target epithelial cells (31). If it is produced in cells that have the receptor expressed (see below), it is transforming. The specific isoform of FGF receptor that binds FGF7, however, is highly restricted in distribution and found almost exclusively on epithelial cells. Aberrant expression leads to proliferative disease.

There is a canonical TATA box 30 nucleotides upstream of its transcription initiation site. Physiological FGF7 transcription is regulated developmentally by an enhancer element located in the 3′ untranslated region of exon 3. This enhancer contains a series of consensus binding sites for a number of known transcription factors, including SP1 and AP1. Although it is probable that the specific trans factors that control the FGF7 gene will belong to the family of octamer-binding proteins (some known to be developmentally regulated), there is no evidence to date for a posttranscriptional or even posttranslational control of FGF7.

G. FGF8

When the activity induced by androgen-treated Shionogi carcinoma cells was isolated and characterized, the molecule was found to have 30–40% homology with the other members of the FGF family and called FGF8 (32,33). The gene, located on human chromosome 10 (Table 1), encodes a protein that has a leader peptide, and for this reason the molecule is found in conditioned media of FGF8 producing cells. While relatively little is known about FGF8, it is found during embryonic development and appears to play a key role in pattern development. Because of the spatiotemporal pattern of gene expression and the distribution of FGF8 protein immunoreactivity during development, it is likely that FGF8 is the true natural growth factor that is responsible for some of the activities first ascribed to FGF1 and FGF2 in embryonic development. Its distribution and expression patterns suggest a role in outgrowth and patterning of limb development (34,35).

H. FGF9

First identified in cell culture supernatants as a glial activating factor (36), FGF9 has 30% homology with the other FGFs and, with the exception that it stimulates the proliferation and activation of glial cells and other cells that express FGF receptors, relatively little is known regarding its physiological function. Because FGF9 has relatively higher specificity for FGFR3 than the other FGFs (37), it is likely to be the ligand responsible for the numerous skeletal disorders associated with the FGF axis. Its ability to transform cells when overexpressed (38) supports the hypothesis that it may also be involved in the development of human malignancies.

I. FGF10

Emoto et al. (39) isolated a cDNA encoding a novel member of the FGF family that is most related to FGF7 (KGF). Like FGF7, FGF10 is mitogenic for keratinizing epidermal cells and essentially inactive on fibroblasts. It is located on chromosome 5p13-p12 little is known regarding its activities, but recent studies have established that, like FGF7, it is involved in wound healing (40) and embryonic development (41) but appears differentially regulated and, like other FGFs, spatially restricted in its expression (42).

J. FGF11, 12, 13, and 14

Using degenerate PCR and random DNA sequencing, Smallwood (43) identified four additional members of the FGF family referred to as fibroblast growth factor homologous factors (FHFs). The genes encoded proteins that were 30% homologous to other FGFs, although they have as much as 70% homology between themselves. Their physiological functions are not known, but they all lack signal peptides and contain nuclear localization sequences. They are expressed primarily in the nervous system, although a comprehensive analysis of protein distribution and gene expression has not been reported. It is assumed that they play a role in nervous system development and function, but their presence in heart and connective tissue implies a broader possibly intracellular function (44).

K. FGF15

When McWirter et al. characterized the downstream targets of the homeodomain oncoprotein E2A-Pbx1 (45), they identified a new member of the FGF family, FGF-15. When the deduced sequence of FGF15 was compared to the other FGF family members, 30% homology was observed, with the greatest similarity being between FGF15 and FGF7 and the least between FGF15 and FGF8. Like the other FGFs, FGF15 also shares a common gene structure: three exons separated by two introns located at exactly the same homologous sequence. Like many of the other FGFs but unlike FGF1, FGF2 and the FHFs, FGF15 has a signal sequence for targeting to the ER-Golgi and secretion. Remarkably, FGF15 is not transforming under these circumstances, suggesting that it does not mediate the effects of E2A-Pbx1 on 3T3 cells. In view of its similarity to FGF7 and the fact that 3T3 cells have no receptor that cross-reacts with FGF7, the findings also suggest that 3T3 cells do not possess a receptor for FGF15, perhaps the same FGF7 receptor, FGFR2.

Although little is known regarding the physiological functions of FGF15, E2A-Pbx1 binds directly to a site proximal to its promoter and the Pbx1 homeodomain is required for induction. The expression pattern of FGF15 suggests that it may play a specialized role in the nervous system. It first appears in the neuroectoderm soon after neurulation and remains restricted throughout development but never overlapping with the two other FGFs that are similarly restricted, FGF3 and FGF8.

L. FGF16

A cDNA encoding a novel member (207 amino acids) of the FGF family from the rat heart was identified by homology-based PCR (46). FGF16 is most similar (73% amino acid identity) to FGF9 and, like FGF9, lacks a typical signal sequence yet is efficiently secreted by Sf9 insect cells infected with recombinant baculovirus containing the FGF16 cDNA. FGF16 mRNA is predominantly expressed in the rat heart among adult tissues. In rat embryos, FGF16 mRNA is predominantly expressed in the brown adipose tissue, but expression decreases greatly after birth. Whether FGF-16 might be a novel secreted FGF that plays a role in development of the brown adipose tissue is unknown.

M. FGF17

A cDNA encoding a novel member (216 amino acids) of the fibroblast growth factor (FGF) family was identified from rat embryos by Ito's laboratory soon after their characterization of FGF16 (47). Among FGF family members, FGF17 is most similar (53.7% amino acid identity) to FGF8. FGF17 has a typical signal sequence at its amino terminus. FGF17 mRNA of approximately 2.1 kb is detected in rat embryos at E14.5 but not at E10.5, and E19.5 and mRNA is found to be preferentially expressed in the neuroepithelia of the isthmus and septum of the rat embryonic brain at E14. Whether FGF17 might be a novel secreted molecule in the induction and patterning of the embryonic brain is unknown.

N. FGF18

Using homology cloning strategies, Itoh's laboratory has also characterized the 18th member of the FGF family (74). It has a typical signal sequence and as such is a secreted protein. When the growth factor is expressed in insect cells and tested on FGF target cells, it is also biologically active. In this instance, FGF18 has neurite-promoting activity on PC12 cells. No information is yet available as to whether it is angiogenic. It is most homologous to FGFs 8 and 17, with which it has almost 53% amino acid identity. Like most FGFs, its physiological function remains largely unknown. Northern blotting and in situ hybridization point to gene expression in adult lungs and not in brain, heart, liver, or kidney. In spite of homology to FGF8, its localization is very distinct. Hence, in tissues like gonads, where FGF8 is found, there is no FGF18. The temporal and spatial patterns of expression of FGF18 are also dramatically different from FGFs 8 and 17.

III. HIGH-AFFINITY FGF RECEPTORS

The FGF receptor family is one of several subclasses of tyrosine kinase receptors that are represented by four distinct genes: FGFR1 (sometimes called flg or fms-like gene), FGFR2 (sometimes called bek or bacterial expressed kinase), FGFR3, and FGFR4 (1).

The FGFRs contain a single membrane-spanning domain, an extracellular domain, and an intracellular domain that has tyrosine kinase activity and is itself phosphorylated. The extracellular domain is characterized by three immunoglobulin-like domains, although each receptor gene can produce mRNA encoding smaller receptors with two or even one Ig domain.

The complexity of the receptor family is increased by a differential RNA splicing that is dictated by the exon structures of each gene. For example, no fewer than six different classes of FGFR1s that can be produced from the FGFR1 gene. They either have no secretory signal sequence, deletion of the first Ig loop, alternative splicing into the third Ig loop, deletions prior to the third Ig loop, or truncation of the transmembrane domain to generate a soluble FGFR1. Together with the isoforms generated by FGFR2 and FGFR3 genes, there are over 100 different combinations possible at the cell surface.

The various isoforms of FGF receptors have different binding affinities for different FGF ligands (48–50). Even the secreted form of FGFR1 is capable of binding FGF2 but not FGF1. Because the third Ig loop appears to confer selectivity, FGF1, FGF2, or sometimes both will interact with FGFR1. When FGFR2 is spliced in the third Ig loop, it generates a receptor that recognizes FGF7 and FGF1 but not the other FGF2–6 and FGF8–9. Similarly, FGF8 and FGF9 appear to have increased selectivity for FGFR3 (51). The isoforms of receptors that are truncated can act as binding proteins secreted from cells in the absence of a transmembrane domain or as dominant negative receptors when trapped at the cell surface. Data are only beginning to surface suggesting that different FGFRs might have different signaling roles. FGFR1 is often associated with malignancies and transformation, the other receptors somewhat less. In any event, because signaling requires receptor dimerization, the complexity of the FGF-FGFR response is compounded by the possibility that heterogeneous dimers have different ligand specificity and altered intracellular substrates for tyrosine phosphorylation than the homodimers.

IV. LOW-AFFINITY RECEPTORS

If the number of FGF ligands and the number of high-affinity receptor isoforms were not enough to create significant complexity, the low-affinity receptors that interact with FGFs at the cell surface add yet more. First, the low-affinity receptors bind the FGFs with relative high affinity (1–20 nM), so that they are thought to play an important role in controlling their activities (1). It is only when they are compared to the very high affinity of FGF high-affinity receptors (kd 1–10 pM) that FGF binding to all heparan sulfate proteoglycans (HSPGs) can be considered low. These HSPGs can be at the cell surface, in the extracellular matrix, or found soluble in biological fluids. They consist of a core protein to which a variable number of glycosaminoglycan chains are linked. Depending on the core protein, HSPGs can be syndecans, glypicans, perlecans, or betaglycans, each having its own particular spectrum of FGF binding, localization in tissues, developmental expression pattern, and ability to control FGF activity. Located in biological fluids, on the cell surface, and at the extracellular matrix, they can sequester FGFs so they either block interactions with their receptors, deliver FGFs to their high-affinity receptors to form a receptor complex, or simply stabilize the mitogens so that they are protease-resistant and acid-stable in a wound healing milieu.

The sequestering function of HSPGs is probably most described for FGF2. Analyses of cells in culture, tissue sections of developing and adult tissues, and biochemical studies have all pointed to the localization of FGF2 in basement membranes, where it is bound to HSPGs (52). Here there are a large number of binding sites that serve to restrict the diffusion off FGF2 to other tissues and cellular compartments. There may also be a regulatory role for the HSPG interaction since the depot of mitogen allows a slow and gradual release to the high-affinity receptor. Accordingly, degradation of basement membranes during wound healing and tissue repair, embryonic development, tumor growth, and metastasis result in the release of FGFs from the basement membranes and their availability to cells expressing high-affinity receptors (12,13). Because the binding of FGFs to HSPGs is sequence-specific, their interaction is also independently controlled from other heparin-binding proteins like hepatocyte growth factor, antithrombin, and TGFβs to name a few.

In addition to the storage and sequestration functions of HSPGs in the control of FGF activities, there is compelling evidence that the HSPGs play a role, with the high-affinity receptor, in the signal transduction cascade (53). When cells are chemically, enzymatically, or genetically stripped of their cell-surface HSPGs, they are no longer responsive to FGFs unless the heparin GAG is added with FGFs to cells. Accordingly, the formation of a heteromultimeric receptor complex that consists of possibly several high- and low-affinity receptors and ligand appears to be responsible for binding, internalization, and cell responsiveness to FGFs. Based on the need for high-affinity receptors to dimerize and to elicit a mitogenic response, the signaling complex most likely includes at least two high-affinity receptors, two HSPGs, and two ligands. Moreover, with the possibility of heterodimerization, specificity and responsiveness could be altered when FGFR1 dimerizes with another, like FGFR3, for example, at the cell surface. The generation of numerous molecular isoforms of each gene product underscores the complexity of the FGF-FGFR axis.

V. SUMMARY AND CONCLUSIONS

The bulk of the evidence surrounding FGFs implicates them in a role in tumor-cell growth and development (54,55,57–60). Yet this complex family of 18 (and probably more) ligands also appears to play important roles in normal cell homeostasis. As their identities have been uncovered, the outstanding questions have turned to understanding their function. The use of genetically altered mice with knocked out genes is pointing to subtle roles for these FGFs and pivotal functions for their receptors. Clearly, research over the next few years will point to a better understanding of what role these molecules play in such processes as angiogenesis in adult and developing tissues.

REFERENCES

1. Galzie Z, Kinsella AR, Smith JA. Fibroblast growth factors and their receptors. Biochem Cell Biol 75:669–685, 1997.
2. Coulier F, Pontarotti P, Roubin R, Hartung H, Goldfarb M, Birnbaum D. Of worms and men: an evolutionary perspective on the fibroblast growth factor (FGF) and FGF receptor families. J Mol Evol 44:43–56, 1997.
3. Burgess WH, Maciag T. The heparin-binding (fibroblast) growth factor family of proteins. Annu Rev Biochem 58:575–606, 1989.
4. Baird A, Klagsbrun M. The fibroblast growth factor family. Cancer Cells 3:239–243, 1991.
5. Baird A, Klagsbrun M. The fibroblast growth factor family: an overview. Ann NY Acad Sci 638:xi–xii, 1991.
6. Finch PW, Rubin JS, Miki T, Ron D, Aaronson SA. Human KGF is FGF-related with properties of a paracrine effector of epithelial cell growth. Science 245:752–755, 1989.
7. Miki T, Fleming TP, Bottaro DP, Rubin JS, Ron D, Aaronson SA. Expression cDNA cloning of the KGF receptor by creation of a transforming autocrine loop. Science 251:72–75, 1991.

8. Shiang R, Thompson LM, Zhu Y-Z, et al. Mutations in the transmembrane domain of FGFR3, cause the most common genetic form of dwarfism, achondroplasia. Cell 78:335–342, 1994.
9. Bonaventure J, Rousseau F, Legeai-Mallet L, Le Merrer M, Munnich A, Maroteaux P. Common mutations in the fibroblast growth factor receptor 3 (FGFR 3) gene account for achondroplasia, hypochondroplasia, and thanatophoric dwarfism. Am J Med Genet 63:148–154, 1996.
10. Prestrelski SJ, Fox GM, Arakawa T. Binding of heparin to basic fibroblast growth factor induces a conformational change. Arch Biochem Biophys 293:314–319, 1992.
11. Gospodarowicz D, Cheng J. Heparin protects basic and acidic FGF from inactivation. J Cell Physiol 128:475–484, 1986.
12. Vlodavsky I, Bar-Shavit R, Ishai-Michaeli R, Bashkin P, Fuks Z. Extracellular sequestration and release of fibroblast growth factor: a regulatory mechanism? Trends Biochem Sci 16:268–271, 1991.
13. Bashkin P, Doctrow S, Klagsbrun M, Svahn CM, Folkman J, Vlodavsky I. Basic fibroblast growth factor binds to subendothelial extracellular matrix and is released by heparitinase and heparin-like molecules. Biochemistry 28:1737–1743, 1989.
14. Renaud F, el Yazidi I, Boilly-Marer Y, Courtois Y, Laurent M. Expression and regulation by serum of multiple FGF1 mRNA in normal transformed, and malignant human mammary epithelial cells. Biochem Biophys Res Commun 219:679–685, 1996.
15. Abraham JA, Mergia A, Whang JL, et al. Nucleotide sequence of a bovine clone encoding the angiogenic protein, basic fibroblast growth factor. Science 233:545–548, 1986.
16. Florkiewicz RZ, Sommer A. Human basic fibroblast growth factor gene encodes four polypeptides: Three initiate translation from non-AUG codons. Proc Natl Acad Sci USA 86:3978–3981, 1989.
17. Florkiewicz RZ, Majack RA, Buechler RD, Florkiewicz E. Quantitative export of FGF-2 occurs through an alternative, energy-dependent, non-ER/Golgi pathway. J Cell Physiol 162: 388–399, 1995.
18. Mignatti P, Rifkin DB. Release of basic fibroblast growth factor, an angiogenic factor devoid of secretory signal sequence: a trivial phenomenon or a novel secretion mechanism? J Cell Biochem 47:201–207, 1991.
19. Baird A. Fibroblast growth factors: What's in a name? Endocrinology 132:487–488, 1993.
20. Yoshida T, Miyagawa K, Odagiri H, et al. Genomic sequence of hst, a transforming gene encoding a protein homologous to fibroblast growth factors and the int-2-encoded protein. Proc Natl Acad Sci USA 84:7305–7309, 1987.
21. Wilkinson DG, Peters G, Dickson C, McMahon AP. Expression of the FGF-related proto-oncogene int-2 during gastrulation and neurulation in the mouse. EMBO J 7:691–695, 1988.
22. Wilkinson DG, Bhatt S, McMahon AP. Expression pattern of the FGF-related proto-oncogene int-2 suggests multiple roles in fetal development. Development 105:131–136, 1989.
23. Delli Bovi P, Curatola AM, Kern FG, Greco A, Ittman M, Basilico C. An oncogene isolated by transfection of Kaposi's sarcoma DNA encodes a growth factor that is a member of the FGF family. Cell 50:729–737, 1987.
24. Niswander L, Tickle C. Vogel A, Martin G. Function of FGF-4 in limb development. Mol Reprod Dev 39:83–89, 1994.
25. Grass S, Arnold HH, Braun T. Alterations in somite patterning of Myf-5-deficient mice: A possible role for FGF-4 and FGF-6. Development 122:141–150, 1996.
26. Zhan X, Bates B, Hu X, Goldfarb M. The human FGF-5 oncogene encodes a novel protein related to fibroblast growth factors. Mol Cell Biol 8:3487–3497, 1988.
27. Iida S, Yoshida T, Naito K, et al. Human hst-2 (FGF-6) oncogene: cDNA cloning and characterization. Oncogene 7:303–309, 1992.
28. Marics I, Adelaide J, Raybaud F, et al. Characterization of the HST-related FGF.6 gene, a new member of the fibroblast growth factor gene family. Oncogene 4:335–340, 1989.
29. Han J-K, Martin GR. Embryonic expression of Fgf-6 is restricted to the skeletal muscle lineage. Dev Biol 158:549–554, 1993.

30. Floss T, Arnold HH, Braun T. A role for FGF-6 in skeletal muscle regeneration. Genes Dev 11:2040–2051, 1997.
31. Aaronson SA, Bottaro DP, Miki T, et al. Keratinocyte growth factor: A fibroblast growth factor family member with unusual target cell specificity. Ann NY Acad Sci 638:62–77, 1991.
32. Gemel J, Gorry M, Ehrlich GD, MacArthur CA. Structure and sequence of human FGF8. Genomics 35:253–257, 1996.
33. Tanaka A, Miyamoto K, Minamino N, et al. Cloning and characterization of an androgen-induced growth factor essential for the androgen-dependent growth of mouse mammary carcinoma cells. Proc Natl Acad Sci USA 89:8928–8932, 1992.
34. Vogel A, Rodriguez C, Izpisúa-Belmonte JC. Involvement of FGF-8 in initiation, outgrowth and patterning of the vertebrate limb. Development 122:1737–1750, 1996.
35. White RA, Dowler LL, Angeloni SV, Pasztor LM, MacArthur CA. Assignment of FGF8 to human chromosome 10q25-q26: mutations in FGF8 may be responsible for some types of acrocephalosyndactyly linked to this region. Genomics 30:109–111, 1995.
36. Miyamoto M, Naruo K-I, Seko C, Matsumoto S, Kondo T, Kurokawa T. Molecular cloning of a novel cytokine cDNA encoding the ninth member of the fibroblast growth factor family, which has a unique secretion property. Mol Cell Biol 13:4251–4259, 1993.
37. Hecht D, Zimmerman N, Bedford M, Avivi A, Yayon A. Identification of fibroblast growth factor 9 (FGF9) as a high affinity, heparin dependent ligand for FGF receptors 3 and 2 but not for FGF receptors 1 and 4. Growth Factors 12:223–233, 1995.
38. Matsumoto-Yoshitomi S, Habashita J, Nomura C, Kuroshima K, Kurokawa T. Autocrine transformation by fibroblast growth factor 9 (FGF-9) and its possible participation in human oncogenesis. Int J Cancer 71:442–450, 1997.
39. Emoto H, Tagashira S, Mattei MG, et al. Structure and expression of human fibroblast growth factor-10. J Biol Chem 272:23191–23194, 1997.
40. Tagashira S, Harada H, Katsumata T, Itoh N, Nakatsuka M. Cloning of mouse FGF10 and up-regulation of its gene expression during wound healing. Gene 197:399–404, 1997.
41. Beer HD, Florence C, Dammeier J, McGuire L, Werner S, Duan DR. Mouse fibroblast growth factor 10: cDNA cloning, protein characterization, and regulation of mRNA expression. Oncogene 15:2211–2218, 1997.
42. Hattori Y, Yamasaki M, Konishi M, Itoh N. Spatially restricted expression of fibroblast growth factor-10 mRNA in the rat brain. Mol Brain Res 47:139–146, 1997.
43. Smallwood PM, Munoz-Sanjuan I, Tong P, et al. Fibroblast growth factor (FGF) homologous factors: new members of the FGF family implicated in nervous system development. Proc Natl Acad Sci USA 93:9850–9857, 1996.
44. Hartung H, Feldman B, Lovec H, Coulier F, Birnbaum D, Goldfarb M. Murine FGF-12 and FGF-13: expression in embryonic nervous system, connective tissue and heart. Mech Dev 64: 31–39, 1997.
45. McWhirter JR, Goulding M, Weiner JA, Chun J, Murre C. A novel fibroblast growth factor gene expressed in the developing nervous system is a downstream target of the chimeric homeodomain oncoprotein E2A-Pbx1. Development 124:3221–3232, 1997.
46. Miyake A, Konishi M, Martin FH, et al. Structure and expression of a novel member FGF-16, of the fibroblast growth factor family. Biochem Biophys Res Commun 243:148–152, 1998.
47. Hoshikawa M, Ohbayashi N, Yonamine A, Konishi M, Ozaki K, Fukui S, Itoh N. Structure and expression of a novel fibroblast growth factor, FGF17, preferentially expressed in the embryonic brain. Biochem Biophys Res Commun 244:187–191, 1998.
48. Ornitz DM, Leder P. Ligand specificity and heparin dependence of fibroblast growth factor receptors 1 and 3. J Biol Chem 267:16305–16311, 1992.
49. Ornitz DM, Xu JS, Colvin JS, et al. Receptor specificity of the fibroblast growth factor family. J Biol Chem 271:15292–15297, 1996.
50. Blunt AG, Lawshé A, Cunningham ML, Seto ML, Ornitz DM, MacArthur CA. Overlapping

expression and redundant activation of mesenchymal fibroblast growth factor (FGF) receptors by alternatively spliced FGF-8 ligands. J Biol Chem 272:3733–3738, 1997.

51. Chellaiah AT, McEwen DG, Werner S, Xu J, Ornitz DM. Fibroblast growth factor receptor (FGFR) 3. Alternative splicing in immunoglobulin-like domain III creates a receptor highly specific for acidic FGF/FGF-1. J Biol Chem 269:11620–11627, 1994.
52. Vlodavsky I, Folkman J, Sullivan R, et al. Endothelial cell-derived basic fibroblast growth factor: synthesis and deposition into subendothelial extracellular matrix. Proc Natl Acad Sci USA 84:2292–2296, 1987.
53. Klagsbrun M, Baird A. A dual receptor system is required for basic fibroblast growth factor activity. Cell 67:229–231, 1991.
54. Halaban R, Kwon B, Ghosh S, Delli Bovi P, Baird A. bFGF as an autocrine growth factor in human melanomas. Mol Cell Biol 8:2933–2941, 1988.
55. Peters G, Brookes S, Smith R, Placzek M, Dickson C. The mouse homolog of the hst/k-FGF gene is adjacent to int-2 and is activated by proviral insertion in some virally induced mammary tumors. Proc Natl Acad Sci USA 86:5678–5682, 1989.
56. Clausse N, Baines D, Moore R, Brookes S, Dickson C, Peters G. Activation of both Wnt-1 and Fgf-3 by insertion of mouse mammary tumor virus downstream in the reverse orientation: A reappraisal of the enhancer insertion model. Virology 194:157–165, 1993.
57. Rosen A, Sevelda P, Klein M, et al. First experience with FGF-3 (INT-2) amplification in women with epithelial ovarian cancer. Br J Cancer 67:1122–1125, 1993.
58. Galdemard C, Brison O, Lavialle C. The proto-oncogene FGF-3 is constitutively expressed in tumorigenic, but not in non-tumorigenic, clones of a human colon carcinoma cell line. Oncogene 10:2331–2342, 1995.
59. Schmitt JF, Susil BJ, Hearn MTW. Aberrant FGF-2, FGF-3, FGF-4 and C-ERB-B2 gene copy number in human ovarian, breast and endometrial tumours. Growth Factors 13:19–35, 1996.
60. Talarico D, Ittmann M, Balsari A, Delli-Bovi P, Basch RS, Basilico C. Protection of mice against tumor growth by immunization with an oncogene-encoded growth factor. Proc Natl Acad Sci USA 87:4222–4225, 1990.
61. Jaye M, Howk R, Burgess W, et al. Human endothelial cell growth factor: cloning, nucleotide sequence, and chromosome localization. Science 233:541–545, 1986.
62. Mergia A, Eddy R, Abraham JA, Fiddes JC, Shows TB. The genes for basic and acidic fibroblast growth factors are on different human chromosomes. Biochem Biophys Res Commun 138:644–651, 1986.
63. Nguyen C, Roux D, Mattei M-G, et al. The FGF-related oncogenes hst and int.2, and the bcl.1 locus are contained within one megabase in band q13 of chromosome 11, while the fgf.5 oncogene maps to 4q21. Oncogene 3:703–708, 1988.
64. Dionne CA, Kaplan R, Seuánez H, O'Brien SJ, Jaye M. Chromosome assignment by polymerase chain reaction techniques: assignment of the oncogene FGF-5 to human chromosome 4. BioTechniques 8:190–194, 1990.
65. De Lapeyriere O, Rosnet O, Benharroch D, et al. Structure, chromosome mapping and expression of the murine Fgf-6 gene. Oncogene 5:823–832, 1990.
66. Mattei MG, DeLapeyriere O, Bresnick J, Dickson C, Birnbaum D, Mason I. Mouse Fgf7 (fibroblast growth factor 7 and Fgf8 (fibroblast growth factor 8) genes map to chromosomes 2 and 19 respectively. Mammalian Genome 6:196–197, 1995.
67. Verdier AS, Mattei MG, Lovec H, et al. Chromosomal mapping of two novel human FGF genes, FGF11 and FGF12. Genomics 40:151–154, 1997.
68. Lovec H, Hartung H, Verdier AS, et al. Assignment of FGF13 to human chromosome band Xq21 by in situ hybridization. Cytogenet Cell Genet 76:183–184, 1997.
69. Wood S, Schertzer M, Yaremko ML. Sequence identity locates CEBPD and FGFR1 to mapped human loci within proximal 8p. Cytogenet Cell Genet 70:188–191, 1995.
70. Dionne CA, Modi WS, Crumley G, O'Brien SJ, Schlessinger J, Jaye M. BEK, a receptor for

multiple members of the fibroblast growth factor (FGF) family, maps to human chromosome 10q25.3↓q26. Cytogenet Cell Genet 60:34–36, 1992.
71. Keegan K, Rooke L, Hayman M, Spurr NK. The fibroblast growth factor receptor 3 gene (FGFR3) is assigned to human chromosome 4. Cytogenet Cell Genet 62:172–175, 1993.
72. Warrington JA, Bailey SK, Armstrong E, et al. A radiation hybrid map of 18 growth factor, growth factor receptor, hormone receptor, or neurotransmitter receptor genes on the distal region of the long arm of chromosome 5. Genomics 13:803–808, 1992.
74. Ohbayashi N, Hoshikawa M, Kimura S, Yamasaki M, Fukui S, Itoh N. Structure and expression of the mRNA encoding a novel fibroblast growth factor, FGF-18 J Biol Chem 273:18161–18164, 1998.

5

TIE-2 Signaling and Venous Malformations

Leena Karttunen
Harvard Medical School, Boston, Massachusetts

Miikka Vikkula
Christian de Duve Institute of Cellular Pathology and Université Catholique de Louvain, Brussels, Belgium

Laurence M. Boon
Christian de Duve Institute of Cellular Pathology and Cliniques Universitaires St-Luc, Université Catholique de Louvain, Brussels, Belgium

Bjorn R. Olsen
Harvard Medical School and Harvard School of Dental Medicine, Boston, Massachusetts

I. INTRODUCTION

Several receptor kinases and their ligands play critical roles in vasculogenesis and angiogenesis. These include the receptor tyrosine kinases for vascular endothelial growth factors; fibroblast growth factors; platelet-derived growth factors; the receptor tyrosine kinase TIE-2, which binds angiopoietins 1 and 2; an orphan receptor TIE-1; the EPH-B4 receptor for ephrin-B2; as well as serine/threonine kinase receptors for members of the TGF-β family of cytokines (for references, see other chapters in this volume). Insights into the signaling pathways initiated by the interaction of these receptors with their ligands have come both from studies in mice and analyses of inherited abnormalities in humans caused by errors in local vascular morphogenesis. For example, mice that are homozygous for inactivated Tie-2 alleles die in utero at a stage when the primary capillary plexus is normally undergoing remodeling to a mature vascular network with arteries and veins (1). A similar phenotype is seen in mice with inactivated angiopoietin-1 alleles or in transgenic mice that overexpress the antagonist ligand angiopoietin-2 (2,3). In humans, activating mutations in TIE-2 are associated with inherited localized mucocutaneous venous malformations. Within the lesions, large, convoluted vascular channels are surrounded by fewer smooth muscle cells than are found in normal veins. This suggests that increased signaling through the TIE-2 receptor pathway negatively affects recruitment and/or differentiation of smooth muscle cells around endothelial tubes in developing veins.

II. VASCULAR MALFORMATIONS

Vascular malformations are a heterogeneous group of congenital blood vessel anomalies (4). Venous malformations differ from vascular tumors, hemangiomas, although they are often erroneously called hemangiomas or cavernous hemangiomas (4,5). Hemangiomas are usually not present at birth but appear within the first 2 weeks after birth, grow rapidly over few months, and then undergo a natural involution within 5–10 years. In contrast, vascular malformations may be present at birth and continue to grow over time without signs of regression. Vascular malformations include arterial, venous, capillary, and lymphatic malformations as well as combined lesions. Capillary malformations are probably the most common, but venous malformations (VMs) are most often seen in vascular centers because they are severe and difficult to treat and more common than, for example, lymphatic malformations. VMs are usually sporadic but several families with dominantly inherited forms have been described and used for molecular genetic studies. Most lesions are located in skin or mucosal membranes of the mouth (Figure 1), although they can also be found within the gastrointestinal tract, as in the case of a family reported by Bean and now referred to as Bean syndrome or ''blue rubber bleb nevus syndrome'' (5).

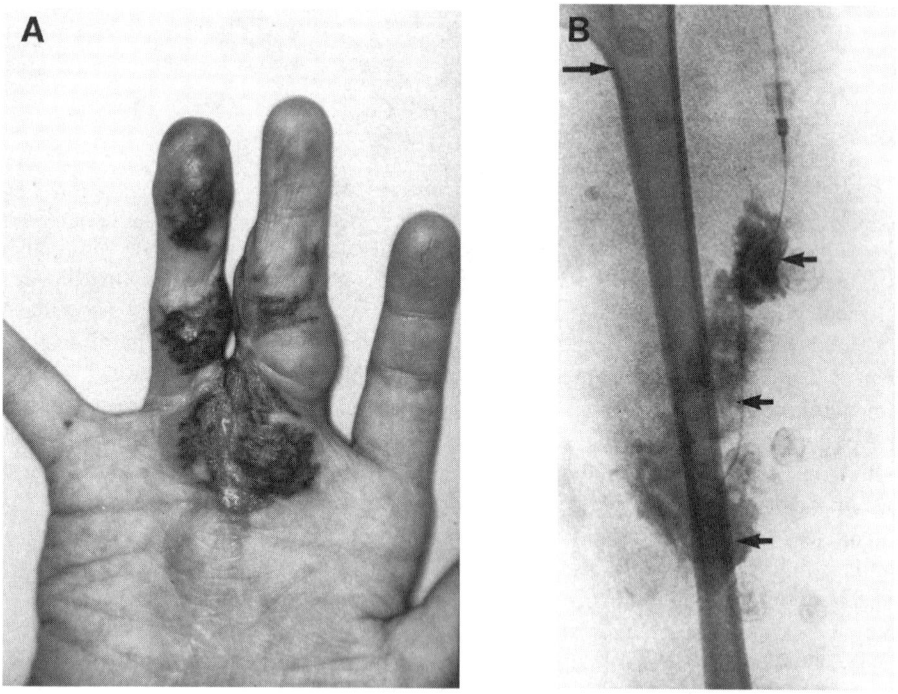

Figure 1 A. Venous malformation of the hand. B. Direct venography in a 16-year-old female patient showing deep venous malformation (short arrows) along the femur (long arrow).

III. VENOUS MALFORMATIONS—HISTOLOGICAL FINDINGS

Cutaneous VM lesions are characterized by a number of saccular channels lined by a flattened endothelium and having deficient smooth muscle cell layer (4,6) (Figure 2). Immunohistochemical studies with antibodies against E-selectin and a general cell proliferation marker Ki-67 indicate that the endothelial cells are not in a proliferative phase (7). Staining of the endothelial cells with antibodies against von Willebrand factor shows that they appear to be firmly attached to the underlying extracellular matrix (6). This histological finding is in contrast to findings in TIE-2 knockout mice, where endothelial cells seem to be rounded and detached from the underlying matrix (1,8).

IV. VENOUS MALFORMATIONS—MUTATION IN TIE-2

The information gained from knockout studies cannot reveal the whole variety of TIE-2 functions because the mice die early in development, and signaling studies using cell cultures only partially reflect the in vivo functions. In addition to these model systems, venous malformations in humans can be used to specifically study the malfunctions in the development and maintenance of veins. Using a genetic approach in two unrelated families with dominantly inherited mucocutaneous venous malformations, the disease locus was assigned to chromosome 9p, and subsequent analysis of the genes in that chromosomal region revealed an arginine-to-tryptophan (R849W) substitution in the conserved region of the kinase domain of the TIE-2 receptor tyrosine kinase (6,9). A second mutation in the same kinase domain, Y897S, in a third family has also been reported (10). In vitro, both of these mutations result in increased autophosphorylation of TIE-2. The R849W mutation is located six amino acid residues from a highly conserved lysine in the kinase

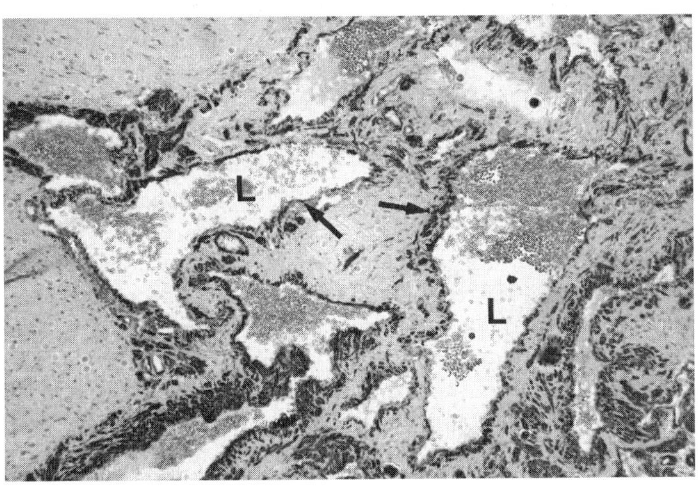

Figure 2 Immunohistochemistry of venous malformation with antibody against smooth muscle α-actin (seen in black). Variable numbers of smooth muscle cells surrounding venous lumens (L) can be seen.

domain. When the TIE-2 kinase domain is compared with the crystallographic structure of the insulin receptor, the arginyl residue corresponding to a glutamyl residue is located in a loop pointing away from the catalytic site (6). The mutation in the TIE-2, replacing a positive charge with a hydrophobic one, could change the tyrosine phosphorylation by either inducing receptor dimerization or causing increased stability of the kinase domain, leading to increased autophosphorylation.

V. DISEASE MECHANISM—CONSEQUENCES OF THE TIE-2 MUTATION

Signaling events involving TIE-2 have only recently begun to be understood, and the downstream target genes regulated by the receptor still remain to be identified. Two ligands, ANG-1 and ANG-2, have been described (3,11). ANG-1 has been shown to induce phosphorylation of TIE-2 and to cause sprouting of endothelial cells (12,13). In addition, when expressed in keratinocytes in transgenic mice, ANG-1 causes increased number and branching of blood vessels in skin (12). ANG-2 seems to act as a natural antagonist to ANG-1 (3). Some molecules associating with the cytosolic kinase domains of TIE-2 have also recently been identified. These include GRB2, SH-PTP2 (14), Dok-R (15), and the p85 subunit of P13-kinase (16). TIE-2 has also been shown to activate two transcription activators, STAT3 and STAT5 (17). These findings suggest that TIE-2 is involved in several different signaling pathways in endothelial cells (Figure 3).

How does the R849W mutation disturb TIE-2 function and cause VMs? In vitro experiments of the R849W mutant TIE-2 show a 6- to 10-fold increase in autophosphorylation of the TIE-2 kinase domain (6). In addition to activating STAT3 and STAT5, expression of mutant TIE-2 in 293 cells also activates STAT1, which regulates the cell cycle through p21 (17). Since increased levels of p21 inhibit cell proliferation, this may explain why VMs grow slowly (often over many years) and why histological analyses show no evidence of increased cell proliferation within VM lesions (7). Activation of several STATs could also lead to transcriptional activation of genes encoding matrix receptors on endothelial cells or molecules secreted by endothelial cells into their surrounding matrix. Such molecules could be important in the adhesion of endothelial cells to the matrix and in regulating the recruitment of smooth muscle cells to the endothelial lining in developing veins. In fact, the histology of VM lesions shows flattened endothelial cells lining the blood vessels without disruptions and fewer than normal smooth muscle cells surrounding the vessel walls (6). These findings suggest that TIE-2 is involved in signaling between endothelial and smooth muscle cells and/or extracellular matrix and that this signaling is impaired in VMs. In support of this suggestion is the finding that lack of Tie-2 or Ang-1 in knockout mice results in detachment of endocardium from myocardium (1,2) and that hematopoietic cells expressing TIE-2 attach more tightly to collagen matrices than cells that do not express TIE-2 (8). A mutation that leads to increased TIE-2 phosphorylation could result in overexpression of genes involved in the attachment and sprouting of endothelial cells normally induced by ANG-1 and inhibited by ANG-2. Constitutive TIE-2 activation could also prevent expression of genes involved in smooth muscle cell recruitment.

TIE-2 is not the only locus associated with VMs. Using a genetic approach, we have recently identified a second locus (17a). As the lesions clinically resemble those caused by TIE-2 mutations, the locus, *VMGLOM*, most probably contains a gene encoding a

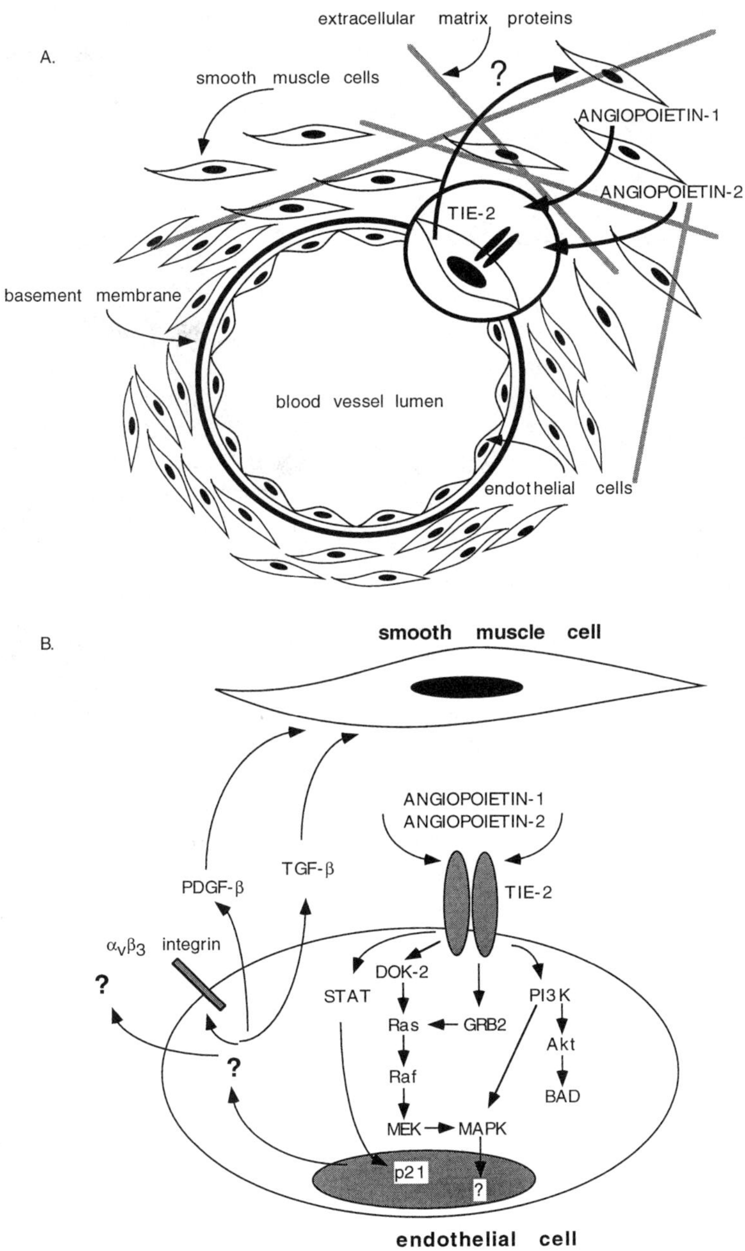

Figure 3 A. Hypothetical TIE-2 signaling pathway in endothelial-smooth muscle cell interactions. B. Possible TIE-2 signaling pathways in endothelial cells. DOK-2 is the human homologue of mouse Dok-R (24).

protein that affects the TIE-2 pathway. The histology of the lesions of the patients linked to the *VMGLOM* locus is slightly different from that of individuals with the TIE-2 mutation. The TIE-2 mutation seems to result in a relative deficiency of smooth muscle cells, whereas lesions from patients linked to the *VMGLOM* show a variable number of poorly differentiated smooth muscle cell–like cells, or ''glomus cells.'' The mutated gene in this locus is likely to provide additional important clues about TIE-2 signaling and how veins develop and are maintained.

VI. DISEASE MECHANISM—LOCALIZED DEFECT

Why do the TIE-2 mutations cause defects specifically in veins, although TIE-2 is expressed in endothelial cells lining most blood vessels? Although the answer to this question is not known, it is possible that proteins involved in TIE-2 signaling pathways are likely to have different expression patterns in arteries and veins because of differential shear stresses and responses to such stresses. Also, genes related to the TIE-2 signaling pathway may be differentially switched ''on'' or ''off'' during vasculogenesis, when blood vessels differentiate to form arteries and veins. At least one receptor ligand pair, ephrin-B2, expressed in arteries, and its receptor EPH-B4, expressed in veins, has already been shown to be involved in the patterning of vasculature (18). As discussed above, TIE-2 is most probably involved in several different pathways to regulate expression of a variety of genes. It has already been shown that TIE-2 associates with Dok-R (15), the p85 subunit of P13-kinase (16), Grb2, and SH-PTP2 (14), which all have different downstream effects. There may also well be differences in the signaling from different tyrosine residues (19), and the R849W may have an effect on only one of these pathways important in veins.

Why do heterozygous patients get only localized lesions and why do not all individuals carrying the mutation get lesions? These questions remain to be answered. It is possible that patients with lesions may have a second, somatic mutation in some other locus or in the TIE-2 locus, and the combination of the inherited TIE-2 mutation and the ''second hit'' could cause a local malformation. Environmental factors such as microtrauma during development and a resulting incorrect healing response of mutant cells may also cause a local effect.

VII. SUMMARY AND FUTURE PERSPECTIVES

Several genes are involved in causing vascular anomalies. For example, hereditary hemorrhagic telangiectasia type 1 (HHT1) is caused by mutations in endoglin, a TGF-β binding protein (20), and HHT2 is caused by mutations in activin-like receptor also involved in TGF-β binding (21). Cerebral cavernous malformations have been linked to chromosome 7q11.23-22, but the mutation has not yet been identified (22,23). Activating mutations in the TIE-2 kinase domain are one cause of venous malformations and another locus has already been identified (6,17a).

For most genetic diseases, identification of the causative mutation does not fully explain the phenotype, especially when penetrance and expressivity may vary. Mutations in TIE-2 result in ligand-independent activation of the receptor kinase and activation of STAT1 in cell culture. Whether these are the consequences in vivo, however, is not known; other molecules playing a role in the TIE-2 pathway may be necessary for malformations to form. We are currently creating a transgenic mouse expressing mutant Tie-2 (R846W corresponding to the human R849W) in endothelial cells. With this model, we will be able to address questions about a second mutational hit involved in VM formation, to find out whether any endothelial cell components are up- or downregulated in vivo as a result of the mutation, and whether there are any potential therapies that could be developed to correct VMs in light of the molecular pathological findings.

ACKNOWLEDGMENTS

Studies from the authors' laboratories were supported in part by the Academy of Finland, Alfred Kordelin Foundation, Finnish Cultural Foundation (to L. K.), Human Frontier Science Program, Fonds de Developpement Scientifique-Université Catholique de Louvain and the Belgian Federal Service for Scientific, Technical and Cultural Affairs (to M. V.), and the National Institutes of Health (AR36820) (to B. R. O.).

REFERENCES

1. DJ Dumont, G Gradwohl, GH Fong, MC Puri, M Gertsenstein, A Auerbach, ML Breitman. Dominant-negative and targeted null mutations in the endothelial receptor tyrosine kinase, tek, reveal a critical role in vasculogenesis of the embryo. Genes Dev 8:1897–1909, 1994.
2. C Suri, PF Jones, S Patan, S Bartunkova, PC Maisonpierre, S Davis, TN Sato, GD Yancopoulos. Requisite role of angiopoietin-1, a ligand for the TIE2 receptor, during embryonic angiogenesis. Cell 87:1171–1180, 1996.
3. PC Maisonpierre, C Suri, PF Jones, S Bartunkova, S Wiegand, C Radziejewski, D Compton, J McClain, TH Aldrich, N Papadopoulos, TJ Daly, S Davis, TN Sato, GD Yancopoulos. Angiopoietin 2, a natural antagonist for Tie2 that disrupts in vivo angiogenesis. Science 277:55–60, 1997.
4. M Vikkula, LM Boon, JB Mulliken, BR Olsen. Molecular basis of vascular anomalies. Trends Cardiovasc Med 8:281–292, 1998.
5. J Mulliken, A Young. Vascular Birthmarks: Hemangiomas and Malformations. Philadelphia: Saunders, 1988.
6. M Vikkula, LM Boon, KL Carraway, JT Calvert, AJ Diamonti, B Goumnerov, KA Pasyk, DA Marchuk, ML Warman, LC Cantley, JB Mulliken, BR Olsen. Vascular dysmorphogenesis caused by an activating mutation in the receptor tyrosine kinase TIE2. Cell 87:1181–1190, 1996.
7. BM Kraling, MJ Razon, LM Boon, D Zurakowski, C Seachord, RP Darveau, JB Mulliken, CL Corless, J Bischoff. E-selectin is present in proliferating endothelial cells in human hemangiomas. Am J Pathol 148:1181–1191, 1996.
8. A Sato, A Iwama, N Takakura, H Nishio, GD Yancopoulos, T Suda. Characterization of TEK receptor tyrosine kinase and its ligands, angiopoietins, in human hematopoietic progenitor cells. Int Immunol 10:1217–1227, 1998.
9. LM Boon, JB Mulliken, M Vikkula, H Watkins, J Seidman, BR Olsen, ML Warman. Assignment of a locus for dominantly inherited venous malformations to chromosome 9p. Hum Mol Genet 3:1583–1587, 1994.
10. JT Calvert, J Berg, NC Nevin, KA Pasyk, SA Simpson, KG Peters, DA Marchuk. Allelic and locus heterogeneity in familial venous malformations syndrome (FVM) (abstr). Am J Hum Genet 63: abstr 2049, 1998.
11. S Davis, TH Aldrich, PF Jones, A Acheson, DL Compton, V Jain, TE Ryan, J Bruno, C

Radziejewski, PC Maisonpierre, GD Yancopoulos. Isolation of angiopoietin-1, a ligand for the TIE2 receptor, by secretion-trap expression cloning. Cell 87:1161–1169, 1996.

12. C Suri, J McClain, G Thurston, DM McDonald, H Zhou, EH Oldmixon, TN Sato, GD Yancopoulos. Increased vascularization in mice overexpressing angiopoietin-1. Science 282:468–471, 1998.
13. TI Koblizek, C Weiss, GD Yancopoulos, U Deutsch, W Risau. Angiopoietin-1 induces sprouting angiogenesis in vitro. Curr Biol 8:529–532, 1998.
14. L Huang, CW Turck, P Rao, KG Peters. GRB2 and SH-PTP2: potentially important endothelial signaling molecules downstream of the TEK/TIE2 receptor tyrosine kinase. Oncogene 11: 2097–2103, 1995.
15. N Jones, DJ Dumont. The Tek/Tie2 receptor signals through a novel Dok-related docking protein, Dok-R. Oncogene 17:1097–1108, 1998.
16. CD Kontos, TP Stauffer, WP Yang, JD York, L Huang, MA Blanar, T Meyer, KG Peters. Tyrosine 1101 of Tie2 is the major site of association of p85 and is required for activation of phosphatidylinositol 3-kinase and Akt. Mol Cell Biol 18:4131–4140, 1998.
17. EI Korpelainen, M Kärkkäinen, Y Gunji, M Vikkula, K Alitalo. Endothelial receptor tyrosine kinases activate the STAT signaling pathway: mutant TIE-2 causing venous malformations signals a distinct STAT activation response. Oncogene, 18:1–8, 1999.

17a. LM Boon, P Brouillard, A Irrthum, L Karttunen, ML Warman, R Rudolph, JB Mulliken, BR Olsen, M Vikkula. A gene for inherited cutaneous venous anomalies (‘‘Glomangiomas’’) localizes to chromosome 1p21–22. Am J Hum Genet, 1999. In press.

18. HU Wang, ZF Chen, DJ Anderson. Molecular distinction and angiogenic interaction between embryonic arteries and veins revealed by ephrin-B2 and its receptor Eph-B4. Cell 93:741–753, 1998.
19. Z Songyang, KL Carraway, MJ Eck, SC Harrison, RA Feldman, M Mohammad, J Schlessinger, SR Hubbard, DP Smith, C Eng, MJ Lorenzo, BAJ Ponder, BJ Mayer, LC Cantley. Catalytic specificity of protein tyrosine kinases is critical for selective signaling. Nature 373: 536–539, 1995.
20. KA McAllister, KM Grogg, DW Johnson, CJ Gallione, MA Baldwin, CE Jackson, EA Helmbold, DS Markel, WC McKinnon, J Murrell, et al. Endoglin, a TGF-beta binding protein of endothelial cells, is the gene for hereditary haemorrhagic telangiectasia type 1. Nature Genet 8:345–351, 1994.
21. DW Johnson, JN Berg, MA Baldwin, CJ Gallione, I Marondel, SJ Yoon, TT Stenzel, M Speer, MA Pericak-Vance, A Diamond, AE Guttmacher, CE Jackson, L Attisano, R Kucherlapati, ME Porteous, DA Marchuk. Mutations in the activin receptor-like kinase 1 gene in hereditary haemorrhagic telangiectasia type 2. Nature Genet 13:189–195, 1996.
22. J Dubovsky, JM Zabramski, J Kurth, RF Spetzler, SS Rich, HT Orr, JL Weber. A gene responsible for cavernous malformations of the brain maps to chromosome 7q. Hum Mol Genet 4: 453–458, 1995.
23. M Gunel, IA Awad, K Finberg, JA Anson, GK Steinberg, HH Batjer, TA Kopitnik, L Morrison, SL Giannotta, C Nelson-Williams, RP Lifton. A founder mutation as a cause of cerebral cavernous malformation in Hispanic Americans. N Engl J Med 334:946–951, 1996.
24. A Di Cristofano, N Carpino, N Dunant, G Friedland, R Kobayashi, A Strife, D Wisniewski, B Clarkson, PP Pandolfi, MD Resh. Molecular cloning and characterization of p56dok-2 defines a new family of RasGAP-binding proteins. J Bio Chem 273:4827–4830, 1998.

6

Novel Autocrine and Paracrine Regulators of the Angiogenic Cascade

Hellmut G. Augustin
University of Göttingen Medical School, Göttingen, Germany

I. INTRODUCTION

The angiogenic cascade requires endothelial cells from small vessels, mostly postcapillary venules, to break away from their quiescent monolayer, proteolytically degrade their underlying basement membrane and the surrounding extracellular matrix, directionally migrate toward the angiogenic stimulus, proliferate, and eventually enter a complex morphogenetic cascade that leads to the three-dimensional organization of new capillaries. These complex microenvironmental interactions require endothelial cells to express a specific repertoire of molecules that shape their angiogenic phenotype. Endothelial cells during angiogenesis engage a number of different molecular systems, such as the proliferation machinery, growth factor receptors, adhesion molecules, cytoskeletal rearrangements, a shift of the proteolytic balance toward an invasive phenotype, and the expression of extracellular matrix molecules (1). A number of molecules specifically or preferentially expressed by angiogenic endothelial cells have been identified in recent years. These include the integrin heterodimers $\alpha_v\beta_3$ and $\alpha_v\beta_5$ as well as the VEGF and Tie receptors [VEGF-R1 (Flt-1), VEGF-R2 (Flk-1), Tie-1, Tie-2] (1). Other known and novel molecules are still to be identified. It is obvious that the angiogenic endothelial cell phenotype is much more complex, as suggested by the relatively few molecules of angiogenic endothelial cells been identified so far. The regulatory forces exerted on endothelial cells during angiogenesis are depicted in Fig. 1. Exogenous cytokines are classically considered as paracrine inducers of angiogenesis (Fig. 1, paracrine in). VEGF is probably the most important angiogenic factor to act in this manner, although endothelial cells themselves synthesize VEGF. Endothelial cells can also be activated by engaging autocrine loops (Fig. 1, autocrine). The heparin-binding growth factor FGF-2 has been analyzed most extensively for its ability to stimulate endothelial cells in an autocrine manner. Last, angiogenically activated endothelial cells may themselves be a source of paracrine-acting cytokines that may act as secondary enhancers of the angiogenic cascade—e.g., by recruiting inflammatory cells (Figure 1, paracrine out).

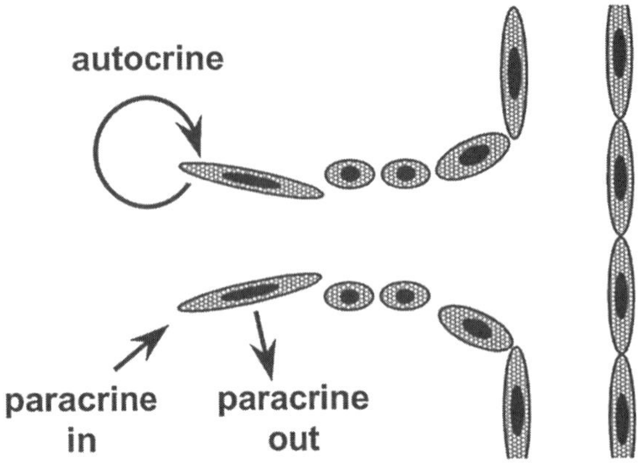

Figure 1 Characterization of the principal forces that act on endothelial cells during angiogenesis. Endothelial cells acquire an invasive proliferating phenotype. They are classically activated by angiogenic growth factors in a paracrine manner (paracrine in). They may activate themselves in autocrine growth regulating loops (autocrine). Last, they themselves can modulate the angiogenic cascade by acting on other cells in a paracrine manner (paracrine out).

II. COMPARATIVE GENE EXPRESSION ANALYSIS OF ACTIVATED AND QUIESCENT ENDOTHELIAL CELLS

A. Rationale

Comparative gene expression analyses of cultured endothelial cells have been extremely useful in the identification of molecules that shape the phenotype of endothelial cells during inflammation. The comparison of resting versus cytokine-activated endothelial cells (e.g., TNF-α stimulation) has led to the identification of a number of biologically relevant surface molecules, such as the selectins (2). Similar approaches have been employed toward the identification of genes that shape the angiogenic endothelial cell phenotype. These experiments involve either two- or three-dimensional angiogenesis assays. We have utilized a two-dimensional lateral sheet migration assay to study comparatively growth-arrested and migrating endothelial cells (Fig. 2) (3). The rationale of these experiments was directed by the hypothesis that laterally sheet migrating endothelial cells, respectively subconfluent endothelial cells, represent some of the molecular properties of angiogenic endothelial cells in vivo, whereas quiescent growth-arrested endothelial cells in culture reflect the resting organ endothelium in vivo. It is important to note that this experimental approach did not involve the stimulation of the cultured endothelial cells with angiogenic growth factors, which may be a useful strategy to identify the response program of endothelial cells mediated by these cytokines. Instead, our working hypothesis was aimed at identifying endothelial cell genes that are expressed under autocrine control just by releasing the cells from growth arrest. Taken together, this hypothesis has held up. Most of the identified differentially expressed genes of subconfluent, migrating endothelial cells could be demonstrated as being involved in shaping the activated endothelial cell phenotype in vivo as it is associated with angiogenesis.

B. Differential RNA Display

Differential RNA display has been developed to rapidly identify differentially expressed 3’UTR sequence tags (4,5). The technique has been applied to the analysis of monospecific

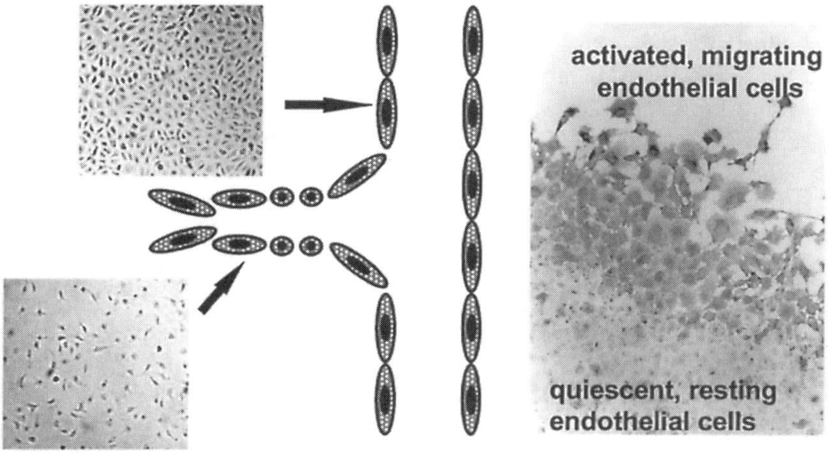

Figure 2 Experimental approach toward the differential gene expression analysis of resting versus activated endothelial cells. Subconfluent, migrating endothelial cells (bottom left) were analyzed as representatives of the invasive, proliferating endothelial cell phenotype during angiogenesis. Confluent monolayers (top left) were comparatively considered as representatives of the quiescent, resting phenotype in vivo. The phenotypic difference between migrating and resting endothelial cells is also demonstrated by the preferential binding of a monoclonal antibody to migrating endothelial cells (right). In vivo, this migrating endothelial cell detecting antibody binds preferentially to angiogenic endothelial cells.

cell populations as well as mRNA from complex tissues (6–8). As a PCR-based technique, RNA display is particularly appealing when the availability of RNA is limiting—for example in the analysis of developmental processes.

RNA display banding patterns of migrating, subconfluent endothelial cells were compared with the banding patterns of quiescent, resting endothelial cells (9). For comparative purposes, it was thus a necessary prerequisite to use pure, monospecific-RNA populations without cross-contaminations of the two functional endothelial cell states to be compared. Subconfluent endothelial cell monolayers that are devoid of confluent islands can be harvested if care is taken to disperse the cells optimally when they are seeded. Fully confluent monolayers were harvested 3 days after growing to confluence. This ensured that the monolayer had reached a state of relative quiescence. It is, however, important to note that even a postconfluent endothelial cell monolayer is not fully quiescent. BrDU labeling experiments of postconfluent populations of several types of endothelial cells [bovine aortic endothelial cells (BAEC), human umbilical vein endothelial cells (HUVEC), human dermal microvascular endothelial cells (HDMVEC)] demonstrated that these cells still have labeling rates between 5 and 10%, which is at least an order of magnitude higher than in vivo (T. Korff and H. G. Augustin, unpublished observation). This has prompted us to develop an endothelial cell spheroid cell culture model that more realistically reflects the quiescent endothelial cell monolayer in vivo as standard two-dimensional tissue culture techniques (10). For the identification of the differentially expressed genes outlined in this chapter, however, RNA populations of two-dimensionally cultured cells were compared.

More than 100 differentially expressed sequence tags of migrating BAEC were isolated from the differential RNA display gels, reamplified, cloned, and used for comparative Northern blots of migrating versus resting endothelial cells. These screening experiments revealed a false-positive rate of close to 50% of genes that could not be confirmed as being differentially expressed by Northern blotting. Of the approximately 50 genes whose expression was truly upregulated in migrating endothelial cells (at least threefold upregulation), close to 50% were known genes, whereas the unknown genes had either sequence homology to expressed sequence tags (ESTs) deposited in publicly available databanks or showed no sequence homology to any known sequence. Among the known molecules whose expression was upregulated in migrating endothelial cells were endothelin (GenBank Y00749), metallothionein-II (GB J00271), follistatin (GB L21716), G(i) protein α-subunit (GB X03642), caveolin 1 (GB U86639), monocyte chemoattractant protein-1 (GB L32659), acetoacetyl-coenzyme A thiolase (GB D90228), annexin I (GB X56649), cyclophilin (GB Y00052), lactate dehydrogenase-A (GB D90141), ras-like protein TC21 (GB M31468), proteasome subunit HC8 (GB D00762), BiP protein (GB X87949), thrombospondin 1 (GB AB005287), ras-like protein TC4 (GB M31469), annexin V-binding protein (GB D64061), ribophorin I (GB X05300), 70-KDa heat-shock cognate protein (GB X53827), chaperonin protein (Tcp20) (GB L27706), lysosomal α-mannosidase (GB U97688), and the KIAA0002 gene (GB D13627). Of these molecules, follistatin and monocyte chemoattractant protein-1 (MCP-1) were selected as candidate molecules for subsequent functional studies (see below). The characterization of novel differentially expressed endothelial cell molecules is the focus of still ongoing experiments.

III. REGULATION OF AUTOCRINE ENDOTHELIAL CELL ACTIVITY THROUGH THE FOLLISTATIN/ACTIVIN-SYSTEM

Follistatin is a member of a family of functionally unrelated cysteine-rich secreted glycoproteins (11). Among these molecules are the endothelial cell glycoproteins SPARC (12) and hevin (13). Originally identified as an FSH-suppressing molecule follistatin was found to bind and inhibit the functions of activins (14,15). As such it received considerable attention as a neutralizing molecule during development by supposedly interfering with the activin-mediated induction of mesoderm (16). Surprisingly, targeted disruption of the follistatin gene revealed that homozygous follistatin-null mice do not exhibit a neural phenotype (17). Instead, follistatin-deficient mice develop only few specific developmental abnormalities, such as defective formation of whiskers. The most prominent feature of follistatin-deficient mice is their pronounced growth retardation, leading to perinatal death of newborn pups (17).

The differential RNA display approach identified follistatin as a differentially expressed gene of migrating endothelial cells (18). Follistatin had previously not been studied as a vascular function-regulating molecule. What sparked our interest in this secreted glycoprotein was the fact that its binding partner, the TGFβ-family member activin-A, had been reported to act as an autocrine and paracrine inhibitor of endothelial cell proliferation (19). Thus, the expression of an inhibitor (follistatin) of a proliferation inhibiting molecule (activin-A) by activated endothelial cells suggested a possible autocrine growth-regulatory loop, which could be conceptualized as the removal of an endogenous growth-inhibiting brake.

Northern blot experiments identified both follistatin transcripts (2.4 kb, 1.7 kb) as prominently expressed in subconfluent, migrating endothelial cells and a downregulation of follistatin expression in quiescent, resting endothelial cells. Subsequent experiments revealed that endothelial cell follistatin expression is under the control of FGF-2. Correspondingly, elevated levels of follistatin protein could be identified in the supernatants of FGF-2–stimulated endothelial cells (18). In contrast to the regulated expression of follistatin in endothelial cells, all endothelial cell populations analyzed constitutively expressed high levels of activin-A, suggesting a role of activin-A in maintaining the quiescent endothelial cell phenotype.

A weak mitogenic effect of follistatin was observed when highly endogenous follistatin-expressing BAE cells were stimulated with exogenous follistatin. In contrast, a prominent proliferation-inducing effect of exogenous follistatin was detected when endothelial cells expressing low endogenous follistatin, such as human umbilical vein endothelial cells (HUVEC), were stimulated with follistatin (18). Based on these observations, we performed rabbit cornea angiogenesis experiments in order to determine whether the proliferation-inducing capacity of follistatin translates into angiogenesis-inducing capacity in vivo. To this end, slow-release Elvax pellets containing low concentrations of recombinant follistatin (20–100 ng) were implanted into the avascular rabbit cornea and studied for their ability to induce sprouting angiogenesis originating from the limbal blood vessels. These experiments revealed that follistatin acts as a moderate specific inducer of angiogenesis in vivo. Coimplantation experiments with FGF-2 demonstrated synergistic angiogenesis-inducing activity of follistatin with this angiogenic factor (18).

The data obtained support a model of autocrine endothelial cell growth regulation through the follistatin/activin system. Consequently, ongoing experiments are aimed at elucidating the possible role of this antagonistic system in the regulation of tumor growth and tumor-associated angiogenesis. In fact, growth regulation through the follistatin/activin system is not just restricted to endothelial cells: Activin-A has been shown to act as an autocrine inhibitor of hepatocyte proliferation (20). Correspondingly, follistatin could be shown to accelerate liver regeneration after partial hepatectomy, supposedly by inhibiting activin-mediated growth arrest of hepatocytes (21).

IV. INDIRECT INDUCTION OF ANGIOGENESIS BY MONOCYTE-CHEMOATTRACTANT MOLECULE (MCP-1)

The CC chemokine MCP-1 is induced by inflammatory cytokines and acts as a potent regulator of monocyte trafficking. Transgenic mice overexpressing MCP-1 have perivascular cuffs of monocytes (22), whereas MCP-1–deficient mice have an inability to recruit macrophages upon challenge (23). Together, these experiments strongly suggest that monocytes are the primary target-cell population of MCP-1.

MCP-1 is expressed by a wide variety of normal and malignant cells, including endothelial cells, monocytes, vascular smooth muscle cells, and fibroblasts as well as glioma, sarcoma, and melanoma cells (24). Originally identified as a PDGF-inducible gene (25), MCP-1 was soon characterized as an inflammatory chemokine that is primarily regulated by inflammatory cytokines such as TNF-α and IL-1 (26–28). Of the different cell types that express MCP-1 upon stimulation by inflammatory cytokines induced expression of MCP-1 by vascular endothelial cells is probably most important for the initial recruitment of monocytes to sites of inflammation. The monocytic cell line U937 has previously been shown to preferentially adhere to migrating endothelial cells that had not been stimulated by inflammatory cytokines (29). Corresponding to these in vitro findings, we could show that monocytic cells adhere preferentially to migrating endothelial cells in vivo following aortic balloon denudation (30). This preferential adhesion of monocytic cells was associated with a massive upregulation of MCP-1 expression at the front of regenerating endothelial cell migration (30).

A detailed analysis of endothelial cell MCP-1 expression in vitro revealed that MCP-1 is a differentially expressed gene of autocrine activated endothelial cells. It is prominently expressed by subconfluent, migrating endothelial cells and downregulated once endothelial cells reach confluence (30). Stimulation as well as antibody inhibition experiments demonstrated that endothelial cell MCP-1 expression is under the control of endogenous FGF-2. Upon stimulation with FGF-2, elevated levels of MCP-1 could be detected within 30 min, suggesting a rapid and direct stimulatory effect of FGF-2.

The regulation of endothelial cell MCP-1 expression through the angiogenic growth factor FGF-2 suggested a link between angiogenesis induction and inflammatory cell trafficking. MCP-1 implantation into the avascular cornea led to the rapid recruitment of macrophages and the induction of corneal angiogenesis (M. Ziche and H.G. Augustin, unpublished observation). Thus, in the context of the rabbit cornea, ectopic recruitment of monocytic cells is sufficient to induce an inflammatory reaction that, in turn, activates angiogenesis. Corresponding to these findings MCP-1 was recently shown to accelerate collateral formation following femoral artery occlusion (31,32).

Tumor cell–derived and/or endothelial cell-derived MCP-1 may be an important mediator of tumor associated inflammatory angiogenesis. This notion is supported by the recently demonstrated association between high macrophage counts in mammary tumors and the degree of vascularization of these tumors as determined by microvessel density counts (33). Similarly, MCP-1 expression and monocyte recruitment has also been associated in a number of other tumor types, including ovarian (34) and prostate carcinomas (35), melanomas (36) and vascular tumors (37) as well as experimental rat tumors (38). Corresponding to these findings, the macrophage function inhibiting substance Linomide exerts its antitumorigenic and antimetastatic activity at least in part by inhibiting tumor associated angiogenesis (39). Taken together, the in vitro findings on the expression of MCP-1 by endothelial cells in association with the in vivo data on the association between MCP-1 expression in tumors, inflammatory cell recruitment, and tumor angiogenesis strongly suggest distinct mechanistic differences between inflammatory tumor angiogenic processes and noninflammatory physiological angiogenic processes as they occur during development and in the female reproductive system. It appears that MCP-1 is a key regulatory molecule of monocyte trafficking that contributes to the distinct qualitative differences between these processes.

V. ENDOTHELIAL CELL ACTIVATION THROUGH THE SDF-1/CXC-R4 LIGAND/RECEPTOR INTERACTION

The experiments outlined above had indicated that endothelial cell–derived MCP-1 contributes to the inflammatory cell trafficking during tumor angiogenesis. In order to exclude direct endotheliotropic functions of MCP-1, we performed proliferation and chemotaxis assays on a variety of different endothelial cell populations. These experiments gave no indication of a direct effect of MCP-1 on endothelial cells. To further confirm these findings, we assessed endothelial cell MCP-1 receptor expression on human umbilical vein endothelial cells (HUVEC) and human dermal microvascular endothelial cells

(HDMVEC). Based on this RT-PCR analysis, it was confirmed that endothelial cells do not express MCP-1 receptors. In fact the experiments showed that both endothelial cell populations did not express any of the hitherto known CC chemokine receptors (CCR-1 through CCR-6).

To complete these experiments, we also analyzed endothelial cell CXC chemokine receptor expression. Several CXC chemokine have been shown to exert angiogenesis-modulating activity, supposedly by directly acting on endothelial cells. Interleukin-8 (IL-8) induces EC proliferation and chemotaxis in vitro and angiogenesis in vivo and appears to be a primary mediator of macrophage-induced angiogenesis (40). In contrast, platelet factor-4 (PF-4) was reported to inhibit EC proliferation in vitro (41) and angiogenesis in vivo (42). Antiangiogenic activities have also been described for the CXC chemokines interferon γ–inducible protein 10 (IP-10) (43) and growth-related oncogene β (GRO-β) (44). In a very elegant-study, Strieter et al. showed that the pro- and antiangiogenic activities of CXC-chemokines are dependent on an amino-terminal ELR motif expressed by the stimulators and not by the inhibitors (45).

Analysis of chemokine receptor expression by RT-PCR revealed that endothelial cells essentially express all known CXC chemokine receptors. As outlined above, endotheliotropic functions of ligands for CXCR-1 (IL-8), CXCR-2 (IL-8) and CXCR-3 (IP-10) have previously been reported. We have consequently performed a detailed analysis of endothelial CXCR-4 expression (46). CXCR-4 is constitutively expressed by quiescent, resting endothelial cells. Cytokine stimulation revealed that bFGF upregulated endothelial CXCR-4 expression, whereas TNF-α downregulates endothelial CXCR-4. Expression of CXCR-4 mRNA as well as protein is also upregulated in autocrine-activated migrating aortic endothelial cells. Furthermore, migrating endothelial cells preferentially present CXCR-4 on the cell surface, as evidenced by cytochemistry and FACS analysis. Similar results have recently been reported (47,48).

In order to elucidate a possible functional role of the interaction of CXCR-4 with its monospecific ligand SDF-1, we performed Boyden chamber chemotaxis experiments with HUVE cells. These experiments showed that SDF-1 can act as a potent inducer of endothelial cell migration (46,48). Checkerboard analyses revealed that the effect of SDF-1 on endothelial cells is chemotactic and not chemokinetic. It would be expected that an in vitro endothelial cell chemotaxis–inducing molecule may be capable of inducing angiogenesis in vivo. This conclusion was recently confirmed with the generation of CXCR-4–deficient mice (49). These mice have an organ-specific angiogenic phenotype characterized by a disturbance of gastrointestinal vascularization. The subsequently performed reanalysis of the phenotype of SDF-1–deficient mice (50) revealed that these mice similarly have a disturbance of gastrointestinal angiogenesis, which further supported the notion that the CXCR-4/SDF-1 interaction is a monospecific receptor/ligand-interaction. Future work will have to reveal the molecular mechanism underlying the organ specific angiogenic phenotype of CXCR-4–or SDF-1–deficient mice and the role of SDF-1 in mediating tumor angiogenesis.

VI. CONCLUSION AND PERSPECTIVE

A plethora of molecules regulates the complex interactions of endothelial cells with their microenvironment during angiogenesis. Paracrine activities can act on endothelial cells, endothelial cells can act on other cells in a paracrine manner, and endothelial cells can

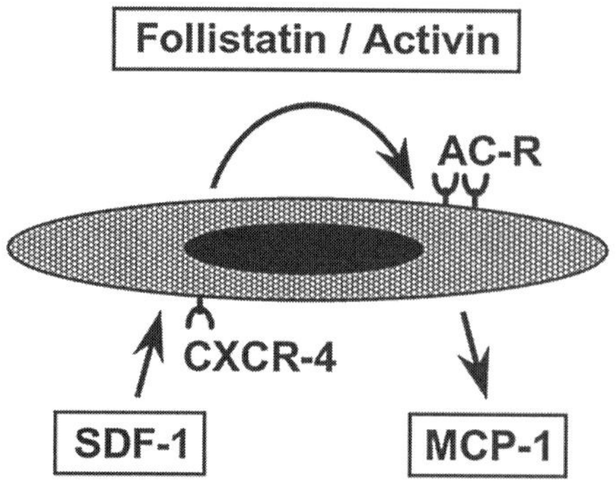

Figure 3 Schematic summary of the identified novel autocrine and paracrine regulators of the angiogenic cascade. The CXCR-4/SDF-1 receptor/ligand interaction stimulates endothelial cells in a paracrine manner. The follistatin/activin system operates as an autocrine growth-regulating system of activated endothelial cells. The differential production of MCP-1 by activated endothelial cells contributes to the induction of angiogenesis in a paracrine manner through the recruitment of monocytes.

act on themselves in autocrine loops (Fig. 1). An example of each of these types of cellular activation is presented in this chapter (Fig. 3). The follistatin/activin system was characterized as an autocrine growth-regulating system of endothelial cells. SDF-1 was identified as a paracrine activator of endothelial cells. Last, endothelial cells act on monocytes in a paracrine manner by secreting MCP-1. Together, the three reported molecular systems contribute to shaping the complex endothelial-cell phenotype during angiogenesis. Other molecules similarly act on endothelial cells. Some of these have already been identified. The results of our differential RNA display experiments suggest that numerous others are still to be identified. These ongoing experiments will contribute to further understanding of the complexity of endothelial cells and eventually establish the hierarchical order in which angiogenesis-regulating molecules act in different situations of angiogenesis.

REFERENCES

1. HG Augustin. Antiangiogenic tumor therapies—will it work? Trends Pharmacol Sci 19:216–222, 1998.
2. TA Springer. Traffic signals for lymphocyte recirculation and leukocyte emigration: the multistep paradigm. Cell 76:301–314, 1994.
3. HG Augustin, BU Pauli. Migrating endothelial cells are distinctly hyper-glycosylated and express specific migration-associated cell surface glycoproteins. J Cell Biol 119:483–491, 1992.
4. P Liang P, AB Pardee. Differential display of eukaryotic messenger RNA by means of the polymerase chain reaction. Science 257:967–971, 1992.
5. P Liang, D Bauer, L Averboukh, P Warthoe, M Rohrwild, H Muller, M Strauss, AB Pardee. Analysis of altered gene expression by differential display. Methods Enzymol 254:304–321, 1995.
6. LP Aiello, GS Robinson, YW Lin, Y Nishio, GL King. Identification of multiple genes in bovine retinal pericytes altered by exposure to elevated levels of glucose by using mRNA differential display. Proc Natl Acad Sci USA 91:6231–6235, 1994.
7. Y Sun, G Hegamyer, NH Colburn. Molecular cloning of five messenger RNAs differentially

expressed in preneoplastic or neoplastic JB6 mouse epidermal cells: one is homologous to human tissue inhibitor of metalloproteinases-3. Cancer Res 54:1139–1144, 1994.
8. ME Russell, U Utans, AF Wallace, P Liang, RJ Arceci, MJ Karnovsky, LR Wyner, Y Yamashita, C Tarn. Identification and upregulation of galactose/N-acetyl-galactosamine macrophage lectin in rat cardiac allografts with arteriosclerosis. J Clin Invest 94:722–730, 1994.
9. DH Kozian, HG Augustin. Rapid identification of differentially expressed endothelial cell genes by RNA-display. Biochem Biophys Res Commun 209:1068–1075, 1995.
10. T Korff, HG Augustin. Integration of endothelial cells in multicellular spheroids prevents apoptosis and induces differentiation. J Cell Biol 143:1341–1352, 1998.
11. L Patthy. Modular exchange principles in proteins. Curr Opin Struct Biol 1:351–361, 1991.
12. TF Lane, EH Sage. The biology of SPARC, a protein that modulates cell-matrix interactions. FASEB J. 8:163–173, 1994.
13. JP Girard, TA Springer. Cloning from purified high endothelial venue cells of hevin, a close relative of the antiadhesive extracellular matrix protein SPARC. Immunity 2:113–123, 1995.
14. U Michel, P Farnworth, JK Findlay. Follistatins: more than follicle-stimulating hormone suppressing proteins. Mol Cell Endocrinol 91:1–11, 1993.
15. T Nakamura, K Takio, Y Eto, H Shibai, K Titani, H Sugino. Activin-binding protein from rat ovary is follistatin. Science 247:836–838, 1990.
16. A Hemmati-Brivanlou, OG Kelly, DA Melton. Follistatin, an antagonist of activin, is expressed in the Spemann organizer and displays direct neuralizing activity. Cell 77:283–295, 1994.
17. MM Matzuk, N Lu, H Vogel, K Sellheyer, DR Roop, A Bradley. Multiple defects and perinatal death in mice deficient in follistatin. Nature 374:360–363, 1995.
18. DH Kozian, M Ziche, HG Augustin. The activin-binding protein follistatin regulates autocrine endothelial cell activity and induces angiogenesis. Lab Invest 76:267–276, 1997.
19. SA McGarthy, R Bicknell. Inhibition of vascular endothelial cell growth by activin-A. J Biol Chem 268:23066–23071, 1993.
20. H Yasuda, T Mine, H Shibata, Y Eto, Y Hasegawa, T Takeuchi, S Asano, I Kojima. Activin A: an autocrine inhibitor of initiation of DNA synthesis in rat hepatocytes. J Clin Invest 92: 1491–1496, 1993.
21. K Kogure, W Omata, M Kanzaki, Y-Q Zhang, H Yasuda, T Mine, I Kojima. A single intraportal administration of follistatin accelerates liver regeneration in partially hepatectomized rats. Gastroenterology 108:1136–1142, 1995.
22. ME Fuentes, SK Durham, MR Swerdel, AC Lewin, DS Barton, JR Megill, R Bravo, SA Lira. Controlled recruitment of monocytes and macrophages to specific organs through transgenic expression of monocyte chemoattractant protein-1. J Immunol 155:5769–5776, 1995.
23. B Lu, BJ Rutledge, L Gu, J Fiorillo, NW Lukacs, SL Kunkel, R North, C Gerard, BJ Rollins. Abnormalities in monocyte recruitment and cytokine expression in monocyte chemoattractant protein 1-deficient mice. J Exp Med 187:601–608, 1998.
24. BJ Rollins. JE/MGP-1: an early-response gene encodes a monocyte-specific cytokine. Cancer Cells 12:517–524, 1991.
25. BJ Rollins, ED Morrison, CD Stiles. Cloning and expression of JE, a gene inducible by platelet-derived growth factor and whose product has cytokine-like properties. Proc Natl Acad Sci USA 85:3738–3742, 1988.
26. JK Brieland, CM Flory, ML Jones, GR Miller, DG Remick, JS Warren, JC Fantone. Regulation of monocyte chemoattractant protein-1 gene expression and secretion in rat pulmonary alveolar macrophages by lipopolysaccharide, tumor necrosis factor-alpha, and interluekin-1 beta. Am J Respir Cell Mol Biol 12:104–109, 1995.
27. Z Brown, ME Gerritsen, WW Carley, RM Strieter, SL Kunkel, J Westwick. Chemokine gene expression and secretion by cytokine-activated human microvascular endothelial cells: differential regulation of monocyte chemo-attractant protein-1 and interleukin-8 in response to interferon-gamma. Am J Pathol 145:913–921, 1994.

28. A Sica, JM Wang, F Colotta, E Dejana, A Mantovani, JJ Oppenhelm, CG Larsen, COC Zachariae, K Matsushima. Monocyte chemotactic and activating factor gene expression induced in endothelial cells by IL-2 and tumor necrosis factor. J Immunol 144:3034–3038, 1990.
29. IA Hauser, E Setter, L Bell, JA Madri. Fibronectin expression correlates with U937 cell adhesion to migrating bovine aortic endothelial cells in vitro. Am J Pathol 143:173–180, 1993.
30. F Wempe, V Lindner, HG Augustin. Basic fibroblast growth factor (bFGF) regulates the expression of the CC chemokine monocyte chemoattractant protein-1 (MCP-1) in autocrine-activated endothelial cells. Arterioscl Thromb Vasc Biol 17:2471–2478, 1997.
31. WD Ito, M Arras, B Winkler, D Scholz, J Schaper, W Schaper. Monocyte chemotactic protein-1 increases collateral and peripheral conductance after femoral artery occlusion. Circ Res 80: 829–837, 1997.
32. M Arras, WD Ito, D Scholz, B Winkler, J Schaper, W Schaper. Monocyte activation in angiogenesis and collateral growth in the rabbit hindlimb. J Clin Invest 101:40–50, 1998.
33. RD Leek, CE Lewis, R Whitehouse, M Greenall, J Clarke, AL Harris. Association of macrophage infiltration with anglogenesis and prognosis in invasive breast carcinoma. Cancer Res 56:4625–4629, 1996.
34. RP Negus, GW Stamp, J Hadley, FR Balkwill. Quantitative assessment of the leukocyte infiltrate in ovarian cancer and its relationship to the expression of C-C chemokines. Am J Pathol 50:1723–1734, 1997.
35. L Mazzucchelli, P Loetscher, A Kappeler, M Uguccioni, M Baggiolini, JA Laissue, C Mueller. Monocyte chemoattractant protein-1 gene expression in prostatic hyperplasia and prostate adenocarcinoma. Am J Pathol 149:501–509, 1996.
36. DT Graves, R Barnhill, T Galanopoulos, HN Antoniades. Expression of monocyte chemotactic protein-1 in human melanoma in vivo. Am J Pathol 140:9–14, 1992.
37. FF Isik, RP Rand, JS Gruss, D Benjamin, CE Alpers. Monocyte chemoattractant protein-1 mRNA expression in hemangiomas and vascular malformations. J Surg Res 61:71–76, 1996.
38. S Yamashiro, M Takeya, T Nishi, J Kuratsu, T Yoshimura, Y Ushio, K Takahashi. Tumor-derived monocyte chemoattractant protein-1 induces intratumoral infiltration of monocyte-derived macrophage subpopulation in transplanted rat tumors. Am J Pathol 145:856–867, 1994.
39. J Vukanovic, JT Isaacs. Linomide inhibits angiogenesis, growth, metastasis, and macrophage infiltration within rat prostatic cancers. Cancer Res 55:1499–1504, 1995.
40. AE Koch, PJ Polverini, SL Kunkel LA Harlow, LA DiPietro, VM Elner, SG Elner, RM Strieter. Interleukin-8 as a macrophage-derived mediator of angiogenesis. Science 258:1798–1801, 1992
41. SK Gupta, JP Singh. Inhibition of endothelial cell proliferation by platelet factor-4 involves a unique action on S phase progression. J Cell Biol 127:1121–1127, 1994.
42. TE Maione, GS Gray, J Petro, AJ Hunt, AL Donner, SI Bauer, HF Carson, RJ Sharpe. Inhibition of angiogenesis by recombinant human platelet factor-4 and related peptides. Science 247:77–79, 1990.
43. DA Arenberg, SL Kunkel, PJ Polverini, SB Morris, MD Burdick, MC Glass, DT Taub, MD Iannettoni, RI Whyte, RM Strieter. Interferon-gamma-inducible protein 10 (IP-10) is an angiostatic factor that inhibits human non-small cell lung cancer (NSCLC) tumorigenesis and spontaneous metastases. J Exp Med 184:981–992, 1996.
44. Y Cao, C Chen, JA Weatherbee, M Tsang, J Folkman. Gro-beta, a -C-X-C-chemokine, is an angiogenesis inhibitor that suppresses the growth of Lewis lung carcinoma in mice. J Exp Med 182:2069–2077, 1995
45. RM Strieter, PJ Polverini, SL Kunkel, DA Arenberg, MD Burdick, J Kasper, J Dzuiba, J Van Damme, A Walz, D Marriott, SY Chan, S Roczniak, AB Shanafelt. The functional role of the ELR motif in CXC chemokine-mediated angiogenesis. J Biol Chem 270:27348–27357, 1995.
46. C Feil, HG Augustin. Endothelial cells differentially express the CXC-chemokine receptor-4

(CXCR-4/fusin) and respond chemotactically to the CXCR-4 ligand SDF-1. Biochem Biophys Res Comm 247:3845, 1998.

47. MV Volin, L Joseph, MS Shockley, PF Davies. Chemokine receptor CXCR4 expression in endothelium. Biochem Biophys Res Commun 242:46–53, 1998.
48. SK Gupta, PG Lysko, K Pillarisetti, E Ohlstein, JM Stadel Chemokine receptors in human endothelial cells: functional expression of CXCR4 and its transcriptional regulation by inflammatory cytokines. J Biol Chem 273:4282–4287, 1998.
49. K Tachibana, S Hirota, H Iizasa, H Yoshida, K Kawabeta, Y Kataoka, Y Kitamura, K Matsushima, N Yoshida, S Nishikawa, T Kishimoto, T Nagasawa. The chemokine receptor CXCR4 is essential for vascularization of the gastrointestinal tract. Nature 393:591–594, 1998.
50. T Nagasawa, S Hirota, K Tachibana, N Takakura, S Nishikawa, Y Kitamura, N Yoshida, H Kikutani, T Kishimoto. Defects of B-cell lymphopoiesis and bone-marrow myelopoiesis in mice lacking the CXC chemokine PBSF/SDF-1. Nature 382:635–638, 1996.

7
Thymosin β4 in Endothelial Cell Differentiation

Derrick S. Grant and R. Wesley Rose
Thomas Jefferson University, Philadelphia, Pennsylvania

I. INTRODUCTION

Angiogenesis, the process of vessel growth or renewal, is required for embryonic development, tissue expansion, reproduction, wound healing, and tumor growth. Normally, in adult animals, blood vessels are stable and angiogenesis is well regulated. Under some pathological conditions in which vessel stability is perturbed, such as diabetic retinopathy and psoriasis, and rapid initiation of the angiogenic process can lead to abnormal angiogenesis and tissue dysfunction. Although several extravascular growth factors are known to stimulate angiogenesis, the specific molecular agents (expressed by endothelial cells) that initiate or stabilize angiogenesis remain unclear. It has been our goal to identify and characterize factors expressed in the endothelial cell that regulate angiogenesis.

The maintenance and integrity of the blood vessel wall depends primarily on the stability of endothelial cells and associated vascular cells. Several in vivo and in vitro models have been developed that mimic at least some aspects of the angiogenic process, and these models provide insight into the steps and initiating factors involved in stimulating endothelial cells to form and maintain structurally intact vessels (1–7). For example, Maciag et al. (8) have shown that when confluent cultured endothelial cells are treated with phorbol esters, approximately 50% of the cells form tube networks in the culture dish. Using this model of endothelial cell differentiation with subtractive hybridization, a novel protein, EDG-1, was identified and shown to increase during tube formation (9,10). This study indicated that this molecular biology approach could shed some light on the genes involved in endothelial cell stabilization tube morphogenesis and activation (as in angiogenesis).

II. THE ROLE OF MATRIX IN ANGIOGENESIS

A major part of vessel regulation depends on the interaction between the cells and the surrounding extracellular matrix. One of these specialized matrices, the basement membrane, is in direct contact with the endothelial and smooth muscle cells that make up the vessel wall. Basement membranes are composed of type IV collagen, the glycoproteins laminin, entactin, and fibronectin, and the proteoglycans (perlecan and chondroitin sulfate proteoglycan). Laminin and collagen type IV are the most abundant components in the matrix and have been shown to be potent regulators of angiogenesis (1,5,11,12). The importance of type IV collagen in angiogenesis (in vivo) is well documented, as shown by experiments in which an inhibitor of collagen synthesis blocked angiogenesis in the growing chorioallantoic membrane (CAM) of the chicken embryo (13,14). Laminin, the major glycoprotein of basement membranes, possesses several cell-binding domains, and these domains promote endothelial cell activities in vitro, including spreading, migration, and differentiation (4,15–18). These proteins within the basement membrane regulate endothelial-cell behavior by exhibiting many transmembrane signals that help to maintain the endothelium in a differentiated state in vivo (1,3,4). During vasculogenesis in the embryo, the presence of basement membrane components usually correlates with the completion of the mature and maturing vessels. Basement membrane also serves as a reservoir (for angiogenic agents) or protective meshwork, since it has the ability to block the degradation of growth factors such as bFGF and TGF-β (19,20). Protease digestion of the basement membrane by endothelial cells can release sequestered growth factors from the ECM which can subsequently stimulate a proliferative or migratory response in the cells (19).

III. THYMOSIN β4 EXPRESSION IN ENDOTHELIAL CELLS ON MATRIGEL

Most of the information regarding the characterization of molecules comprising the basement membrane has been derived from extraction of these components from a matrix-rich murine sarcoma known as EHS (21). This tumor produces a large quantity of basement membrane from which a proteinaceous substance, Matrigel, can be extracted. Matrigel contains laminin, collagen type IV, heparan sulfate proteoglycan, and several growth factors [i.e., TGF-β, EGF, insulin-like growth factor 1, basic fibroblast growth factor (bFGF), and platelet-derived growth factor (PDGF)] (20). Matrigel has biochemical and morphological features similar to those of common basement membrane. This extracted matrix is biologically active and is known to promote the differentiation of a variety of cells (21). Matrigel has been used by several investigators to induce cultured cells (including endothelial cells) to differentiate in vitro (22). Since vascular endothelial cells are in direct contact with basement membranes, we have used Matrigel to evaluate the biological activity of basement membrane components in endothelial differentiation.

When human umbilical vein endothelial cells (HUVEC) are seeded onto Matrigel, the cells align and differentiate into capillary-like structures with lumens within 24 hr (4,18). The formation of capillary-like tubes by HUVEC cultured on Matrigel is a complex process that involves several steps and the biosynthesis of numerous gene products (23). We have shown previously (11) that the addition of cycloheximide or actinomycin D to this model inhibits tube formation. Thus, protein synthesis is essential to tube formation on Matrigel. Furthermore, using antibodies to a laminin receptor (LBP32/67) and integrin receptors (β1, α5, β4, α6, α2), we demonstrated that blocking these receptors on HUVEC inhibits tube formation on Matrigel. This would indicate that some of the differentiation (and synthesis of proteins required for tube formation) may be mediated through receptor signaling.

We have investigated the early genes that are induced during endothelial cell differentiation on Matrigel (23) by determining the identity of some of the early genes expressed by the cells during the process. A differential cDNA cloning technique performed with the RNA from endothelial cells cultured for 4 hr on plastic or 4 hr on Matrigel showed that several genes (35 total) were induced by Matrigel. Some of these genes included the messages for calmodulin, tubulin, a collagen-like gene, ribosomal proteins, human elongation and splicing factors, several novel genes, and thymosin β4. We have focused on thymosin β4, which is found in several vascular cell types but was not previously shown to be associated with endothelial cell differentiation. The thymosin β4 cDNA clone isolated from endothelial cells recognized a message of 800 bp, slightly larger than found in thymocytes due to an extended 3′ region. Cells cultured on plastic do not significantly change their expression of thymosin β4 over a 6-hr period, whereas cells cultured on Matrigel show a fivefold increase at 4 through 6 hr. This demonstrates that, on Matrigel, there is an increase in the synthesis of thymosin β4 mRNA, 4–6 hr into the tube formation process.

IV. BACKGROUND INFORMATION ON THYMOSIN β4

Thymosin β4, a small polypeptide (4.9 kDa), was first identified and isolated from the calf thymus (24–28). Thymosin β4 is a member of a large group of thymosin proteins (both alpha and beta), which were first isolated from fraction 5 (F5) of calf thymus (29). The F5 preparation contained proteins ranging from 1000–15,000 Mr and the component peptides were named with respect to their migration in isoelectric focusing gels. Peptides with pI values below 5.0 were named alpha 1, 2, 3, etc.; those with pI values between 5.0 and 7.0 were designated beta 1, 2, 3, etc.; and those with pI values above 7.0 were designated gamma. The sequence of thymosin β4 is almost identical to that of β8, β9, β10, and, more recently, β15 (30).

V. MOLECULAR AND STRUCTURAL BIOLOGY OF THYMOSIN β4

The thymosin β4 gene has been described in the rat (31) and was found to consist of multiple intron-containing genes that have 2 kb-pair-long transcription units containing two introns. The human thymosin β4 gene, which is identical to the human interferon-inducible gene 6–26, is localized to seven distinct chromosomes (1, 2, 4, 9, 11, 20, and X), some of which are pseudogenes (32). This protein is conserved, since a cDNA encoding human thymosin-β4 has been isolated and found to be 84% homologous to rat thymosin β4 (the homology between their coding regions is 93%), and was over 95% homologous in the human, mouse, and cow (27,33). At the transcriptional level, a thymosin β4 cDNA clone has been isolated from a mouse pre–B cell line, and it was shown by RNase protection assays that two sizes of thymosin β4 mRNA existed: a long form containing a 98-nucleotide insertion and a short form that corresponds to the known rat and human mRNA (34). Southern blotting revealed that both forms were encoded by a single gene in the mouse and that the two forms may arise by differential RNA splicing (34). Analysis of the amino acid and cDNA nucleotide sequences of rat thymosin β4 revealed that it is synthesized as a 44–amino acid peptide, which is processed into a 43–amino acid peptide by removal of the first methionyl residue and acetylation of the exposed NH_2-terminal seryl residue (32). Thymosin β4 has no signal peptide, suggesting that it is not secreted; however, it is present in human serum and may be secreted by some cell types (32). Thymosin β4 is an abundant protein found in circulating platelets, and the presence of this protein increases in wounds and pathologies where platelets are lysed. This may explain, in part, the increase in local or circulating levels of thymosin β4 in pathological conditions.

VI. PREDICTED STRUCTURE OF Tβ4 PROTEIN

Thymosin β4, a 43–amino acid protein, contains two alpha-helical regions; the remaining regions appear to be highly mobile (Fig. 1). Recent NMR study shows that thymosin β4 is mostly unfolded in aqueous solution (35), whereas it adopts a specific conformation when it binds to actin and may extend over widely separated regions of the actin monomer.

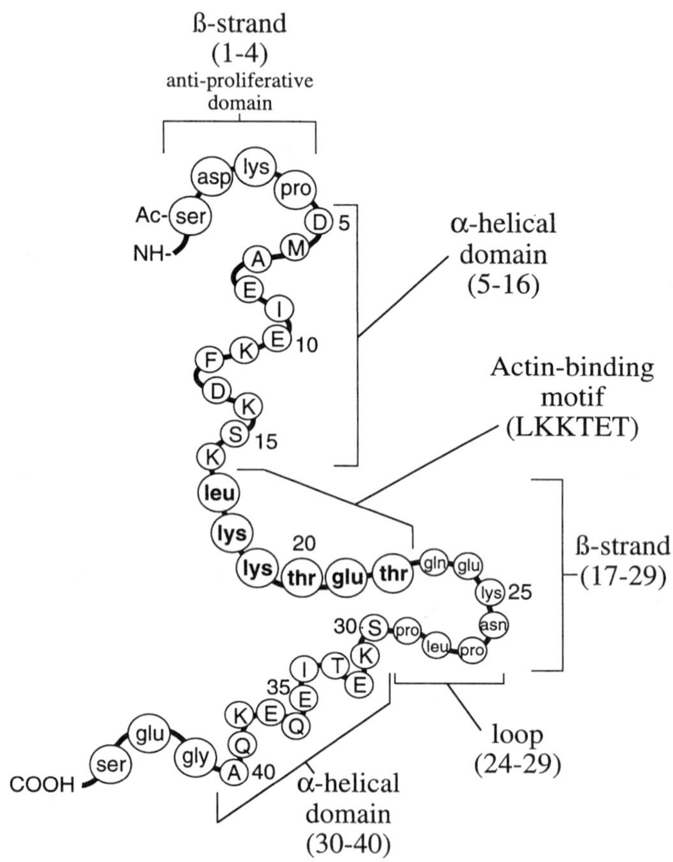

Figure 1 Proposed structure of monomeric thymosin β4. Shown in the schematic diagram are the first β-strand, residues 1–4, containing the hypothesized antiproliferative domain. Residues 5–16 contain the first α-helical domain; the second β-strand (residues 17–29) contains the actin-binding region LKKTET (residues 17–22) as well as a loop (residues 24–29). The second α-helical domain is located at the carboxy-terminal end, residues 30–40 (35,37,57,58).

In an alcohol/water environment, thymosin β4 contains two alpha-helical regions that encompass residues 4–6 and 30–40. The remaining regions of the molecule appear to be highly flexible and a unique structure for these regions cannot be defined (see Fig. 1). Two subdomains or regions in Tβ4 have been identified that function as either an actin-binding domain (residues 17–22) or as an antiproliferative ''differentiation'' domain (residues 1–4). The actin-binding region, a 9-residue sequence, is present in all thymosin β4 (designated LKKTET, residues 17–22) and is highly homologous to the actin-binding sequence of *Acanthamoeba* actobindin (36). The site for thymosin β4 may, however, extend beyond subdomain 1, since Dnase I which binds to acitin subdomain 2 can also block binding of Tβ4 to actin (37). The second subdomain of thymosin β4 is found in the N-terminus, Acetyl-N-Ser-Asp-Lys-Pro (AcSDKP), and has been shown to be a negative regulator of hematopoiesis, specifically with respect to pluripotent stem cell proliferation.

VII. KNOWN THYMOSIN β4 FUNCTIONS

Not much is known about the expression or the regulation of thymosin β4 in endothelial cells. The major biological property of thymosins from F5 is the induction of differentiation of specific subclasses of T lymphocytes (26). Thymosin β4, which was first thought to be a very specific thymic hormone, is now believed to be responsible for a much more general though still obscure function in the cell owing to its wide distribution in many cells and tissues (38). Since thymosin β4 is one of the most abundant thymosins, a variety of regulatory or signaling functions have been attributed to it, including the induction of T-cell differentiation and stimulation of the secretion of lutenizing hormone releasing factor (reviewed in Safer, 1992) (37). However many questions exist as to whether thymosin β4 is a true hormone on the basis of its widespread distribution and abundance within cells, low circulating levels (under normal physiological states), and lack of direct evidence for its secretion. The exact role of thymosin β4 is still controversial due to its relatively low level of circulation. Normal human circulating levels of thymosin β4 range from 3 to 83 ng/ml in serum (39). Others have determined that thymosin β4 concentration in whole blood ranges from 12 to 19.5 μg/mL and 0.387 μg/mL in plasma (40). The tissue distribution of thymosin β4 varies and is highly concentrated (50 μg/mL) in some tumors (41) and is concentrated in human wound and blister fluids (42). All of these sites usually involve increased local angiogenesis. It is possible that thymosin β4 may play an active role in these processes by increasing the vessels' response to angiogenic factors.

Thymosin β4 is also an actin binding protein. It is responsible for sequestering the major part of the unpolymerized pool of actin (43). Thus, in the past year thymosin β4's role has shifted from that of a hormone to an intracellular regulator of cytoskeletal elements (44). It has been shown to inhibit G actin polymerization by forming a 1:1 complex with the monomer, thus leading to the hypothesis that thymosin β4 is involved in the regulation of actin polymerization in the cell. The binding of thymosin β4 to actin is mainly mediated by a hexamotif, LKKTET (36). Vancompernolle et al. (36) further showed that if the lysine residues 18 and 19 were replaced by glutamic acid, the thymosin β4 variant showed no binding to actin. Moreover it has been shown that transfection and overexpression of thymosin β4 in CV1 fibroblasts induces extensive loss of phalloidin-stained actin stress fibers, and a decrease in the number of focal adhesions, as evidenced by a decrease in discrete vinculin staining (45). These data strongly support the hypothesis that thymosin β4 also functions intracellularly as an actin monomer sequestering protein, but does not rule out a possible extracellular role for thymosin β4.

VIII. RECENT DISCOVERIES REGARDING THYMOSIN β4 FUNCTION AND RELATIONSHIP TO VESSELS

Thymosin β4 has been proposed to have several functions and is present in several organ systems. For instance, this protein is ubiquitously expressed in many cells and tissues (46,47), including cells in the developing human kidney (33), and is differentially expressed in developing brain tissues during embryogenesis. In the rat, it is present in lower amounts in embryonic lung, spleen, adrenal glands, heart, and liver (47). One important thymosin β4 activity includes its ability to partially or fully reconstitute immune functions in thymus-deprived or immunodeprived animals (26,27). Other hormone-like functions include the modulation of cell proliferation. Thymosin β4 expression has been shown to be directly associated with cell cycle regulation (48). A related form of thymosin β4, β15, has been shown to be upregulated in the highly mobile and metastatic Dunning cancer cell lines (30). Thymosin β4 is also increased in the serum of patients over 60, in newborns, and in patients suffering from AIDS, hepatitis, and inflammatory bowel disease (49). There are a few other studies suggesting that exogenous thymosin β4 may play a regulatory role. For instance, when authentic thymosin β4 was incubated with bone marrow cells, a tetrapeptide was formed from the N-terminal subdomain, designated AcSDKP. When the synthetic form of this peptide was injected into rats, it inhibited the action of a stimulator of hematopoietic stem cell proliferation, preventing the switching of quiescent cells into the cell cycle (50). More recently it has been shown that thymosin β4 itself can also inhibit the cell proliferation of $CD34^{2+}$ bone marrow cells directly (51). Thymosin β10 however, which has an alanine in place of the serine, could not form this tetrapeptide (52), which suggest some specificity for β4. Increased thymosin β4 expression has been shown to have a strong correlation with developing blood vessels and endocardial cushions (53). Thymosin β4 is highly expressed in developing brain, where an extensive blood vessel network is rapidly forming (54). Others have also indicated that the increase in thymosin β4 levels in growing tumors and malignant tissue should be attributed to the vascularization process rather than the tumor gene activation (33,53). More recently Malinda et al. (55) have shown that the exogenous addition of thymosin β4 increases migration in human umbilical vein endothelial cells. Exogenous treatment also increased the production of matrix metalloproteinase, which may degrade the basement membrane during angiogenesis. These results provide the first direct evidence that Tβ4 has chemoattractive activity and promotes angiogenesis at least in part by stimulating the migration of endothelial cells.

Therefore, thymosin β4 appears to play an important role in normal vessel development, differentiation, and maintenance. Our study (23) was the first showing that the expression of thymosin β4 is involved in endothelial cell differentiation, and now this observation is supported by other investigators. Additionally, recent data have shown that thymosin β4 acts as a specific chemoattractant for EC as well as stimulating increased matrix and metalloproteinase production. Further mechanistic investigations are needed, however, to establish whether thymosin β4 is differentially expressed in differentiating vessels ubiquitously—also whether it directly induces differentiation in angiogenic endothelial cells or if it is a product of the differentiation process.

IX. EXPRESSION OF THYMOSIN β4 IN TRANSFECTED HUVEC

We have recently investigated the relationship between the expression of thymosin β4 in endothelial cells and their ability to form capillary-like tubes in vitro. Thymosin β4 full-length cDNA was recently transfected into endothelial cells and the presence of the protein was detected by FITC immunocytochemistry. HUVEC transfected with the thymosin β4 cDNA were strongly stained within 24 hr with 50–60% of the cells, showing strong positive staining (Fig. 2A). We also compared the expression of von Willebrand factor in endothelial cells transfected with thymosin β4 cDNA and observed that there was a posi-

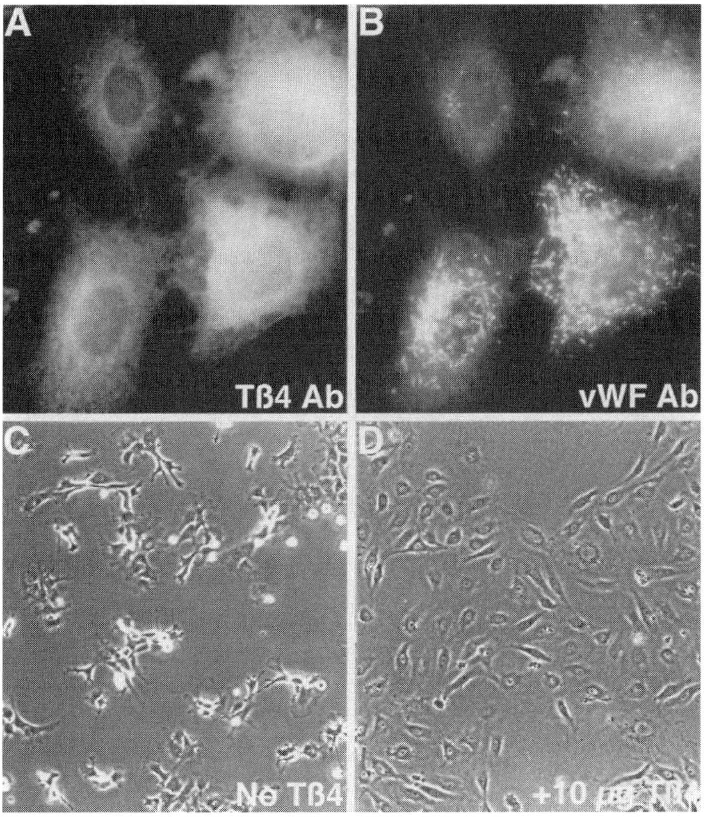

Figure 2 Transfection of HUVEC with thymosin β4. A and B. Immunofluorescent staining of transfected HUVEC shows strong positive staining for (A) thymosin β4, and (B) von Willebrand factor (vWF). C and D. Effect of Tβ4 transfection on HUVEC attachment to tissue culture plastic at 30 mins. Shown are phase-contrast micrographs of (C) nontransfected cells and (D) cells transfected with 10 μg Tβ4.

tive correlation between increased expression (over basal levels) of von Willebrand factor and thymosin β4 protein expression in transfected cells (Fig. 2B). Endothelial cells transfected with thymosin β4 mRNA appear to induce an increase in vWF expression, as observed with differentiated endothelial cells in vivo.

X. BIOLOGICAL BEHAVIOR OF HUVEC TRANSFECTED WITH THYMOSIN β4

Since endothelial cells expressed increased levels of thymosin β4 during differentiation, we examined the biological behavior of thymosin β4–transfected HUVEC. Endothelial cells transfected with thymosin β4 expression vector were examined for their ability to attach and spread on tissue culture plastic. Like normal cells, the mock-transfected cells were not fully adherent, nor did they all spread within 30 min (cells usually take about 60 min to attach and spread) (Fig. 2C). In contrast, transfected cells became rapidly adherent and were almost completely spread within 30 min (Fig. 2D).

The ability of thymosin β4–transfected endothelial cells to attach to matrix proteins was also examined. In attachment assays, normal HUVEC showed a dose-dependent increase in attachment to laminin within the short 30-min time span as compared to mock-transfected cells. The thymosin β4–transfected cells also demonstrated an increase in rate of attachment to laminin (at dosages of 2–10 μg) as compared to the normal or mock-transfected cells. Transfected cells also demonstrated an attachment to collagen IV as compared to mock-transfected or normal cells. Thus, cells transfected with thymosin β4 are more adherent and spread more quickly than control cells, suggesting that this protein has a biological function in promoting adhesion and spreading. At times greater than 30 min, no difference in the total number of attached cells was observed, demonstrating that thymosin β4 expression affected only the rate and not the total number of attached cells (data not shown).

XI. THE EFFECT OF Tβ4 TRANSFECTION ON HUVEC TUBE FORMATION

Induction of tubes by thymosin β4 was also examined in transfected and mock-transfected cells plated on Matrigel. No differences in adhesion to Matrigel were observed at any time point between transfected and nontransfected cells. However, differences in the rate of tube formation were seen between transfected and mock-transfected cells when the cells were examined at 2, 4, and 6 hr after seeding on Matrigel (23). The mock-transfected cells demonstrated a pattern of adhesion and alignment similar to that found in normal HUVEC, and cells formed fairly flattened monolayers (similar to HUVEC on plastic) in 2–6 hr. At these same time points, the thymosin β4–transfected cells appeared more elongated, and structures resembling normal tube formation, usually observed at approximately 10 hr, were now observed at 4 hr. There was no difference in tube formation between transfected and mock-transfected cell types and nontransfected cells following a 24-hr incubation on Matrigel, when complete tubes were observed. Thus, the cells transfected with thymosin β4 formed tubes more quickly on Matrigel than the control cells and the final tube formation was similar to the normal. Therefore it appears that thymosin β4 accelerates tube formation. These changes may be due in part to the cells' increased ability to spread and migrate on Matrigel or the effect of increased thymosin β4 on the cells' actin cytoskeleton.

XII. DISTRIBUTION OF THYMOSIN β4 IN CAPILLARY-LIKE STRUCTURES IN VITRO AND IN VESSELS IN VIVO

A. In Vitro

The incubation of HUVEC on Matrigel results in the elongation and reorganization of the cells into tubelike structures. This network of tubes is attached firmly to the Matrigel via regions of flattened endothelial cells. As discussed above, extensive reorganization of the cells occurs at 4–6 hr during tube formation. Tubes immunostained with thymosin β4 antibody and viewed by confocal microscopy demonstrated positive immunofluorescent staining in most of the cells of the tube structures (23). Actin colocalization showed some filaments running within the axis of the tube in the periphery of the cell cytoplasm. Staining for actin in the flattened junctional areas appeared somewhat more diffuse. Therefore, areas where tubes were actively forming expressed an increase in thymosin β4 protein.

B. In Vivo

When bFGF-supplemented Matrigel (bFGF is added to increase angiogenesis in the Matrigel in vivo angiogenesis model) is implanted subcutaneously in mice, many regions of the implants contain distinct blood vessels (56). When paraffin sections of these implants were immunostained with an antibody to thymosin β4, numerous cells within the Matrigel were stained. In particular, cells organized into linear vessel-like structures were all strongly stained, whereas single cells did not appear to be stained. Other areas of the tissue sections were also examined; staining was observed in the basal layer of the epithelium of the skin and in some cells of the connective tissue. These data indicate that thymosin β4 is present in forming vessels in the in vitro and in vivo angiogenesis assays as well as in mature vessels but do not tell us whether the message for thymosin β4 is increased.

XIII. THYMOSIN β4 ANTISENSE AND TUBE FORMATION

We have tried to block the effect of thymosin β4 in endothelial cells by inhibiting the expression of this protein. For instance, when antisense oligonucleotides (made to the first 15 amino acids in TB4) to thymosin β4 were added to the tube-forming assay, an inhibition of tube formation was observed. The tube area was not significantly affected by the sense oligo between 0–10 μg/mL (0–2 μM). At 10 and 20 μg/mL of the antisense oligo, tube formation was inhibited by approximately 50% (Fig. 3). A scrambled nonsense oligo did not have an effect on tube formation at any of the concentrations (data not shown). At concentrations of 20 μg/mL sense oligo (4 μM), there was some inhibition of tube formation as compared to controls lacking oligonucleotides. In contrast, the antisense oligo elicited an inhibition (36%) in tube formation with only 5 μg/mL (1 μM). As a control experiment, we performed Western blots on whole-cell lysate from cells incubated in the presence of sense and antisense oligos and found that protein synthesis for thymosin β4 was blocked only with the antisense oligo [data not shown (23)]. In addition, we immuno-

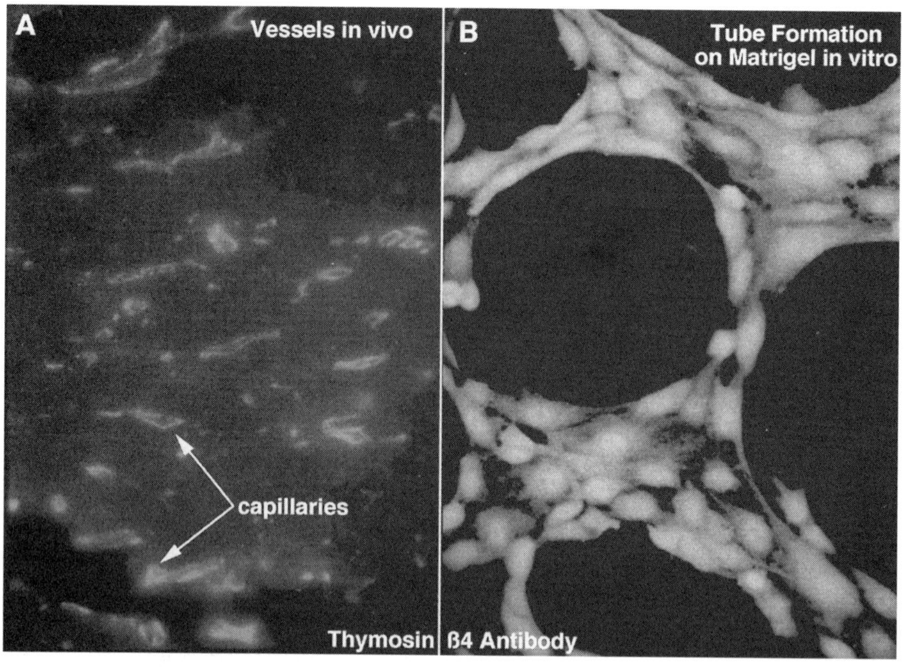

Figure 3 A. Immunofluorescent staining of Tβ4 expression in blood vessels in vivo. B. Expression of Tβ4 during HUVEC differentiation on Matrigel in vitro as detected by immunohistochemistry.

stained HUVEC after a 4-hr incubation with the thymosin β4 antisense oligo and showed a reduction in the intracellular staining for the protein. These data demonstrate that blocking endogenous production of thymosin β4 inhibits tube formation on Matrigel.

XIV. PARACRINE EFFECT OF THYMOSIN β4 ON HUVEC

Thymosin β4 is present in the circulation at levels of 1–50 μg/mL; however, this concentration varies depending on the physiological or pathological condition in tissues. For instance clot lysis has been shown to release thymosin β4 from platelets rich in this protein. Other groups have shown that thymosin β4, either injected or added to cells in vitro, could affect cell behavior, including migration and proliferation (37). To determine whether thymosin β4 acts in an autocrine or paracrine manner to regulate angiogenesis, we added different amounts of purified thymosin β4 protein as prepared according to Low et al. (26,29) to our endothelial cell in vitro systems. We used the purified protein to determine if direct addition of thymosin β4 to the HUVEC could elicit the same biological response as the transfection and increased cytoplasmic expression described above.

XV. EFFECT OF EXOGENOUS THYMOSIN β4 ON HUVEC PROLIFERATION

The following experiments examined the effect of thymosin β4 on endothelial cells. HUVEC were seeded in normal endothelial cell medium and purified thymosin β4 protein was added to each well at 0, 5, 10, and 20 μg. The cell number per 10 fields was counted using a computer-assisted morphometric analysis program or by Promega's CellTiter 96Aq assay, which is based on the conversion of MTS to formazan (whose absorbance at 490 nm is measured) by live cells. The results demonstrated that the presence of thymosin β4 ≥ 1 μg/mL caused an increase in cell proliferation over 4 days.

XVI. EFFECT OF EXOGENOUS THYMOSIN β4 ON HUVEC ATTACHMENT

We tested the effect of thymosin β4 on HUVEC attachment. This was done by first incubating a plate of subconfluent HUVEC overnight with 20 μg/mL of thymosin β4. The cells were then washed and lifted with EDTA (at 37°C), then 200,000 cells were added to duplicate 16-mm wells (24-well plate) uncoated or coated with laminin or collagen IV. After a 30-min incubation at 37°C, the wells were washed and stained and the number of attached cells were quantitated using computer-assisted image analysis.

The results demonstrated that the presence of the thymosin β4 protein increased the attachment of HUVEC to laminin and collagen IV (Fig. 4). This effect was reversible; for example, if cells were washed after incubation with thymosin β4 and the assay was done 24 hr later, there was no increase in HUVEC attachment (data not shown).

XVII. THYMOSIN β4 AND ANGIOGENIC SPROUTING EX VIVO

All of the processes examined above are exhibited during normal angiogenesis and seen during capillary sprouting in rat aortic rings (7). We have examined the effect of thymosin β4 on these sprouting vessels ex vivo. By adding as little as 5 μg/mL of synthetic thymosin β4, we have observed changes in the vascular sprouts from the rat aorta. Normally, after incubating 1-mm pieces of rat aortic rings in a collagen gel for 24 hr, cells extend from the rim of the ring and migrate out into the collagen I gel. In the presence of 5 μg of thymosin β4, the cells migrate further and begin to assemble into a tube network (see Fig.

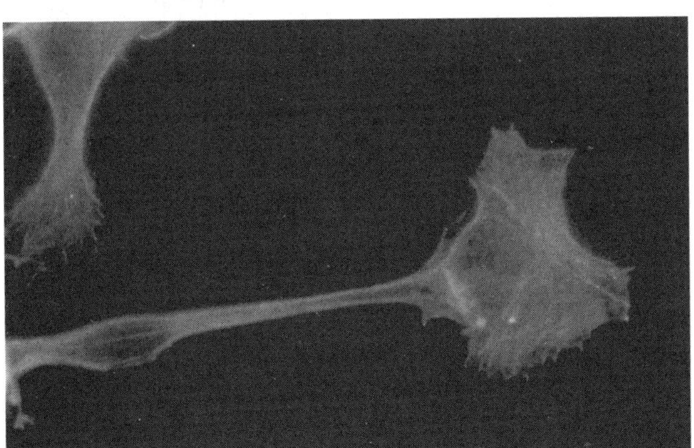

Figure 4 Effect of Tβ4 transfection on HUVEC morphology. As shown, transfected cells assume a more elongated morphology, with cytoplasmic processes contacting neighboring cells.

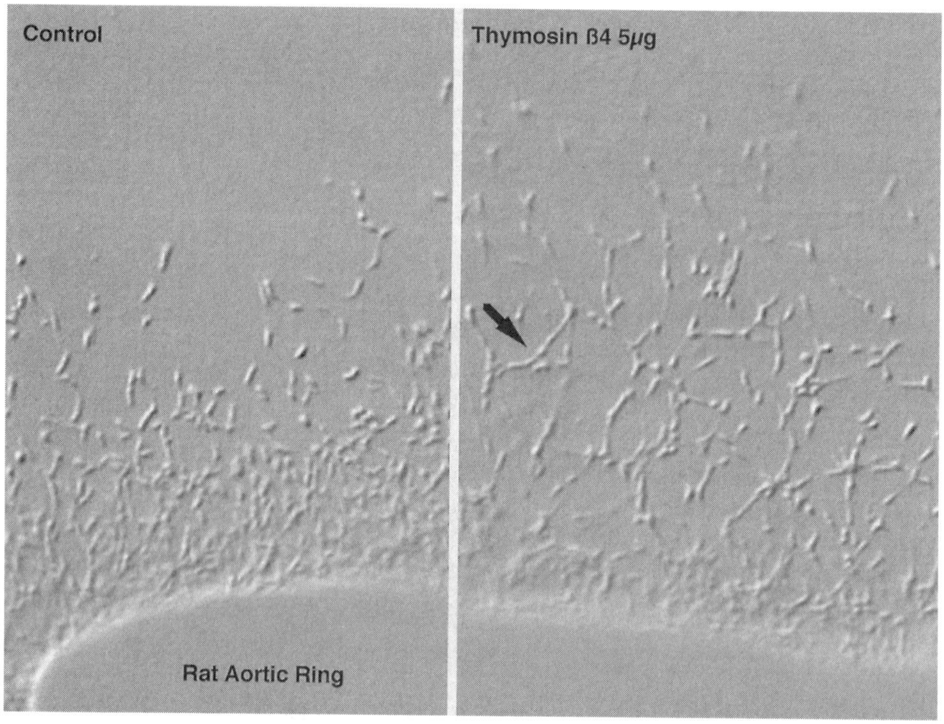

Figure 5 Sprouting of endothelial cells from rat aortic rings incubated in collagen I gel. Panel on left shows normal response and on the right in the presence of 5 μg of thymosin β4. Arrow (right panel) indicates the organization of endothelial cells into tubelike structures.

5). These networks maintained stable vessels over a period of 2 to 3 days. Therefore it appears that thymosin β4 may play an important role in the formation of endothelial cells sprouting from the aortic rings.

In summary, we have shown that thymosin β4 has a paracrine effect on differentiating thymocytes as well as several other functions on several cell types. The majority of thymosin β4 is bound intracellularly to actin; however, it has also been shown to be present in the circulation and is increased during development or diseased states. Others have suggested that there is an association of thymosin β4 expression with angiogenesis in vivo, and we have strong evidence that this protein is involved in the regulation of EC differentiation in vitro. Our studies have defined an endogenous and exogenous role of this inducible gene product in vessel formation, which should provide a better understanding of its ability as a angiogenic modulator with potential clinical applications.

REFERENCES

1. Ingber DE, Folkman J. How does extracellular matrix control capillary morphogenesis? Cell 58:803–805, 1989.
2. Jaffe EA, Nachman RL, Becker CG, Minick CR. Culture of human endothelial cells derived from umbilical veins-identification by morphological and immunological criteria. J Clin Invest 52:2745–2756, 1973.

3. Kramer RH, Fuh GM. Type IV collagen synthesis by cultured human microvascular endothelial cells and its deposition in the subendothelial basement membrane. Biochemistry 24:7423–7430, 1985.
4. Kubota Y, Kleinman HK, Martin GR, Lawley TJ. Role of laminin and basement membrane in the morphological differentiation of human endothelial cells into capillary-like structures. J Cell Biol 107:1589–1598, 1988.
5. Madri JA, Pratt BM. Endothelial cell-matrix interactions: in vitro models of angiogenesis. J Histochem Cytochem 34:85–91, 1986.
6. Nicosia RF, McCormick JF, Bielunas J. The formation of endothelial webs and channels in plasma clot culure. Scan Elect Microsc 2:793–799, 1984.
7. Nicosia RF, Ottinetti A. Growth of microvessels in serum-free matrix culture of rat aorta. Lab Invest 63:115–122, 1990.
8. Maciag T. Molecular and cellular mechanisms of angiogenesis. Import Adv Oncol 85–98, 1990.
9. Hla T, Maciag T. An abundant transcript induced in differentiating human endothelial cells encodes a polypeptide with structural similarities to G-protein-coupled receptors. J Biol Chem 265:9308–9313, 1990.
10. Hla T, Maciag T. Isolation of immediate-early differentiation mRNAs by enzymatic amplification of subtracted cDNA from human endothelial cells. Biochem Biophys Res Commun 167:637–643, 1990.
11. Grant DS, Lelkes PI, Fukuda K, Kleinman HK. Intracellular mechanisms involved in basement membrane induced blood vessel differentiation in vitro. In Vitro Cell Dev Biol 21A:327–336, 1991.
12. Grant DS, Kleinman HK, Martin GR. The role of basement membranes in vascular development. Ann NY Acad Sci 588:61–72, 1990.
13. Maragoudakis ME, Panoutsacopoulou M, Sarmonika M. Rate of basement membrane biosynthesis as an index to angiogenesis. Tissue Cell 20:531–539, 1988.
14. Maragoudakis ME, Missirlis E, Karakiulakis GD, Sarmonica M, Bastakis M, Tsopanoglou N. Basement membrane biosynthesis as a target for developing inhibitors of angiogenesis with anti-tumor properties. Kidney Int 43:147–150, 1993.
15. Kleinman HK, Sephel GC, Tashiro K, Weeks BS, Burrous BA, Adler SH, et al. Laminin in neuronal development. Ann NY Acad Sci 580:302–310, 1990.
16. Kleinman HK, Weeks BS, Schnaper HW, Kibbey MC, Yamamura K, Grant DS. The laminins: a family of basement membrane glycoproteins important in cell differentiation and tumor metastases. Vitam Horm 47:161–186, 1993.
17. Kibbey MC, Grant DS, Kleinman HK. Role of the SIKVAV site of laminin in promotion of angiogenesis and tumor growth: an in vivo Matrigel model. J Natl Cancer Inst 84:1633–1638, 1992.
18. Grant DS, Tashiro K-I, Segui-Real B, Yamada Y, Martin GR, Kleinman HK. Two different laminin domains mediate the differentiation of human endothelial cells into capillary-like structures in vitro. Cell 58:933–943, 1989.
19. Roberts JM, Forrester JV. Factors affecting the migration and growth of endothelial cells from microvessels of bovine retina. Exp Eye Res 50:165–172, 1990.
20. Vukicevic S, Kleinman HK, Luyten FP, Roberts AB, Roche NS, Reddi AH. Identification of multiple active growth factors in basement membrane Matrigel suggests caution in interpretation of cellular activity related to extracellular matrix components. Exp Cell Res 202:1–8, 1992.
21. Kleinman HK, McGarvey ML, Liotta LA, Gehron-Robbey P, Tryggvasson K, Martin GR. Isolation and characterization of type IV procollagen, laminin and heparan sulfate proteoglycan from the EHS sarcoma. Biochemistry 24:6188, 1986.
22. Kleinman HK, Schnaper HW. Basement membrane matrices in tissue development (comment). Am J Respir Cell Mol Biol 8:238–239, 1993.

23. Grant DS, Kinsella JL, Kibbey MC, LaFlamme S, Burbelo PD, Goldstein AL, et al. Matrigel induces thymosin beta 4 gene in differentiating endothelial cells. J Cell Sci 108:3685–3694, 1995.
24. Badamchian M, Strickler MP, Stone MJ, Goldstein AL. Rapid isolation of thymosin beta 4 from thymosin fraction 5 by preparative high-performance liquid chromatography. J Chromatogr 459:291–300, 1988.
25. Gomez-Marquez J, Dosil M, Segade F, Bustelo X, Pichel J, Dominguez F, et al. Thymosin-β4 gene: preliminary characterisation and expression in tissues, thymic cells, and lymphocytes. J Immunol 143:2740–2744, 1989.
26. Low T, Goldstein A. Thymosins: structure, function and therapeutic application. Thymus 6: 27–42, 1984.
27. Low TL, Goldstein AL. Thymosin beta 4. Methods Enzymol 116:248–255, 1985.
28. Low TL, Lin CY, Pan TL, Chiou AJ, Tsugita A. Structure and immunological properties of thymosin beta 9 Met, a new analog of thymosin beta 4 isolated from porcine thymus. Int J Pept Protein Res 36:481–488, 1990.
29. Low TLK, Goldstein AL. Chemical characterization of thymosin β4. J Biol Chem 257:1000–1006, 1981.
30. Bao L, Loda M, Janmey PA, Stewart R, Anand AB, Zetter BR. Thymosin beta 15: a novel regulator of tumor cell motility upregulated in metastatic prostate cancer (see comments) Nature Med 2:1322–1328, 1996.
31. Varghese S, Kronenberg HM. Rat thymosin beta 4 gene: intron-containing gene and multiple retroposons. J Biol Chem 266:14256–14261, 1991.
32. Clauss IM, Wathelet MG, Szpirer J, Islam MQ, Levan G, Szpirer C, et al. Human thymosin-beta 4/6-26 gene is part of a multigene family composed of seven members located on seven different chromosomes. Genomics 9:174–180, 1991.
33. Hall A. Differential expression of thymosin genes in human tumors and in the developing human kidney. Int J Cancer 48:672–677, 1991.
34. Rudin C, Engler P, Storb U. Differential splicing of thymosin β4 mRNA. J Immunol 144: 4857–4862, 1990.
35. Zarbock J, Oschkinat H, Hannappel E, Kalbacher H, Voelter W, Holak TA. Solution conformation of thymosin beta 4: a nuclear magnetic resonance and simulated annealing study. Biochemistry 29:7814–7821, 1990.
36. Vancompernolle K, Goethals M, Huet C, Louvard D, Vandekerckhove J. G- to F-actin modulation by a single amino acid substitution in the actin binding site of actobindin and thymosin beta 4. Embo J 11:4739–4749, 1992.
37. Safer D. The interaction of actin with thymosin beta 4 (news). J Muscle Res Cell Motil 13: 269–271, 1992.
38. Livaniou E, Mihelic M, Evangelatos GP, Haritos A, Voelter W. A thymosin beta 4 ELISA using an antibody against the N terminal fragment thymosin beta 4 (1–14). J Immunol Meth 148:9–14, 1992.
39. Weller F, Mutchnik M. Enzyme immunoassay measurement of thymosin β4. J Immunoassay 8:203–217, 1987.
40. Hannappel E, Kampen MV. Determination of thymosin β4 in human blood cells and serum. J Chrom 397:279–285, 1987.
41. Gold JS, Bao L, Ghoussoub RAD, Zetter BR, Rimm DL. Localization and quantitation of expression of the cell motility-related protein thymosin β15 in human breast tissue. Mod Pathol 11:1106–1012, 1997.
42. Frohm M, Gunne H, Bergman AC, Agerberth B, Bergman T, Boman A, et al. Biochemical and antibacterial analysis of human wound and blister fluid. Eur J Biochem 237:86–92, 1996.
43. Safer D, Elzinga M, Nachmias VT. Thymosin beta 4 and Fx, an actin-sequestering peptide, are indistinguishable. J Biol Chem 266:4029–4032, 1991.
44. Sanger JM, Golla R, Safer D, Choi JK, Yu KR, Sanger JW, et al. Increasing intracellular

concentrations of thymosin beta 4 in PtK2 cells: effects on stress fibers, cytokinesis, and cell spreading. Cell Motil Cytoskel 31:307–322, 1995.
45. Yu FX, Lin SC, Morrison BM, Yin HL. Effects of thymosin beta 4 and thymosin beta 10 on actin structures in living cells. Cell Motil Cytoskel 27:13–25, 1994.
46. Condon MR, Hall AK. Expression of thymosin beta-4 and related genes in developing human brain. J Mol Neurosci 3:165–170, 1992.
47. Lin S, Morrison-Bogorad M. Developmental expression of mRNAs encoding thymosins β4 and β10 in rat brain and other tissues. J Mol Neurosci 2:35–44, 1990.
48. Schöbitz B, Netzker R, Hannappel E, Brand K. Cell-cycle-regulated expression of thymosin beta 4 in thymocytes. Eur J Biochem 199:257–262, 1991.
49. Naylor PH, Friedman KA, Hersh E, Erdos M, Goldstein AL. Thymosin alpha 1 and thymosin beta 4 in serum: comparison of normal, cord, homosexual and AIDS serum. Int J Immunopharmacol 8:667–676, 1986.
50. Safer D, Nachmias VT. Beta thymosins as actin binding peptides. Bioessays 16(8):590, 1994.
51. Bonnet D, Lemoine FM, Frobert Y, Bonnet ML, Baillou C, Najman A, et al. Thymosin beta4, inhibitor for normal hematopoietic progenitor cells. Exp Hematol 24:776–782, 1996.
52. Safer D, Nachmias VT. Beta thymosins as actin binding peptides [published erratum appears in Bioessays 16:590, 1994.] Bioessays 16:473–479, 1994.
53. Gomez-Marquez J, Franco dAF, Carpintero P, Anadon R. High levels of mouse thymosin beta4 mRNA in differentiating P19 embryonic cells and during development of cardiovascular tissues. Biochim Biophys Acta 1306:187–193, 1996.
54. Carpinterio P, Anadon R, del AF, Gomez MJ. The thymosin beta 4 gene is strongly activated in neural tissues during early postimplantation mouse development. Neurosci Lett 184:63–66, 1995.
55. Malinda KM, Goldstein AL, Kleinman HK. Thymosin β4 stimulates directional migration of human umbilical vein endothelial cells. FASEB J 11:474–481, 1997.
56. Passaniti A, Taylor RM, Pili R, Guo Y, Long PV, Haney JA, et al. A simple, quantitative method for assessing angiogenesis and antiangiogenic agents using reconstituted basement membrane, heparin, and fibroblast growth factor. Lab Invest 67:519–528, 1992.
57. Cheviron N, Grillon C, Carlier MF, Wdzieczak BJ. The antiproliferative activity of the tetrapeptide Acetyl-N-SerAspLysPro, an inhibitor of haematopoietic stem cell proliferation, is not mediated by a thymosin beta 4-like effect on actin assembly. Cell Prolif 29:437–446, 1996.
58. Becker S, Armbruster FP, Muller B, Echner H, Kapurnotu A, Livaniou E, et al. Theoretical and experimental epitope mapping of thymosin beta 4. J Immunol Meth 177:131–137, 1994.

8

EDG-1, an Endothelial-Derived G Protein–Coupled Receptor for Bioactive Lipids, Regulates Morphogenetic Differentiation

Menq-Jer Lee, Shobha Thangada, Catherine H. Liu, Brian Thompson, Nicolas Ancellin, Michael Kluk, and Timothy Hla
University of Connecticut School of Medicine, Farmington, Connecticut

Angiogenesis, the process of new blood vessel formation, is important in embryonic development and normal physiological events, such as fetal vasculogenesis, wound healing, and corpus luteum formation (1,2). Dysregulated angiogenesis is involved in pathological conditions, such as solid tumor growth and diabetic retinopathy (3). The process of angiogenesis is regulated by soluble growth factors (4) and extracellular matrix–derived signals (5), among others. Moreover, it is generally thought that angiogenesis in vivo is achieved by balance between positive (angiogenic) and negative (antiangiogenic) factors (1,6). For example, it has been known for some time that the tumors can stimulate the process of angiogenesis. Ferrara's group (7) and others demonstrated that growing tumors secrete vascular endothelial growth factor (VEGF), a specific angiogenic factor, which stimulated angiogenesis. Also, Isner's group (8), using gene transfer techniques, directly showed that VEGF induced collateral artery formation in the ischemic hindlimb.

However, the molecular mechanisms underlying angiogenesis have remained elusive due to the complexity of in vivo phenomenon. In 1980, Folkman and Haudenschild (9) showed that isolated endothelial cells in culture can form a complete vascular structure. Following this striking discovery, cultured endothelial cells have become a widely used in vitro model system to study angiogenesis. Furthermore, Montesano and Orci (10) showed that the tumor promotor phorbol 12-myristate 13-acetate (PMA) was a potential angiogenic factor in vitro, inducing tubule formation of endothelial cells cultured in three-dimensional collagen gel.

We employed a cDNA subtraction technique to isolate PMA-induced genes from endothelial cells in order to study genes involved in the differentiative aspects of angiogenesis. One such isolated gene was named endothelial differentiation gene 1 (*EDG-1*) (11). Northern blotting analysis indicated that *EDG-1* was rapidly and transiently induced by PMA in endothelial cells and was superinduced in the presence of the protein synthesis inhibitor cycloheximide. Thus, *EDG-1* is an immediate-early gene. Since immediate-early genes are important for cell growth and differentiation (12), we hypothesized that PMA-induced EDG-1 may regulate endothelial differentiation.

I. G_i IS A MAJOR G_α SUBUNIT COUPLED TO EDG-1 G PROTEIN–COUPLED RECEPTOR

Sequence analysis indicated that EDG-1 encodes a novel G protein–coupled receptor (GPCR) with seven hydrophobic transmembrane domains (11). This raised several questions, such as which heterotrimeric G proteins are coupled to EDG-1, what signals are mediated by EDG-1, what is the ligand for EDG-1, and what the physiological functions of EDG-1 are in vivo.

It has been well documented that the third intracellular loop (i_3) of a GPCR is important for interacting with its cognate G proteins (13). We therefore constructed a fusion protein in which the intracellular loop 3 (i_3) domain of EDG-1 was linked to the C-terminus of glutathione S-transferase (GST) (14). By affinity chromatography, we observed that the pertussis toxin–sensitive but not cholera toxin–sensitive G proteins specifically interacted with GST-EDG-1-i_3. This interaction was inhibited in the presence of GTP$_\gamma$S, suggesting that this in vitro interaction is physiologically relevant. The G proteins bound to EDG-1-i_3 immunoreacted with anti-$G_{i\alpha}$ antibodies. The G_α specificity was further confirmed by the facts that (a) the $G_{i\alpha}$ polypeptides were coimmunoprecipitated with intact EDG-1 GPCR (14) and (b) yeast two hybrid interaction system (14). Also, we demonstrated that EDG-1 GPCR was capable of activating $G_{i\alpha}$-mediated signals, including mitogen-activated protein (MAP) kinase (14), phospholipase A_2 (14), inhibition of adenylate cyclase (15), and so on.

II. EDG-1 IS A HIGH-AFFINITY RECEPTOR FOR THE BIOACTIVE LIPID SPP

Identifying of the physiological ligand and signaling relevant for a orphan receptor is a formidable task. Phylogram analysis indicated that EDG-1 is most closely related to the cannabinoid receptor family, suggesting that EDG-1 ligand might be lipid. Initially, we utilized the G_i/MAP kinase pathway to screen the ligand for EDG-1 by using EDG-1 transiently overexpressed cells. However, despite this labor-intensive and tedious approach, the results obtained were difficult to interpret due to the presence of endogenous EDG-1 and other G_i-coupled receptors in many cell types. Therefore, we examined a variety of cells to find cell lines that express extremely low levels of *EDG-1* mRNA. Those lines were then utilized to establish EDG-1 stable transfectants, which were used to screen for EDG-1 ligand. Among the lines screened, we found that human embryonic kidney 293 (HEK293) fibroblast-like cells have barely detectable *EDG-1* mRNA. Interestingly, all the EDG-1 stably overexpressed HEK293 clones (HEK293EDG-1) exhibited a distinct morphology (16). The control vector transfected cells (HEK293pCDNA) were evenly distributed at day 1 and reached confluency at day 3 after plating. In contrast, HEK293EDG-1 cells were tightly aggregated at day 1 and formed network-like structures at day 3 after plating. This distinct culture morphology of HEK293EDG-1 cells is similar to the tubular structure in differentiated endothelial cells and thus was termed *morphogenetic differentiation*. The differentiation was inhibited by anti-M2 antibody, which recognizes the FLAG-epitope at the N-terminus of EDG-1, indicating that the differentiation was EDG-1–specific.

One of the major questions is why EDG-1 overexpression results in morphogenetic differentiation. Since EDG-1 is derived from endothelial cells exposed to the bloodstream in vivo, we hypothesized that the EDG-1 ligand is present in serum and that ligand/EDG-1 interaction results in morphogenetic differentiation. To test this hypothesis, fetal bovine serum (FBS), which is routinely used for cell culture, was subjected to various manipulations (16). Heat-denatured FBS (95° for 1 hr) still contains the differentiation-inducing ability, indicating that serum-borne EDG-1 ligand is very heat-stable. Interestingly, the differentiation-inducing ability of FBS was abolished when serum was treated with charcoal to remove serum-borne bioactive lipids (16). In addition, we observed that the acid eluant of charcoal-bound serum lipids was capable of inducing differentiation specifically in HEK293EDG-1 cells but not in control cells. Collectively, these data suggest that EDG-1 ligand is a serum-borne bioactive lipid.

Subsequently, we employed the differentiation assay in HEK293EDG-1 cells as a method to screen for the EDG-1 ligand. Among all available lipids, we found that only sphingosine-1-phosphate (SPP) and lysophosphatidic acid (LPA) were able to induce EDG-1 cell differentiation (16,17). However, SPP was more potent than LPA; significant differentiation was observed at 1 μM of SPP, whereas LPA was effective at 10–20 μM. Other lipids—such as sphingosine, sphingomyelin, ceramide, ceramide-1-phosphate, lysophosphatidylserine, lysophosphatidyl ethanolamine, lysophosphatidyl inositol, lysophosphatidyl choline, leukotriene B4 and C4, platelet-activating factor, anandamide, 12-hydroxyeicosatetraenoic acid (12-HETE), 5-HETE, 15-HETE, di-HETEs, and 13-hydroxydodecanoic acid—were ineffective at concentrations as high as 50 μM. Although EDG-1 signals to activate MAP kinase via $G_{i\alpha}$ subunit, PD98059 and pertussis toxin, which inhibit ERK-2 signaling pathway and $G_{i\alpha}$, respectively, did not inhibit EDG-1–mediated differentiation. However, C3 exotoxin, an inhibitor of Rho small GTPase, completely prevented this morphogenesis, suggesting a requirement for the small GTPase Rho (16).

Several lines of evidence supported the notion that SPP is an important physiological ligand for EDG-1. First, radiolabeled SPP binding was observed only in HEK293EDG-1 cells, whereas binding was negligible in vector-only transfected control cells (16). This binding was inhibited by cold SPP but not by other lipid analogues, indicating that the binding was highly specific. Scatchard plot analysis showed that the apparent Kd of EDG-1 for SPP is 8.1 nM. SPP is present in serum with concentration much greater than the Kd of EDG-1 (18). Second, SPP was shown to be able to activate MAP kinase directly via the EDG-1 receptor and $G_{i\alpha}$ subunit. Third, SPP induced EDG-1 receptor phosphorylation. Fourth, EDG-1 receptor was internalized into intracellular vesicles specifically by incubating with SPP at 37°C. Together, these results establish formally that SPP is a high-affinity ligand for EDG-1 and suggest that SPP/EDG-1 interaction may be of physiological significance.

Similarly, radiolabeled LPA binding was significantly enhanced in HEK293EDG-1 (17). This binding is specific, since unlabeled LPA but not other lipid analogues competed for the binding. Scatchard plot analysis showed that there were two distinct LPA binding sites present in HEK293EDG-1 cells (the Kds were 178 nM and 2.3 μM, respectively), whereas the high-affinity binding site was readily detected in HEK293pCDNA cells. This suggested that EDG-1 was the low-affinity binding site for LPA. Nevertheless, LPA was found to activate MAP kinase through EDG-1 in a pertussis toxin–sensitive manner (17).

In addition, EDG-1 receptor was shown to be rapidly phosphorylated in response to LPA (17). However, consistent with the morphogenetic differentiation and binding affinity, both MAP kinase activation and receptor phosphorylation required much higher concentration of LPA than SPP. This supported the notion that LPA is a low-affinity ligand for EDG-1.

Intriguingly, the binding of radiolabeled SPP was poorly competed by the presence of excess amount of cold LPA, and vice versa (17). However, following treatment of HEK293EDG-1 cells with SPP at 37°C, which induced EDG-1 receptor internalization into intracellular compartments, the subsequent cell-surface binding of radiolabeled LPA was diminished (17). A likely explanation for these observations is that SPP and LPA bind to distinct sites on EDG-1 receptor. This notion awaits definitive assignment of binding sites by site-directed mutagenesis, which is currently under way in our laboratory.

SPP and LPA are two structurally and functionally similar serum-borne bioactive lipids. They are released from activated platelets (18,19) and are responsible for many biological activities of serum. Also, SPP and LPA can regulate a wide range of cellular responses, including proliferation, platelet aggregation, stress fiber formation, neurite retraction, cell rounding, and tumor cell migration, among others (20,21). Moreover, it has been demonstrated that SPP and LPA regulate most or all of these responses by activating specific G protein–coupled receptors present on the surface of numerous cell types (20,21). Our data collectively show that EDG-1 is a high-affinity receptor for SPP and a low-affinity receptor for LPA.

III. EDG-1 REGULATES MORPHOGENESIS BY INDUCING THE FORMATION OF ADHERENS JUNCTION VIA THE RHO PATHWAY

HEK293EDG-1 cells aggregated strongly in suspension, whereas HEK293pCDNA cells did not (16). This aggregation was Ca^{2+}-dependent and was completely prevented by EGTA. Because the cadherins mediate calcium-dependent homotypic adhesion mechanisms (22), we analyzed the level of cadherin family polypeptides in HEK293EDG-1 and HEK293pCDNA cells. Indeed, the expression of both P- and E-cadherins was increased in HEK293EDG-1 cells. However, expression of cytoplasmic cadherin-associated proteins—such as α-, β-, and γ-catenin—was not altered. Electron microscopy analysis showed that HEK293EDG-1 cells had abundant well-developed adherens junction-like structures. In addition, consistent with the morphogenetic differentiation, expression of P-cadherin in HEK293EDG-1 cells was enhanced by SPP and, to a lesser extent by LPA. The induction of P-cadherin expression was inhibited by C3 exotoxin but not by pertussis toxin. Together, these data suggest that SPP/LPA can signal through EDG-1 to regulate the biogenesis or maintenance of the adherens junctional complexes and morphogenetic differentiation. Rho was required for both EDG-1–induced cadherin expression and the formation of adherens junctions, consistent with early observation that the small guanosine triphosphatases (GTPase) Rho and Rac are required for the establishment of cadherin-dependent cell-cell contacts (23).

IV. SUMMARY

Morphogenetic differentiation of cells is a fundamental event in embryonic development. It is dysregulated in angiogenesis and pathological conditions such as tumorigenesis. Morphogenetic differentiation is affected by cell-cell as well as cell-matrix adhesion molecules, which are, in turn, controlled by extracellular factors (24). A huge body of evidence indicated that factors acting through receptor tyrosine kinase and G protein–coupled receptors (GPCR) regulate the processes of morphogenetic differentiation. Examples for factors acting on receptor tyrosine kinases are hepatocyte growth factor/scatter factor (HGF/SF) (25), fibroblast growth factor (FGF) (26), and vascular endothelial growth factor (VEGF) (7). On the other hand, cAMP receptor in *Dictyostelium* (27) and Wnt/Frizzled signaling (28) are examples of factors acting through GPCR to regulate morphogenetic differentiation. Our studies establish that bioactive lipids, SPP and LPA, also serve as morphogenetic factors by activating EDG-1. Also, in agreement with previous reports (23), small guanine nucleotide-binding protein Rho is required for the establishment of EDG-1-mediated, cadherin-dependent cell-cell contacts.

Other members of EDG family GPCRs were recently isolated from different laboratories, including EDG-2 (VZG-1) (29), EDG-3 (30), EDG-4 (31), and EDG-5 (H218/AGR16) (30). All the members were shown to be the receptors for lysophospholipids and/or lysosphingolipids. EDG-1 is a high-affinity receptor for SPP and a low-affinity receptor for LPA. EDG-3 and EDG-5 have been shown to be the SPP receptors (30), whereas EDG-2 and EDG-4 are LPA receptors (29,31). Interestingly, EDG-3 and EDG-5 also responded significantly to other sphingolipids, including sphingosine, dihydrosphingosine, dihydrosphingosine-1-phosphate, and sphingosylphosphorylcholine (SPC) in Jurkat T cells; but they responded to SPP and SPC only in *Xenopus* oocytes (30). Thus, it is possible that one of the mechanisms to regulate the responsiveness and discriminate between EDG family sphingolipid receptors is at the level of G-protein coupling. Together, these results imply that the variable expression and distinct signaling pathways may be responsible for different physiological functions in vivo of individual EDG family members.

In conclusion, EDG-1 is a receptor for platelet-derived serum-borne bioactive lipids. At least two distinct signalings are mediated by EDG-1 receptor. EDG-1 signals via the G_i to regulate MAK kinase, phospholipase A_2 and adenyl cyclase. In addition, EDG-1

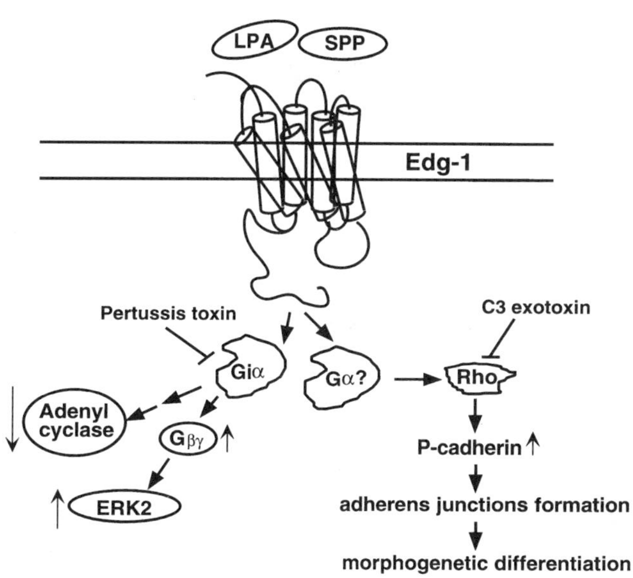

Figure 1 The signaling pathways regulated by EDG-1.

signals via an unidentified G_α subunit to stimulate the adherens junctions formation, which requires the small GTP-binding protein Rho (Fig. 1). Modulation of EDG-1 function may offer a novel target for angiogenic therapeutics.

REFERENCES

1. J Folkman. Tumor angiogenesis. Adv Cancer Res 43:175–203, 1985.
2. W Risau, I Flamme. Vasculogenesis. Annu Rev Cell Dev Biol 11:73–91, 1995.
3. J Folkman. Angiogenesis in cancer, vascular, rheumatoid and other disease. Nature Med 1: 27–31, 1995.
4. KH Plate, G Breier, HA Weich, W Risau. Vascular endothelial growth factor is a potential tumor angiogenesis factor in human gliomas in vivo. Nature 359:845–848, 1992.
5. S Stromblad, DA Cheresh. The role of cell adhesion molecules in angiogenesis. Trends Cell Biol 6:462–468, 1996.
6. MS O'Reilly, L Holmgren, Y Shing, C Chen, RA Rosenthal, M Moses, WS Lane, Y Cao, EH Sage, J Folkman. Angiostatin: a novel angiogenesis inhibitor that mediates the suppression of metastases by a Lewis lung carcinoma. Cell 79:315–328, 1994.
7. KJ Kim, B Li, J Winer, M Armanini, N Gillett, HS Philips, N Ferrara. Inhibition of vascular endothelial growth factor–induced angiogenesis suppresses tumour growth in vitro. Nature 362:671–674, 1993.
8. S Takeshita, Y Tsurumi, T. Couffinahl, T Asahara, C Bauters, J Symes, N Ferrara, JM Isner. Gene transfer of naked DNA encoding for three isoforms of vascular endothelial growth factor stimulates collateral development in vivo. Lab Invest 75:487–501, 1996.
9. J Folkman, C Haudenschild. Angiogenesis in vitro. Nature 288:551–556, 1980.
10. R Montesano, L Orci. Phorbol esters induce angiogenesis in vitro from large-vessel endothelial cells. J Cell Physiol 130:284–291, 1987.
11. T Hla, T Maciag. An abundant transcript induced in differentiating human endothelial cells encodes a polypeptide with structural similarities to G-protein–coupled receptor. J Biol Chem 265:9308–9313, 1990.
12. J Tooze. DNA Tumor Viruses. Cold Spring Harbor, New York: Cold Spring Harbor Laboratory, 1980.
13. BR Conklin, HR Bourne. Structural elements of G alpha subunits that interact with G beta gamma, receptors, and effectors. Cell 73:631–641, 1993.
14. MJ Lee, M Evans, T Hla. The inducible G Protein-coupled receptor edg-1 signals via the Gi/ Mitogen-activated protein kinase pathway. J Biol Chem 271:11272–11279, 1996.
15. JR VanBrocklyn, MJ Lee, R Menzeleev, A Olivera, L Edsall, O Cuvillier, DM Thomas, PJP Coopman, S Thangade, CH Liu, T Hla, S Spiegel. Dual actions of sphingosine-1-phosphate: extracellular through the Gi-coupled receptor Edg-1 and intracellular to regulate proliferation and survival. J Cell Biol 142:229–240, 1998.
16. MJ Lee, JR VanBrocklyn, S Thangada, CH Liu, AR Hand, R Menzeleev, S Spiegel, T Hla. Sphingosine-1-phosphate as a ligand for the G-protein–coupled receptor EDG-1. Science 279: 1552–1555, 1998.
17. MJ Lee, S Thangada, CH Liu, BD Thompson, T Hla. Lysophosphatidic acid stimulate the G-protein–coupled receptor EDG-1 as a low affinity agonist. J Biol Chem 273:22105–22112, 1998.
18. Y Yatomi, Y Igarashi, L Yang, N Hisano, R Qi, N Asazuma, K Satoh, Y Ozaki, S Kume. Sphingosine 1-phosphate, a bioactive sphingolipid abundantly stored in platelets, is a normal constituent of human plasma and serum. J Biochem 121:969–973, 1997.
19. T Eichholtz, K Jalink, I Fahrenfort, WH Moolenaar. The bioactive phospholipid lysophosphatidic acid is released from activated platelets. Biochem J 291:677–680, 1993.

20. WH Moolenaar, O Kranenburg, FR Postma, GC Zondag. Lysophosphatidic acid: G-protein signalling and cellular responses. Curr Opin Cell Biol 9:168–173, 1997.
21. S Spiegel, O Cuvillier, L Edsall, T Kohama, R Menzeleev, A Olivera, D Thomas, Z Tu, J VanBrocklyn, Wang F. Roles of sphingosine-1-phosphate in cell growth, differentiation, and death. Biochemistry 63:69–73, 1998.
22. S Hirano, N Kimoto, Y Shimoyama, S Hirohashi, M Takeichi. Identification of a neural alpha-catenin as a key regulator of cadherin function and multicellular organization. Cell 20:293–301, 1992.
23. A Hall. Rho GTPases and the actin cytoskeleton. Science 279:509–514, 1998.
24. BM Gumbiner. Cell adhesion: the molecular basis of tissue architecture and morphogenesis. Cell 84:345–357, 1996.
25. R Montesano, K Matsumoto, T Nakamura, L Orci. Identification of a fibroblast-derived epithelial morphogen as hepatocyte growth factor. Cell 67:901–908, 1991.
26. JM Slack, BG Darlington, JK Heath, SF Godsave. Mesoderm induction in early Xenopus embryos by heparin-binding growth factors. Nature 326:197–200, 1987.
27. P Lilly, L Wu, DL Welker, PN Devreotes. A G-protein β-subnit is essential for Dictyostelium development. Genes Dev 7:986–995, 1993.
28. X He, JP Saint-Jeannet, Y Wang, J Nathans, I Dawid, H Varmus. A member of the frizzled protein family mediating axis induction by Wnt-5A. Science 275:1652–1654, 1997.
29. JH Hecht, JA Weiner, SR Post, J Chun. Ventricular zone gene-1 (vzg-1) encodes a lysophospatidic acid receptor expressed in neurogenic region of the developing cerebral cortex. J Cell Biol 135:1071–1083, 1997.
30. S An, T Bleu, W Huang, OG Hallmark, SR Coughlin, EJ Goetzl. Identification of cDNAs encoding two G protein-coupled receptors for lysosphingolipids. FEBS Lett 417:279–282, 1997.
31. S An, T Bleu, W Huang, OG Hallmark, EJ Goetzl. Characterization of a novel subtype of human G protein-coupled receptor for lysophosphatidic acid. J Biol Chem 273:7906–7910, 1998.

9
The Role of HIF-1α and ARNT Proteins in Blood Vessel Development

Emin Maltepe and M. Celeste Simon
University of Chicago, Chicago, Illinois

I. INTRODUCTION

The recent excitement over angiogenesis is in part due to the critical role played by this process in tumorigenesis. The dependence of growing tumor masses on neovascularization is a well-established fact. A tumor cannot grow beyond a diameter of approximately 2 mm without vascularizing, owing to the limitations of diffusion as a means of nutrient and oxygen delivery (1). This requires that a tumor be perfused by a network of capillaries distributed approximately every three cell layers to meet the metabolic demands of rapid growth (2). The more vascularized a tumor is, the better it is able to grow and the worse the prognosis it carries for the patient (3,4). An important therapeutic modality for inhibiting solid tumor growth may thus involve the ability to specifically target tumor blood vessels. It is for this reason that the recently discovered endogenous inhibitors of angiogenesis have drawn the attention not only of the scientific community but also of the public at large. Study of the underlying mechanisms regulating the vascularization of growing tumors, however, requires an understanding of the fundamental physiological functions provided by blood vessels. As these include oxygen and nutrient delivery, it would seem logical that the supply of such agents be linked to demand via an intersection of physiological and genetic processes. Recent studies indicate that this is indeed the case and that these processes require the activities of a rapidly growing family of transcription factors characterized by the presence of basic helix-loop-helix (bHLH) as well as PAS domains.

II. bHLH-PAS PROTEINS

Spatially and temporally coordinated gene expression is key to all biological processes. Structural analysis of transcription factors has resulted in their distribution into multiple classes depending on the nature of the modular motifs responsible for their specific activities. The PAS domain was originally described as a region of homology identified within the three founding members of this family (5): the *Drosophila* period (*Per*) (6), the mammalian arylhydrocarbon nuclear translocator (Arnt), and the *Drosophila* single-minded (*Sim*) (6) genes. The PAS protein family now contains 11 members in the mouse with highly diverse functions, ranging from mediating the transcriptional response to xenobiotic exposure via the arylhydrocarbon receptor (Ahr) (7,8) to regulating circadian rhythms via *Per* and the recently described CLOCK protein (9–11). Basic helix-loop-helix (bHLH) proteins function as heterodimeric complexes, with the basic domains of each being responsible for DNA binding and the HLH domains serving as primary heterodimerization interfaces. Unlike classic bHLH proteins, the recognition sequences for bHLH-PAS transcription factors consist of asymmetrical half sites, each being recognized by one member of the heterodimeric complex. This ability to form heterodimers allows the formation of multiple combinations to help mediate the requirement for binding-site specificity. In general, a limited number of family members exhibit broad expression profiles, with the multiple heterodimerization partners for each introducing one level of specificity by displaying a much more restricted expression profile (12). Additionally, multiple heterodimerization partners for a specific family member can be expressed within the same cell, generating the potential for functional interference between different functional pathways (13). Alternatively, similar to the ability of Id, an HLH protein that lacks a basic domain, to heterodimerize with and inhibit the transcriptional activities of other bHLH proteins (14), *Per* lacks a bHLH domain but can still homodimerize (15) or heterodimerize with other bHLH-PAS proteins and act in a dominant negative fashion by inhibiting their DNA-binding capabilities (16). Thus, the combinatorial complexity of the interactions within the bHLH-PAS transcription factor family can be quite extensive.

PAS domains are interesting in that they have been shown to have multiple functions including potential cofactor recruitment, the regulation of heterodimerization specificity, as well as ligand or Hsp90 binding. For example, replacement of the PAS domain of the *Drosophila* Trachealess (Trh) protein with the analogous region of the Sim protein has been shown to be sufficient to convert it into a functional Sim protein without altering its target-gene specificity (17). Thus, it has been proposed that the Trh PAS domain is responsible for recruiting a different set of cofactors or coactivators important for its role in tracheal development, whereas Sim's PAS domain may recruit factors involved in central nervous system (CNS) midline development. In addition, while the bHLH domain of the dioxin receptor has the intrinsic ability to homodimerize, addition of the PAS domain dramatically reduces this ability, thereby promoting heterodimerization, but only with other bHLH-PAS proteins (18), thus regulating its heterodimerization specificity. Activation of the Ah receptor via dioxin or other polycylic aromatic hydrocarbons has been shown to occur via binding of these compounds to its PAS domain (19,20). In the unactivated state, the Ahr is located within the cytoplasm complexed with two molecules of Hsp90 (21,22), which interact with it via the PAS domain. Hsp90 also interacts with the PAS domain of mammalian Sim (23).

Database searches have identified PAS-containing proteins in highly diverse organisms including eubacteria, archaea, cyanobacteria, fungi, plants, insects, and mammals (24). It has recently been postulated that the photoactive yellow protein (PYP) of the bacterial blue-light photoreceptor, which also displays ligand/cofactor binding and signal transduction properties governed by heterodimer formation, represents a structural prototype of the PAS domain found in higher organisms (25). bHLH-PAS proteins, however, seem to be restricted to multicellular organisms, as no homologues have been identified in yeast, while a variety of bHLH-PAS family members are found in *Caenorhabtidis elegans*, *Drosophila*, and higher organisms. Whatever the organism involved, a prominent theme within this family tends to involve the sensing of environmental conditions and regulation of the appropriate cellular and organismal responses.

III. THE TRANSCRIPTIONAL RESPONSE TO HYPOXIA

Oxygen (O_2) deprivation, or hypoxia, has been shown to be a strong stimulant of erythropoietin (Epo) production by the kidney in an attempt to increase the blood's oxygen-carrying capacity (26). A search for factors responsible for this action resulted in the discovery of another bHLH-PAS family member, the hypoxia-inducible factor-1α (HIF-1α). Biochemical purification of HIF-1, as the active transcription factor is known, revealed that it was composed of the previously described ARNT protein complexed with HIF-1α (27). While both genes are broadly expressed at the mRNA level, in general, only ARNT protein is readily detectable in cells, with HIF-1α protein levels increasing upon exposure to lowered oxygen tensions (28). This posttranslational regulation has been shown to be attributable to the rapid ubiquitin-mediated degradation of HIF-1α under normoxic conditions (29). Epo secretion represents only a small component of the response to hypoxia, however, which ranges from the organismal to the cellular and includes alterations of respiratory and cardiovascular rates, neovascularization, and metabolic adaptation (30,31). A significant component of each of these responses is mediated by the activities of proteins encoded by genes that are upregulated in response to hypoxia. Binding sites for the HIF-1 complex, known as hypoxia response elements (HREs), have been identified within the promoters or enhancers of numerous hypoxia-responsive genes including glycolytic enzymes, glucose transporters, and angiogenic and hematopoietic factors, to name a few (32). As HIF-1α and ARNT are widely expressed and as many of these responses can be noted to some extent in every cell type studied, the HIF-1 complex may constitute a master regulator of the transcriptional response to hypoxia.

The ability to mount a transcriptional response to hypoxia implies the existence of an O_2-sensor. Since the HIF-1 complex is activated at O_2 partial pressures prior to the point where O_2 becomes rate-limiting for respiration, it seems that molecular O_2 plays a unique dual role acting not only as the terminal electron acceptor of the mitochondrial respiratory chain but also as a ligand affecting transcription factor activity. While much progress has been made delineating the factors involved in hypoxic gene expression, relatively little is known regarding the manner in which cells actually sense O_2 concentration. Mitochondria represent obvious candidates as sites of O_2 sensing given that they are responsible for 90–95% of the O_2 consumed by a cell. Although mitochondria are in an ideal position to respond to changes in O_2 partial pressures, they have historically been discounted as sites of O_2 sensing for two principal reasons: (a) the low apparent K_m of mitochondria for O_2 (33) and (b) the inability to stimulate Epo secretion with inhibitors of mitochondrial electron transport (34,35). It has thus been argued that the transcriptional response to hypoxia, which is triggered at O_2 concentrations that are not rate-limiting for electron transport, cannot be transduced by mitochondrial processes that do not alter their activities unless exposed to near anoxic conditions.

Recent studies challenge this conventional view of O_2 sensing by indicating that cytochrome *c* oxidase kinetic activity is responsive to changes in O_2 tension within the physiological range (36,37). Additionally, it has recently been shown that cells depleted of mitochondrial DNA, and hence mitochondrial DNA-derived protein, are unable to activate the HIF-1 complex and thus cannot mount a transcriptional response to hypoxia (38).

Furthermore, inhibitors of mitochondrial electron transport that act upstream of complex III are able to inhibit hypoxia-induced transcription, indicating that an intact respiratory chain is required for this response. The formation of reactive oxygen species (ROS) via single electron reduction of O_2 by ubisemiquinone (39) most likely plays an important signaling role, as scavengers of ROS are also able to block hypoxia-induced HIF-1 activation. The ability of iron to be incorporated into newly translated HIF-1α protein (40) suggests a mechanism whereby hypoxia-induced mitochondrial ROS formation may directly modulate HIF-1α activity via redox chemistry, inhibiting ubiquitin-mediated degradation and leading to nuclear accumulation and complex formation with the constitutively nuclear ARNT protein.

IV. OXYGEN, TUMOR ANGIOGENESIS, AND GROWTH

Tumor-cell hypoxia has been proposed to be an important factor governing the tumor's vascularity (41). Cells within solid tumors are exposed to extremely heterogeneous microenvironments displaying wide variations in O_2 distribution and pH due to rapid growth and disorganized vascularization (42). Vascular endothelial growth factor (VEGF), a potent endothelial cell–specific mitogen and attractant (43), is thought to play an important role in tumor angiogenesis. Rapidly growing tumors display robust levels of VEGF expression, especially in cells adjacent to regions of necrosis, presumably representing the most hypoxic though still viable cell populations (41). Additionally, embryonic stem (ES) cells rendered deficient for VEGF via gene targeting are unable to form tumors in mice owing to impaired vascularity (44). As the VEGF gene is exquisitely sensitive to hypoxia, because of the presence of a functional HRE within its promoter (45), HIF-1 mediated VEGF expression may be an important component regulating the vascularization of growing tumors.

While the described transcriptional response to hypoxia clearly plays an adaptive role, enabling cells to endure exposure to diminished O_2 supply while promoting improved vascularization and increases in the blood's O_2-carrying capacity, an additional function of the HIF-1 complex has been described which may ensure that this response remains self-limiting. Multiple investigators have previously reported that exposure of cells to severe O_2 deprivation, or anoxia, leads to growth arrest and/or apoptosis (46–49). Additionally, these stimuli have been shown to induce increases in p53 protein levels and pRb hypophosphorylation, both of which are involved in cell-cycle inhibition. The necrotic cores of tumors are thought to arise because of such anoxia-induced apoptosis. It has also been shown that the hypoxic nature of tumors may select for mutant p53-expressing cells in an effort to overcome this stress-induced cell-cycle arrest (50). Recent reports from multiple laboratories suggest that the HIF-1 complex is also responsible for mediating this growth-inhibitory function. It has been shown that the p53 tumor suppressor protein is stabilized under exposure to hypoxic conditions or to chemical agents that mimic physiological hypoxia via a direct interaction with the HIF-1 complex (51). Furthermore, the antiapoptotic protein Bcl-2 is downregulated during hypoxia via a process that is also HIF-1 dependent (52). Thus, the cellular response to hypoxia may include a component that prevents further growth until the normalization of environmental conditions. This may play an important role during embryogenesis, where continued growth during limb formation, for example, would be limited so as not to outpace the developing vascular supply.

V. ARNT PROTEINS, OXYGEN REGULATION, AND EMBRYOGENESIS

As stated previously, ARNT forms heterodimeric complexes with a number of bHLH-PAS family members, including HIF-1α, the arylhydrocarbon receptor (AHR), Per, and murine homologs of *Drosophila* single-minded, SIM1 and SIM2 (27,53,54). In response to environmental toxins, AHR-ARNT heterodimers regulate genes required for the metabolism of xenobiotics (7,55,56), whereas HIF-1α-ARNT heterodimers regulate genes involved in O_2 homeostasis (45,57,58). Adaptive responses to hypoxia include the transcriptional activation of genes such as transferrin, VEGF, inducible nitric oxide synthase (iNOS), erythropoietin, glucose transporters, and the glycolytic enzymes aldolase A (ALDA), enolase 1, lactate dehydrogenase A, phosphofructokinase L (PFK-L), and phosphoglycerate kinase 1 (PGK-1) (see Ref. 32). Each of these proteins is critical to intracellular, local, or systemic physiological responses to O_2 deprivation. For example, glucose transporters (GLUT1, GLUT3, etc.) and glycolytic enzymes are essential for anaerobic ATP synthesis, erythropoietin stimulates red cell production and the O_2-carrying capacity of blood, transferrin delivers iron to the bone marrow for incorporation into hemoglobin, VEGF mediates vascularization, and iNOS modulates vascular tone. Hypoxia-response DNA elements (5′-ACGTG-3′) that bind HIF-1α-ARNT have been identified in each of these genes. Therefore, substantial evidence supporting a role for HIF-1α and ARNT in hypoxia-inducible gene regulation exists. However, related proteins such as EPAS-1/HIF-2α/MOP2 (32,59,60), ARNT2 (61), and MOP3 (62) have recently been cloned. Furthermore, ARNT has multiple dimerization partners including HIF-1α, EPAS-1, AHR, SIM1, SIM2, and Per. To definitively establish the role of ARNT and HIF-1α in O_2 gene regulation and mammalian development, $Arnt^{-/-}$ (63,64) and $Hif1\alpha^{-/-}$ (52,65,66), embryonic stem (ES) cells, and mice that lack each protein have been generated.

By analyzing $Arnt^{-/-}$ ES cells, we showed that genes normally activated by low O_2 tension failed to respond to hypoxia in the absence of ARNT (63). ARNT and HIF-1α are expressed in ES cells, allowing a role for ARNT in the regulation of these genes to be directly analyzed. The glycolytic enzymes PGK-1, ALDA, PFK-L; the glucose transporters GLUT-1 and GLUT-3; and the endothelial mitogen VEGF were activated by hypoxia (1.5% O_2) in $Arnt^{+/+}$ ES cells. However, two independently derived $Arnt^{-/-}$ ES cell lines did not increase production of PGK-1, ALDA, PFK-1, GLUT-1, GLUT-3, or VEGF mRNAs during growth under hypoxic conditions. Furthermore, two reagents that mimic physiological hypoxia, cobalt chloride and desferrioxamine, increased transcription of these genes in $Arnt^{+/+}$ cells but not in $Arnt^{-/-}$ cells. GLUT-1 and VEGF mRNA expression has been shown to increase in response to glucose deprivation in glioma cells (67,68). We therefore tested to see whether these O_2-regulated genes still responded to hypoglycemia in $Arnt^{-/-}$ ES cells. $Arnt^{+/+}$ ES cells grown in low-glucose medium showed elevated levels of PGK-1, ALDA, VEGF, and GLUT-1, with kinetics similar to those observed in hypoxic culture conditions. In direct contrast, $Arnt^{-/-}$ ES cells failed to induce the expression of these genes in response to hypoglycemia. Of note, the $Arnt^{-/-}$ cells exhibited a 60% reduction in viability compared to $Arnt^{+/+}$ cells after 72 hr of glucose deprivation. Thus, a failure to stimulate genes like GLUT-1 during glucose starvation is correlated with decreased cell survival. Taken together, these results demonstrate that ARNT is essential for responses to both O_2 and glucose deficiency. Furthermore, hypoxia and hypoglycemia signaling pathways may converge to activate a common group of genes in a process that is mediated by ARNT.

The role of ARNT in vivo was assessed by producing *Arnt*$^{-/-}$ mice. *Arnt*$^{-/-}$ embryos died in utero by day 10.5 of gestation with multiple defects in vascular development. Although vasculogenesis (the de novo formation of blood vessels) appeared to be normal, angiogenesis (the subsequent maturation of the vascular plexus) in the yolk sac, branchial arches, and placenta was deficient (63) (E. Maltepe and C. Simon, unpublished observations). For example, histological staining revealed the presence of fused endothelial cells in the yolk sacs of mutant embryos, but the vasculature was highly irregular. In addition, the vessels of *Arnt*$^{-/-}$ yolk sacs had fewer surrounding smooth-muscle cells, reminiscent of tissue factor mutant (*TF*$^{-/-}$) yolk sac vessels. When dissected free of the yolk sac, *Arnt*$^{-/-}$ embryos were generally wasted and necrotic, exhibiting stunted development. Despite abnormal extraembryonic vasoformation, the heart, dorsal aortas, and intersomitic blood vessels appeared to be intact. These results contrast with *Vegf*$^{+/-}$ and *Vegf*$^{-/-}$ embryos, which displayed a more severe reduction in blood vessels throughout the conceptus and incorrect cardiac development (44,69). Nevertheless, *Arnt*$^{-/-}$ embryos transcribed less VEGF mRNA in the yolk sac and many areas within the embryo proper. High levels of VEGF were detected in the neural tube and primitive gut, which may explain the differences between *Vegf*$^{-/-}$ and *Arnt*$^{-/-}$ embryos, given the close proximity of these structures to the heart, dorsal aortas, and somitic vessels. Similar defects were observed in *Arnt*$^{-/-}$ embryos independently derived by Kozak et al. (1997). These investigators observed neural tube closure defects, forebrain hypoplasia, delayed rotation of the embryo, and a failure of the placenta to vascularize. However, they did not detect any defects in the yolk sac circulation. The reasons for this discrepancy between the two mutant strains of mice are not clear, but it could be due to slightly different 129 substrains (SvJ versus Sv-CP) employed by each group. The conclusions from both studies are that ARNT is required for proper vascularization and survival of the day-10.5 mouse embryo.

VI. HIF-1α, HYPOXIA RESPONSES, AND DEVELOPMENT

Null mutations in *HIF-1α* have now been created by three independent research groups (52,65,66). The results with *Hif1α*$^{-/-}$ ES cells confirm the importance of HIF-1 in hypoxic gene induction. All three reports demonstrate that like ARNT, HIF-1α is essential for upregulation of a number of hypoxically responsive genes (e.g., VEGF, PGK-1, ALDA, and GLUT-1). However, a requirement for HIF-1α in the basal expression of many O_2-regulated genes (e.g., VEGF, PFK-L, GLUT-3, and GAPDH) in ES cells may be more stringent than for ARNT (52,65). Furthermore, two reports (52,66) conclude that HIF-1α is also essential for responses to glucose deprivation, while Iyer et al. maintain that hypoglycemic gene induction is not HIF-1α–dependent (65). Their data is consistent with the findings that HIF-1 DNA- binding activity is not induced by glucose deprivation (E. Maltepe and C. Simon, unpublished observations) (65). Therefore, the precise requirement of HIF-1α or the hypoxia response element in glucose gene regulation remains unclear. It remains a formal possibility that ARNT and/or HIF-1α regulate hypoglycemia responses via alternate DNA consensus elements or a posttranscriptional mechanism, such as a hypoglycemia-induced mRNA binding protein.

Analysis of *Hif-1α*$^{-/-}$ embryos revealed multiple defects in cephalic vascularization, neural fold formation, and cardiovascular development (65,66). Cardiovascular abnormalities included pericardial effusion, myocardial hyperplasia, and ventricular outflow obstruction. Mutant embryos also exhibited enlarged vascular structures in the cephalic region and poorly developed dorsal aortas. Of note, *Hif-1α*$^{-/-}$ embryos demonstrated a high number of hypoxic and dead cells because of inadequate perfusion (65,66). Collectively, these defects led to fetal death between days 8.5 and 10.5 of gestation. Although decreased vascularization was detected in the branchial arches of *Arnt*$^{-/-}$ mice, the enlarged cephalic vascular structures and dorsal aortic abnormalities were not observed (63). This may be due to the presence of HIF-1α-ARNT2 heterodimers in the nervous system. A defect in placental development has not been reported for *Hif-1α*$^{-/-}$ embryos. It seems possible that ARNT may dimerize with another bHLH protein (*HIF-2α*) during placentation. It will be of interest to examine placental development in the absence of this bHLH transcription factor.

VII. THE ROLE OF HIF-1 DURING TUMOR GROWTH

ES cells form teratocarcinomas when injected into immunocompromised mice. The role of hypoxia in tumors in the induction of angiogenesis and selection of p53 mutant cells has been described above. Unlike their wild-type counterparts, *Vegf*$^{-/-}$ ES cells fail to generate tumors upon subcutaneous introduction into the backs of beige/nude mice (44). As VEGF is clearly a target of HIF-1, we analyzed the ability of *Arnt*$^{-/-}$ ES cells to form teratocarcinomas. Surprisingly, *Arnt*$^{-/-}$ ES cells produced tumors of comparable mass to *Arnt*$^{+/+}$ cells over a 4-week period (E. Maltepe and C. Simon, manuscript in preparation). However, these results can be explained by the presence of both ARNT2 and VEGF expression in the mutant tumors and subsequent vascularization. Importantly, ARNT2 "rescues" the defect in *Arnt*$^{-/-}$ ES cells to form a HIF-1 complex and induce the expression of VEGF and GLUT1 in response to hypoxia. We propose that both ARNT and ARNT2 can heterodimerize with HIF-1α, stimulate VEGF production, and promote tumor angiogenesis. This hypothesis can be directly tested by generating tumors that lack both ARNT and ARNT2.

The role of HIF-1α in tumor growth has also been tested using *Hif1α*$^{-/-}$ ES cells (65,66). Ryan et al. reported that *Hif1α*$^{-/-}$ teratocarcinomas exhibited a substantial decrease (75%) in tumor mass relative to wild type tumors after 3 weeks (1998), suggesting that HIF-1 is critical for rapid cell growth in the tumor microenvironment. This was accompanied by decreased VEGF expression and a 40% reduction in tumor vessel density. These results strongly suggest that hypoxic regulation of angiogenic agents like VEGF by HIF-1 is critical to tumor vascularization. Carmeliet et al. also found that loss of HIF-1α reduced hypoxic induction of VEGF and prevented the formation of large blood vessels in mutant teratocarcinomas (52). However, the growth of ES cell–derived tumors was not retarded over a 4-week period in their experiment. In fact, from 5–8 weeks, tumor growth was actually accelerated. They concluded that the increase in tumor size was due to decreased apoptosis in the *Hif1α*$^{-/-}$ teratocarcinomas. Hypoxia and hypoglycemia reduced proliferation and increased apoptosis in wild-type but not *Hif1α*$^{-/-}$ ES cells. HIF-1 has been shown to stabilize wild-type p53 via the formation of a physical complex with this cell-cycle regulator (51). Carmeliet et al. demonstrated that the accumulation of p53 and p21, reduction in Bcl-2, and induction of apoptosis during hypoxia and hypoglycemia is dependent on HIF-1α (52). Therefore, *Hif1α*$^{-/-}$ tumors may be able to outpace vascular supply due to insufficient hypoxia/hypoglycemia–induced growth arrest or apoptosis. It is difficult to reconcile the results obtained by both studies; however, the tumors were generated in different strains of immunocompromised mice. Ryan et al. (66) utilized *Rag-1*$^{-/-}$ mice, while Carmeliet et al. injected ES cells into *Nu/Nu* mice. The *Rag-1*$^{-/-}$ mice may be more immunodeficient, allowing tumors to grow more rapidly in the Ryan analysis (5 g versus 2 g after 3 weeks), and thereby allowing this group to complete their study in less time than the Carmeliet group (3 weeks versus 8 weeks). It seems probable, however, that hypoxia-induced transcriptional responses play an important role within solid tumors. Hepatomas deficient for the ARNT protein have been shown to exhibit reduced vascularity and slower growth (70), while our own results suggest that ARNT-deficient ES cells are not impaired in their ability to form tumors, owing to compensation by other family members that do not exist in hepatocytes. A precise assessment of the requirement for various bHLH-PAS proteins during tumor growth clearly requires additional study.

VIII. SUMMARY AND CONCLUSIONS

The expression of HIF-1 during embryogenesis and in solid tumors suggests a model whereby increasing tissue mass leads to formation of O_2- and nutrient-deprived cells, HIF-1 activation, and the subsequent induction of a program of target genes that regulate glycolysis, vascularization, erythropoiesis, cellular proliferation, and apoptosis. Delivery of O_2 and glucose to ischemic tissues is critical in a number of pathophysiological situations, including heart attack, stroke, and cancer. Gene-targeting experiments that eliminate the HIF-1α and ARNT proteins have conclusively shown that these bHLH-PAS transcription factors play a critical role in the development of blood vessels during fetal life and tumorigenesis. However, many questions remain, including the precise mechanisms whereby cells sense O_2 and glucose starvation, how this signal is propagated, how HIF-1 DNA binding activity is induced, and how additional related proteins (HIF-2α, ARNT2, MOP3) fit into this picture. Finally, the molecular basis of how cells control proliferation, growth arrest, or apoptosis during hypoxia or hypoglycemia remains unclear. We are only beginning to understand the complex mechanisms allowing for a proper interaction between these fundamental physiological and genetic phenomena. Analysis of the bHLH-PAS protein family in general and the HIF-1 complex in particular should yield further insights into this intriguing area of study.

REFERENCES

1. J Folkman, R Cotran. Relation of vascular proliferation to tumor growth. Int Rev Exp Pathol 16:207–248, 1976.
2. IF Tannock. The relation between cell proliferation and the vascular system in a transplanted mouse mammary tumor. Br J Cancer 22:258–273, 1968.
3. N Weidner, JP Semple, WR Welch, J Folkman. Tumor angiogenesis and metastasis—correlation in invasive breast carcinoma. N Engl J Med 324:1–8, 1991.
4. N Weidner, PR Carroll, J Flax, W Blumenfeld, J Folkman. Tumor angiogenesis correlates with metastasis in invasive prostate carcinoma. Am J Pathol 143:401–409, 1993.
5. JR Nambu, JO Lewis, KA Wharton Jr, ST Crews. The Drosophila single-minded gene encodes a helix-loop-helix protein that acts as a master regulator of CNS midline development. Cell 67:1157–1167, 1991.
6. ST Crews, JB Thomas, CS Goodman. The Drosophila single-minded gene encodes a nuclear protein with sequence similarity to the per gene product. Cell 52:143–151, 1988.

7. O Hankinson. The aryl hydrocarbon receptor complex. Annu Rev Pharmacol Toxicol 35:307–340, 1995.
8. JV Schmidt, CA Bradfield. Ah receptor signaling pathways. Annu Rev Cell Dev Biol 12:55–89, 1996.
9. DP King, Y Zhao, AM Sangoram, LD Wilsbacher, M Tanaka, MP Antoch, TD Steeves, MH Vitaterna, JM Kornhauser, PL Lowrey, FW Turek, JS Takahashi. Positional cloning of the mouse circadian clock gene. Cell 89:641–653, 1997.
10. N Gekakis, D Staknis, HB Nguyen, FC Davis, LD Wilsbacher, DP King, JS Takahashi, CJ Weitz. Role of the CLOCK protein in the mammalian circadian mechanism (see comments). Science 280:1564–1569, 1998.
11. TK Darlington, K Wager-Smith, MF Ceriani, D Staknis, N Gekakis, TDL Steeves, CJ Weitz, JS Takahashi, SA Kay. Closing the circadian loop: CLOCK-induced transcription of its own inhibitors per and tim (see comments). Science 280:1599–1603, 1998.
12. S Jain, E Maltepe, MM Lu, C Simon, CA Bradfield. Expression of ARNT, ARNT2, HIF1alpha, HIF2alpha and Ah receptor mRNAs in the developing mouse. Mech Dev 73:117–123, 1998.
13. K Gradin, J McGuire, RH Wenger, I Kvietikova, ML Whitelaw, R Toftgard, L Tora, M Gassmann, L Poellinger. Functional interference between hypoxia and dioxin signal transduction pathways: competition for recruitment of the Arnt transcription factor. Mol Cell Biol 16:5221–5231, 1996.
14. R Benezra, RL Davis, D Lockshon, DL Turner, H Weintraub. The protein Id: a negative regulator of helix-loop-helix DNA binding proteins. Cell 61:49–59, 1990.
15. LE Huang, Z Arany, DM Livingston, HF Bunn. Activation of hypoxia-inducible transcription factor depends primarily upon redox-sensitive stabilization of its alpha subunit. J Biol Chem 271:32253–32259, 1996.
16. MC Lindebro, L Poellinger, ML Whitelaw. Protein-protein interaction via PAS domains: role of the PAS domain in positive and negative regulation of the bHLH/PAS dioxin receptor-Arnt transcription factor complex. EMBO J 14:3528–3539, 1995.
17. E Zelzer, P Wappner, BZ Shilo. The PAS domain confers target gene specificity of Drosophila bHLH/PAS proteins. Genes Dev 11:2079–2089, 1997.
18. I Pongratz, C Antonsson, ML Whitelaw, L Poellinger. Role of the PAS domain in regulation of dimerization and DNA binding specificity of the dioxin receptor. Mol Cell Biol 18:4079–4088, 1998.
19. ML Whitelaw, M Gottlicher, JA Gustafsson, L Poellinger. Definition of a novel ligand binding domain of a nuclear bHLH receptor: co-localization of ligand and hsp90 binding activities within the regulable inactivation domain of the dioxin receptor. Embo J 12:4169–4179, 1993.
20. KM Dolwick, HI Swanson, CA Bradfield. In vitro analysis of Ah receptor domains involved in ligand-activated DNA recognition. Proc Natl Acad Sci USA 90:8566–8570, 1993.
21. LA Carver, V Jackiw, CA Bradfield. The 90-kDa heat shock protein is essential for Ah receptor signaling in a yeast expression system. J Biol Chem 269:30109–30112, 1994.
22. C Antonsson, B Arulampalam, ML Whitelaw, S Pettersson, L Poellinger. Constitutive function of the basic-loop-helix/PAS factor Arnt. J Biol Chem 270:13968–13972, 1995.
23. J McGuire, P Coumailleau, ML Whitelaw, JA Gustafsson, L Poellinger. The basic helix-loop-helix/PAS factor Sim is associated with hsp90: implications for regulation by interaction with partner factors. J Biol Chem 270:31353–31357, 1995.
24. IB Zhulin, BL Taylor, R Dixon. PAS domain S-boxes in Archaea, Bacteria and sensors for oxygen and redox. Trends Biochem Sci 22:331–333, 1997.
25. JL Pellequer, KA Wager-Smith, SA Kay, ED Getzoff. Photoactive yellow protein: a structural prototype for the three-dimensional fold of the PAS domain superfamily. Proc Natl Acad Sci USA 95:5884–5890, 1998.
26. W Jelkmann. Erythropoietin: structure, control of production, and function. Physiol Rev 72: 449–489, 1992.
27. GL Wang, B-H Jiang, EA Rue, GL Semenza. Hypoxia-inducible factor 1 is a basic helix-

loop-helix-PAS heterodimer regulated by cellular O_2 tension. Proc Natl Acad Sci USA 92: 5510–5514, 1995.
28. BH Jiang, GL Semenza, C Bauer, HH Marti. Hypoxia-inducible factor 1 levels vary exponentially over a physiologically relevant range of O2 tension. Am J Physiol 271:C1172–1180, 1996.
29. S Salceda, J Caro. Hypoxia-inducible factor lalpha (HIF-1alpha) protein is rapidly degraded by the ubiquitin-proteasome system under normoxic conditions: its stabilization by hypoxia depends on redox-induced changes. J Biol Chem 272:22642–22647, 1997.
30. K Guillemin, MA Krasnow. The hypoxic response: huffing and HIFing. Cell 89:9–12, 1997.
31. E Maltepe, MC Simon. Oxygen, genes, and development: an analysis of the role of hypoxic gene regulation during murine vascular development (in process citation). J Mol Med 76:391–401, 1998.
32. RH Wenger, M Gassmann. Oxygen(es) and the hypoxia-inducible factor-1. Biol Chem 378: 609–616, 1997.
33. KM Sinjorgo, OM Steinebach, HL Dekker, AO Muijsers. The effects of pH and ionic strength on cytochrome c oxidase steady-state kinetics reveal a catalytic and a noncatalytic interaction domain for cytochrome c [published erratum appears in Biochim Biophys Acta 851:491 1986]. Biochim Biophys Acta 850:108–115, 1986.
34. E Necas, J Neuwirt. The effect of inhibitors of energy metabolism on erythropoietin production. J Lab Clin Med 79:388–396, 1972.
35. CW Pugh, CC Tan, RW Jones, PJ Ratcliffe. Functional analysis of an oxygen-regulated transcriptional enhancer lying 3′ to the mouse erythropoietin gene. Proc Natl Acad Sci USA 88: 10553–10557, 1991.
36. NS Chandel, GRS Budinger, PT Schumacker. Molecular oxygen modulates cytochrome c oxidase function. J Biol Chem 271:18672–18677, 1996.
37. NS Chandel, GR Budinger, SH Choe, PT Schumacker. Cellular respiration during hypoxia: role of cytochrome oxidase as the oxygen sensor in hepatocytes. J Biol Chem 272:18808–18816, 1997.
38. NS Chandel, E Maltepe, C Mathieu, E Goldwasser, MC Simon, PT Schumacker. Mitochondrial reactive oxygen species trigger hypoxia induced transcription. Proc Natl Acad Sci USA 95:11715–11720, 1998.
39. JF Turrens, A Alexandre, AL Lehninger. Ubisemiquinone is the electron donor for superoxide formation by complex III of heart mitochondria. Arch Biochem Biophys 237:408–414, 1985.
40. V Srinivas, X Zhu, S Salceda, R Nakamura, J Caro. Hypoxia-inducible Factor 1 α (HIF-1α) is a non-heme iron protein. J Biol Chem 273:18019–18022, 1998.
41. D Shweiki, A Itin, D Soffer, E Keshet. Vascular endothelial growth factor induced by hypoxia may mediate hypoxia-initiated angiogenesis. Nature 359:843–845, 1992.
42. G Helmlinger, F Yuan, M Dellian, RK Jain. Interstitial pH and pO2 gradients in solid tumors in vivo: high-resolution measurements reveal a lack of correlation. Nature Med 3:177–182, 1997.
43. N Ferrara, T Davis-Smyth. The biology of vascular endothelial growth factor. Endocrinol Rev 18:4–25, 1997.
44. N Ferrara, K Carver-Moore, H Chen, M Dowd, L Lu, KS O'Shea, L Powell-Braxton, KJ Hillan, MW Moore. Heterozygous embryonic lethality induced by targeted inactivation of the VEGF gene. Nature 380:439–442, 1996.
45. JA Forsythe, BH Jiang, NV Iyer, F Agani, SW Leung, RD Koos, GL Semenza. Activation of vascular endothelial growth factor gene transcription by hypoxia-inducible factor 1. Mol Cell Biol 16:4604–4613, 1996.
46. TG Graeber, JF Peterson, M Tsai, K Monica, AJ Fornace Jr, AJ Giaccia. Hypoxia induces accumulation of p53 protein, but activation of a G1-phase checkpoint by low-oxygen conditions is independent of p53 status. Mol Cell Biol 14:6264–6277, 1994.

47. A Krtolica, JW Ludlow. Hypoxia arrests ovarian carcinoma cell cycle progression, but invasion is unaffected. Cancer Res 56:1168–1173, 1996.
48. RJ Muschel, EJ Bernhard, L Garza, WG McKenna, CJ Koch. Induction of apoptosis at different oxygen tensions: evidence that oxygen radicals do not mediate apoptotic signaling. Cancer Res 55:995–998, 1995.
49. C Schmaltz, PH Hardenbergh, A Wells, DE Fisher. Regulation of proliferation-survival decisions during tumor cell hypoxia. Mol Cell Biol 18:2845–2854, 1998.
50. TG Graeber, C Osmanian, T Jacks, DE Housman, CJ Koch, SW Lowe, AJ Giaccia. Hypoxia-mediated selection of cells with diminished apoptotic potential in solid tumours (see comments). Nature 379:88–91, 1996.
51. WG An, M Kanekal, MC Simon, E Maltepe, MV Blagosklonny, LM Neckers. Stabilization of wild-type p53 by hypoxia-inducible factor 1alpha. Nature 392:405–408, 1998.
52. P Carmeliet, Y Dor, JM Herbert, D Fukumura, K Brusselmans, M Dewerchin, M Neeman, F Bono, R Abramovitch, P Maxwell, CJ Koch, P Ratcliffe, L Moons, RK Jain, D Collen, E Keshet. Role of HIF-1alpha in hypoxia-mediated apoptosis, cell proliferation and tumour angiogenesis. Nature 394:485–490, 1998.
53. HI Swanson, WK Chan, CA Bradfield. DNA binding specificities and pairing rules of the Ah receptor, ARNT, and SIM proteins. J Biol Chem 270:26292–26302, 1995.
54. EC Hoffman, H Reyes, FF Chu, F Sander, LH Conley, BA Brooks, O Hankinson. Cloning of a factor required for activity of the Ah (dioxin) receptor. Science 252:954–958, 1991.
55. DW Nebert, FJ Gonzales. P450 genes: structure, evolution, and regulation. Annu Rev Biochem 56:945–993, 1987.
56. C Telakowski-Hopkins, R King, C Pickett. Glutathione S-transferase Ya subunit gene: identification of regulatory elements required for basal level and inducible expression. Proc Natl Acad Sci USA 85:1000–1004, 1988.
57. GL Wang, GL Semenza. General involvement of hypoxia-inducible factor in transcriptional response to hypoxia. Proc Natl Acad Sci USA 90:4304–4308, 1993.
58. HF Bunn, RO Poyton. Oxygen sensing and molecular adaptation to hypoxia. Physiol Rev 76: 839–885, 1996.
59. H Tian, SL McKnight, DW Russell. Endothelial PAS domain protein 1 (EPAS1), a transcription factor selectively expressed in endothelial cells. Genes Dev 11:72–82, 1997.
60. JB Hogenesch, WK Chan, VH Jackiw, RC Brown, YZ Gu, M Pray-Grant, GH Perdew, CA Bradfield. Characterization of a subset of the basic-helix-loop-helix-PAS superfamily that interacts with components of the dioxin signaling pathway. J Biol Chem 272:8581–8593, 1997.
61. K Hirose, M Morita, M Ema, J Mimura, H Hamada, H Fujii, Y Saijo, O Gotoh, K Sogawa, Y Fujii-Kuriyama. cDNA cloning and tissue-specific expression of a novel basic helix-loop-helix/PAS factor (Arnt2) with close sequence similarity to the aryl hydrocarbon receptor nuclear translocator (Arnt). Mol Cell Biol 16:1706–1713, 1996.
62. JB Hogenesch, YZ Gu, S Jain, CA Bradfield. The basic-helix-loop-helix-PAS orphan MOP3 forms transcriptionally active complexes with circadian and hypoxia factors. Proc Natl Acad Sci USA 95:5474–5479, 1998.
63. E Maltepe, JV Schmidt, D Baunoch, CA Bradfield, MC Simon. Abnormal angiogenesis and responses to glucose and oxygen deprivation in mice lacking the protein ARNT. Nature 386: 403–407, 1997.
64. KR Kozak, B Abbott, O Hankinson. ARNT-deficient mice and placental differentiation. Dev Biol 191:297–305, 1997.
65. NV Iyer, LE Kotch, F Agani, SW Leung, E Laughner, RH Wenger, M Gassmann, JD Gearhart, AM Lawler, AY Yu, GL Semenza. Cellular and developmental control of O2 homeostasis by hypoxia-inducible factor 1 alpha. Genes Dev 12:149–162, 1998.
66. HE Ryan, J Lo, RS Johnson. HIF-1a is required for solid tumor formation and embryonic vascularization. EMBO J. 17:3005–3015, 1998.
67. D Shweiki, M Neeman, A Itin, E Keshet. Induction of vascular endothelial growth factor

expression by hypoxia and by glucose deficiency in multicell spheroids: implications for tumor angiogenesis. Proc Natl Acad Sci USA 92:768–772, 1995.
68. I Stein, M Neeman, D Shweiki, A Itin, E Keshet. Stabilization of vascular endothelial growth factor mRNA by hypoxia and hypoglycemia and coregulation with other ischemia-induced genes. Mol Cell Biol 15:5363–5368, 1995.
69. P Carmeliet, V Ferreira, G Breir, S Pollefeyt, L Kieckens, M Gertenstein, M Fahrig, A Vandenhoeck, K Harpal, C Eberhard, C Declercq, J Pawling, ML, D Collen, W Risau, A Nagy. Abnormal blood vessel development and lethality in embryos lacking a single VEGF allele. Nature 380:435–439, 1996.
70. PH Maxwell, GU Dachs, JM Gleadle, LG Nicholls, AL Harris, IJ Stratford, O Hankinson, CW Pugh, PJ Ratcliffe. Hypoxia-inducible factor-1 modulates gene expression in solid tumors and influences both angiogenesis and tumor growth. Proc Natl Acad Sci USA 94:8104–8109, 1997.

10
Scatter Factor as a Tumor Angiogenesis Factor

Eliot M. Rosen
Long Island Jewish Medical Center, New Hyde Park, and Albert Einstein College of Medicine, Bronx, New York

Katrin Lamszus, Saijun Fan, and Itzhak D. Goldberg
Long Island Jewish Medical Center, New Hyde Park, New York

John Laterra
The Johns Hopkins School of Medicine and Kennedy-Krieger Research Institute, Baltimore, Maryland

Peter J. Polverini
University of Michigan School of Dentistry, Ann Arbor, Michigan

I. INTRODUCTION

Scatter factor (SF) is a heparin-binding glycoprotein that dissociates (''scatters'') sheets of epithelium (Stoker et al., 1987; Rosen et al., 1989). SF is identical to hepatocyte growth factor (HGF) (Weidner et al., 1991a; Bhargava et al., 1992), a serum mitogen that stimulates the proliferation of primary hepatocyte cell cultures and functions as an hepatotrophic factor for liver repair (Miyazawa et al., 1989; Nakamura et al., 1989). SF is a heterodimeric protein consisting of a 60-kDa α-chain containing four kringle domains and a 30-kDa trypsin-like β-chain (Gherardi et al., 1989; Rosen et al., 1990b; Weidner et al., 1990). SF has 38% amino acid sequence identity to the proenzyme plasminogen (Nakamura et al., 1989) but lacks protease activity (Rosen et al., 1990b) due to replacement of two essential amino acids at the catalytic triad of the β-chain. SF is synthesized as a 728–amino acid precursor (preproSF); intracellular cleavage of a 31–amino acid signal peptide results in its secreted single-chain form (proSF), which is biologically inactive (Lokker et al., 1992). Extracellular cleavage of proSF at ^{494}arg-^{495}val yields active two-chain SF. HGF activator, a novel serine protease homologous to coagulation factor XII (Hagemann factor), may be a physiological cleavage enzyme for proSF (Miyazawa et al., 1993). This enzyme is produced in zymogen form; it may be activated by a proteolytic cascade initiated by tissue injury (Miyazawa et al., 1994). ProSF is also cleaved by the plasminogen activators [urokinase (uPA) and tissue plasminogen activator (tPA)], but at stoichiometric rather than enzymatic quantities. The SF receptor is the protein product of the c-Met proto-oncogene (Bottaro et al., 1991), a transmembrane tyrosine kinase (TK) expressed predominantly by epithelia (Gonzatti-Haces et al., 1988).

SF transduces three major classes of cellular actions in vitro: motility, growth, and morphogenesis (reviewed in Rosen et al., 1994b). In addition to cell dissociation, SF induces random movement of isolated epithelial cells, chemotactic (gradient-directed) migration, migration from carrier beads to flat surfaces, and invasion through extracellular matrix proteins (Rosen et al., 1990a–c, 1991a; Weidner et al., 1990; Bhargava et al., 1992; Li et al., 1994). SF stimulates the mRNA and protein expression of both urokinase (uPA) and uPA receptor (uPAR) (Pepper et al., 1992; Grant et al., 1993; Rosen et al., 1994a). The net effect is to increase the amount of uPA bound to uPAR on the cell surface. Receptor-bound uPA on the cell surface is catalytically active and is thought to mediate focal degradation of the extracellular matrix necessary to clear a path for invading cells (Saksela and Rifkin, 1988). Thus, SF appears to be able to "switch on" a program of cell activities for invasion. SF is mitogenic for various normal cell types, including epithelial cells, vascular endothelial cells, and melanocytes (Kan et al., 1991; Rubin et al., 1991; Halaban et al., 1992). SF is also a potent morphogen. SF induces Madin-Darby canine kidney (MDCK) epithelial cells incubated in collagen I gels to organize into a network of branching tubules with the proper apical-basolateral polarity (Montesano et al., 1991; Santos et al., 1993). Similarly, SF induces mammary epithelial cells to form ductlike structures (Tsarfaty et al., 1992). Thus, SF can activate specific programs of cell differentiation depending upon the cell type and environment.

II. BLOOD VESSEL WALL CELLS AS TARGETS FOR SCATTER FACTOR: IN VITRO ANGIOGENIC ACTIVITIES

A. Vascular Endothelial Cells (ECs)

During the early stages of angiogenesis in vivo, ECs from pre-existing small vessels (usually venules that lack a smooth muscle covering) focally degrade the subendothelial basement membrane, migrate into the interstitium toward an angiogenic stimulus, and form capillary sprouts (Folkman, 1985). Sprouting ECs proximal to the migrating tip proliferate; and the EC sprouts organize into an anastamosing network of capillary tubes. Finally, these ECs synthesize new basement membrane. Adhesion of smooth muscle cells (SMCs) and pericytes and synthesis of new basement membrane are associated with the termination of the angiogenic response (Antonelli-Orlidge et al., 1989). Stimulation of EC motility, proliferation, and capillary-like tube formation in vitro correlate with the induction of angiogenesis in vivo (Folkman, 1985).

Both large-vessel and microvessel ECs express the c-Met receptor and are biologically responsive to SF (Rosen et al., 1990b,c, 1991b; Bussolino et al., 1992; Grant et al., 1993; Naidu et al., 1994). SF is chemotactic to ECs and stimulates random motility, as demonstrated in assays using microwell modified Boyden chambers (Rosen et al., 1990b, 1991b). In addition, SF induces the migration of ECs cultured on microcarrier beads from the beads to flat culture surfaces (Rosen et al., 1990b,c). In chemoinvasion assays, SF induces penetration of ECs through porous filters coated with Matrigel, a reconstituted matrix of basement membrane (Rosen et al., 1991b). Maximal stimulation of EC motility and invasion is typically observed at SF concentrations of 2–20 ng/mL.

In addition to motility, SF stimulates DNA synthesis and proliferation of some EC types, including HUVEC and human omental microvessel ECs (Rubin et al., 1991; Morimoto et al., 1991). Capillary-like tube formation is an independent property of ECs, not directly related to motility or proliferation (Grant et al., 1989). When ECs are plated onto a surface of Matrigel, they cease proliferation, extend long cytoplasmic processes, and begin to organize into capillary-like tubes. SF stimulates capillary-like tube formation in various EC lines by up to five to tenfold, as determined by computerized digital image analysis of stained cultures (Rosen et al., 1991b; Grant et al., 1993). SF also induces increased expression of uPA activity by EC cultures (Rosen et al., 1991b; Grant et al., 1993). Most of the SF-induced urokinase activity is cell-associated rather than secreted. The majority of cell-associated urokinase is bound to uPAR on the cell surface, where it is well positioned to mediate focal degradation of extracellular matrix proteins, a prerequisite for invasion. Taken together, these findings indicate that SF can induce most or all of the phenotypic characteristics expected of ECs undergoing angiogenesis.

B. Vascular Smooth Muscle Cells (SMCs) and Pericytes

In vitro, various types of SMCs produce SF at rates comparable to those of high-producer human lung fibroblasts (Rosen et al., 1989, 1990a). The biological and chemical properties of SMC-derived SF are very similar to those of fibroblast-derived SF, and it is likely that these molecules are identical (Rosen et al., 1990a,b).

In addition to producing SF, SMCs also express the c-Met receptor and are responsive to SF. Thus, SMCs within microvessel walls in psoriatic plaques and in tumor microvasculature stained positively for c-Met protein (Grant et al., 1993; Jin et al., 1997; Joseph et al., 1995), suggesting that they are potential target cells for SF. Recruitment of SMCs and pericytes (which are generally regarded as microvascular SMCs) is an essential component of angiogenesis. These cells stabilize newly formed vessels, thus contributing to termination of the angiogenic process (Antonelli-Orlidge et al., 1989). Cultured pericytes from bovine retina express c-Met mRNA, as determined by reverse-transcriptase PCR analysis (unpublished results), consistent with the finding that pericytes express c-Met protein in vivo. Moreover, we found that SF stimulates the proliferation of bovine aortic SMC and bovine retinal pericytes in vitro (Rosen and Goldberg, 1995), consistent with the putative role of SMCs in angiogenesis and the presence of SMCs in new microvessels induced by SF (see below).

C. Cytoprotective Activity of Scatter Factor

We have reported that SF protects various lines of epithelial and carcinoma cells against apoptosis (a form of programmed cell death) induced by cytotoxic agents, including DNA-damaging agents doxoruxicin (adriamycin), ionizing radiation (6 MeV therapeutic x-rays), and ultraviolet (UV-C) radiation (Fan et al., 1998). The cytoprotective activity of SF for these cell types is due in part to upregulation of cell-survival pathways, including enhanced expression of Bcl-X_L, an antiapoptotic protein closely related to Bcl-2. In agreement with these findings, SF also protected bovine aortic endothelial cells (BAECs) against cytotoxicity caused by Adriamycin (a DNA topoisomerase IIα inhibitor) and nitrogen mustard (a DNA alkylating agent) (see Table 1A); and SF blocked apoptosis induced by adriamycin (data not shown). In contrast to these two DNA-damaging agents, SF gave little or no protection of ECs against taxol and vincristine, microtubule poisons that arrest cells in mitosis (Table 1A).

In another set of experiments, we examined the ability of SF to protect BAECs against Adriamycin in the present of thrombospondin-1 (TSP-1), an endothelial cell inhibitor and anti-angiogenic protein. At a dose of 4 nM, TSP-1 alone did not affect cell viability, although TSP-1 slightly enhanced the toxicity of Adriamycin. However, TSP1 markedly

Table 1 Effect of Scatter Factor (SF) on Cell Viability of Bovine Aortic Endothelial Cells Treated with Cytotoxic Agents Using the MTT Assay[a]

A. SF protects endothelial cells against DNA-damaging agents but not mitotic spindle poisons

		Cell viability (%)	
Agent	Dose (μM)	−SF	+SF
DNA-damaging agents			
Adriamycin (ADR)	5	40 ± 2	95 ± 2
Nitrogen mustard	15	12 ± 3	60 ± 2
Mitotic spindle poisons			
Taxol	10	43 ± 2	44 ± 3
Vincristine	10	51 ± 2	57 ± 1

B. Thrombospondin-1 (TSP1) reverses SF-mediated protection of endothelial cells

Additions SF (ng/mL)	TSP-1 (ng/mL)	ADR (μM)	Cell viability (%)
0	0	0	100
100	0	0	102 ± 3
0	4	0	96 ± 3
0	0	5	40 ± 2
0	4	5	33 ± 1
100	0	5	90 ± 2
100	4	5	60 ± 3

[a] Subconfluent proliferating cells in 96-well dishes were preincubated without (−) or with (+) SF (100 ng/mL × 48 hr) and ± TSP-1 in cell culture medium (DMEM plus 5% fetal calf serum). Cells were then sham-treated or treated with agent (2 hr for Adriamycin and nitrogen mustard, 24 hr for Taxol and vincristine); washed twice to remove drug; incubated for another 72 hr in fresh drug-free culture medium; and assayed for MTT dye conversion by spectrophotometry (ref). Values represent cell viability relative to sham-treated control cells and are means ± SEMs of N = 10 replicate wells.

inhibited SF-mediated protection against ADR, resulting in a large decrease in cell viability of cells treated with (SF + Adriamycin + TSP-1) as compared with (SF + Adriamycin) (Table 1B). At > 4 nM, TSP1 alone caused decreases in cell viability of BAECs. Corresponding results were obtained in assays of apoptosis using agarose gel electrophoresis (data not shown).

These findings are interesting for two reasons. First, they suggest another mechanism by which SF may promote angiogenesis: i.e., by blocking endothelial cell apoptosis, a mechanism by which microvessels within granulation tissue regress during the later phases of wound healing (ref). Second, the findings may have therapeutic implications for cancer treatment. Thus, SF present in tumors (see below) may protect not only tumor cells but also the tumor neovasculature against the cytotoxic actions of some types of chemotherapy and radiation therapy. If our findings hold true for ECs in general, then mitotic spindle poisons may be more effective in targeting the vasculature of tumors with a high endogenous SF content.

III. IN VIVO ANGIOGENIC ACTIVITY OF SCATTER FACTOR

We used two different assays, the mouse Matrigel assay and the rat cornea assay, to demonstrate the ability of SF to induce new blood vessel formation in vivo (Grant et al., 1993; Naidu et al., 1994). In the former, different doses of SF were mixed with 0.5 mL of Matrigel in the liquid state at 4°C. The Matrigel was injected subcutaneously into either XID nude beige mice or C57/BL mice, allowing the formation of a solid gel that retained SF and permitted prolonged exposure of the surrounding tissues to it. Animals were sacrificed after 10 days, and the ingrowth of blood vessels into the Matrigel plugs was quantitated by computerized image analysis of tissue sections stained with Masson's trichrome. Angiogenesis assessed at day 10 increased in a dose-dependent manner from 2–200 ng/mL of SF, up to four to five times control values. Responses were similar in nude mice

Table 2 Inhibition of Scatter Factor (SF)–Induced Angiogenesis in the Rat Corneal Neovascularization Assay by SF α-Chain Peptides NK1 and NK2 and by c-Met IgG

Agent(s) added	Positive assays (%)
A. Inhibition of SF-mediated angiogenesis by NK1 and NK2 peptides	
Control assays	
Hydron + PBS	0/2 (0)
bFGF (50 ng)	2/2 (100)
rhSF (50 ng)	3/3 (100)
NK1 (25 ng)	0/2 (0)
NK2 (25 ng)	0/2 (0)
rhSF ± peptide	
bFGF (50 ng) + NK1 (25 ng)	2/2 (100)
bFGF (50 ng) + NK2 (25 ng)	2/2 (100)
rhSF (50 ng) + NK1 (25 ng)	0/2 (0)
rhSF (50 ng) + NK2 (25 ng)	0/2 (0)
B. Inhibition of SF-mediated angiogenesis by c-Met IgG	
Control assays	
Hydron + PBS	0/2 (0)
bFGF (100 ng)	2/2 (100)
rhSF (100 ng)	3/3 (100)
c-Met IgG (1 μg)	0/4 (0)
rhSF ± c-Met IgG	
rhSF (100 ng)	3/3 (100)
rhSF (100 ng) + c-Met IgG (250 ng)	3/6 (50)
rhSF (100 ng) + c-Met IgG (500 ng)	1/5 (20)
rhSF (100 ng) + c-Met IgG (1 μg)	0/5 (0)

Abbreviations: bFGF, basic fibroblast growth factor; PBS, phosphate-buffered saline; rhSF, recombinant human scatter factor. c-Met IgG is a chimeric antibody consisting of the ligand-binding domain of the c-Met receptor fused to a human IgG heavy chain backbone. Assays were performed and responses scored as described before (Tolsma et al., 1993).

and C57/BL mice. Inflammatory responses were not observed in nude mice and were found only at supramaximal SF doses (2000 ng/mL) in C57/BL mice.

In the second assay, SF was dissolved in Hydron polymer and dried Hydron pellets were placed in surgically created pockets about 1.5 mm from the limbus of the avascular rat cornea. Animals were perfused with colloidal carbon and sacrified after 7 days; the growth of new vessels from the limbus toward the pellet was them assessed. SF induced dose-dependent corneal neovascularization similar to that observed in the murine assay (Grant et al., 1993). The maximal SF-induced responses were similar in intensity to those induced by basic fibroblast growth factor (bFGF) or vascular endothelial growth factor (VEGF). Antibodies against SF blocked SF-induced angiogenesis but did not affect bFGF-induced angiogenesis. Inflammatory responses (monocyte/macrophage infiltration) were observed only at supramaximal doses of SF. Therefore, SF appears to be at least as potent an inducer of angiogenesis as bFGF or VEGF.

Our findings indicate that recruitment of inflammatory cells does not play a major role in SF-induced angiogenesis. However, in the mouse Matrigel assay, histologic sections prepared at early times (days 2–3), revealed many SMC/pericyte-like cells present in the Matrigel. Moreover, at higher doses of SF, histological sections prepared on day 10 revealed SMCs in some of the newly formed vessels in the Matrigel (Grant et al.,

Table 3 Inhibition of Chemotaxis of Vascular Endothelial Cells Toward Scatter Factor (SF) by Recombinant Human SF α-Chain Peptides NK1 and NK2[a]

Agent(s) added (ng/ml)	Cells per ten 40 grids (N)	% Inhibition
None	18 ± 3 (4)	—
rhSF (20)	136 ± 12 (4)	0
NK1 (20)	21 ± 4 (3)	—
NK1 (50)	26 ± 11 (3)	—
NK1 (200)	12 ± 4 (3)	—
rhSF (20) + NK1 (20)	130 ± 10 (3)	8
rhSF (20) + NK1 (50)	101 ± 4 (3)	36
rhSF (20) + NK1 (200)	56 ± 6 (3)	63
NK2 (20)	15 ± 3 (3)	—
NK2 (50)	9 ± 3 (3)	—
NK2 (200)	17 ± 10 (3)	—
rhSF (20) + NK2 (20)	94 ± 1 (3)	33
rhSF (20) + NK2 (50)	100 ± 10 (3)	23
rhSF (20) + NK2 (200)	54 ± 5 (3)	68

[a] Chemotactic migration of human dermal microvascular endothelial cells toward different combinations of rhSF and NK1 or NK2 was assayed using 48-well microwell-modified Boyden chambers, as described before (see Lamszus et al., 1997). Migration values (means ± standard errors) represent the number of cells that had migrated across porous collagen-coated filters over a 4-hr assay incubation period. Values in parentheses represent the number of replicate wells tested.

1993). Thus, SF-induced angiogenesis appears to be mediated by direct effects on ECs and, in addition, by direct or indirect effects on SMCs.

Several peptide fragments of the SF α-chain, designated NK1 and NK2, are generated by alternative mRNA processing. These peptides consist of the N-terminal prekringle region plus the first kringle domain (NK1) or the first two kringle domains (NK2) of the SF α-chain (Chan et al., 1991; Lokker et al., 1992; Cioce et al., 1996). Recombinant human NK1 and NK2 peptides have been found to bind with high affinity to the c-Met receptor and to exert partial agonistic and partial antagonistic activity compared with full-length heterodimeric SF (Cioce et al., 1996; Lokker et al., 1992). Utilizing the rat corneal neovascularization assay, we found that neither NK1 nor NK2 were capable of inducing angiogenesis, but each of these peptides inhibited angiogenesis induced by full-length SF (Table 2A). A chimeric antibody consisting of a human IgG heavy chain fused to bivalent c-Met receptor extracellular binding domains [designated ''*c-Met* IgG'' (Mark et al., 1992)] also blocked SF-mediated angiogenesis in the rat cornea (Table 2B). One microgram of *c-Met* IgG ablated the angiogenic action of 100 ng/mL of rhSF, a dose sufficient to induce a maximal angiogenic response. NK1 and NK2 also inhibited chemotactic migration of human microvascular endothelial cells toward SF (Table 3), indicating that these may be useful reagents for specifically blocking SF-induced angiogenesis.

IV. ROLE OF SCATTER FACTOR AS A TUMOR ANGIOGENESIS FACTOR

A. Clinical Studies

Recent studies suggest that tumor angiogenesis, indicated by large numbers of microvessels in tumor stroma, is an independent indicator of poor prognosis in patients with invasive breast cancer and other tumor types (Weidner et al., 1991b, 1992; Bosari et al., 1992). Consistent with the idea that SF may function as a tumor angiogenesis factor, we have found that SF and c-Met expression are both upregulated during the transition from low-grade noninvasive to high-grade invasive tumors in breast cancer, urothelial bladder cancer, and gliomas (Jin et al., 1997a, b; Joseph et al., 1995; Lamszus et al., 1998); Rosen et al., 1996; Rosen et al., 1997; Yao et al., 1996; Ole Schmidt et al., in press). In addition, other investigators have reported that a high SF content of invasive breast cancers is a powerful and independent predictor of tumor relapse and death (Yamashita et al., 1994).

To investigate the potential role of SF as a tumor angiogen, we measured the SF content and the content of von Willebrand's factor (VWF), an endothelial-specific marker protein, in extracts of invasive human breast cancers. Studies of over 200 tumor specimens revealed a strong and highly significant correlation between the SF content and the VWF content of the tumor extract (Jin et al., 1997a, b) (Table 4). These results suggest that the overall level of tumor vascularity, as indicated by the VWF content, is related to the SF content in biopsy specimens of breast cancers. It was initially though that tumor-derived SF was produced by tumor-associated stromal cells (e.g., fibroblasts, SMCs, macrophages), since the vast majority of carcinoma cell lines, including a number of human breast cancer cell lines, do not produce any detectable SF in vitro. However, more recent studies indicate that breast carcinoma cells express both SF protein and SF mRNA in vivo, suggesting that loss of the ability to produce SF may be an artifact of cell culture (Jin et al., 1997; Tuck et al., 1996; Wang et al., 1994).

Table 4 Relationship Between Context of von Willebrand factor (VWF) and Scatter Factor (SF) in Extracts of Invasive Human Breast Cancers[a]

SF content range (ng/mg)	VWF content (ng/mg protein)
I. 0.00–0.50	26 ± 7 (50)
II. 0.51–1.00	49 ± 8 (62)
III. 1.01–2.00	93 ± 27 (57)
IV. 2.01–up	203 ± 51 (55)

[a] Human breast cancer tissue specimens frozen in OCT compound were thawed; and tissue extracts were prepared as described before (ref). Values of SF and VWF content were determined by two-antibody ELISAs and were normalized to the protein content of the extract (ref). Values represent means ± standard errors (number of individual tumors assayed). Statistical comparisons of VWF vs. SF content range were made using two-tailed Student's t-tests: 0–1.00 vs. 1.01–up, $p < 0.001$; 0–0.50 vs. 0.51–1.00, $p = 0.040$; 0–0.50 vs. 2.01–up, $p < 0.001$; 1.01–2.00 vs. 2.01–up, $p = 0.061$.

B. Kaposi's Sarcoma (KS)

KS is a multicellular neoplasm characterized by a major component of EC proliferation and angiogenesis. The epidemic form of KS occurs frequently in homosexual males with acquired immunodeficiency syndrome (AIDS). AIDS-KS is usually regarded as a multifocal cytokine-dependent tumor rather than a metastasizing cancer. Recently, SF was found to mediate several functions that may be related to the pathogenesis of AIDS-related KS: (a) conversion of normal human ECs into a KS tumor cell-like phenotype; (b) autocrine stimulation of KS tumor cell proliferation; and (c) angiogenic activity of conditioned medium from retrovirally infected human T lymphocyte cultures (Naidu et al., 1994). Furthermore, immunoreactive SF and c-Met receptor were detected in a variety of cell types within KS tumor sections. These findings suggest that SF may function as a mediator of EC proliferation and angiogenesis in AIDS-related KS lesions.

C. Experimental Animal Models

We have examined the ability of SF to stimulate angiogenesis in an animal model of human breast cancer. We transfected human SF cDNA into MDAMB231 human breast cancer cells so that they overexpressed SF. Tumors generated from SF-transfected (SF+) cell clones grew much more rapidly in the mammary fat pads of nude mice than did tumors generated control-transfected (SF−) clones (Lamszus et al., 1997). Extracts from the SF+ tumors contained 50-fold a higher SF content than did extracts prepared from SF− control tumors. In contrast to the in vivo results, in vitro proliferation rates and cloning efficiencies were no greater in SF+ vs. SF− clones. We investigated the discrepancy between the in vivo and in vitro growth results and found that (a) microvessel densities were significantly higher in SF+ vs. SF− tumors; (b) chemotactic activity for capillary endothelial cells was greater in conditioned media and primary tumor extracts from SF+ vs SF− clones; (c) angiogenic activity in the rat cornea assay was much higher in SF+ vs. SF− tumor extracts; (d) chemotactic and angiogenic activity in SF+ tumor extracts was not explained by secondary changes in levels of other regulators; and (e) those activities were neutralized by an anti-SF monoclonal (Lamszus et al., 1997). Thus, the increased tumor growth rates of SF+ MDAMB231 cells appears to be due in part to an angiogenic response induced by SF.

In addition to the breast cancer model, we have developed several rodent models for the orthotopic growth of glial tumors. We found that 9L rat glioma cells transfected with human SF cDNA grew much more rapidly than control-transfected cells both as brain tumors and as subcutaneous tumors in syngeneic host rats (Laterra et al., 1997). Analysis of brain tumor sections using a digital image analysis system revealed significantly higher microvessel densities in SF+ tumors than in SF− control tumors. SF+ tumors also showed significantly higher expression of urokinase, an enzyme required for proteolytic degradation of extracellular matrix and invasion of migrating ECs during the angiogenic process. In parallel studies, SF also induced increased growth and microvessel formation in brain tumors from SF-transfected human glioblastoma cells (U373MG) grown as brain tumors in immunocompromised (SCID) mice, as compared with control U373MG cells (Laterra et al., 1997). These findings suggest that overexpression of SF stimulates in vivo tumor growth, in part by conferring an angiogenic tumor phenotype.

V. CONCLUSIONS

The cytokine scatter factor is potent inducer of endothelial cell activation and of angiogenesis. Studies of clinical tumor specimens and experimental animal models of human tumors suggest that the expression of SF confers a more invasive and angiogenic phenotype.

ACKNOWLEDGMENTS

This work was supported in part by grants from the U.S. Public Health Service (CA-64869 and CA-64416). We are indebted to Dr. Ralph Schwall, Genentech, Inc. (South San Francisco, CA) for providing sheep antiserum against SF, recombinant human two-chain human SF, and c-Met IgG. We are grateful to Drs. Jeffrey Rubin, Donald Bottaro, Stephen Stahl, and Paul Wingfield, Laboratory of Cellular and Molecular Biology, National Cancer Institute (Bethesda, MD) for providing recombinant human truncated SF peptides (NK1 and NK2).

REFERENCES

Antonelli-Orlidge A, Smith SR, D'Amore PA. Influence of pericytes on capillary endothelial cell growth. Am Rev Resp Dis 140:1129–1131, 1989.

Bhargava M, Joseph A, Knesel J, Halaban R, Li Y, Pang S, Goldberg I, Setter E, Donovan MA, Zarnegar R, Michalopoulos GA, Nakamura T, Faletto D, Rosen EM. Scatter factor and hepatocyte growth factor: activities, properties, and mechanism. Cell Growth Differ 3:11–20, 1992.

Bosari S, Lee AK, DeLellis RA, Wiley BD, Heatley GJ, Silverman ML. Microvessel quantitation and prognosis in invasive breast carcinoma. Hum Pathol 23:755–761, 1992.

Bottaro DP, Rubin JS, Faletto DL, Chan AM-L, Kmiecik TE, Vande Woude GF, Aaronson SA. Identification of the hepatocyte growth factor receptor as the c-met proto-oncogene product. Science 251:802–804, 1991.

Bussolino F, DiRenzo MF, Ziche M, Bocchieto E, Olivero M, Naldini L, Gaudino G, Tamagnone L, Coffer A, Comoglio PM. Hepatocyte growth factor is a potent angiogenic factor which stimulates endothelial cell motility and growth. J Cell Biol 119:629–641, 1992.

Chan AM-L Rubin JS, Bottaro DP, Hirschfield DW, Chedid M, Aaronson SA: Identification of a competitive antagonist encoded by an alternative transcript. Science 254:1382–1385, 1991.

Cioce V, Csaky KG, Chan AM-L, Bottaro DP, Taylor WG, Jemsen R, Aaronson SA, Rubin JS. HGF/NK1 is a naturally occurring HGF/SF variant with partial agonist/antagonist activity. J Biol Chem 271:13110–13115, 1996.

Fan S, Wang J-A, Yuan R-Q, Rockwell S, Andres J, Zlatapolskiy A, Goldberg ID, Rosen EM. Scatter factor protects epithelial and carcinoma cells against apoptosis induced by DNA-damaging agents. Oncogene 17:131–141, 1998.

Folkman J. Tumor angiogenesis. Adv Cancer Res 43:175–203, 1985.

Gherardi E, Gray J, Stoker M, Perryman M, Furlong R. Purification of scatter factor, a fibroblast-derived basic protein which modulates epithelial interactions and movement. Proc Natl Acad Sci USA 86:5844–5848, 1989.

Gohda E, Matsunaga T, Kataoka H, Yamamoto I. TGF-β is a potent inhibitor of hepatocyte growth factor secretion by human fibroblasts. Cell Biol Int Reports 16:917–926, 1992.

Gonzatti-Haces M, Seth A, Park M, Copeland T, Oroszlan S, Vande Woude GF. Characterization of the TPR-MET oncogene p65 and the MET protooncogene p140 protein tyrosine kinases. Proc Natl Acad Sci USA 85:21–25, 1988.

Grant DS, Tashiro KI, Segui-Real B, Yamada Y, Martin GR, Kleinman HK. Two different laminin domains mediate the differentiation of human endothelial cells into capillary-like structures in vitro. Cell 58:933–943, 1989.

Grant DS, Kleinman HK, Goldberg ID, Bhargava M, Nickoloff BJ, Polverini P, Rosen EM. Scatter factor induces blood vessel formation in vivo. Proc Natl Acad Sci USA 90:1937–1941, 1993.

Halaban R, Rubin J, Funasaka Y, Cobb M, Boulton T, Faletto D, Rosen E, Chan A, Yoko K, White W, Cook C, Moellmann G. Met and hepatocyte growth factor/scatter factor signal transduction in normal melanocytes and melanoma cells. Oncogene 7:2195–2206, 1992.

Jin L, Fuchs A, Schnitt SS, Yao Y, Joseph A, Lamszus K, Park M, Goldberg ID, Rosen EM. Expression of SF and c-met receptor in benign and malignant breast tissue. Cancer 79:749–760, 1997a.

Jin L, Yuan R-Q, Fuchs A, Yao Y, Joseph A, Schwall R, Guida A, Schnitt SJ, Hastings HM, Andres J, Turkel G, Polverini PJ, Goldberg ID, Rosen EM. Expression of IL-1β in human breast cancer. Cancer 80:421–434, 1997b.

Joseph A, Weiss GH, Jin L, Fuchs A, Chowdhury S, O'Shaughnessy P, Goldberg ID, Rosen EM. Expression of scatter factor in human bladder carcinoma. J Natl Cancer Inst 87:372–377, 1995.

Kan M, Zhang GH, Zarnegar R, Michalapoulos G, Myoken Y, McKeehan WL, Stevens JL. Hepatocyte growth factor-hepatopoietin A stimulates the growth of rat proximal tubule epithelial cells (rpte), rat non-parenchymal liver cells, human melanoma cells, mouse keratinocytes, and stimulates anchorage-independent growth of SV40-transformed rpte. Biochem Biophys Res Commun 174:331–337, 1991.

Lamszus K, Jin L, Fuchs A, Shi YE, Chowdhury S, Yao Y, Polverini PJ, Goldberg ID, Rosen EM. Scatter factor stimulates tumor growth and tumor angiogenesis in human breast cancers in the mammary fat pads of nude mice. Lab Invest 76:339–353, 1997.

Lamszus K, Schmidt NO, Jin L, Laterra J, Zagzag D, Way D, Witte M, Weinand M, Goldberg ID, Westfal M, Rosen EM. Scatter factor promotes motility of human glioma and neuromicrovascular endothelial cells. Int J Cancer 75:19–28, 1998.

Laterra J, Rosen E, Nam M, Rangathan S, Fielding K, Johnston P. Scatter factor/hepatocyte growth factor expression enhances human glioblastoma tumorigenicity and growth. Biochem Biophys Res Commun 235:743–747, 1997.

Laterra J, Nam M, Rosen EM, Rao JS, Lamszus K, Johnston P. Scatter factor/hepatocyte growth factor gene transfer enhances glioma growth and angiogenesis in vivo. Lab Invest 76:565–577, 1997.

Li Y, Bhargava MM, Joseph A, Jin L, Rosen EM, Goldberg ID. The effect of scatter factor and hepatocyte growth factor on motility and morphology of non-tumorigenic and tumor cells. In Vitro Cell Dev Biol 30A:105–110, 1994.

Lokker NA, Mark MR, Luis EA, Bennett GL, Robbins KA, Baker JB, Godowski PJ. Structure-function analysis of hepatocyte growth factor: identification of variants that lack mitogenic activity yet retain high affinity receptor binding. EMBO J 11:2403–2410, 1992.

Mark MR, Lokker NA, Zioncheck TF, Luis EA, Godowski PJ. Expression and characterization of hepatocyte growth factor receptor-IgG fusion proteins: effects of mutations in the potential proteolytic cleavage sites on processing and ligand binding. J Biol Chem 267:26166–26171, 1992.

Miyazawa K, Tsubouchi H, Naka D, Takahashi K, Okigaki M, Arakaki N, Nakayama S, Hirono S, Sakiyama O, Gohda E, Daikuhara Y, Kitamura N. Molecular cloning and sequence analysis of cDNA for human hepatocyte growth factor. Biochem Biophys Res Commun 163:967–973, 1989.

Miyazawa K, Shimomura T, Kitamura A, Kondo J, Morimoto Y, Kitamura N. Molecular cloning and sequence analysis of the cDNA for a human serine protease responsible for activation of hepatocyte growth factor: structural similarity of protease precursor to blood coagulation factor XII. J Biol Chem 268:10024–10028, 1993.

Miyazawa K, Shimomura T, Naka D, Kitamura N. Proteolytic activation of hepatocyte growth factor in response to tissue injury. J Biol Chem 269:8966–8970, 1994.

Montesano R, Matsumoto K, Nakamura T, Orci L. Identification of a fibroblast-derived epithelial morphogen as hepatocyte growth factor. Cell 67:901–908, 1991.

Morimoto, A., Okamura, K., Hamanaka, R., Sato, Y., Shima, N., Higashio, K., Kuwano, M. Hepatocyte growth factor modulates migration and proliferation of microvascular endothelial cells in culture. Biochem Biophys Res Commun 179:1042–1049, 1991.

Naidu YM, Rosen EM, Zitnik R, Goldberg I, Park M, Naujokas M, Polverini PJ, Nickoloff BJ. Role of scatter factor in the pathogenesis of AIDS-related Kaposi's sarcoma. Proc Natl Acad Sci USA 91:5281–5285, 1994.

Nakamura T, Nishizawa T, Hagiya M, Seki T, Shimonishi M, Sugimura A, Shimizu S. Molecular cloning and expression of human hepatocyte growth factor. Nature 342:440–443, 1989.

Ole Schmidt N, Westfal M, Hagel C, Stavrou D, Ergun S, Rosen EM, Lamszus K. Expression of vascular endothelial growth factor, hepatocyte growth factor/scatter factor, and basic fibroblast growth factor in human gliomas and their relation to angiogenesis. Int J Cancer. In press.

Pepper MS, Matsumoto K, Nakamura T, Orci L, Montesano R. Hepatocyte growth factor increases urokinase-type plasminogen activator (u-PA) and u-PA receptor expression in Madin-Darby canine kidney epithelial cells. J Biol Chem 267:20493–20496, 1992.

Rosen EM, Goldberg ID, Kacinski BM, Buckholz T, Vinter DW. Smooth muscle releases an epithelial cell scatter factor which binds to heparin. In Vitro Cell Dev Biol 25:163–173, 1989.

Rosen EM, Laterra J, Joseph A, Jin L, Fuchs A, Way D, Witte M, Weinand M, Goldberg ID. Scatter factor expression and regulation in human glial tumors. Int J Cancer 67:248–255, 1996.

Rosen EM, Meromsky L, Setter E, Vinter DW, Goldberg ID. Smooth muscle-derived factor stimulates mobility of human tumor cells. Invasion Metastasis 10:49–64, 1990a.

Rosen, EM, Meromsky L, Setter E, Vinter DW, Goldberg ID. Purification and migration-stimulating activities of scatter factor. Proc Soc Exp Biol Med 195:34–43, 1990b.

Rosen EM, Meromsky L, Setter E, Vinter DW, Goldberg ID. Quantitation of cytokine-stimulated migration of endothelium and epithelium by a new assay using microcarrier beads. Exp Cell Res 186:22–31, 1990c.

Rosen EM, Goldberg ID: Scatter factor and angiogenesis. Adv Cancer Res 67:257–279. 1995.

Rosen EM, Goldberg ID, Liu D, Setter E, Donovan MA, Bhargava M, Reiss M, Kacinski BM.

Tumor necrosis factor stimulates epithelial tumor cell motility. Cancer Res 57:5315–5321, 1991a.

Rosen EM, Grant D, Kleinman H, Jaken S, Donovan MA, Setter E, Luckett PM, Carley W. Scatter factor stimulates migration of vascular endothelium and capillary-like tube formation. In: Goldberg ID, Rosen EM, eds. Cell Motility Factors. Basel: Birkhauser-Verlag, 1991b, pp. 76–88.

Rosen EM, Jaken S, Carley W, Setter E, Bhargava M, Goldberg ID. Regulation of motility in bovine brain endothelial cells. J Cell Physiol 146:325–335, 1991c.

Rosen EM, Joseph A, Jin L, Yao Y, Chau M-H, Fuchs A, Gomella L, Hasings H, Goldberg ID, Weiss GH. Urinary and tissue levels of scatter factor in transitional cell carcinoma of bladder. J Urol 157:72–78, 1997.

Rosen EM, Knesel J, Goldberg ID, Bhargava M, Joseph A, Zitnik R, Wines J, Kelley M, Rockwell S. Scatter factor modulates the metastatic phenotype of the EMT6 mouse mammary tumor. Int J Cancer 57:706–714, 1994a.

Rosen EM, Nigam SK, Goldberg ID. Mini-Review: Scatter factor and the c-met receptor: a paradigm for mesenchymal: epithelial interaction. J Cell Biol 127:1783–1787, 1994b.

Rubin JS, Chan AM-L, Bottaro DP, Burgess WH, Taylor WG, Cech AC, Hirschfield DW, Wong J, Miki T, Finch PW, Aaronson SA. A broad spectrum human lung fibroblast-derived mitogen is a variant of hepatocyte growth factor. Proc Natl Acad Sci USA 88:415–419, 1991.

Saksela O, Rifkin DB. Cell-associated plasminogen activation: regulation and physiologic functions. Annu Rev Cell Biol 4:93–126, 1988.

Santos OFP, Moira LA, Rosen EM, Nigam SK. Modulation of HGF-induced tubulogenesis and branching by multiple phosphorylation mechanisms. Dev Biol 159:538–545, 1993.

Stoker M, Gherardi E, Perryman M, Gray J. Scatter factor is a fibroblast-derived modulator of epithelial cell mobility. Nature 327:238–242, 1987.

Tolsma SS, Volpert OV, Good DJ, Frazier WA, Polverini PJ, Bouck N. Peptides derived from two separate domains of the matrix protein thrombospondin-1 have antiangiogenic activity. J Cell Biol 122:497–511, 1993.

Tsarfaty I, Resau JH, Rulong S, Keydar I, Faletto D, Vande Woude GF. The met proto-oncogene receptor and lumen formation. Science 257:1258–1261, 1992.

Tuck AB, Park M, Sterns EE, Boag A, Elliott BE. Coexpression of hepatocyte growth factor and receptor (Met) in human breast carcinoma. Am J Pathol 148:225–232, 1996.

Wang Y, Selden AC, Morgan N, Stamp GW, Hodgson HJ. Hepatocyte growth factor/scatter factor expression in human mammary epithelium. Am J Pathol 144:675–682, 1994.

Weidner KM, Behrens J, Vandekerckhove J, Birchmeier W. Scatter factor: molecular characteristics and effect on invasiveness of epithelial cells. J Cell Biol 111:2097–2108, 1990.

Weidner KM, Arakaki N, Vandekereckhove J, Weingart S, Hartmann G, Rieder H, Fonatsch C, Tsubouchi H, Hishida T, Daikuhara Y, Birchmeier W. Evidence for the identity of human scatter factor and human hepatocyte growth factor. Proc Natl Acad Sci USA 88:7001–7005, 1991a.

Weidner N, Semple JP, Welch WR, Folkman J. Tumor angiogenesis and metastasis—correlation in invasive breast carcinoma. New Engl J Med 324:1–8, 1991b.

Weidner N, Folkman J, Pozza F, Bevilaqua P, Allred EN, Moore DH, Meli S, Gasparini G. Tumor angiogenesis: a new significant and independent prognostic indicator in early stage breast carcinoma. J Natl Cancer Inst 84:1875–1887, 1992.

Yamashita J, Ogawa M, Yamashita S, Nomura K, Kuramoto M, Saishoji T, Sadahito S. Immunoreactive hepatocyte growth factor is a strong and independent predictor of recurrence and survival in human breast cancer. *Cancer Res* 54:1630–1633, 1994.

Yao Y, Jin L, Fuchs A, Joseph A, Hastings H, Goldberg ID, Rosen EM. Scatter factor protein levels in human breast cancers: Clinicopathologic and biologic correlations. *Am J Pathol* 149:1707–1717, 1996.

11
Thromboxane A_2 as an Antiangiogenic Factor

Anthony W. Ashton
Albert Einstein College of Medicine, Bronx, New York

J. Anthony Ware
Albert Einstein College of Medicine and Montefiore Medical Center, Bronx, New York

I. INTRODUCTION

Thromboxane A_2 was first identified as a vasospastic substance produced during anaphylaxis and was named rabbit aorta-contracting substance (RCS). RCS was subsequently shown to be a bioactive metabolite of PGG_2, released during arachidonic acid induced platelet aggregation (Ogletree, 1987). Hamberg et al. (1975) renamed RCS thromboxane A_2; this eicosanoid is metabolized from PGH_2, as are many other PG products, but is distinguished by its short biological half life and potent vasoconstrictive properties. As such, TXA_2 has an autocoid/paracrine role in regulating cellular processes in clinical conditions in which it is produced and released. TXA_2 is synthesized primarily by activated platelets; however, during certain kinds of stimulation, TXA_2 is also synthesized by endothelial cells and leukocytes. The biological effects of TXA_2 are manifest through binding to a cell-surface receptor that is a member of the G protein–linked receptor family. Most of the attention devoted to TXA_2 and its receptors has focused on its role in promoting platelet aggregation, initiating thrombosis, and inducing vasoconstriction. The recognition that TXA_2 receptors (TP) are expressed on endothelial cells (Clesham et al., 1992; Kent et al., 1993) has raised speculation that TXA_2 might influence vascular growth. This possibility has important clinical implications, as many conditions in which TXA_2 is elevated are also those in which new vessel formation is an essential component of the disease's pathophysiology. Our laboratory has been interested in the possibility that manipulation of TXA_2 or its receptor (TP) can affect angiogenesis. To devise strategies to test this possibility and to develop therapeutics aimed at exploiting TXA_2 for the purposes of altering angiogenesis, it is necessary to consider the pathophysiological conditions in which TXA_2 is elevated and the biochemical and molecular mechanisms by which the eicosanoid, its receptor, and their interaction are regulated.

II. PATHOPHYSIOLOGICAL CONDITIONS INVOLVING TXA_2 AND ITS RECEPTOR

As noted above, the eicosanoid TXA_2 is synthesized and released by activated platelets, monocytes, and cells of the vessel wall. Both local and systemic elevations of TXA_2 are important mediators of several vascular and thrombotic diseases, including myocardial infarction (Fitzgerald et al., 1986), reocclusion following coronary thrombosis (Fitzgerald et al., 1988), pregnancy-induced hypertension (Meagher et al., 1993), and ischemia of several organs. TXA_2 is an important mediator of bronchial asthma and pulmonary hypertension (Mikhail et al., 1998) and increases following angioplasty in peripheral vascular atherosclerosis (Rossi et al., 1997). Antagonism of TXA_2's action lessens reperfusion-mediated injury of several organs, including the liver (Suehiro et al., 1996) and the heart. Thus, pharmacological manipulation of TXA_2 and its receptor has been a central part of the therapeutic regimen for ischemic and thrombotic diseases. One of the mechanisms by which TXA_2 mediates unstable coronary artery disease, acute myocardial infarction, and hypertensive diseases is that it causes platelet aggregation and release of vasoconstrictive and promitogenic substances from platelet granules. In stimulated human monocytes, TXA_2 has an autocrine function. It is not only produced by stimulated monocytes but also facilitates TNF-α and IL-β production by these cells (Kameda et al., 1995). TXA_2 also has several effects on endothelial cells, including induction of prostacyclin release (Clesham et al., 1992; Kent et al., 1993), and, as discussed below, inhibition of endothelial cell migration and tube formation.

Regulation of the TXA_2 receptor (TP) is a consequence of several disease states as well; increased numbers of TPs have been detected on the platelets of patients with pregnancy-induced hypertension (Liel et al., 1993), acute ischemia (Dorn, et al., 1990; Modesti, et al., 1995), and postoperative graft thrombosis (Brothers et al., 1998). Thus, it is clear that not only TXA_2 is increased during ischemia and other vascular conditions but that the number of TP receptors is increased as well. Because of this increase, it is of particular relevance from a clinical standpoint to determine if the TXA_2 produced and released during ischemia affects formation of new vessels by either angiogenesis or collateral recruitment.

III. BIOCHEMISTRY/ENZYMOLOGY OF COX-1 AND -2 AND TXA_2 SYNTHASE

Arachidonic acid is the primary source of lipid-based biological mediators in all mammals. Arachidonic acid is liberated from phospholipids by the action of phospholipases (Figure 1) and is metabolized by three independent metabolic pathways:

1. The cyclooxygenase pathway, resulting in prostaglandins and thromboxanes
2. The lipoxygenase pathway, resulting in leukotrienes, lipoxins, hydroxyeicosatetraenoic, and hydroperoxyeicosatetraenoic acids
3. The cytochrome P-450 monooxygenase pathway, resulting in epoxides and hydroxyeicosatetraenoic acids

The action of cyclooxygenase on arachidonic acid produces the intermediate metabolite PGH_2, the central prostanoid from which all other prostaglandins and thromboxanes are enzymatically derived. Under the influence of prostaglandin endoperoxide synthase, arachidonic acid undergoes oxygenation and cyclization of a pentane ring, leading to the formation of the unstable metabolite PGG_2 (cyclooxygenase activity). The peroxidase reaction then catalyzes the reduction of PGG_2 and the endoperoxide PGH_2 is formed (Miyamoto et al., 1976; Ohki et al., 1979). The prostaglandins PGD_2, PGE_2, and PGI_2 are derived from PGH_2 by their respective synthases, with PGD_2 being cytosolic and PGE_2 and PGI_2 synthases membrane-bound. PGD_2 and PGE_2 synthases require glutathione as a cofactor (Hecker et al., 1989).

Thromboxanes are derived from PGH_2 by the action of thromboxane synthase (Figure 1), a membrane-resident cytochrome P-450–like heme-thiolate enzyme, that adds an oxan:oxetane ring to the structure of PGH_2, producing equimolar quantities of hydroxyheptadecatrienoic acid and malondialdehyde in the process. The rate-limiting step in the production of TXA_2 is the induction of COX and the production of PGH_2, as TXA_2 production is dependent on the kinetics of COX induction in monocytes (Orlandi et al., 1994). TXA_2 spontaneously hydrolyzes to the dormant hemiacetal TXB_2 (with a $t_{1/2} = 30$ sec) (Hecker et al., 1989). The chemical inactivation is irreversible. TXA_2 synthase is not constitutive but is rapidly induced. TXA_2 synthase has a K_m for PGH_2 of 10 μM and can synthesize up to 3198 TXA_2 molecules per min (Hecker et al., 1987; Xia et al., 1993). Studies using inhibitors of protein synthesis show that the protein has a half-life of 16.4 h; however, the biological half-life is only 10.5 hr (Nusing et al., 1993). This supports the findings of Jones and Fitzpatrick (1991), who found that TXA_2 synthase undergoes inactivation during the course of normal enzyme function, most likely due to tight, nondestructive association of substrate or product with the heme prosthetic group. Inactivation by destructive radicals also occurs; however, this reduces only the apoenzyme function of the protein. During the formation of TXA_2, PGH_2 binds with the 9 oxygen atom oriented toward the heme group of TXA_2 synthase. The heme group must be in the ferric state to cleave the endoperoxide bond. Cleavage of the endoperoxide bond and immediate formation of the heme thiolate ligand complex gives rise to the alkoxy radical. The C11-C12 bond is then broken by ''β-scission'' leading to formation of the allylic radical. At this point the metabolite has two possible fates: if the radical goes through Fe(IV)-oxidation to the carbocation followed by ionic rearrangement then TXA_2 is formed. If the radical is allowed to decompose, then HHT and malondialdehyde are formed (Hecker and Ullrich, 1989). This mechanism also explains the formation of HHT and malondialdehyde with TXA_2 synthase in vitro.

The determining factor of the specific form of prostanoid/thromboxane synthesized is the relative amount of the necessary enzymes, thus leading to tissue specific expression patterns for many of these products; e.g., TXA_2 is synthesized mainly by platelets and macrophages while PGI_2 is synthesized predominantly by vascular smooth muscle and endothelial cells (Hecker et al., 1989).

IV. STRUCTURE/GENOMICS OF TP AND TXA_2 SYNTHASE

A. TXA_2 Synthase

Thromboxane A_2 is synthesized from the precursor PGH_2 via the action of TXA_2 synthase (Fig. 1). The TXA_2 synthase enzyme is a membrane-bound protein monomeric hemoprotein containing one heme group per molecule of enzyme. The mRNA for TXA_2 synthase yields a protein composed of 534 amino acids (M_r 58–60 kDa), which corresponds to the observed size of the protein isolated from platelets and lung (Yokoyama et al., 1991;

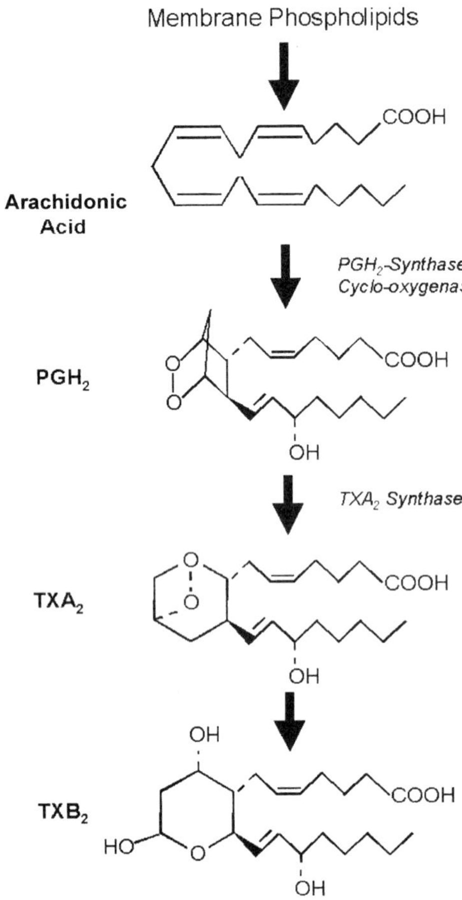

Figure 1 Biochemical synthesis of TXA_2 from membrane phospholipids via the sequential action of cyclooxygenase and TXA_2 synthase.

Haurand and Ullrich, 1985). TXA_2 synthase is constitutively expressed in monocytes and platelets, cells in which the highest expression is seen (Orlandi et al., 1994; Ullrich and Nusing, 1990). The enzyme is closely analogous to cytochrome P-450 enzymes as determined by optical and EPR spectroscopy (Haurand and Ullrich, 1985; Nusing et al., 1990). Indeed, the enzyme is 34–36% homologous to members of the cytochrome P-450 III gene family and contains a conserved Cys (Cys^{480}). In TXA_2 synthase, this sequence is in the C-terminus (residues 472–492) (Yokoyama et al., 1991). Cys^{480} along with Arg^{478}, Phe^{127}, and Asn^{110} are predicted to be involved in binding the heme prosthetic group, including formation of the heme thiolate ligand in the enzyme's active site (Wang et al., 1996; Ullrich and Nusing, 1990). Indeed, mutants with Cys^{480} mutated to Ser are devoid of function (Xia et al., 1993). The Arg^{408} provides a hydrophobic environment for substrate binding and Arg^{413} is thought to form part of the active site (Wang et al., 1996). Other residues important for enzyme function are Arg^{410} and Glu^{433} and Ile^{346} (Wang et al., 1996). TXA_2 synthase contains five potential asparagine-linked glycosylation sites; however, the enzyme has never been found in a glycosylated state (Yokoyama et al., 1991). According to hydropathy plots of the protein's primary structure, there are 10 hydrophobic regions in the protein (residues 8–46, 73–84, 86–90, 139–164, 209–239, 278–283, 331–364, 388–398, 425–438 and 480–498) with residues 8–46 comprising the membrane resident regions of the protein.

The gene for TXA_2 synthase (TBXAS1) was isolated from an embryonic cDNA library using TXA_2 synthase cDNA as a probe (Miyata et al., 1994). The human gene spans more than 75 kb and consists of 13 exons and 12 introns with the splice donor/acceptor sites conforming to the GT/AG rule. Exons vary in size from 53–315 bp. Analysis of the 5′ flanking region (1.2 kb) revealed consensus binding sites for the transcription factors AP-1, AP-2, GATA-1, PEA-3, LF-A1, myb, xenobiotic- and cAMP-response elements, a basic transcription element and a CCAAT box. Multiple transcription start sites exist, with the major initiation site 142 bp upstream of the translational start site. Gene transcription is negatively regulated by *cis*-acting elements between bases −1562 and −248 in the 5′ flanking region and it is believed that regions outside 1.6 kb in the 5′ region result in tissue-specific expression of TXA_2 synthase (Tazawa et al., 1996). The gene has been assigned to chromosome 7 (q33–q34) by in situ hybridization.

The cDNA was first cloned in 1991 by RT-PCR from human platelets and was found to be 1.9 kb in length, which contained 80 bases of 5′ sequence with a 1602 base open reading frame (Yokoyama et al., 1991). The second ATG sequence has been determined as the translational initiation codon, based on the surrounding Kozak sequence (1984). A polyadenylation signal AATAAA appears at residues 1761–1766. A second cDNA was later cloned from lung; however, this encoded an inactive enzyme devoid of a catalytic site (Xia et al., 1993). Northern blotting revealed ubiquitous expression that was particularly high in leukocytes, spleen, lung, and liver (Miyata et al., 1994). Two TXA_2 synthase mRNA species, 2.2 and 3.7 kb, were found in RNA from platelets and lung, with the 3.7-kb band thought to represent an immature form of the mRNA. TXA_2 synthase mRNA is induced twofold by treatment with phorbol esters (Yokoyama et al., 1991). Restriction mapping and reporter studies with the TXA_2 synthase promotor reveal that bases −1 to −286 are essential for gene transcription and sequences within the region −365 to −665 suppress transcription (Lee et al., 1996). This repressive sequence functions independently of orientation and position and was defined as a ''silencer sequence.'' The AP-1 binding site was shown to be critical for promotor function, although AP-1 is not currently believed to be the transcription factor responsible. Binding of undefined *cis*-acting elements was identified for the repression of promotor activity within 5.5 kb of the transcriptional start site.

B. TXA_2 Receptor

The biological activity of TXA_2 is mediated through binding to a cell surface–associated receptor. The thromboxane receptor (TP), like other prostaglandin receptors, is a ''serpentine'' or seven-membrane-spanning G protein–linked receptor. The expression of the receptor is widespread in that it is expressed in leukocytes, macrophages, and vascular cells, suggesting a role for TXA_2 as a potential autocrine factor in inflammatory and ischemic/thrombotic conditions; however, it is also expressed in neural tissue, nephrons, and epithelium (Raychowdhury et al., 1994; Hirata et al., 1991; Blackman et al., 1998; Asano et al., 1996; Simmons et al., 1993; Chiyotani et al., 1992). In humans, two isoforms exist, which have been designated TP-α and -β. Both receptor isoforms have been identified on platelets and endothelial cells (Hirata et al., 1996; Yokota et al., 1998), while the TP-α isoform only has also been identified in smooth muscle cells, monocytes, immature thymocytes, monocytes, nephrons, and oligodendrocytes (Simmons et al., 1993; Ushikubi et al.,

1993; Blackman et al., 1998; Asano et al.; 1996, Chiyotani et al., 1992). In other species studied—including rat, mouse and bovine—only one isoform, homologous mostly to TP-α, has been identified thus far (Dorn II et al., 1997; Kitanaka et al., 1995; Namba et al., 1992; Muck et al., 1998).

The two receptor isoforms, TP-α and -β are derived from alternate splicing of a single gene (Raychowdhury et al., 1994) (Figure 2). The TP-α gene product was first cloned from megakaryocytic and placental tissue, and the TP-β isoform was isolated from a human endothelial cell library (Hirata et al., 1991; Raychowdhury et al., 1994). The two isoforms are identical in structure except for the C-terminal domain, which diverges after residue 328. TP-α cDNA from the placental cDNA library was 2.9 kb in length, with an open reading frame of 1029 bp, and encodes a protein 343 residues in length (37.5 kDa). Northern analysis showed that the major mRNA species for TP-α was 2.8 kb, with a lesser amount of a 3.5 kb transcript also present. The TP-β cDNA from the human endothelial cell cDNA library contained an open reading frame of 1224 bp, which encoded a protein 407 amino acids in length (approximately 44 kDa) (Raychowdhury et al., 1994). The transcript was shown to be 2.8 kb and was downregulated with stimulation by agonist. The receptor presents a number of sites for N-linked glycosylation (Hirata et al., 1991) that could result in a mass for the TP-α receptor closer to 55 kDa on SDS-PAGE analysis, which is that suggested by protein detection (Borg et al., 1994).

The TXA_2 receptor is encoded by a single gene (TBXA2R) located on chromosome 19p13.3 (Schwengel et al., 1993; Nusing et al., 1993b). TBXA2R spans over 15 kb and consists of three exons and two introns. The first intron, intron 1, exists in the 5′-noncoding region, 83 bp upstream from the ATG start site, and is 6.3 kb long. Intron 2, with a length of 4.3 kb, is located at the end of the sixth transmembrane region, thereby separating it from the downstream coding sequences, including the seventh transmembrane region and the 3′-untranslated region (Nusing et al., 1993b). PCR-based analysis revealed that the gene contained an additional 115-bp-long noncoding exon, exon 1b, located in intron 1. The existence of two receptor isoforms in humans is a product of a novel splicing event within exon 1b. This splicing event is the result of the retention of the intron 4 in TP-β (Nusing et al., 1993b). Residues 982–985 conform to a consensus intron donor-acceptor sequence. However, TP-α results from the failure to utilize this potential splicing site, resulting in encoding of the cytoplasmic domain that is an intron in TP-β. The two transcripts reconverge at 1641–1643 before the polyadenylation site, which is shared by both transcripts.

Regulation of receptor expression is also a complex process. The 5′ flanking region site of TBXA2R, within 2.1 kb of the transcription start site, contains two transcription initiation sites starting in two different putative promoter regions, one upstream of each of the two exons (D'Angelo et al., 1995). The gene does not contain a typical TATA or CAAT consensus sequence. Primer extension analysis identified multiple alternate transcription initiation sites that mapped to a novel region in intron 1 of the TP gene (termed exon 1B). The two promotors in the TP gene, P1 and P2, each contain a specific complement of transcription factor binding sites. The major promoter, P1, positioned 5′ to exon 1, contains two AP2 and four Sp1 elements and a single TRE site. Promotor P2 (located downstream of exon 1B) contains two AP2, two Sp1, three APRRE, one GRET, and one c-Myc site. The different complements of transcription factor binding sites may be the basis for the differential cell- and tissue-specific expression of TP-α and β. Polymorphisms have also been found in the TP gene, especially in the P2 promotor region. Gene transcription is upregulated by phorbol esters with the phorbol ester response element −1.91 to −1.95 kb upstream from the transcriptional start site, in a cysteine-rich region among a cluster of AP-2 binding sites. The transcription factor involved has been identified as Sp1, which binds to a consensus sequence in the opposite strand of DNA. Gene transcription is inhibited by exposure to lipopolysaccharide although the transcription factors involved have not been identified (Ishiguro et al., 1994).

V. STRUCTURAL DETERMINANTS OF TP FOR ITS LIGANDS

The transmembrane domains of the TXA_2 receptor (TP) enclose a region that forms the binding site for TXA_2 and its analogues. The ligand-binding site of the TP is composed of a serine residue from transmembrane domain V, an arginine residue from domain VII and the intervening hydrophobic pocket (Yamamoto et al., 1993). This is consistent with the known properties of TXA_2 and TXA_2 antagonists, having a hydrogen-bonding group such as hydroxyl, a carboxyl group, and a hydrophobic moiety. The residues $Cys^{102,105,183}$ are all important for ligand binding, but not residues $Cys^{11,68,257}$ (Fig. 2) (Chiang et al.,

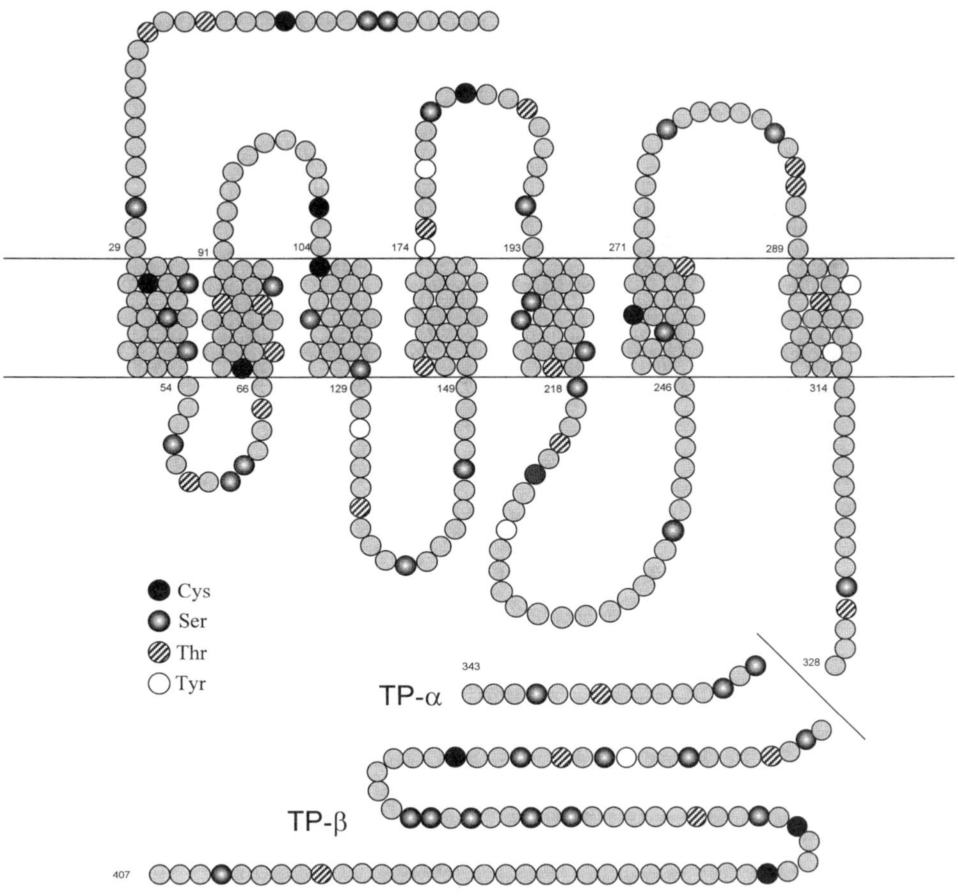

Figure 2 Structure of the TXA_2/PGH_2 receptor: This is a representative model of the TXA_2/PGH_2 receptor deduced from the amino acid sequence. The positions of all cysteine, serine, threonine, and tyrosine residues are indicated.

1996; D'Angelo et al., 1996). Cys^{105} (first extracellular loop) and Cys^{183} (second extracellular loop) are conserved among all prostanoid receptors and are presumed to form a disulfide bond important for receptor stability (D'Angelo et al., 1996). Cys^{102} must play an essential role in ligand-receptor interaction as mutation of this residue results in a decreased binding capacity and low amplitude Ca^{2+} influx. The putative seventh transmembrane segment is highly related in all cloned members of the eicosanoid receptor family and forms a critical portion of the ligand binding pocket for G protein–coupled receptors. Multiple mutations in this region ablate ligand binding to the receptor. Of particular note is the role of Trp^{299} in discrimination of receptor ligands. Mutation of Trp^{299} to leucine allowed a definite discrimination between agonist and antagonist binding sites in competition and saturation binding experiments (Funk et al., 1993). The residues Ser^{201} and Ser^{255} (Fig. 2) distinguish between ligand binding, with mutants decreasing binding of the TXA_2 mimetic IBOP {[1S-(1alpha, 2beta(5Z), 3alpha(1E, 3R), 4alpha]-7-[3-(3-hydroxy-4-(4′-iodophenoxy)-1-butenyl)-7-oxabicyclo-(2.2.1)heptan-2yl]-5′-heptenoic acid} but not the receptor antagonist SQ29548 (Chiang et al., 1996).

TPs from different tissues vary greatly in their affinity for various ligands. Endothelial cells from different vascular beds in different species have dissociation constants (K_D) in the range of 2–10 nM, while rat smooth muscle cells have a K_D of 0.5 nM and human platelet receptors a K_D of 20 nM (Hanasaki et al., 1988; Lasserre et al.; 1992, Hunt et al.; 1992, D'Angelo et al., 1996). The α and β isoforms of the TXA_2 receptor have identical K_D for the same ligand, which is consistent with mutational analysis showing the cytoplasmic tail has little to do with ligand binding (Habib et al., 1997). Although highly homologous, TP-α receptors from human platelets bind IBOP with lower affinity than those from rat vascular smooth muscle cells (D'Angelo et al., 1996). Dorn and colleagues (1997) found that in chimeric receptors, in which the first transmembrane domain of the human receptor was replaced with the rat equivalent, binding of IBOP, but not SQ29548, was enhanced. Detailed analysis has found that the substitution $Val^{37} \rightarrow$ Ala, in combination with either a $Val^{36} \rightarrow$ Leu or $Ala^{40} \rightarrow$ Gly, in the human receptor was responsible for the greater affinity for IBOP. Increased affinity is thought to result from lengthening of the amino acid side chain at position 37 extending the isopropyl group further into the putative binding pocket. The co-substitutions result in shortening of the side chains of adjacent residues making room for the extended side chain at position 37 (Sato et al., 1994)

A series of 3-benzoyl or 3-phenylsulfonyl-2-substituted thiazolidine derivatives have been synthesized and evaluated for their TP receptor-antagonizing effect on (15S)-15-hydroxy-11 alpha,9 alpha-(epoxymethano)prosta-5(Z),13(E)-dienoic acid (U-46619)-induced aggregation of rabbit platelet-rich plasma (PRP). A simple 2-aryl-thiazolidine derivative, 3-benzoyl-2-(4-hydroxy-3-methoxyphenyl)thiazolidine (5a), showed mild TP antagonist activity. Modification of 5a led to 2-chloro-4-[3-(4-chlorophenylsulfonyl)thiazolidin-2-ylmet hyl]phenoxyacetic acid (29d), which showed 10 times more potent TP antagonist activity than did 5a.

VI. TXA_2 MANIPULATION OF INTRACELLULAR SIGNAL TRANSDUCTION PATHWAYS

The downstream signaling processes of the TXA_2 receptor are still largely not understood. Analogizing between the TXA_2 and adrenergic receptor systems permits speculation that the difference in the length of the TP intracellular tails may mediate differential regulation of receptor metabolism and intracellular signaling pathways. Furthermore, the TP-β receptor contains an additional eight serine, three threonine, and a single tyrosine residue not present in the TP-α receptor (Fig. 2). Many residues present in both receptors are currently known to be important for receptor-mediated intracellular signal transduction. Cys^{223} and Thr^{221}, located in intracellular loop three (Fig. 2), are important for transduction of calcium signals (D'Angelo et al., 1996). Mutation of other residues in this region (His^{224} and His^{227}) result in a receptor with a normal phenotype. Hirata and colleagues have found that a mutation in the first cytoplasmic loop of TP-α ($Arg^{60} \rightarrow$ Leu) impairs agonist-induced activation of phospholipase C and adenylyl cyclase (Okuma et al., 1996; Hirata et al., 1994; Hirata et al., 1996).

The TP receptor has seven transmembrane domains and extracellular N and intracellular C termini. The two isoforms differ in the length of their intracellular domains; the cytoplasmic tail of TP-α is 15 amino acids in length, while TP-β has a 79–amino acid tail (Raychowdhury et al., 1994). The differences in the length of the cytoplasmic tails confer associations with different G proteins. Both isoforms are linked to G_q/G_{11} (Kinsella et al., 1997) with TP-α specifically linked to $G_{\alpha s}$ while TP-β is linked to $G_{\alpha i}$ (Hirata et al., 1996). Thus ligand binding to TP-α also induces increases in cAMP and PKA activity, while TP-β inhibits cAMP accumulation. Binding of TXA_2 to its receptor induces phospholipase C-catalyzed hydrolysis of phosphoinositides (Baldassare et al., 1993; Hanasaki et al., 1990), PKC activation (Allison et al., 1986) and induction of cytoskeletal change (Shock et al., 1997). TXA_2 also induces MAP kinase activity and DNA synthesis in vascular smooth muscle (Hanasaki et al., 1990). TXA_2 also upregulates the expression of CD18 on peripheral blood neutrophils and causes adhesion of PMN to endothelium from aorta and pulmonary artery (Wiles et al., 1991). Exposure of thymocytes to TXA_2 mimetics induces the disappearance of the CD4+8+ thymocytes via induction of apoptosis (Ushikubi et al., 1993).

Stimulation of TP receptors leads to receptor phosphorylation and desensitization (Habib et al., 1997). The TP receptors present multiple phosphorylation sites in their cytoplasmic domains and have been shown to become phosphorylated by the ser-thr kinases PKA, PKC and PKG in vitro (Kinsella, et al., 1994; Habib et al., 1997). Prolonged stimulation of TP-α and -β has different effects on desensitization and receptor expression. Stimulation of TP-β upregulates cell surface expression of the receptor (Yukawa et al., 1997) while expression of TP-α remains unchanged. A PKC-dependent mechanism is involved in the desensitization of the response of both receptor subtypes; however, TP-β is more sensitive than TP-α to the activation of PKC (Yukawa, et al., 1997). Phosphorylation of TP by PKA does not influence receptor function markedly (Kinsella, 1994). In platelets, TP phosphorylation by PKG abrogates TXA_2 induced activation (Wang, 1998). However, stimulation of heterologous G protein–linked receptors that stimulate PKC, such as the thrombin receptor, failed to phosphorylate either of the TP isoforms (Habib et al., 1997).

VII. EFFECT OF INHIBITING TXA_2 OR COX ON ANGIOGENESIS

In contrast to the extensive studies to assess whether blocking the actions of TXA_2 or of the TP receptor had an antithrombotic effect, relatively few studies have assessed whether such interference promotes function of ischemic myocardium, and none have assessed whether TXA_2 affects angiogenesis in vivo. A few studies from the 1970s and early 1980s tested the effect of aspirin on coronary collateral flow in animal models of ischemia (Shaub et al., 1982; Capurro et al., 1979; Davenport et al., 1981). These studies found that aspirin enhanced such flow at late periods after infarction but did not differentiate between vasodilation of the collateral vessels and creation of new vessels by angiogenesis or other mechanisms. Since the predominant prostaglandin produced by EC is the vasodilator PGI_2, aspirin-induced vasodilation would seem unlikely; in fact, others have found that aspirin caused collateral vessel constriction (Altman et al., 1993). Thus, the possibility that TXA_2 inhibits angiogenesis and that its selective inhibition might promote new vessel formation is suggested. Very recent studies have determined that prostaglandins produced by tumor cells by a COX-2-dependent mechanism stimulate angiogenesis, and that COX-1 regulates angiogenesis within the endothelial cell (Tsujii et al., 1998). As noted earlier, the COX enzymes mediate TXA_2 formation but also that of many other prostaglandins; TXA_2 is not a major product of EC under most circumstances, implicating PGI_2 or another COX product in promotion of angiogenesis. Some of those, including PGI_2, have recently been shown to be mediators of angiogenesis induced by cytokines such as VEGF (Murohara et al., 1998). Taken together with our preliminary results (see below), these results suggest the possibility that these effects of COX-2 inhibition result from effects on prostaglandins other than TXA_2 and that selective inhibition of TXA_2 may instead facilitate development of new vessels.

A. Effect of TXA_2 and TP Antagonism on Endothelial Migration, Cell Survival, and Tube Formation

Recently, we have performed several studies indicating that the addition of a TXA_2 mimetic (IBOP) to human endothelial cells (HEC) prevents their ability to migrate, as assessed by either the standard Boyden chamber assay and a wound-healing assay. Stimulation of human endothelial cells with IBOP inhibited their migration by 50% and abrogated capillary formation in a three-dimensional culture model of angiogenesis. The effects of IBOP were time- and concentration-dependent, with an IC_{50} of 25 nM (Ashton et al., submitted). The inhibition of tube formation by IBOP was not due to any marked effect on the number of integrins expressed on the surface of human endothelial cells or the cytoskeletal structure. Since gap junction–mediated intercellular communication (GJIC) is required for the migration of a variety of cells, we examined the effects of IBOP on GJIC in HEC. IBOP reduced the propagation of calcium waves and the passage of microinjected dyes between human endothelial cells (Fig. 3). The effects of IBOP on migration and communication are mimicked by the GJIC inhibitor, 18 β-glycerrhetinic acid, with a similar time course. Exposure to IBOP for 24 hr did not reduce the expression of connexin 43, a major endothelial gap junction protein in HEC lysates, nor did it affect its phosphorylation, but it did alter the histological appearance of the gap junctions. We concluded from these observations that TXA_2 prevented HEC migration and angiogenesis in vitro by functioning as a "molecular brake" and that at least one of the mechanisms by which this occurred was by reduced GJIC. Agents that increase cytoplasmic levels of cAMP have previously been shown to induce aggregation of hemichannels into larger-sized plaques (Wang and Rose, 1997; Wang and Rose, 1995); such cAMP-induced enlargement of gap junctions has been associated with decreased GJIC in myocytes (Sakai et al., 1992). However, only TP-α has thus far been shown to increase intracellular cAMP levels (Hirata et al., 1996).

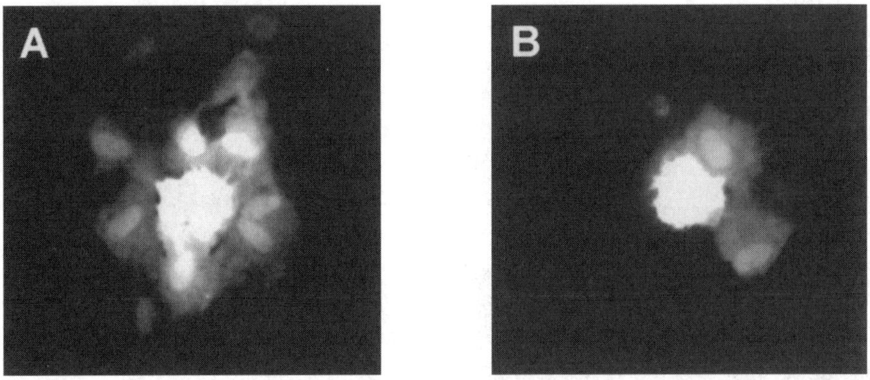

Figure 3 Characterization of GJIC in HEC monolayers during TXA_2 stimulation. Pictorial representation showing the spread of the fluorescent dye, Lucifer yellow, through a confluent monolayer of untreated (A) or IBOP-treated (B) endothelial cells. Photographs were taken using FITC filters (400x).

In further preliminary studies, we discovered that both TP-α and TP-β subtypes of the TXA_2 receptor are expressed in low-passage HEC (Yokota et al., submitted). Thus, we investigated the effect of selectively activating either TP-α or TP-β on the migration of endothelial cells. Rat microvascular endothelial cells, which have no native TP expression, were transfected with TP-α and TP-β cDNAs and were studied in the same model of migration involving a scrape wound to investigate the role of different TP subtypes in the inhibition of endothelial cell migration (Fig. 4). Migration of rat endothelial cells transfected with either TP-α or TP-β were similarly reduced by the addition of IBOP (30 nM). IBOP treatment of the rat cells transfected with the vector alone did not demonstrate altered migration, and TP-α and -β clones migrated similarly to vector only cells in the absence of ligand. Thus, the reduction in migration mediated by IBOP could result from activation of either or both TP subtypes in HEC. This indicates that alternate pathways are used by the two isoforms to achieve the same effects.

We also examined the hypothesis that the antiangiogenic effects of TXA_2 might be mediated through the induction of EC apoptosis. Indeed, we found that the TXA_2 mimetic IBOP increases HEC apoptosis and destroys EC networks formed in vitro on a basement membrane matrix gel (Fig. 5). Initial investigations showed a fivefold increase in the hypodiploid population of IBOP treated HEC from matrix gel cultures as determined by flow cytometry. Morphologically, IBOP-treated cells had condensed or fragmented nuclei characteristic of cells undergoing apoptosis (Fig. 5). To distinguish between the role of the two TP subtypes in EC apoptosis we employed autologous endothelial cells overexpressing the isoforms TP-α and TP-β, as described elsewhere. IBOP caused apoptosis in both TP-α and TP-β as shown by DNA fragmentation (so-called laddering) and TUNEL staining.

In TP-β, both TUNEL staining and DNA fragmentation were suppressed by the adenylyl cyclase activator forskolin, while that of the TP-α EC was not affected by forskolin or by the PKA inhibitor H-89. Thus, we believe that apoptosis of EC may be an important consequence of TXA_2 exposure and that either TP subtype could mediate apoptosis in EC and have suggested that such apoptosis may cause TXA_2-mediated destruction of vascular structures. Apoptotic signaling differs in each subtype in that deacti-

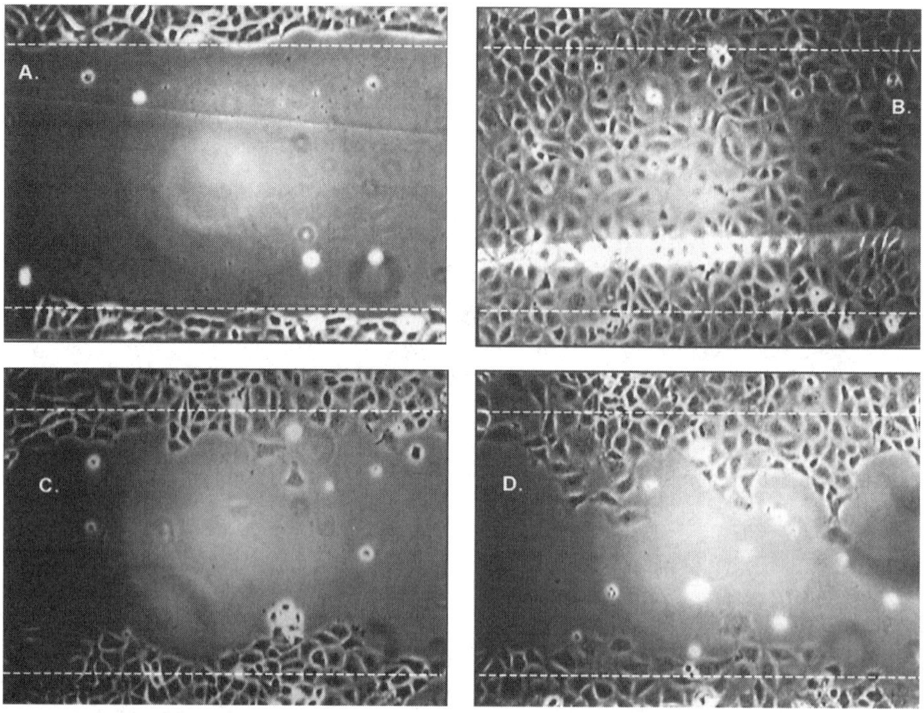

Figure 4 Inhibition of endothelial cell migration by ligand binding to individual TP isoforms. Stable clones of capillary endothelial cells transfected with either TP-α or TP-β cDNAs or empty vector were grown to confluence and wounded by scraping (A). Cells were washed and incubated with 30 nM IBOP for 24 hr. The rate of wound closure for TP-α–(C) and -β–(D) transfected clones was compared to that of clones containing empty vector (B).

vation of the adenylyl cyclase pathway prevents apoptosis caused by TP-β but not by TP-α stimulation.

We considered the possibility that some of the inhibitory effects of TXA_2 on migration, angiogenesis, or apoptosis result from antagonism of the effect of known angiogenic growth factors, especially FGF-2 and VEGF. The angiogenic growth factor FGF-2 enhances migration of the endothelial cells as well as tube formation on matrix gel and protects endothelial cells from apoptosis. TXA_2 strongly antagonized the proangiogenic effects of FGF-2 on endothelial cells in the tube-forming assay on matrix gel, with 50 nM IBOP retarding the proangiogenic effects of FGF-2 at concentrations of FGF-2 greater than 20 ng/mL. IBOP also inhibited the augmentation of HEC migration in the presence of FGF-2. As noted above, IBOP (50 nM) inhibited wound healing by 56%. FGF-2 (20 ng/mL) increased the migration rate of HEC twofold over control, cells exposed to FGF-2 and IBOP together migrated at the same rate as control cells. IBOP prevented the promigratory effects of FGF-2 concentrations as high as 40 ng/mL.

The mechanism by which TXA_2 opposes the effects of FGF-2 on endothelial cell migration and angiogenesis is yet to be determined. The induction of intercellular communication by FGF-2 in migrating endothelial cells is not opposed by costimulation with TXA_2; however, FGF-2 reversed the inhibition of calcium wave propagation by TXA_2 and returned the number of responding cells to that of control cultures. Thus the antimigra-

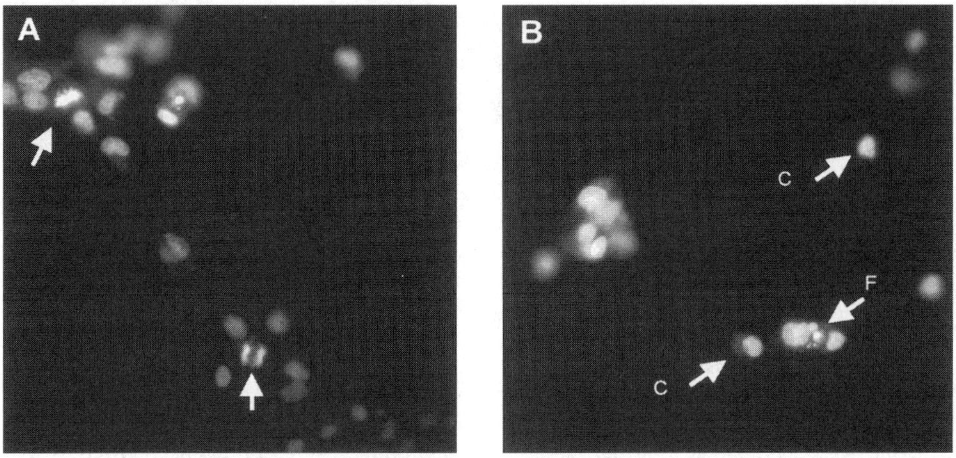

Figure 5 Morphology of HEC on matrix gel cultures. HEC were allowed to form tubes on matrix gel for 16 h, were stained with Hoechst 33342, and fixed. Panel A, control cultures forming tubes between islands of monolayered HEC, arrows denote proliferating cells at ends of the tube. Panel B, morphology of IBOP treated HEC from matrigel cultures *in situ*. Notice the absence of tube formation and random arrangement of cells. Arrows denote cells with either condensed (C) or fragmented nuclei (F), which were considered apoptotic.

tory and antiangiogenic properties of TXA_2 are reversed by FGF-2, possibly through the restoration of GJIC. The ability of TXA_2 to antagonize FGF induced HEC migration does not depend entirely on its ability to prevent its enhancement of GJIC. Addition of IBOP does not decrease the number of FGF-2 high-affinity receptor sites (flg); however, possible pathways distal to cell surface binding and receptor expression are yet to be studied.

VIII. CONCLUSIONS

We have found activation of the TXA_2 receptor to be a powerful antiangiogenic stimulus. The ability to prevent new vessel formation and induce endothelial cell apoptosis in existing vessels makes antagonism of this system an attractive target for intervention and promotion of angiogenesis. Furthermore, the antagonism of this system may enhance the response of endothelium to the proangiogenic stimuli from endogenous growth factors. Understanding the biochemical basis of the effects of TXA_2 on endothelial cell biology and angiogenesis and producing strategies to antagonize them may enhance reendothelialization and vascularization in pathologies in which platelet activation is a hallmark and defects in angiogenesis exist.

REFERENCES

Allison, AC, WJ Kowalski, and B Strulovici. Effects of enprostil on platelets, endothelial cells, and other cell types, and second messenger systems by which these effects are mediated. Am J Med 1986; 81(2A):34–39.

Altman, JD, D Dulas, T Pavek, and RJ Bache. Effect of aspirin on coronary collateral blood flow. Circulation 1993; 87:583–589.

Asano, K, S Taniguchi, A Nakao, T Maruyama, T Watanabe, and K Kurokawa. Distribution of thromboxane A2 receptor mRNA along the mouse nephron segments. Biochem Biophys Res Commun 1993; 226:613–617.

Ashton, AW, R Yokota, G John, XM Zhu, SO Suadicani, DC Spray, and JA Ware. Inhibition of endothelial cell migration, intercellular communication and vascular tube formation by thromboxane A_2 (TXA_2). 1999. Submitted.

Baldassare, JJ, AP Tarver, PA Henderson, WM Mackin, B Sahagan, and GJ Fisher. Reconstitution of thromboxane A2 receptor-stimulated phosphoinositide hydrolysis in isolated platelet membranes: involvement of phosphoinositide-specific phospholipase C-beta and GTP-binding protein Gq. Biochem J 1993; 291(pt 1):235–240.

Blackman, SC, G Dawson, K Antonakis, and GC Le Breton. The identification and characterization of oligodendrocyte thromboxane A2 receptors. J Biol Chem 1998; 273:475–483.

Borg, C, CT Lim, DC Yeomans, JP Dieter, D Komiotis, EG Anderson, and GC Le Breton. Purification of rat brain, rabbit aorta, and human platelet thromboxane A2/prostaglandin H2 receptors by immunoaffinity chromatography employing anti-peptide and anti-receptor antibodies. J Biol Chem 1994; 269:6109–6116.

Brothers, TE, JG Robison, BM Elliott, and JM Boggs. Preoperative thromboxane A2/prostaglandin H2 receptor activity predicts early graft thrombosis. J Vasc Surg 1998; 27:317–325. Discussion, 326–328.

Capurro, NL, KC Marr, R Aamodt, RE Goldstein, and SE Epstein. Aspirin-induced increase in collateral flow after acute coronary occlusion in dogs. Circulation 1979; 59:744–747.

Chiang, N, WM Kan, and HH Tai. Site-directed mutagenesis of cysteinyl and serine residues of human thromboxane A2 receptor in insect cells. Arch Biochem Biophys 1996; 334:9–17.

Chiyotani, A, J Tamaoki, N Sakai, K Isono, M Kondo, and K Konno. Thromboxane A_2 mimetic U46619 stimulates ciliary motility of rabbit tracheal epithelial cells. Prostaglandins 1992; 43: 111–120.

Clesham, G, H Parsaee, S Joseph, JR McEwan, and J MacDermot. Activation of bovine endothelial thromboxane receptors triggers release of prostacyclin but not EDRF. Cardiovasc Res 1992; 26:513–517.

D' Angelo, DD, MG Davis, WA Houser, JJ Eubank, ME Ritchie, and GW Dorn II. Characterization of 5′ end of human thromboxane receptor gene: organizational analysis and mapping of protein kinase C—responsive elements regulating expression in platelets. Circ Res 1995; 77: 466–474.

D' Angelo, DD, JJ Eubank, MG Davis, and GW Dorn II. Mutagenic analysis of platelet thromboxane receptor cysteines: roles in ligand binding and receptor-effector coupling. J Biol Chem 1996; 271:6233–6240.

Davenport, N, RE Goldstein, N Capurro, LC Lipson, RO Bonow, NR Shulman, and SW Epstein. Sulfinpyrazone and aspirin increase epicardial coronary collateral flow in dogs. Am J Cardiol 1981; 47:848–854.

Dorn, GWD, N Liel, JL Trask, DE Mais, ME Assey, and PV Halushka. Increased platelet thromboxane A2/prostaglandin H2 receptors in patients with acute myocardial infarction. Circulation 1990; 81:212–218.

Dorn, GW II, MG Davis, and DD D'Angelo. Structural determinants for agonist binding affinity to thromboxane/prostaglandin endoperoxide (TP) receptors: analysis of chimeric rat/human TP receptors. J Biol Chem 1997; 272:12399–12405.

Fitzgerald, DJ, L Roy, F Catella, and GA FitzGerald. Platelet activation in unstable coronary disease. N Engl J Med 1986; 315:983–989.

Fitzgerald, DJ, J Fragetta, and GA FitzGerald. Prostaglandin endoperoxides modulate the response to thromboxane synthase inhibition during coronary thrombosis. J Clin Invest 1988; 82:1708–1713.

Funk, CD, L Furci, N Moran, and GA Fitzgerald. Point mutation in the seventh hydrophobic domain of the human thromboxane A2 receptor allows discrimination between agonist and antagonist binding sites. Mol Pharmacol 1993; 44:934–939.

Habib, A, R Vezza, C Creminon, J Maclouf, and GA FitzGerald. Rapid, agonist-dependent phosphorylation in vivo of human thromboxane receptor isoforms: minimal involvement of protein kinase C. J Biol Chem 1997; 272:7191–7200.

Hamberg, M, J Svensson, and B Samuelsson. Thromboxanes: a new group of biologically active compounds derived from prostaglandin endoperoxides. Proc Natl Acad Sci USA 1975; 72: 2994–2998.

Hanasaki, K, K Nakano, H Kasai, H Kurihara, and H Arita. Identification of thromboxane A2 receptor in cultured vascular endothelial cells of rat aorta. Biochem Biophys Res Commun 1988; 151:1352–1357.

Hanasaki, K, T Nakano, and H Arita. Receptor-mediated mitogenic effect of thromboxane A2 in vascular smooth muscle cells. Biochem Pharmacol 1990; 40:2535–2542.

Haurand, M, and V Ullrich. Isolation and characterization of thromboxane synthase from human platelets as a cytochrome P-450 enzyme. J Biol Chem 1985; 260:15059–15067.

Hecker, M, M Haurand, V Ullrich, U Diczfalusy, and S Hammarstrom. Products, kinetics, and substrate specificity of homogeneous thromboxane synthase from human platelets: development of a novel enzyme assay. Arch Biochem Biophys 1987; 254:124–135.

Hecker, M, ML Foegh, and PW Ramwell. The Eicosanoids: Prostagladins, thromboxanes, leukotrienes and related compounds. In: B.G. Katzung, ed. Basic and Clinical Pharmacology. Englewood Cliffs, N.J.: Prentice-Hall, 1989, pp. 228–241.

Hecker, M, and V Ullrich. On the mechanism of prostacyclin and thromboxane A2 biosynthesis. J Biol Chem 1989; 264:141–150.

Hirata, M, Y Hayashi, F Ushikubi, Y Yokota, R Kageyama, S Nakanishi, and S Narumiya. Cloning and expression of cDNA for a human thromboxane A2 receptor. Nature 1991; 349:617–620.

Hirata, T, A Kakizuka, F Ushikubi, I Fuse, M Okuma, and S Narumiya. Arg60 to Leu mutation of the human thromboxane A2 receptor in a dominantly inherited bleeding disorder. J Clin Invest 1994; 94:1662–1667.

Hirata, T, F Ushikubi, A Kakizuka, M Okuma, and S Narumiya. Two thromboxane A2 receptor isoforms in human platelets: opposite coupling to adenylyl cyclase with different sensitivity to Arg60 to Leu mutation. J Clin Invest 1996; 97:949–956.

Hunt, JA, JE Merritt, J MacDermot, and M Keen. Characterization of the thromboxane receptor mediating prostacyclin release from cultured endothelial cells. Biochem Pharmacol 1992; 43: 1747–1752.

Ishiguro, S, S Arii, K Monden, Y Adachi, N Funaki, H Higashitsuji, S Fujita, M Furutani, M Mise, T Kitao, et al. Identification of the thromboxane A2 receptor in hepatic sinusoidal endothelial cells and its role in endotoxin-induced liver injury in rats. Hepatology 1994; 20:1281–1286.

Jones, DA, and FA Fitzpatrick. Thromboxane A2 synthase: modification during ''suicide'' inactivation. J Biol Chem 1991; 266:23510–23514.

Kameda, H, T Yoshida, Y Ichikawa, and M Homma. Thromboxane A2 synthase inhibitor, DP-1904, decreases TNF alpha secretion from monocytes and inhibits E-selectin and ICAM-1 expression on the endothelial cell surfaces. Adv Prostaglandin Thromboxane Leukotriene Res 1995; 23:423–425.

Kent, KC, LJ Collins, FT Schwerin, MK Raychowdhury, and JA Ware. Identification of functional PGH2/TxA2 receptors on human endothelial cells. Circ Res 1993; 72:958–965.

Kinsella, BT, OM DJ, and GA FitzGerald. Phosphorylation and regulated expression of the human thromboxane A2 receptor. J Biol Chem 1994; 269:29914–29919.

Kinsella, BT, OM DJ, and GA Fitzgerald. The human thromboxane A2 receptor alpha isoform (TP alpha) functionally couples to the G proteins Gq and G11 in vivo and is activated by the isoprostane 8-epi prostaglandin F2 alpha. J Pharmacol Exp Ther 1997; 281:957–964.

Kitanaka, J, H Hashimoto, Y Sugimoto, M Sawada, M Negishi, A Suzumura, T Marunouchi, A Ichikawa, and A Baba. cDna cloning of a thromboxane A2 receptor from rat astrocytes. Biochim Biophys Acta 1995; 1265:220–223.

Lasserre, B, AP Huu, C Navarro-Delmasure, and V Dossou-Gbete. Binding of a thromboxane A2 (TXA2)/prostaglandin H2 (PGH2) receptor antagonist to rabbit and pig heart membrane protein. Prostaglandins Leukot Essent Fatty Acids 1992; 47:153–157.

Lee, KD, SJ Baek, and RF Shen. Multiple factors regulating the expression of human thromboxane synthase gene. Biochem J 1996; 319 (pt 3):783–791.

Liel, N, I Nathan, T Yermiyahu, Z Zolotov, JR Lieberman, A Dvilansky, and PV Halushka. Increased platelet thromboxane A2/prostaglandin H2 receptors in patients with pregnancy induced hypertension. Thromb Res 1993; 70:205–210.

Meagher, EA, and GA FitzGerald. Disordered eicosanoid formation in pregnancy-induced hypertension. Circulation 1993; 88:1324–1333.

Mikhail, G, AH Chester, JS Gibbs, JA Borland, NR Banner, and MH Yacoub. Role of vasoactive mediators in primary and secondary pulmonary hypertension. Am J Cardiol 1998; 82:254–255.

Miyamoto, T, N Ogino, S Yamamoto, and O Hayaishi. Purification of prostaglandin endoperoxide synthetase from bovine vesicular gland microsomes. J Biol Chem 251:2629–2636.

Miyata, A, C Yokoyama, H Ihara, S Bandoh, O Takeda, E Takahashi, and T Tanabe. Characterization of the human gene (TBXAS1) encoding thromboxane synthase. Eur J Biochem 1994; 224:273–279.

Modesti, PA, A Colella, I Cecioni, A Costoli, D Biagini, A Migliorini, and GG Neri Serneri. Increased number of thromboxane A2-prostaglandin H2 platelet receptors in active unstable angina and causative role of enhanced thrombin formation. Am Heart J 1995; 129:873–879.

Muck, S, AA Weber, J Meyer-Kirchrath, and K Schror. The bovine thromboxane A2 receptor: molecular cloning, expression, and functional characterization. Naunyn-Schmiedebergs Arch Pharmacol 1998; 357:10–16.

Murohara, T, JR Horowitz, M Silver, Y Tsurumi, D Chen, A Sullivan, and JM Isner. Vascular endothelial growth factor/vascular permeability factor enhances vascular permeability via nitric oxide and prostacyclin. Circulation 1998; 97:99–107.

Namba, T, Y Sugimoto, M Hirata, Y Hayashi, A Honda, A Watabe, M Negishi, A Ichikawa, and S Narumiya. Mouse thromboxane A2 receptor: cDNA cloning, expression and northern blot analysis. Biochem Biophys Res Commun 1992; 184:1197–203.

Nusing, R, S Schneider-Voss, and V Ullrich. Immunoaffinity purification of human thromboxane synthase. Arch Biochem Biophys 1990; 280:325–330.

Nusing, R, M Goerig, AJ Habenicht, and V Ullrich. Selective eicosanoid formation during HL-60 macrophage differentiation: regulation of thromboxane synthase. Eur J Biochem 1993; 212: 371–376.

Nusing, RM, M Hirata, A Kakizuka, T Eki, K Ozawa, and S Narumiya. Characterization and chromosomal mapping of the human thromboxane A2 receptor gene. J Biol Chem 1993b; 268: 25253–25259.

Ohki, S, N Ogino, S Yamamoto, and O Hayaishi. Prostaglandin hydroperoxidase, an integral part of prostaglandin endoperoxide synthetase from bovine vesicular gland microsomes. J Biol Chem 1997; 254:829–836.

Okuma, M, T Hirata, F Ushikubi, A Kakizuka, and S Narumiya. Molecular characterization of a dominantly inherited bleeding disorder with impaired platelet responses to thromboxane A2. Polish J Pharmacol 1996; 48:77–82.

Orlandi, M, G Bartolini, B Belletti, E Spisni, and V Tomasi. Thromboxane A2 synthase activity in platelet free human monocytes. Biochim Biophys Acta 1994; 1215:285–290.
Raychowdhury, MK, M Yukawa, LJ Collins, SH McGrail, KC Kent, and JA Ware. Alternative splicing produces a divergent cytoplasmic tail in the human endothelial thromboxane A2 receptor. J Biol Chem 1994; 270:7011.
Rossi, P, P Kuukasjarvi, JP Salenius, M Tarkka, T Kerttula, J Alanko, I Mucha, and A Riutta. Percutaneous transluminal angioplasty increases thromboxane A2 production in claudicants. Prostaglandins Leukot Essent Fatty Acids 1997; 56:369–372.
Sakai, N, MG Blennerhassett, and RE Garfield. Intracellular cyclic AMP concentration modulates gap junction permeability in parturient rat myometrium. Can J Physiol Pharmacol 1992; 70: 358–364.
Sato, M, Y Kawashima, J Goto, Y Yamane, Y Chiba, S Jinno, M Satake, T Imanishi, and C Iwata. Synthesis and evaluation of novel thiazolidine derivatives as thromboxane A2 receptor antagonists. Chem Pharm Bull 1994; 42:521–529.
Schaub, RG, KA Gates, and RE Roberts. Effect of aspirin on collateral blood flow after experimental thrombosis of the feline aorta. Am J Vet Res 1982; 43:1647–1650.
Schwengel, DA, N Nouri, DA Meyers, and RC Levitt. Linkage mapping of the human thromboxane A2 receptor (TBXA2R) to chromosome 19p13.3 using transcribed 3′ untranslated DNA sequence polymorphisms. Genomics 1993; 18:212–215.
Shock, DD, K He, JD Wencel-Drake, and LV Parise. Ras activation in platelets after stimulation of the thrombin receptor, thromboxane A2 receptor or protein kinase C. Biochem J 1997; 321(pt 2):525–530.
Simmons, TR, JA Cook, JN Moore, and PV Halushka. Thromboxane A2 receptors in equine monocytes: identification of a new subclass of TXA2 receptors. J Leukoc Biol 1993; 53:173–178.
Suehiro, T, K Yanaga, H Itasaka, K Kishikawa, K Shirbe, M Shimada, and K Sugimachi. Thromboxane A2 in preservation-reperfusion injury: the effect of thromboxane A2 synthetase inhibitor. J Surg Res 1996; 62:216–223.
Tazawa, R, ED Green, K Ohashi, KK Wu, and LH Wang. Characterization of the complete genomic structure of human thromboxane synthase gene and functional analysis of its promoter. Arch Biochem Biophys 1996; 334:349–356.
Tsujii, M, S Kawano, S Tsuji, H Sawaoka, M Hori, and RN Dubois. Cyclooxygenase regulates angiogenesis induced by colon cancer cells. Cell 1998; 93:705–716.
Ullrich, V, and R Nusing. Thromboxane synthase: from isolation to function (review). Stroke 1990; 21(12 suppl):IV134–IV138.
Ushikubi, F, Y Aiba, K Nakamura, T Namba, M Hirata, O Mazda, Y Katsura, and S Narumiya. Thromboxane A2 receptor is highly expressed in mouse immature thymocytes and mediates DNA fragmentation and apoptosis. J Exp Med 1993; 178:1825–1830.
Wang, Y, and B Rose. Clustering of C × 43 cell-to-cell channels into gap junction plaques: regulation by cAMP and microfilaments. J Cell Sci 1995; 108:3501–3508.
Wang, LH, N Matijevic-Aleksic, PY Hsu, KH Ruan, KK Wu, and RJ Kulmacz. Identification of thromboxane A2 synthase active site residues by molecular modeling-guided site-directed mutagenesis. J Biol Chem 1996; 271:19970–19975.
Wang, Y, and B Rose. An inhibition of gap-junctional communication by cadherins. J Cell Sci 1997; 110:301–309.
Wang, G-R, Y Zhu, PV Halushka, TM Lincon and ME Mendelsohn. Mechanism of platelet inhibition by nitric oxide: In vivo phosphorylation of thromboxane receptor by cyclic GMP-dependent protein kinase. Proc Natl Acad Sci USA 1998; 95:1–6.
Wiles, ME, R Welbourn, G Goldman, HB Hechtman, and D Shepro. Thromboxane-induced neutrophil adhesion to pulmonary microvascular and aortic endothelium is regulated by CD18. Inflammation 1991; 15:181–199.
Xia, Z, RF Shen, SJ Baek, and HH Tai. Expression of two different forms of cDNA for thromboxane

synthase in insect cells and site-directed mutagenesis of a critical cysteine residue. Biochem J 1993; 295 (pt 2):457–461.
Yamamoto, Y, K Kamiya, and S Terao. Modeling of human thromboxane A2 receptor and analysis of the receptor-ligand interaction. J Med Chem 1993; 36:820–825.
Yokota, R, AW Ashton, Y Gao, Y Shizukuda, T Michaeli, and JA Ware. Induction of endothelial cell apoptosis and reversal of angiogenesis in vitro by thromboxane A2. 1999. Submitted.
Yokoyama, C, A Miyata, H Ihara, V Ullrich, and T Tanabe. Molecular cloning of human platelet thromboxane A synthase. Biochem Biophys Res Commun 1991; 178:1479–1484.
Yukawa, M, R Yokota, RT Eberhardt, L von Andrian, and JA Ware. Differential desensitization of thromboxane A2 receptor subtypes. Circ Res 1997; 80:551–556.

12

Modulation of Endothelial Cell Function by Estrogen

H. William Schnaper, Maria Luiza C. Albuquerque, Constance Runyan, and Susan C. Hubchak
Northwestern University Medical School, Chicago, Illinois

John McGuire
Washington University Medical School, St. Louis, Missouri

Seunghee Kim-Schulze
Mount Sinai School of Medicine, New York, New York

I. INTRODUCTION

Estrogen is one of a group of feminizing gonadal steroid hormones comprising estrogens, progesterone, and related molecules. It is produced from testosterone by the action of aromatase, also called estrogen synthase, an enzyme expressed in numerous tissues including adipose tissue and muscle cells (1). The most active estrogen in the circulation is 17β-estradiol. Estrogen metabolites, including estrone, estriol and other, less common forms (see Table 1), vary based upon the oxidation status of the side group attached to the C-17 atom, shown in Fig. 1, or other side groups. Estrogen is responsible in part for the development of secondary female sex characteristics such as the adult mammary gland and, along with progestins, is a critical regulator of the menstrual cycle and an important factor in maintaining pregnancy. As has been the case with other steroid hormones, the availability of more sophisticated approaches to understanding estrogen function has revealed additional roles that are not classically associated with gonadal function.

One important role of estrogen is in the regulation of vascular biology. In this chapter, we consider the effects of estrogen on the blood vessel wall, particularly as it relates to the issue of angiogenesis. Data suggesting that estrogen plays a role in angiogenesis can be grouped into several categories. First, indirect evidence supports a role for estrogen in modulating vascular function and, by extension, suggests that estrogen affects the activity of the cells in the vessel wall. Second, estrogen-stimulated events during the menstrual cycle are associated with angiogenesis. Third, the role of estrogen in modulating endothelial biology has been more directly assessed in a variety of experimental studies evaluating how estrogen influences angiogenesis or activities of endothelial cells that contribute to angiogenesis. Here, we examine the evidence indicating that estrogen modulates endothe-

Table 1 Common Metabolites of Estrogen

Estradiol (17α, -β)
Estriol
Estrone
Conjugates (examples):
16-Epiestriol
16α-Hydroxyestrone
16β-Hydroxyestrone
16-Keto-estradiol
2-Methoxyestrone
2-Methoxyestriol
15α-Hydroxyestrone
15β-Hydroxyestrone

lial cell function and angiogenesis. We then consider potential mechanisms by which such modulation may occur.

II. ESTROGEN AND ATHEROSCLEROSIS: EVIDENCE FOR EFFECTS ON THE VESSEL WALL

It has long been known that premenopausal women enjoy significant protection from cardiovascular disease (2). Estrogen treatment has been thought to improve morbidity and mortality from stroke (see Ref. 3 and references contained therein), although this assertion has proven to be controversial (4). Evaluating beneficial effects of estrogen on the cerebral blood circulation is difficult, since estrogens may have direct, salient effects on central nervous system function (5). In contrast, studies of heart disease seem more straightforward. Most studies of postmenopausal women who have received hormone replacement therapy (HRT) (6–9) show a beneficial effect on cardiovascular status, with mortality or disease incidence in women receiving estrogen therapy being two-thirds that of women who have not received exogenous estrogen. The indications for HRT and an analysis of the risk of breast (10) or uterine (11) cancer in women who are receiving such treatment

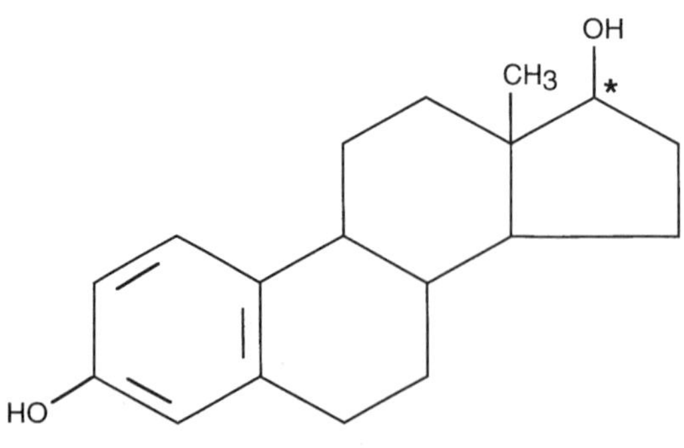

Figure 1 Chemical structure of 17β-estradiol. The carbon-17 is indicated by an asterisk.

are beyond the scope of this chapter. The reader is referred to the references cited here for additional information.

Estrogen may improve the cardiovascular status of premenopausal women in part through effects on vascular lipid metabolism (12,13). Estrogen decreases the influx of cholesterol into the vessel wall (14), suggesting that either permeability (passive mechanism) or transport (active mechanism) through the endothelial barrier is affected. The plasma lipoprotein profile in the presence of feminizing concentrations of circulating estrogen is less atherogenic than in controls, with increased production of high-density lipoproteins and decreased low-density lipoproteins (LDL) (15). Oxidation of LDL, a critical component of the atherogenic process, is decreased as well (16). Although this finding could represent an antioxidant effect of estrogen rather than a receptor-mediated effect on endothelial cells, together these data suggest that estrogen alters atherogenesis through effects on the vessel wall.

Estrogen also modifies vessel wall function as it pertains to regulation of vascular tone. Postmenopausal estrogen replacement therapy is associated with increased coronary artery caliber (11). Vascular diameter may be so sensitive to estrogen that it varies throughout the menstrual cycle (17). Even short-term treatment with estrogen leads to vasodilation (18). One important mechanism of action is to counter vasoconstricting signals, such as that of acetylcholine (19). Estradiol appears to act by stimulating endothelial production of nitric oxide (20–22), opening of calcium-dependent potassium channels (23) and cyclic GMP–dependent relaxation of vascular smooth muscle cells (VSMC) (24).

Finally, an effect of estradiol on endothelial cell function may reduce arterial wall thickening in response to internal vascular injury (25,26). When the endothelial lining of the blood vessel is stripped away, shear forces on the denuded intima stimulate VSMC to migrate into the lesion, proliferate, and secrete extracellular matrix. The extent of the thickening reaction in this model thus represents the balance between activation of VSMC to proliferate and secrete connective tissue proteins and stimulation of endothelial cells to heal intimal wounds. Accelerated re-endothelialization is accompanied by decreased vessel wall thickness as well as more rapid recovery of endothelial function (as demonstrated by nitric oxide production) (26). Decreased fibrous plaque formation suggests a shift in the balance between endothelial and smooth muscle cell activities, supporting a role for estrogen in modulating vascular cell function.

III. ANGIOGENESIS RELATED TO OVULATION AND PREGNANCY

A role for gonadal hormones in angiogenesis has been suggested by studies of the reproductive system. In the adult mammal, this system is the most significant site of physiological tissue regeneration and remodeling. Rapid growth of decidual tissue requires generation of a supporting vascular system. During cyclical growth, increased estrogen production and angiogenesis are temporally related (27). In the cycling human female, preovulatory levels of estradiol approximate those in men and are in the range of 25–50 pg/mL. Similar levels are found postmenopause (28,29). At the time of ovulation, there is a sharp spike in levels (as high as 500 pg/mL in humans), followed by a lower but still elevated level (200–300 pg/mL) during the luteal phase of the menstrual cycle (30). During this part of the cycle, critical events occur to support a potential pregnancy, with development of the corpus luteum and preparation of decidual tissue in the endometrium

(31). These events are accompanied by a high degree of angiogenic activity. Blood flow to critical organs increases just after ovulation (32). In the rat (33) or ewe (34) ovary, rapidly dividing endothelial cells produce vessels to support the developing corpus luteum. Simultaneously, rapid angiogenesis in the time just after the estrogen surge produces the coiled arteries of the endometrium (35). Conversely, estrogen withdrawal in cattle induces rapid vessel regression accompanied by rounded endothelial cells suggestive of apoptosis (36). Similar changes to those seen in the maturing ovarian follicle and uterus may occur in vascularization of the mammary gland during pregnancy and lactation (37).

Multiple cytokines and growth factors have been implicated in reproductive angiogenesis including basic fibroblast growth factor (bFGF), vascular endothelial cell growth factor (VEGF), and tumor necrosis factor α (TNF-α). It is noteworthy that mice null for bFGF remain fertile (38). In contrast, inhibition of VEGF activity using soluble Flt-2 (VEGF receptor) leads to failure of corpus luteum development and lack of endometrial maturation, including decreased endometrial angiogenesis (39). These data do not clarify whether VEGF alone, as opposed to multiple effects of a functioning corpus luteum, is responsible for subsequent angiogenesis. Nonetheless, a critical role is suggested for VEGF. The most direct effect of estrogen on this system may be stimulation of VEGF production by relevant tissues, as described in the rat (40), monkey (41), and human both in vivo and in vitro (42). These studies indicate that at least one mechanism by which estrogen may stimulate endothelial cell function in decidual tissue is through VEGF-mediated activity. In subsequent discussion, we explore other, more complex mechanisms by which estrogen affects endothelial cell function.

IV. EVIDENCE THAT ESTROGEN DIRECTLY MODIFIES ENDOTHELIAL CELL FUNCTION

Angiogenic effects of estrogen are not limited to the tissues important for reproductive biology, nor are they mediated only by VEGF. Schechter and colleagues reported (43) that estrogen induces bFGF production by gonadotropic cells in the periphery of the pituitary in Fischer 344 but not Sprague-Dawley rats. These peripheral sites were adjacent to areas of increased capillary formation required to support ongoing glandular hypertrophy. Estrogen also stimulates bFGF production by postmenopausal endometrium (44). In addition, estrogen alters the endothelial cell response to bFGF. Morales and colleagues employed a model in which bFGF, mixed into the basement membrane extract Matrigel, is injected subcutaneously in mice. Matrigel is a liquid at 4°C but polymerizes at 37°C to form a scaffolding that is subject to vascular invasion and can be removed subsequently for analysis. Without bFGF, little response is observed after 7 days (Fig. 2). However, when bFGF is added to the Matrigel, there is marked invasion of cellular elements that are almost entirely von Willebrand factor–positive. Organized vessels are present and contain erythrocytes (Figure 2). If the mice are subjected to ovariectomy 2 weeks before implanting the Matrigel, there is a marked (65–70%) reduction in cellular elements, which is reconstituted by subcutaneous implantation of a sustained-release estrogen pellet (45). Johns and colleagues further refined this assay by injecting fluorescein-dextran intravenously to permit measurement of the plasma volume in the vascularized Matrigel plug, based on fluorescence activity (46). Evaluating normal and estrogen receptor (ER)α-null (ERKO) mice, they determined that bFGF-enhanced angiogenesis was abrogated in the ERKO mice (Fig. 3). This result indicates that estrogenic enhancement of angiogenesis requires the intact

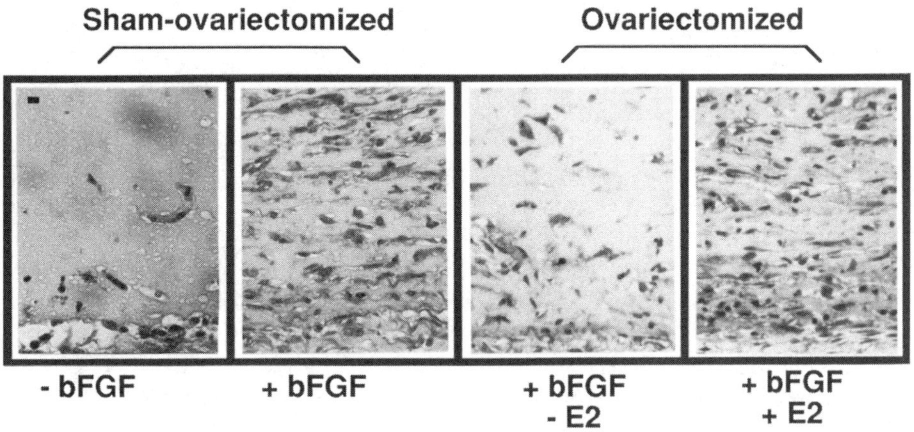

Figure 2 Effect of ovariectomy on vascularization of a Matrigel plug. Ovariectomy leads to a 65% decrease in cellular elements invading the Matrigel; the response is reconstituted by implantation of a sustained-release estrogen pellet. (From Ref. 98.)

estrogen receptor. Indeed, the findings suggest that ERα action is essential for bFGF-induced angiogenic activity.

Further indication of a role for ER in angiogenesis is derived from studies investigating the effect of specific ER antagonists on angiogenesis in vitro. Angiogenesis induced by bFGF or by VEGF in the chick chorioallantoic membrane (CAM) assay is inhibited by the ER antagonists tamoxifen, 4-hydroxy-tamoxifen, nafoxidene, clomiphene, and ICI182, 780 (47,48). The results suggest either that these antagonists have their own, direct effects on endothelial cell function or that endogenous estrogen production in these assays supports the response to the growth factors. In accord with the latter possibility, regression of breast cancers after tamoxifen therapy is accompanied by vessel regression (49).

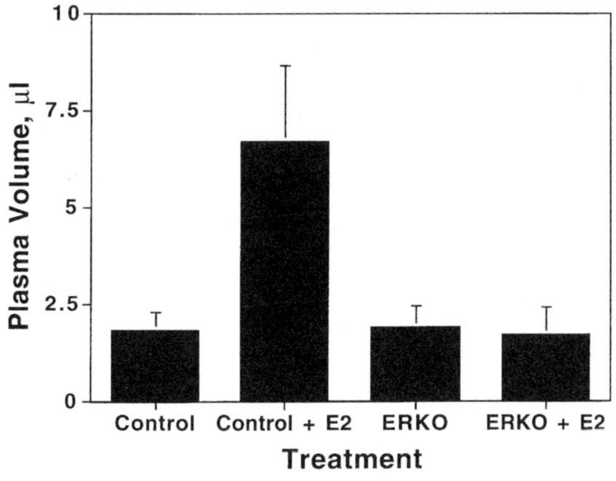

Figure 3 Vascularization of a Matrigel plug in wild-type and ERα-deficient (ERKO) mice that have been ovariectomized. Matrigel and bFGF were injected subcutaneously in the mice and plasma volume of the Matrigel was determined 10 days later, as described in Ref. 46. Estrogen replacement enhances angiogenesis in the wild-type but not ERKO, mice. (From Ref. 46.)

V. IN VITRO ASSAYS OF ENDOTHELIAL CELL FUNCTION RELATED TO ANGIOGENESIS

The hypothesis that estrogen is important in angiogenesis has been further tested by in vitro studies of the endothelial cell response to estrogen. Angiogenesis can be broken down into a number of component functions (50). Injury to the vessel wall or disruption of the basement membrane that surrounds the capillary leads to activation of endothelial cells. The cells secrete extracellular matrix proteases, permitting them to migrate into the stromal space and attach to other matrices. Proliferation yields a sufficient mass of cells to permit organization into new tubular structures. With secretion and remodeling of a new basement membrane, a mature capillary is formed. These functions of protease production, attachment, migration, and organization do not occur in discrete steps but rather are invoked as necessary in order to complete the angiogenic process. Each function can be modeled in vitro to test for effects of estrogen.

Endothelial cells can be depleted of estrogen by culture for 48 hr in phenol red-free medium (51) containing charcoal-stripped serum. When human umbilical vein endothelial cells (HUVEC) are hormone-depleted and then treated with 3 nM estradiol for 3 hr, they show accelerated rates of attachment to plastic, collagens, fibronectin, laminin, and other substrates, although there is no difference in the number of cells that eventually attach (45). Endothelial cells also show increased rates of proliferation in the presence of estrogen (45). Interestingly, when estrogen is added to hormone-depleted cells, a lag period of 24 to 48 hr passes before the cells manifest growth curves that parallel those of estrogen-replete cells (52), raising the possibility that the effect of estradiol on cell growth requires the activation of intermediate mechanisms.

When migration is evaluated either by a Boyden chamber assay or by the ability of the cells to migrate into a ''wound'' scraped in an endothelial cell monolayer, estradiol enhances rates of migration. Organization of cultured endothelial cells into capillary-like structures has been evaluated using an assay in which endothelial cells are plated on a substratum of Matrigel in culture. This causes the cells to migrate, align, and form tubes whose central lumens are surrounded by cells having appropriate apical-basal orientation. This assay is inhibited by inhibitors of cell attachment, migration, or protease production, further supporting its relevance to angiogenesis in vivo (53–56). When estrogen-depleted HUVEC are plated on Matrigel, they show thin, poorly formed structures in the tube-forming assay. In contrast, estradiol causes a dose-dependent increase in the number of tubes that are formed, peaking at 1–2 ng/mL (45). Figure 4 shows an example of estradiol-enhanced tube formation by cultured human coronary artery endothelial cells (HCAEC).

Without the appropriate combination of signals (e.g., in the absence of integrin binding), stimuli that normally induce angiogenesis may instead induce apoptosis (57,58). Several groups have determined that estradiol decreases rates of endothelial cell apoptosis (59,60). In an in vitro model, endothelial cells treated with TNF-α undergo apoptosis at rates as high as 50%, accompanied by increased expression of the interleukin-1β converting enzyme (ICE) that has been associated with this process. These events are inhibited by estradiol at nanomolar concentrations. This protective effect of estradiol is abrogated by simultaneous treatment with the specific estrogen receptor antagonist, ICI182,780 (59). These results indicate that estrogen may prevent apoptosis, suggesting that estrogen is affecting the balance achieved among the multiple signals that regulate programmed cell death.

Another critical balance, that between extracellular matrix proteases and their inhibitors, is an essential determinant of the cells' ability to break out of their initial location (50), migrate (61), and organize (55,56). This balance is tightly regulated, and endothelial

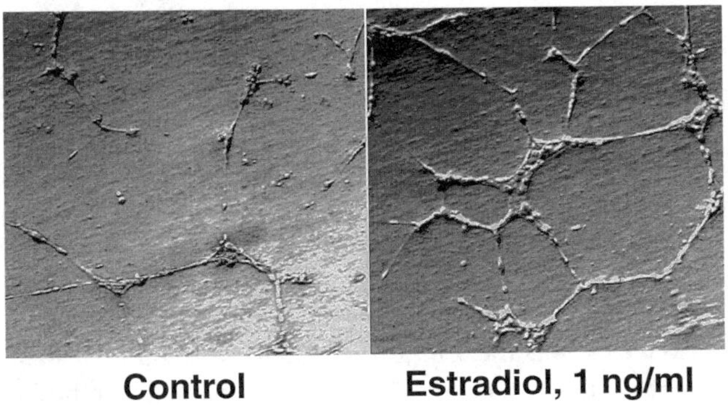

Figure 4 Enhancement of human coronary artery endothelial cell (HCAEC) tube formation in vitro by estradiol. Hormone-depleted HCAEC (4×10^4) were plated on Matrigel in 24-well plates. The structures formed on Matrigel appear thicker and more complete. When the structures were quantified by image analysis, the surface area of the culture covered with tubes was approximately doubled. See text and Ref. 45 for details of the methods.

cells under different conditions may show different levels of protease or protease inhibitor expression. For example, proliferating HUVEC express less extracellular matrix proteinases on a molar basis than they do the protease inhibitors, whereas in human renal microvascular endothelial cells, the molar ratio is approximately even (62). Thus, cells from different sources show different relative expression of proteases and protease inhibitors. The status of cellular activation is also important. When cells from either of these tissue sources reach quiescence, the balance shifts in favor of protease inhibitor. In the case of mRNA for plasminogen activators and their inhibitors, estradiol exaggerates this tendency, so that active endothelial cells treated with estradiol express even more plasminogen activator mRNA, and estrogen-treated quiescent endothelial cells express even more of the inhibitor, PAI-1 mRNA (63) (depicted schematically in Fig. 5). PAI-1 expression is important, since it plays a role in fibrinolysis as well as regulating matrix degradation. In vivo, high concentrations of estradiol may be thrombogenic (64), a phenomenon felt to contribute to the hypercoagulability that may complicate pregnancy. However, at least one study suggests that estradiol reduces PAI-1 protein concentrations when endothelial cells are

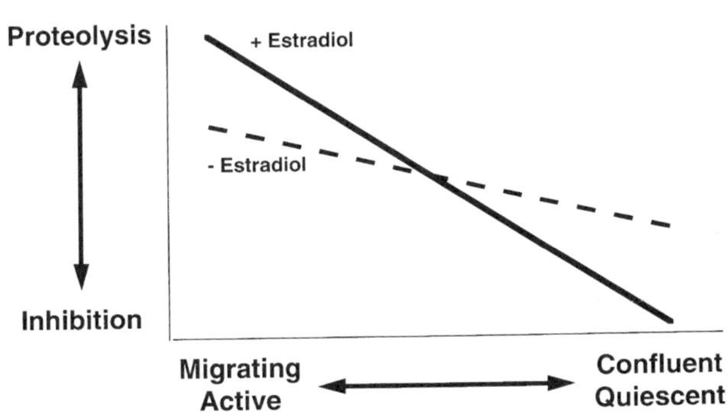

Figure 5 Depiction of the effect of estradiol on relative production of plasminogen activators and plasminogen activator inhibitors. Both the tendency of active cells to express more proteases and of quiescent cells to produce more inhibitor mRNA are exaggerated by estradiol.

cultured in vitro (65). Thus, in the case of the proteolytic and coagulation cascades, the effects of estradiol are complex. They may be dose-dependent, or estradiol may amplify established activities rather than stimulating new ones.

VI. ROLE OF AN ESTROGEN RECEPTOR IN ESTROGENIC MODULATION OF ENDOTHELIAL CELL FUNCTION

In many of the assays described above, a specific ER antagonist inhibits the effect of estrogen. In addition, mice that are ERα-null do not show an angiogenic response to bFGF in vivo. Strikingly, positive effects of estradiol on intimal healing may not be abrogated in ERα-null mice (66), suggesting either that ER-mediated effects are specific for bFGF or that ERα plays a special role in angiogenesis. However, an endothelial cell ER has been difficult to define. Classically, the ER is a cytoplasmic protein that, when bound by estrogen, undergoes phosphorylation, dimerization, and translocation to the nucleus. There, it participates in transcription regulatory complexes (67). A ''binding site'' for estrogen in endothelial cell cultures was described as early as 1978 (68) and a putative receptor was described in bone endothelial cells in 1993 (69). A more functionally defined receptor was described in 1996 by two groups (52,70). Specific nuclear staining was accompanied by Scatchard analysis suggesting a dissociation constant of 5 nM and 2- to 8 $\times 10^4$ receptors per cell. We are concerned that these estimates may be somewhat high in view of their value relative to that for MCF-7 breast cancer cells, which show more intense staining but similar dissociation constant and receptor number estimations. However, these results have been obtained consistently and are similar to results obtained for human VSMC (71,72).

In endothelial cells, unlike the response of MCF-7 cells, estradiol stimulates expression of mRNA for ERα. Similar stimulation has been found in cultured pituitary cells (73). This difference could reflect the fact that endothelial cells are not ''classical'' estrogen targets, that MCF-7 cells are transformed whereas HUVEC are not, or another, as yet unknown explanation. The endothelial ERα shows specific DNA binding, since in electrophoretic mobility shift assays a shifted complex is competed by excess unlabeled estrogen response element (ERE) and is supershifted by antibody against ERα (52). Recently, we have confirmed the functional nature of this ER by demonstrating that an ERE-luciferase construct, transiently expressed in HUVEC, is activated by estradiol treatment (S. Kim-Schulze and H.W. Schnaper, unpublished). Thus, the endothelial ER is induced to act as a transcriptional regulator by estradiol and is therefore a classical, functional ER. High estradiol concentrations can induce gene repressor activity by ER, either directly or indirectly (74). It is therefore not surprising that, when estrogen concentrations become significantly higher than the physiological range, they may inhibit responses (45,75).

Considerable attention has been focused on the recently described alternative estrogen receptor, the ERβ (76). Originally detected in transformed gonadal tissues, this isoform was found to be expressed at high levels in an endothelial cell line as determined by reverse transcription-polymerase chain reaction. However, several lines of evidence suggest that the ERα is the predominant receptor mediating the responses described above in HUVEC and HCAEC. First, the antibodies used to detect ER in the cell immunostaining and gel-shift assays do not cross-react with ERβ. Second, we and others have found it difficult to demonstrate convincingly the presence of ERβ protein in early-passage, primary cell cultures or in endothelial cells in vivo. Third, ERβ is able to affect the AP-1 promoter via indirect binding (77); we have been unable to reproduce this finding in primary endothelial cells (S. Kim-Schulze S.C. Hubchak, and H.W. Schnaper, unpublished). While the finding that ERα-null mice still respond to E2 in preventing intimal thickening after injury (66) is suggestive of a role for ERβ, the data are consistent with alternative interpretations, such as a non-receptor-mediated mechanism of estrogen action. These issues will be resolved only when an ERβ-deficient mouse becomes available for study.

VII. ACTIVATION OF THE EXTRACELLULAR SIGNAL-REGULATED PROTEIN KINASE (ERK) MAP KINASE BY ESTRADIOL

ERKs are mitogen-associated protein (MAP) kinases that have been found to play a role in proliferation and other cellular processes (78). They are a critical part of the cascade by which growth-factor receptors transmit signals through protein tyrosine kinases (79), $p21^{ras}$ (80) and lipid-mediated signaling pathways (81). Estrogen-stimulated endothelial cells show an unusual ERK MAP kinase response, with activity peaking at 3 hr rather than at 5–15 min, as is usually the case with this signaling pathway (82). The delay in activation likely results from the time required to synthesize new protein(s), since induction of activity is inhibited by blocking gene transcription with actinomycin D or translation with cycloheximide. Activation is also ER-mediated, since the specific ER antagonist ICI182,780 inhibits it. MAP kinase activation may result from secretion of the newly synthesized protein, since media conditioned by estrogen-treated cells stimulate more rapid (5-min) activity in naive cells. This capacity is decreased by prior incubation of the conditioned medium with neutralizing anti-bFGF antibody, and activation of ERKs by estradiol or by conditioned medium is temporally associated with phosphorylation of the bFGF receptor. Further, bFGF mRNA expression is induced within 50–90 min of estrogen stimulation (81). These data describe an autocrine loop for ERK MAP kinase activation that involves bFGF.

VIII. OTHER SIGNAL TRANSDUCTION MECHANISMS AFFECTED BY ESTRADIOL

Activation of ERKs by this pathway likely represents a classical mechanism of action for estradiol: stimulating ER to function as a transcription factor, possibly activating (among others) the *fgf2* gene. For other signaling pathways, the mechanism is less clear. Estrogen-stimulated signaling is mediated in part by induction of calcium flux in endothelial cells (83). Estrogen-stimulated calcium flux could play a role in the NO production that mediates reflex vessel relaxation and could also transduce additional signals, possibly by activating protein kinases (83). Since the relaxation response is very rapid (18), there may be insufficient time for new protein synthesis to occur. Certainly, it is possible that signaling is mediated by the ER without the induction of new gene transcription. Endothelial cell release (as opposed to synthesis) of bFGF, which occurs by a process that is not well understood, also may be affected by estradiol. Endothelial cells cultured on dishes coated with laminin, fibronectin, or type IV collagen, but not with type I collagen or gelatin, release bFGF rapidly (within 2 hr) after treatment with estradiol. This release is not blocked by inhibitors of protein synthesis or of calcium flux but is inhibited by calphostin C, an inhibitor of protein kinase C. Since this response is specific for cells cultured on basement membrane proteins, it suggests an interaction among estradiol, integrin binding, and protein kinase C that leads to a novel system for estrogen-mediated cell signal transduction (84). Additional data suggest that estradiol enhances endothelial cell focal adhesion kinase (FAK) activity (60). Thus, the ER could mediate rapid signaling through nongenomic mechanisms.

IX. ESTROGEN IS MORE THAN ONE COMPOUND

In evaluating an effect of ''estrogen,'' it is important to consider which estrogen is involved. As noted previously, there are many metabolites of estrogen. Whereas estradiol is proangiogenic, 2-methoxy-estradiol is clearly antiangiogenic (85), possibly inducing endothelial cell apoptosis (86). Similarly, antiestrogens were initially defined as such by their ability to inhibit estrogen effects in specific assays relative to gonadal function. Since these compounds bind to the ER, they may induce various activities in different tissues. For example, tamoxifen is an antiestrogen used to treat breast cancer. It is a partial agonist in some in vitro angiogenesis assays (H.W. Schnaper, unpublished), and has been suggested to reduce cardiac mortality in at least one study (87). Recently, other ER ''antagonists'' have been found to be partial agonists that may have salient effects on tissues such as bone or cardiovascular system but no gonadal stimulatory effects (88,89).

X. SUMMARY AND CONCLUSION

The ability of estrogen to promote angiogenesis appears to be well supported by the scientific literature, although the mechanism for this effect is less certain. Figure 6 diagrams

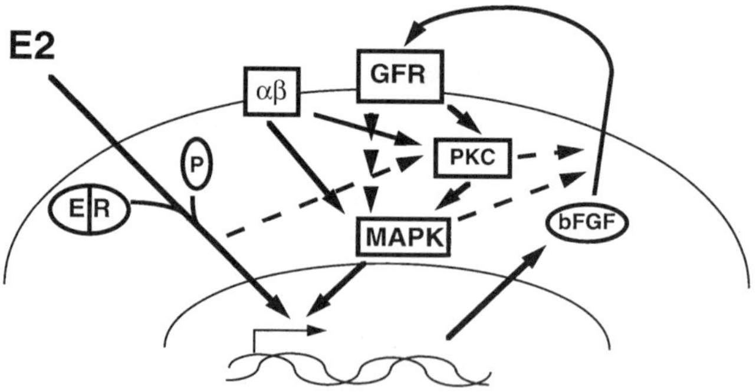

Figure 6 Potential effects of estrogen on endothelial cell signal transduction. Estradiol (E2) binds to the estrogen receptor (ER) and initiates gene transcription. One of the transcribed genes (bFGF) stimulates an autocrine, growth factor receptor (GFR)–mediated autocrine loop that leads to activation of MAP kinases (MAPK), which also have transcription regulatory activity. GFR binding may directly or indirectly stimulate protein kinase C (PKC), which also stimulates MAPK and may facilitate bFGF release. Binding of integrin receptors (αβ) may enable activated ER to stimulate PKC and, in turn, to facilitate bFGF release. Solid lines, known pathways; broken lines, proposed pathways.

several potential effects of estradiol on cell signaling. The ER acts as a transcription factor in endothelial cells and can stimulate an autocrine (bFGF) loop to activate ERK MAP kinases, which themselves stimulate gene transcription. Growth factors may also activate the ER in the absence of estradiol (90), although the role of MAP kinases in this ER activation remains controversial (91). If such activation does occur, it defines an amplification loop for signal transduction between the nucleus and the cytoplasm. Estradiol may have additional effects on the cell membrane or other signaling pathways (PKC, NO), further amplifying the cellular response. Recent data indicating that vascular smooth muscle cells may produce aromatase (92) support the notion that estrogen production represents a local system for autocrine and paracrine modulation of vascular function.

Finally, the ability of estradiol to interact with and amplify signals initiated by growth-factor stimulation provides insight into a basic mechanism of the cellular response to stimulation. Katzenellenbogen and colleagues have described systems in which growth factors prolong or exaggerate activation of steroid hormone receptors, enhancing their function (93,94). Additional studies from the groups of Brown (95), O'Malley (96), and Kadonaga (97) have suggested a molecular mechanism for this enhancement. These findings established the concept of "cross-talk" in which growth factors affect steroid hormone signaling. The data presented here suggest that similar cross-talk may occur in the opposite direction as well, with estrogen-ER interaction enhancing the efficacy of other signals transduced across the cell membrane.

REFERENCES

1. A Brodie, Q Lu, and J Nakamura. Aromatase in the normal breast and breast cancer. J Steroid Biochem Mol Biol 61:281–286, 1997.
2. MJ Stampfer, GA Colditz, and WC Willett. Menopause and heart disease: a review. Ann NY Acad Sci 592:193–203, 1990.
3. JW Simpkins, G Rajakumar, YQ Zhang, CE Simpkins, D Greenwald, CJ Yu, N Bodor, and AL Day. Estrogens may reduce mortality and ischemic damage caused by middle cerebral artery occlusion in the female rat. J Neurosurg 87:724–730, 1997.
4. DB Petitti, S Sidney, CP Quesenberry, and A Bernstein. Ischemic stroke and use of estrogen and estrogen/progestogen as hormone replacement therapy. Stroke 29:23–28, 1998.
5. I Wickelgrin. Estrogen stakes claim to cognition. Science 276:675–678, 1997.
6. BM Psaty, SR Heckbert, D Atkins, R Lemaitre, TD Koepsell, PW Wahl, DA Siscovick, and EH Wagner. The risk of myocardial infarction associated with the combined use of estrogens and progestins in postmenopausal women. Arch Intern Med 154:1333–1339, 1994.
7. D Grady, SM Rubin, DB Petitti, CS Fox, D Black, B Ettinger, VL Ernster, and SR Cummings. Hormone therapy to prevent disease and prolong life in postmenopausal women. Ann Intern Med 117:1016–1037, 1992.
8. AR Folsom, PJ Mink, TA Sellers, C-P Hong, W Zheng, and JD Potter. Hormonal replacement therapy and morbidity and mortality in a prospective study of postmenopausal women. Am J Public Health 85:1128–1132, 1995.
9. MH Criqui, L Suarez, E Barrett-Connor, J McPhillips, DL Wingard, and C Garland. Postmenopausal estrogen use and mortality: results from a prospective study in a defined, homogeneous community. Am J Epidemiol 128:606–614, 1988.
10. GA Colditz, SE Hankinson, DJ Hunter, WC Willett, JE Manson, MJ Stampfer, C Hennekens, B Rosner, and FE Speizer. The use of estrogens and progestins and the risk of breast cancer in postmenopausal women. N Engl J Med 332:1589–1593, 1995.
11. JM Sullivan. Estrogen replacement. Circulation 94:2699–2702, 1996.

12. J Haarbo, P Leth-Espensen, S Stender, and C Christiansen. Estrogen monotherapy and combined estrogen-progestogen replacement therapy attenuate aortic accumulation of cholseterol in ovariectomized cholesterol-fed rabbits. J Clin Invest 87:1274–1279, 1991.
13. J Haarbo, LB Nielsen, S Stender, and C Christiansen. Aortic permeability to LDL during estrogen therapy. Arterioscler Thromb 14:243–247, 1994.
14. JL Hough and DB Zilversmit. Effect of 17 beta estradiol on aortic cholesterol content and metabolism in cholesterol-fed rabbits. Arteriosclerosis 6:57–63, 1986.
15. BW Walsh, I Schiff, B Rosner, L Greenberg, V Ravnikar, and FM Sacks. Effects of postmenopausal estrogen replacement on the concentrations and metabolism of plasma lipoproteins. N Engl J Med 325:1196–1204, 1991.
16. MN Sack, DJ Rader, and RO Cannon III. Oestrogen and inhibition of oxidation of low-density lipoproteins in postmenopausal women. Lancet 343:269–270, 1994.
17. M Hashimoto, M Akishita, M Eto, M Ishikawa, K Kozaki, K Toba, Y Sagara, Y Taketani, H Orimo, and Y Ouchi. Modulation of endothelium-dependent flow-mediated dilation of the brachial artery by sex and menstrual cycle. Circulation 92:3431–3435, 1996.
18. JK Williams, MR Adams, DM Herrington, and TB Clarkson. Short-term administration of estrogen and vascular responses of atherosclerotic coronary arteries. J Am Coll Cardiol 20: 452–457, 1992.
19. DM Gilligan, DM Badar, JA Panza, AA Quyyumi, and RO Cannon III. Acute vascular effects of estrogen in postmenopausal women. Circulation 90:786–791, 1994.
20. GM Rubanyi, AD Freay, K Kauser, D Sukovich, G Burton, DB Lubahn, JF Couse, SW Curtis, and KS Korach. Vascular estrogen receptors and endothelium-derived nitric oxide production in the mouse aorta. J Clin Invest 99:2429–2437, 1997.
21. U Lang, RS Baker, and KE Clark. Estrogen-induced increases in coronary blood flow are antagonized by inhibitors of nitric oxide synthesis. Eur J Obstet Gynecol 74:229–235, 1997.
22. P Collins, J Shay, C Jiang, and J Moss. Nitric oxide accounts for dose-dependent estrogen-mediated coronary relaxation after acute estrogen withdrawal. Circulation 90:1964–1968, 1994.
23. GC Wellman, AD Bonev, MT Nelson, and JE Brayden. Gender differences in coronary artery diameter involve estrogen, nitric oxide, and Ca^{2+}-dependent K^+ channels. Circ Res 79:1024–1030, 1996.
24. RE White, DJ Darkow, and JL Falvo Lang. Estrogen relaxes coronary arteries by opening BKCa channels through a cGMP-dependent mechanism. Circ Res 77:936–942, 1995.
25. S-J Chen, H Li, J Durand, S Oparil, and Y-F Chen. Estrogen reduces myointimal proliferation after balloon injury of rat carotid artery. Circulation 93:577–584, 1996.
26. K Krasinski, I Spyridopoulos, T Asahara, R van der Zee, JM Isner, and DW Losodo. Estradiol accelerates functional endothelial recovery after arterial injury. Circulation 95:1768–1772, 1997.
27. FR Rodger, FM Young, HM Fraser, and PJ Illingworth. Endothelial cell proliferation follows the mid-cycle luteinizing hormone surge, but not human chorionic gonadotropin rescue, in the human corpus luteum. Hum Reprod 12:1723–1729, 1997.
28. C Castelo-Branco, MJ Martinez de Osaba, A Fortuny, X Iglesias, and J Gonzalez-Merlo. Circulating hormone levels in menopausal women receiving different hormone replacement therapy regimens: a comparison. J Reprod Med 40:556–560, 1995.
29. S Lum, EA Woltering, WS Fletcher, and RF Pommier. Changes in serum estrogen levels in women during tamoxifen therapy. Am J Surg 173:399–402, 1997.
30. F Naftolin and G Tolis. Neuroendocrine regulation of the menstrual cycle. Clin Obstet Gynecol 21:17–29, 1978.
31. GS Jones. Luteal phase defect. A review of pathophysiology. Curr Opin Obstet Gynecol 3: 641–648, 1991.
32. W Collins, D Jurkovic, T Bourne, A Kurjak, and S Campbell. Ovarian morphology, endocrine function and intra-follicular blood flow during the peri-ovulatory period. Hum Reprod 6:319–324, 1991.

33. GT Meyer and JK McGeachie. Angiogenesis in the developing corpus luteum of pregnant rats: a stereologic and autoradiographic study. Anat Rec 222:18–25, 1988.
34. JL Cavender and WJ Murdoch. Morphological studies of the microcirculatory system of periovulatory ovine follicles. Biol Reprod 39:989–997, 1988.
35. IR Kaiserman-Abramof and HA Padykula. Angiogenesis in the postovulatory primate endometrium: the coiled arteriolar system. Anat Rec 224:479–489, 1989.
36. U Modlich, F-J Kaup, and HG Augustin. Cyclic angiogenesis and blood vessel regression in the ovary: blood vessel regression during luteolysis involves endothelial cell detachment and vessel occlusion. Lab Invest 74:771–780, 1996.
37. M Matsumoto, H Nishinakagawa, M Kurohmaru, Y Hayashi, and J Otsuka. Pregnancy and lactation affect the microvasculature of the mammary gland in mice. J Vet Med Sci 54:937–943, 1992.
38. M Zhou, RL Sutliff, RJ Paul, JN Lorenz, JB Hoying, CC Haudenschild, M Yin, JD Coffin, L Kong, EG Kranias, W Luo, GP Boivin, JJ Duffy, SA Pawlowski, and T Doetschman. Fibroblast growth factor 2 control of vascular tone. Nature Med 4:201–207, 1998.
39. N Ferrara, H Chen, T Davis-Smith, H-P Gerber, T-N Nguyen, D Peers, V Chisholm, KJ Hillan, and RH Schwall. Vascular endothelial growth factor is essential for corpus luteum angiogenesis. Nature Med 4:336–340, 1998.
40. SM Hyder, GM Stancel, C Chiapetta, L Murthy, HL Boetter-Tong, and S Makela. Uterine expression of vascular endothelial growth factor is increased by estradiol and tamoxifen. Cancer Res 56:3954–3960, 1996.
41. N Ravindrath, L Little-Ihrig, HS Phillips, N Ferrara, and AJ Zeleznik. Vascular endothelial growth factor messenger ribonucleic acid expression in the primate ovary. Endocrinology 131: 254–260, 1992.
42. JL Shifrin, JF Tseng, CJ Zaloudek, IP Ryan, YG Meng, N Ferrara, RB Jaffe, and RN Taylor. Ovarian steroid regulation of vascular endothelial growth factor in the human endometrium: implications for angiogenesis during the menstrual cycle and in the pathogenesis of endometriosis. J Clin Endocrinol Metab 81:3112–3118, 1996.
43. J Schechter and R Weiner. Changes in basic fibroblast growth factor coincident with estradiol-induced hyperplasia of the anterior pituitaries of Fisher 344 and Sprague-Dawley rats. Endocrinology 129:2400–2408, 1991.
44. M Rusnati, G Casarotti, S Pecorelli, G Ragnotti, and M Presta. Estro-progestinic replacement therapy modulates the levels of basic fibroblast growth factor (bFGF) in postmenopausal endometrium. Gynecol Oncol 48:88–93, 1993.
45. DE Morales, KA McGowan, DS Grant, S Maheshwari, D Bhartiya, MC Cid, HK Kleinman, and HW Schnaper. Estrogens promote angiogenic activity in human umbilical vein endothelial cells in vitro and in a murine model. Circulation 91:755–763, 1995.
46. A Johns, AD Freay, W Fraser, KS Korach, and GM Rubanyi. Disruption of estrogen receptor gene prevents 17 beta estradiol-induced angiogenesis in transgenic mice. Endocrinology 137: 4511–4513, 1996.
47. AR Gagliardi, B Hennig, and DC Collins. Antiestrogens inhibit endothelial cell growth stimulated by angiogenic growth factors. Anticancer Res 16:1101–1106, 1996.
48. A Gagliardi and DC Collins. Inhibition of angiogenesis by antiestrogens. Cancer Res 53:533–535, 1993.
49. E Furman Haran, AF Maretzek, I Goldberg, A Horowitz, and H Degani. Tamoxifen enhances cell death in implanted MCF7 breast cancer by inhbiting endothelium growth. Cancer Res 54: 5511–5514, 1994.
50. J Folkman and C Haudenschild. Angiogenesis in vitro. Nature 288:551–556, 1980.
51. Y Berthois, JA Katzenellenbogen, and BS Katzenellenbogen. Phenol red in tissue culture media is a weak estrogen: implications concerning the study of estrogen-responsive cells in culture. Proc Natl Acad Sci USA 83:2496–2500, 1986.
52. S Kim-Schulze, KA McGowan, SC Hubchak, MC Cid, MB Martin, HK Kleinman, GL Greene,

and HW Schnaper. Expression of an estrogen receptor by human coronary artery and umbilical vein endothelial cells. Circulation 94:1402–1407, 1996.
53. Y Kubota, HK Kleinman, GR Martin, and TJ Lawley. Role of laminin and basement membrane in the morphological differentiation of human endothelial cells into capillary-like structures. J Cell Biol 107:1589–1598, 1988.
54. HW Schnaper, HK Kleinman, and DS Grant. Role of laminin in endothelial cell recognition and differentiation. Kidney Int 43:20–25, 1993.
55. HW Schnaper, DS Grant, WG Stetler-Stevenson, R Fridman, G D'Orazi, RE Bird, M Hoyhtya, TR Fuerst, J Quigley, D French, and HK Kleinman. Type IV collagenases and TIMPs modulate endothelial cell morphogenesis in vitro. J Cell Physiol 156:235–246, 1993.
56. HW Schnaper, ES Barnathan, A Mazar, S Ellis, S Maheshwari, SL Cortez, WH Baricos, and HK Kleinman. Plasminogen activators augment endothelial cell organization in vitro by two distinct pathways. J Cell Physiol 165:107–118, 1995.
57. FF Isik, NS Gibran, YC Jang, L Sandell, and SM Schwartz. Vitronectin decreases microvascular endothelial cell apoptosis. J Cell Physiol 175:149–155, 1998.
58. S Stromblad, JC Becker, M Yebra, PC Brooks, and DA Cheresh. Suppression of p53 activity and p21WAF1/CIP1 expression by vascular cell integrin $\alpha_V\beta_3$ during angiogenesis. J Clin Invest 98:426–433, 1996.
59. I Spyridopoulos, AB Sullivan, M Kearney, JM Isner, and DW Losordo. Estrogen receptor–mediated inhibition of human endothelial cell apoptosis: estradiol as a survival factor. Circulation 95:1505–1514, 1997.
60. RJ Alvarez, SJ Gips, N Moldovan, CC Wilhide, EE Milliken, AT Hoang, RH Hruban, HS Silverman, CV Dang, and PJ Goldschmidt-Clermont. 17β-estradiol inhibits apoptosis of endothelial cells. Biochem Biophys Res Commun 237:372–381, 1997.
61. R Tsuboi, Y Sato, and DB Rifkin. Correlation of cell migration, cell invasion, receptor number, proteinase production and basic fibroblast growth factor levels in endothelial cells. J Cell Biol 110:511–517, 1990.
62. CB Louise and TG Obrig. Human renal microvascular endothelial cells as a potential target in the development of the hemolytic uremic syndrome as related to fibrinolysis factor expression in vitro. Microvasc Res 47:377–387, 1994.
63. HW Schnaper, KA McGowan, SC Hubchak, MC Cid, and S Kim-Schulze. Endothelial cells, estrogen and angiogenesis. In: U-F Habenicht and JR Lancaster, eds. Ernst Schering Research Foundation Workshop 21: Nitric Oxide, Cytochromes P450 and Sexual Steroids. Berlin: Springer-Verlag, 1997, pp. 283–301.
64. F Grodstein, MJ Stampfer, SZ Goldhaber, JE Manson, GA Colditz, FE Speizer, WC Willett, and CH Hennekens. Prospective study of exogenous hormones and risk of pulmonary embolism in women. Lancet 348:983–987, 1996.
65. KK Koh, R Mincemoyer, MN Bui, G Csako, F Pucino, V Guetta, M Waclawiw, and RO Cannon. Effects of hormone-replacement therapy on fibrinolysis in postmenopausal women. N Engl J Med 336:683–690, 1997.
66. MD Iafrati, RH Karas, M Aronovitz, M King, TR Sullivan, Jr., DB Lubahn, TF O'Donnell, Jr., KS Korach, and ME Mendelsohn. Estrogen inhibits the vascular injury response in estrogen receptor α-deficient mice. Nature Med 3:545–548, 1997.
67. MG Parker. Structure and function of the oestrogen receptor. J Neuroencrinol 5:223–228, 1993.
68. P Colburn and V Buonassisi. Estrogen-binding sites in endothelial cell cultures. Science 201: 817–819, 1978.
69. ML Brandi, C Crescioli, A Tanini, U Frediani, D Agnusdei, and C Gennari. Bone endothelial cells as estrogen targets. Calcif Tissue Int 53:312–317, 1993.
70. CD Venkov, AB Rankin, and DE Vaughan. Identification of authentic estrogen receptor in cultured endothelial cells: a potential mechanism for steroid hormone regulation of endothelial function. Circulation 94:727–733, 1996.

71. DW Losordo, M Kearney, EA Kim, J Jekanowski, and JM Isner. Variable expression of the estrogen receptor in normal and atherosclerotic coronary arteries of premenopausal women. Circulation 89:1501–1510, 1994.
72. RH Karas, BL Patterson, and ME Mendelsohn. Human vascular smooth muscle cells contain functional estrogen receptor. Circulation 89:1943–1950, 1994.
73. KE Friend, LW Ang, and MA Shupnik. Estrogen regulates the expression of several different estrogen receptor mRNA isoforms in rat pituitary. Proc Natl Acad Sci USA 92:4367–4371, 1995.
74. ST Pottraz, T Bellido, H Mocharia, D Crabb, and SC Manolagas. 17β-estradiol inhibits expression of human interleukin-6 promoter-reporter constructs by a receptor-dependent mechanism. J Clin Invest 93:944–950, 1994.
75. SK Banerjee, DR Campbell, AP Weston, and DK Banerjee. Biphasic estrogen response on bovine adrenal medulla capillary endothelial cell adhesion, proliferation and tube formation. Mol Cell Biochem 177:97–105, 1997.
76. GGMJ Kuiper, E Enmark, M Pelto-Huikko, S Nilsson, and J-A Gustafsson. Cloning of a novel estrogen receptor expressed in rat prostate and ovary. Proc Natl Acad Sci USA 93:5925–5930, 1996.
77. K Paech, P Webb, GG Kuiper, S Nilsson, J Gustafsson, PJ Kushner, and TS Scanlan. Differential ligand activation of estrogen receptors ERα and ERβ at AP1 sites. Science 277:1508–1510, 1997.
78. G Pages, P Lenormond, G L'Allemain, et al. Mitogen-activated protein kinases p42mpk and p44mapk are required for fibroblast proliferation. Proc Natl Acad Sci USA 90:8319, 1993.
79. D Bokemeyer, A Sorokin, and MJ Dunn. Multiple intracellular MAP kinase signaling cascades. Kidney Int 49:1187–1198, 1995.
80. J Xing, DD Ginty, and ME Greenberg. Coupling of the RAS-MAPK pathway to gene activation by RSK2, a growth factor-regulated CREB kinase. Science 273:959–963, 1996.
81. BC Duckworth and LC Cantley. Conditional inhibition of the mitogen-activated protein kinase cascade by wortmannin: dependence on signal strength. J Biol Chem 272:27665–27670, 1997.
82. S Kim-Schulze, WL Lowe Jr, and HW Schnaper. Estrogen induces mitogen-activated protein kinase (ERK 1/2) activity in endothelial cells via an autocrine loop involving basic fibroblast growth factor. Circulation 98:413–421, 1998.
83. I Fleming, B Fisslthaler, and R Busse. Calcium signaling in endothelial cells involves activation of tyrosine kinases and leads to activation of mitogen-activated protein kinases. Circ Res 76:522–529, 1995.
84. MLC Albuquerque, SK Akiyama, and HW Schnaper. Human coronary artery endothelial cell secretion of basic fibroblast growth factor is enhanced by matrix proteins, 17β-estradiol and a PKC signaling pathway. Exp Cell Res 245:163–169, 1998.
85. T Fotsis, Y Zhang, MS Pepper, H Adlercreutz, R Montesano, PP Nawroth, and L Schweigerer. The endogenous oestrogen metabolite 2-methoxyoestradiol inhibits angiogenesis and suppresses tumour growth. Nature 368:237–239, 1994.
86. T-L Yue, X Wang, CS Louden, S Gupta, K Pillarisetti, J-L Gu, TK Hart, PG Lysko, and GZ Feuerstein. 2-methoxyestradiol, an endogenous estrogen metabolite, induces apoptosis in endothelial cells and inhibits angiogenesis: possible role for stress-activated preotein kinase signaling pathway and Fas expression. Mol Pharmacol 51:951–962, 1997.
87. LE Rutqvist and A Mattson. Cardiac and thrombolic morbidity among postmenopausal women with early-stage breast cancer in a randomized trial of adjuvant tamoxifen. J Natl Cancer Inst 85:1398–1406, 1993.
88. VL Baker, M Draper, S Paul, S Allerheiligen, M Glant, J Shifren, and RB Jaffe. Reproductive endocrine and endometrial effects of raloxifene hydrochloride, a selective estrogen receptor modulator, in women with regular menstrual cycles. J Clin Endocrinol Metab 83:6–13, 1998.
89. KA Treinen, S Rehm, and PJ Wier. An evaluation of the novel selective estrogen receptor

modulator, idoxifene, for effects on reproduction in rats and rabbits. Toxicol Sci 41:199–207, 1998.

90. S Kato, H Endoh, Y Masuhiro, T Kitamoto, S Uchiyama, H Sasaki, S Masushige, Y Gotoh, E Nishida, H Kawashima, D Metzger, and P Chambon. Activation of the estrogen receptor through phosphorylation by mitogen-activated protein kinase. Science 270:1491–1494, 1995.
91. PB Joel, AM Traish, and DA Lannigan. Estradiol-induced phosphorylation of serine 118 in the estrogen receptor is independent of p42/p44 mitogen-activated protein kinase. J Biol Chem 273:13317–13323, 1998.
92. F Bayard, S Clemens, F Meggetto, N Blaes, G Delsol, and JC Faye. Estrogen synthesis, estrogen metabolism, and functional estrogen receptors in rat arterial smooth muscle cells in culture. Endocrinology 136:1523–1529, 1995.
93. SM Aronica, WL Kraus, and BS Katzenellenbogen. Estrogen action via the cAMP signaling pathway: stimulation of adenylate cyclase and cAMP-regulated gene transcription. Proc Natl Acad Sci USA 91:8517–8521, 1994.
94. H Cho and BS Katzenellenbogen. Synergistic activation of estrogen receptor-mediated transcription by estradiol and protein kinase activators. Mol Endocrinol 7:441–452, 1993.
95. B Hanstein, R Eckner, J DiRenzo, S Halachmi, H Liu, B Searcy, R Kurokawa, and M Brown. p300 is a component of an estrogen receptor coactivator complex. Proc Natl Acad Sci USA 93:11540–11545, 1996.
96. CL Smith, SA Onate, M-J Tsai, and BW O'Malley. CREB binding to protein acts synergistically with steroid receptor coactivator-1 to enhance steroid receptor-dependent transcription. Proc Natl Acad Sci USA 93:8884–8888, 1996.
97. WL Kraus and JT Kadonaga. p300 and estrogen receptor cooperatively activate transcription via differential enhancement of initiation and reinitiation. Genes Dev 12:331–342, 1998.
98. T Gura. Estrogen: key player in heart disease among women. Science 269:771–773, 1995.

13
Nitric Oxide and Angiogenesis

Anthony Johns and Stefan Zöllner
Berlex Biosciences, Richmond, California

Angiogenesis primarily occurs during the development of the organism and during embryogenesis; it is essential for normal development and survival. The extent of angiogenesis in the adult organism is almost completely downregulated, although it has been demonstrated that there is limited angiogenesis in the postnatal period (1,2). There is almost complete downregulation of angiogenesis in the adult with the exception of the female reproductive tract (3); however, angiogenesis does occur in pathological conditions where it may be desirable (wound healing and cardiac or limb ischemia) or undesirable (diabetic retinopathy or tumor growth).

In the past decade a plethora of angiogenic stimulators and inhibitors have been described. The endogenous agents that stimulate angiogenesis include peptide growth factors, cytokines, hormones, chemokines, and hematopoietic growth factors. A list of these agents is given in Table 1. There have also been a number of reports of endogenous inhibitors of angiogenesis, shown in Table 2. The balance between the many endogenous activators and inhibitors can therefore provide very precise control of angiogenesis in any one region of the whole organism. Local demand for new blood flow to any one region of the body, owing either to a physiological demand (i.e., exercise or pregnancy) or a pathological condition (ischemia due to peripheral occlusive vascular disease) has also been demonstrated to increase angiogenesis and therefore blood vessel density. These known physiological stimuli all involve conditioning of the vascular bed to profound vasodilation.

Furchgott and Zawadzki first described the common vasodilator mechanism in all blood vessels to be due to the release of an endothelium-derived relaxing factor (EDRF), which was later identified as nitric oxide (NO), a molecule with profound vasomotor regulatory effects (4). Subsequent investigations established a plethora of diverse regulatory activities for NO, including inhibition of platelet aggregation, leukocyte adherence, and smooth muscle proliferation (5). Studies on endothelial cells in culture have also revealed an inhibition of proliferation of these cells by compounds known to release NO (SIN-1); however, in these experiments, high concentrations of NO donors were used (1 mM) and the NO concentration was probably not physiological (6). NO synthesis has been demonstrated to be tightly regulated by a family of isoenzymes (7,8) with different cell distribution patterns. All the isoforms share the common property of catalyzing the conversion of L-arginine to L-citrulline, yielding free NO (9,10). All the NO synthase

Table 1 Endogenous Promotors of Angiogenesis

Peptide growth factors
Acidic fibroblast growth factor (aFGF)
Basic fibroblast growth factor (bFGF)
Vascular edothelial growth factors A, B, C, D (VEGF-A, -B, -C, -D)
Placental growth factor (PIGF)
Platelet-derived growth factor (PDGF)
Hepatocyte growth factor (HGF)
Transforming growth factors α, β (TGF-α, -β)
Insulin-like growth factor 1 (IGF-1)
Hormones
Estrogens
Follistatin
Proliferin
Prostaglandin E1, E2
Cytokines, chemokines, and immune mediators
Tumor necrosis factor α (TNF-α)
Monocyte chemoattractant protein 1 (MCP-1)
Interleukin 8 (IL-8)
Hemotopoietic growth factors
Erythropoietin
Granulocyte colony-stimulating factor (G-CSF)
Granulocyte-macrophage colony-stimulating factor (GM-CSF)
Enzymes
Platelet-derived endothelial cell growth factor (PD-ECGF; thymidine phosphorylase)
Angiogenin (ribonuclease A homologue)

Table 2 Endogenous Inhibitors of Angiogenesis

Peptides
Angiotensin
Endostatin
Laminin peptides
Fibronectin peptides
Hormones
2-Methoxyestradiol (2-ME)
Proliferin-related protein
Cytokines, chemokines, and immune mediators
Tumor necrosis factor α (TNF-α)
Interleukin 12 (IL-12)
Interferons
Platelet factor 4 (PF-4)
Interferon γ–inducible protein (IP-10)
Enzymes inhibitors
Plasminogen activator inhibitors 1, 2 (PAI-1, -2)
Tissue metalloproteinase inhibitors 1, 2, 3, 4 (TIMP-1, -2, -3, -4)

isoforms share a common substrate, and this fact has been exploited in both in vitro and in vivo studies. Administration of L-arginine supplements has been shown to reduce vasomotor tone (11), whereas deletion of the gene encoding for the endothelial cell specific NO synthase (NOS 3, eNOS) has been shown to result in systemic hypertension (12) and pulmonary vasoconstriction (13).

There are many published reports implying that NO may be involved in angiogenesis. These papers include reports that the endothelial cell–specific growth factor VEGF can increase cytosolic free calcium in cultured endothelial cells (14) and that, in isolated canine coronary artery, VEGF causes an endothelial-dependent, dose-dependent relaxation that could be attenuated by pretreatment with N^G-monomethyl-L-arginine (L-NMMA) (15). Isner's group also confirmed the release of NO from normal arteries by treatment with VEGF (16) and extended these studies to demonstrate that VEGF promoted recovery of the ischemia-induced endothelial-dependent relaxation in the rabbit hindlimb. Subsequent work by the same group demonstrated that a hypotensive effect of infusion of VEGF could be demonstrated in animals (17) as well as humans (18) and that this treatment inhibited neointimal thickening in two different animal restenosis models where arterial injury had been induced by balloon angioplasty (19,20).

Indirect evidence that NO can be involved in an angiogenic response is available from the studies in oncology. Increased levels of NO have been reported in human tumors (21,22), and it has been demonstrated that transfection of the inducible NO synthase into an adenocarcinoma cell line gave rise to more vascularized tumors than the wild-type cells when injected into animals (23). Other experiments using inhibitors of NO synthase also strongly support the importance of NO in xenografted tumor vascularization (24,25). Ziche's group has also demonstrated that vascularization of tumors that have been initiated with breast cancer cells overexpressing VEGF can be attenuated by treatment with NO synthase inhibitors (26).

The link between the release of NO from endothelial cells and blood vessel formation by the known angiogenic factor VEGF has been extended to determine if there was a more direct link between the production of NO induced by VEGF and angiogenesis. Papapetropoulos et al. (27,28) and Morbidelli et al. (29) demonstrated that many of the angiogenic responses to VEGF in vitro could be correlated to increases in cGMP in the cultured endothelial cells and that the angiogenic responses elicited by VEGF could be attenuated by inhibitors of NO synthesis. These angiogenic responses included not only endothelial cell proliferation but also migration and tube formation. There is also evidence that similar results can be demonstrated in vivo, since Ziche and coworkers have demonstrated that, in rabbits, corneal angiogenesis induced by VEGF but not bFGF can be inhibited by treatment of the animals with inhibitors of NO production (26,30).

There are also a number of papers suggesting that NO has a negative influence on the angiogenic process, making the role of NO in angiogenesis controversial. Lau and Ma (31), for example, reported that NO inhibits migration of cultured endothelial cells, an essential step for angiogenesis (32). Pipili-Synetos et al. (33) reported that NO donors inhibit angiogenesis in the chick chorioallantoic membrane and tube formation in the Matrigel assay (34). In contrast, Guo et al. (35) showed that exogenous administration of a novel NO donor stimulated proliferation of cultured rat aortic endothelial cells, and Ziche et al. suggested that NO may play a role in angiogenesis elicited not only by VEGF (26,29) but also by substance P (30). They also reported that the other known angiogenesis promotor, bFGF (26), was not mediated by NO, although, like VEGF, it has been demonstrated to cause endothelial-dependent vasodilation and NO release from endothelial cells.

In order to elucidate more fully the role of NO in ischemia-induced angiogenesis, Isner's group has performed important experiments in both rabbits and mice (36).

In the rabbit ischemic hindlimb model (36), they found that dietary L-arginine supplementation significantly decreased blood pressure, reduced arterial resistance, increased flow at rest, and increased flow reserve in the ischemic limb of L-arginine-treated animals as compared with untreated controls. This improvement in the functional indexes persisted for 3 days following discontinuation of the L-arginine, suggesting a structural change in the blood flow to the ischemic limb. Physical evidence for an increased collateral blood flow to the ischemic limb in the L-arginine–treated animals was also obtained by angiography performed in vivo and by histological examination performed at necropsy.

Isner's group also investigated the effects of ischemia on angiogenesis in both normal mice and in eNOS-knock-out (KO) mice (36). The degree of angiogenesis was determined using both laser Doppler imaging in situ and by measuring capillary density, histologically, following necropsy. In the eNOS-KO mice the spontaneous angiogenic response to ischemia was severely attenuated as compared with the normal mice. The authors also showed that the levels of VEGF were comparable in the ischemic limbs of both the control and KO mice, suggesting that the reason for the attenuated response to ischemia was the lack of NO release from the endothelium of the eNOS-KO mice. In order to rule out any involvement of VEGF, Isner's group also transfected a vector containing VEGF into the mice. This vector had already been shown to compensate for the low angiogenic response in Apo E-KO mice (37). This vector, or injection of recombinant VEGF protein, failed to improve the angiogenic response in the eNOS-KO animals, confirming that VEGF was not the causative agent in the lack of angiogenesis in these mice. The authors argue that the attenuation of angiogenesis in these mice cannot be accounted for by the hypertension in these animals (11,38), since other hypertensive animals show normal angiogenic responses to ischemia (39,40).

These results from Isner's group support the findings of Papapetropoulos et al., who showed that NO produced by the endothelial cells in response to the addition of VEGF was required for the maximum action of this growth factor on both capillary tube formation in vitro and on the development of vascular networks (27,28). These results are also consistent with the in vivo studies that first demonstrated a general role of NO in angiogenesis performed by Ziche et al. (26) and the specific role of NO in being the downstream mediator of the action of VEGF in vivo (26). Ziche et al. were also the first to demonstrate a specific effect of the non-isozyme-specific NO synthesis inhibitor L-NAME on the angiogenesis produced by VEGF but not bFGF (26). The effective blockade of the VEGF action in the eNOS-KO mice (36) would therefore suggest that the L-NAME effect seen by Ziche et al. (26) was due to inhibition of the eNOS isozyme.

The ability to restore the effects of ischemia on angiogenesis in the eNOS-KO mice with a NO donor was not successful (36), so a positive role of exogenously applied NO on angiogenesis still has to be demonstrated. This would suggest that either the amount of NO or the time at which the NO is given is critical to observing a positive effect of exogenous NO on angiogenesis.

In conclusion, the evidence would suggest that endogenous NO can act as a modulator of angiogenesis in a number of clinically important situations and that endogenous NO is one of the main mediators of the action of VEGF as an angiogenic agent by a mechanism yet to be elucidated. The evidence for NO being an important regulator in the angiogenic response to ischemia is the most convincing. The implications of NO being an important angiogenic factor would also suggest that any disturbance of the endothelial function of releasing NO would have detrimental effects on processes that require angiogenesis for the normal functioning of the organism. NO may be required for endothelial cell proliferation and migration (29,41,42), and since inhibition of NO synthesis has multiple other effects, alternative mechanisms, including increased leukocyte-endothelial interaction (43), increased perimicrovascular fibrosis, hyperreactivity to serotonin (44), and increased activity of tissue angiotensin-converting enzyme (45) cannot be excluded. However, in vivo studies have established that endothelial dysfunction (measured as the release of NO) can be partially reversed in response to L-arginine administration independent of the presence or absence of global L-arginine deficiency (46–52). Also, Weyrich et al. (53) have shown that L-arginine could ameliorate myocardial ischemia-reperfusion injury in a cat model and that this effect was associated with the reestablishment of a more normal endothelial function.

REFERENCES

1. W Risau, I Flamme. Vasculogenesis. Annu Rev Cell Dev Biol 11:73–91, 1995.
2. L Beck Jr, PA D'Amore. Vascular development: cellular and molecular regulation. FASEB J 11:365–373, 1997.
3. LP Reynolds, SD Killilea, DA Redmer. Angiogenesis in the female reproductive system. FASEB J 6:886–892, 1992.
4. RF Furchgott, JV Zawadzki. The obligatory role of endothelial cells in the relaxation of arterial smooth muscle by acetylcholine. Nature 288:373–376, 1980.
5. HH Schmidt, U Walter. NO at work. Cell 78:919–925, 1994.
6. A Lopez-Farre, L Sanchez de Miguel, C Caramelo, J Gomez-Macias, R Garcia, JR Mosquera, T de Frutos, I Millas, F Rivas, G Echezarreta, S Casado. Role of nitric oxide in autocrine control of growth and apoptosis of endothelial cells. Am J Physiol 272:H760–H768, 1997.
7. U Forstermann, II Closs, JS Pollock, M Nakane, P Schwarz, I Gath, H Kleinert. Nitric oxide synthase isoenzymes: characterization, purification, molecular cloning, and functions. Hypertension 23:1121–1131, 1994.
8. WC Sessa. The nitric oxide synthase family of proteins. J Vasc Res. 31:131–143, 1994.
9. S Moncada, EA Higgs. The L-arginine-nitric oxide pathway. N Engl J Med 329:2002–2012, 1993.
10. LJ Ignarro. Biosynthesis and metabolism of endothelium-derived nitric oxide. Annu Rev Pharmacol Toxicol 30:535–560, 1990.
11. MA Creager, SJ Gallagher, XJ Girerd, SM Coleman, VJ Dzau, JP Cooke. L-arginine improves endothelium-dependent vasodilation in hypercholesterolemic humans. J Clin Invest 90:1248–1252, 1992.
12. PL Huang, Z Huang, H Mashimo, KD Bloch, MA Moskowitz, JA Bevan, MC Fishman. Hypertension in mice lacking the gene for endothelial nitric oxide synthase. Nature 377:239–242, 1995.
13. W Steudel, F Ichinose, P Huang, WE Hurford, RC Jones, JA Bevan, MC Fishman, WM Zapol. Pulmonary vasoconstriction and hypertension in mice with targeted disruption of the endothelial nitric oxide synthase (NOS3) gene. Circ Res 81:34–41, 1997.
14. TA Brock, HF Dvorak, DR Senger. Tumor-secreted vascular permeability factor increases cytosolic Ca^{2+} and von Willebrand factor release in human endothelial cells. Am J Pathol 138:213–221, 1991.
15. DD Ku, JK Zaleski, S Liu, TA Brock. Vascular endothelial growth factor induces EDRF-dependent relaxation in coronary arteries. Am J Physiol 265:H586–H592, 1993.

16. R van der Zee, T Murohara, Z Luo, F Zollmann, J Passeri, C Lekutat, JM Isner. Vascular endothelial growth factor (VEGF)/vascular permeability factor (VPF) augments nitric oxide release from quiescent rabbit and human vascular endothelium. Circulation 95:1030–1037, 1997.
17. JR Horowitz, A Rivard, R van der Zee, MD Hariawala, DD Sheriff, DD Esakof, GM Chaudhry, JF Symes, JM Isner. Vascular endothelial growth factor/vascular permeability factor produces nitric oxide–dependent hypotension. Arterioscler Thromb Vasc Biol 17:2793–2800, 1997.
18. JM Isner. Angiogenesis. In: EJ Topol, ed. Textbook of Cardiovascular Medicine. Philadelphia: Lippincott-Raven, pp. 2491–2518, 1998.
19. T Asahara, C Bauters, CJ Pastore, M Kearney, S Rossow, S Bunting, N Ferrara, JF Symes, JM Isner. Local delivery of vascular endothelial growth factor accelerates reendothelialization and attenuates intimal hyperplasia in balloon-injured rat carotid artery. Circulation 91:2793–2801, 1995.
20. T Asahara, D Chen, Y Tsurumi, M Kearney, S Rossow, J Passeri, J Symes, JM Isner. Accelerated restitution of endothelial integrity and endothelium-dependent function following phVEGF165 gene transfer. Circulation 94:3291–3302, 1996.
21. LL Thomsen, FG Lawton, RG Knowles, JE Beesley, V RiverosMoreno, S Moncada. Nitric oxide synthase activity in human gynecological cancer. Cancer Res 54:1352–1354, 1994.
22. CS Cobbs, JE Brenman, KD Alpade, DS Bredt, MA Israel. Expression of nitric oxide synthase in human central nervous system tumors. Cancer Res 55:727–730, 1995.
23. DC Jenkins, IG Charles, LL Thomsen, DW Moss, LS Holmes, SA Baylis, P Rhodes, K Westmore, PC Emson, S Moncada. Role of nitric oxide in tumor angiogenesis. Proc Natl Acad Sci USA 92:4392–4396, 1995.
24. GD Kennovin, DG Hirst, MRL Stratford, FW Flitney. Inducible nitric oxide synthase is expressed in tumor associated vasculature; inhibition retards tumor growth in vivo. In: S Moncada, M Feelish, R Busse, and EA Higgs, eds. Biology of Nitric Oxide: Vol. 4. Enzymology, Biochemistry and Immunology. London: Portland Press, 1994, pp. 473–479.
25. A Orucevic, PK Lala. N^G-nitro-L-arginine methyl ester, an inhibitor of nitric oxide synthesis, ameliorates interleukin 2-induced capillary leakage and reduces tumor growth in adenocarcinoma-bearing mice. Br J Cancer 73:189–196, 1996.
26. M Ziche, L Morbidelli, R Choudhuri, H-T Zhang, S Donnini, HJ Granger, R Bicknell. Nitric oxide synthase lies downstream from vascular endothelial growth factor-induced but not fibroblast growth factor-induced angiogenesis. J Clin Invest 99:2625–2634, 1997.
27. A Papapetropoulos, KM Desai, RD Rudic, B Mayer, R Zhang, MP Ruiz-Torres, G Garcia-Cardena, JA Madri, WC Sessa. Nitric oxide synthase inhibitors attenuate transforming-growth-factor-b1-stimulated capillary organization in vitro. Am J Pathol 150:1835–1844, 1997.
28. A Papapetropoulos, G Garcia-Cardena, JA Madri, WC Sessa. Nitric oxide production contributes to the angiogenic properties of vascular endothelial growth factor in human endothelial cells. J Clin Invest 100:3131–3139, 1997.
29. L Morbidelli, C-H Chang, JG Douglas, HJ Granger, F Ledda, M Ziche. Nitric oxide mediates mitogenic effect of VEGF on coronary venular endothelium. Am J Physiol 270:H411–H415, 1995.
30. M Ziche, L Morbidelli, E Masini, S Amerini, HJ Granger, CA Maggi, P Geppetti, F Ledda. Nitric oxide mediates angiogenesis in vivo and endothelial cell growth and migration in vitro promoted by substance P. J Clin Invest 94:2036–2044, 1994.
31. Y-T Lau, W-C Ma. Nitric oxide inhibits migration of cultured endothelial cells. Biochem Biophys Res Commun 221:670–674, 1996.
32. MM Sholley, GP Ferguson, HR Seibel, JL Montour, JD Wilson. Mechanisms of neovascularization: vascular sprouting can occur without proliferation of endothelial cells. Lab Invest 51: 624–634, 1984.
33. E Pipili-Synetos, E Sakkoula, ME Maragoudakis. Nitric oxide is involved in the regulation of angiogenesis. Br J Pharmacol 108:855–857, 1993.

34. E Pipili-Synetos, E Sakkoula, G Haralabopoulos, P Andriopoulou, P Peristeris, ME Maragoudakis. Evidence that nitric oxide is an endogenous antiangiogenic mediator. Br J Pharmacol 111:894–902, 1994.
35. J-P Guo, MM Panday, PM Consigny, AM Lefer. Mechanisms of vascular preservation by a novel NO donor following rat carotid artery intimal injury. Am J Physiol 269:H1122–H1131, 1995.
36. T Murohara, T Asahara, M Silver, C Bauters, H Masuda, C Kalka, M Kearney, D Chen, JF Symes, MC Fishman, PL Huang, JM Isner. Nitric oxide synthase modulates angiogenesis in response to tissue ischemia. J Clin Invest 101:2567–2578, 1998.
37. T Couffinhal, M Silver, LP Zheng, M Kearney, B Witzenbichler, JM Isner. Angiogenesis is impaired in ApoE Knock out mice due to a reduced expression of vascular endothelial growth factor (abstr). Circulation 94:1–102, 1996.
38. EG Shesely, N Maeda, H-S Kim, KM Desai, JH Krege, VE Laubach, PA Sherman, WC Sessa, O Smithies. Elevated blood pressures in mice lacking endothelial nitric oxide synthase. Proc Natl Acad Sci USA 93:13176–1318, 1996.
39. RJ Tomanek, JC Searls, PA Lachenbruch. Quantitative changes in the capillary bed during developing peak and stabilized cardiac hypertrophy in the spontaneously hypertensive rat. Circ Res 51:295–304, 1982.
40. P Anversa, AV Loud, F Giacomelli, J Weiner. Absolute morphometric study of myocardial hypertrophy in experimental hypertension: II. Ultrastructure of myocardium and interstitium. Lab Invest 38:597–609, 1978.
41. J-P Guo, T Murohara, MM Panday, AM Lefer. Nitric oxide promotes endothelial cell proliferation: role in inhibiting restenosis (abstr). Circulation 92:1–750, 1996.
42. JF Arnal, T Munzel, RC Venema, NL James, CL Bai, WE Mitch, DG Harrison. Interaction between L-arginine and L-glutamine change endothelial NO production: an effect independent of NO synthase substrate availability. J Clin Invest 95:2565–2572, 1995.
43. P Kubes, M Suzuki, DN Granger. Nitric oxide:an endogenous modulator of leukocyte-adhesion. Proc Natl Acad Sci USA 88:4651–4655, 1991.
44. A Ito, K Egashira, T Kadokami, Y Fukumoto, T Takayanagi, R Nakaike, T Kuga, K Sueishi, H Shimokawa, A Takeshita. Chronic inhibition of endothelium-derived nitric oxide synthesis causes coronary microvascular structural changes and hyperreactivity to serotonin in pigs. Circulation 92:2636–2644, 1995.
45. M Takemoto, K Egashira, M Usui, K Numaguchi, H Tomita, H Tsutsui, H Shimokawa, K Sueishi, A Takeshita. Important role of tissue angiotensin-converting enzyme activity in the pathogenesis of coronary vascular and myocardial structural changes induced by long-term blockade of nitric oxide synthesis in rats. J Clin Invest 99:278–287, 1997.
46. Y Tsurumi, T Murohara, K Krasinski, C Dongfen, B Witzenbichler, M Kearney, T Couffinhal, JM Isner. Reciprocal relationship between VEGF and NO in the regulation of endothelial integrity. Nature Med 3:879–886, 1997.
47. RH Boger, SM Bode-Boger, W Thiele, W Junker, K Alexander, JC Frolich. Biochemical evidence for impaired nitric oxide synthesis in patients with peripheral arterial occlusive disease. Circulation 95:2068–2074, 1997.
48. B-Y Wang, RC Candipan, M Arjomandi, PT Hsium, PS Tsao, JP Cooke. Arginine restores nitric oxide activity and inhibits monocyte accumulation after vascular injury in hypercholesterolemic rabbits. J Am Coll Cardiol 28:1573–1579, 1996.
49. A Chauhan, RS More, PA Mullins, G Taylor, MC Petch, PM Schofield. Aging-associated endothelial dysfunction in humans is reversed by L-arginine. J Am Coll Cardiol 28:1796–1804, 1996.
50. M Hamon, B Vallet, C Bauters, N Wernert, EP McFadden, J-M Lablanche, B Dupuis, ME Bertrand. Long-term oral administration of L-arginine reduces intimal thickening and enhances neoendothelium-dependent acetylcholine-induced relaxation after arterial injury. Circulation 90:1357–1362, 1994.

51. SM Bode-Boger, RH Boger, H Alfke, D Heinzel, D Tsikas, A Creutzig, K Alexander, J Frolich. L-arginine induces nitric-oxide dependent vasodilation in patients with critical limb ischemia: a randomized controlled study. Circulation 93:85–90, 1996.
52. XJ Girerd, AT Hirsch, JP Cooke, VJ Dzau, MA Creager. L-arginine augments endothelium-dependent vasodilation in cholesterol-fed rabbits. Circ Res 67:1301–1308, 1990.
53. AS Weyrich, X-L Ma, AM Lefer. The role of L-arginine in ameliorating reperfusion injury after myocardial ischemia in the cat. Circulation 86:279–288, 1992.

14
Reactive Oxygen Species and Angiogenesis

Peter I. Lelkes and Christopher R. Waters
University of Wisconsin Medical School, Milwaukee, Wisconsin

I. INTRODUCTION

Oxygen homeostasis is of critical importance for maintaining the viability of all tissues. Attempts to restore normal oxygen levels after episodes of hypoxia and/or ischemia paradoxically result in further damage through the generation of free radicals, such as superoxide ion (O_2^-), hydrogen peroxide (H_2O_2), and hydroxyl radical ($\cdot OH$), cumulatively termed reactive oxygen species (ROS). Excess ROS production results in extensive ''oxidative injury,'' which has been implicated in the pathogenesis of numerous diseases, prominently in ischemia/reperfusion injuries to the brain or to the cardiovascular system—e.g. after myocardial infarction (1,2). Clinically, prevention of ROS generation by using antioxidants or other ROS scavengers holds great promise for mitigating hypoxia/reperfusion–related injuries.

Lack of oxygen, generally associated with the initiation and progression of diverse disease states, can also trigger unique ''repair'' mechanisms: hypoxia per se might act as the crucial inducer of angiogenesis/neovascularization,—namely, the very same mechanism that is aimed at restoring and securing adequate oxygen supply, via hypoxia-sensitive promoter elements in a number of angiogenesis-related genes, such as growth factors and their cognate receptors. For example, hypoxia enhances expression/production of vascular endothelial cell growth factor (VEGF) in vascular smooth muscle cells or fibroblasts, concomitant with the upregulation of cognate VEGF receptors, such as KDR/flk, and flt on endothelial cells (3–8). As these genes are upregulated in response to hypoxia, the gene products initiate or contribute to the formation of collateral blood vessels (9). Indeed, these recent discoveries form the basis for novel gene therapy based approaches to ''therapeutic'' angiogenesis (10).

These studies clearly imply a stimulating role for hypoxia in angiogenesis in vivo, which is indirect; that is, it requires paracrine signaling. Previous studies have shown that (mostly long-term) exposure to hypoxia affects physiological responses of cultured EC, such as proliferation, expression of adhesion molecules, and secretion of fibrinolytic enzymes and cytokines (11–14). However, it remains unclear what if any role hypoxia per se plays in activating the angiogenic endothelial cell phenotype. This issue is currently being addressed using several in vitro model systems of angiogenesis (15). For example,

Phillips and coworkers have recently shown that chronic exposure to hypoxia enhances the migration of calf pulmonary microvascular EC into collagen gels and the formation of capillary-like tubes. In this model, hypoxia-induced tubulogenesis involves autocrine mechanisms, such as an increase in the expression/release of angiogenic growth factors (presumably bFGF) or cytokines, concomitant with enhanced secretion of fibrinolytic/matrix-degrading enzymes (16).

In contrast, when endothelial cells are plated (in the absence of hypoxia) on the complex, laminin-rich reconstituted basement membrane (Matrigel) extracted from the murine Englebreth-Holm-Swarm (EHS) tumor, angiogenic factors contained in this three-dimensional (3D) matrix will stimulate the cells to rapidly (within hours) form a capillary-like network of tubular structures. For the sake of simplicity we will refer to this process as *tube-formation* or *tubular morphogenesis* (17,18). We recently reported that a brief (<60 min) exposure to hypoxia, followed by reoxygenation, enhanced "spontaneous" tubular morphogenesis of human microvascular endothelial cells (HMVEC) cultured on Matrigel (19). These conditions of hypoxia/reoxygenation (H/R) are clinically relevant in that they are reminiscent of transient ischemia, in which tissues experience brief episodes of ischemia/hypoxia followed by prolonged reoxygenation. Previous studies have demonstrated that exposure of endothelial cells to H/R leads to the intracellular generation of ROS (20,21).

We hypothesized that these endogenous ROS might directly and rapidly enhance the angiogenic properties of the cells cultured on a permissive 3D extracellular matrix, such as Matrigel (22). In support of our hypothesis, we found that exposure of human microvascular endothelial cells grown on Matrigel to a brief episode of hypoxia (40–60 min) and subsequent reoxygenation significantly accelerated (up to threefold) the rate of tubular morphogenesis as determined by computer-aided morphometry. This effect was not dependent on activation of PKC or upregulation/release of angiogenic growth factors. Rather, H/R but not hypoxia alone caused the formation of ROS and the activation of the nuclear transcription factor NF6B, both of which were inhibited by diverse ROS scavengers. Furthermore, tube formation was inhibited, also under normoxic conditions, by the same antagonists in a dose-dependent fashion, suggesting a correlation between ROS activation of NF6B and the angiogenic process (19).

In this chapter we review and further explore how ROS might affect some of the distinct steps in the angiogenic cascade, such as EC migration, proliferation, and initiation of matrix remodeling. Our data suggest that ROS generated by exposure to H/R might be a necessary and perhaps sufficient second-messenger system for the initiation of the cellular signaling cascade that leads to tubular morphogenesis. We hypothesize that, in the clinical setting of hypoxia/reoxygenation—e.g., during ischemic preconditioning, enhanced activation of ROS-dependent intracellular signaling may accelerate the rate of neovascularization also in vivo, thus contributing to the alleviation of certain ischemic lesions.

II. MATERIALS AND METHODS

A. Cell Culture

Normal human dermal microvascular endothelial cells (HMVEC, from Clonetics Corporation, San Diego, CA) and pooled human aortic endothelial cells (pHAEC, pooled from three different donors) were grown in an optimized cell culture medium (1:1 mixture of MCDB 131/M199 with supplements), as previously described (23–25). Most of the phenotypic traits of these EC, including their capability to rapidly form tubes on Matrigel (in vitro angiogenesis), were maintained through the tenth passage, corresponding to about 30 population doublings (26,27).

B. In Vitro Tube Formation on Matrigel

Matrigel was prepared from EHS tumor according to published procedures (18,28). Twelve-well cell culture plates (surface area = 3.5 cm^2/well, from Costar) were coated with 150 μL/cm^2 ice-cold Matrigel and incubated overnight at 37°C. Following two washes with phosphate-buffered saline (PBS), EC were seeded at the densities indicated below; the ensuing tubes were evaluated 2–24 h after the initiation of the experiments. The total length of the tubes formed was quantitated by computer-aided image analysis using established morphometric methods (19,29,30). As inhibitors of reactive oxygen species (ROS), we used several ''free radical'' scavengers, pyrollidine dithiocarbamate (PDTC), N-acetyl cysteine (NAC), superoxide dismutase (SOD), and catalase.

C. Generation of Hypoxia/Reoxygenation (H/R)

Hypoxia was generated by flooding a conventional CO_2 incubator with a mixture of 95% N_2/5% CO_2. Dissolved oxygen in the media was measured on line with a calibrated oxygen electrode (from ORION). The cells, plated on Matrigel, were exposed for the indicated time periods to hypoxia and then transferred into normoxia (19,30). For some of the biochemical experiments detailed below, confluent EC monolayers were exposed to the same regimen of H/R. As a positive control for ROS generation, confluent EC monolayers were exposed to H_2O_2 at the concentrations and for the times indicated below. The generation of ROS was verified using 2,7-dichlorodihydrofluorescein diacetate (DCF, from Molecular Probes) as an indicator, as previously described (19,30).

D. Cell Proliferation

The effect of H/R and ROS inhibitors on EC proliferation was assessed using a fluorimetric assay, based on measuring the amount of nuclear staining with DNA-specific stain, Hoechst 22508 (31). Briefly, 24 hr after seeding the cells in 96 wells at low densities (5000 cells per well), the cells were exposed to various concentrations of the ROS antagonists described above. Twenty four hours later, the cell numbers were assayed, as previously described. Parenthetically, in the presence of ROS inhibitors, other established cell-enumeration assays [MTT or Alamar Blue (32,33)], which rely on processing of the indicator molecules by mitochondrial enzymes, returned invalid results.

E. Cell Migration

The effect of H/R and of various ROS inhibitors on EC migration was assessed using established techniques. In brief, confluent EC monolayers were ''wounded'' with a homemade plastic comb (prepared from a cell scraper) and, after washing to remove the debris, exposed to various ROS inhibitors at the doses indicated. The extent of cell migration was evaluated after 24 hr using computer-aided morphometry.

F. Zymographic Detection of Metalloproteinase Production

The effects of H/R on the secretion of matrix degrading metalloproteinases (MMP) was evaluated by conventional zymography (34). In brief, EC were plated on Matrigel under conditions of H/R as well as appropriate normoxic/hypoxic controls, as detailed above. Eight hours after plating, the supernatant was collected and analyzed by SDS-PAGE, using precast 10% Tris-Glycine gels containing 0.1% gelatine (Novex, San Diego, CA). Following electrophoresis, gels were washed for 1 hr at RT in distilled water containing 2.5% (v/v) Triton X-100. The gels were then incubated overnight at 37°C in developing buffer [1 mM Tris-Base, 40 mM Tris-HCl, 200 mM NaCl, 7 mM CaCl2, and 0.2% (v/v) Nonidet P-40 in distilled water], stained for 4 hr with Coommassie Blue, and then destained in distilled water. The gels were photographed and the bands evaluated densitometrically.

G. H/R-Induced Signalization

In order to analyze H/R-activated intracellular signaling pathways, confluent pHAEC were exposed to various conditions of H/R for the times indicated below. Activation of the mitogen-activated protein kinase (MAPK), also known as extracellular signal-regulated kinase (ERK), was assessed by Western blotting, using specific antibodies (from Santa Cruz) against ERK and its phosphorylated form (pERK), as previously described (35).

H. Statistical Analysis

All experiments were carried out in triplicate and repeated at least three times unless indicated otherwise. The results, presented as means ± standard error of the means (SEM), were evaluated by ANOVA and Bonferroni's posttest analysis using SigmaStat (Jandel), with $p < 0.05$ being considered as statistically significant.

III. RESULTS AND DISCUSSION

The Matrigel assay is a particularly useful in vitro angiogenesis system in that it adequately models certain dynamic stages in the angiogenic cascade, such as EC migration, elongation, and assembly of tubes (36). One of the main advantages of the Matrigel system over other 3D gel assemblies, such as collagen or fibrin gels, is that tubular morphogenesis on Matrigel occurs relatively fast (approximately 8–24 h), depending on the seeding density and the particular cell type used (27). A potential disadvantage is the fact that unless one deals with growth factor–depleted Matrigel, tube formation in this matrix is a "spontaneous" event, presumably initiated by a combination of environmental cues contained in this complex mixture of extracellular matrix proteins, growth factors, and fibrinolytic/matrix–degrading enzymes (36). Inhibition of "spontaneous" tube formation by anti-angiogenic compounds is rather straightforward to assess (18,28). Detecting the effects of proangiogenic compounds/conditions, such as the tumor-promoting agent phorbol myristoyl acetate (PMA), thrombin, or H/R, requires more sophisticated approaches (19,30).

Recently, Philips and coworkers have shown that chronic exposure to hypoxia (at least 3 days) enhances tubular morphogenesis of endothelial cells growing on collagen gels (37). In our system, however, exposure to hypoxia (P_{O_2}, 7.5 mmHg) for up to 24 hr did not affect the rate or extent of capillary-like tube formation by HMVEC. Rather, in the Matrigel system, exposure to 40–60 min of hypoxia with subsequent reoxygenation for 8 hr (H/R) led to a two- to threefold acceleration/enhancement of capillary tube formation (38). Subsequently, we demonstrated that the acceleration of tubular morphogenesis by H/R did not involve some enhanced expression/activity of the ''classical'' promoters of angiogenesis, such as upregulation/release of angiogenic growth factors or activation of PKC (19,30). We therefore hypothesized (and demonstrated) that H/R enhances the production of reactive oxygen species (ROS) and thus may accelerate tubular morphogenesis. Indeed, established ROS scavengers, such as pyrollidine dithiocarbamate (PDTC), superoxide dismutase (SOD), catalase, but not the nitroxyl-radical based ROS antagonist 4-hydroxy-2,2,6,6-tetramethyl piperidine-1-N-oxyl (TPL, from Sigma), inhibited tube formation in a dose-dependent manner. Importantly, inhibition of tubular morphogenesis by these ROS scavengers was not limited to H/R-enhanced tube formation. Rather, SOD, catalase, and, most prominently, PDTC (but not TPL) were also effective under normoxic conditions, suggesting that ROS are vital ingredients of the angiogenic process per se.

To ''positively'' demonstrate a role of ROS, specifically of hydrogen peroxide, in enhanced tubular morphogenesis, we plated HMVEC on Matrigel and exposed the cells for various periods of time and at different doses of H_2O_2. As seen in Fig. 1, both the concentration and time of exposure to H_2O_2 are critical. Maximal enhancement of tube formation is observed when the cells are exposed to 1 mM H_2O_2 at the time of plating

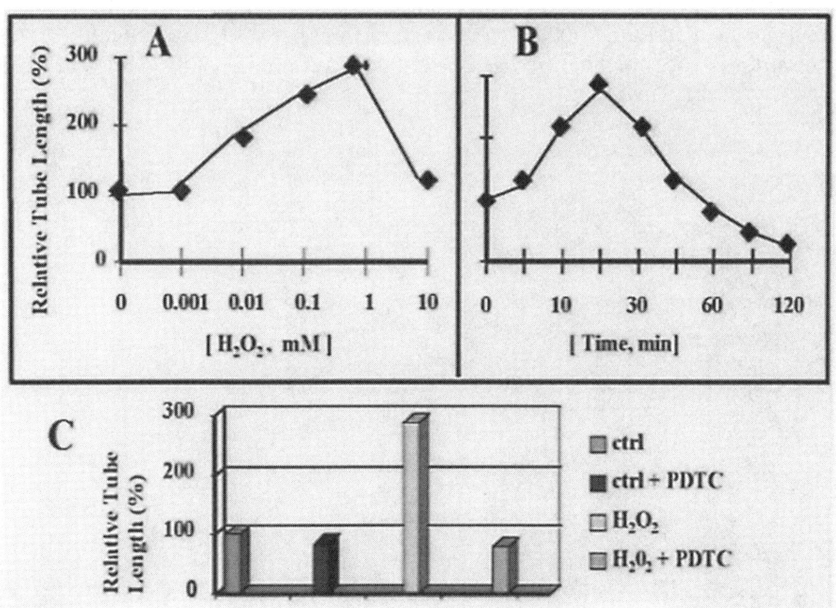

Figure 1 Effect of hydrogen peroxide on tube formation on matrigel. Human microvascular endothelial cells (HMVEC) were plated at 50,000 cells per well into Matrigel-coated 12-well culture plates and exposed to H_2O_2 for the times and the at the doses indicated. The extent of tube formation was determined after 8 hr morphometrically, as previously described (19,30). A. Dose response: the cells were incubated for 15 min with H_2O_2 at the doses indicated. B. Time course: the cells were incubated with 1 mM H_2O_2 for the times indicated. C. Reversibility of H_2O_2 (1 mM, 15 min) induced enhanced tubulogenesis by the ROS antagonist PDTC (10 μM).

for no longer than 15 min. Exposure of the cells for 15 min to higher concentrations of peroxide (we tested up to 10 mM) will reverse the enhancement of tube formation (panel A). Similarly, prolonged exposure of the cells to 1 mM H_2O_2 will abrogate the acceleration/enhancement of tube formation seen under optimized conditions. For incubations exceeding 1 hr, it will altogether inhibit tubular morphogenesis (panel B). Finally, H_2O_2-induced enhancement of tube formation is inhibited by concomitant incubation of the cells with 10 μM PDTC (panel C).

The angiogenic cascade is a complex, multistep process that involves a tightly regulated sequence of individual steps, including endothelial cell proliferation, migration, and remodeling of the basement membrane (39). To investigate which of these pertinent steps might be affected by H/R-generated ROS, we tested the effects of ROS antagonists on EC proliferation and migration. As seen in Figs. 2 and 3, PDTC inhibited both the proliferation and the migration of HMVEC in a dose-dependent fashion. Similar results were observed with other ROS inhibitors used (not shown). These findings are of relevance for all in vitro models of angiogenesis, suggesting that two of the fundamental mechanisms of the angiogenic cascade critically depend on the availability of ROS. Specifically for the Matrigel system, where tubular morphogenesis occurs through migration but not through cell proliferation (40), our results suggest that the enhancement of tube formation might be due to an enhanced rate of endothelial cell migration of this permissive substrate. Currently studies are under way to test this hypothesis directly using computer-aided image analysis and modeling.

The formation of capillary networks on Matrigel requires activation of fibrinolytic enzymes and of matrix-degrading metalloproteinases (MMPs) (40,41). Specifically, endothelial cell migration is accompanied by the induction/secretion of both MMPs and plas-

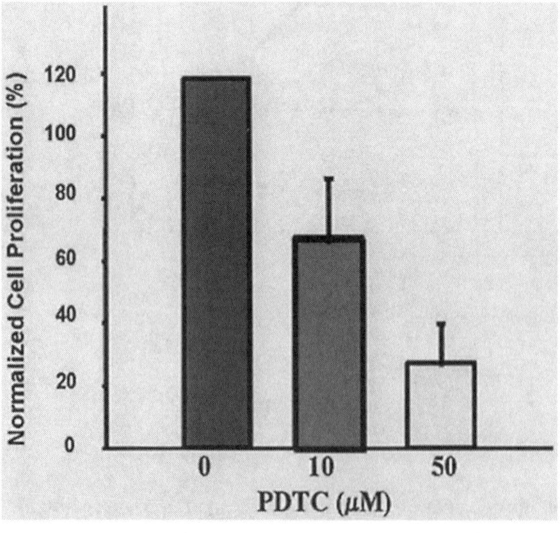

Figure 2 Inhibition of EC proliferation by the ROS antagonist pyrollidine dithiocarbamate (PDTC). Human microvascular EC (5000 per well) were plated onto 96-well plates and incubated with PDTC at the doses indicated. EC numbers were determined fluorimetrically (for details, see "Materials and Methods"). The data, normalized to the proliferation of untreated controls, represent mean ± SEM from three experiments, each performed in triplicate.

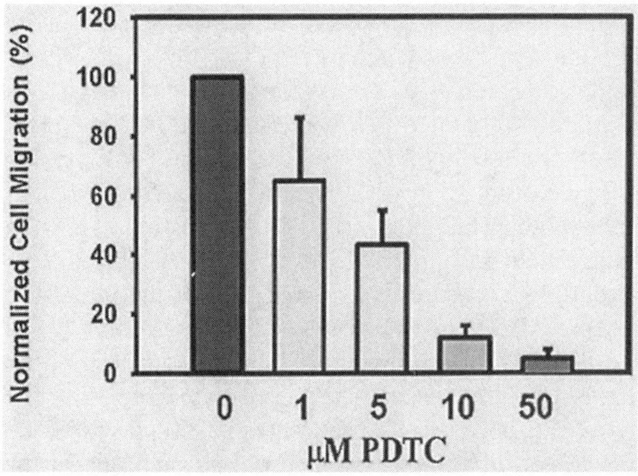

Figure 3 Inhibition of EC migration by PDTC. Confluent monolayers of HMVEC were wounded by scraping and exposed to the concentrations of PDTC as indicated. Migration of the cells from the wounded edges was determined morphometrically (for details, see ''Materials and Methods''). The data (mean ± SEM from three independent experiments, each in duplicate) are normalized to the migration of control, untreated EC (165.7 ± 24 µm/24 hr).

minogen activators (42,43). We hypothesized that exposure to H/R might enhance the production of MMPs in EC plated on Matrigel. To test this hypothesis, we analyzed the supernatant of test cultures by zymography. As see in Fig. 4, Matrigel alone does contain a leachable MMP of 72 kDa (MMP2, also known as gelatinase A). When EC are plated onto Matrigel under normoxic conditions, MMP production does not seem to be enhanced (lane B), while there is an apparent increase in MMP secretion when the cells are exposed

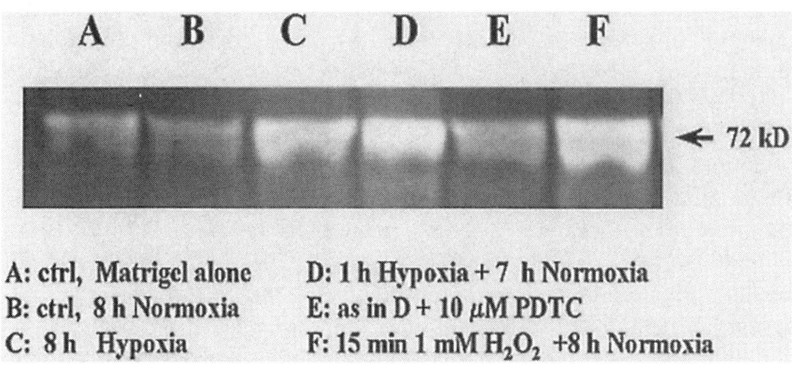

Figure 4 Effect of hypoxia/reoxygenation on the production of matrix-degrading metalloproteinases (MMP). HMVEC (2500,000 per dish) were seeded onto Matrigel-coated 35-mm Petri dishes and exposed to the various experimental conditions. After 8 hr the supernatant was analyzed by zymography (for details, see ''Materials and Methods''). A. Matrigel alone, incubated for 8 hr in the culture medium w/o cells; B. Cells exposed to 8 hr of normoxia. C. Cells exposed to 8 hr of hypoxia. D. Cells exposed to 1 hr of hypoxia and then to 7 hr of normoxia. E. As in D, in the presence of 10 µM PDTC. F. Cells exposed for 15 min to hydrogen peroxide and then, after washing, to normoxia.

to hypoxia for 8 hr (lane C). By contrast, exposing the cells for 1 hr to hypoxia with subsequent reoxygenation for 7 hr yields significantly enhanced levels of MMP, which, according to computer-aided image analysis, are five- to sixfold higher than normoxic controls (lane D). This effect is essentially abrogated in the presence of 10 μM PDTC (lane E). Finally, a brief exposure of the cells to 1 mM H_2O_2 at the time of plating also yields a significant upregulation of MMP production (lane F). Taken together with our previous results, these results suggest that the availability of ROS is necessary at least at the beginning of the process that leads to tubular morphogenesis.

We previously reported that extended exposure to hypoxia, beyond the initial 60-min period, will reverse the ''angiogenic'' effect of H/R (19,30). It remains to be tested whether the amount of ROS formed depends on the length of the exposure to hypoxia and whether it is crucial for distinguishing between beneficial and damaging levels of ROS. Alternatively, the reversal of H/R enhanced tube formation after >60 min exposure to hypoxia might indicate that the ROS-sensitive signaling step occurs only during the initial stages of the angiogenic cascade. Recent studies from several laboratories have suggested that the ROS-inducible nuclear transcription factor NF6B is a central switchboard in endothelial cell activation (44,45). In testing the hypothesis that H/R might enhance the nuclear translocation NF6B complex, we demonstrated that NF6B binding was significantly enhanced in nuclear extracts from cells exposed to 40 min of hypoxia followed by 30 min of normoxia. Moreover, H/R-induced nuclear translocation NF6B was abrogated by PDTC (10 μM). Furthermore, we confirmed the identity of the transcription factor translocated into the nucleus upon activation by H/R as p65, using a supershift assay (30). The activation of NF6B in response to angiogenic stimuli might not be restricted to H/R. Other promoters of angiogenesis that presumably induce oxidative stress seem to act through the same pathway. For example, recent studies have demonstrated a role for NF6B in the angiogenic response of coronary microvessel endothelial cells in response to a particular arachidonic acid metabolite 12(R)-Hydroxy-5,8,14(Z,Z,Z)-eicosatrienoic acid but not to other eicosanoids (46,47).

Previous studies have presented ample evidence that the initiation of the angiogenic cascade process via growth factors and other receptor agonists requires phosphorylation-dependent cellular signaling (4,48). Specifically, inhibitors of tyrosine kinases abrogate angiogenesis (49,50). We therefore hypothesized that H/R might affect angiogenesis by upregulating mitogen-activated protein kinase (MAPK, also known as ERK), a central, tyrosine phosphorylation–dependent cellular signaling pathway. To test our hypothesis, we exposed confluent monolayers of EC to H/R and determined the activation of ERK by Western blotting, using specific antibodies against the phosphorylated form of ERK. As seen in Fig. 5, baseline activity (phosphorylation) of ERK is small (lane 1) and not further enhanced by exposing the cells to 30 min of hypoxia (lane 2). If anything, pERK is decreased in hypoxia. By contrast, reoxygenation of the cells results in a transient increase in pERK activity, which is pronounced after 15 min (lane 3) and returns to baseline after 30 min (lane 4). These results, together with the work of others, suggest that ROS and ROS-mediated MAPK-dependent signaling might be involved in the angiogenic cascade.

Taken together, these results clearly indicate that the angiogenic process, at least as far as it is represented in this in vitro model by the formation of tubelike structures, requires the production of ROS. ROS seem to affect all facets of the angiogenic cascade, such as EC proliferation, migration, and remodeling of the basement membrane. Moreover, our experiments with hydrogen peroxide suggest that excess ROS formation will abrogate tube formation. In the context of the ''oxygen paradox,'' it is important to note that main-

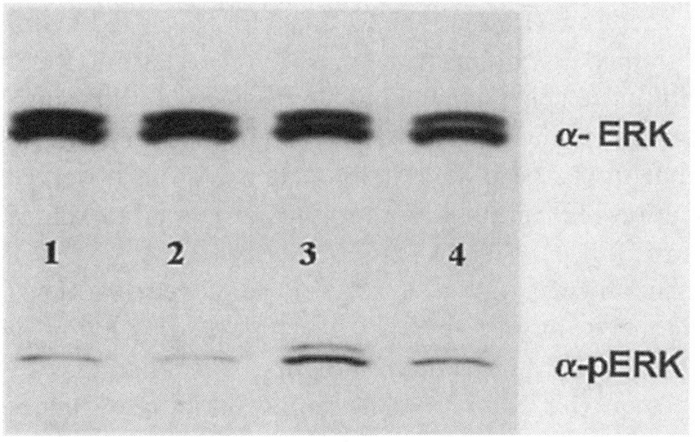

Figure 5 Activation of MAPK by hypoxia/reoxygenation. Confluent monolayers of pooled human aortic endothelial cells (pHAEC) grown in 100-mm dishes were exposed to either normoxia, hypoxia, or H/R as indicated. MAPK activity was determined by Western blotting (for details, see ''Materials and Methods''). Top row: blots were probed with anti-ERK (to show even loading of the gels). Bottom row: anti-pERK, using an antibody specifically directed against the phosphorylated form of ERK. Lane 1: Normoxia, baseline control; Lane 2: 30 min of hypoxia. Lane 3: 30 min of hypoxia, 15 min of normoxia. Lane 4: 30 min of hypoxia, 30 min of normoxia.

taining a basal (low) level of endogenous ROS as a by-product of various fundamental cell-physiological processes, such as the metabolism of cytochrome P-450, or production of NADPH, is obviously necessary for the normal functioning of any cell (51). Our results suggest that the transient generation of ROS, above and beyond constitutively produced baseline levels, may be a pivotal mechanism that is required for the initiation/maintenance of the angiogenic cascade. As such, we surmise that angiogenesis is yet another form of cellular activation, in which the generation of ROS constitutes a unique signaling mechanism using free radicals as second messengers. Furthermore, the angiogenic process can be viewed as yet another process of endothelial cell activation, in which the tight regulation of ROS formation is pivotal for determining the outcome (52,53). Within an angiogenic environment (such as provided by a 3D scaffold made of ECM), stimulation of ''appropriate'' ROS production will promote angiogenesis, while excess ROS production will be inhibitory. As such, this process resembles activation of endothelial cells by other agonists, such as growth factors, cytokines, and/or mechanical forces, in which the extent and duration of ROS production is commensurate with the beneficial and/or harmful action of these agonists. For example, exposure of cultured EC to, respectively, laminar and pulsatile/oscillatory/perturbed flow results in differential activation of ROS. In the case of laminar shear stress, ROS production is moderate and transient, while larger amounts of ROS are produced for prolonged periods of time when the cells are exposed to pulsatile/oscillatory flow (54). It is believed that laminar shear stress acts as a vasoprotective mechanism—e.g., by activating nitric oxide production and downregulating the expression of vascular cell adhesion molecule-1 (VCAM-1) (55). In contrast, perturbed, flow patterns, such as are encountered in areas of atherosclerotic predilections, will upregulate VCAM-1 expression, an early hallmark of predispositions to atherosclerotic lesions (56).

Our studies and those of others have not as yet identified the exact nature of ROS generated by H/R, which are important for the acceleration of in vitro tubular morphogenesis. Recent studies have presented evidence for the formation of both H_2O_2 and O_2 in pulmonary artery endothelial cells exposed to H/R (21). The ROS antagonists used in our study, PDTC, SOD, and catalase, are rather ubiquitous scavengers of a variety of ROS. Therefore, further experimentation with more specific ROS antagonists or with inhibitors of specific metabolic pathways known to produce unique ROS are needed in order to pinpoint a (or the) particular ROS that might be directly responsible for the observed effects. The lack of an inhibitory effect of TPL on tubular morphogenesis might exclude the involvement of nitroxide radicals (57).

At present, there is a persistent controversy about the role in angiogenesis of yet another group of oxygen radicals produced during generated during the production of nitric oxide (NO). Based on the effects of NO donors and inhibitors, several studies by the group of Maragoudakis seem to indicate that NO is an endogenous inhibitor of angiogenesis both in vivo in the chick chorioallantoic membrane model as well as in vitro using the Matrigel model (58–60). Ziche and coworkers, on the other hand, report that nitric oxide donors activate angiogenic events (61–65). Based on the inhibitory action of the NOS-antagonist L-NAME, these authors suggested that angiogenesis, stimulated by VEGF but not by bFGF, might be critically dependent on NO action, and that NOS might lie downstream of the VEGF signaling pathway and upstream for the activation of MAPK (65–67). More recently, it was shown that NOS and NO production in several human endothelial cell types is also activated by bFGF and that L-NAME inhibited bFGF-induced tubular morphogenesis in the fibrin-gel model (68).

Possible explanations for these discrepancies invoke endothelial cell heterogeneity—namely, differential sensitivities of various cell types used in vitro and/or differential effects of NO antagonists and mimics in the various in vivo models of angiogenesis or differences in the doses and the duration of NO present in the various systems. In the context of our discussion, we suggest that the dichotomy in the action of NO on angiogenesis might be due to the different redox states of the cells at the onset of the experiments. NO can increase the amount of available ROS, for example, by reacting with superoxide anions to form highly reactive peroxynitrite ($ONOO^-$) (69). On the other hand, NO production can also alter the balance between other, endogenous ROS (70) and thus either increase or decrease their availability/potential for angiogenesis.

IV. CONCLUSIONS AND FUTURE DIRECTIONS

Our studies provide further evidence that brief exposure to hypoxia followed by reoxygenation enhances/accelerates the formation of a network of capillary-like tubes by HMVEC cultured on Matrigel. Importantly, the mechanism, of H/R-enhanced tube formation does not involve upregulation of PKC or the secretion/release of angiogenic growht factors but rather the generation of ROS and the subsequent activation of ROS-inducible signaling pathways, such as the activation of MAPK and the nuclear translocation of NF6B. Thus, the effects described in this chapter are quite distinct from those triggered by prolonged exposure to hypoxia alone.

Mechanistically, our studies clearly implicate ROS signaling as an important angiogenic second-messenger system that affects cellular signaling mechanisms involved in several steps of the angiogenic cascade, such as EC migration, proliferation, and matrix remodeling (Fig. 6). In this context, initiation of angiogenesis may be taken as a transient, temporally limited sequela of injury-like activation signals, causing the EC to abandon

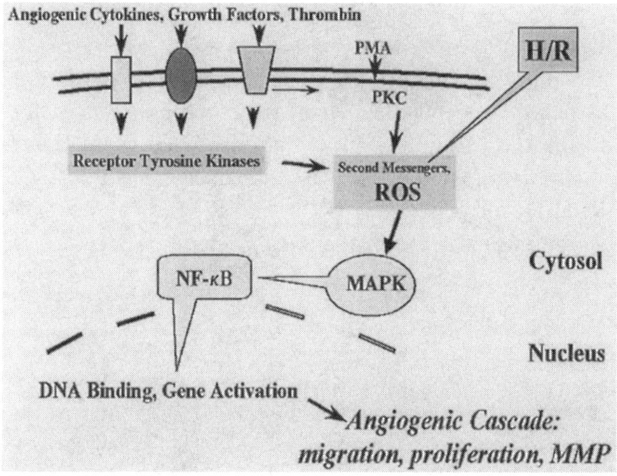

Figure 6 Model for a possible role of ROS in the initiation of the angiogenesis. Upon exposure to hypoxia/reoxygenation, ROS-sensitive signaling cascades and transcription factors as exemplified, but not limited to the MAPK pathway and NF6B, are mobilized, which, in turn, led to the activation of a distinct set of genes that will initiate specific events in the angiogenic cascade, such EC migration, proliferation, and the production of matrix degrading/fibrinolytic enzymes.

the quiescent phenotype in favor of an activated one. As shown in the past, ROS activation might serve as a central switchboard which is turned on by numerous ''events'' known to activate endothelial cells, including exposure to inflammatory cytokines or to hemodynamic forces (44,71,72). As depicted in our model (Figure 6), angiogenesis shares some of the common pathways of other EC ''injuries'' that activate endothelial cells, notably the activation of MAPK-dependent signaling and of ROS-dependent transcription factors, such as NF6B.

The role of ROS in angiogenesis might be either direct or indirect. A direct role of ROS would require that H/R-generated ROS activate unique angiogenic signaling cascades. Our experiments indicate the ROS activate cellular signaling via ERK and NF6B. Presently, we are carrying out experiments to evaluate the role of these pathways in tubular morphogenesis. However, we suspect that these pathways are far too ubiquitous and nonspecific to be uniquely angiogenic. Therefore we and others are striving for a more detailed understanding of the distinct ROS signaling mechanisms that will induce the angiogenic response. Alternatively, transient elevation of ROS by H/R may induce the rapid production of angiogenic cytokines, such as IL-1, IL-6 and IL-8 (37,73), leading to the upregulation of fibrinolytic/matrix-degrading enzymes, which, in turn, will enhance tube formation by facilitating endothelial migration. Experiments are on the way to test for these possibilities.

Clinically, the observed enhancement of angiogenesis by H/R may be relevant for recent attempts at mitigating the severity of ischemic lesions to the myocardium by ''ischemic preconditioning''—that is, by brief, repetitive episodes of occlusion(hypoxia/ischemia) followed by reperfusion/reoxygenation (74–76). Based on our studies, we hypothesize that one of the reasons this approach is apparently quite successful might be because of enhanced generation of ROS and hence of angiogenesis in the wake of the repetitive regimen of hypoxia and subsequent reoxygenation.

ACKNOWLEDGMENTS

This study was supported in part by a grant in aid from the American Heart Association, National Center, as well as by grants from the Milwaukee Heart Research Foundation and Berlex Biosciences. We would like to acknowledge the contribution of Drs. K.A. Hahn, J. Wigboldus, and D.A. Sukovich, who participated in earlier stages of these studies. We thank Drs. Hynda Kleinman, NIDR, NIH, for generously providing the EHS tumor, and M.M. Samet for his help in the design and operation of our H/R system.

REFERENCES

1. AM Lefer, PS Tsao, DJ Lefer, and X-L Ma. Role of endothelial dysfunction in the pathogenesis of reperfusion injury after myocardial ischemia. FASEB J 5:2029–2034, 1991.
2. BC White, LI Grossman, and GS Krause. Brain injury by global ischemia and reperfusion. Neurology 43:1656–1665, 1993.
3. D Shweiki, A Itin, D Soffer, and E Keshet. Vascular endothelial growth factor induced by hypoxia may mediate hypoxia-initiated angiogenesis. Nature 359:843–845, 1992.
4. B Millauer, S Wizigmann-Voos, H Schnürch, R Martinez, NPH Moller, W Risau, and A Ullrich. High affinity VEGF binding and developmental expression suggest Flk-1 as a major regulator of vasculogenesis and angiogenesis. Cell 72:835–846, 1993.
5. L Seetharam, N Gotoh, Y Maru, G Neufeld, S Yamaguchi, and M Shibuya. A unique signal transduction from FLT tyrosine kinase, a receptor for vascular endothelial growth factor VEGF. Oncogene 10:135–147, 1995.
6. H Takagi, GL King, N Ferrara, and LP Aiello. Hypoxia regulates vascular endothelial growth factor receptor KDR/Flk gene expression through adenosine A_2 receptors in retinal capillary endothelial cells. Invest Ophthalmol Vis Sci 37:1311–1321, 1996.
7. H Tian, SL McKnight, and DW Russell. Endothelial PAS domain protein 1 (EPASI), a transcription factor selectively expressed in endothelial cells. Genes Dev 11:72–82, 1997.
8. E Brogi, G Schatteman, T Wu, EA Kim, L Varticovski, B Keyt, and JM Isner. Hypoxia-induced paracrine regulation of vascular endothelial growth factor receptor expression. J Clin Invest 97:469–476, 1996.
9. JA Ware and M Simons. Angiogenesis in ischemic heart disease. Nature Med 3:158–164, 1997.
10. MS Pepper. Manipulating angiogenesis—from basic science to the bedside. Arterioscler Thromb Vasc Biol 17:605–619, 1997.
11. CL Klein, H Köhler, F Bittinger, M. Otto, I Hermanns, and CJ Kirkpatrick. Comparative studies on vascular endothelium in vitro: 2. Hypoxia: its influences on endothelial cell proliferation and expression of cell adhesion molecules. Pathobiology 63:1–8, 1995.
12. J Wojta, RL Jones, BR Binder, and RL Hoover. Reduction in pO2 decreases the fibrinolytic potential of cultured bovine endothelial cells derived from pulmonary arteries and lung microvasculature. Blood 71:1703–1706, 1988.
13. M Tucci, SI Hammerman, S Furfaro, JJ Saukonnen, TJ Conca, and HW Farber. Distinct effect of hypoxia on endothelial cell proliferation and cycling. Am J Physiol 272:C1700–C1708, 1997.
14. G Zünd, DP Nelson, EJ Neufeld, AL Dzus, J Bischoff, JE Mayer, and SP Colgan. Hypoxia enhances stimulus-dependent induction of E-selectin on aortic endothelial cells. Proc Natl Acad Sci USA 93:7075–7080, 1996.
15. R Auerbach, W Auerbach, and I Polakowski. Assays for angiogenesis: a review. Pharmacol Ther 51:1–11, 1991.

16. PG Phillips, LM Birnby, and A Narendran. Hypoxia induces capillary network formation in cultured bovine pulmonary microvessel endothelial cells. Am J Physiol Lung Cell Mol Physiol 268:L789–L800, 1995.
17. Y Kubota, HK Kleinman, GR Martin, and TJ Lawley. Role of laminin and basement membrane in the morphological differentiation of human endothelial cells into capillary-like structures. J Cell Biol 107:1589–1598, 1988.
18. DS Grant, PI Lelkes, K Fukuda, and HK Kleinman. Intracellular mechanisms involved in basement membrane induced blood vessel differentiation in vitro. In Vitro Cell Dev Biol Anim 27A:327–336, 1991.
19. PI Lelkes, KA Hahn, S Karmiol, and DH Schmidt. Hypoxia/reoxygenation enhances tube formation of cultured human micro-vascular endothelial cells: the role of reactive oxygen species. In: ME Maragoudakis, ed. Angiogenesis: Models, Modulators, and Clinical Applications. New York: Plenum Press, 1998, pp. 321–336.
20. W Yang and ER Block. Effect of hypoxia and reoxygenation on the formation and release of reactive oxygen species by porcine pulmonary artery endothelial cells. J Cell Physiol 164: 414–423, 1995.
21. JJ Zulueta, R Sawheny, FS Yu, CC Cote, and PM Hassoun. Intracellular generation of reactive oxygen species in endothelial cells exposed to anoxia-reoxygenation. Am J Physiol 272:L897–L902, 1997.
22. KA Hahn, DH Schmidt, and PI Lelkes. Hypoxia enhances in vitro angiogenesis of human microvascular endothelial cells cultured on Matrigel (abstr). NATO-ASI on Molecular, Cellular and Clinical Aspects of Angiogenesis. New York: Plenum Press, 1995.
23. VG Manolopoulos, MM Samet, and PI Lelkes. Regulation of the adenylyl cyclase signalling system in various types of cultured endothelial cells. J Cell Biochem 57:590–598, 1995.
24. MD Silverman, VG Manolopoulos, BR Unsworth, and PI Lelkes. Tissue factor expression is differentially modulated by cyclic mechanical strain in various human endothelial cells. Blood Coagul Fibrinolysis 7:281–288, 1996.
25. K Kanda, GT Hayman, M. D. Silverman, and PI Lelkes. Distinct expression of ICAM-1 and VCAM-1 in various human endothelial cell types and smooth muscle cells. Endothelium 6: 33–44, 1998.
26. PI Lelkes, M Silverman, DM Wankowski, S Zhang, J Varani, M Dame, J Shen, and S Karmiol. A comparison of umbilical vein endothelial cells (HUVEC) and dermal microvascular endothelial cells (HMVEC) with respect to adhesion molecule expression, tissue factor activity, and neutrophil killing (abstr). ASCB Thirty-fourth Annual Meeting: Molecular Biology of the Cell. 1994.
27. PI Lelkes, VG Manolopoulos, M Silverman, S Zhang, S Karmiol, and BR Unsworth, On the possible role of endothelial cell heterogeneity in angiogenesis. In: ME Maragoudakis, ed. Molecular, Cellular, and Clinical Aspects of Angiogenesis. New York: Plenum Press, 1996, pp. 1–18.
28. GC Haralabopoulos, DS Grant, HK Kleinman, PI Lelkes, SP Papaioannou, and ME Maragoudakis. Inhibitors of basement membrane collagen synthesis prevent endothelial cell alignment in Matrigel in vitro and angiogenesis in vivo. Lab Invest 71:575–582, 1994.
29. MM Samet and PI Lelkes. Flow patterns and endothelial cell morphology in a simplified model of an artificial ventricle. Cell Biophys 23:139–163, 1993.
30. PI Lelkes, KL Hahn, DA Sukovich, S Karmiol, and DH Schmidt. On the possible role of reactive oxygen species in angiogenesis. In: A Hudetz and D Bruleyeds, eds. Oxygen Transport to Tissue XX. New York: Plenum Press, 1998, pp. 295–310.
31. E Papadimitriou and PI Lelkes. Measurement of cell numbers in microtiter culture plates using the fluorescent dye Hoechst 33258. J Immunol Meth 162:41–45, 1993.
32. PI Lelkes, E Ramos, VV Nikolaychik, DM Wankowski, BR Unsworth, and TJ Goodwin. GSTF-2: a new, versatile cell culture medium for diverse normal and transformed mammalian cells. In Vitro Cell Dev Biol Anim. 33:344–351, 1997.

33. VV Nikolaychik, MM Samet, and PI Lelkes. A New method for continual quantitation of viable cells on endothelialized polyurethanes. J Biomater Sci Polymer Ed 7:881–891, 1996.
34. P Zaoui, C Barro, and F Morel. Differential expression and secretion of gelatinases and tissue inhibitor of metalloproteinase-1 during neutrophil adhesion. Biochim Biophys Acta Gen Subj 1290:101–112, 1996.
35. D Rasouly, D Shavit, R Zuniga, RB Elejalde, BR Unsworth, A Yayon, P Lazarovici, and PI Lelkes. Staurosporine induces neurite outgrowth in neuronal hybrids (PC12EN) lacking NGF receptors. J Cell Biochem 62:356–371, 1996.
36. S Baatout. Endothelial differentiation using Matrigel. Anticancer Res 17:451–455, 1997.
37. T Shono, M Ono, H Izumi, S Jimi, K Matsushima, T Okamoto, K Kohno, and M Kuwano. Involvement of the transcription factor NF-kappaB in tubular morphogenesis of human microvascular endothelial cells by oxidative stress. Mol Cell Biol 16:4231–4239, 1996.
38. J Teissie, M Prats, P Soucaille, and JF Tocanne. Evidence for conduction of protons along the interface between water and a polar lipid monolayer. Proc Natl Acad Sci USA 82:3217–3221, 1985.
39. ME Maragoudakis. Angiogenesis. Ann Cardiac Surg 8:13–19, 1994.
40. AB Zimrin, B Villeponteau, and T Maciag. Models of in vitro angiogenesis: Endothelial cell differentiation on fibrin but not Matrigel is transcriptionally dependent. Biochem Biophys Res Commun 213:630–638, 1995.
41. K Ito, M Ryuto, S Ushiro, M Ono, A Sugenoya, A Kuraoka, Y Shibata, and M Kuwano. Expression of tissue-type plasminogen activator and its inhibitor couples with development of capillary network by human microvascular endothelial cells on Matrigel. J Cell Physiol 162:213–224, 1995.
42. LA Cornelius, LC Nehring, JD Roby, WC Parks, and HG Welgus. Human dermal microvascular endothelial cells produce matrix metalloproteinases in response to angiogenic factors and migration. J Invest Dermatol 105:170–176, 1995.
43. C Iwasaka, K Tanaka, M Abe, and Y Sato. Ets-1 regulates angiogenesis by inducing the expression of urokinase-type plasminogen activator and matrix metalloproteinase-1 and the migration of vascular endothelial cells. J Cell Physiol 169:522–531, 1996.
44. T Collins. Biology of disease: endothelial nuclear factor-kB and the initiation of the atherosclerotic lesion. Lab Invest 68:499–508, 1993.
45. T Collins, MA Read, AS Neish, MZ Whitely, D Thanos, and T Maniatis. Transcriptional regulation of endothelial cell adhesion molecules:NF-kB and cytokine-inducible enhancers. FASEB J 9:899–909, 1995.
46. M Laniado-Schwartzman, Y Lavrovsky, RA Stoltz, MS Conners, JR Falck, K Chauhan, and NG Abraham. Activation of nuclear factor kB and oncogene expression by 12(R)-hydroxyeicosatrienoic acid, an angiogenic factor in microvessel endothelial cells. J Biol Chem 269:24321–24327, 1994.
47. RA Stoltz, NG Abraham, and ML Schwartzman. The role of NF-kB in the angiogenic response of coronary microvessel endothelial cells. Proc Natl Acad Sci USA 93:2832–2837, 1996.
48. JL Kinsella, DS Grant, BS Weeks, and HK Kleinman. Protein kinase C regulates endothelial cell tube formation on basement membrane matrix, matrigel. Exp Cell Res 199:56–62, 1992.
49. T Oikawa, H Ashino, M Shimamura, M Hasegawa, I Morita, S Murota, M Ishizuka, and T Takeuchi. Inhibition of angiogenesis by erbstatin, an inhibitor of tyrosine kinase. J Antibiot 46:785–790, 1993.
50. T Mustonen and K Alitalo. Endothelial receptor tyrosine kinases involved in angiogenesis. J Cell Biol 129:895–898, 1995.
51. HM Lander. An essential role for free radicals and derived species in signal transduction. FASEB J 11:118–124, 1997.
52. J Abe and BC Berk. Reactive oxygen species as mediators of signal transduction in cardiovascular disease. Trends Cardiovasc Med 8:59–64, 1998.

53. RW Alexander. Atherosclerosis as disease of redox-sensitive genes. Trans Am Clin Climatol Assoc 109:129–145, 1998.
54. GW De Keulenaer, DC Chappell, N Ishizaka, RM Nerem, RW Alexander, and KK Griendling. Oscillatory and steady laminar shear stress differentially affect human endothelial redox state: Role of a superoxide-producing NADH oxidase. Circ Res 82:1094–1101, 1998.
55. O Traub and BC Berk. Laminar shear stress: mechanisms by which endothelial cells transduce an atheroprotective force. Arterioscler Thromb Vasc Biol 18:677–685, 1998.
56. DC Chappell, SE Varner, RM Nerem, RM Medford, and RW Alexander. Oscillatory shear stress stimulates adhesion molecule expression in cultured human endothelium. Circ Res 82: 532–539, 1998.
57. E Beit-Yannai, R Zhang, V Trembovler, A Samuni, and E Shohami. Cerebroprotective effect of stable nitrofxide radicals in closed head injury in the rat. Brain Res 717:22–28, 1996.
58. E Pipili-Synetos, E Sakkoula, and ME Maragoudakis. Nitric oxide is involved in the regulation of angiogenesis. Br J Pharmacol 108:855–857, 1993.
59. E Pipili-Synetos, E Sakkoula, G Haralabopoulos, P Andriopoulou, P Peristeris, and ME Maragoudakis. Evidence that nitric oxide is an endogenous antiangiogenic mediator. Br J Pharmacol 111:894–902, 1994.
60. E Pipili-Synetos, A Papageorgiou, E Sakkoula, G Sotiropoulou, T Fotsis, G Karakiulakis, and ME Maragoudakis. Inhibition of angiogenesis, tumour growth and metasis by the NO-releasing vasodilators, isosorbide mononitrate and dinitrate. Br J Pharmacol 116:1829–1834, 1995.
61. M Ziche, L Morbidelli, E Masini, S Amerini, HJ Granger, CA Maggi, P Geppetti, and F Ledda. Nitric oxide mediates angiogenesis in vivo and endothelial cell growth and migration in vitro promoted by substance P. J Clin Invest 94:2036–2044, 1994.
62. M Ziche, L Morbidelli, A Parenti, and F Ledda. Nitric oxide modulates angiogenesis elicited by prostaglandin E1 in rabbit cornea. Adv Prostaglandin Thrombox Leukot Res 23:495–497, 1995.
63. L Morbidelli, CH Chang, JG Douglas, HJ Granger, F Ledda, and M Ziche. Nitric oxide mediates mitogenic effect of VEGF on coronary venular endothelium. Am J Physiol 270:H411–H415, 1996.
64. M Ziche, A Parenti, F Ledda, P Dell'Era, HJ Granger, CA Maggi, and M Presta. Nitric oxide promotes proliferation and plasminogen activator production by coronary venular endothelium through endogenous bFGF. Circ Res 80:845–852, 1997.
65. M Ziche, L Morbidelli, R Choudhuri, H-T Zhang, S Donnini, and HT Granger. Nitric oxide synthase lies downstream from vascular endothelial growth factor-induced but not basic fibroblast growth factor-induced angiogenesis. J Clin Invest 99:2625–2634, 1997.
66. JD Hood, CJ Meininger, M Ziche, and HJ Granger. VEGF upregulates ecNOS message, protein, and NO production in human endothelial cells. Am J Physiol 274:H1054–H1058, 1998.
67. A Parenti, L Morbidelli, X-L Cui, JG Douglas, JD Hood, HJ Granger, F Ledda, and M Ziche. Nitric oxide is an upstream signal of vascular endothelial growth factor-induced extracellular signal-regulated kinase$_{1/2}$ activation in postcapillary endothelium. J Biol Chem 273:4220–4226, 1998.
68. S Babaei, K Teichert-Kuliszewska, J-C Monge, F Mohamed, MP Bendeck, and DJ Stewart. Role of nitric oxide in the angiogenic response in vitro to basic fibroblast growth factor. Circ Res 82:1007–1015, 1998.
69. R Foresti, JE Clark, CJ Green, and R Motterlini. Thiol compounds interact with nitric oxide in regulating heme oxygenase-1 induction in endothelial cells–Involvement of superoxide and peroxynitrite anions. J Biol Chem 272:18411–18417, 1997.
70. MP Gupta, V Evanoff, and CM Hart. Nitric oxide attenuates hydrogen peroxide-mediated injury to porcine pulmonary artery endothelial cells. Am J Physiol 272:L1133–L1141, 1997.
71. LM Khachigian, V Lindner, AJ Williams, and T Collins. Egr-1-induced endothelial gene expression: a common theme in vascular injury. Science 271:1427–1431, 1996.

72. MA Read, MZ Whitley, AJ Williams, and T Collins. NF-kappaB and IkappaBalpha: an inducible regulatory system in endothelial activation. J Exp Med 179:503–512, 1994.
73. Y Ala, O Palluy, J Favero, C Bonne, G Modat, and J Dornand. Hypoxia/reoxygenation stimulates endothelial cells to promote interleukin-1 and interleukin-6 production: effects of free radical scavengers. Agents Actions 37:134–139, 1992.
74. R Strasser, P Htun, and W Schaper. Salvage of jeopardized myocardium by ischemic preconditioning: Is the quest over? Mol Cell Biochem 161:209–215, 1996.
75. N Maulik, M Watanabe, YL Zu, CK Huang, GA Cordis, JA Schley, and DK Das. Ischemic preconditioning triggers the activation of MAP kinases and MAPKAP kinase 2 in rat hearts. FEBS Lett 396:233–237, 1996.
76. K Przyklenk and RA Kloner. Role of protein kinase C in ischemic preconditioning: in search of the ''pure and simple truth''. Basic Res Cardiol 91:41–43, 1996.

15

Modulators of Angiogenesis
Hemodynamic Forces

Olga Hudlicka and Margaret D. Brown
University of Birmingham, Birmingham, England

I. INTRODUCTION

The role of hemodynamic factors in the growth of blood vessels during development was pointed out over a hundred years ago by Thoma (1). Clark (2) later demonstrated in tadpole tails that growth of capillaries occurred by sprouting in those vessels with high flow, whereas they regressed if flow stopped. Furthermore, in the rabbit ear chamber capillaries with faster flow developed over several weeks into venules (3). Despite these early findings, the importance of hemodynamic factors as regulators of angiogenesis has long been overshadowed by the discovery of growth factors, whose involvement particularly in vessel growth under pathological circumstances such as in tumors or wounds is well established. The significance of hemodynamic forces is, however, becoming more and more apparent with the detection of genes in endothelial or smooth muscle cells that can be activated in vitro by fluid shear stress, stretch, or cyclic strain, as discussed in other chapters. Whether such findings are applicable in the context of understanding mechanisms of vessel growth in vivo remains to be fully clarified. Skalak and Price (4) recently reviewed the role of certain hemodynamic forces in vascular remodeling, indicating the importance of circumferential wall stress in formation of new arterioles. While its involvement in growth of capillaries is much less clear, it is possible that increased capillary wall tension could decrease the thickness of endothelial cells (5) and thereby make them more susceptible to the action of growth factors (6).

One problem with extrapolations from findings in vitro to the in vivo situation is that, with few exceptions, tissue cultures involve only one type of cell, which can display a different phenotype from that in vivo (7). For example, primary cultures of rat myocardial capillaries lose alkaline phosphatase, an enzyme well expressed in capillary endothelium in this species in vivo (8). Knowledge of the involvement of different hemodynamic factors in angiogenesis in vivo is based on assessment of circulatory parameters that enable the calculation of shear stress within vessels and their circumferential wall stress (wall tension). Shear stress (τ) can be estimated in capillaries, arterioles, and larger vessels, since it is a function of velocity of flow (Vrbc), viscosity (η), and vessel radius (r) [$\tau = \eta (4\text{Vrbc}/r)$]. Stretch experienced by vessels can be expressed as circumferential wall stress (σ) or wall tension (T) and can be calculated in capillaries from diameter and pressure

(P) by T = P · r and in larger vessels by incorporation of a value for wall thickness (h) as well ($\sigma = P \cdot r/h$). Cyclic strain is more difficult to assess but changes in pulse pressure are its closest approximation. These are most evident in larger arteries, and although there are some data available on their involvement in vascular remodeling, very little is known about their role in angiogenesis. Since angiogenesis usually begins by growth of capillaries, which could later change into arterioles, this chapter briefly reviews the most relevant studies on the effect of shear stress, stretch, and cyclic strain on cultured endothelial cells with respect to proliferation and gives examples to indicate whether and to what extent these three components may be involved in angiogenesis in vivo.

II. SHEER STRESS, CYCLIC STRAIN AND STRETCH IN RELATION TO ENDOTHELIAL CELL PROLIFERATION IN VITRO

A. Shear Stress

Although the fact that shear stress results in proliferation of endothelial cells in culture was established more than 15 years ago (9–11), many studies aimed at elucidating the underlying signal transduction mechanisms have not provided a complete understanding, and very probably there is more than one mechanism involved. Laminar shear stress induces synthesis of nitric oxide, NO (12,13) which promotes DNA synthesis in endothelial cells (14). Shear stress also triggers an inward rectifying K^+ selective current (15) which increases Ca^{2+} entry into endothelial cells and hence intracellular Ca^{2+} concentration, a condition necessary for nuclear division. It causes perturbation of the endothelial cell membrane and phospholipids (16,17) and activates phosphatidyl inositol–specific phospholipase C with subsequent hydrolysis of phosphatidylinositol 4,5 diphosphate into inositol triphosphate (IP3) and diacyl glycerol (18). It also upregulates a number of genes—e.g., for metalloproteinases (19), for platelet-derived growth factor and basic fibroblast growth factor (20), or immediate early genes such as c-fos and c-jun. Resnick and Gimbrone (21) summarized the effect of shear stress on changes in endothelial cells which take either seconds or minutes (activation of ion channels, release of NO and prostaglandins, increases in cytosolic Ca^{2+}), minutes to hours (increase in eNOS, immediate early genes—e.g., c-fos, some growth factors, and cell replication) or hours to days (changes in cell shape and extracellular matrix remodeling). Changes in cell shape lead to reorganization of the cytoskeleton, which could affect the extracellular matrix and create local adhesion sites (22), thereby modifying endothelial cell migration and proliferation. Finally, perturbation of the endothelial cell surface in tissue cultures by flow, and thus shear stress, inhibited synthesis of fibronectin (23), which is the first extracellular matrix component to provide an environment facilitatory for migration of endothelial cell precursors in chick chorioallantoic membrane (CAM) (24) and in developing heart (25).

B. Stretch and Cyclic Strain

Stretch in vivo can be due to increased vessel diameter and/or pressure and hence wall tension. It can also arise from extravascular stretch imposed on vessels by growth and expansion of the surrounding tissue, which is certainly important during development and probably also in situations such as heart or skeletal muscle hypertrophy. In tissue cultures, stretch has been shown to increase DNA synthesis and endothelial cell division within 5–7 days of exposure (26), possibly via the inositol phosphatide pathway, since levels of IP_3 were elevated (27). It also unregulated the expression of certain growth factors in endothelial cells (28), together with protooncogene expression and phosphoinositide turnover in smooth muscle cells (29). When endothelial cells were grown on hydrogel, they spread out gradually in relation to its concentration, the highest incorporation of ^{3}H-thymidine being found in the thinnest cells (30). Incorporation of ^{3}H-thymidine was also higher in endothelial cells grown on substrata such as collagen IV, which encouraged a greater degree of spreading (31). Cells exposed to cyclic strain had increased production of PGI_2 (32) and tPA (33), both of which are involved in angiogenesis (34,35). Stretch may increase intracellular calcium ion concentration, a prerequisite for cell proliferation, via stretch-activated ion channels (36) and, since it inhibits collagen production (37), may weaken the basement membrane. Stretch due to increased pressure in vivo stimulates hypertrophy and hyperplasia of vascular smooth muscle in resistance arteries of differing sizes (38,39) and in veins when exposed to arterial levels of pressure (40). However, Dethlefsen et al. (41) recently demonstrated that this depends more on the phenotype of smooth muscle cells than a generalized effect of stretch, since smooth muscle cells isolated from human saphenous arteries proliferated in response to stretch while those from the aorta did not.

The most recent findings on the effects of mechanical forces on signal transduction and gene expression in endothelial cells have been reviewed by Chien and Shyy (42). Shear stress activates small GTPases, guanine nucleotide binding proteins with an intrinsic GTPase activity, which are associated with the cell membrane, such as Ras. These in turn activate MAPkinases (Mitogen activated protein kinases), ERK (extracellular signal-regulated kinase) and JNK (c-Jun N-terminal kinase) which are involved in cell growth (ERK) and apoptosis (JNK). MAPkinases are also activated by protein tyrosine kinases, which are present in focal adhesion sites on the abluminal side of the endothelial cell and, in turn, activate genes c-jun and c-fos. Genes regulated by shear stress have a shear stress-responsive element in their promoter region. Cyclic strain or stretch also induce many of the signalling events elicited by shear stress, and impose oxidative stress on endothelial cells. Its important role is also in the reorganisation of $\alpha5$, $\alpha2$, and $\alpha1$ integrins, which assist in mediating the adhesion of cells to extacellular matrix as a condition of growth.

C. Extracellular Matrix (ECM)

Endothelial cells secrete ECM–related proteins as well as metalloproteinases and tissue inhibitors of metalloproteinases (43,44), thus keeping in balance enzymes capable of disrupting and maintaining the basement membrane. ECM molecules determine, in tissue culture at least, whether endothelial cells proliferate, differentiate, or undergo involution. When cells start migrating, they express mRNA for urokinase (uPA), a proteolytic enzyme which helps to disrupt the extracellular matrix to create space for newly formed capillaries (35,45). Matrix protease activity can be blocked by TIMP-1 (tissue inhibitor of metalloproteinases), thereby inhibiting migration (46). ECM also regulates endothelial cell growth by changing cell response to growth factors. Furthermore, binding of certain ECM molecules to transmembrane integrin receptors activates the $Na+/H+$ antiport, which stimulates DNA synthesis (47).

The role of integrins in angiogenesis has recently been reviewed by Stromblad and Cheresh (48), who demonstrated that disruption of $\alpha v\beta3$ binding to ECM prevented vessel formation on CAM, while both $\alpha v\beta 3$ and $\alpha v\beta 5$ were involved in basic fibroblast growth factor- and vascular endothelial growth factor-induced angiogenesis in the cornea (49). Moreover, integrins caused induction of the early growth genes such as c-jun and c-myc. The influence of different growth factors on ECM has been established by several groups (47,48) and the importance of physical factors in ECM remodelling has been emphasized (50), yet exact mechanisms have not been fully clarified.

The role of extracellular matrix during vasculogenesis and angiogenesis in vivo has recently been demonstrated by Rongish et al. (25). They used concomitant staining for lectins and CD31 (PECAM1) in the heart to identify both vessels and extracellular matrix components such as fibronectin, laminin, and collagens I, III, and IV, enabling the sequence of vessel development and the role of the individual ECM components in migration, proliferation, and tube formation to be followed. The appearance of fibronectin preceded vasculogenesis and provided an environment to facilitate cell migration while laminin production coincided with tube formation and maturation, closely followed by appearance of type IV collagen and the newly forming basement membrane. Collagen I and III were not related to tube formation but were an important component of arteriolar and venular adventitia. A similar sequence was shown previously in the chicken chorioallantoic membrane (24).

III. HEMODYNAMIC FACTORS IN THE REGULATION OF ANGIOGENESIS UNDER PHYSIOLOGICAL CIRCUMSTANCES

A. Female Reproductive Organs

Reynolds et al. (51,52) recently reviewed the changes occurring during angiogenesis in the ovary, uterus, and placenta with respect to changes in blood flow and growth factors. Although the latter, together with estrogen, progesterone, and gonadotrophin, are very important in the control of vascular growth, data were also presented showing the importance of blood flow in vascular growth and regression. Ovarian blood flow is increased before ovulation, and follicles about to ovulate (only a small percentage of those present) show increased vascularity, which decreases in those undergoing involution. Mature corpus luteum receives most of the ovarian blood flow and the proportion of endothelial cells within it increases by 50%. Decreased ovarian flow, on the other hand, may play a role in luteal regression. Endometrial proliferation in the uterus also occurs in parallel with increased blood flow. Despite this, microvascular volume density remains constant throughout the estrous cycle, indicating that there is a coordinated growth of vascular and extravascular tissue.

A 50% increase in the vascular density in the endometrium in ewes 24 days after mating was also linked to increased blood flow, although it is not clear to what extent the latter, which is regulated by the levels of estrogen and progesterone, preceded vascular growth. Increased blood flow is also necessary for placental development. The role of blood flow is also demonstrated by the vessels supplying the uterus, which undergo marked changes after delivery as blood flow through them is drastically reduced. Arteries and veins constrict and longitudinal muscle bundles in veins bulge into the lumen, leading to complete collapse of the vessel wall. Endothelial cells in the arteries swell and start to fuse about 12 hr postpartum, ending 6 days after delivery, with complete involution of the vessels to the prepregnancy state in the fundus of the uterus within 40–44 days (53).

B. Exercise Training

Growth of capillaries and expansion of the whole vascular bed can be induced by exercise training in skeletal and cardiac muscle. Increased labeling index for ^{3}H-thymidine in myocardial endothelial cells in rats trained by swimming confirms cellular proliferation in the heart, but there is no comparable direct evidence of proliferation in skeletal muscles. Despite this, specific histochemical methods or low-power electron microscopy used to visualize capillaries has established that an increase in capillary-per-fiber ratio (C:F) or capillary density (CD) occurring without any change in fiber area is due to the appearance of new vessels. There are, however, differences in the time sequence and pattern of growth depending on the type of training, tissue and age (see Refs. 54–56 for reviews).

The evidence that changes in blood flow may induce capillary growth during exercise training remains circumstantial. Ingjer (57) showed increased capillary supply after 6 months of endurance training in athletes around fast oxidative and slow muscle fibers, which are the first to be recruited in endurance exercise, and blood flow is preferentially increased in muscles or regions of muscles with a high proportion of these fibers (58,59). The effects on capillary supply of high-intensity sprint or low-intensity endurance training have been studied in white (fast glycolytic fibers), mixed (fast glycolytic and fast oxidative), and red (fast and slow oxidative) gastrocnemius muscles in rats. C:F was increased by high-intensity sprint training in white gastrocnemius in red parts of gastrocnemius by low-intensity endurance training (60). Changes in blood flow measured during different running speeds (58) correspond with these changes in capillary supply. As a result of high-intensity training, resting blood flow and C:F were higher only in white gastrocnemius, whereas low-intensity endurance training increased blood flow in the mixed and red parts of the muscle (61) and C:F increased in the red part only (Fig. 1). The correlation between the changes in blood flow and capillary supply therefore indicate a possible role for blood flow in capillary growth in trained skeletal muscles, but there are no data on parameters such as shear stress or wall tension in individual capillaries with different types of training to indicate a more direct involvement of hemodynamic factors.

The situation is even less clear in the heart. As in skeletal muscle, coronary blood flow is elevated during each bout of exercise, implying the likelihood of increased perfusion through individual capillaries. Endurance exercise training also results in bradycardia, and it has been established that ''athletic'' animals like hares or wild rats have a lower heart rate and higher capillary density than analogous, more sedentary species of similar body weight, such as rabbits or laboratory rats (62). Exercise, particularly if performed vigorously over a long period of time, could therefore be expected to result in an increase in myocardial capillary density. Surprisingly, this happens only to a limited extent and mostly in young animals. Increased myocardial CD (63), C:F (64) and total capillary length (65) have all been reported in young but not adult rats trained by running. Capillary proliferation in the heart was demonstrated by incorporation of ^{3}H-thymidine into nuclei of endothelial cells using electron microscopy with autoradiography (66) in young and adult but not old rats trained by swimming (67) (Fig. 2).

Laughlin and McAllister (56) concluded that capillary growth is more easily elicited by swimming than treadmill endurance training, whereas training by weight lifting did not have any effect on the size of the coronary vascular bed in cats (68). Capillary growth in the heart is thus dependent on age, type of training, and possibly species, but its relationship to the intensity of training is more controversial. Vigorous training has been shown to reduce myocardial capillarity in rats (69,70), pigs (71), and possibly dogs (72). By contrast, increased C:F ratio was found in young rats that swam 1 hr daily over 10 weeks

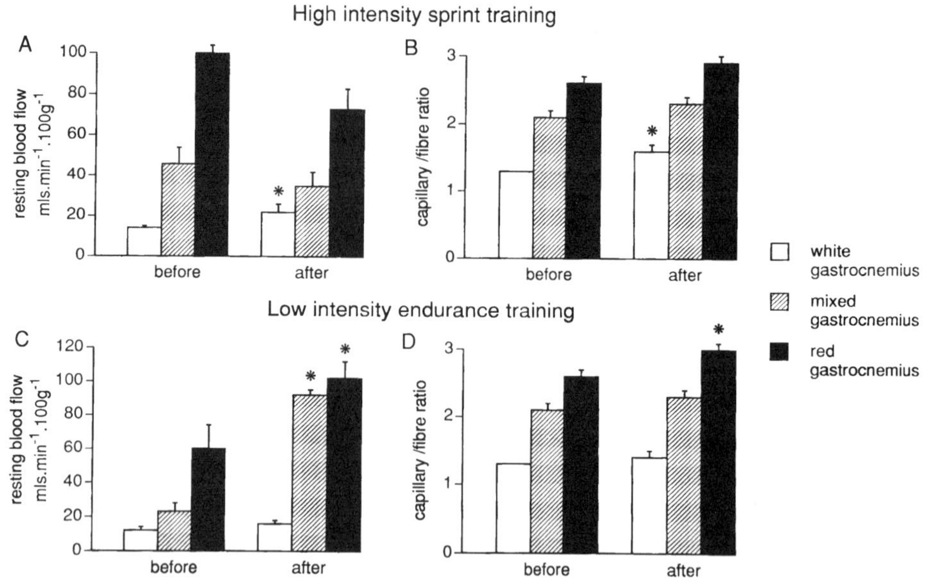

Figure 1 Resting blood flow (A and C) in three different parts of rat gastrocnemius muscle before and after high-intensity spring training (top) and low-intensity endurance training (bottom). B and D show capillary supply (capillary-per-fiber ratio C:F) for the same muscle regions after the same training regimes. Asterisks indicate significant differences after training versus before training. (From Refs. 59, 60, 176, and 177.)

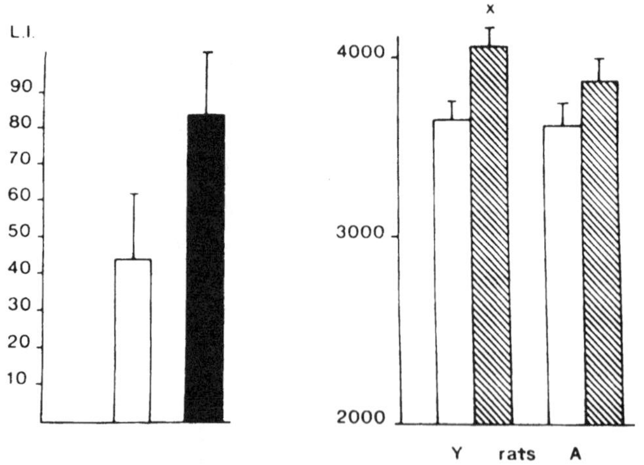

Figure 2 Left panel: Growth of capillaries assessed as labeling index for ^{3}H-thymidine (LI in counts per minute per mg DNA) in hearts of rats trained by swimming. (From Ref. 178.) Right panel: Capillary density (ordinate) expressed as number of capillary profiles per square millimeter in hearts of rats trained by treadmill running Y = young animals, A = adult animals. (From Ref. 63.)

but not those trained only twice a week (73). If the workload was increased by weights added to the animals, however, increased C:F ratio was found even in rats that swam only twice weekly (74). Such differences in response could depend upon the magnitude, duration, and frequency of increase in coronary blood flow during repetitive exercise bouts.

A lack of changes in capillary density does not necessarily indicate a complete lack of capillary growth, since training results in a mild heart hypertrophy and thus larger fibers. Maintenance of the same CD must therefore imply that some capillary growth has occurred, else CD would tend to decrease as in pathological heart hypertrophy. On the other hand, it is possible that a relative lack of capillary growth in the heart due to training is seen because CD is already very high; a similar lack of capillary growth in response to training was observed in skeletal muscles with a high initial capillary supply in species such as greyhounds (see Ref. 54). However, White et al. (180) recently showed that capillaries growing during early stages of training are transformed into arterioles so no increase in CD is found later.

Hemodynamic factors that could induce capillary growth in the hearts of moderately endurance-trained young animals could arise from the increased coronary blood flow (and hence shear stress). This may improve perfusion of capillaries, since the maximal transcapillary exchange capacity of trained hearts is higher (75). However, direct measurements of flow parameters at the level of individual capillaries in hearts of trained animals are not available owing to the technical difficulties in obtaining them. In addition to blood flow, stretch and relaxation of skeletal muscles and stretch of cardiac myocytes in conjunction with bradycardia and prolongation of diastole may disturb the configuration of the extracellular matrix, giving rise to mechanical factors that act on capillaries from the abluminal side, disturbing the basement membrane and release of growth factors.

C. Exposure to Hypoxia

Capillary growth in response to exposure to hypoxia has been studied in animals and humans, with conflicting results (for review, see Ref. 54). Relatively few studies report increased capillarity in skeletal muscles, and it is now generally considered that earlier reports of increases were misleading, since high-altitude hypoxia, particularly during longer exposure, leads to muscle fiber atrophy, giving rise to an apparent higher CD without actual capillary growth. A recent report on relatively short exposure to hypoxia, which did not alter fiber area, showed increased C:F in oxidative muscles (76). It remains to be determined to what extent capillary growth under these condition can be due to increased blood flow or upregulation of vascular endothelial growth factor (VEGF) by hypoxia. In the heart, increased capillarization in the right, but seldom the left, ventricle can be explained by the fact that the right ventricle works harder to overcome high resistance in the pulmonary bed and has higher blood flow in comparison with hearts of animals not exposed to hypoxia (77,78). Although acute hypoxia does not lead to perfusion of a greater proportion of capillaries (79), this does not exclude the possibility that velocity of flow in vessels which are perfused is increased.

D. Cold Exposure

It has been suggested (80) that the increased CD reported by some authors in skeletal muscle of animals exposed to high-altitude hypoxia may be due to the accompanying low temperature. Indeed, long-term exposure to cold leads to increased capillarity, particularly in red muscles, in a variety of animals (see Ref. 54). Recent studies on cold-acclimated rats and hamsters have shown that increased C:F in skeletal muscles was accompanied by increased blood flow, while an increase in capillary supply in the heart, which was of greater magnitude, was not (81–83).

IV. HEMODYNAMIC FACTORS IN MODELS OF PHYSIOLOGICAL ANGIOGENESIS

While an association between increased blood flow and the role of hemodynamic factors as modulators of capillary growth can be inferred from studying situations of physiological angiogenesis, the hypothesis needs further support. With this is mind, we have attempted to estimate shear stress and wall tension in capillaries in models of angiogenesis that would mimic to some extent situations occurring during exercise training. In skeletal muscles, this can be achieved by increasing activity on a chronic basis by electrical stimulation; in hearts, exercise training can be modeled by long-term bradycardia; and in both tissues, the role of increased blood flow can be replicated by prolonged pharmacological vasodilation.

A. Chronic Electrical Stimulation of Skeletal Muscles

It has long been known (84) that postural muscles, which are almost permanently active, have a considerably higher capillary density than muscles used for rapid movements. When Salmons and Vrbova (85) imposed the pattern of nerve activity occurring in postural muscles, such as soleus (continuous low-frequency firing), on fast-contracting muscles, the latter became slower. Using this approach, we established that 5–10 Hz continuous stimulation of the nerves supplying lower limb flexors [tibialis anterior (TA), extensor digitorum longus (EDL), extensor hallucis proprius (EHP)] resulted in rapid capillary growth in rats (Fig. 3) and rabbits, and in gracilis muscle in cats. The incorporation of the thymidine analogue bromodeoxyuridine (BrdU) into nuclei colocalized with capillaries was trebled in muscles stimulated for 2 days, and capillary supply, assessed as C:F, increased after 4 days, reaching values 40% higher after 7 days and doubling in muscles stimulated for 28 days (see Ref. 86 and 87). When blood supply to the stimulated muscles was limited by ligation of the common iliac artery, there was no change in C:F (88), thereby emphasizing the importance of blood flow in the process of capillary growth.

Measurements of red blood cell velocity (Vrbc) and diameters in capillaries during acute contractions and after 7 days of stimulation allowed calculation of shear stress. Assuming that viscosity was unchanged, shear stress was elevated 75% by acute contractions (89) and increased about 50% even at rest in muscles stimulated for 7 days (90). Capillaries were slightly wider, and the diameters of arterioles were increased in the first few days of stimulation, implying that capillary pressure and thus wall tension could also be increased.

There are several possible mechanisms by which increased shear stress and/or capillary wall tension could stimulate capillary proliferation. Shear stress results in release of NO and prostaglandins from endothelium both in vivo and in isolated microvessels (91). It upregulates endothelial nitric oxide synthase (eNOS) in vitro (92), an effect which can

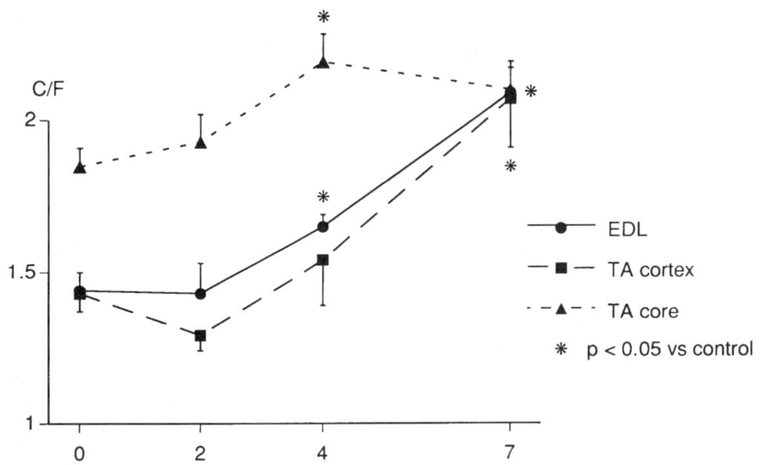

Figure 3 Capillary-per-fiber ratio (C:F) in rat tibialis anterior (TA, oxidative core and glycolytic cortex) and extensor digitorum longus (EDL, approximately half glycolytic and half oxidative) rat muscles, stimulated at 10 Hz, 8 h per day for 2, 4, or 7 days (abscissa). Day 0 indicates values in control animals.

also be produced by cyclic strain (93). Whereas the role of NO in angiogenesis is still controversial (94,95), prostaglandins induce angiogenesis in a variety of circumstances (96–98). In fact, there is evidence that both of these substances could be involved in the induction of capillary proliferation and increased C:F in stimulated skeletal muscle, since these were significantly attenuated by concurrent administration of the cyclooxygenase inhibitor indomethacin (89,87) and abolished by treatment with the NO synthase inhibitor L-NNA (99,87). Whether increased shear stress or capillary wall tension leads to activation of angiogenic growth factors in stimulated muscles is somewhat contradictory. There was no evidence for the involvement of basic fibroblast growth factor (FGF-2) (100), but the levels of a small molecular angiogenic factor, endothelial cell–stimulating angiogenic factor (ESAF) were increased in muscles stimulated for 7 days (101). Messenger RNA for VEGF was reported to be increased transiently in muscles stimulated for 1 day with a secondary increase after 21 days (102), but it was not elevated in muscles stimulated for a longer period of time (103). VEGF protein expression was greater after 3, 21, and 56 days in muscles stimulated for 24 hr per day (104), and even with 8 hr per day stimulation there was a higher proportion of capillaries colocalized with VEGF immunoexpression in muscles stimulated for 4, but not 2–3, days than in controls (105). VEGF has been shown to stimulate NO production in vitro (106,107) and in vivo (108), but the possibility of a reciprocal relationship has not been explored. As referred to above, acute muscle contractions increase shear stress, which results in immediate release of NO. Thus, although VEGF mRNA is increased after a single bout of exercise (109), the possibility of a more rapid release of NO, which could activate the VEGF gene, cannot be excluded.

Actively contracting muscles are also exposed to cyclic shortening and relaxation, which imposes mechanical stresses on the extracelluar matrix and capillaries tethered to it. To explore the involvement of mechanical factors acting on the capillary wall from outside, permanent stretch of the EDL muscle was produced by extirpation of its agonist,

TA. This resulted in capillary growth without changes in muscle blood flow or upregulation of FGF-2, but disturbance of the extracellular matrix did activate the growth factor ESAF (110).

B. Long-Term Bradycardia and Capillary Angiogenesis

Low heart rate in endurance-trained animals does not necessarily induce capillary growth (71), but data from comparative physiology indicate that very active animals have a lower heart rate and higher capillary density than sedentary animals of related species (62). A drastic reduction in heart rate by electrical bradycardial pacing for 4 weeks indeed resulted in increased capillary density in rabbits (111) and pigs (112). Pharmacological bradycardia by treatment with the sinus-acting drug alinidine also increased capillary density in rats (113). The evidence for capillary growth was an increase in CD occurring without changes in myocyte diameters after 3–4 weeks of pacing, estimated from staining of vessels for endothelial alkaline phosphatase (rabbits and rats) or lectin *Griffonia simplicifolia I* (pigs) or from low-power electron microscopy. The greater the bradycardia, the larger was the increase in CD. Heart rate reduction by nearly 50% in rabbits resulted in 25–30% higher capillary density than in the animals with a comparable heart weight, while a decrease of about 30% in the heart rate in pigs or rats led to a 20% increase in capillarity (114) (Table 1).

The mechanism by which bradycardia stimulates capillary growth could relate to changes in perfusion characteristics. Heart rate reduction induced by alinidine (115) or acute bradycardial pacing (116) improved subendocardial perfusion in ischemic hearts, and a similar effect in normal hearts would increase wall shear stress. Although coronary flow per beat was found to be higher in paced hearts (117), it is unlikely that increased shear stress could explain capillary growth. Vrbc is much higher during systole than during diastole (118); even with a drastic reduction in heart rate, the relative durations of systole and diastole do not change sufficiently for Vrbc to be significantly altered overall (119). Capillary diameters, however, are wider during diastole than systole (118), and even if capillary pressures do not seem to be different, capillary wall tension will be higher (119,120). During bradycardia, this will occur for relatively longer periods of time and may cause disruption of the capillary basement membrane and enable endothelial cell migration and capillary elongation.

Yet another mechanical stimulus for vessel growth resulting from bradycardia may

Table 1 Degree of Bradycardia[a] and Increase in Capillary Supply[b] in Experimental Animals

	Resting heart rate beats per minute	Heart rate during alinidine or pacing, beats per minute	Capillary supply Pretreatment	Posttreatment
Rats	380 ± 20	288 ± 10	C:F 1.30 ± 0.04	1.54 ± 0.02
Pigs	117 ± 7	77 ± 2	CD 1445 ± 67	1709 ± 104
Rabbits	277 ± 5	151 ± 3	CD 1963 ± 168	2456 ± 194

[a] Produced in anesthetized rats by alinidine (3 $mg.kg^{-1}$ i.p) and in anesthetized pigs and rabbits by right atrial pacing.

[b] C:F (capillary-per fiber-ratio) or CD (capillary density) after 5 weeks of treatment with alinidine in rats and 4–5 weeks of pacing in pigs and rabbits.

be linked with the mechanical distortion of the whole cardiac structure due either to the elongation of myocytes during prolonged diastole or to increased contraction force consequent to the positive inotropic effect of chamber stretch. Since capillaries are tethered to myocytes, exaggerated changes in myocyte contraction could disturb their basement membrane and/or extracellular matrix and initiate the angiogenesis cascade, with endothelial cell migration and proliferation. The importance of the latter in capillary growth was studied using a drug with positive inotropic action, the β-agonist dobutamine. Long-term administration of dobutamine increased stroke volume, with a concomitant decrease in coronary blood flow in dogs (121), and simulated some effects of exercise training in rats (122). In rabbits, it had a positive inotropic and minimal chronotropic effect when given by continuous intravenous infusion for 2 weeks. It did not cause heart hypertrophy but increased maximal stroke work and capillary density without affecting maximal coronary blood flow, thus demonstrating an effect on the growth of capillaries that involved mechanical but not hemodynamic factors (123).

The mechanical factors arising from changes in heart rate—elongation of myocytes during prolonged diastole or increased force of contraction—could potentially lead to activation of growth factors. In stretched perfused hearts, Li et al. (124) showed a sixfold increase in VEGF expression, which was mediated by TGF-β. Although mRNA for VEGF was not examined in bradycardially paced hearts, TGF-β mRNA was found to be elevated (120). Increased mRNA for VEGF was, however, found in rat hearts where chronic bradycardia was elicited by treatment with alinidine (125), and Hudlicka et al. (120) showed a good correlation between capillary density and levels of ESAF in bradycardially paced hearts, whereas mRNA for FGF-2 did not seem to be related to capillary supply.

C. Long-Term Pharmacological Dilation

The preceding sections suggest a relationship between blood flow and vascular growth under a variety of physiological circumstances, but the role of increased flow in angiogenesis has been more clearly defined by the use of vasodilating drugs (see Ref. 54 for review). In skeletal muscles, C:F was increased by administration of adenosine, dipyridamole, propentofylline, or prazosin, all vasodilators with completely different mechanisms of action. Prazosin did not increase flow in the heart and therefore did not increase myocardial capillary density, while adenosine or propentofylline did (Fig. 4). Dipyridamole caused capillary proliferation in both heart and skeletal muscle, evident as increased labeling index for ^{3}H-thymidine (126,127) whereas prazosin increased C:F by capillary elongation and splitting (128) without proliferation (129). Increased total capillary length and increased numbers of capillary endothelial nuclei were reported in the hearts of rats and rabbits fed for several weeks with ethanol as a vasodilator (130,131). Although some of these drugs, such as adenosine, can stimulate endothelial cell proliferation directly (132), others, such as dipyridamole, have no similar effect in tissue cultures (133).

The hemodynamic effects of dipyridamole in coronary capillaries have been assessed by intravital microscopy, showing increased Vrbc without any change in diameters (134). This would increase shear stress in these vessels (120), a similar effect to that produced in the capillary bed of skeletal muscles by prazosin (90). Some drugs may increase capillary wall tension without altering capillary diameter. For example, dipyridamole increased pressure in venules in the heart (135), which would result in increased capillary pressure, and prazosin could have a similar effect in skeletal muscles by dilating arterioles.

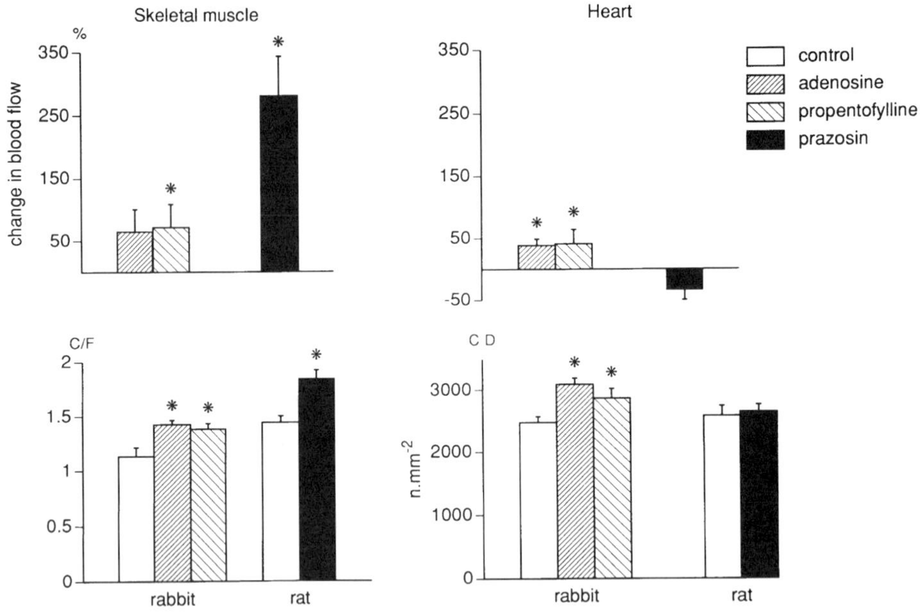

Figure 4 Top: changes in blood flow as percent of resting values (ordinate) in skeletal muscle (left) and heart (right) at the end of 3- to 4-hr intravenous infusion of adenosine, propentofylline or prazosin. Bottom: C:F in extensor digitorum longus muscle (left) and capillary density (CD) in the hearts (right) of animals which had been treated chronically with adenosine or propentofylline (continuous intravenous infusion 24 hr per day for 3–5 weeks) or prazosin (in drinking water for 5 weeks). Asterisks indicate significance against control values. (From Refs. 174 and 179.)

V. INVOLVEMENT OF HEMODYNAMIC FACTORS IN PATHOLOGICAL ANGIOGENESIS

Although growth of vessels in pathological situations such as tumors, healing wounds, or retinopathies is regulated mainly by growth factors, some involvement of hemodynamic factors should also be considered. Clark et al. (3) demonstrated the importance of blood flow in capillary growth and transformation of capillaries into large microvessels in the rabbit ear chamber preparation. They also showed that cold, which caused vasoconstriction, slowed down growth of vessels, while increased temperature, causing dilation, accelerated it. Very recently, Ichioka et al. (136) accelerated the growth of vasculature in the same preparation by long-term administration of prazosin, linked with elevation of shear stress in venules lasting up to 13 days postoperatively. More rapid healing of dermal wounds by electrical stimulation has been attributed to increased blood flow (137). In tumors, blood flow is sluggish and the failure of vasodilating drugs such as papaverine or isoprenaline to increase flow indicates that vessels are already maximally dilated (138). In this condition, however, wall tension would be increased, which could enhance the susceptibility of endothelial cells to growth factors (6). Stefansson et al. (139) have also presented evidence supporting the importance of vasodilation in different forms of retinopathies. Even though the evidence for involvement of hemodynamic factors in vessel growth in pathological situations is rather anecdotal, it should be acknowledged that changes in blood flow, shear stress, or wall tension have the potential to modify the effect of other factors involved.

VI. MODULATION OF ANGIOGENIC PROCESSES BY HEMODYNAMIC AND MECHANICAL FACTORS

The accepted view of angiogenesis is that it starts by breakage of capillary basement membrane, followed by endothelial cell migration outside the capillary, with mitotic activity occurring at the base of the migrating cells, making sprouts, which later connect to other capillaries. This process requires activation of proteolytic enzymes to disintegrate the extracellular matrix in order for endothelial cell migration and sprout formation to occur, and it is thought to be initiated by the presence of growth factors toward which the capillaries start to grow (140). Although it is now known that growth factors such as FGF-2 or VEGF can cause arteriolar dilation at lower concentrations than those required for endothelial cell proliferation (141), and hence could generate hemodynamic forces in the existing capillaries, such a mechanism has not been experimentally investigated.

Another type of angiogenesis called intussusceptive growth has been described by Burri and Tarek (142). This starts by narrowing of capillaries and fusion of their opposing walls, followed by ingrowth of surrounding mesenchymal tissue, eventually dividing the

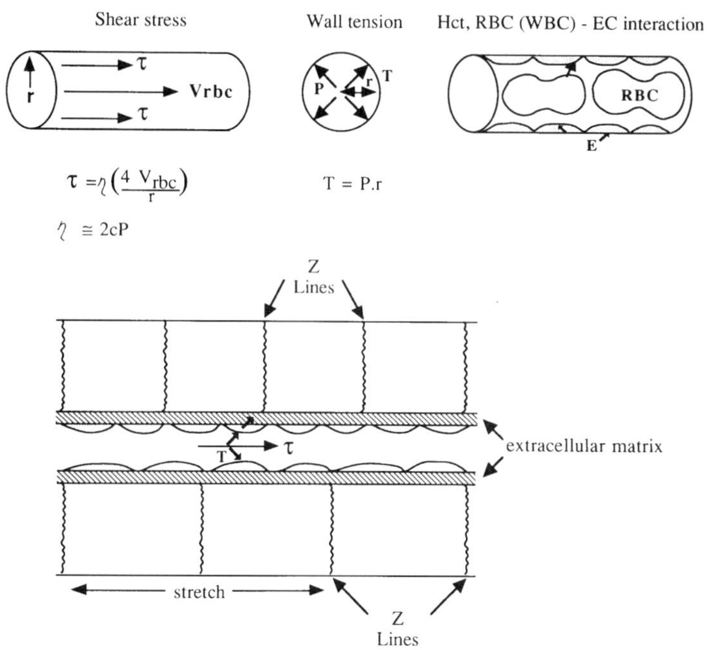

Figure 5 Scheme of mechanical factors with the potential to influence microvascular growth. Vrbc = velocity of red blood cells, τ = shear stress, η = viscosity, cP = centipoise, P = pressure, r = radius, T = tension, E = endothelial cell, RBC = red blood cell, WBC = white blood cell. Interval between Z lines indicates sarcomere length. (From Ref. 87.)

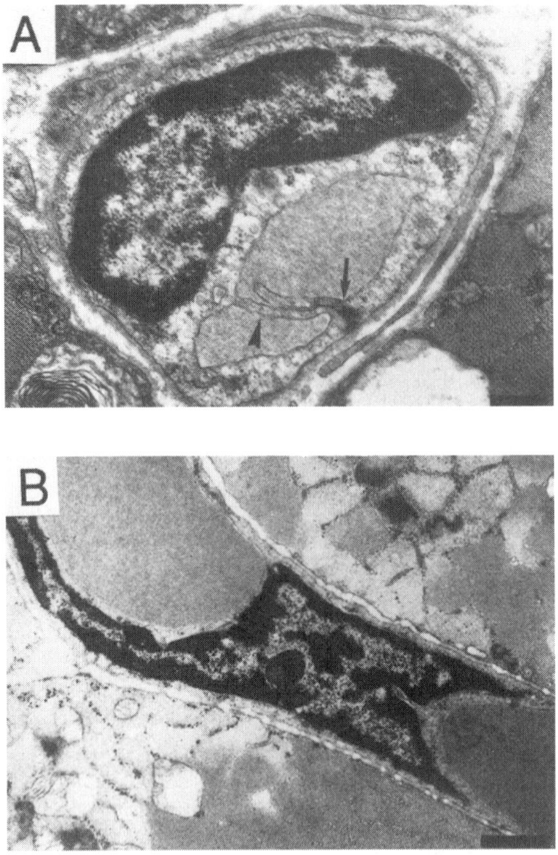

Figure 6 A. Electron micrograph of a capillary from rat extensor hallucis proprius (EHP) muscle after 1 week of treatment with prazosin. Thin luminal processes (arrowheads) traverse the lumen to form a complete septum. B. Electron micrograph of twin capillary lumens sharing a single nucleus (in the middle) from rat EHP after 2 weeks prazosin treatment. Bars = 0.5 μm. (From Ref. 128, by permission of Springer Verlag.)

capillary in two. The factors that promote this type of growth are still not clear but, as it occurs mainly during development in many organs, it could be explained by a faster rate of growth of surrounding tissue than of endothelial cells, exerting pressure that causes the initial capillary collapse. Yet another type of capillary growth is by longitudinal splitting, described only during development in the heart (143).

To what extent can hemodynamic factors influence the way in which capillary growth takes place? The effect of shear stress or wall tension/circumferential wall stress, which act mainly on the luminal surface of a vessel, on growth of endothelial cells in vitro and growth of capillaries in vivo has already been referred to. In addition, the extracellular matrix and surrounding tissue can influence stress on the capillary wall (Fig. 5). Our own models of physiological angiogenesis separate, to a certain degree, the role of the factors acting predominantly from the luminal surface (administration of vasodilators and increased shear stress) from those acting predominantly on the abluminal side (stretch), with a combination of both factors present during chronic electrical stimulation.

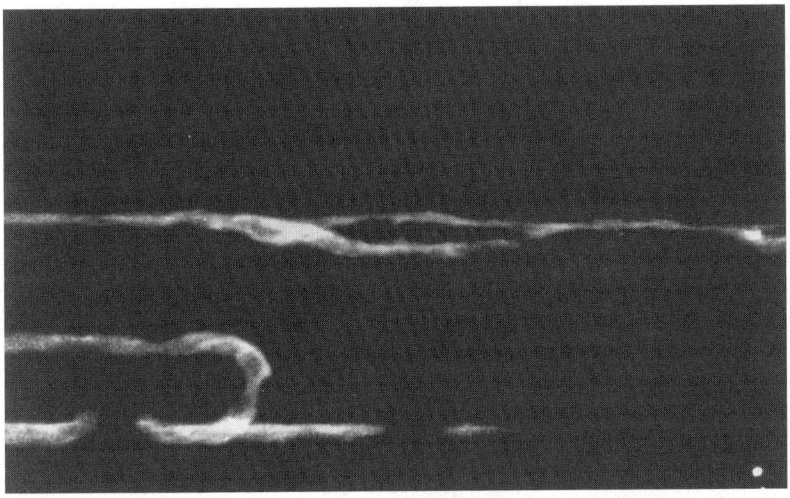

Figure 7 Confocal microscopic image of capillaries, visualised by the fluorescent lectin method (Ref. 147) in a whole mount preparation of rat extensor hallucis proprius muscle from a 2-week prazosin-treated animal. The central capillary shows a region of divergence and convergence, consistent with the proposed mechanism of growth by longitudinal splitting. (Courtesy of Dr. F. Hansen-Smith.) (From Ref. 128, by permission of Springer Verlag.)

The form of capillary growth differed in these models accordingly. While administration of prazosin caused increased C:F without proliferation (129), proliferation of endothelial cells occurred concurrently with increased capillarization during long-term stretch (144) and preceded it during chronic stimulation (87,89).

These different patterns of angiogenesis were demonstrated using electron and confocal microscopy techniques. In capillaries of skeletal muscles from prazosin-treated animals, there were signs of neither basement membrane breakage nor abluminal capillary sprouting. However, numerous protrusions into the capillary lumen (Fig. 6A) ultimately led to division of capillaries (Fig. 6B) by longitudinal splitting (Fig. 7) (128). By contrast, increased C:F in stretched muscles was achieved by transient proliferation, with an increased labeling index for BrdU in capillary-linked nuclei and mitotic figures in endothelial cells, increased numbers of abluminal processes with focal breakdown of the basement membrane at the tips of some, and signs of sprout formation (144). These findings were also confirmed by confocal microscopy (145,146). In stimulated muscles, where increased shear stress is combined with changes in the configuration of the extracellular matrix, there were occurrences of basement membrane disruption (Fig. 8) and formation of both luminal and abluminal processes with a rich formation of sprouts within 4 days and an expanded capillary bed within a week (Fig. 9) (147). Capillary sprouts in stimulated muscles were also seen by intravital microscopy (148). Increased adhesion of white blood cells in stimulated muscles to the walls of confluent venules (149) could contribute to an increase in capillary pressure and thus increased capillary wall tension (Fig. 10).

It appears, therefore, that hemodynamic and mechanical factors can induce capillary growth, which takes different forms depending on whether the factors act primarily upon the luminal or abluminal vessel surface. Where shear stress is involved, the growth could be mediated by nitric oxide and/or prostaglandins. Where mechanical factors act predomi-

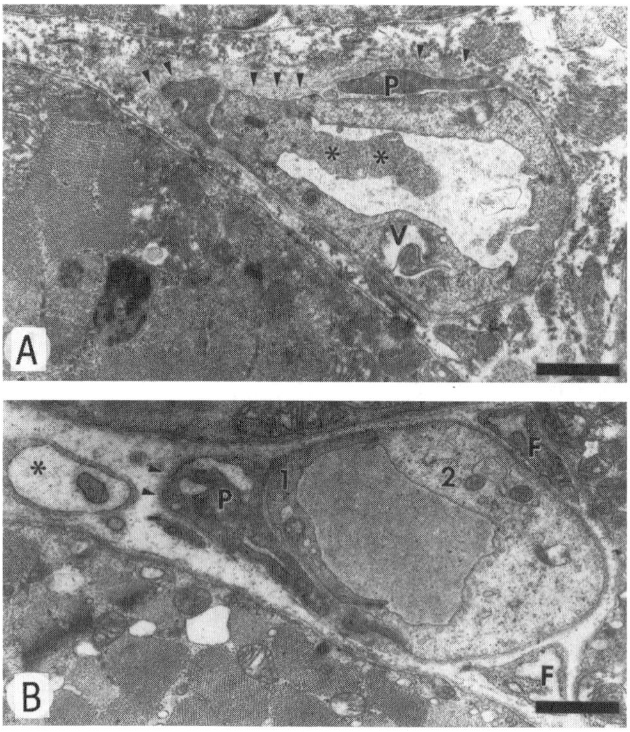

Figure 8 Electron micrographs of capillaries from extensor digitorum longus muscles stimulated for 4 days. A. Arrowheads indicate regions of very amorphous capillary basement membrane, asterisks identify a luminal protrusion, V indicates a vesicle with a myelin figure. B. A capillary composed of two endothelial cells of different electron density and content of mitochondria and pinocytic vesicles, indicating differing metabolic activity, is surrounded by a pericyte (P) and a fibroblast-like cell (F). A structure marked by * may represent the tip of a capillary sprout. Bar = 1 μm. (From Ref. 147, by permission of Springer Verlag.)

nantly from the abluminal side, disturbance of the extracellular matrix is required, possibly involving growth factors and/or integrins.

VII. GROWTH OF ARTERIOLES AND LARGER VESSELS

Whereas endurance training undoubtedly leads to capillary growth in skeletal muscle, this may or may not be accompanied by growth of arterioles. In rats trained by treadmill running, an increased number of arterioles with a very small increase in C:F was seen in spinotrapezius muscle; in gracilis, C:F but not the number of arterioles was higher (150). Although the resting diameter of arterioles of all categories was similar in sedentary and trained animals, dilation of the larger arterioles during muscle contractions was greater after training (151). This could explain the findings of higher maximal blood flow, an indicator of the size of the whole vascular bed, in the limbs of trained humans (152,153) (Fig. 11).

When muscle activity was increased by chronic electrical stimulation, the size of

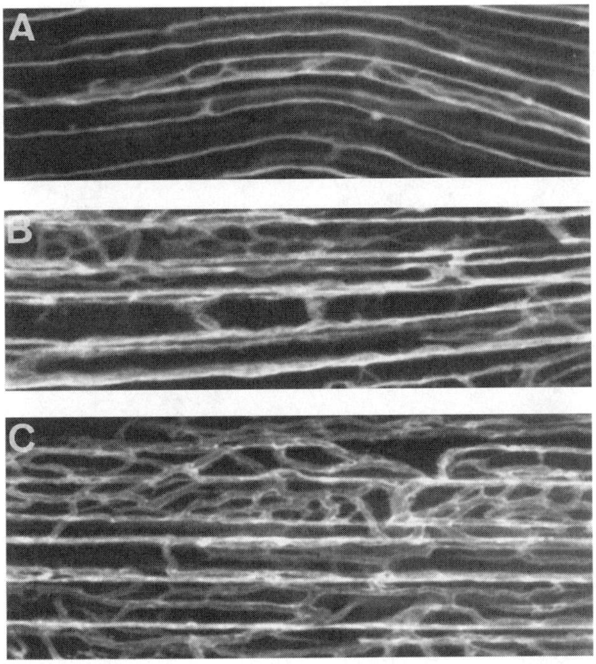

Figure 9 Capillaries in a whole mount of rat extensor hallucis proprius visualised by laser scanning confocal microscopy after staining for lectin *Griffonia simplicifolia I.* A. Control muscle, B and C. Muscles stimulated for 4 and 7 days respectively. (From Ref. 147, by permission of Springer Verlag.)

the whole vascular bed, estimated by casting, was enlarged after only 1 week of stimulation (154). This was most likely due to an increased density of the small arterioles, shown using immunostaining for alpha smooth muscle actin (αSMA) (155). There was little increase in the number of arterioles >20 μm in diameter or in vessels categorized as mature by the presence of smooth muscle myosin heavy chains. There was, however, a considerable increase in the number of smaller vessels (<10 μm diameter) stained for αSMA, particularly those that were not completely encircled by αSMA-positive staining, which may have been capillaries in the process of transforming into arterioles, a process suggested to occur in developing muscles (156). The size of feed arteries was increased after 4 weeks of stimulation (157) and maximal conductance increased after 2 weeks (158). These changes are unlikely to have occurred any earlier, since blood flow in contracting muscles was only marginally higher after 7 days of stimulation than in control muscles (159) (Fig. 12). A similar increase in the arteriolar network was described in muscles of prazosin-treated animals (160) and after treatment with the β_2-agonist salbutamol (161).

The stimulus for growth of arterioles is not clear. Hester and Duling (162) observed only a small increase in Vrbc in arterioles in contracting muscles and since their diameters were larger than at rest, calculated shear stress decreased rather than increased. Nevertheless, wider diameters would result in increased circumferential wall stress, which is known to be an important factor initiating growth of arterioles (4). Increased circumferential wall stress could also play a role in ''arteriolarization'' of capillaries in stimulated muscles, since the increase in arteriolar density seen after 7 days was preceded, in muscles stimu-

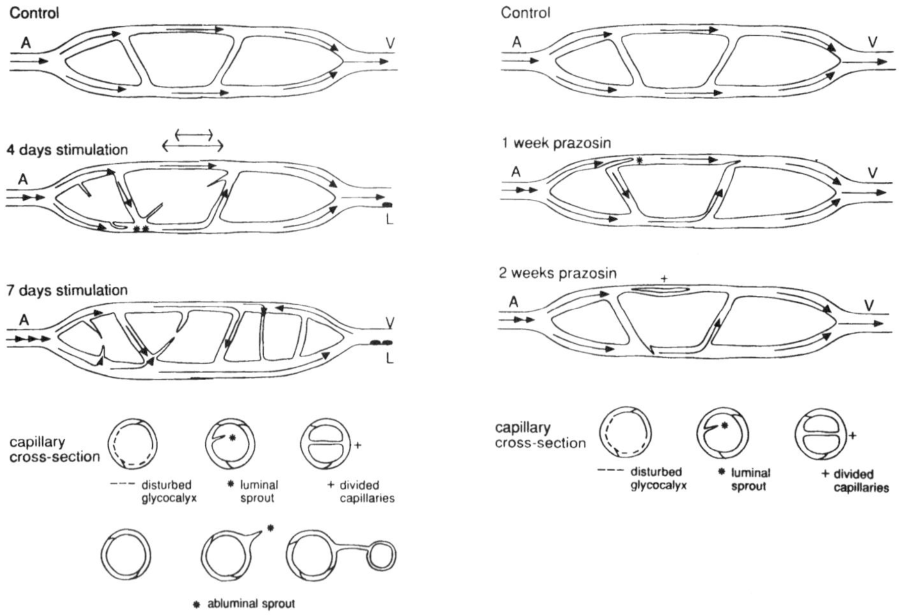

Figure 10 Scheme showing capillary growth in rat muscles during chronic electrical stimulation (left side) and prazosin treatment (right side). Arrows inside capillaries indicate direction of flow; double or triple arrows in arterioles (A) represent faster flow. V = venule, L = leukocytes, which adhere to venular endothelium more in chronically stimulated muscles and may contribute to increased capillary pressure. Asterisks inside capillary cross sections indicate luminal sprouting; asterisks outside capillary cross section indicate abluminal sprouting. Lines with double arrowheads outside capillaries in 4-day stimulated muscles indicate the changes in sarcomere length with muscle contraction and relaxation, also applicable to 7-day stimulated muscles. Interrupted lines in the capillary cross section indicate possible disturbance of glycocalyx, which could be necessary for luminal sprouting to occur. (From Ref. 87.)

lated for 2 days, by wider arteriolar diameters (89). It is therefore probable that the mechanical stimuli for capillary and arteriolar growth are different, with shear stress being an important stimulus in capillaries and circumferential wall stress in arterioles. Whereas capillaries in stimulated muscles grow by sprouting (148) and/or by longitudinal splitting (147), Skalak and Price (4) suggest three possible mechanisms to explain the growth of arterioles. Capillary ''arteriolarization'' could occur either by migration of smooth muscles cells towards capillaries, by pericytes transforming into smooth muscle cells or by fibroblasts migrating toward capillaries and changing into smooth muscle cells. There is some support for the role of fibroblasts in this process. Zhou and Egginton (163) described a close apposition of proliferating fibroblasts to capillary endothelial cells in stimulated muscles on the basis of immunogold labeling for BrdU, and Price and Skalak (164) demonstrated a great increase in the number of proliferating fibroblasts in the adventitia of arterioles in prazosin-treated animals. The great increase in the numbers of capillaries partially encircled by αSMA in stimulated muscles indicates either the migration of smooth muscle cells or increased coverage by pericytes. However, electron microscopic studies (165) showed pericyte withdrawal rather than increased coverage of the capillary circumference in stimulated muscles, so this issue needs to be further investigated. Whether the proposed

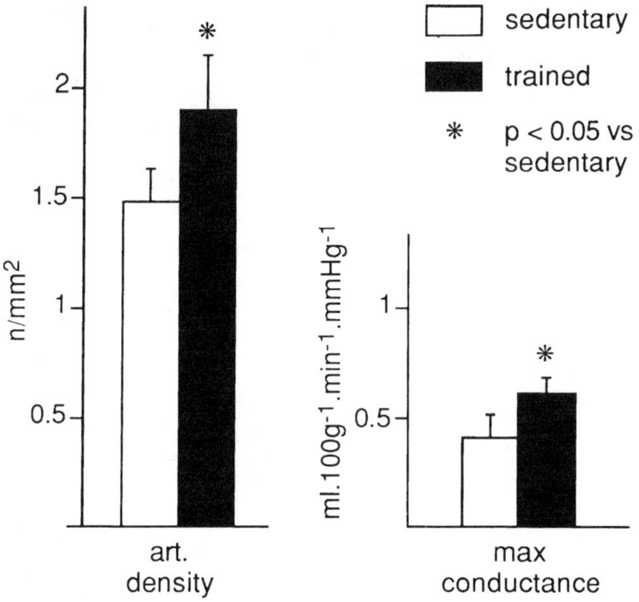

Figure 11 Arteriolar density expressed as number per square millimeter in skeletal muscles from sedentary and treadmill-trained rats and maximal limb vascular conductance in sedentary and trained individuals. (From Refs. 150 and 153.)

explanation of arteriolar growth in stimulated muscles is applicable to muscles exposed to exercise training is again a matter for further study.

A higher maximal conductance and greater vasodilator capacity in endurance trained or long-term stimulated muscles may also be explained by an increased activity of eNOS, observed in the aorta (166) and in skeletal muscle arterioles from trained rats (167). A generalized increase in NO production and a greater dilation of different vessel categories would probably increase the circumferential wall stress and could lead to a permanent increase the vessel wall diameters.

In contrast to skeletal muscle, exercise training does not inevitably lead to capillary growth in the heart, but growth of arterioles and large conduit vessels has frequently been reported. Evidence for growth of the larger coronary vessels with training includes increased weight of vascular casts relative to heart weight in rats trained by swimming (168) and increased size of the coronary arteries at postmortem in marathon runners (169). Higher maximal coronary blood flow (170), conductance (171), and maximum capillary diffusion capacity (72) in hearts of trained compared with sedentary animals might be due to these changes in the resistance and conduit vasculature but do not appear to be linked with capillary growth (Fig. 13). A study of training effects in pig hearts found coronary blood flow to be higher in subepicardial layers and unchanged in subendocardial layers with a lower CD in both regions, but a significant increase in the density of arterioles (71). White and Bloor (172) recently demonstrated an increase in the number of coronary arterioles starting after 3 weeks of training in pigs. After 16 weeks, however, the increase in arteriolar numerical density only accounted for 11% of the 50% increase in total cross-sectional area of the vascular bed, the rest being due to an increase in diameter of existing

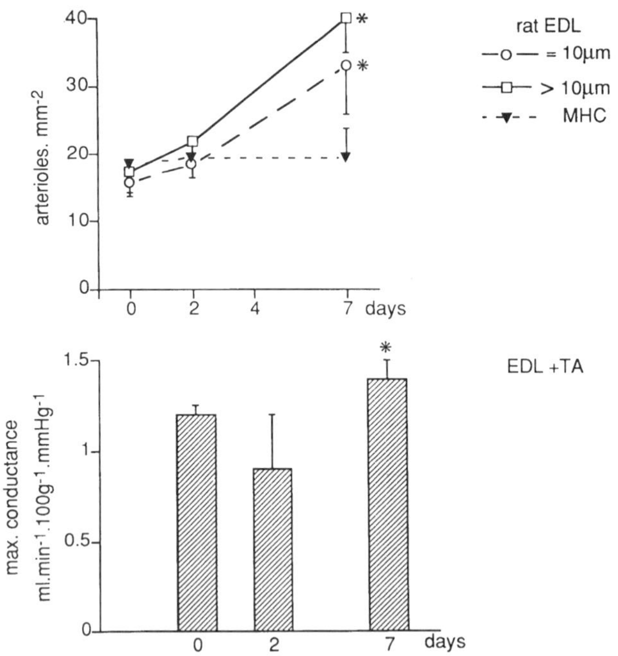

Figure 12 Arteriolar density (top) and maximal vascular conductance (bottom) in control (0), 2-, and 7-day stimulated rat EDL muscles. Classification of arterioles was based on staining for α smooth muscle actin (diameter > 10 μm) and myosin heavy chains (MHC), the latter representing mature arterioles. Maximal conductance was measured in contracting muscles using radioactive microspheres. Data are shown as means ± SEM, with * representing values significantly different from controls. (From Refs. 155 and 159.)

vessels. This complements the previous findings of an increase in the area of both left anterior descending and left circumflex coronary arteries in trained dogs (173) (Fig. 13).

The factors involved in the growth of arterioles or enlargement of conduit vessels in the heart are not fully understood. Higher blood flow is clearly not important, since neither long-term administration of vasodilating drugs (174) nor sympathectomy (175) led to increased maximal coronary blood flow. Furthermore, maximal coronary blood flow was not increased in bradycardially paced hearts or in animals treated with dobutamine, two interventions that stimulated significant growth of capillaries. It therefore remains a challenge to establish why procedures that readily stimulate capillary growth in the heart do not affect growth of larger vessels, while those that do not lead to capillary proliferation can induce growth of the arterial vascular tree.

VIII. CONCLUSIONS

Hemodynamic forces connected with changes in blood flow or blood pressure result in changes in shear stress or circumferential wall stress/wall tension. In vitro, fluid shear stress and cyclic strain, or stretch, induce endothelial proliferation by a variety transduction signals that include activation of phospholipases, protein kinases, and mitogen-activated

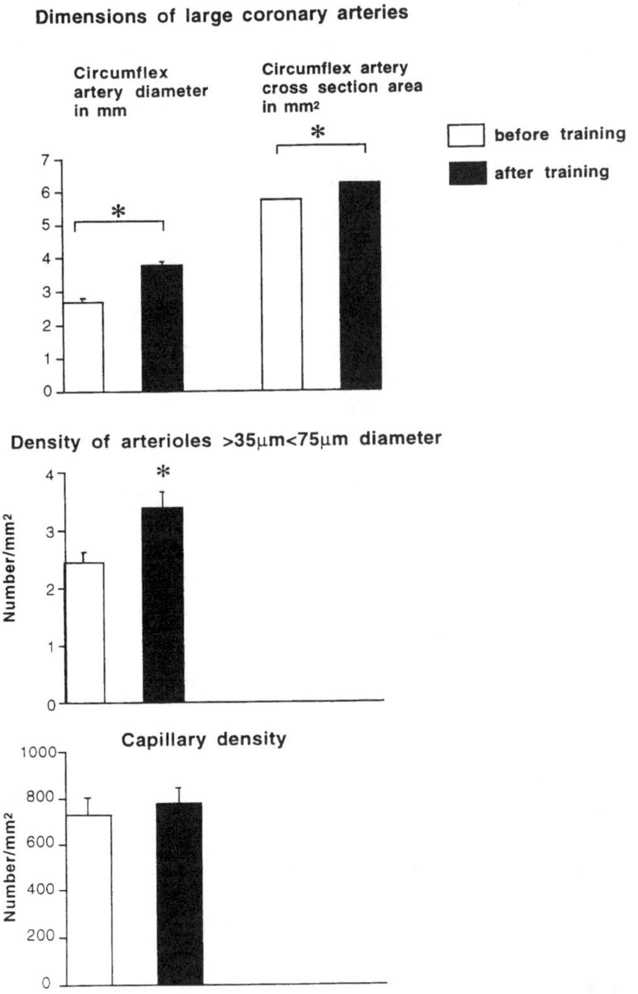

Figure 13 Changes in the coronary vascular bed in dogs (circumflex artery and capillary density) and pigs (density of arterioles) after endurance training. Data are shown as means ± SEM, * indicating values significantly different from those before training. (From Refs. 71 and 173.)

kinases, leading to upregulation of genes for growth factors or genes directly connected with cellular growth. Although there is sufficient evidence showing parallel changes in blood flow and capillary growth under physiological (female reproductive organs, long-term exposure to hypoxia or cold, endurance exercise) and pathological (wounds, tumors, retinopathies) conditions, there are very few direct studies providing data on hemodynamic forces in association with capillary growth.

Shear stress has been measured in only a few experiments, showing that it is increased in capillaries or venules in animals treated with vasodilating drugs (in skeletal muscles, hearts, and wounds), and in muscles subjected to a long-term activity, in all cases being linked with increased capillary supply. To date there are no data on changes in shear stress in capillaries in animals exposed to any type of training that leads to capillary growth. Capillary wall tension can be increased in skeletal muscles exposed to high levels of contractile activity and in hearts with a drastically reduced heart rate, situations resulting in capillary growth. Wall tension/circumferential wall stress is a very important factor for the growth of arterioles, which is a feature of muscles exposed to high levels of activity due to either training or electrical stimulation and the hearts of trained animals. Different hemodynamic forces can result in differing forms of angiogenesis, with capillary bed expansion due to endothelial cell migration and longitudinal splitting in situations where the forces are acting on the luminal side of the vessel and to endothelial cell proliferation and abluminal capillary sprouting when the mechanical forces act mainly from the abluminal side.

REFERENCES

1. R Thoma. Untersuchungen uber die Histogenese und Histomechanik des Gefassystems. Stuttgart: Enkeverlag, 1893.
2. ER Clark. Studies on the growth of blood vessels in the tail of the frog. Am J Anat 23:37–88, 1918.
3. ER Clark, WJ Hitchler, HT Kirby-Smith, RO Rex, JH Smith. General observations on the ingrowth of new blood vessels into standardized chamber in the rabbit's ear and the subsequent changes in the newly grown vessels over a period of months. Anat Rec 50:129–168, 1931.
4. TC Skalak, RJ Price. Mechanical stresses in microvascular remodelling. Microcirculation 3: 143–165, 1996.
5. J Lee, GW Schmid-Schonbein. Biomechanics of skeletal muscle capillaries: hemodynamic resistance, endothelial distensibility, and pseudopod formation. Ann Biomed Eng 23:226–246, 1995.
6. J Folkman, HP Greenspan. Influence of geometry on control of cell growth. Biochem Biophys Acta 34:905–910, 1975.
7. DT Shima, KB Saunders, A Gougos, PA D'Amore. Alterations in gene expression associated with changes in the state of endothelial differentiation. Differentiation 58:217–226, 1995.
8. H Lindner, U Behrends, Y Misumi, E Kremmer, G Eissner, S Schultz-Hector. Loss of myocardial capillary endothelial cell alkaline phosphatase (ALP) activity in primary endothelial cell culture. Cell Tissue Res 291:497–506, 1998.
9. CF Jr Dewey, SR Bussolari, MA Gimbrone Jr, PF Davies. The dynamic response of vascular endothelial cells to fluid shear stress. J Biochem Eng 103:177–185, 1981.
10. PF Davies, A Remuzzi, EJ Gordon, CF Dewey, MA Gimbrone. Turbulent fluid shear stress induces vascular endothelial cell turnover in vitro. Proc Nat Acad Sci USA 83:2114–2117, 1986.
11. J Ando, H Nomura, A Kamiya. The effect of fluid shear stress on the migration and proliferation of cultured endothelial cells. Microvasc Res 33:62–70, 1987.
12. M Hecker, AM Isch, E Bassenge, R Busse. Vasoconstriction and increased flow: two principal mechanisms of shear stress-dependent endothelial autacoid release. Am J Physiol 265: H828–H833, 1993.
13. M Noris, M Morigi, R Donadelli, S Aillo, M Foppolo, M Todeschini, S Orisio, G Remuzzi, A Remuzzi. Nitric oxide synthesis of cultured endothelial cells is modified by flow conditions. Circ Res 76:536–543, 1995.
14. M Ziche, L Morbidelli, E Masini, H Granger, P Geppetti, F Ledda. Nitric oxide promotes DNA synthesis and cyclic GMP formation in endothelial cells from postcapillary venules. Biochem Biophys Res Commun 192:1198–1203, 1993.

15. SP Olesen, DE Clapham, PF Davies. Haemodynamic shear stress activate a K+ current in vascular endothelial cells. Nature 331:168–170, 1988.
16. A Bhagyalakshmi, F Berthiaume, KM Reich, JA Frangos. Fluid shear stress stimulates membrane phospholipid metabolism in cultured endothelial cells. J Vasc Res 29:443–449, 1992.
17. F Berthiaume, JA Frangos. Fluid flow increases membrane permeability to merocyanine 540 in human endothelial cells. Biochim Biophys Acta 1192:209–218, 1994.
18. MU Nollert, SG Eskin, LV McIntire. Shear stress increases inositol triphosphate levels in human endothelial cells. Biochem Biophys Res Commun 170:281–282, 1990.
19. LC Brown, FC Messick, MP Koki, IG Hamilton, PR Giraud. Fluid flow stimulates metalloproteinase production and deposition into extracellular matrix of endothelial cells (abstr). FASEB J 9:A617, 1995.
20. AM Malek, GH Gibbons, VJ Dzau, S Izumo. Fluid shear stress differentially modulates expression of genes encoding basic fibroblast growth factor and platelet derived growth factor-B chain in vascular endothelium. J Clin Invest 92:2013–2201, 1993.
21. N Resnick, MA Gimbrone. Hemodynamic forces are complex regulators of endothelial gene expression. FASEB J 9:874–882, 1995.
22. PF Davies. Flow-mediated endothelial mechanotransduction. Physiol Rev 75:519–560, 1995.
23. A Gupte, JA Frangos. Effects of flow on the synthesis and release of fibronectin by endothelial cells. In vitro 26:57–60, 1990.
24. DH Ausprunck, SM Dethlefsen, ER Higgins. Distribution of fibronectin, calcium and type IV collagen during development of blood vessels in the chick chorioallantoic membrane. In: RN Feinberg, G-K Schere, R Auerbach, eds. The Development of the Vascular System. Basel: Karger, 1991, pp. 93–108.
25. BJ Rongish, G Hinchman, MK Doty, S Baldwin, RJ Tomanek. Relationship of the extracellular matrix to coronary neovascularization during development. J Mol Cell Cardiol 28:2203–2215, 1996.
26. BE Sumpio, AJ Banes, AG Levin, G Johnson. Mechanical stress stimulates aortic endothelial cells to proliferate. J Vasc Surg 6:252–256, 1987.
27. BE Sumpio. Haemodynamic forces and the biology of endothelium—signal transduction pathways in endothelial cells subjected to physical forces in vitro. J Vasc Surg 13:744–746, 1991.
28. BE Sumpio, W Du, G Galagher, X Wang, LM Khachigian, T Collins, MA Gimbrone, N Resnick. Regulation of PDGF β in endothelial cells exposed to cyclic strain. Arteriosclerosis Thromb Vasc Biol 18:349–355, 1998.
29. F Lyall, MR Deehan, IA Greer, F Boswell, WC Brown, GT McInnes. Mechanical stretch increases proto-oncogene expression and phosphoinositide turnover in vascular smooth muscle cells. J Hypertens 12:1139–1145, 1994.
30. CC Haudenschild. Growth control of endothelial cells in atherogenesis and tumour angiogenesis. In: BM Altura, ed. Advances in Microcirculation, Vol. 9. Basel: Karger, 1980, pp. 226–251.
31. DE Ingber, J Folkman. Regulation of endothelial growth factor action—solid state control of extracellular matrix. Prog Clin Biol Res 249:273–282, 1987.
32. T Iba, S Maitz, T Furbert, O Rosales, MD Widmann, B Spillane, T Shin, T Sonoda, BE Sumpio. Effect of cyclic stretch on endothelial cells from different vascular beds. Circ Shock 35:193–198, 1991.
33. T Iba, T Sain, T Sonoda, O Rosales, BE Sumpio. Stimulation of endothelial secretion of tissue-type plasminogen activator by repetitive stretch. J Surg Res 50:457–460, 1991.
34. PM Gullino, G Alessandri, M Ziche. Regulatory events in angiogenesis. Front Diabetes 9: 175–182, 1990.
35. MS Pepper, R Montesano, SJ Mandriota, L Orci, JD Vassalli. Angiogenesis: a paradigm for balanced extracellular proteolysis during cell migration and morphogenesis. Enzyme Protein 49:138–162, 1996.

36. JB Lansman, TJ Hallam, TJ Rink. Single stretch-activated ion channels in vascular endothelial cells as mechanotransducers? Nature 325:811–812, 1987.
37. BE Sumpio, AJ Banes, GW Link, T Iba. Modulation of endothelial cell phenotype by cyclic stretch: inhibition of collagen production. J Surg Res 48:415–420, 1990.
38. B Folkow. Physiological aspects of primary hypertension. Physiol Rev 62:347–504, 1982.
39. MJ Mulvany, C Aaalkjaer. Structure and function of small arteries. Physiol Rev 70:921–962, 1990.
40. AA Liebow. Situations which lead to changes in vascular patterns. In: Handbook of Physiology: Circulation, Vol. 2. Bethesda, MD: American Physiological Society, 1963, pp. 1258–1259.
41. SM Dethlefsen, D Shepro, PA D'Amore. Comparison of the effects of mechanical stimulation on venous and arterial smooth muscle cells. J Vasc Res 33:405–413, 1996.
42. S Chien, YJ Shyy. Effects of mechanical forces on signal transduction and gene expression in endothelial cells. Hypertension 32:162–169, 1998.
43. EN Unemori, KS Bouhana, E Werb. Vectorial secretion of extracellular matrix proteins, matrix degrading proteinases and tissue inhibitor of metalloproteinases by endothelial cells. J Biol Chem 265:445–451, 1990.
44. D Moscatelli, DB Rifkin. Membrane and matrix localisation of proteases: a common theme in tumour invasion and angiogenesis. Biochem Biophys Acta 948:67–85, 1988.
45. MS Pepper, JD Vassalli, R Montesano, L Orci. Urokinase-type plasminogen activator is induced in migrating capillary endothelial cells. J Cell Biol 105:2535–2541, 1987.
46. MD Johnson, HRC Kim, L Chesler, G Tsaowu, N Bouck, PJ Polverini. Inhibition of angiogenesis by tissue inhibitor of metalloproteinase. J Cell Physiol 160:194–202, 1994.
47. D Ingber. Extracellular matrix and cell shape: potential control points for inhibition of angiogenesis. J Cell Biochem 47:236–241, 1991.
48. A Stromblad, DA Cheresh. Cell adhesion and angiogenesis. Trends Cell Biol 6:462–467, 1996.
49. M Friedlander, PC Brooks, RW Shaffer, CM Kincaid, JA Varner, DA Cheresh. Definition of two angiogenic pathways by distinct αv integrins. Science 270:1500–1502, 1995.
50. JF Woessner Jr. Matrix metalloproteinases and their inhibitors in connective tissue remodeling. FASEB J 5:2145–2154, 1991.
51. LP Reynolds, SD Killilea, DA Redmer. Angiogenesis in the female reproductive system. FASEB J 6:886–692, 1992.
52. LP Reynolds, DA Redmer, AT Grazul-Bilska, SD Killilea, RM Moor. Angiogenesis in the female reproductive organs. In: M Margoudakis, ed. Molecular, Cellular and Clinical Aspects of Angiogenesis. New York: Plenum Press, 1996, pp. 125–139.
53. JS Maher. Morphologic and histochemical changes in postpartum uterine blood vessels. Arch Pathol 67:175–180, 1959.
54. O Hudlicka, MD Brown, S Egginton. Angiogenesis in skeletal and cardiac muscle. Physiol Rev 72:369–417, 1992.
55. O Hudlicka, MD Brown. Postnatal growth of the heart and its blood vessels. J Vasc Res 33: 266–287, 1996.
56. MH Laughlin, RM McAllister. Exercise training-induced coronary vascular adaptation. J Appl Physiol 73:2209–2225, 1992.
57. F Ingjer. Effects of endurance training on muscle fibre ATPase activity, capillary supply and mitochondria in man. J Physiol (Lond) 294:419–432, 1979.
58. MH Laughlin, RB Armstrong. Muscle blood flow distribution patterns as a function of running speed in rats. Am J Physiol 243:H296–H306, 1982.
59. RB Armstrong, MH Laughlin. Exercise blood flow patterns within and among rat muscles after training. Am J Physiol 246:H59–H68, 1984.
60. D Gute, MH Laughlin, JF Amann. Regional changes in capillary supply in skeletal muscle

of interval-trained-sprint and low-intensity, endurance-trained rats. Microcirculation 1:183–193, 1994.

61. WL Sexton, MH Laughlin. Influence of endurance exercise training on distribution of vascular adaptations in rat skeletal muscle. Am J Physiol 266:H483–H490, 1994.
62. O Poupa, K Rakusan, B Ostadal. The effect of physical activity upon the heart of the vertebrates. Med Sport 4:202–235, 1970.
63. RJ Tomanek. Effects of age on the extent of the myocardial capillary bed. Anat Rec 167: 55–62, 1970.
64. T Mattfeldt, KL Kramer, R Zeitz, G Mall. Stereology of myocardial hypertrophy induced by physical exercise. Virchow's Arch Pathol Anat Histopathol 409:473–484, 1986.
65. P Anversa, V Levicky, C Beghi, SL McDonald, Y Kikkawa. Morphometry of exercise-induced right ventricular hypertrophy in the rat. Circ Res 52:57–64, 1983.
66. E Mandache, G Unge, LE Appelgren, A Ljungqvist. The proliferation activity of the heart tissues in various forms of experimental cardiac hypertrophy studied by electron microscope autoradiography. Virchow's Arch Cell Pathol 12:112–122, 1973.
67. G Unge, S Carlsson, A Ljungqvist, G Tornling, J Adolfsson. The proliferative activity of myocardial capillary wall cells in variously aged swimming exercised rats. Acta Pathol Microbiol Scand Sec A 87:15–17, 1979.
68. KW Ho, RR Roy, JF Taylor, WW Heusner, WD VanHuss. Differential effects of running and weight lifting on the rat coronary arterial tree. Med Sci Sports Exerc 15:472–477, 1983.
69. GD Tharp, CT Wagner. Chronic exercise and cardiac vascularisation. Eur J Appl Physiol 48:97–104, 1982.
70. P Anversa, V Beghi, V Levicky, SL McDonald, Y Kikkawa. Morphometry of right ventricular hypertrophy induced by strenuous exercise in rat. Am J Physiol 243:H850–H861, 1982.
71. EA Breisch, FC White, LE Nimmo, CM Bloor. Cardiac vasculature and flow during pressure-overload hypertrophy. Am J Physiol 251:H1031–1037, 1986.
72. MH Laughlin, RJ Tomanek. Myocardial capillarity and maximal capillary diffusion capacity in exercise-trained dogs. J Appl Physiol 63:1481–1486, 1987.
73. AS Leon, CM Bloor. Effects of exercise and its cessation on the heart and its blood supply. J Appl Physiol 24:485–490, 1968.
74. RD Bell, RL Rasmussen. Exercise and myocardial capillary-fiber ratio during growth. Growth 38:237–244, 1974.
75. MH Laughlin, KA Overholser, MY Bhatte. Exercise training increases coronary transport reserve in minature swine. J Appl Physiol 67:1140–1149, 1989.
76. D Deveci, P Bryan, S Egginton, JM Marshall. Chronic hypoxia induces angiogenesis in striated muscle of rat (abstr). J Physiol (Lond) 513P:86P, 1998.
77. K Kasalicky, J Ressl, D Urbanova, J Widimsky, B Ostadal, V Pelouch, M Week, J Prochazka. Relative organ blood flow in rats exposed to intermittent high altitude hypoxia. Pflugers Arch 368:111–115, 1977.
78. Z Turek, H Turek-Maischeider, RA Claessens, BEM Ringnalda, F Kreuzer. Coronary blood flow in rats native to simulated high altitude and in rats exposed to it later in life. Pflugers Arch 355:49–62, 1975.
79. M Steinhausen, H Tillmanns, H Thederan. Microcirculation of the epimyocardial layer of the heart. I. A method for in vivo observation of microcirculation of the superficial ventricular myocardium of the heart and capillary flow pattern under normal and hypoxic conditions. Pflugers Arch 378:9–14, 1978.
80. N Banchero, SR Kayar, AJ Lechner. Increased capillarity in skeletal muscle of growing guinea pigs acclimated to cold and hypoxia. Respir Physiol 62:245–256, 1985.
81. D Deveci, S Egginton. The effect of chronic cold exposure on muscle capillarity and composition in hamsters and rats (abstr). Int J Microcirc Clin Exp 16:205, 1996.
82. D Deveci, S Egginton. Effects of acute and chronic cold exposure on tissue blood flow (abstr). Int J Microcirc Clin Exp 17: 200, 1997.

83. D Deveci, S Egginton. A whole animal study of regional blood flow during normothermia and hypothermia using fluorescent labelled microspheres (abstr). Physiologist 41:266, 1998.
84. L Ranvier. Note sur les vaisseaux sanguins et la circulation dans les muscles rouges. C R Seance Soc Biol 26:28–31, 1874.
85. S Salmons, G Vrbova. The influence of activity on some contractile characteristics of mammalian fast and slow muscles. J Physiol (Lond) 210:535–549, 1969.
86. O Hudlicka. What makes blood vessels grow? J Physiol (Lond) 444:1–24, 1991.
87. O Hudlicka. Is physiological angiogenesis in skeletal muscle regulated by changes in microcirculation? Microcirculation 5:7–23, 1998.
88. O Hudlicka, S Price. The role of blood flow and/or muscle hypoxia in capillary growth in chronically stimulated fast muscles. Pflugers Arch 417:62–72, 1990.
89. O Hudlicka, MD Brown, S Pearce, H Silgram. Prostaglandins (PG) and nitric oxide (NO): role in regulation of microcirculation and capillary growth in stimulated skeletal muscles (abstr). J Vasc Surg 35:380.
90. JM Dawson, O Hudlicka. Can changes in microcirculation explain capillary growth in skeletal muscle? Int J Exp Pathol 74:65–71, 1993.
91. A Koller, G Kaley. Shear stress dependent regulation of vascular resistance in health and disease: role of endothelium. Endothelium 4:247–272, 1996.
92. Z Xiao, Z Zhang, SL Diamond. Shear stress induction of the endothelial nitric oxide synthase gene is calcium dependent but not calcium activated. J Cell Physiol 171:205–211, 1997.
93. MA Awolesi, WC Sessa, BE Sumpio. Cyclic strain upregulates nitric oxide synthase in cultured bovine aortic endothelial cells. J Clin Invest 96:1449–1454, 1995.
94. JP Guo, T Murohara, MM Panday, AM Lefer. Nitric-oxide promotes endothelial-cell proliferation—role in inhibiting restenosis. Circulation 92:855–606, 1995.
95. E Pipili-Synetos, A Papageorgiou, E Sakkoula, G Sotiropoulou, T Fotsis, G Karakiulakis, ME Maragoudakis. Inhibition of angiogenesis, tumor growth and metastasis by the NO-releasing vasodilators, isosorbide mononitrate and dinitrate. Br J Pharmacol 116:1829–1834, 1995.
96. D Ben Ezra. Neovasculogenic ability of prostaglandins, growth factors and synthetic chemoattractants. Am J Ophthalmol 86:455–461, 1978.
97. DH Form, R Auerbach. PGE2 and angiogenesis. Proc Soc Exp Biol Med 172:214–218, 1983.
98. M Ziche, G Alessandri, PM Gullino. Gangliosides promote the angiogenic response. Lab Invest 61:629–634, 1989.
99. O Hudlicka, MD Brown, H Silgram. Role of nitric oxide in capillary proliferation in chronically stimulated rat skeletal muscle (abstr). Int J Microcirc Clin Exp 16(suppl 1):92, 1996.
100. MD Brown, H Walter, FM Hansen-Smith, O Hudlicka, S Egginton. Lack of involvement of basic fibroblast growth factor (FGF-2) in capillary growth in skeletal muscles exposed to long-term contractile activity. Angiogenesis 2:81–91, 1998.
101. MD Brown, O Hudlicka, RF Makki, JB Weiss. Low-molecular-mass endothelial cell–stimulating angiogenic factor in relation to capillary growth induced in rat skeletal muscle by low-frequency electrical stimulation. Int J Microcirc Clin Exp 15:111–116, 1995.
102. D Skorjanc, FJ Jaschinski, G Heine, D Pette. Sequential increases in capillarization and mitochondrial enzymes in low-frequency stimulated rabbit muscles. Am J Physiol 274:C810–C818, 1998.
103. J Hang, L Kong, J-W Gu, TH Adair. VEGF gene expression is upregulated in electrically stimulated rat skeletal muscle. Am J Physiol 269:H1827–H1831, 1995.
104. BH Annex, CE Torgan, P Lin, DA Taylor, MA Thompson, KG Peters, WE Kraus. Induction and maintenance of increased VEGF protein by chronic motor nerve stimulation in skeletal muscle. Am J Physiol 274:H860–867, 1998.
105. S Jeal, MD Brown, O Hudlicka, S Egginton. Involvement of vascular endothelial growth factor in capillary growth as a result of chronic electrical stimulation of skeletal muscle in conscious rats (abstr). J Physiol (Lond) 499P:39P, 1997.
106. L Morbidelli, CH Chang, JG Douglas, HJ Granger, F Ledda, M Ziche. Nitric oxide mediates

mitogenic effect of VEGF on coronary venular endothelium. Am J Physiol 270:H4111–H415, 1996.
107. JD Hood, CJ Meininger, M Ziche, HJ Granger. VEGF upregulates ecNOS message, protein and NO production in human endothelial cells. Am J Physiol 274:H1054–H1058, 1998.
108. JR Horowitz, A Rivard, R van der Zee, M Hariawala, DD Sheriff, DD Esakof, GM Chaudry, JF Symes, JM Isner. Vascular endothelial growth factor/vascular permeability factor produces nitric oxide-dependent hypotension: evidence for a maintenance role in quiescent adult endothelium. Arterioscler Thromb Vasc Biol 17:2793–2800, 1997.
109. EC Breen, EC Johnson, H Wagner, HM Tseng, LA Sung, PD Wagner. Angiogenic growth factor mRNA responses in muscle to a single bout of exercise. J Appl Physiol 81:355–361, 1996.
110. S Egginton, O Hudlicka, H Walter, MD Brown, JB Weiss, A Bate. Capillary growth in relation to blood flow and performance in overloaded rat skeletal muscle. J Appl Physiol 85:2025–2032, 1998.
111. AJA Wright, O Hudlicka. Capillary growth and changes in heart performance induced by chronic bradycardial pacing in the rabbit. Circ Res 49:469–478, 1981.
112. MD Brown, MK Davies, O Hudlicka, P Townsend. Long-term bradycardia by electrical pacing: a new method for studying heart rate reduction. Cardiovasc Res 28:1774–1779, 1994.
113. MD Brown, MJ Cleasby, O Hudlicka. Capillary supply of hypertrophied rat hearts after chronic treatment with the bradycardiac alinidine (abstr). J Physiol (Lond) 427:40P, 1990.
114. MD Brown, MK Davies, O Hudlicka. The effect of long-term bradycardia on heart microvascular supply and performance. Cell Mol Biol Res 40:137–142, 1994.
115. HC Schamhardt, PD Verdouw, PR Saxena. Improvement of perfusion and function of ischemic porcine myocardium after reduction of heart rate by alinidine. Cardiovasc Pharmacol 3:728–738, 1981.
116. C Indolfi, BD Guth, T Miura, S Miyazaki, R Schulz, J Ross Jr. Heart rate reduction improves myocardial ischemia in swine; role of interventricular blood flow redistribution. Am J Physiol 261:H910–H917, 1991.
117. O Hudlicka, D West, S Kumar, F El Khelly, AJA Wright. Can growth of capillaries in the heart and skeletal muscle be explained by the presence of an angiogenic factor? Br J Exp Pathol 70:237–246, 1989.
118. H Tillmanns, S Ikeda S, H Hansen, JS Sarma, JH Fauvel, RJ Bing. Microcirculation in the ventricle of the dog and turtle. Circ Res 34:561–569, 1974.
119. O Hudlicka. Mechanical factors involved in the growth of the heart and its blood vessels. Cell Mol Biol Res 40:143–152, 1994.
120. O Hudlicka, MD Brown, H Walter, JB Weiss, A Bate. Factors involved in capillary growth in the heart. Mol Cell Biochem 147:57–68, 1995.
121. CS Liang, RR Tuttle, WB Hood Jr, H Gavras. Conditioning effects of chronic infusion of dobutamine: comparison with exercise training. J Clin Invest 64:613–619, 1979.
122. P Buttrick, A Malhotra, S Factor, D Geenen, J Scheuer. Effect of chronic dobutamine administration on hearts of normal and hypertensive rats. Circ Res 63:173–181, 1988.
123. MD Brown, O Hudlicka. Capillary supply and cardiac performance in the rabbit after chronic dobutamine treatment. Cardiovasc Res 25:909–915, 1991.
124. J Li, T Hampton, JP Morgan, M Simons. Stretch-induced VEGF expression in the heart. J Clin Invest 100:18–24, 1997.
125. W Zheng, RJ Tomanek. VEGF is upregulated in bradycardia-induced coronary angiogenesis in rat (abstr). FASEB J 12:A71, 1998.
126. G Tornling, G Unge, J Adolfsson, A Ljungqvist, S Carlsson. Proliferative activity of capillary wall cells in skeletal muscle of rats during long-term treatment with dipyridamole. Arzneim Forsch 30:622–623, 1980.
127. G Tornling. Capillary neoformation in the heart of dipyridamole treated rats. Acta Pathol Microbiol Immunol Scand 90 (sect A):269–271, 1982.

128. A-I Zhou, S Egginton, O Hudlicka, MD Brown. Internal division of capillaries in rat skeletal muscle in response to chronic vasodilator treatment with al antagonist prazosin. Cell Tissue Res 293:293–303, 1998.
129. MD Brown, O Hudlicka, JB Weiss, A Bate, H Silgram. Prazosin-induced capillary growth in rat skeletal muscle: link with endothelial-cell-stimulating angiogenic factor (abstr). Int J Microcirc Clin Exp 16:207, 1996.
130. G Mall, T Mattfeldt, B Volk. Ultrastructural morphometric study on the rat heart after chronic ethanol feeding. Virchows Arch A Pathol Anat 389:59–77, 1981.
131. G Mall, T Mattfeldt, P Reiger, B Volk, VA Frolov. Morphometric analysis of the rabbit myocardium after chronic ethanol feeding—early capillary changes. Basic Res Cardiol 77: 57–67, 1982.
132. CJ Meininger, ME Schelling, HJ Granger. Adenosine and hypoxia stimulate proliferation and migration of endothelial cells. Am J Physiol 255:H554–H562, 1988.
133. W Jakob, J Zipper, SB Savoly, WE Siems, KD Jentzsch. Is dipyridamole an angiogenic agent? Exp Pathol 22:217–224, 1982.
134. H Tillmanns, M Steinhausen, H Leinberger, H Thederan, W Kubler. The effect of coronary vasodilators on the microcirculation of the ventricular myocardium. In: H Tillmanns, W Kubler, H Zebe, eds. Microcirculation of the Heart. Berlin: Springer Verlag, 1982, pp. 305–312.
135. H Tillmanns, M Steinhausen, H Leinberger, H Thederan, W Kubler. Pressure measurement in the terminal vascular bed of the epimyocardium of rats and cats. Circ Res 49:1201–1211, 1981.
136. S Ichioka, M Shibata, K Kosaki, Y Sato, K Harii, A Kamiya. In vivo measurement of morphometric and hemodynamic changes in microcirculation during angiogenesis under chronic alpha-1 adrenergic blocker treatment. Microvasc Res 55:165–174, 1998.
137. GD Gentzkow. Electrical stimulation to heal dermal wounds. J Dermatol Surg Oncol 19: 753–758, 1993.
138. H-I Peterson. Innervation von Tumorgefassen, Einfluss vasoaktiver Substanzen und Regulation der Tumordurchblutung. Prog Appl Microcir 2:69–77, 1983.
139. E Stefansson, M Landers, ML Wolbarsht. The role of oxygen and vasodilation in diabetic retinopathy. In: EA Friedman, FA L'Esperance, eds. Diabetic-Neural-Retinal Syndrome Retinopathy, vol 2. New York: Grune & Stratton, 1982, pp. 117–151.
140. DH Ausprunck, J Folkman. Migration and proliferation of endothelial cells in preformed and newly formed blood vessels during tumor angiogenesis. Microvasc Res 14:53–66, 1977.
141. MD Brown, O Hudlicka, D Damon, BR Duling. Vasoactive effects of basic and acidic fibroblast growth factors in hamster cheek pouch arterioles. Int J Microcirc Clin Exp 16:308–312, 1995.
142. PH Burri, MR Tarek. A novel mechanism of capillary growth in the rat pulmonary microcirculation. Anat Rec 228:35–45, 1990.
143. JP van Groningen, AC Weninck, LH Testers. Myocardial capillaries increase in number by splitting of existing vessels. Acta Embryol 184:65–70, 1991.
144. A-I Zhou, S Egginton, MD Brown, O Hudlicka. Capillary growth in overloaded hypertrophic adult skeletal muscle. Anat Rec 252:49–63, 1998.
145. FM Hansen-Smith, S Egginton, O Hudlicka. Topographic changes in microvessels during early stages of angiogenesis in adult rat extensor hallucis proprius (EHP) muscle (abstr). Int J Microcirc Clin Exp 16(S1):93, 1996.
146. FM Hansen-Smith, O Hudlicka, S Egginton. Growth of terminal arterioles in overloaded extensor muscles (abstr). FASEB J 12:A20, 1998.
147. FM Hansen-Smith, O Hudlicka, S Egginton. In vivo angiogenesis in adult rat skeletal muscle: early changes in capillary network architecture and ultrastructure. Cell Tiss Res 286:123–136.
148. R Myrhage, O Hudlicka. Capillary growth in chronically stimulated adult skeletal muscle

as studied by intravital microscopy and histological methods in rabbits and rats. Microvasc Res 16:73–90, 1978.
149. JM Dawson, O Hudlicka. Inhibition of capillary growth in skeletal muscle with GPA 1734 (abstr). Int J Micorcirc 9:134, 1990.
150. JM Lash, HG Bohlen. Functional adaptations of rat skeletal muscle arterioles to aerobic exercise training. J Appl Physiol 72:2052–2062, 1992.
151. JM Lash. Contribution of arterial feed vessels to skeletal muscle functional hyperemia. J Appl Physiol 76:1512–1519, 1994.
152. LI Sinoway, J Shenberger, J Wilson, D McLaughlin, T Musch, R Zelis. A 30-day forearm work protocol increases maximal forearm blood flow. J Appl Physiol 62:1063–1067, 1987.
153. PG Snell, WH Martin, JC Buckley, CG Blomqvist. Maximal vascular leg conductance in trained and untrained men. J Appl Physiol 62:606–610, 1987.
154. JD Dawson, O Hudlicka. The effect of long-term activity on the microvasculature of rat glycolytic skeletal muscle. Int J Microcirc Clin Exp 8:53–69, 1989.
155. FM Hansen-Smith, S Egginton, O Hudlicka. Growth of arterioles in chronically stimulated adult skeletal muscle. Microcirculation 5:49–59, 1998.
156. RJ Price, GK Owens, TC Skalak. Immunohistochemical identification of arteriolar development using markers of smooth muscle differentiation—evidence that capillary arterialization proceeds from terminal arterioles. Circ Res 75:520–527, 1994.
157. TH Adair, J Hang, ML Wells, FD Magee, JP Montani. Long-term electrical stimulation of rabbit skeletal muscle increases growth of paired arteries and veins. Am J Physiol 269:H717–H724, 1995.
158. MD Brown, O Hudlicka. Vascular conductance and capillary supply in heart and skeletal muscle in response to altered activity (abstr). Microcirculation 2:98, 1995.
159. S Egginton, O Hudlicka. Early changes in performance, blood flow and capillary fine structure in rat fast muscles induced by electrical stimulation. J Physiol (Lond) 515:265–275.
160. RJ Price, TC Skalak. Chronic alpha 1-adrenergic blockade stimulates terminal and arcade arteriolar development. Am J Physiol 271:H752–H759, 1996.
161. JW Dusseau, PM Hutchins. Stimulation of arteriolar number by salbutamol in spontaneously hypertensive rats. Am J Physiol 236:H134–H140, 1979.
162. RL Hester, B Duling. Red cell velocity during functional hyperemia: implications for rheology and oxygen transport. Am J Physiol 255:H236–244, 1988.
163. A-L Zhou, S Egginton. Immunogold labelling of proliferating cells during skeletal muscle angiogenesis (abstr). J Vasc Res 35:386, 1998.
164. RJ Price, TC Skalak. Prazosin administration enhances proliferation of arteriolar adventitial fibroblasts. Microvasc Res 55:138–145, 1998.
165. S Egginton, O Hudlicka, MD Brown, L Graciotti, A-L Granata. In vivo pericyte-endothelial cell interaction during angiogenesis in adult cardiac and skeletal muscle. Microvasc Res 51:213–228, 1996.
166. MM Delp, MH Laughlin. Time course of enhanced endothelium-mediated dilation in the aorta of trained rats. Med Sci Sports Exerc 29:1454–1461, 1997.
167. D Sun, A Huang, A Koller, G Kaley. Short term daily exercise activity enhances endothelial NO synthesis in skeletal muscle arterioles of rats. J Appl Physiol 76:2241–2247, 1994.
168. J Tepperman, D Pearlman. Effects of exercise and anaemia on coronary arteries of small animals as revealed by the corrosion-cast technique. Circ Res 9:576–584, 1961.
169. MD McKirnan, CM Bloor. Clinical significance of coronary vascular adaptations to exercise training. Med Sci Sports Exerc 26:1262–1268, 1994.
170. P Buttrick, HA Levte, TF Schaible, G Clambrone, J Scheuer. Early increases in coronary vascular response in exercised rats are independent of cardiac hypertrophy. J Appl Physiol 59:1861–1985, 1985.
171. KL Spear, JE Koerner, RL Terjung. Coronary blood flow in physically trained rats. Cardiovasc Res 12:135–143, 1978.

172. FC White, CM Bloor. Coronary growth, not angiogenesis, is the dominant response increasing coronary vascular bed cross-sectional area induced by exercise training (abstr). FASEB J 9:A909, 1995.
173. HL Wyatt, J Mitchell. Influence of physical conditioning and deconditioning on coronary vasculature of dogs. J Appl Physiol 215:619–625, 1978.
174. AMAR Ziada, O Hudlicka, KR Tyler, AJA Wright. The effect of long term vasodilation on capillary growth and performance in rabbit heart and skeletal muscle. Cardiovasc Res 18: 724–732, 1984.
175. RJ Torry, PM Connell, DM O'Brien, WM Chilian, RJ Tomanek. Sympathectomy stimulates capillary but not precapillary growth in hypertrophic hearts. Am J Physiol 260:H1515–H1521, 1991.
176. MH Laughlin, RJ Korthuis, WL Sexton, RB Armstrong. Regional muscle blood flow capacity and exercise hyperemia in high-intensity trained rats. J Appl Physiol 64:2420–2427, 1988.
177. RB Armstrong, MH Laughlin. Blood flow within and among rat muscles as a function of time during high speed treadmill exercise. J Physiol (Lond) 344:189–208, 1983.
178. A Ljungqvist, G Unge, S Carlsson. The myocardial capillary vasculature in exercising rats with increased cardiac pressure load. Acta Pathol Microbiol Scand A Pathol 84:244–246, 1976.
179. AMAR Ziada, O Hudlicka, KR Tyler. The effect of long-term administration of alpha1 blocker prazosin on capillary growth and performance in rabbit heart and skeletal muscle. Pflugers Arch 415:355–360, 1989.
180. FC White, CM Bloor, MD McKirnan, SM Caroll. Exercise training in swine promotes growth of arteriolar bed and capillary angiogenesis in heart. J Appl Physiol 85:1160–1168, 1998.

16
Oxygen and Tumor Angiogenesis

Yuval Dor and Eli Keshet
The Hebrew University, Jerusalem, Israel

I. INTRODUCTION

Cell survival depends on a continuous supply of oxygen and nutrients carried by the blood. Therefore, every cell of the body must by sufficiently close to a blood capillary to allow for efficient nutrient diffusion. A tissue cannot grow beyond ~1 mm in diameter before requiting new blood vessels to invade and nourish it. Tumor cells are no exception to that rule, and a vast body of evidence indeed shows that continuous blood vessel formation is obigatory for tumor growth and metastasis.

What drives tumor vascularization? One important mechanism is the ''angiogenic switch'': a stochastic, mutational event in the series of genetic hits required for tumor progression. The angiogenic switch is proposed to endow the tumor with the ability to recruit blood vessels, either by downregulation of natural angiogenesis inhibitors or by inducing angiogenic factors. The angiogenic switch may explain phenomena such as a tumor dormancy for an extended period of time (i.e., when the tumor remains in its avascular phase) or the variable timing for the emergence of the angiogenic tumor phenotype in a mouse model of multistage tumorigenesis (1).

A second mechanism of tumor vascularization is ischemia-induced angiogenesis. Ischemia-driven angiogenesis is a fundamental homeostatic feedback response by which a normal tissue matches its energy supply to its metabolic needs. The ischemic tissue detects insufficient oxygen (hypoxia) or insufficient glucose (hypoglycemia) and responds by inducing the production of angiogenic factors that attract new blood vessels to the ischemic area. Tumor cells may retain the capacity for sensing ischemia and respond, like their normal counterparts, by upregulating angiogenic factor production. Normal endothelial cells in the surrounding tissue expressing cognate receptors then respond by sprouting in a mechanism fundamentally similar to physiological angiogenesis (2).

Recently, genetic studies in mice have enabled us to better assess the relative contribution of ischemia-induced angiogenesis and of a genetic angiogenic switch to overall tumor angiogenesis. From these studies it has become apparent that tumor ischemia is a major contributor to tumor neovascularization.

This review focuses on the adaptive, homeostatic nature of tumor angiogenesis. Hypoxia-induced angiogenesis is discussed with respect to molecular events in the pathway initiated with oxygen sensing and culminating in increased vessel formation. Section II deals with the molecular mechanism of hypoxia/hypoglycemia sensing by tumor cells that culminates in increased levels of the transcription factor hypoxia-inducible factor 1 (HIF1). Section III discusses ischemia-induced tumor angiogenesis, highlighting the role of the HIF1 target gene vascular endothelial growth factor (VEGF). Section IV introduces the concept of vessel maturation with respect to regulation by oxygen. We conclude by emphasizing some differences between stress-induced angiogenesis in normal tissue and in tumors, which may be of clinical importance.

II. OXYGEN SENSING BY TUMOR CELLS

A. HIF1

Ischemia is manifest at the cellular level mainly as lack of oxygen and glucose. Hypoxia-inducible factor 1 (HIF1) has emerged as a key mediator of the mammalian response to hypoxia and probably also to hypoglycemia. It can be detected in virtually every cell exposed to hypoxic stress, including transformed cells. HIF1 is a heterodimeric transcription factor composed of two bHLH-PAS subunits, both of which participate in recognizing the hypoxia response element (HRE) on target genes. The β subunit (known also as ARNT, for aryl hydrocarbon receptor nuclear translocator) is a dimerization partner for several other transcription factors and is found in large excess in the cell. The α subunit seems to participate solely in the HIF1 complex, and its level determines HIF1 transcriptional activity. Evolutionary conservation of HIF1α suggests a fundamental common role in highly diverse organisms. For example, human and mouse HIF1α subunits share 95% identity in amino acid sequence and an HIF1α homologue is also found in *Drosophila* (3–5).

B. Oxygen Sensing and HIF1α Activation

Low oxygen concentrations lead to a posttranslational regulation of HIF1α by a multistep process that includes stabilization of the protein, nuclear translocation, and activation of its transcriptional activation domain (TAD). Under normoxia, HIF1α is rapidly degraded (half-life <5 min) via the ubiquitin/proteasome system. Upon exposure to hypoxia, degradation is markedly inhibited, resulting in accumulation of HIF1α to levels sufficient to activate its respective target genes. Mutational analysis has defined a 200–amino acid domain [oxygen-dependent degradation domain (ODD), amino acids 401–603] that is required for oxygen-dependent ubiquitination/degradation of HIF1α. The mechanism by which HIF1α evades proteasome-mediated degradation under hypoxic conditions is not known (6–8). Following its accumulation, HIF1α translocates to the nucleus in a hypoxia-dependent manner, mediated by a hypoxia-inducible nuclear translocation signal located in its carboxy terminus (9). Once in the nucleus, HIF1α binds HIF1β and forms the heterodimer that recognizes HRE on target genes.

Hypoxia also increases the transcriptional activity of HIF1α independently of the changes in steady-state level of the protein. This effect is mediated by two TADs found at the carboxy terminus of HIF1α (amino acids 531–575 and 786–826), separated by a domain (amino acids 576–785) that inhibits transactivation under normoxia (10–12). The enhancement (or derepression) of HIF1α transcriptional activity under hypoxia involves the recruitment of the transcriptional coactivator CREB-binding protein CBP/p300, which has histone acetyltransferase activity and is thought to modify chromatin structure to enable transcription (9,13) (Fig. 1).

Induction of HIF1α increases exponentially at oxygen concentrations below 6%, with maximal induction at 0.5% O_2. This range of oxygen concentrations is comparable to oxygen concentrations measured in vivo within ischemic/hypoxic tissues, including tumors (14,15).

The molecular mechanism of oxygen sensing and the signal transduction pathway that leads to HIF1α induction are not understood. Pharmacological data have suggested that oxygen sensing may involve a heme-containing protein that undergoes a conformational change between the oxy and deoxy states of the heme-bound iron (16). Other studies have suggested the involvement of tyrosine kinases/phosphatases in the transduction pathway (17) or a role for reactive oxygen species such as hydrogen peroxide (H_2O_2) (8). It seems that better insights into the pathway of oxygen sensing upstream of HIF1 might require a genetic approach.

Importantly, conditions other than hypoxia are also conductive to activation of HIF1 target genes. Notably, hypoglycemia, another consequence of ischemia, also leads to HIF1 activation (18). The mechanism of HIF1 activation by hypoglycemia is not known. Insulin and IGF-1 activate HIF1, probably by stabilizing HIF1α protein (19); V-SRC can activate

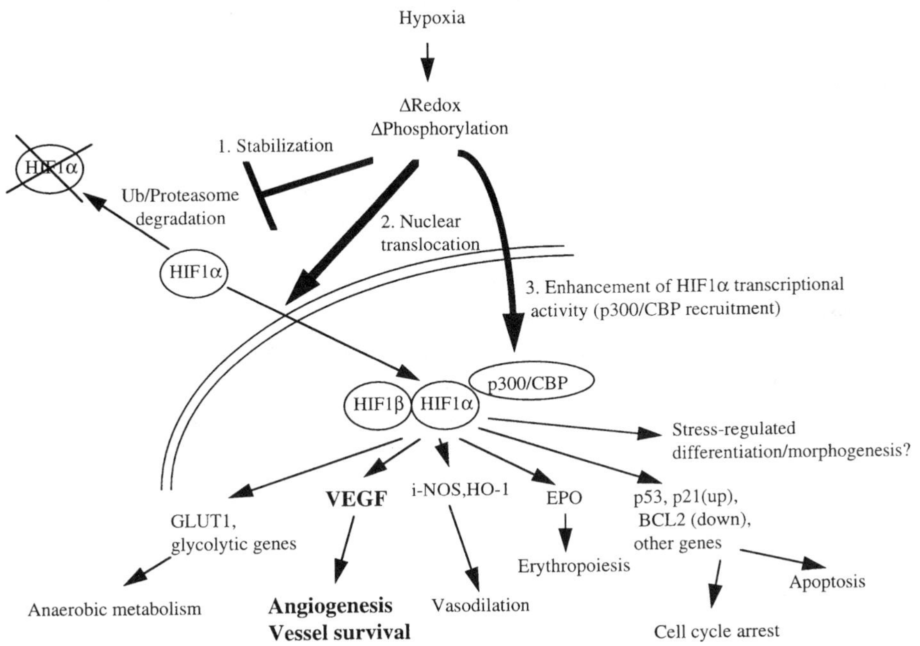

Figure 1 Current view of hypoxia signal transduction. Low oxygen levels, sensed by an unknown oxygen sensor (possibly a heme protein) activate HIF1α via three independent mechanisms, indicated by bold arrows: stabilization of the protein, translocating it to the nucleus, and enhancing its transcriptional activity by recruitment of p300/CBP. The active HIF1 complex then up regulates its target genes, containing HRE. The VEGF subpathway affecting angiogenesis is indicated in bold letters.

HIF1 in the absence of hypoxia or lead to superinduction under hypoxia (20,21); RAS/PI3-kinase were reported to modulate the HIF1-dependent hypoxic response (22,23); and p53 may inhibit HIF1 transcriptional activity [although this requires very high levels of p53 (24)]. The modulation of HIF1 activity by oncogenes on the one hand and by environmental stress on the other is of interest, as it suggests HIF1 as an integrator of both genetic and epigenetic influences on tumor angiogenesis (see below).

C. HIF1 Function and Target Genes

Once activated, HIF1 induces transcription of a wide range of target genes, all of which contain at least one HRE [core sequence 5′-(A/G)CGTG-3′] at the promoter or 3′ enhancer regions (25). The known target genes of HIF1 are thought to play a role in the adaptive response to oxygen/glucose shortage and can be categorized according to their level of action within the organism:

1. Hypoxia-induced expression of erythropoietin (10) and transferrin (26) improves systemic oxygen delivery by increasing erythropoiesis.
2. Inducible nitric oxide synthase (i-NOS) (27) and hemoxygenase 1 (HO-1) (28) cause local vasodilatation through production of the vasoactive molecules NO and CO, respectively; VEGF initiates local compensatory angiogenesis (29) (see below).
3. Glucose transporter 1 (Glut1) and most glycolytic enzymes allow cellular metabolic adaptation to anaerobic conditions, as well as to shortage of glucose, by streamlining glucose transport and usage (30).
4. A novel role for HIF1 is the control of cellular proliferation and apoptosis in response to hypoxia. Exposure of cells to severe hypoxic stress leads to either cell-cycle arrest or apoptosis, and it appears that HIF1 is necessary for the induction of these responses, although its direct target genes are not yet identified. The physiological rationale of the hypoxia-induced cell-cycle arrest and apoptosis is not well understood. One possible explanation is that hypoxia causes genetic instability (31–35), thus raising the need for programmed cell death analogous to DNA damage–induced cell cycle arrest/apoptosis. The responses related to HIF1-dependent proliferation/apoptosis include the downregulation of the anti-apoptotic protein Bc12, induction of p53, and induction of the cyclin kinase inhibitor p21/WAF1 (36). Surprisingly, HIF1 upregulates p53 in hypoxia by a novel, nontranscriptional mechanism—namely, stabilization of p53 through direct protein-protein interaction of HIF1α and p53 (37). Tumor cells may escape from hypoxia-mediated cell-cycle arrest/apoptosis by mutating p53, thus gaining growth advantage in hypoxic niches of the tumor (38).
5. Analysis of HIF1-deficient mice showed that the gene is essential for embryonic development past embryonic day 10.5 (E10.5); in addition to vascular defects that could be attributed to loss of VEGF induction, the mutant embryos displayed various other defects such as failure of neural tube closure (18,39). These defects suggest new roles for HIF1 in development and differentiation that still await molecular characterization.

D. HIF1-Independent Hypoxic Responses

Not all hypoxic responses are mediated by HIF1. Certain genes, including the molecular chaperone GRP78, the DNA damage-induced GADD153, and the cyclin kinase inhibitor p27, are induced by hypoxia in a HIF1-independent manner (36). One candidate regulator of these responses is HIF2, a transcription factor composed of ARNT and HIF2α, a close homologue of HIF1α (40–43). HIF2 is also induced by hypoxia (44) and activates the same HRE on its respective target genes. During embryogenesis, HIF2 appears to be essential for the production of catecholamines (45). Despite the fact that both transcription factors bind to the same DNA sequence, they appear to activate a nonoverlapping repertoire of target genes. It is not yet clear what determines the target gene specificity of HIF1/HIF2.

In summary, the cellular hypoxic response appears to be mediated largely by the transcription factor HIF1. In Section III, we discuss the angiogenic aspect of this response and its role in tumor angiogenesis.

III. ISCHEMIA-REGULATED TUMOR ANGIOGENESIS

A. Modes of Angiogenesis Regulation

Like other processes of neovascularization in the adult organism, tumor neovascularization takes place by sprouting angiogenesis. Quiescent endothelial cells in nearby blood vessels respond to an angiogenic stimulus and sprout in the direction from which the angiogenic stimulus emanates. The complex process of sprouting angiogenesis begins with local destabilization of the vessel structure by loosening endothelial-periendothelial cell contacts and degrading the basement membrane. The endothelial cells in this activated region then change shape, invade the surrounding stroma, and form a migrating column with a proliferating leading edge. Endothelial cell cords then undergo morphogenic changes and form the lumen of a new capillary tube. New capillary tubes fuse into loops and blood flow starts. The newly made network undergoes extensive remodeling to form larger vessels, circulating blood in a directional manner. The recruitment of pericytes stabilizes vessels and renders them "mature" (see Section IV below) (46).

The trigger for angiogenesis in the fully developed organism may be hormonal stimulation, as in the female reproductive system and in certain hormone-dependent tumors. In other physiological settings, such as increased mass of muscle tissue or increased activity of neuronal tissues, the trigger appears to be microenvironmental stress. As mentioned above, in the case of tumors, the angiogenic switch has been proposed to be a stochastic genetic event taking place in tumor cells (1,2).

In all cases, the tissue responds by either upregulating the production of an angiogenic factor or downregulating the production of an angiogenesis inhibitor. The growing list of angiogenic factors includes VEGF, bFGF, aFGF, TGF-β, PDGF-B, and others. The list of angiogenesis inhibitors includes, among others, thrombospondin-1, endostatin, and angiostatin. While the mechanism of action of natural angiogenesis inhibitors is not known, a recurrent theme appears to be their generation as cleavage products of larger proteins.

The endothelial decision to sprout or not to sprout is proposed to be based on the net balance of positive and negative regulators of angiogenesis (the balance hypothesis).

Tilting the balance toward tumor neovascularization may well depend on the tumor microenvironment. Here we highlight evidence for ischemic triggering of tumor angiogenesis via the HIF1 pathway and its mediation by ischemia-induced VEGF as well as by additional hypoxia-induced angiogenic factors.

B. Genetic Evidence for the Role of HIF1 in Tumor Angiogenesis

Direct examination of the role of HIF1 in tumor angiogenesis became possible with the availability of HIF1-deficient tumorigenic cell lines. HIF1α-deficient embryonic stem (ES) cells as well as a mouse hepatoma line deficient in the HIF1β subunit were xenografted into nude mice and assayed for tumor growth and angiogenesis (36,47,48). HIF1-deficient tumors showed a significant reduction in vascularity. Notably, HIF1α-deficient tumors were deficient in medium- and large-sized vessels, while capillary density was comparable with that of the corresponding wild-type tumors (36). This deficit in blood vessels was reflected in a 50% reduction in blood flow rate and a 90% reduction in overall oxygen delivery to HIF1α-null tumors. The result of impaired vascularity was that cells within the mass of HIF1-null tumors were under constant metabolic stress, as demonstrated by augmented expression of HIF1-independent stress-induced genes (like GRP78). These results attest to the key role of HIF1 in tumor angiogenesis and highlight hypoxia-induced angiogenesis as a major contributor to tumor neovascularization. As discussed below, the HIF1 target gene that mediates hypoxia-driven angiogenesis is VEGF, whose expression is greatly compromised (albeit still detectable) in HIF1-null tumors.

HIF1 target genes fulfill multiple roles in the feedback response to hypoxia; accordingly, its inactivation might affect the tumor phenotype through other than direct effects on the vasculature. As noted above, yet unidentified HIF1 target genes appear to control the protective mechanism of growth arrest/apoptosis under hypoxia, which prevents cells experiencing severe hypoxia from replicating their DNA under circumstances prone to mutations or chromosomal aberrations (31–35). Importantly, HIF1 null tumor cells fail to exert this control when exposed to hypoxia while fully maintaining the capacity of cytokine-induced apoptosis (36). The effect of HIF1 deficiency on the overall tumor phenotype is thus a complex trait that may depend on the different growth and metabolic parameters of different tumors. Indeed, the effect of HIF1 deficiency on tumor growth rate was found to depend on the experimental tumor system employed. In two cases, HIF1-deficient tumors [both HIF1β-defective hepatoma (47) and HIF1α-defective ES-derived teratocarcinoma (48)] had lower growth rates, whereas, in another study, HIF1α deficiency led to accelerated tumor growth, likely from diminished growth arrest and apoptosis in hypoxic tumor microenvironments (36). An intriguing extrapolation of these findings is that the bulk of oxygen delivered to the tumor is required for allowing cells to pass the cell-cycle checkpoints, while for cell viability per se, lower oxygen levels are likely to be sufficient.

Another important question is whether HIF1 target genes can be activated in tumors by factors other than ischemia. As mentioned above, some oncogenes are able to activate HIF1 target genes in the absence of or by synergy with hypoxic stress. This may cause hypoxia-independent HIf1-induced tumor angiogenesis. A clear definition of the actual inducers of HIF1 in tumors may come from isolation and manipulation of the oxygen-sensing components upstream of HIF1 or from correlating HIF1 protein abundance/activity with a tumor hypoxic microenvironment.

C. VEGF and Tumor Angiogenesis

VEGF has emerged in recent years as a major tumor angiogenic factor. Reducing VEGF production or antagonizing its biological activity using different experimental modalities have all resulted in a markedly reduced tumor vascularization and, consequently, in retarded tumor growth and also often in tumor regression. These include the use of neutralizing VEGF antibodies (49–51), expression of antisense VEGF mRNA (52–55), use of a dominant-negative form of VEGF receptor (56), and conditional switching of VEGF expression (57). Similarly, VEGF-null tumorigenic cells are significantly less angiogenic than their normal counterparts (36,58). Elevated levels of VEGF are found in the serum and urine of tumor-bearing patients (59). The mechanism of VEGF action on endothelial cells, although extensively studied, is not fully understood. Secreted VEGF acts in a paracrine fashion, through binding the endothelial cell–specific tyrosine kinase receptors VEGFR1 (Flt1) and VEGFR2 (Flk1/KDR) and the recently discovered neuropilins (60). The signal-transduction pathway of VEGF is only partially characterized. Ex vivo, VEGF was found active in promoting endothelial cell proliferation, endothelial cell chemotaxis, capillary morphogenesis, and capillary sprouting. In vivo, VEGF activities include, in addition to promoting overall neovascularization, induction of vascular permeability (61,62), promotion of vessel fusions (63), enhancement of pericyte recruitment (64), and a vascular survival function (65). It remains to be determined whether this multiplicity of biological activities by a single growth factor represents multiple responses in a single transduction pathway or divergent transduction pathways. Also, different threshold levels of VEGF might be required to initiate the different responses. For example, formation of large vessels might require more persistent receptor activation (57,63,64), explaining the finding that HIF1-null tumors, which express lower levels of VEGF, show normal capillary density but lack large vessels (36).

What drives VEGF expression in tumors? Initial in situ observations revealed elevated levels of VEGF expression in presumed hypoxic tumor micoenvironments, primarily around necrotic areas, where extreme hypoxia was not compatible with cell survival (66,67). The hypoxic status of the VEGF-expressing cells was confirmed using the in situ hypoxia marker EF5 in both multicell spheroids (68) and xenografted gliomas (69). In addition to hypoxic tumor cells, macrophages and lymphocytes recruited to the hypoxic niches are also abundant sources of VEGF. Yet the trigger for VEGF expression by those cells appears also to be hypoxic stress (70,71).

In vitro studies have uncovered three levels of control on hypoxia-induced VEGF. First, VEGF transcription is induced three- to fourfold during hypoxia (and hypoglycemia) by direct binding of HIF1 to the HRE present in the VEGF promoter (30). Second, hypoxia as well as hypoglycemia leads to a posttranscriptional stabilization of VEGF mRNA. The intrinsically short half-life of VEGF mRNA ($\sim$30 min in normoxia) is extended during stress, accounting for a further three- to fourfold increase in steady-state message levels. The stabilization is presumed to occur through hypoxia-augmented binding of protein(s) to the 3′-untranslated region (UTR) (72–75). The relative contribution of transcription and stability to VEGF levels in tumors was elegantly dissected in a xenografted glioma system using a β-gal reporter hooked to either HRE or 3′-UTR sequences of VEGF. As expected, the two elements were found to contribute independently to hypoxic induction of the reporter (69). Third, VEGF production is also regulated at the translational level: The 5′-UTR of VEGF contains an internal ribosome entry site (IRES) that allows efficient translation during stress conditions, where conventional cap-dependent translation is compromised (but VEGF is most needed) (76).

In conclusion, VEGF is a major contributor to tumor angiogenesis, and microenvironmental stress appears to be the main driving force for its expression in tumors. Yet VEGF expression in tumors may also be controlled genetically (see below).

D. Other Hypoxia-Regulated Angiogenic Factors

In addition to VEGF, other factors with angiogenic activity are hypoxia-inducible. Cyclooxygenase-2 (COX-2), the inducible prostaglandin synthesizing enzyme, is upregulated by hypoxia (77,78) and was recently shown to modulate colon cancer angiogenesis (79). Platelet-derived growth factor B (PDGF-B) (80) and tissue factor (TF) (81) are also induced by hypoxia. These two proteins, however, are thought to play a role at later stages of vessel formation—namely, remodeling of the capillary plexus and vessel maturation (64,82,83). Interestingly, VEGF receptors 1 and 2 are also induced by hypoxia, presumably reflecting a positive feedback loop to accelerate angiogenesis during hypoxia (84–86). The molecular mechanism of hypoxic induction in all these cases is not well defined. For COX-2, a dependency on NFκB was reported; VEGF-R1 hypoxic induction depends on an HRE element, suggesting that it is a direct target of HIF2. Hypoxic induction of TF is not dependent on HIF1 (36). Taken together, these data suggest additional roles for hypoxia in angiogenesis beyond direct induction of VEGF.

E. Relation of Hypoxic HIF1/VEGF Induction to Oncogenes

The hypoxia/HIF1/VEGF axis in tumors may have additional important connections. As noted above, several oncogenes can activate HIF1 under normoxic conditions or augment its expression during hypoxia. In addition, in situ hybridization analysis has shown that HIF1α mRNA expression is subject to various nonhypoxic controls during embryonic development (our unpublished observations). However, it remains to be seen whether elevated levels of HIF1α mRNA under normoxia are accompanied by accumulation of a bioactive protein.

Downstream of HIF1, VEGF expression is clearly subjected to regulation by other factors. HIF1α-deficient teratocarcinomas often differentiate into cell lineages that constitutively express VEGF irrespective of hypoxia (our unpublished observations). A prime example for hypoxia-independent expression of VEGF, directly related to tumor angiogenesis, is the Von-Hippel Lindau (VHL) tumor suppressor gene. Mutations in VHL give rise to constitutive high levels of VEGF due to increased stability of its mRNA (87). This leads to hemangioblastomas, one of the hallmarks of VHL disease. Thus, an oncogenic mutation can disrupt the fine hypoxic control of VEGF expression (88). A genetic view also argues for HIF1-independent mechanisms for VEGF-induced angiogenesis: VEGF deficiency has a more severe effect on tumor angiogenesis than HIF1 deficiency (37). Therefore, HIF1 cannot account for all VEGF-induced angiogenesis.

Recently, tumor cells were shown to induce VEGF transcription in nontransformed neighboring stromal cells (89). Thus, the tumor microenvironment affects not only tumor cells but also stromal cells surrounding it. The mechanism of this induction and its significance are not known.

Although not fully understood, stress-independent VEGF expression may provide an interesting link between stress-induced angiogenesis and a mutational angiogenic switch: oncogenic mutations may exploit the existing physiological pathway for stress-driven, VEGF-induced angiogenesis by short-cutting its negative controls and rendering it constitutive.

Another potential link between hypoxia and the angiogenic switch is the genomic instability induced by hypoxia. Hypoxic stress may not only select tumor clones with growth advantage (as was shown for p53 loss) but also increase genetic variability within the tumor, thereby increasing the probability for oncogenic/angiogenic mutations.

IV. VESSEL MATURATION

A. Oxygen-Regulated VEGF as a Survival Factor for Newly Formed Vessels

Following formation of a primary vascular plexus, oxygen tension fine tunes the density of blood vessels to the metabolic needs of the tissue. That is, hypoxia induces the formation of more sprouts, while excess of oxygen (hyperoxia) causes vascular regression, or pruning. Pruning is initiated by shedding of endothelial cells from the vessel wall; these subsequently die by apoptosis (57,65).

The same oxygen-regulated pathway—namely, oxygen sensing/HIF1/VEGF—is thought to regulate both hypoxia-induced angiogenesis and hyperoxia-induced vascular regression. This was shown most clearly in the context of the developing retina, where experimental manipulation of VEGF levels had a profound effect on the vasculature: increase in VEGF led to formation of more vessels, and suppression of VEGF led to vessel regression and exaggeration of the otherwise normal process of vascular pruning (65). This activity of VEGF uncovered its important role as a survival factor for newly formed vessels. Like axons competing for NGF during neural networking, newly formed vessels compete for a limited amount of VEGF for their survival. VEGF level is determined by HIF1 activity, reflecting the metabolic needs of the tissue.

The same mechanism works in tumor vessels: when VEGF in a growing tumor is downregulated experimentally through the use of a switchable promoter, expression of antisense RNA, or a neutralizing antibody, the newly formed tumor vasculature collapses. As in the developmental vascular pruning, endothelial cell shedding from the vessel wall appears to be an early step in tumor vessel regression (57).

B. Selective Dependence of Immature Tumor Vessels on VEGF

While blood vessels in the adult organism retain the capacity for VEGF-induced angiogenesis, they are independent of VEGF for survival. This is in sharp contrast to the newly formed immature vasculature. Thus, the transition from a VEGF-dependent phase to a VEGF-independent phase is a hallmark of vessel maturation. The physiological rationale for vessel maturation, in the sense of rendering vessels refractory to VEGF depletion, is clear: following a plasticity time window for the fine adjustment of vascular density to the metabolic needs of the tissue, stochastic fluctuations in oxygen delivery should not lead to collapse of major vessels (64,65).

The driving force for vessel maturation appears to be the acquisition of a pericyte/smooth muscle cell coating by the preformed endothelial plexus. A programmed delay in periendothelial cell recruitment allows for this plasticity window; upon its completion, vessels become refractory to VEGF withdrawal. Indeed, only pericyte-covered vessels survive natural or experimental VEGF withdrawal. Moreover, when pericyte coverage in the retina is disrupted, the natural trimming of vessels (caused by reduced VEGF due to sufficient oxygenation) is exaggerated. Pericytes are thought to provide the nearby endothelial cells with alternative survival factors, perhaps by depositing those factors in the shared extracellular matrix (64). The molecular mechanism of periendothelial cell recruitment is not well understood. Among the factors implicated in this process is PDGF-B expressed by endothelial cells and thought to attract PDGFβ receptor–positive pericytes. Interestingly, PDGF-B is upregulated by hypoxia (80). Pericyte coverage is also enhanced by VEGF itself by an unknown mechanism (64). Other factors that affect pericyte recruitment and vascular remodeling were suggested on the basis of gene knockout studies. These include the endothelial cell–specific receptors TIE1 (90) and TIE2 (91), TIE2 ligands angiopoietin 1 (Ang1) (92) and 2 (93,94), and the tissue factor (TF) system (82).

The principles of vascular remodeling and maturation outlined above seem to hold in tumors as well. As mentioned before, many tumor vessels collapse following experimental VEGF withdrawal. A follow-up study has demonstrated that only immature tumor vessels are obliterated. Analysis of natural human tumors for the status of "vessel maturation" (determined by association with smooth muscle actin–positive cells) has shown that, unlike the respective normal tissue, tumors contain a significant fraction of immature vessels. This difference should render the tumor vasculature more vulnerable than the surrounding normal tissue to VEGF withdrawal. Indeed, suppression of VEGF production in human prostate cancer through androgen ablation therapy leads to selective obliteration of immature tumor vessels (95).

The basis for this phenomenon, which has important clinical implications, is not known. We speculate that it has to do with the more dynamic metabolism of the tumors, where fast turnover of cells [even within "dormant metastases" (96)] may lead to rapid fluctuations in oxygen demand at different microenvironments. The resultant fluctuating levels of VEGF will enforce a continuous vascular remodeling, preventing the slow and steady establishment of pericyte coverage. Indeed, as many as 4.5% of endothelial cells in human glioblastomas are engaged in cell division, as judged by BrdU labeling of primary human glioblastoma (97). This is in sharp contrast to the estimated 0.01% of dividing cells in a normal adult vessel (1).

V. SUMMARY

The concept highlighted in this review is that ischemic stress is a major contributor to tumor angiogenesis as it is for physiological angiogenesis. Oxygen and glucose levels are sensed by tumor cells, which integrate this information and translate it to the activation of the transcription factor HIF1. HIF1 controls a wide array of target genes that participate in energy homeostasis, by both matching nutrient supply to the metabolic needs of the tumor and adjusting tumor metabolism to the available nutrients. Here we have focused on one critical facet of this homeostatic response, namely, formation and regression of blood vessels. The main effector of HIF1 in matching vascular supply to metabolic needs is VEGF, acting as both an angiogenic factor and a survival factor for newly formed vessels. Ischemia increases VEGF level, thus promoting angiogenesis; excess oxygen reduces VEGF, thus promoting regression of surplus vessels. Evidence from clinical as well as from mouse models suggests that the HIF1-VEGF axis, controlled by oxygen/glucose levels, plays an important role in regulating tumor angiogenesis. This may mean that the angiogenic stage in tumor progression may not always require specific genetic hits, as suggested by the original angiogenic switch hypothesis. A possible link between ischemia-induced angiogenesis and the angiogenic switch hypothesis is the exploitation of the HIF1-VEGF angiogenic pathway by oncogenes, modulating it independently of microenvironmental stress.

During development, oxygen tension fine tunes vascular density for a defined period of time, during which excess vessels regress (due to a limited amount of VEGF). This plasticity window ends when vessels reach maturity, marked by pericyte coverage. Tumor vasculature, composed of normal endothelium, retains the machinery for remodeling the immature vascular tree. However, a significant fraction of tumor vessels appears to persist in the immature state, thereby remaining vulnerable to VEGF withdrawal. This may result from the highly dynamic tumor metabolism, changing its demands rapidly and preventing formation of a stable vascular architecture. This difference between tumor and normal tissue vasculature may be of a significant clinical importance.

REFERENCES

1. D Hanahan, J Folkman. Patterns and emerging mechanisms of the angiogenic switch during tumorigenesis. Cell 86:353–364, 1996.
2. Y Dor, E Keshet. Ischemia-driven angiogenesis. Trends Cardiovasc Med 7:289–294, 1997.
3. GL Wang, BH Jiang, EA Rue, GL Semenza. Hypoxia-inducible factor 1 is a basic-helix-loop-helix-PAS heterodimer regulated by cellular O2 tension. Proc Natl Acad Sci USA 92:5510–5514, 1995.
4. G Semenza. Hypoxia-inducible factor 1: master regulator of O2 homeostasis. Curr Opin Genet Dev 8:588–594, 1998.
5. NC Bacon, P Wappner, JF O'Rourke, SM Bartlett, B Shilo, CW Pugh, PJ Ratcliffe. Regulation of the Drosophila bHLH-PAS protein Sima by hypoxia: functional evidence for homology with mammalian HIF-1 alpha. Biochem Biophys Res Commun 249:811–816, 1998.
6. LE Huang, Z Arany, DM Livingston, HF Bunn. Activation of hypoxia-inducible transcription factor depends primarily upon redox-sensitive stabilization of its alpha subunit. J Biol Chem 271:32253–32259, 1996.
7. LE Huang, J Gu, M Schau, HF Bunn. Regulation of hypoxia-inducible factor 1alpha is mediated by an O2-dependent degradation domain via the ubiquitin-proteasome pathway. Proc Natl Acad Sci USA 95:7987–7992, 1998.
8. S Salceda, J Caro. Hypoxia-inducible factor 1alpha (HIF-1alpha) protein is rapidly degraded by the ubiquitin-proteasome system under normoxic conditions: its stabilization by hypoxia depends on redox-induced changes. J Biol Chem 272:22642–22647, 1997.
9. POK Kallio, O'Brien S, Carrero P, Makino Y, Tanaka H, Poellinger L. Signal transduction in hypoxic cells: inducible nuclear translocation and recruitment of the CBP/p300 coactivator by the hypoxia-inducible factor-1alpha. EMBO J 17:6573–6586, 1998.
10. BH Jiang, E Rue, GL Wang, R Roe, GL Semenza. Dimerization, DNA binding, and transactivation properties of hypoxia-inducible factor 1. J Biol Chem 271:17771–17778, 1996.
11. BH Jiang, JZ Zheng, SW Leung, R Roe, GL Semenza. Transactivation and inhibitory domains

of hypoxia-inducible factor 1alpha: modulation of transcriptional activity by oxygen tension. J Biol Chem 272:19253–19260, 1997.

12. CW Pugh, JF O'Rourke, M Nagao, JM Gleadle, PJ Ratcliffe. Activation of hypoxia-inducible factor-1: definition of regulatory domains within the alpha subunit. J Biol Chem 272:11205–11214, 1997.
13. Z Arany, LE Huang, R Eckner, S Bhattacharya, C Jiang, MA Goldberg, HF Bunn, DM Livingston. An essential role for p300/CBP in the cellular response to hypoxia. Proc Natl Acad Sci USA 93:12969–12973, 1996.
14. BH Jiang, GL Semenza, C Bauer, HH Marti. Hypoxia-inducible factor 1 levels vary exponentially over a physiologically relevant range of O2 tension. Am J Physiol 271 (4 pt 1):C1172–1180, 1996.
15. G Helmlinger, F Yuan, M Dellian, RK Jain. Interstitial pH and pO2 gradients in solid tumors in vivo: high-resolution measurements reveal a lack of correlation. Nature Med 3:177–182, 1997.
16. MA Goldberg, SP Dunning, HF Bunn. Regulation of the erythropoietin gene: evidence that the oxygen sensor is a heme protein. Science 242:1412–1415, 1988.
17. GL Wang, BH Jiang, GL Semenza. Effect of protein kinase and phosphatase inhibitors on expression of hypoxia-inducible factor 1. Biochem Biophys Res Commun 216:669–675, 1995.
18. E Maltepe, JV Schmidt, D Baunoch, CA Bradfield, MC Simon. Abnormal angiogenesis and responses to glucose and oxygen deprivation in mice lacking the protein ARNT. Nature 386: 403–407, 1997.
19. ELY Zelzer, Kahana C, Shilo BZ, Rubinstein M, Cohen B. Insulin induces transcription of target genes through the hypoxia-inducible factor HIF-1alpha/ARNT. EMBO J 17:5085–5094, 1998.
20. D Mukhopadhyay, L Tsiokas, XM Zhou, D Foster, JS Brugge, VP Sukhatme. Hypoxic induction of human vascular endothelial growth factor expression through c-Src activation. Nature 375:577–581, 1995.
21. BH Jiang, F Agani, A Passaniti, GL Semenza. V-SRC induces expression of hypoxia-inducible factor 1 (HIF-1) and transcription of genes encoding vascular endothelial growth factor and enolase 1: involvement of HIF-1 in tumor progression. Cancer Res 57:5328–5335, 1997.
22. NM Mazure, EY Chen, P Yeh, KR Laderoute, AJ Giaccia. Oncogenic transformation and hypoxia synergistically act to modulate vascular endothelial growth factor expression. Cancer Res 56:3436–3440, 1996.
23. NM Mazure, EY Chen, KR Laderoute, AJ Giaccia. Induction of vascular endothelial growth factor by hypoxia is modulated by a phosphatidylinositol 3-kinase/Akt signaling pathway in Ha-ras-transformed cells through a hypoxia inducible factor-1 transcriptional element. Blood 90:3322–3331, 1997.
24. MV Blagosklonny, WG An, LY Romanova, J Trepel, T Fojo, L Neckers. p53 inhibits hypoxia-inducible factor-stimulated transcription. J Biol Chem 273:11995–11998, 1998.
25. RH Wenger, M Gassmann. Oxygen(es) and the hypoxia-inducible factor-1. Biol Chem 378: 609–616, 1997.
26. A Rolfs, I Kvietikova, M Gassmann, RH Wenger. Oxygen-regulated transferrin expression is mediated by hypoxia-inducible factor-1. J Biol Chem 272:20055–20062, 1997.
27. G Melillo, LS Taylor, A Brooks, T Musso, GW Cox, L Varesio. Functional requirement of the hypoxia-responsive element in the activation of the inducible nitric oxide synthase promoter by the iron chelator desferrioxamine. J Biol Chem 272:12236–12243, 1997.
28. PJ Lee, BH Jiang, BY Chin, NV Iyer, J Alam, GL Semenza, AM Choi. Hypoxia-inducible factor-1 mediates transcriptional activation of the heme oxygenase-1 gene in response to hypoxia. J Biol Chem 272:5375–5381, 1997.
29. JA Forsythe, BH Jiang, NV Iyer, F Agani, SW Leung, RD Koos, GL Semenza. Activation of vascular endothelial growth factor gene transcription by hypoxia-inducible factor 1. Mol Cell Biol 16:4604–4613, 1996.

30. GL Semenza, PH Roth, HM Fang, GL Wang. Transcriptional regulation of genes encoding glycolytic enzymes by hypoxia-inducible factor 1. J Biol Chem 269:23757–23763, 1994.
31. A Coquelle, F Toledo, S Stern, A Bieth, M Debatisse. A new role for hypoxia in tumor progression: induction of fragile site triggering genomic rearrangements and formation of complex DMs and HSRs. Mol Cell 2:259–265, 1998.
32. CK Luk, L Veinot Drebot, E Tjan, IF Tannock. Effect of transient hypoxia on sensitivity to doxorubicin in human and murine cell lines. J Natl Cancer Inst 82:684–692, 1990.
33. SD Young, RS Marshall, RP Hill. Hypoxia induces DNA overreplication and enhances metastatic potential of murine tumor cells. Proc Natl Acad Sci USA 85:9533–9537, 1988.
34. GC Rice, C Hoy, RT Schimke. Transient hypoxia enhances the frequency of dihydrofolate reductase gene amplification in Chinese hamster ovary cells. Proc Natl Acad Sci USA 83: 5978–5982, 1986.
35. TY Reynolds, S Rockwell, PM Glazer. Genetic instability induced by the tumor microenvironment. Cancer Res 56:5754–5757, 1996.
36. P Carmeliet, Y Dor, JM Herbert, D Fukumura, K Brusselmans, M Dewerchin, M Neeman, F Bono, R Abramovitch, P Maxwell, CJ Koch, P Ratcliffe, L Moons, RK Jain, D Collen, E Keshet. Role of HIF-1alpha in hypoxia-mediated apoptosis, cell proliferation and tumour angiogenesis. Nature 394:485–490, 1998.
37. WG An, M Kanekal, MC Simon, E Maltepe, MV Blagosklonny, LM Neckers. Stabilization of wild-type p53 by hypoxia-inducible factor 1alpha. Nature 392:405–408, 1998.
38. TG Graeber, C Osmanian, T Jacks, DE Housman, CJ Koch, SW Lowe, AJ Giaccia. Hypoxia-mediated selection of cells with diminished apoptotic potential in solid tumours (see comments). Nature 379:88–91, 1996.
39. NV Iyer, LE Kotch, F Agani, SW Leung, E Laughner, RH Wenger, M Gassmann, JD Gearhart, AM Lawler, AY Yu, GL Semenza. Cellular and developmental control of O2 homeostasis by hypoxia-inducible factor 1 alpha. Genes Dev 12:149–162, 1998.
40. I Flamme, T Frohlich, M von Reutern, A Kappel, A Damert, W Risau. HRF, a putative basic helix-loop-helix-PAS-domain transcription factor is closely related to hypoxia-inducible factor-1 alpha and developmentally expressed in blood vessels. Mech Dev 63:51–60, 1997.
41. M Ema, S Taya, N Yokotani, K Sogawa, Y Matsuda, Y Fujii Kuriyama. A novel bHLH-PAS factor with close sequence similarity to hypoxia-inducible factor 1alpha regulates the VEGF expression and is potentially involved in lung and vascular development. Proc Natl Acad Sci USA 94:4273–4278, 1997.
42. H Tian, SL McKnight, DW Russell. Endothelial PAS domain protein 1 (EPAS1), a transcription factor selectively expressed in endothelial cells. Genes Dev 11:72–82, 1997.
43. JB Hogenesch, WK Chan, VH Jackiw, RC Brown, YZ Gu, M Pray Grant, GH Perdew, CA Bradfield. Characterization of a subset of the basic-helix-loop-helix-PAS superfamily that interacts with components of the dioxin signaling pathway. J Biol Chem 272:8581–8593, 1997.
44. MS Wiesener, H Turley, WE Allen, C William, KU Eckardt, KL Talks, SM Wood, KC Gatter, AL Harris, CW Pugh, PJ Ratcliffe, PH Maxwell. Induction of endothelial PAS domain protein-1 by hypoxia: characterization and comparison with hypoxia-inducible factor-1 alpha. Blood 92:2260–2268, 1998.
45. H Tian, R Hammer, A Matsumoto, D Russell, S McKnight. The hypoxia-responsive transcription factor EPAS1 is essential for catecholamine homeostasis and protection against heart failure during embryonic development. Genes Dev 12:3320–3324, 1998.
46. W Risau. Mechanisms of angiogenesis. Nature 386:631–742, 1997.
47. PH Maxwell, GU Dachs, JM Gleadle, LG Nicholls, AL Harris, IJ Stratford, O Hankinson, CW Pugh, PJ Ratcliffe. Hypoxia-inducible factor-1 modulates gene expression in solid tumors and influences both angiogenesis and tumor growth. Proc Natl Acad Sci USA 94:8104–8109, 1997.
48. HE Ryan, J Lo, RS Johnson. HIF-1 alpha is required for solid tumor formation and embryonic vascularization. EMBO J 17:3005–3015, 1998.

49. M Asano, A Yukita, T Matsumoto, S Kondo, H Suzuki. Inhibition of tumor growth and metastasis by an immunoneutralizing monoclonal antibody to human vascular endothelial growth factor/vascular permeability factor 121. Cancer Res 55:5296–5301, 1995.
50. P Borgstrom, KJ Hillan, P Sriramarao, N Ferrara. Complete inhibition of angiogenesis and growth of microtumors by anti-vascular endothelial growth factor neutralizing antibody: novel concepts of angiostatic therapy from intravital videomicroscopy. Cancer Res 56:4032–4039, 1996.
51. P Borgstrom, MA Bourdon, KJ Hillan, P Sriramarao, N Ferrara. Neutralizing anti-vascular endothelial growth factor antibody completely inhibits angiogenesis and growth of human prostate carcinoma micro tumors in vivo. Prostate 35:1–10, 1998.
52. KP Claffey, LF Brown, LF del Aguila, K Tognazzi, KT Yeo, EJ Manseau, HF Dvorak. Expression of vascular permeability factor/vascular endothelial growth factor by melanoma cells increases tumor growth, angiogenesis, and experimental metastasis. Cancer Res 56:172–181, 1996.
53. T Oku, JG Tjuvajev, T Miyagawa, T Sasajima, A Joshi, R Joshi, R Finn, KP Claffey, RG Blasberg. Tumor growth modulation by sense and antisense vascular endothelial growth factor gene expression: effects on angiogenesis, vascular permeability, blood volume, blood flow, fluorodeoxyglucose uptake, and proliferation of human melanoma intracerebral xenografts. Cancer Res 58:4185–4192, 1998.
54. SY Cheng, HJ Huang, M Nagane, XD Ji, D Wang, CC Shih, W Arap, CM Huang, WK Cavenee. Suppression of glioblastoma angiogenicity and tumorigenicity by inhibition of endogenous expression of vascular endothelial growth factor. Proc Natl Acad Sci USA 93:8502–8507, 1996.
55. M Saleh, SA Stacker, AF Wilks. Inhibition of growth of C6 glioma cells in vivo by expression of antisense vascular endothelial growth factor sequence. Cancer Res 56:393–401, 1996.
56. B Millauer, LK Shawver, KH Plate, W Risau, A Ullrich. Glioblastoma growth inhibited in vivo by a dominant-negative Flk-1 mutant. Nature 367:576–579, 1994.
57. LE Benjamin, E Keshet. Conditional switching of vascular endothelial growth factor (VEGF) expression in tumors: induction of endothelial cell shedding and regression of hemangioblastoma-like vessels by VEGF withdrawal. Proc Natl Acad Sci USA 94:8761–8766, 1997.
58. N Ferrara, K Carver Moore, H Chen, M Dowd, L Lu, KS O'Shea, L Powell Braxton, KJ Hillan, MW Moore. Heterozygous embryonic lethality induced by targeted inactivation of the VEGF gene. Nature 380:439–442, 1996.
59. J Folkman. Seminars in Medicine of the Beth Israel Hospital, Boston: clinical applications of research on angiogenesis. N Engl J Med 333:1757–1763, 1995.
60. S Soker, S Takashima, HQ Miao, G Neufeld, M Klagsbrun. Neuropilin-1 is expressed by endothelial and tumor cells as an isoform-specific receptor for vascular endothelial growth factor. Cell 92:735–745, 1998.
61. DR Senger, CA Perruzzi, J Feder, HF Dvorak. A highly conserved vascular permeability factor secreted by a variety of human and rodent tumor cell lines. Cancer Res 46:5629–5632, 1986.
62. DR Senger, SJ Galli, AM Dvorak, CA Perruzzi, VS Harvey, HF Dvorak. Tumor cells secrete a vascular permeability factor that promotes accumulation of ascites fluid. Science 219:983–985, 1983.
63. CJ Drake, CD Little. Exogenous vascular endothelial growth factor induces malformed and hyperfused vessels during embryonic neovascularization. Proc Natl Acad Sci USA 92:7657–7661, 1995.
64. L Benjamin, I Hemo, E Keshet. A plasticity window for blood vessel remodelling is defined by pericyte coverage of the preformed endothelial network and is regulated by PDGF-B and VEGF. Development 125:1591–1598, 1998.
65. T Alon, I Hemo, A Itin, J Pe'er, J Stone, E Keshet. Vascular endothelial growth factor acts as a survival factor for newly formed retinal vessels and has implications for retinopathy of prematurity. Nature Med 1:1024–1028, 1995.

66. D Shweiki, A Itin, D Soffer, E Keshet. Vascular endothelial growth factor induced by hypoxia may mediate hypoxia-initiated angiogenesis. Nature 359:843–845, 1992.
67. KH Plate, G Breier, HA Weich, W Risau. Vascular endothelial growth factor is a potential tumour angiogenesis factor in human gliomas in vivo. Nature 359:845–848, 1992.
68. NS Waleh, MD Brody, MA Knapp, HL Mendonca, EM Lord, CJ Koch, KR Laderoute, RM Sutherland. Mapping of the vascular endothelial growth factor–producing hypoxic cells in multicellular tumor spheroids using a hypoxia-specific marker. Cancer Res 55:6222–6226, 1995.
69. A Damert, M Machein, G Breier, MQ Fujita, D Hanahan, W Risau, KH Plate. Up-regulation of vascular endothelial growth factor expression in a rat glioma is conferred by two distinct hypoxia-driven mechanisms. Cancer Res 57:3860–3864, 1997.
70. JH Harmey, E Dimitriadis, E Kay, HP Redmond, D Bouchier Hayes. Regulation of macrophage production of vascular endothelial growth factor (VEGF) by hypoxia and transforming growth factor beta-1. Ann Surg Oncol 5:271–278, 1998.
71. P Salven, P Heikkila, A Anttonen, M Kajanti, H Joensuu. Vascular endothelial growth factor in squamous cell head and neck carcinoma: expression and prognostic significance. Mod Pathol 10:1128–1133, 1997.
72. E Ikeda, MG Achen, G Breier, W Risau. Hypoxia-induced transcriptional activation and increased mRNA stability of vascular endothelial growth factor in C6 glioma cells. J Biol Chem 270:19761–19766, 1995.
73. AP Levy, NS Levy, MA Goldberg. Post-transcriptional regulation of vascular endothelial growth factor by hypoxia. J Biol Chem 271:2746–2753, 1996.
74. AP Levy, NS Levy, MA Goldberg. Hypoxia-inducible protein binding to vascular endothelial growth factor mRNA and its modulation by the von Hippel-Lindau protein. J Biol Chem 271: 25492–25497, 1996.
75. I Stein, M Neeman, D Shweiki, A Itin, E Keshet. Stabilization of vascular endothelial growth factor mRNA by hypoxia and hypoglycemia and coregulation with other ischemia-induced genes. Mol Cell Biol 15:5363–5368, 1995.
76. I Stein, A Itin, P Einat, R Skaliter, Z Grossman, E Keshet. Translation of vascular endothelial growth factor mRNA by internal ribosome entry: implications for translation under hypoxia. Mol Cell Biol 18:3112–3119, 1998.
77. YS Ji, Q Xu, JF Schmedtje Jr. Hypoxia induces high-mobility-group protein I(Y) and transcription of the cyclooxygenase-2 gene in human vascular endothelium. Circ Res 83:295–304, 1998.
78. JF Schmedtje Jr, YS Ji, WL Liu, RN DuBois, MS Runge. Hypoxia induces cyclooxygenase-2 via the NF-kappaB p65 transcription factor in human vascular endothelial cells. J Biol Chem 272:601–608, 1997.
79. M Tsujii, S Kawano, S Tsuji, H Sawaoka, M Hori, RN DuBois. Cyclooxygenase regulates angiogenesis induced by colon cancer cells. Cell 93:705–716, 1998.
80. S Kourembanas, RL Hannan, DV Faller. Oxygen tension regulates the expression of the platelet-derived growth factor-B chain gene in human endothelial cells. J Clin Invest 86:670–674, 1990.
81. JF O'Rourke, CW Pugh, SM Bartlett, PJ Ratcliffe. Identification of hypoxically inducible mRNAs in HeLa cells using differential-display PCR: role of hypoxia-inducible factor-1. Eur J Biochem 241:403–410, 1996.
82. P Carmeliet, N Mackman, L Moons, T Luther, P Gressens, I Van Vlaenderen, H Demunck, M Kasper, G Breier, P Evrard, M Muller, W Risau, T Edgington, D Collen. Role of tissue factor in embryonic blood vessel development. Nature 383:73–75, 1996.
83. P Lindahl, BR Johansson, LP C Betsholtz. Pericyte Loss and Microaneurysm formation in PDGF-B-Deficient Mice. Science 277:242–245, 1997.
84. H Takagi, GL King, N Ferrara, LP Aiello. Hypoxia regulates vascular endothelial growth

factor receptor KDR/Flk gene expression through adenosine A2 receptors in retinal capillary endothelial cells. Invest Ophthalmol Vis Sci 37:1311–1321, 1996.

85. C Kremer, G Breier, W Risau, KH Plate. Up-regulation of flk-1/vascular endothelial growth factor receptor 2 by its ligand in a cerebral slice culture system. Cancer Res 57:3852–3859, 1997.
86. HP Gerber, F Condorelli, J Park, N Ferrara. Differential transcriptional regulation of the two vascular endothelial growth factor receptor genes. Flt-1, but not Flk-1/KDR, is up-regulated by hypoxia. J Biol Chem 272:23659–23667, 1997.
87. JR Gnarra, S Zhou, MJ Merrill, JR Wagner, A Krumm, E Papavassiliou, EH Oldfield, RD Klausner, WM Linehan. Post-transcriptional regulation of vascular endothelial growth factor mRNA by the product of the VHL tumor suppressor gene. Proc Natl Acad Sci USA 93:10589–10594, 1996.
88. O Iliopoulos, AP Levy, C Jiang, WG Kaelin Jr, MA Goldberg. Negative regulation of hypoxia-inducible genes by the von Hippel-Lindau protein. Proc Natl Acad Sci USA 93:10595–10599, 1996.
89. D Fukumura, R Xavier, T Sugiura, Y Chen, EC Park, N Lu, M Selig, G Nielsen, T Taksir, RK Jain, B Seed. Tumor induction of VEGF promoter activity in stromal cells. Cell 94:715–725, 1998.
90. MC Puri, J Rossant, K Alitalo, A Bernstein, J Partanen. The receptor tyrosine kinase TIE is required for integrity and survival of vascular endothelial cells. EMBO J 14:5884–5891, 1995.
91. TN Sato, Y Tozawa, U Deutsch, K Wolburg Buchholz, Y Fujiwara, M Gendron Maguire, T Gridley, H Wolburg, W Risau, Y Qin. Distinct roles of the receptor tyrosine kinases Tie-1 and Tie-2 in blood vessel formation. Nature 376:70–74, 1995.
92. C Suri, PF Jones, S Patan, S Bartunkova, PC Maisonpierre, S Davis, TN Sato, GD Yancopoulos. Requisite role of angiopoietin-1, a ligand for the TIE2 receptor, during embryonic angiogenesis (see comments). Cell 87:1171–1180, 1996.
93. PC Maisonpierre, C Suri, PF Jones, S Bartunkova, SJ Wiegand, C Radziejewski, D Compton, J McClain, TH Aldrich, N Papadopoulos, TJ Daly, S Davis, TN Sato, GD Yancopoulos. Angiopoietin-2, a natural antagonist for Tie2 that disrupts in vivo angiogenesis (see comments). Science 277:55–60, 1997.
94. T Asahara, D Chen, T Takahashi, K Fujikawa, M Kearney, M Magner, GD Yancopoulos, JM Isner. Tie2 receptor ligands, angiopoietin-1 and angiopoietin-2, modulate VEGF-induced postnatal neovascularization (see comments). Circ Res 83:233–240, 1998.
95. LE Benjamin, D Golijanin, A Itin, D Pode, E Keshet. Selective ablation of immature blood vessels in established human tumors follows vascular endothelial growth factor withdrawal. J Clin Invest 103:159–165, 1999.
96. L Holmgren, MS O'Reilly, J Folkman. Dormancy of micrometastases: balanced proliferation and apoptosis in the presence of angiogenesis suppression (see comments). Nature Med 1: 149–153, 1995.
97. T Nagashima, T Hoshino, KG Cho. Proliferative potential of vascular components in human glioblastoma multiforme. Acta Neuropathol 73:301–305, 1987.

17
Fibrin and Proteolysis in Repair-Associated Angiogenesis

Pieter Koolwijk
Gaubius Laboratory TNO-PG, Leiden, The Netherlands

Victor W. M. van Hinsbergh
Gaubius Laboratory TNO-PG, Leiden, and Institute for Cardiovascular Research, Vrije Universiteit, Amsterdam, The Netherlands

I. OVERVIEW

The major difference between the formation of new blood vessels during developmental and under pathological conditions is that pathological or repair-associated angiogenesis is accompanied by vascular leakage (1), the occurrence of inflammatory cells (2), and the presence of a temporary fibrin matrix (3–5). In this chapter, the role of the components of the plasminogen activator system during the invasion of endothelial cells into a fibrin matrix is described. At the basolateral side of the endothelial cell, the urokinase-type plasminogen activator (u-PA) bound to a specific cellular receptor (u-PA receptor) is involved in the proteolytic modulation of matrix proteins and cell-matrix interaction. Inhibition of u-PA activity or prevention of the binding of u-PA to the u-PA receptor reduces the in vitro invasion of human microvascular endothelial cells (hMVEC) and the formation of capillary-like tubular structures of hMVEC in three-dimensional fibrin matrices, induced by a combination of the cytokine tumor necrosis factor-α (TNF-α) and the angiogenic factors' basic fibroblast growth factor (bFGF) or vascular endothelial growth factor (VEGF). The structure of the fibrin fibers in the matrix is of importance during the invasion of endothelial cells into a fibrin matrix and the formation of capillary-like structures in the fibrin matrix in vitro.

II. INTRODUCTION

Angiogenesis, the outgrowth of new blood vessels from existing ones, is an essential process during development; in adults, it occurs in the reproductive system of females (formation of corpus luteum and placenta) (6). In addition, the formation of new blood vessels is also an essential factor during tissue repair (formation and regression of granulation tissue), which is necessary to restore healthy tissue after wounding and/or inflammation. Furthermore, angiogenesis is associated with many pathological conditions, such as chronic inflammation, including rheumatoid arthritis (7,8), malignancies (9), and retinopathy caused by metabolic dysregulation—in particular, diabetes (10,11). What is common to all these pathological conditions is that the angiogenesis is accompanied by vascular leakage of the vessels involved (1), the occurrence of inflammatory cells and mediators (2), and the presence of fibrin (3–5). These latter factors are absent in angiogenesis during embryonic development. Therefore, it is likely that ''developmental angiogenesis'' and adult ''repair-associated or pathological angiogenesis'' are two separate processes with many identical features but with different properties with respect to their regulation (12).

Fibrin, formed after the wounding of a blood vessel and when plasma leaks from blood vessels, forming a fibrous exudate, is a temporary matrix that not only acts as a barrier preventing further blood loss but also provides a structure in which new microvessels can infiltrate during wound healing. Outgrowth of microvessels as well as the subsequent (partial) disappearance of these vessels is essential to ensure adequate wound healing and prevent the formation of scar tissue. A controlled invasion of the temporary fibrin matrix by endothelial cells (but also by inflammatory or tumor cells) is facilitated by components of the plasminogen activator/plasmin system, which allow a spatially controlled lysis of the fibrin matrix at the basolateral side of the cell (13). To do this optimally, the endothelium must respond differently to fibrin depending on whether the fibrin is present at its luminal or basolateral sides (12). Fibrin generated at the luminal side may occlude the vessels, which will cause serious damage to the distal tissues. To prevent this, the endothelium is able instantaneously to increase the fibrinolytic activity in the blood compartment. However, this leads to strong systemic fibrinolytic activity and recurrent bleeding unless the fibrinolysis is restricted in time and to a limited distance from the endothelium. To fulfill this dual role, both in the prevention of local intravascular fibrin accumulation and a differently timed fibrinolysis in angiogenesis and tissue repair, endothelial cells are equipped with a complex regulatory system involving activators/enzymes, inhibitors, cell polarity, and cellular receptors. After introducing the components and regulation of these components of the plasminogen activator/plasmin system, we will focus on the contribution of the urokinase-type plasminogen activator (u-PA) in pericellular proteolytic events associated with endothelial cell migration, the invasion of capillary-like tubular structures into fibrin matrices in vitro and during angiogenesis in vivo.

III. THE PLASMIN/PLASMINOGEN ACTIVATOR SYSTEM

A. Activators/Enzymes

The activation of the plasmin/plasminogen activator system is schematically presented in Fig. 1. The key enzyme, plasmin, is formed from the zymogen plasminogen by proteolytic activation by one of the known mammalian plasminogen activators (PAs): tissue-type plasminogen activator (t-PA) or urokinase-type plasminogen activator (u-PA) (14,15). All three serine proteases of the plasmin/plasminogen activator system—plasminogen, t-PA and u-PA—are synthesized as single polypeptide chains, and each is converted by specific proteolytic cleavage to a molecule with two polypeptide chains connected by a disulfide bond. The A-chain or amino-terminal part of the molecule is built up by domains that determine the interaction of the proteases with matrix proteins and cellular receptors, whereas the B-chain or C-terminal part of the molecule contains the proteolytically active

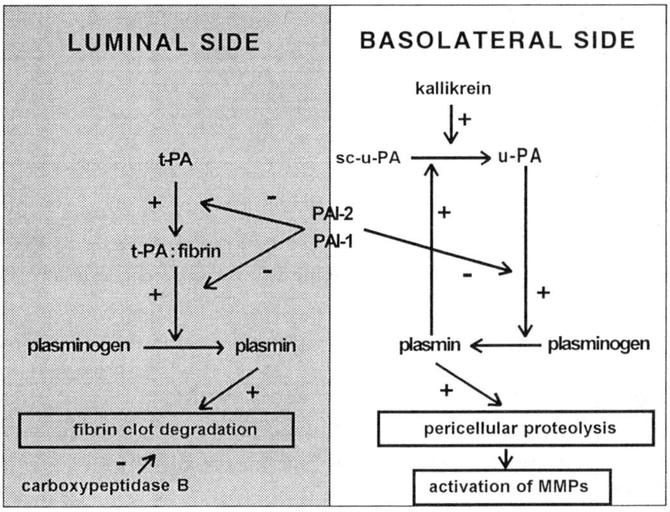

Figure 1 Scheme of the plasminogen activation system. Abbreviations: +, activation; −, inhibition; PA, plasminogen activator; t-PA, tissue-type PA; tcu-PA, two-chain urokinase-type PA; scu-PA; single-chain urokinase-type PA; PAI, PA inhibitor.

site. The proteolytic cleavage of plasminogen and single-chain (sc) u-PA to their respective two-chain (tc) forms is necessary to disclose the proteolytically active site and to activate the molecule. The interaction of plasminogen with fibrin or the cell surface occurs predominantly via binding sites in the kringle structures, which recognize lysine residues of proteins, in particular C-terminal lysines. Because B-type carboxypeptidases remove C-terminal lysine residues from potential binding sites for plasminogen in fibrin or on the cell surface, they can act as negative regulators of the fibrinolytic system (16).

B. Inhibitors

The activities of the proteases of the plasmin/plasminogen activator system are controlled by potent inhibitors, which are members of the serine protease inhibitor (serpin) superfamily. Free circulating plasmin, not bound to fibrin, is instantaneously inhibited by α_2-antiplasmin (17). The interaction of α_2-antiplasmin and plasmin is facilitated by the lysine binding domain of plasmin; it is attenuated when plasmin is bound to fibrin. The predominant regulators of t-PA and u-PA activities are the plasminogen activator inhibitor type 1 (PAI-1) and PAI-2 (18,19). PAI activity in human plasma is normally exclusively PAI-1, present in blood platelets and synthesized by endothelial cells, smooth muscle cells, hepatocytes, and many other cell types in culture (20,21). PAI-1 binds to vitronectin, which stabilizes its inhibitory activity. PAI-2 is produced by monocytes/macrophages, placental trophoblasts, and certain tumor cell lines (19).

C. Receptors

Regulation of fibrinolytic activity also occurs by means of cellular receptors. These receptors direct the action of PAs and plasmin to focal areas on the cell surface or are involved in the clearance of the PAs. High-affinity binding sites for plasminogen (22–24), t-PA (24,25), and u-PA (26–28) are found on various types of cells, including endothelial cells.

Plasminogen binds with a moderate affinity (120 to 340 nM depending on whether the Lys- or Glu-form of plasminogen is used) but with a high capacity (3.9 to 14 $\times$ 10^5 molecules per cell) to endothelial cells (23,29). This binding, which is also observed with many other cell types, is mediated by the lysine binding sites of kringles 1–3 of the plasmin(ogen) molecule. Because these lysine binding sites are also involved in the interaction of plasmin with α_2-antiplasmin, occupation of lysine binding sites protects plasmin from instantaneous inhibition by α_2-antiplasmin not only when plasmin is bound to fibrin (see above) but also when it is bound to the cellular receptors.

The nature of the plasminogen receptors is not fully resolved. In addition to gangliosides, which directly or indirectly contribute to the plasminogen binding (30), at least eight proteins have been reported to be involved in plasminogen binding. Among them are members of the low-density lipoprotein receptor family, such as gp330 and LRP, annexin II, an as yet unidentified 45-kDa protein, GbIIb/IIIa, and α-enolase (see Ref. 24 for review). Lipoprotein Lp(a), which has strong structural homology with a large part of the plasminogen molecule, can compete for plasminogen binding to endothelial cells (31,32). This competition also involves lysine binding sites.

Specific binding of t-PA to human endothelial cells has been reported (33–36). T-PA binds via its growth-factor domain with high affinity to annexin II on human endothelial cells in culture (37). Different epitopes of the annexin II molecule are involved in plasminogen and t-PA binding. It is conceivable that annexin II, like fibrin, forms a ternary complex with t-PA and plasminogen on the endothelial cell surface (24). In addition, t-PA interacts with matrix-bound PAI-1 (35).

Binding of u-PA to cells is mediated via a specific u-PA receptor (CD87), which has been identified and cloned (38,39). It is present on many cell types, including neutrophils, T cells, monocytes/macrophages, fibroblasts, and endothelial cells (40–44). It is a heavily glycosylated glycosylphosphatidylinositol (GPI)-anchored protein (28) made up of three cysteine-rich CD-59-like domains, which binds both single-chain u-PA and two-chain u-PA via their growth-factor domain (45). After synthesis, it is proteolytically processed at its carboxyl terminus and subsequently anchored in the plasma membrane by a GPI-group (46). It comprises three domains, which are structurally homologous to snake venom α-toxins (28). Human endothelial cells in vitro contain about 40,000–140,000 u-PA receptors per cell (47,48).

The u-PA receptor acts not only as a site for focal pericellular proteolysis by u-PA but is also involved in the clearance of the u-PA:PAI-1 complex (49). In contrast to receptor-bound scu-PA or noninhibited tcu-PA, the u-PA:PAI-1 complex is rapidly internalized together with the u-PA receptor, followed by the degradation of the u-PA:PAI-1 complex and the return of the empty u-PA receptor to the plasma membrane (50). Internalization of the GPI-linked u-PA receptor probably occurs after interaction with another receptor, such as the α-macroglobulin/low-density-lipoprotein receptor–related protein (LRP) (51) or the VLDL-receptor (52). VLDL-receptors were demonstrated on capillary and arteriolar endothelial cells in vivo (53). Recently, it has been demonstrated that the cation-independent, mannose 6-phosphate/insulin-like growth factor II receptor is also involved in targeting of the u-PA receptor to intracellular lysosomes (54). The LRP and LRP-like proteins on liver hepatocytes are involved in the clearance of plasmin-α-antiplasmin, PA:PAI-1 complexes (55–58) and probably free PAs from the circulation. In addition, t-PA can also be cleared by mannose receptors present on macrophages and liver endothelial cells (59) and by α-fucose receptors on hepatocytes (37).

IV. MODES OF ACTION BY THE u-PA RECEPTOR

Upon secretion, scu-PA binds to the u-PA receptor, which facilitates a local and controlled accumulation of u-PA and is subsequently converted to the proteolytically active tcu-PA. Since the endothelial cell also contains plasmin(ogen) receptors, an interplay between receptor-bound u-PA and receptor-bound plasmin(ogen) and plasmin formation is likely to occur (Fig. 2). The plasmin generated can degrade a number of matrix proteins and can activate several matrix-degrading metalloproteinases (MMPs), in particular stromelysin-1 (MMP-3), interstitial collagenase-1 (MMP-1), gelatinase-B (MMP-9), and membrane-type MMPs (60–66). Binding of u-PA to u-PA receptors on the cell surface limits plasminogen activation to focal areas, such as the focal attachment sites and cellular protrusions in-

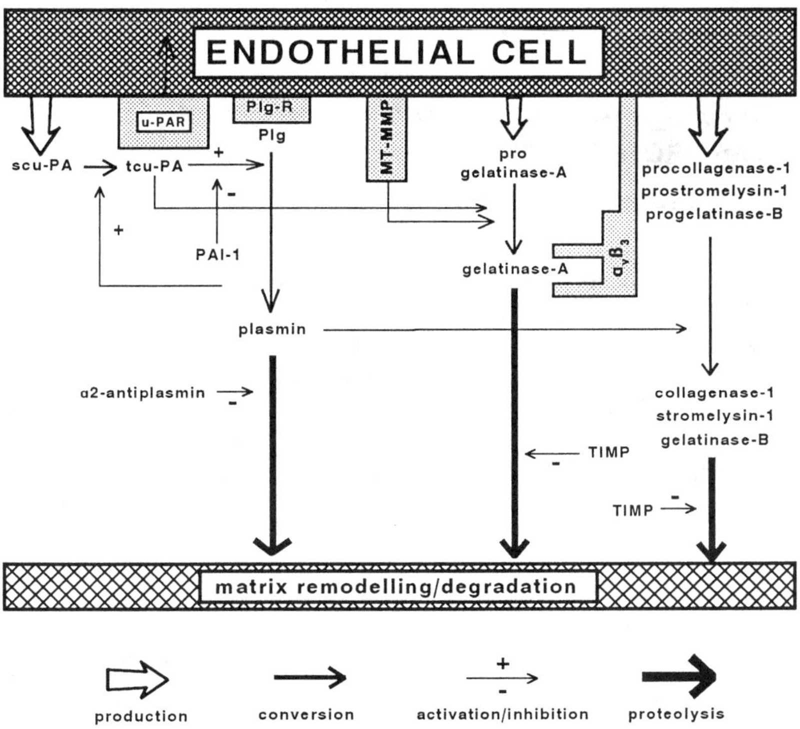

Figure 2 Schematic representation of the interaction between the u-PA/plasmin system and the presumed activation of matrix-degrading metalloproteinases (MMPs). Abbreviations: u-PA, urokinase-type PA; sc-u-PA, single-chain u-PA; tc-u-PA, two-chain u-PA; u-PAR, u-PA receptor; Plg, plasminogen; Plg-R, Plg receptor; PAI-1, PA inhibitor-1; MT-MMP, membrane-bound MMP; TIMP, tissue inhibitor of MMP. +, stimulation; −, inhibition.

volved in cell migration and invasion (42). In addition, a direct plasmin-independent proteolytic action of u-PA on matrix proteins may also occur (67). Like free u-PA activity, receptor-bound tcu-PA is subject to inhibition by PAI-1. As a consequence, u-PA is active for only a short time.

Binding of u-PA to the u-PA receptor also evokes activation of monocytes and neutrophils with regard to enhanced chemotaxis, adhesion, and production of TNF-α and MMPs (68–72). This activation by u-PA is independent of its enzymatic activation and is mediated by an increase of diacylglycerol (DAG) generation (73), the phosphorylation of several proteins of the Jak/Stat pathway (74–76), Erk1 and Erk2 (77) and the induction of c-fos gene expression (78). Binding of u-PA to the u-PA receptor induces a conformational change in the u-PA receptor, by which it can act as a ligand for other cell-surface components. This conformational site is also present on chymotrypsin-cleaved soluble u-PA receptor (79) and lies between domains 1 and 2, as demonstrated by u-PA receptor mutants and synthetic peptides (80).

In addition to the modes of action described above, the u-PA receptor is also involved in the interaction between cells and the extracellular matrix. The u-PA receptor has been found in focal attachment sites, where integrin-matrix interactions occur, and in cell-cell contact areas (81–83). Besides the ability of the u-PA receptor to form complexes with the β_1 and β_2 integrins (84–86), the u-PA receptor occupied by u-PA interacts avidly with vitronectin (26,87).

V. REGULATION OF PLASMINOGEN ACTIVATION AND PERICELLULAR PROTEOLYSIS

A. Regulation of Proteases and Inhibitors

Interleukin-1 (IL-1) and tumor necrosis factor-α (TNF-α) exert many effects on the vascular endothelium, such as the induction or increase of the transcription of the genes of the leukocyte adhesion molecules E-selectin, VCAM-1 and ICAM-1, cyclooxygenase-2, and a number of proteases and protease inhibitors. The inflammatory mediators TNF-α, IL-1, and the bacterial lipopolysaccharide (LPS) induce the synthesis of u-PA in human endothelial cells in vitro (88–90), which occurs entirely toward the basolateral side of the cell, whereas the secretion of t-PA and PAI-1 proceeds equally to the luminal and basolateral sides of the cell (88). The polar secretion of u-PA suggests that u-PA may be involved in local remodeling of the basal membrane of the cell. The activity of u-PA is controlled in space by interaction of u-PA with its cellular receptor and by the inhibitor PAI-1. Whereas u-PA is normally not found in endothelial cells in vivo, association of u-PA with the endothelium was observed in inflammatory tissues during acute appendicitis (91), rheumatoid arthritis (7,92), and tumor development (93).

Simultaneously with the increase in u-PA, TNF-α, IL-1, as well as LPS markedly increase the production of PAI-1 in endothelial cells in vitro (94–97), which was largely inhibited by the isoflavone compound genistein (98). The increase in PAI-1 induced by inflammatory mediators may represent a protective mechanism of the cell against uncontrolled u-PA activity. In vivo, administration of TNF-α, IL-1, or LPS causes an increase in PAI-1 concentration in the circulation and a PAI-1 mRNA increase in the endothelium of various organs (99,100).

The angiogenic growth factors bFGF, aFGF, and VEGF-A have species- and tissue-dependent effects on the production of u-PA and PAI-1 by endothelial cells. Whereas bFGF and VEGF-A are both potent inducers of u-PA in bovine (101–103) and mouse endothelial cells (104), bFGF and VEGF-A do not enhance u-PA production in human endothelial cells (48,105). The PAI-1 expression in human endothelial cells is slightly decreased by the addition of bFGF or VEGF-A (48,106) but stimulated in bovine endothelial cells (102).

B. u-PA Receptor Expression

TNF-α increases u-PA receptor expression in human microvascular endothelial cells (48) and monocytes (107,108) but not in endothelial cells from human umbilical vein or aorta (48). The expression of the u-PA receptor is also enhanced by many angiogenic growth factors, including bFGF, aFGF, and VEGF-A (48,109,110) and by the activation of protein kinase C or the elevation of the cellular cAMP concentration (13,111). The effects of bFGF and VEGF-A on the induction of u-PA receptors by human endothelial cells are regulated independently of their effects on cell proliferation. Similarly, Presta et al. (112) have shown that the bFGF-induced u-PA expression in bovine endothelial cells proceeds independently of the stimulation of mitogenesis by this growth factor.

VI. u-PA/u-PA RECEPTOR AND CELL MIGRATION OF ENDOTHELIAL CELLS

Migrating and invading cells, such as monocytes and tumor cells, express u-PA activity bound to u-PA receptors on their cellular protrusions (42,113) and on focal attachment sites (81,114). Receptor-bound u-PA was also demonstrated in focal adhesion sites of fibroblasts (114) and endothelial cells (83). A continuous activation and removal of u-PA bound to the receptor could contribute to the formation and detachment of focal attachment sites and hence to locomotion of the cell (115). Receptor-bound u-PA activity is also thought to be involved in smooth muscle and endothelial cell migration and in the formation of new blood vessels (angiogenesis). Inhibition of plasminogen activation interferes with smooth muscle cell migration in vitro (116,117) and affects smooth muscle migration and proliferation in vivo (118). Intimal hyperplasia of injured arteries is less pronounced in mice lacking u-PA than in wild-type or t-PA–deficient mice, whereas PAI-1–deficient mice show an increased intimal proliferation (119). Animals made deficient in plasminogen show comparable pathological features to those with a combined deficiency of u-PA and t-PA (120,121). These data suggest that u-PA and plasminogen play a role in cell recruitment.

Several investigators (103,122,123) have demonstrated the involvement of u-PA and the u-PA receptor in the migration of bovine endothelial cells. After a monolayer of these endothelial cells is wounded, the cells that migrate into the wounded area express enhanced u-PA activity (122) and u-PA receptor expression (103) induced by bFGF released from the matrix of the wounded area (123). This growth factor is a potent inducer of plasminogen activator activity, in particular u-PA, in bovine endothelial cells (101,124) but not in human endothelial cells (48). In both species, bFGF increases the number of u-PA receptors on endothelial cells (48,103,109).

VII. THE PLASMIN/PLASMINOGEN ACTIVATOR SYSTEM AND THE FORMATION OF ENDOTHELIAL TUBES IN FIBRIN MATRICES

Angiogenesis in the adult is associated with tissue repair after injury by wounding or inflammation. This repair-associated angiogenesis is usually accompanied by the presence of a temporary matrix of fibrin (3,4,125), which is formed after the wounding of a blood vessel and when plasma leaks from blood vessels. The fibrin matrix provides a structure into which new microvessels can infiltrate during wound healing. Proper timing of the outgrowth of microvessels as well as the subsequent (partial) disappearance of these vessels is essential to ensure adequate wound healing and prevent the formation of scar tissue. Although essential for the formation of granulation tissue and tissue repair, angiogenesis, once under the control of pathological stimuli, can contribute to a number of pathological conditions, such as tumor neovascularization, pannus formation in rheumatoid arthritis, and diabetic retinopathy.

An understanding of the mechanisms involved in angiogenesis may provide clues to preventing pathological angiogenesis without seriously impairing tissue repair. Because of the specific roles of fibrin and inflammatory cells and mediators in pathological angiogenesis in the adult (2–4,126–128), we have focused our studies on the invasion of human endothelial cells into three-dimensional fibrin matrices and the role of endothelial plasminogen activators in this process. This model resembles the recanalization of a fibrin clot by invading endothelial cells. This invasion is usually preceded by the infiltration of inflammatory cells, which interact with the vascular structures from which endothelial cells subsequently migrate into the fibrin clot (129).

Pepper and Montesano (130,131) demonstrated a direct correlation between the bFGF-induced formation of capillary sprouts by bovine microvascular endothelial cells in three-dimensional fibrin matrices and the expression of PA. The extent of tube formation and the diameter of the tubes formed was reduced by the simultaneous presence of TGF-β (131). The latter is a growth factor, which exerts a strong enhancement of PAI-1 synthesis in cultured endothelial cells and thus inhibits PA activity (101). In addition to bFGF, VEGF can also stimulate bovine endothelial cells to form tubular structures and can act cooperatively with bFGF in this induction (132).

In contrast, quiescent monoalyers of human microvascular endothelial cells grown on fibrin matrices exposed to bFGF or VEGF alone do not form tubular structures. However, when bFGF or VEGF are added simultaneously with TNFα, a large number of capillary-like tubular structures are formed (Fig. 3, Ref. 48). The outgrowth of tubular structures requires u-PA receptor–bound u-PA and plasmin activity and is completely reduced by anti-u-PA immunoglobulins, as well as by anti-u-PA receptor antibodies inhibiting the interaction of u-PA with its receptor, but not by anti-t-PA antibodies (Fig. 4). Furthermore, proteolytic activation of plasminogen appears to be involved, because the plasmin inhibitor aprotinin largely inhibits the formation of tubular structures (48,133). Recent immunohistochemical studies performed on cross sections of these invading capillary-like tubular structures showed an enhanced expression of both u-PA and the u-PAR antigen by invading endothelial cells compared with noninvading endothelial cells (133a). These data agree with the data on bovine endothelial cells except that in human endothelial cells, a second mediator is required to induce u-PA synthesis. In animals bFGF and aFGF have been shown to induce neovascularization (134,135). Because it is difficult to rule out inflammation in these experiments in vivo, it remains to be elucidated whether these

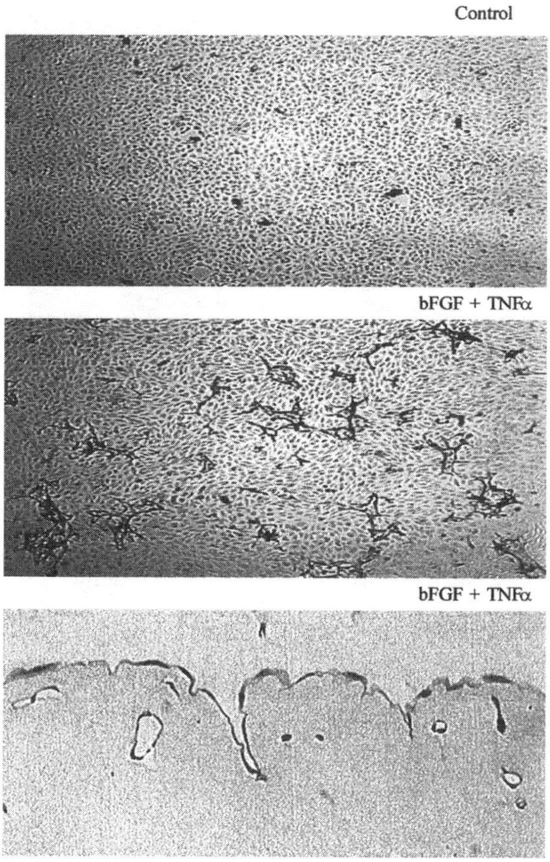

Figure 3 Formation of capillary-like tubular structures in a three-dimensional fibrin matrix by human endothelial cells. Top. Phase contrast microscopy of human microvascular endothelial cells grown under control conditions on top of a three-dimensional fibrin matrix. Middle. Formation of tubular structures is induced by the simultaneous addition of the growth factor bFGF (5 ng/mL) and the cytokine TNF (5 ng/mL). Bottom. Light microscopy of hemotoxylin/phloxin–stained cross section perpendicular to the fibrin matrix surface of capillary-like tubular structures of microvascular endothelial cells.

growth factors act in vivo in conjunction with inflammatory mediators or independently of the latter mediators. Min et al. (136) have recently shown that prevention of the binding of u-PA to the u-PAR by a fusion product of the epidermal growth factor–like domain of murine u-PA and Fc portion of human IgG reduced bFGF-induced angiogenesis in vivo and the growth of a B16 melanoma in syngeneic mice, indicating that u-PA–mediated angiogenesis also occurs in these in vivo systems.

Irrespective of the exact role of fibrin in stimulating angiogenesis (137–139), the fibrin structure has important consequences for wound healing and angiogenesis. The structure of fibrin is determined not only by the concentration of fibrinogen and thrombin present in the tissues but also by the presence of matrix molecules—such as fibronectin, vitronectin, hyaluronic acid, and thrombospondin—and by environmental conditions, such

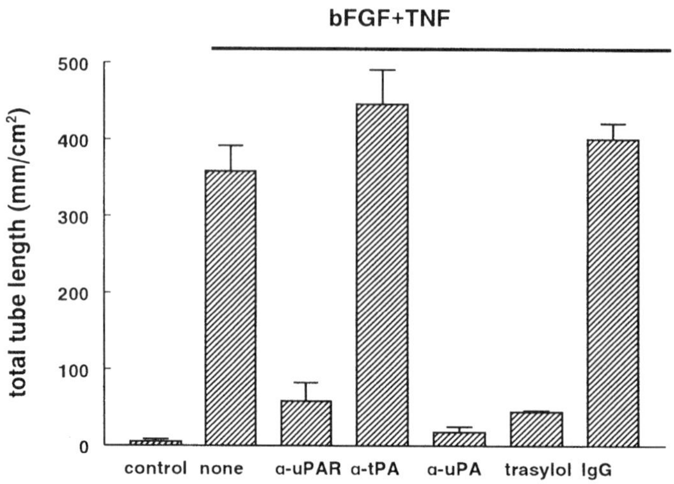

Figure 4 Effect of various inhibitors of the u-PA/plasmin system on the bFGF/TNF-α–induced formation of tubular structures of human microvascular endothelial cells in a three-dimensional fibrin matrix in vitro.

as the pH and ionic strength. The invasion of endothelial cells into a fibrin matrix and the formation of capillary-like structures in the fibrin matrix in vitro is an attractive model with which to study this feature of the angiogenesis process associated with tissue repair. A recent paper of Nehls and Herrmann (140) points to the importance of the fibrin structure in endothelial cell migration and the formation of tubular structures. These authors show that the rigidity of the fibrin matrix has a strong impact on tube formation by bovine endothelial cells in response to bFGF and VEGF. In addition, we have found that this was also the case when human microvascular endothelial cells were used (133). When fibrin matrices were prepared at a pH of 7.0, they consisted of malleable matrices with thick fibrin fibers, which were very sensitive to fibrinolysis. On the other hand, when these fibrin matrices were prepared at pH 7.8, they were rigid, finely dispersed, and more resistant to fibrinolysis. When these matrices were equilibrated to a pH of 7.4 and subsequently covered by human microvascular endothelial cells, which became exposed to bFGF or VEGF in combination with TNF-α, the cells formed tubular structures much more easily and faster on the fibrin matrices prepared at pH 7.0; however, because of excessive fibrinolysis, the stability of these structures was less than that of the structures in the fibrin matrices prepared at pH 7.4 or 7.8.

It remains to be determined whether this effect of the fibrin structure reflects a mechanical barrier of movement of cells by a dense fibrin network or is due to an inadequate spacing of cell-binding epitopes in the fibrin network. Fibrin contains at least two cell-binding domains for endothelial cells: an RGD sequence in its α-chains, which binds to the vitronectin receptor—i.e. the $\alpha_v\beta_3$ integrin (141,142)—and another site in the β-chain, residues 5–42 (143), which binds to a 130-kDa receptor (144). Brooks et al. (145,146) and Hammes et al. (147) have shown that inhibition of the $\alpha_v\beta_3$ integrin reduces angiogenesis in several in vivo models. Friedlander et al. (148) recently reported that both $\alpha_v\beta_3$ and $\alpha_v\beta_5$ integrins are involved in growth factor–stimulated angiogenesis in the rabbit cornea, but via distinct mechanisms. It is of interest to note that bFGF- and TNF-α-induced angiogenesis is inhibited by an antibody against the $\alpha_v\beta_3$ integrin, whereas angiogenesis induced by VEGF or by a protein kinase C activating phorbol ester required $\alpha_v\beta_5$ integrin. It remains to be established whether both $\alpha_v\beta_3$- and $\alpha_v\beta_5$-dependent mechanisms are active in the invasion of endothelial cells into a fibrin matrix. The involvement of $\alpha_v\beta_3$ integrin interaction with the fibrin matrix is likely (149), while the role of $\alpha_v\beta_5$ integrin, which more selectively interacts with vitronectin, has to be evaluated.

ACKNOWLEDGMENTS

The financial support by the Netherlands Heart Foundation (grant 95-193), the Dutch Cancer Society (grant TNOP 97-1511) and the Praeventiefonds (grants 28-2621 and 28-2622) is gratefully acknowledged.

REFERENCES

1. HF Dvorak, LF Brown, M Detmar, AM Dvorak. Vascular permeability factor/vascular endothelial growth factor, microvascular hyperpermeability, and angiogenesis. Am J Pathol 146: 1029–1039, 1995.
2. PJ Polverini. Macrophage-induced angiogenesis—a review. Macrophage-Derived Cell Regulatory Factors 1:54–73, 1989.
3. JA Nagy, LF Brown, DR Senger, N Lanir, L Van De Water, AM Dvorak, HF Dvorak. Pathogenesis of tumor stroma generation: a critical role for leaky blood vessels and fibrin deposition. Biochim Biophys Acta 948:305–326, 1989.
4. HF Dvorak, DR Senger, AM Dvorak. Fibrin as a component of the tumor stroma: origins and biological significance. Cancer Metastasis Rev 2:41–73, 1983.
5. H Bárdos, P Molnár, G Csécsei, R Adány. Fibrin deposition in primary and metastatic human brain tumours. Blood Coagul Fibrinolysis 7:536–548, 1996.
6. E Bacharach, A Itin, E Keshet. Invivo patterns of expression of urokinase and its inhibitor PAI-1 suggest a concerted role in regulating physiological angiogenesis. Proc Natl Acad Sci USA 89:10686–10690, 1992.
7. JB Weinberg, AMM Pippen, CS Greenberg. Extravascular fibrin formation and dissolution in synovial tissue of patients with osteoarthritis and rheumatoid arthritis. Arthritis Rheum 34:996–1005, 1991.
8. AE Koch. Angiogenesis: implications for rheumatoid arthritis. Arthritis Rheum 41:951–962, 1998.
9. J Folkman, Y Shing. Angiogenesis. J Biol Chem 267:10931–10934, 1992.
10. LP Aiello, RL Avery, PG Arrigg, BA Keyt, HD Jampel, ST Shah, LR Pasquale, H Thieme, MA Iwamoto, JE Park, HV Nguyen, LM Aiello, N Ferrara, GL King. Vascular endothelial growth factor in ocular fluid of patients with diabetic retinopathy and other retinal disorders. N Engl J Med 331:1480–1487, 1994.
11. RN Frank. Diabetic retinopathy. Prog Retinal Eye Res 14:361–392, 1995.
12. VWM Van Hinsbergh, P Koolwijk, R Hanemaaijer. Role of fibrin and plasminogen activators in repair-associated angiogenesis: in vitro studies with human endothelial cells—regulation of angiogenesis. EXS 79:391–411, 1997.
13. VWM van Hinsbergh. Impact of endothelial activation on fibrinolysis and local proteolysis in tissue repair. Ann NY Acad Sci 667:151–162, 1992.
14. F Bachmann. Fibrinolysis. In: M Verstraete, J Vermylen, R Lijnen, J Arnout, eds. Thrombosis and Haemostasis 1987. Leuven: Leuven University Press, 1987, pp. 227–265.
15. P Wallén. Structure and function of tissue plasminogen activator and urokinase. In: PJ

Castellino, PJ Gaffney, MM Samama, A Takada, eds. Fundamental and Clinical Fibrinolysis. Amsterdam: Elsevier, 1987, pp. 1–18.
16. A Redlitz, AK Tan, DL Eaton, EF Plow. Plasma carboxypeptidases as regulators of plasminogen system. J Clin Invest 96:2534–2538, 1995.
17. WE Holmes, L Nelles, HR Lijnen, D Collen. Primary structure of human α2-antiplasmin, a serine protease inhibitor (serpin). J Biol Chem 262:1659–1664, 1987.
18. C Fearns, F Samad, DJ Loskutoff. Synthesis and localization of PAI-1 in the vessel wall. In: VWM van Hinsbergh, ed. Vascular Control of Hemostasis. Amsterdam: Harwood Academic Publishers, 1996, pp. 207–226.
19. F Bachmann. The enigma PAI-2: gene expression, evolutionary and functional aspects. Thromb Haemost 74:172–179, 1995.
20. ED Sprengers, C Kluft. Plasminogen activator inhibitors. Blood 69:381–387, 1987.
21. A Wohlwend, D Belin, JD Vassalli. Plasminogen activator-specific inhibitors produced by human monocytes/macrophages. J Exp Med 165:320–339, 1987.
22. LA Miles, EF Plow. Plasminogen receptors: ubiquitous sites for cellular regulation of fibrinolysis. Fibrinolysis 2:61–71, 1988.
23. EF Plow, T Herren, A Redlitz, LA Miles, JL Hoover-Plow. The cell biology of the plasminogen system. FASEB J 9:939–945, 1995.
24. KA Hajjar. Cellular receptors in the regulation of plasmin generation. Thromb Haemostas 74:294–301, 1995.
25. KA Hajjar. The endothelial cell tissue plasminogen activator receptor—specific interaction with plasminogen. J Biol Chem 266:21962–21970, 1991.
26. Y Wei, DA Waltz, N Rao, RJ Drummond, S Rosenberg, HA Chapman. Identification of the urokinase receptor as cell adhesion receptor for vitronectin. J Biol Chem 269:32380–32388, 1994.
27. F Blasi, M Conese, LB Møller, N Pedersen, U Cavallaro, MV Cubellis, F Fazioli, L Hernandez-Marrero, P Limongi, P Munoz-Canoves, M Resnati, L Riittinen, N Sidenius, E Soravia, MR Soria, MP Stoppelli, D Talarico, T Teesalu, S Valcamonica. The urokinase receptor: structure, regulation and inhibitor-mediated internalization. Fibrinolysis 8:182–188, 1994.
28. K Danø, N Behrendt, N Brunner, V Ellis, M Ploug, C Pyke. The urokinase receptor: protein structure and role in plasminogen activation and cancer invasion. Fibrinolysis 8:189–203, 1994.
29. KA Hajjar, RL Nachman. Endothelial cell-mediated conversion of Glu-plasminogen to Lys-plasminogen: further evidence for assembly of the fibrinolytic system on the endothelial cell surface. J Clin Invest 82:1769–1778, 1988.
30. LA Miles, CM Dahlberg, EG Levin, EF Plow. Gangliosides interact directly with plasminogen and urokinase and may mediate binding of these fibrinolytic components to cells. Biochemistry 28:9337–9343, 1989.
31. RL Nachman. Review: Stratton Lecture. Thrombosis and atherogenesis: molecular connections. Blood 79:1897–1906, 1992.
32. LA Miles, GM Fless, EG Levin, AM Scanu, EF Plow. A potential basis for the thrombotic risks associated with lipoprotein(a). Nature 399:301–303, 1989.
33. KA Hajjar, NM Hamel, PC Harpel, RL Nachman. Binding of tissue plasminogen activator to cultured human endothelial cells. J Clin Invest 80:1712–1719, 1987.
34. DP Beebe. Binding of tissue plaminogen activators to human umbilical vein endothelial cells. Thromb Res 46:241–254, 1987.
35. ES Barnathan, A Kuo, H Van der Keyl, KR McCrae, GR Larsen, DB Cines. Tissue-type plasminogen activator binding to human endothelial cells: evidence for two distinct binding sites. J Biol Chem 263:7792–7799, 1988.
36. M Mulder, U Kohnert, S Fischer, VWM van Hinsbergh, JH Verheijen. The interaction of recombinant tissue type plasminogen activator and recombinant plasminogen activator (r-PA/BM 06.022) with human endothelial cells. Blood Coagul Fibrinolysis 8:124–133, 1997.

37. KA Hajjar and CM Reynolds. α-Fucose-mediated binding and degradation of tissue-type plasminogen activator by HepG2 cells. J Clin Invest 93:703–710, 1994.
38. N Behrendt, E Rønne, M Ploug, T Petri, D Lober, LS Nielsen, WD Schleuning, F Blasi, E Appella, K Danø. The human receptor for urokinase plasminogen activator—NH2-terminal amino acid sequence and glycosylation variants. J Biol Chem 265:6453–6460, 1990.
39. H Solberg, D Lober, J Eriksen, M Ploug, E Rønne, N Behrendt, K Danø, G Høyer-Hansen. Identification and characterization of the murine cell surface receptor for the urokinase-type plasminogen activator. Eur J Biochem 205:451–458, 1992.
40. U Zacharias, W Handschack, F Schneider, K Loster, C Kleitke, F Noll, H Will. Inhibition of urokinase activity and prevention of urokinase receptor binding by monoclonal antibodies. J Immunol Methods 130:81–90, 1990.
41. A Nykjaer, B Møler, RF Todd, T Christensen, PA Andreasen, J Gliemann, CM Petersen. Urokinase receptor–An activation antigen in human T lymphocytes. J Immunol 152:505–516, 1994.
42. A Estreicher, J Mühlhauser, JL Carpentier, L Orci, J-D Vassalli. The receptor for urokinase type plasminogen polarizes expression of the protease to the leading edge of migarting monocytes and promotes degradation of enzyme inhibitor complexes. J Cell Biol 111:783–792, 1990.
43. M Gonzalez-Gronow, G Gawdi, SV Pizzo. Plasminogen activation stimulates an increase in intracellular calcium in human synovial fibroblasts. J Biol Chem 268:20791–20795, 1993.
44. ES Barnathan. Characterization and regulation of the urokinase receptor of human endothelial cells. Fibrinolysis 6:1–9, 1992.
45. E Appella, EA Robinson, SJ Ullrich, MP Stoppelli, A Corti, G Cassani, F Blasi. The receptor-binding sequence of urokinase: a biological function for the growth-factor module of proteases. J Biol Chem 262:4437–4440, 1987.
46. M Ploug, N Behrendt, D Lober, K Danø. Protein structure and membrane anchorage of the cellular receptor for urokinase-type plasminogen activator. Semin Thromb Haemost 17:183–193, 1991.
47. RC Haddock, ML Spell, CD Baker, JR Grammer, JM Parks, M Speidel, FM Booyse. Urokinase binding and receptor identification in cultured endothelial cells. J Biol Chem 266: 21466–21473, 1991.
48. P Koolwijk, MGM Van Erck, WJA De Vree, MA Vermeer, HA Weich, R Hanemaaijer, VWM Van Hinsbergh. Cooperative effect of TNFα, bFGF, and VEGF on the formation of tubular structures of human microvascular endothelial cells in a fibrin matrix: role of urokinase activity. J Cell Biol 132:1177–1188, 1996.
49. D Olson, J Pöllönen, G Høyer-Hansen, E Rønne, K Sakaguchi, TC Wun, E Appella, K Danø, F Blasi, Internalization of the urokinase-plasminogen activator inhibitor type-1 complex is mediated by the urokinase receptor. J Biol Chem 267:9129–9133, 1992.
50. A Nykjaer, M Conese, EI Christensen, D Olson, O Cremona, J Gliemann, F Blasi. Recycling of the urokinase receptor upon internalization of the uPA:serpin complexes. EMBO J 16: 2610–2620, 1997.
51. A Nykjaer, CM Petersen, B Møller, PH Jensen, SK Moestrup, TL Holtet, M Etzerodt, HC Thøgersen, M Munch, PA Andreasen, J Gliemann. Purified α2-macroglobulin receptor/LDL receptor-related protein binds urokinase/plasminogen activator inhibitor type-1 complex—evidence that the α2-macroglobulin receptor mediates cellular degradation of urokinase receptor-bound complexes. J Biol Chem 267:14543–14546, 1992.
52. CW Heegaard, ACW Simonsen, K Oka, L Kjoller, A Christensen, B Madsen, L Ellgaard, L Chan, PA Andreasen. Very low density lipoprotein receptor binds and mediates endocytosis of urokinase-type plasminogen activator-type-1 plasminogen activator inhibitor complex. J Biol Chem 270:20855–20861, 1995.
53. KL Wyne, RK Pathak, MC Seabra, HH Hobbs. Expression of the VLDL receptor in endothelial cells. Arterioscler Thromb Vasc Biol 16:407–415, 1996.

54. A Nykjaer, EI Christensen, H Vorum, H Hager, CM Petersen, H Roigaard, HY Min, F Vilhardt, LB Møller, S Kornfeld, J Gliemann. Mannose 6-phosphate/insulin-like growth factor-II receptor targets the urokinase receptor to lysosomes via a novel binding interaction. J Cell Biol 141:815–828, 1998.
55. G Bu, S Williams, DK Strickland, AL Schwartz. Low density lipoprotein receptor-related protein/α2-macroglobulin receptor is an hepatic receptor for tissue-type plasminogen activator. Proc Natl Acad Sci USA 89:7427–7431, 1992.
56. K Orth, EL Madison, MJ Gething, JF Sambrook, J Herz. Complexes of tissue-type plasminogen activator and its serpin inhibitor plasminogen-activator inhibitor type-1 are internalized by means of the low density lipoprotein receptor–related protein/α2-macroglobulin receptor. Proc Natl Acad Sci USA 89:7422–7426, 1992.
57. K Orth, T Willnow, J Herz, MJ Gething, JF Sambrook. Low density lipoprotein receptor-related protein is necessary for the internalization of both tissue-type plasminogen activator-inhibitor complexes and free tissue-type plasminogen activator. J Biol Chem 269:21117–21122, 1994.
58. GJ Bu, I Warshawsky, AI Schwartz. Cellular receptors for the plasminogen activators. Blood 83:3427–3436, 1994.
59. M Otter, MM Barrett-Bergshoeff, DC Rijken. Binding of tissue-type plasminogen activator by the mannose receptor. J Biol Chem 266:13931–13935, 1991.
60. G Murphy and AJP Docherty. The matrix metalloproteinases and their inhibitors. Am J Respir Cell Mol Biol 7:120–125, 1992.
61. H Nagase, K Suzuki, JJ Enghild, G Salvesen. Stepwise activation mechanisms of the precursors of matrix metalloproteinases-1 (tissue collagenase) and 3 (stromelysin). Biomed Biochim Acta 50:749–754, 1991.
62. AY Strongin, IE Collier, PA Krasnov, LT Genrich, BL Marmer, GI Goldberg. Human 92 kDa type-IV collagenase—functional analysis of fibronectin and carboxyl-end domains. Kidney Int 43:158–162, 1993.
63. JF Santibanez and J Martinez. Membrane-associated procollagenase of leukemic cells is activated by urokinase-type plasminogen activator. Leuk Res 17:1057–1062, 1993.
64. Y Okumura, H Sato, M Seiki, H Kido. Proteolytic activation of the precursor of membrane type 1 matrix metalloproteinase by human plasmin—a possible cell surface activator. FEBS Lett 402:181–184, 1997.
65. P Carmeliet, L Moons, R Lijnen, M Baes, V Lemaitre, P Tipping, A Drew, Y Eeckhout, S Shapiro, F Lupu, D Collen. Urokinase-generated plasmin activates matrix metalloproteinases during aneurysm formation. Nature Genet 17:439–444, 1997.
66. V Knauper, H Will, C Lopez-Otin, B Smith, SJ Atkinson, H Stanton, RM Hembry, G Murphy. Cellular mechanisms for human procollagenase-G (MMP-13) activation—evidence that MT1-MMP (MMP-14) and gelatinase A (MMP-2) are able to generate active enzyme. J Biol Chem 271:17124–17131, 1996.
67. JP Quigley, LI Gold, R Schwimmer, LM Sullivan. Limited cleavage of cellular fibronectin by plasminogen activator purified from transformed cells. Proc Natl Acad Sci USA 84:2776–2780, 1987.
68. MR Gyetko, RF Todd III, CC Wilkinson, RG Sitrin. The urokinase receptor is required for human monocyte chemotaxis in vitro. J Clin Invest 93:1380–1387, 1994.
69. RG Sitrin, RF Todd III, E Albrecht, MR Gyetko. The urokinase receptor (CD87) facilitates CD11b/CD18-mediated adhesion of human monocytes. J Clin Invest 97:1942–1951, 1996.
70. AL Kindzelskii, ZO Laska, RF Todd III, HR Petty. Urokinase-type plasminogen activator receptor reversibly dissociates from complement receptor type 3 (αM beta 2′ CD11b/CD18) during neutrophil polarization. J Immunol 156:297–309, 1996.
71. NK Rao, G-P Shi, HA Chapman. Urokinase receptor is a multifunctional protein: influence of receptor occupancy on macrophage gene expression. J Clin Invest 96:465–474, 1995.

72. RG Sitrin, SB Shoilenberger, RM Strieter, MR Gyetko. Endogenously produced urokinase amplifies tumor necrosis factor-α secretion by THP-1 mononuclear phagocytes. J Leukoc Biol 59:302–311, 1996.
73. E Anichini, G Fibbi, M Pucci, R Caldini, M Chevanne, M Delrosso. Production of second messengers following chemotactic and mitogenic urokinase-receptor interaction in human fibroblasts and mouse fibroblasts transfected with human urokinase receptor. Exp Cell Res 213:438–448, 1994.
74. I Dumler, T Petri, WD Schleuning. Interaction of urokinase-type plasminogen activator (u-PA) with its cellular receptor (u-PAR) induces phosphorylation on tyrosine of a 38 kDa protein. FEBS Lett 322:37–40, 1993.
75. I Dumler, A Weis, OA Mayboroda, C Maasch, U Jerke, H Haller, DC Gulba. The Jak/Stat pathway and urokinase receptor signaling inhuman aortic vascular smooth muscle cells. J Biol Chem 273:315–321, 1998.
76. M Konakova, F Hucho, WD Schleuning. Downstream targets of urokinase-type plasminogen-activator-mediated signal transduction. Eur J Biochem 253:421–429, 1998.
77. DHD Nguyen, IM Hussaini, SL Gonias. Binding of urokinase-type plasminogen activator to its receptor in MCF-7 cells activates extracellular signal-regulated kinase 1 and 2 which is required for increased cellular motility. J Biol Chem 273:8502–8507, 1998.
78. I Dumler, T Petri, WD Schleuning. Induction of c-fos gene expression by urokinase-type plasminogen activator in human ovarian cancer cells. FEBS Lett 343:103–106, 1994.
79. AAR Higazi, A Mazar, JY Wang, N Quan, R Griffin, R Reilly, J Henkin, DB Cines. Soluble human urokinase receptor is composed of two active units. J Biol Chem 272:5348–5353, 1997.
80. F Fazioli, M Resnati, N Sidenius, Y Higashimoto, E Appella, F Blasi. A urokinase-sensitive region of the human urokinase receptor is responsible for its chemotactic activity. EMBO J 16:7279–7286, 1997.
81. J Pöllänen, K Hedman, LS Nielsen, K Danø, A Vaheri. Ultrastructural localization of plasma membrane-associated urokinase-type plasminogen activator at focal contacts. J Cell Biol 106: 87–95, 1988.
82. K Takahashi, K Ikeo, T Gojobori, M Tanifuji. Local function of urokinase receptor at the adhesion contact sites of a metastatic tumor cell. Thromb Res 55–61, 1990.
83. G Conforti, C Dominguez-Jimenez, E Rønne, G Høyer-Hansen, E Dejana. Cell-surface plasminogen activation causes a retraction of in vitro cultured human umbilical vein endothelial cell monolayer. Blood 83:994–1005, 1994.
84. J Bohuslav, V Horejsi, C Hansmann, J Stöckl, UH Weidle, O Majdic, I Bartke, W Knapp, H Stockinger. Urokinase plasminogen activator receptor, β2-integrins, and Src-kinases within a single receptor complex of human monocytes. J Exp Med 181:1381–1390, 1995.
85. DI Simon, NK Rao, H Xu, Y Wei, O Majdic, E Rønne, L Kobzik, HA Chapman. Mac-1 (CD11b/CD18) and the urokinase receptor (CD87) form a functional unit on monocytic cells. Blood 88:3185–3194, 1996.
86. Y Wei, M Lukashev, DI Simon, SC Bodary, S Rosenberg, MV Doyle, HA Chapman. Regulation of integrin function by the urokinase receptor. Science 273:1551–1555, 1996.
87. SM Kanse, C Kost, OG Wilhelm, PA Andreasen, KT Preissner. The urokinase receptor is a major vitronectin-binding protein on endothelial cells. Exp Cell Res 224:344–353, 1996.
88. VWM Van Hinsbergh, EA Van den Berg, W Fiers, G Dooijewaard. Tumor necrosis factor induces the production of urokinase-type plasminogen activator by human endothelial cells. Blood 75:1991–1998, 1990.
89. MJ Niedbala, MS Stein-Picarella. Tumor necrosis factor induction of endothelial cell urokinase-type plasminogen activator mediated proteolysis of extracellular matrix and its antagonism by γ-Interferon. Blood 79:678–687, 1992.
90. J Wojta, H Zoellner, M Gallicchio, EL Filonzi, JA Hamilton, K McGrath. Interferon-α2 counteracts interleukin-1 α-stimulated expression of urokinase-type plasminogen activator

in human foreskin microvascular endothelial cells in vitro. Lymphokine Cytokine Res 13: 133–138, 1994.

91. J Grøndahl-Hansen, LT Kirkeby, E Ralfkiaer, P Kristensen, LR Lund, K Danø. Urokinase-type plasminogen activator in endothelial cells during acute inflammation of the appendix. Am J Pathol 135:631–636, 1989.
92. Z Szekanecz, GK Haines, AE Koch. Differential expression of the urokinase receptor (CD87) in arthritic and normal synovial tissues. J Clin Pathol 50:314–319, 1997.
93. YL Xu, J Hagege, J Doublet, P Callard, J Sraer, E Rønne, E Rondeau. Endothelial and macrophage upregulation of urokinase receptor expression in human renal cell carcinoma. Hum Pathol 28:206–213, 1997.
94. RR Schleef, MP Bevilacqua, M Sawdey, MA Gimbrone, DJ Loskutoff. Cytokine activation of vascular endothelium: effect on tissue-type plasminogen activator and type 1 plasminogen activator inhibitor. J Biol Chem 263:5797–5803, 1988.
95. M Colucci, JA Paramo, D Collen. Generation in plasma of a fast-acting inhibitor of plasminogen activator in response to endotoxin stimulation. J Clin Invest 75:818–824, 1985.
96. JJ Emeis and T Kooistra. Interleukin 1 and lipopolysaccharide induce an inhibitor of tissue-type plasminogen activator in vivo and in cultured endothelial cells. J Exp Med 163:1260–1266, 1986.
97. VWM Van Hinsbergh, T Kooistra, EA Van den Berg, HM Princen, W Fiers, JJ Emeis. Tumor necrosis factor increases the production of plasminogen activator inhibitor in human endothelial cells in vitro and in rats in vivo. Blood 72:1467–1473, 1988.
98. VWM Van Hinsbergh, M Vermeer, P Koolwijk, J Grimbergen, T Kooistra. Genistein reduces tumor necrosis factor α-induced plasminogen activator inhibitor-1 transcription but not urokinase expression in human endothelial cells. Blood 84:2984–2991, 1994.
99. PHA Quax, CM Van den Hoogen, JH Verheijen, T Padro, R Zeheb, TD Gelehrter, TJC Van Berkel, J Kuiper, JJ Emeis. Endotoxin induction of plasminogen activator and plasminogen activator inhibitor type 1 mRNA in rat tissues in vivo. J Biol Chem 265:15560–15563, 1990.
100. M Kccton, Y Eguchi, M Sawdey, C Ahn, DJ Loskutoff. Cellular localization of type 1 plasminogen activator inhibitor messenger RNA and protein in murine renal tissue. Am J Pathol 142:59–70, 1993.
101. O Saksela, D Moscatelli, DB Rifkin. The opposing effects of basic fibroblast growth factor and transforming growth factor β on the regulation of plasminogen activator activity in capillary endothelial cells. J Cell Biol 105:957–963, 1987.
102. MS Pepper, N Ferrara, L Orci, R Montesano. Vascular endothelial growth factor (VEGF) induces plasminogen activators and plasminogen activator inhibitor-1 in microvascular endothelial cells. Biochem Biophys Res Commun 181:902–906, 1991.
103. MS Pepper, AP Sappino, R Stocklin, R Montesano, L Orci, J-D Vassalli. Upregulation of urokinase receptor expression on migrating endothelial cells. J Cell Biol 122:673–684, 1993.
104. M Bastaki, EE Nelli, P Dell'Era, M Rusnati, MP Molinari-Tosatti, S Parolini, R Auerbach, LP Ruco, L Possati, M Presta. Basic fibroblast growth factor-induced angiogenic phenotype in mouse endothelium—a study of aortic and microvascular endothelial cell lines. Arterioscler Thromb Vasc Biol 17:454–464, 1997.
105. A Bikfalvi, C Sauzeau, H Moukadiri, J Maclouf, N Busso, M Bryckaert, J Plouët, G Tobelem. Interaction of vasculotropin/vascular endothelial cell growth factor with human umbilical vein endothelial cells: binding, internalization, degradation, and biological effects. J Cell Physiol 149:50–59, 1991.
106. T Kaji, S Hiraga, S Ohkawara, M Inada, C Yamamoto, H Kozuka, F Koizumi. Regulation by basic fibroblast growth factor of glycosaminoglycan biosynthesis in cultured vascular endothelial cells. Microvasc Res 49:268–276, 1995.
107. JC Kirchheimer, Y-H Nong, HG Remold. IFN-gamma, tumor necrosis factor-α, and urokinase regulate the expression of urokinase receptors on human monocytes. J Immunol 141: 4229–4234, 1988.

108. RG Sitrin, RF Todd III, IF Mizukami, TJ Gross, SB Shollenberger, MR Gyetko. Cytokine-specific regulation of urokinase receptor (CD87) expression by U937 mononuclear phagocytes. Blood 84:1286–1275, 1994.
109. P Mignatti, R Mazzieri, DB Rifkin. Expression of the urokinase receptor in vascular endothelial cells is stimulated by basic fibroblast growth factor. J Cell Biol 113:1193–1201, 1991.
110. SJ Mandriota, G Seghezzi, J-D Vassalli, N Ferrara, S Wasi, R Mazzieri, P Mignatti, MS Pepper. Vascular endothelial growth factor increases urokinase receptor expression in vascular endothelial cells. J Biol Chem 270:9709–9716, 1995.
111. DJ Langer, A Kuo, K Karikó, M Ahuja, BD Klugherz, KM Ivanics, JA Hoxie, WV Williams, BT Liang, DB Cines, ES Barnathan. Regulation of the endothelial cell urokinase-type plasminogen activator receptor—evidence for cyclic AMP-dependent and protein kinase-C dependent pathways. Circ Res 72:330–340, 1993.
112. M Presta, JAM Maier, G Ragnotti. The mitogenic signalling pathway but not the plasminogen activator-inducing pathway of basic fibroblast growth factor is mediated through protein kinase C in fetal bovine aortic endothelial cells. J Cell Biol 109:1877–1884, 1989.
113. K Danø, PA Andreasen, J Grøndahl-Hansen, P Kristensen, LS Nielsen, L Skriver. Plasminogen activators, tissue degradation, and cancer. Adv Cancer Res 44:139–266, 1985.
114. CA Hebert, JB Baker. Linkage of extracellular plasminogen activator to fibroblast cytoskeleton: colocalization of cell surface urokinase with vinculin. J Cell Biol 105:1241–1247, 1988.
115. F Blasi. Urokinase and urokinase receptor—a paracrine/autocrine system regulating cell migration and invasiveness. Bioessays 15:105–111, 1993.
116. RR Schleef, CR Birdwell. The effect of proteases on endothelial cell migration in vitro. Exp Cell Res 141:503–508, 1982.
117. P Quax, P Koolwijk, JH Verheijen, VWM Van Hinsbergh. The role of plasminogen activators in vascular pericellular proteolysis. In: VWM Van Hinsbergh, ed. Vascular Control of Hemostasis. Amsterdam: Harwood Academic Publishers, 1996, pp. 227–245.
118. AW Clowes, MM Clowes, YP Au, MA Reidy, D Belin. Smooth muscle cells express urokinase during mitogenesis and tissue-type plasminogen activator during migration in injured rat carotid artery. Circ Res 67:61–67, 1990.
119. P Carmeliet, D Collen. Evaluation of the plasminogen/plasmin system in transgenic mice. Fibrinolysis 8:269–276, 1994.
120. VA Ploplis, P Carmeliet, S Vazirzadeh, I Van Vlaenderen, L Moons, EF Plow, D Collen. Effects of disruption of the plasminogen gene on thrombosis, growth, and health in mice. Circulation 92:2585–2593, 1995.
121. TH Bugge, TT Suh, MJ Flick, CC Daugherty, J Romer, H Solberg, V Ellis, K Danø JL Degen. The receptor for urokinase-type plasminogen activator is not essential for mouse development or fertility. J Biol Chem 270:16886–16894, 1995.
122. MS Pepper, J-D Vassalli, R Montesano, L Orci. Urokinase-type plasminogen activator is induced in migrating capillary endothelial cells. J Cell Biol 105:2535–2541, 1987.
123. Y Sato and DB Rifkin. Autocrine activities of basic fibroblast growth factor: Regulation of endothelial cell movement, plasminogen activator synthesis, and DNA synthesis. J Cell Biol 107:1199–1205, 1988.
124. A Gualandris and M Presta. Transcriptional and posttranscriptional regulation of urokinase-type plasminogen activator expression in endothelial cells by basic fibroblast growth factor. J Cell Physiol 162:400–409, 1995.
125. HF Dvorak, JA Nagy, B Berse, LF Brown, KT Yeo, TK Yeo, AM Dvorak, L Van De Water, TM Sioussat, DR Senger. Vascular permeability factor, fibrin, and the pathogenesis of tumor stroma formation. Ann NY Acad Sci 667:101–111, 1992.
126. SJ Leibovich, PJ Polverini, HM Shepard, DM Wiseman, V Shively, N Nuseir. Macrophage-induced angiogenesis is mediated by tumour necrosis factor-α. Nature 329:630, 1987
127. M Fràter-Schröder, W Risau, R Hallman, P Gautschi, P Böhlen. Tumor necrosis factor

type α, a potent inhibitor of endothelial cell growth in vitro, is angiogenic in vivo. Proc Natl Acad Sci USA 84:5277–5281, 1987.

128. AE Koch, PJ Polverini, SL Kunkel, LA Harlow, LA DiPietro, VM Elner, SG Elner, RM Strieter. Interleukin-8 as a macrophage-derived mediator of angiogenesis. Science 258:1798–1801, 1992.
129. HC Kwaan. Tissue fibrinolytic activity studied by a histochemical method. Fed Proc 25:52–56, 1966.
130. R Montesano. Regulation of angiogenesis Invitro. Eur J Clin Invest 22:504–515, 1992.
131. MS Pepper, D Belin, R Montesano, L Orci, J-D Vassalli. Transforming growth factor-β-1 modulates basic fibroblast growth factor induced proteolytic and angiogenic properties of endothelial cells invitro. J Cell Biol 111:743–755, 1990.
132. MS Pepper, N Ferrara, L Orci, R Montesano. Potent synergism between vascular endothelial growth factor and basic fibroblast growth factor in the induction of angiogenesis invitro. Biochem Biophys Res Commun 189:824–831, 1992.
133. A Collen, P Koolwijk, ME Kroon, VWM Van Hinsbergh. Influence of fibrin structure on the formation and maintenance of capillary-like tubules by human microvascular endothelial cells. Angiogenesis 2:153–165, 1998.
133a. ME Kroon, P Koolwijk, H van Goor, UH Weidle, A Collen, G van der Phim, VWM van Hinsbergh. Role and localization of urokinase receptor in the formation of new microvascular structures in fibrin matrices. Am J Pathol 154:1731–1742, 1999.
134. JA Thompson, KD Anderson, JM DiPietro, JA Zwiebel, M Zametta, WF Anderson, T Maciag. Site-directed neovessel formation in vivo. Science 241:1349–1352, 1988.
135. KN Broadley, AM Aquino, SC Woodward, A Buckley-Sturrock, Y Sato, DB Rifkin, JM Davidson. Monospecific antibodies implicate basic fibroblast growth factor in normal wound repair. J Clin Invest 61:571–575, 1989.
136. HY Min, LV Doyle, CR Vitt, CL Zandonella, JR Stratton-Thomas, MA Shuman, S Rosenberg. Urokinase receptor antagonists inhibit angiogenesis and primary tumor growth in syngeneic mice. Cancer Res 56:2428–2433, 1996.
137. HF Dvorak, S Harvey, P Estrella, LF Brown, J McDonagh, AM Dvorak. Fibrin containing gels induce angiogenesis: implications for tumor stroma generation and wound healing. Lab Invest 57:673–686, 1987.
138. WD Thompson, EB Smith, CM Stirk, FI Marshall, AJ Stout, A Kocchar. Angiogenic activity of fibrin degradation products is located in fibrin fragment-E. J Pathol 168:47–53, 1992.
139. CM Stirk, A Kochhar, EB Smith, WD Thompson. Presence of growth-stimulating fibrin degradation products containing fragment E in human atherosclerotic plaques. Atherosclerosis 103:159–169, 1993.
140. V Nehls and R Herrmann. The configuration of fibrin clots determines capillary morphogenesis and endothelial cell migration. Microvasc Res 51:347–364, 1996.
141. E Dejana, MG Lampugnani, M Giorgi, M Gaboli, PC Marchisio. Fibrinogen induces endothelial cell adhesion and spreading via the release of endogenous matrix proteins and the recruitment of more than one integrin receptor. Blood 75:1509–1517, 1990.
142. P Thiagarajan, AJ Rippon, DH Farrell. Alternative adhesion sites in human fibrinogen for vascular endothelial cells. Biochemistry 35:4169–4175, 1996.
143. LA Bunce, LA Sporn, CW Francis. Endothelial cell spreading on fibrin requires fibrinopeptide B cleavage and amino acid residues 15-42 of the β chain. J Clin Invest 89:842–850, 1992.
144. JK Erban and DD Wagner. A 130-kDa protein on endothelial cells binds to amino acids 15-42 of the B β chain of fibrinogen. J Biol Chem 267:245–2458, 1992.
145. PC Brooks, RAF Clark, DA Cheresh. Requirement of Vascular Integrin $\alpha_v\beta_3$ for angiogenesis. Science 264:569–571, 1994.
146. PC Brooks, AMP Montgomery, M Rosenfeld, RA Reisfeld, TH Hu, G Klier, DA Cheresh.

Integrin $\alpha_v\beta_3$ antagonists promote tumor regression by inducing apoptosis of angiogenic blood vessels. Cell 79:1157–1164, 1994.

147. HP Hammes, M Brownlee, A Jonczyk, A Sutter, KT Preissner. Subcutaneous injection of a cyclic peptide antagonist of vitronectin receptor-type integrins inhibits retinal neovascularization. Nature Med 2:529–533, 1996.
148. M Friedlander, PC Brooks, RW Shaffer, CM Kincaid, JA Varner, DA Cheresh. Definition of two angiogenic pathways by distinct alpha(v) integrins. Science 270:1500–1502, 1995.
149. MC Chang, BR Wang, TF Huang. Characterization of endothelial cell differential attachment to fibrin and fibrinogen and its inhibition by Arg-Gly-Asp-containing peptides. Thromb Haemost 74:764–769, 1995.

18
The Role of E-Selectin in Tumor Angiogenesis and Metastasis

Jing-Liang Wang, James S. Tomlinson, and Mai Nguyen
University of California, Los Angeles, Los Angeles, California

I. INTRODUCTION

The selectins are a family of adhesion molecules consisting of three members: E-, P-, and L-selectins. These molecules share a common structural domain organization and similar carbohydrate recognition properties. There are five discrete domains: a lectin-binding domain, followed by an epidermal growth factor (EGF)–type domain, a variable number of consensus repeats, a transmembrane region, and a short cytoplasmic tail. The selectins bind with varying affinity to a series of carbohydrates including Lewis X (LeX), sialyl-Lewis X (sLeX), sialyl-Lewis A (sLeA), and phosphorylated or sulfated polysaccharides. E-selectin is expressed exclusively by endothelial cells (1). P-selectin is an intracellular glycoprotein found in the α-granules of platelets and in the Weibel-Palade bodies of endothelial cells (2). L-selectin is found in lymphocytes (3). This chapter concentrates on E-selectin and its potential roles in angiogenesis and metastasis. For information on the selectins in general, readers are referred to several recently published reviews (4–6).

E-selectin, also known as endothelial leukocyte adhesion molecule-1 (ELAM-1), is a 115-kDa glycoprotein (7). The molecule was cloned in the late 1980s (8,9). E-selectin expression is induced by interleukin-1 (IL-1), tumor necrosis factor (TNF), or lipopolysaccharide. It can bind to a wide variety of cells in the immune system including neutrophils, lymphocytes, monocytes, eosinophils, and basophils. Extensive evidence supports the hypothesis that E-selectin plays an important role in inflammatory processes by mediating the binding of leukocytes to the endothelium of inflamed sites (10).

A number of E-selectin ligands have recently been isolated from leukocytes of multiple species. A 150-kDa E-selection glycoprotein ligand, ESL-1, has been identified in a mouse neutrophil cell line (11). PSGL-1 (P-selectin glycoprotein ligand) on human neutrophils has also been proposed to bind to E-selectin (12). Some investigators have suggested that L-selectin (100 kDa) on neutrophils also function as ligands for E-selectin (13). In addition, a 250-kDa glycoprotein ligand, identified on bovine γ/δT cells, binds to E-selectin (14).

II. ROLE OF E-SELECTIN IN ANGIOGENESIS

Recently, certain in vitro and in vivo data have suggested that E-selectin may play a role in angiogenesis. Growth of new capillaries—i.e., angiogenesis—is a complex and highly regulated process that normally occurs during specific physiological events and during development. Abnormal capillary growth, however, underlies many disease states, such as diabetic retinopathy, rheumatoid arthritis, and hemangiomas (15). In addition, neovascularization of tumors occurs by an aberrant stimulation of normally quiescent endothelial cells to migrate, proliferate, and form new capillary blood vessels. To identify cell adhesion molecules required for angiogenesis, we used an in vitro model in which bovine capillary endothelial cells can be induced to form capillary-like tubes (16). Monoclonal antibodies directed against the carbohydrate epitopes sialyl Lewis-X and sialyl Lewis-A inhibited capillary formation. We postulated that a member of the selectin family of adhesion molecules may be involved in capillary formation because these proteins bind to sialyl Lewis-X/A–containing ligands. We isolated a 2.8-kb cDNA from a bovine capillary endothelial cell cDNA library that encodes a polypeptide with a 71% identity to human E-selectin. Antibodies directed against the bovine E-selectin inhibited capillary formation, suggesting that, in addition to its role in leukocyte adhesion to endothelium, a form of E-selectin is involved in capillary morphogenesis (17). Furthermore, structural analysis of the asparagine-linked oligosaccharides synthesized by bovine capillary endothelial cells induced to form capillary tubes revealed that these cells are enriched in sialylated, fucosylated glycans, providing biochemical evidence for the presence of potential ligands for E-selectin (18).

In 1995, Koch et al. (19) showed that human recombinant soluble E-selectin induced chemotaxis of human endothelial cells in vitro and was angiogenic in rat cornea. Soluble E-selectin acted on endothelial cells in part through a sialyl Lewis-X–dependent mechanism. The chemotactic activity of rheumatoid synovial fluid for endothelial cells as well as its angiogenic activity were blocked by antibodies to soluble E-selectin.

In contrast to the above supportive in vitro data, the studies of selectin knockout mice have raised questions about the postulated role of E-selectin in angiogenesis. Neither the E-selectin knockout nor the P/E-selectin knockout mice demonstrated any vascular or developmental abnormalities (20). Furthermore, endothelial cells isolated from these knockout mice were able to form capillary tubes in vitro (21). These mice did nonetheless exhibit deficiencies in leukocyte homeostasis and trafficking. There are some possible explanations for this phenomenon. First, compensatory molecules or redundant pathways may be upregulated in E-selectin-deficient mice. Since the P/E-selectin knockout mice are also viable, it appears that P-selectin does not compensate for E-selectin in angiogenesis. Second, E-selectin may not play a central role in murine angiogenesis. Wild-type mice express very low levels of E-selectin in the microvascular endothelium (21). Third, E-selectin may not play a crucial role in fetal and developmental angiogenesis but may still be important in adult angiogenesis, particularly in disease states where angiogenesis is undesirable. Although knockout mice are powerful scientific tools, there have been instances where they fail to recapitulate the functional significance of even established and well-studied molecules (22,23).

In humans, there exist in vivo data that support a role for E-selectin in angiogesis. The rare patients with leukocyte adhesion deficiency type 2 lack sLeX epitopes, which are carbohydrate ligands for E-selectin. These patients suffer from mental retardation, facial abnormalities, and dwarfism (24–26). Kraling et al. showed that normal human newborn foreskin and placenta stained positive for E-selectin, whereas adult skin did not, suggesting that E-selectin is preferentially expressed in angiogenic and/or proliferating tissues (27).

Immunohistochemical studies from different laboratories have lent support to the hypothesis that E-selectin is upregulated in the tumor vasculature. We have found that E-selectin's expression was increased in malignant breast tumors as opposed to their benign counterparts (28) (Fig. 1). Furthermore, E-selectin staining was found to be significantly increased in the estrogen receptor–negative carcinomas as opposed to the estrogen receptor–positive ones. The results indicated that endothelial E-selectin expression during angiogenesis is related to breast carcinoma progression in vivo. Other investigators have also observed similar increased expression of E-selectin in human tumor specimens. Kraling et al. showed that E-selectin is present in dividing endothelial cells in biopsy specimens of proliferating human hemangioma by immunohistochemical analysis of frozen sections (27). Also, Fox et al. (29) detected the presence of selectins in the human breast. These authors reported that 21% of benign breast specimens and 52% of breast tumor specimens stained positive for E-selectin. Charpin et al. further confirmed E-selectin expression in breast carcinomas (30). Also recently, Ye et al. (31) and Suzuki et al. (32) reported that E-selectin is present in vessels of colorectal cancer specimens. Groves et al. (33) and Staal-van den Brekel et al. (34) were also able to detect E-selectin in blood vessels of squamous cell carcinoma and in lung cancer, respectively.

Furthermore, E-selectin levels in the serum have been reported to be elevated in patients with cancer (35–39). In one study, soluble E-selectin levels in cancer patients were used as a marker of the therapeutic efficacy of CM-101, an antiangiogenic drug currently in clinical trials (40).

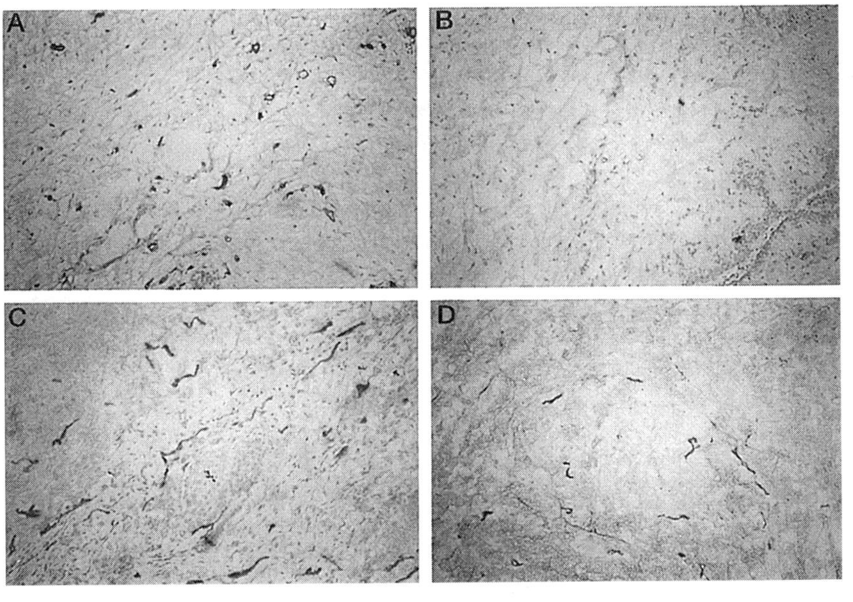

Figure 1 Immunohistochemical staining of (A) normal breast with vWf (von Willebrand factor) antibody; (B) normal breast with E-selectin antibody; (C) breast cancer with vWf antibody; and (D) breast cancer with E-selectin antibody. (Magnification ×280.)

III. ROLE OF E-SELECTIN IN METASTASIS

The process of tumor metastasis is complicated and involves multiple steps (41). One of the first steps of metastasis involves the adhesion of tumor cells to the endothelium. This interaction occurs as the tumor cells attempt to extravasate through the endothelial cells lining the tumor microvessels or those lining the lymphatic channels. In the bloodstream, the tumor cells must adhere to the luminal surface of the endothelial cells lining the blood vessels in order to again extravasate and seed metastatic colonies. Many investigators have hypothesized that selectin-ligand interactions play an important role in the adhesion processes between tumor cells and the endothelial cells (42). More specifically, this adhesion process involves the interaction between E-selectin on endothelial cells and its glycoconjugate ligands sialyl Lewis X (sLeX) and sialyl Lewis A (sLeA) on tumor cells (43).

Much indirect evidence supports the hypothesis that the glycoconjugate ligands sLeX and sLeA play a role in human cancer metastasis. It has been shown by immunohistochemistry that increased expression of sLeX and/or sLeA in human cancer specimens correlates with metastasis and poor survival. This observation applies to a number of gastrointestinal tumors, including colon, rectal, and gastric (44–48). In addition, the more malignant skin cancers, such as melanomas and squamous cell carcinomas, express more sLeX and/or sLeA than the less malignant basal cell carcinomas (49–52). A similar phenomenon has also been reported in breast cancer (53), prostate cancer (54), and lung cancer (55).

In in vitro studies, many researchers have shown that sLeX and/or sLeA glycoconjugates on the tumor cell surface bind to the E-selection molecules on the surface of endothelial cells (56). This phenomenon of E-selectin–ligand interaction appears to apply to the majority of gastrointestinal cancer cell lines tested, particularly those derived from colon tumors (57,58). The same phenomenon was observed in vitro in only a fraction of the melanoma and lung cancer cell lines. Furthermore, highly metastatic colonic as well as lung carcinoma cells, as opposed to their less metastatic counterparts, bind more efficiently to activated endothelial cells that express E-selection [59,60]. More recently, an in vivo study has provided direct evidence that expression of E-selection can direct metastasis of tumor cells expressing appropriate ligands (61). Two other in vivo studies reported that tumor metastasis can be suppressed with E-selectin inhibition (62,63).

Various E-selectin ligands have recently been identified from human colon cancer cells. Two lysosomal membrane glycoproteins, lamp-1 and lamp-2, have been shown to mediate the binding of colon cancer cells to E-selectin (64). Colon cancer cells are also known to express a high level of mucins. Three of these mucins, leukosialin, MUC-1, and MUC-2, are capable of binding E-selectin (65–67). We have identified a 400-kDa glycoprotein ligand for E-selectin from human colon cancer cells by using human E-selectin-lg affinity chromatography (68).

IV. CLINICAL IMPLICATIONS

E-selectin is one of the few adhesion molecules that is truly restricted to the endothelium. Therefore, the inflammatory and angiogenic functions of this molecule can be inhibited with antibodies (69,70) and adenoviral vectors with transcriptional specificity carried by the E-selectin promoter (71,72). Furthermore, the potential exists to target E-selectin in order to influence tumor metastasis. This goal may be achieved by competitive inhibition

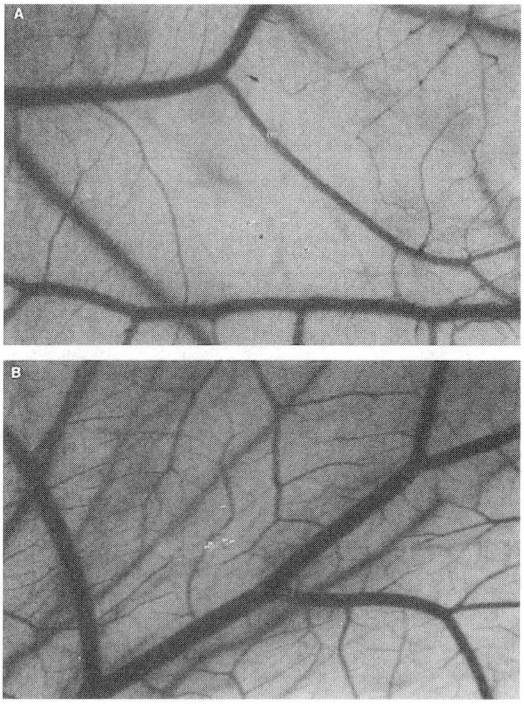

Figure 2 The chick chorioallantoic membrane with (A) a pellet containing 2 mg of the analogue of sialyl Lewis X CY-1503 and (B) a pellet containing 2 mg of the Lewis X carbohydrates.

of E-selection binding with soluble E-selectin (62); oligosaccharide analogues (73,74) (Fig. 2); and drugs that inhibit E-selectin binding, such as prostacyclins (75).

ACKNOWLEDGMENTS

We would like to thank Dr. Joyce Bischoff for insightful advice and Dr. Frederick Eilber for valuable guidance and support. Work performed in the author's laboratories have been supported by NIH CA69433, the American Cancer Society, the UCLA Jonsson Comprehensive Cancer Center, and the University of California Cancer Research Coordinating Committee. Dr. James Tomlinson is a recipient of NIH T32 CA09535-13.

REFERENCES

1. J Bischoff. Cell adhesion in vascular biology. J Clin Invest 99:373–376, 1997.
2. A Weller, S Isenmann, D Vestweber. Cloning of the mouse endothelial selectins. Expression of both E- and P-selectin is inducible by tumor necrosis factor alpha. J Biol Chem 267:15176–15183, 1992.
3. LM Stoolman, SD Rosen. Possible role for cell-surface carbohydrate-binding molecules in lymphocyte recirculation. J Cell Biol 96:722–729, 1983.

4. TF Tedder, DA Steeber, A Chen, P Engel. The selectins: vascular adhesion molecules. FASEB J 9:866–873, 1995.
5. GS Kansas. Selectins and their ligands: current concepts and controversies. Blood, 88:3259–3287, 1996.
6. PM Sass. The involvement of selectins in cell adhesion, tumor progression, and metastasis. Cancer Invest 16:322–328, 1998.
7. J Bischoff. Functions of E-selectin. Trends Glycosci Glycotech 6:351–365, 1994.
8. C Hession, L Osborn, D Goff, G Chi-Rosso, C Vassallo, M Pasek, C Pittaack, R Tizard, S Goelz, K McCarthy, S Hopple, R Lobb. Endothelial leukocyte adhesion molecule 1: direct expression cloning and functional interactions. Proc Natl Acad Sci USA 87:1673–1677, 1987.
9. MP Bevilacqua, S Stengelin, MA Gimbrone Jr, B Seed. Endothelial leukocyte adhesion molecule 1: an inducible receptor for neutrophils related to complement regulatory proteins and lectins. Science 243:1160–1164, 1989.
10. LA Lasky. Selectin-carbohydrate interactions and the initiation of the inflammatory response. Annu Rev Biochem 64:113–139, 1995.
11. M Steegmaler, A Levinovitz, S Isenmann, E Borges, M Lenter, HP Kocher, B Kleuser, D Vestweber. The E-selectin-ligand ESL-1 is a variant of a receptor for fibroblast growth factor. Nature 373:615–620, 1995.
12. DJ Goetz, DM Greif, H Ding, RT Camphausen, S Howes, KM Comess, KR Snapp, GS Kansas, FW Luscinskas. Isolated P-selectin glycoprotein ligand-1 dynamic adhesion to P- and E-selectin. J Cell Biol 137:509–519, 1997.
13. O Zollner, MC Lenter, JE Blanks, E Borges, M Steegmaler, H Zerwes, D Vestweber. L-Selectin from human, but not from mouse neutrophils, binds directly to E-selectin. J Cell Biol 136:707–716, 1997.
14. B Walcheck, G Watts, MA Jutila. Bovine (gamma/delta) T cells bind E-selectin via a novel glycoprotein receptor: first characterization of a lymphocyte/E-selectin interaction in an animal model. J Exp Med 178:853–863, 1993.
15. J Folkman. Angiogenesis in cancer, vascular, rheumatoid and other disease. Nature Med 1: 27–31, 1995.
16. D Ingber, J Folkman. Mechanochemical switching between growth and differentiation during fibroblast growth factor stimulated angiogenesis in vitro: role of extracellular matrix. J Cell Biol 109:317–330, 1989.
17. M Nguyen, NA Strubel, J Bischoff. A role for sialyl Lewis-X/A glycoconjugates in capillary morphogenesis. Nature 365:267–269, 1993.
18. M Nguyen, J Folkman, J Bischoff. 1-Deoxymannojirimycin inhibits capillary tube formation in vitro. J Biol Chem 267:26157–26165, 1992.
19. AE Koch, MM Halloran, CJ Haskell, MR Shah, PJ Polverini. Angiogenesis mediated by soluble forms of E-selection and vascular cell adhesion molecule-1. Nature 376:517–519, 1995.
20. PS Frenette, DD Wagner. Insights into selectin function from knockout mice. Thromb Haemostas 78:60–64, 1997.
21. ME Gerritsen, CP Shen, WJ Atkinson, RC Padgett, MA Gimbrone Jr, DS Milstone. Microvascular endothelial cells from E-selectin-deficient mice form tubes in vitro. Lab Invest 75:175–184, 1996.
22. S Jiralerspong, PI Patel. Regulation of the hypoxanthine phosphoribosyl transferase gene: in vitro and in vivo approaches. Proc Soc Exp Biol Med 212:116–127, 1996.
23. MA Blasco, HW Lee, MP Hande, E Samper, PM Lansdorp, RA DePinho, CW Greider. Telomere shortening and tumor formation by mouse cells lacking telomerase RNA. Cell 91:25–34, 1997.
24. A Etzioni, M Frydman, S Pollack, I Avidor, M So, ML Phillips, JO Paulson, R Gershoni-Baruch. Brief report: recurrent severe infections caused by a novel leukocyte adhesion deficiency. Res Hematol 327:1789–1792, 1992.

25. UH von Andrian, EM Berger, L Ramezani, JD Chambers, HD Ochs, JM Harlan, JC Paulson, A Etzioni, KE Arfors. In vivo behavior of neutrophils from two patients with distinct inherited leukocyte adhesion deficiency syndromes. J Clin Invest 91:2893–2897, 1993.
26. A Etzioni. Adhesion molecule deficiencies and their clinical significance. Cell Adh Comm 2: 257–260, 1994.
27. BM Kraling, MJ Razon, LM Boon, D Zurakowski, C Seachord, RP Darveau, JB Mulliken, CL Corless, J Bischoff. E-selectin is present in proliferating endothelial cells in human hemangiomas. Am J Pathol 148:1181–1191, 1996.
28. M Nguyen, CL Corless, BM Kraling, C Tran, T Atha, J Bischoff, SH Barsky. Vascular expression of E-selectin is increased in estrogen-receptor-negative breast cancer: a role for tumor-cell-secreted interleukin-1α. Am J Pathol 150:1307–1314, 1997.
29. SB Fox, DH Turner, KC Gatter, AL Harris. The increased expression of adhesion molecules ICAM-3, E- and P-selectins on breast cancer endothelium. J Pathol 177:369–376, 1995.
30. C Charpin, D Bergeret, S Garcia, L Andrac, F Martini, N Horschowski, R Choux, MN Lavaut. ELAM selectin expression in breast carcinomas detected by automated and quantitative immunohistochemical assays. Int J Oncol 12:1041–1048, 1998.
31. C Ye, K Kiriyama, C Mistuoka, R Kannagi, K Ito, T Watanabe, K Kondo, K Akiyama, H Takagi. Expression of E-selectin on endothelial cells of small veins in human colorectal cancer. Int J Cancer 61:455–460, 1995.
32. Y Suzuki, H Ohtani, T Mizoi, S Takeha, K Shiiba, S Matsuno, H Nagura. Cell adhesion molecule expression by vascular endothelial cells as an immune/inflammatory reaction in human colon carcinoma. Jpn J Cancer Res 86:585–593, 1995.
33. RW Groves, MH Allen, EL Ross, G Ahsan, JNWN Barker, DM MacDonald. Expression of selectin ligands by cutaneous squamous cell carcinoma. Am J Pathol 143:1220–1225, 1993.
34. AJ Staal-van den Brekel, FBJM Thunnissen, WA Buurman, EFM Wouters. Expression of E-selectin, intercellular adhesion molecule (ICAM)-1 and vascular cell adhesion molecule (VCAM)-1 in non-small-cell lung carcinoma. Virchows Arch 428:21–27, 1996.
35. RE Banks, AJH Gearing, IK Hemingway, DR Norfolk, TJ Perren, PJ Selby. Circulating intercellular adhesion molecule-1 (ICAM-1), E-selectin and vascular cell adhesion molecule-1 (VCAM-1) in human malignancies. Br J Cancer 68:122–124, 1993.
36. N Matsuura, T Narita, C Mitsuoka, N Kimura, R Kannagi, T Imai, H Funahashi, H Takagi. Increased concentration of soluble E-selectin in the sera of breast cancer patients. Anticancer Res 17:1367–1372, 1997.
37. G Velikova, RE Banks, A Gearing, I Hemingway, MA Forbes, SR Preston, M Jones, J Wyatt, K Miller, U Ward, J Al-Maskatti, SM Singh, NS Ambrose, JN Primrose, PJ Selby. Circulating soluble adhesion molecules E-cadherin, E-selectin, intercellular adhesion molecule-1 (ICAM-1) and vascular cell adhesion molecule-1 (VCAM-1) in patients with gastric cancer. Br J Cancer 76:1398–1404, 1997.
38. M Benekli, IH Gullu, G Tekuzman, MC Savas, M Hayran, G Hascelik, D Firat. Circulating intercellular adhesion molecule-1 and E-selectin levels in gastric cancer. Br J Cancer 78:267–271, 1998.
39. M Hebbar, F Revillion, MM Louchez, MO Vilain, C Fournier, J Bonneterre, JP Peyrat. The relationship between concentrations of circulating soluble E-selectin and clinical, pathological, and biological features in patients with breast cancer. Clin Cancer Res 5:373–380, 1998.
40. BD Wamil, GB Thurman, HW Sundell, RF DeVore, G Wakefield, DH Johnson, YF Wang, CG Hellerqvist. Soluble E-selectin in cancer patients as a marker of the therapeutic efficacy of CM101, a tumor-inhibiting antineovascularization agent, evaluated in phase I clinical trial. J Cancer Res Clin Oncol 123:173–179, 1997.
41. IJ Fidler. Molecular biology of cancer: Invasion and metastasis. In: VT DeVita Jr, S Hellman, SA Rosenberg, eds. Cancer: Principles and Practice of Oncology. Philadelphia: Lippincott-Raven, 1997, pp. 135–152.

42. KV Honn, DG Tang. Adhesion molecules and tumor cell interaction with endothelium and subendothelial matrix. Cancer Metast Rev 11:353–375, 1992.
43. M Fukuda. Possible roles of tumor-associated carbohydrate antigens. Cancer Res 56:2237–2244, 1996.
44. S Itai, J Nishikata, T Yoneda, K Ohmori, H Yamabe, S Arii, T Tobe, R Kannagi. Tissue distribution of 2-3 and 2-6 sialyl Lewis A antigens and significance of the ratio of two antigens for the differential diagnosis of malignant and benign disorders of the digestive tract. Cancer 67:1576–1587, 1991.
45. S Nakamori, M Kameyama, S Imaoka, H Furukawa, O Ishikawa, Y Sasaki, T Kabuto, T Iwanaga, Y Matsushita, T Irimura. Increased expression of sialyl Lewis x antigen correlates with poor survival in patients with colorectal carcinoma: clinicopathological and immunohistochemical study. Cancer Res 53:3632–3637, 1993.
46. T Nakayama, M Watanabe, T Katsumata, T Teramoto, T Kitajima. Expression of sialyl Lewis A as a new prognostic factor for patients with advanced colorectal carcinoma. Cancer 75: 2051–2056, 1995.
47. H Ura, R Denno, K Hirata, K Yamaguchi, T Yasoshima, T Shishido. Close correlation between increased sialyl-Lewis x expression and metastasis in human gastric carcinoma. World J Surg 21:773–776, 1997.
48. Y Ikeda, M Mori, T Kamakura, M Saku, K Sugimachi. Immunohistochemical expression of sialyl Tn and sialyl Lewis antigens in stromal tissue correlates with peritoneal dissemination in stage IV human gastric cancer. Eur J Surg Oncol 21:168–175, 1995.
49. RW Groves, MH Allen, EL Ross, G Ahsan, JNWNBarker, DM MacDonald. Expression of selectin ligands by cutaneous squamous cell carcinoma. Am J Pathol 143:1220–1225, 1993.
50. T Kageshita, SS Hirai, T Kimmura, N Hanai, S Ohta, T Ono. Association between sialyl Lewis[a] expression and tumor progression in melanoma. Cancer Res 55:1748–1751, 1995.
51. N Miller, RG Vile, IR Hart. Selectin ligands on human melanoma cells. Glycoconj J 13:33–43, 1996.
52. MH Ravindranath, AA Amiri, PM Bauer, MC Kelley, R Essner, DL Morton. Endothelial-selectin ligands sialyl Lewis X and sialyl Lewis A are differentiation antigens immunogenic in human melanoma. Cancer 79:1686–1697, 1997.
53. J Renkonen, T Paavonen, R Renkonen. Endothelial and epithelial expression of sialyl Lewis x and sialyl Lewis a in lesions of breast carcinoma. Int J Cancer 74:296–300, 1997.
54. S Martensson, SA Bigler, M Brown, PH Lange, MK Brawer, S Hakomori. Sialyl-Lewis x related carbohydrate antigens in the prostate. Hum Pathol 26:735–739, 1995.
55. J Ogawa, A Sano, H Inoue, S Koide. Expression of Lewis-related antigen and prognosis in stage I non-small-cell lung cancer. Ann Thorac Surg 59:412–415, 1995.
56. R Renkonen, P Mattila, ML Majuri, J Rabina, S Toppila, J Renkonen, L Hirvas, J Niittymaki, JP Turunen, O Renkonen, T Paavonen. In vitro experimental studies of sialyl Lewis X and sialyl Lewis A on endothelial and carcinoma cells: crucial glycans on selectin ligands. Glycoconj J 14:593–600, 1997.
57. D Lauri, L Needham, I Martin-Padura, E Dejana. Tumor cell adhesion to endothelial cells: endothelial leukocyte adhesion molecule-1 as an inducible adhesive receptor specific for colon carcinoma cells. J Natl Cancer Inst 83:1321–1324, 1991.
58. A Takada, K Ohmori, T Yoneda, K Tsuyuoka, A Hasegawa, M Kiso, R Kannagi. Contribution of carbohydrate antigens sialyl Lewis A and sialyl Lewis X to adhesion of human cancer cells to vascular endothelium. Cancer Res 53:354–361, 1993.
59. M Martin-Satue, R Marrugat, JA Cancelas, J Blanco. Enhanced expression of $\alpha(1,3)$-fucosyltransferase genes correlates with E-selectin–mediated adhesion and metastatic potential of human lung adenocarcinoma cells. Cancer Res 58:1544–1550, 1998.
60. R Sawada, S Tsuboi, M Fukuda. Differential E-selectin–dependent adhesion efficiency in sublines of a human colon cancer exhibiting distinct metastatic potentials. J Biol Chem 269:1425–1431, 1994.

61. L Biancone, M Araki, K Araki, P Vassalli, I Stamenkovic. Redirection of tumor metastasis by expression of E-selectin in vivo. J Exp Med 183:581–587, 1996.
62. G Mannori, D Santoro, L Carter, C Corless, RM Nelson, MP Bevilacqua. Inhibition of colon carcinoma cell lung colony formation by a soluble form of E-selectin. Am J Pathol 151:233–243, 1997.
63. P Brodt, L Fallavollita, RS Bresalier, S Meterissian, CR Norton, BA Wolitsky. Liver endothelial E-selectin mediates carcinoma cell adhesion and promotes liver metastasis. Int J Cancer 71:612–619, 1997.
64. R Sawada, JB Lowe, M Fukuda. E-selectin-dependent adhesion efficiency of colonic carcinoma cells is increased by genetic manipulation of their cell surface lysosomal membrane glycoprotein-1 expression levels. J Biol Chem 268:12675–12681, 1993.
65. D Baeckstrom, K Zhang, N Asker, U Ruetschi, M Ek, GC Hansson. Expression of the leukocyte-associated sialoglycoprotein CD43 by colon carcinoma cell line. J Biol Chem 270:13688–13692, 1995.
66. K Zhang, D Baeckstrom, H Brevinge, GC Hansson. Comparison of sialyl-Lewis a-carrying CD43 and MUC1 mucins secreted from a colon carcinoma cell line for E-selectin binding and inhibition of leukocyte adhesion. Tumor Biol 18:175–187, 1996.
67. G Mannori, P Crottet, O Cecconi, K Hanasaki, A Aruffo, RM Nelson, A Varki, MP Bevilacqua. Differential colon cancer cell adhesion to E-, P-, and L-selectin: role of mucin-type glycoproteins. Cancer Res 55:4425–4431, 1995.
68. J Tomlinson, JL Wang, T Walton, MC Lee, J Bischoff, M Nguyen. Identification of a glycoprotein ligand for E-selectin on human colon cancer cells. Manuscript submitted.
69. E Dejana, I Martin-Padura, D Lauri, S Bernasconi, MR Bani, A Garofalo, R Giavazzsi, J Magnani, A Mantovani, S Menard. Endothelial leukocyte adhesion molecule-1-dependent adhesion of colon carcinoma cells to vascular endothelium is inhibited by an antibody to Lewis fucosylated type I carbohydrate chain. Lab Invest 66:324–330, 1992.
70. U Srinivas, P Pahlsson, A Lundblad. E-selectin-sialyl Lewis a dependent adhesion of colon cancer cells, is inhibited differently by antibodies against E-selectin ligands. Scand J Immunol 44:197–203, 1996.
71. RT Jaggar, HY Chan, AL Harris, R Bicknell. Endothelial cell-specific expression of tumor necrosis factor-α from the KDR or E-selectin promoters following retroviral delivery. Hum Gene Ther 8:2239–2247, 1997.
72. T Walton, JL Wang, A Ribas, SH Barsky, J Economou, M Nguyen. Endothelium-specific expression of an E-selectin promoter recombinant adenoviral vector. Anticancer Res 18:1357–1360, 1998.
73. MS Mulligan, JB Lowe, RD Larsen, J Paulson, Z Zheng, S DeFrees, K Maemura, M Fukuda, PA Ward. Protective effects of sialylated oligosaccharides in immune complex-induced acute lung injury. J Exp Med 178:623–631, 1993.
74. M Nguyen, FR Eilber, S Defrees. Novel synthetic analogs of sialyl Lewis X can inhibit angiogenesis in vitro and in vivo. Biochem Biophys Res Comm 228:716–723, 1996.
75. GW Daneker, SA Lund, SW Caughman, CA Staley, WC Wood. Antimetastatic prostacyclins inhibit the adhesion of colon carcinoma to endothelial cells by blocking E-selectin expression. Clin Exp Metast 14:230–238, 1996.

19
In Vitro and In Vivo Models of Angiogenesis

Hynda K. Kleinman, Katherine M. Malinda, Yong Song Gho, and M. Lourdes Ponce
National Institutes of Health, Bethesda, Maryland

I. INTRODUCTION

The basal surfaces of endothelial cells are in contact with a basement membrane, the thin extracellular matrix that separates the endothelium from the underlying stroma (1,2). This matrix serves as a selective barrier to cells and to macromolecules and also as a storage depot for cytokines and growth factors. The basement membrane functions to maintain the endothelial cell phenotype and integrity of the endothelium. The first step in angiogenesis is the breakdown of the basement membrane followed by cell migration, tube formation, proliferation, and, finally, resynthesis of the basement membrane (3). The major components of the basement membrane include collagen IV, perlecan, laminin, entactin, and numerous growth factors and cytokines, all of which interact in a highly specific manner (2). Several in vitro and in vivo assays for angiogenesis have been described (Tables 1 and 2). It is generally considered important to perform at least two of these assays to confirm angiogenic or antiangiogenic activity, since many of the assays do not encompass all of the steps in the angiogenic process. Many of these assays are in wide use, but several are difficult to perform and quantitate. This chapter focuses on three assays that employ basement membrane matrix (Matrigel) (4), including an in vitro tube assay (5), sprouting from aortic ring explants (6), and an in vivo subcutaneous implant model (7). All three assays are quick and reliable and can measure both stimulators and inhibitors of angiogenesis. Matrigel is commercially available, which saves time and makes the assays even easier to perform. Because of the variations in batches of Matrigel, it is best to work with the same lot of material as much as possible and one must always include positive (10% serum or bFGF) and negative controls (no additions).

II. ENDOTHELIAL CELL TUBE FORMATION ON BASEMENT MEMBRANE

When endothelial cells are plated on a gelled layer of basement membrane, the cells rapidly attach, migrate, and form capillary-like structures with a lumen (5,8) (Fig. 1). This assay

Table 1 Angiogenesis Assays In Vitro

Substratum	Time
Plastic with low serum	3–6 weeks
Gelatin or fibronectin	4–8 weeks
Collagen I gel	2–4 weeks
Fibrin clot	4–7 days
Aortic ring on Matrigel/fibrin/collagen I	5–11 days
Adipose tissue in collagen gel	5–12 days
Collagen IV	4 days
Matrigel	18 hr

Table 2 Angiogenesis Assays In Vivo

Tissue	Time
Chick chroioallantoic membrane	4–8 days
Sponge implants	1–2 weeks
Rabbit or rodent cornea	2–5 days
Polyvinyl disk implant	14 days
Matrigel	3–10 days

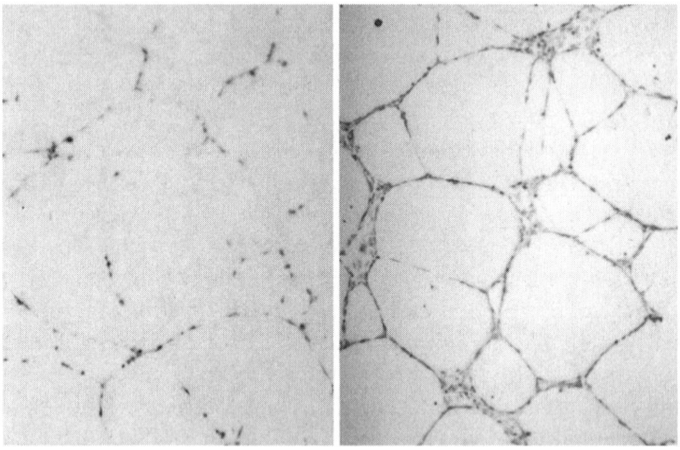

Figure 1 Formation of capillary-like tubes on Matrigel by human umbilical vein endothelial cells. Approximately 330 μL of Matrigel is allowed to gel for 30 min in each well of a 24-well plate. Then cells (40,000/mL) in media containing 10% serum (for testing inhibitors of angiogenesis) or without serum (for testing stimulators of angiogenesis) are added and incubated overnight. With some batches of Matrigel, a low level of serum (2–5%) gives partial tubes and can be used to test for possible stimulators of angiogenic activity. Since there are batch-to-batch variations in Matrigel, it is important to pretest each batch in the absence of serum and in the presence of various levels of serum to identify the optimal conditions. Control (left) and ECGS-treated cells (right) are shown.

mimics many of the steps in angiogenesis, including cell migration, tube formation, and basement membrane degradation and synthesis. The assay, however, does not require cell proliferation and, in fact, no cell proliferation is observed for several days. The assay has a number of advantages over other assays (Table 3). The tube assay is rapid, is usually complete within 12 to 18 hr, and can be performed in 96-well plates. It is easy to quantitate, and compounds show similar activity in other assays (9,10). The assay does require primary endothelial cells, although some established cell lines are beginning to show promise. Despite this limitation, this assay is a good, quick first screen for endothelial cell activity. It should be noted that the ability to form capillary-like tube structures on Matrigel supports the role of the basement membrane as a potent regulator of vessel differentiation in vivo.

A variety of factors influence tube formation in this assay, many of which are known to be important or to influence steps in angiogenesis (Table 4). For example, protease activity is necessary for the initial breakdown of the basement membrane in the first step of angiogenesis (11,12). Exogenous proteases, such as recombinant 72-kDa collagenase (13), stimulate tube formation, while inhibitors, such as TIMP, inhibit tube formation (14). Likewise, collagen synthesis is important in the final step of tube formation, when the basement membrane is resynthesized. Stimulators of collagen synthesis, such as ascorbic acid, increase tube formation, whereas collagen synthesis inhibitors, such as tricyclodecan-9-yl xanthanate (D609), block tube formation (9,15). A number of other growth factors known to promote angiogenesis—including bFGF, TGFβ, HGF, etc.—also promote tube formation (reviewed in Ref. 10). Thus, the tube assay appears to be a reliable model for testing angiogenic and antiangiogenic factors.

An example of a factor identified using the tube assay is thymosin β4. Thymosin β4 is a 43–amino acid molecule initially described in the thymus. Using subtractive cDNA cloning of endothelial cells on plastic and on Matrigel, the mRNA levels of thymosin β4 were found to increase four-to sixfold (16). Thymosin β4 has potent and specific cell migration activity for endothelial cells and was found to promote angiogenesis and to be localized in vessels in vivo (16,17). It is active at the nanogram level and is probably an important mediator of endothelial cell behavior. A related family member, thymosin β15, has been found to be an important migration factor in prostate cancer (18). The discovery of thymosin β4 and its angiogenic activity furhter point to the utility of the in vitro tube assay on Matrigel.

Table 3 Characteristics of Tube Assay on Matrigel

Advantages
Quick—18 hr
Can perform multiple samples—i.e., 96 well plates
Reliable—compounds show similar activity in ''established'' assays
Easy to quantitate
Can measure inhibitors and stimulators of angiogenesis
Disadvantages
Does not require cell proliferation, an important step in angiogenesis
Dependent on primary cell cultures of endothelial cells
Variations in Matrigel preparations may effect baseline

Table 4 Activity of Various Factors in Tube Assay on Matrigel

Factor	Activity
Laminin peptide SIKVAV	Stimulates
Laminin peptide YIGSR or RGD	Inhibits
Thymosin β4	Stimulates
bFGF	Stimulates
TGF beta	Stimulates
Alpha interferon	Stimulates
Gamma interferon	Inhibits
HGF/SF	Stimulates
Haptoglobin	Stimulates
Phorbol esters	Stimulates
Estrogen	Stimulates
IP-10	Inhibits
Inhibitors of collagen synthesis	Inhibits
Stimulators of collagen synthesis	Stimulates
Recombinant 72-kDa gelatinase	Stimulates
TIMP-1 or TIMP-2	Inhibits
PAI-1	Inhibits
Amino-terminal fragment of uPA	Stimulates
Thrombin	Stimulates

The in vitro tube assay has also been used in large screens for angiogenic activity. For example, vasculitis patients were found to have angiogenic activity in their sera. Using differential ammonium sulfate precipitation and column chromatography, the active angiogenic factor was identified with only the tube assay as the initial screen (19). The active factor, haptoglobin, was confirmed later, using two in vivo assays, as the angiogenic factor. Furthermore, antibodies to haptoglobin abolished the angiogenic activity in the patients' sera. Successful isolation of active factors from serum is generally difficult, but the tube assay demonstrated reliable activity at all levels, confirming the utility of the assay. Since the basement membrane is highly important in regulating angiogenic activity, we decided to look for endothelial cell binding sites on laminin-1, which has been found to be very biologically active. Recently we have completed a large screen of over 700 synthetic peptides from laminin-1 using the tube assay as our initial screen. Several active peptides were identified, including four that appear to be endothelial cell–specific (20). The use of the tube assay as a high-throughput screen has made these large searches for angiogenic activity possible.

III. SPROUTING FROM AORTIC RINGS ON BASEMENT MEMBRANE

The ability of cultured pieces of aorta to permit endothelial cell outgrowth has been exploited as a method to test for angiogenic and antiangiogenic factors (6,21,22). New vessels actually grow out from the cut ends of the tissue (Fig. 2). The assay was initially

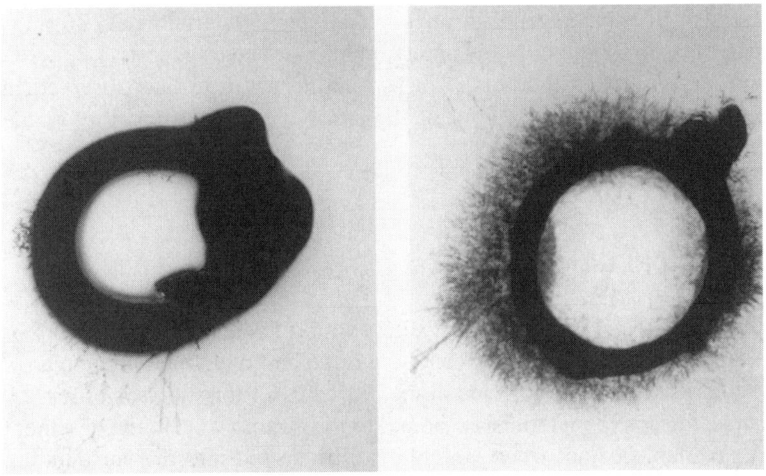

Figure 2 Sprouting of capillaries from pieces of aorta placed on gelled Matrigel. Approximately 110 μL of Matrigel is allowed to gel for 30 min in each well of a 48-well plate. A thin slice of aorta from a 6-week-old rat is placed in the center of the well, 50 μL of Matrigel is added on top of the tissue, and the dish is placed back into the incubator for 30 min more to allow the Matrigel to gel. Serum-free endothelial cell-growth medium is added together with the test factors (no additions to test stimulators and bFGF to test inhibitors) and incubated for 5–7 days. Control (left) and ECGS-treated aortic rings (right) are shown.

performed with the slices of aorta placed on either gelled collagen I or fibrin (21,22), but it is equally effective on Matrigel (6). Commercial preparations of collagen I are not concentrated enough to form solid gels. Thus, laboratory preparations of collagen I are necessary. The assay is performed with aortas from young (about 6 weeks old) rats, but human aortas and other vessels have also been used successfully. Serum-free medium is used to prevent fibroblastic outgrowth and thus allows for testing in the absence of serum factors. This assay appears to involve all of the steps involved in angiogenesis, with pericytes even observed migrating along the vessel sprouts.

A major advantage to the assay is that it is generally complete within 5 days (Table 5). Some sprouting is observed on day 4, by day 6–7, there are generally too many vessels

Table 5 Characteristics of Aortic Ring Sprouting Assay on Matrigel

Advantages
Quick—5 days
Can measure 30 samples per aorta
Can measure inhibitors and stimulators
Reliable—compounds show similar activity in ''established assays''
Results are clearly visible
Disavantages
Not easy to quantitiate
Each test must be performed in sextuplets due to variations
Uses animals

to quantitate. The assay is also advantageous, since one can obtain approximately 30 rings per rat aorta, and it can measure both inhibitors and stimulators of angiogenesis. Likewise, compounds show similar activity in the aortic ring-sprouting assay to that observed in established angiogenesis assays. A nice feature of the assay is the easy visibiltiy of the sprouts.

IV. IN VIVO MATRIGEL PLUG ASSAY

Subcutaneous injection of Matrigel provides an excellent support vehicle for testing angiogenic and antiangiogenic factors (7) (Fig. 3). Even though the basement membrane matrix is rich in angiogenic factors, when it is injected alone, little vessel formation is observed. This is probably because the active materials are bound to the components in the basement membrane and are unavailable or not freely soluble to diffuse out into the surrounding tissues and attract endothelial cells. Active fractions to be tested for angiogenic activity are mixed with the Matrigel prior to injection. Antiangiogenic factors are tested in the presence of bFGF; in some cases, heparin is included with the bFGF to increase the angiogenic response. The assay is easy to perform and does not even require that the animals be anesthetized. It appears that all of the steps in angiogenesis occur during the assay, and the results are highly reliable. There are several advantages to this assay over other in vivo assays (Table 6). The use of subcutaneous plugs of Matrigel, formed by injection of the liquid—which gels inside the animals—is very easy and quick. Many samples can be tested at one time. The assay is easily quantitated after histology and one can test for either an inhibitor or a stimulator of angiogenesis.

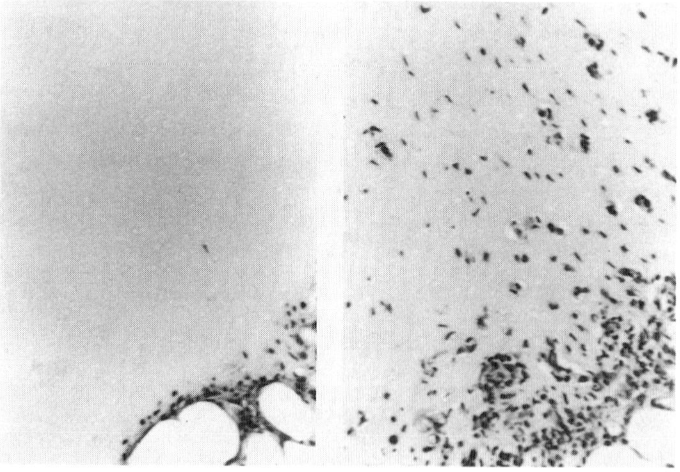

Figure 3 In vivo angiogenesis into Matrigel subcutaneously injected with angiogenic factors. Cold (liquid) Matrigel is injected subcutaneously into C57 Bl 6 mice either alone or containing test factors (factor alone for testing stimulators and bFGF plus factor for testing inhibitors). After 5–7 days, the mice are sacrificed and the Matrigel plugs are removed and examined for vessel density. Control (left) and ECGS-containing plugs (right) are shown.

Table 6 Characteristics of the In Vivo Matrigel Plus Assay

Advantages
Nonsurgical, easy to do
Can test up to two samples per animal
Assay time is 4–7 days plus time for histological staining
Reliable—compounds show similar activity in established assays
Easy to quantitate
Can measure either inhibitors or stimulators of angiogenesis
Disadvantages
Dependent on histology
Baseline variations in different batches of Matrigel
Variations in animal response
Uses animals

V. USE OF ASSAYS TO IDENTIFY ANGIOGENIC FACTORS

Through use of both the in vitro tube assay and in vivo basement membrane plug assay, a number of known angiogenic factors have been found to be active and several new factors have been identified (Table 4). Perhaps one of the most interesting angiogenic factors is estrogen, which was known to be important in reproduction and to have cardioprotective effects in women (23–25). It should also be noted that premenopausal women have more vasculitic and autoimmune diseases than postmenopausal women and men (26,27). Estradiol was found to greatly increase tube formation on Matrigel in vitro (25). Endothelial cell adhesion, migration, and proliferation were also promoted by estradiol in vitro via an estrogen receptor (28,29). The in vitro assay required the use of hormone-free serum and phenol-free medium. Using the in vivo Matrigel plug assay, estrogen was found to support bFGF-induced angiogenesis on ovariectomized animals receiving slow-release estrogen. Control ovariectomized animals not receiving estrogen showed approximately 50% less angiogenesis, supporting a major role for estrogen in vessel formation. Finally, the expression of a potent angiogenic factor, VEGF, has been found to be increased by estrogen (30), and antiestrogens have been shown to inhibit angiogenesis (31). These results may in part explain some of the cardioprotective effects of estrogen, in that repair of damaged vessels may occur more quickly in premenopausal women.

Using the tube and plug assays, interferon-inducible protein (IP-10) was found to be a potent inhibitor of angiogenesis (32,33). IP-10 had no effect on endothelial cell attachment, growth, or migration. Subsequent studies have shown the importance of IP-10 in reducing tumorigenesis (34). Thus, the use of these two assays has defined a number of important physiological regulators of angiogenesis.

VI. SUMMARY

Several reliable assays that use basement membrane to assess angiogenesis have been described. These assays are quick and relatively easy to perform and have been used to define the steps in angiogenesis and to identify new modulators. The importance of the basement membrane in angiogenesis is confirmed by the in vitro assays described here. Furthermore, when endothelial cells are plated on collagen I gels in tube-forming assays, a matrix accumulates inside the tubes, suggesting that the tubes are inside out (35–37). Thus, there is specificity in the signals provided by the basement membrane. Given the potent therapeutic uses of angiogenic and antiangiogenic factors in wound repair and cancer, there is a need for reliable, quantitative angiogenic assays to screen for activity and to determine the mechanisms of action.

REFERENCES

1. GR Martin, R Timpl, K Kuhn. Basement membrane proteins: molecular structure and function. 39:1–50, 1988.
2. R Timpl, JC Brown. Supramolecular assembly of basement membranes. BioEssays 18:123–132, 1996.
3. W Risau. Mechanisms of angiogenesis. Nature 386:671–674, 1997.
4. HK Kleinman, ML McGarvey, JR Hassell, V Star, F Cannon, GW Laurie, GR Martin. Basement membrane complexes with biological activity. Biochemistry 25:312–318, 1986.
5. Y Kubota, HK Kleinman, GR Martin, TJ Lawley. Role of laminin and basement membrane in the morphological differentiation of human endothelial cells into capillary-like structures. J Cell Biol 107:1589–1598, 1988.
6. RF Nicosia, A Ottinetti. Modulation of microvascular growth and morphogenesis by reconsituted basement membrane gel in three-dimensional cultures of rat aorta: a comparative study of angiogenesis in matrigel, collagen, fibrin, and plasma clot. In Vitro Cell Dev Biol 26:119–128, 1990.
7. A Passanitti, R Taylor, R Pili, Y Guo, P Long, J Haney, R Pauley, DS Grant, GR Martin. A simple, quantitative method for assessing angiogeneic and antiangiogenic agents using reconstituted basement membrane, heparin, and fibroblast growth factor. Lab Invest 67:519–528, 1992.
8. DS Grant, KI Tashiro, B Segui-Real, Y Yamada, GR Martin, HK Kleinman. Two different laminin domains mediate the differentiation of human endothelial cells into capillary-like structures in vitro. Cell 58:933–943, 1989.
9. DS Grant, PI Lelkes, K Fukuda, HK Kleinman. Intracellular mechanisms involved in basement membrane induced blood vessel differentiation in vitro. In Vitro Cell Dev Biol 27A:327–336, 1991.
10. DS Grant, D Morales, MC Cid, HK Kleinman. Angiogenesis models identify factors which regulate endothelial cell differentiation. In: ME Maragoudakis, ed: Angiogenesis: Molecular Biology, Clinical Aspects. New York: Plenum Press, 1994, pp. 51–60.
11. W Risau, V Lemmon. Changes in the vascular extracellular matrix during embryonic vasculogenesis and angiogenesis. Dev Biol 125:441–450, 1988.
12. HW Schnaper, DS Grant, WG Stetler-Stevenson, R Fridman, G D'Orazi, A Murphy, RE Bird, M Hoyhtya, TR Feurst, J Quigley, D French, HK Kleinman. Type IV collagenases and TIMPs modulate endothelial cell morphogenesis in vitro. J Cell Physiol 156:235–246, 1993.
13. HW Schnaper, ES Barnathan, A Mazar, S Maheshwari, S Ellis, SL Cortez, WH Baricos, HK Kleinman. Plasminogen activators augment endothelial cell organization in vitro by two distinct pathways. J Cell Physiol 165:107–118, 1995.
14. PD Brown. Matrix metalloproteinase inhibitors. Angiogenesis 1:142–154, 1997.
15. GC Haralabopoulos, DS Grant, HK Kleinman, PI Lelkes, SP Papaioannou, ME Maragoudakis. Inhibitors of basement membrane collagen synthesis prevent endothelial cell alignment in matrigel in vitro and angiogenesis in vivo. Lab Invest 71:575–582, 1994.
16. DS Grant, JL Kinsella, MC Kibbey, S LaFlamme, PD Burbelo, AL Goldstein, HK Kleinman.

Matrigel induces thymosin β4 in differentiating endothelial cells. J Cell Sci 108:3685–3694, 1995.
17. KM Malinda, AL Goldstein, HK Kleinman. Thymosin β4 stimulates directional migration of human umbilical vein endothelial cells. FASEB J 11:474–481, 1997.
18. L Bao, M Loda, PA Janmey, R Stewart, B Anand-Apte, BR Zetter. Thymosin β15: a novel regulator of tumor cells motility upregulated in metastatic prostate cancer. Nature Med 2: 1322–1328, 1996.
19. MC Cid, DS Grant, GS Hoffman, R Auerbach, AS Fauci, HK Kleinman. Identification of haptoglobin as an angiogenic factor in sera from patients with systemic vasculitis. J Clin Invest 91:977–985, 1993.
20. ML Ponce, KM Malinda, MC Delgado, M Nomizu, Y Yamada, HK Kleinman. Peptide sequences from laminin γ1 chain promote angiogenesis. Mol Biol Cell 180a:1044, 1997.
21. RF Nicosia, T Tchao, J Leighton. Histotypic angiogenesis in vitro: light microscopic, ultrastuctural, and radio-autographic studies. In Vitro 18:538–549, 1982.
22. RF Nicosia, JA Madri. The microvascular extracellular matrix: developmental changes during angiogenesis in the aortic ring plasma clot model. Am J Pathol 128:78–90, 1987.
23. MY Farhat, MC Lavigne, PW Ramwell. The vascular protective effects of estrogen. FASEB J 10:615–624, 1996.
24. JK Findlay. Angiogenesis in reproductive tissues. J Endocrinol 111:357–366, 1986.
25. DE Morales, KA McGowan, DS Grant, S Maheshwari, D Bhartiya, MC Cid, HK Kleinman, HW Schnaper. Estrogen promotes angiogenic activity in human umbilical vein endothelial cells in vitro and in a murine model. Circulation 91:755–763, 1995.
26. MJ Stampfer, GA Golditz, C Willett, JE Mauson, B Rosner, FE Speizer, CH Henneken. Postmenopausal estrogen therapy and cardiovascular disease: ten year follow-up from the nurses' health study. N Engl J Med 325:756–762, 1991.
27. NK Wenger, L Speroff, B Packard. Cardiovascular health and disease in women. N Engl J Med 329:247–256, 1993.
28. S Kim-Schulze, KA McGowan, SC Hubchak, MC Cid, MB Martin, HK Kleinman, GL Greene, HW Schnaper. Expression of an estrogen receptor by human coronary artery and umbilical vein endothelial cells. Circulation 94:1402–1407, 1996.
29. A Johns, AD Freay, W Fraser, KS Korach, GM Rubanyi. Disruption of estrogen receptor gene prevents 17 beta estradiol-induced angiogenesis in transgenic mice. Endocrinology 137:4511–4513, 1996.
30. J Nakamur, A Savinov, Q Lu, A Brodie. Estrogen regulates vascular endothelial growth/permeability factor expression in 7,12-dimethlybenz(a)anthracene-induced rat mammary tumors. Endocrinology 137:5589–5596, 1996.
31. A Gagliardi, DC Collins. Inhibition of angiogenesis by antiestrogens. Cancer Res 53:533–535, 1993.
32. AL Angiolillo, C Sgadari, DD Taub, F Lioa, JM Farber, S Maheshwari, HK Kleinman, GH Reaman, G Tosato. Human interferon-inducible protein 10 is potent inhibitor of angiogenesis. J Exp Med 182:155–162, 1995.
33. RM Streiter, SL Kunkel, DA Arenberg, MD Burdick, PJ Poverini. Interferon gamma–inducible protein 10 (IP-10), a member of the C-X-C chemokine family, is an inhibitor of angiogenesis. Biochem Biophys Res Commun 210:51–57, 1995.
34. DA Arenberg, SL Kunkel, PJ Polverini, SB Morris, MD Burdick, MC Glass, DT Taub, MD lannettoni, RI Whyte, RM Streiter. Interferon gamma-inducible protein 10 (IP-10) is an angiostatic factor that inhibits human non-small cell lung cancer (NSCLC) tumorigenesis and spontaneous metastases. J Exp Med 184:981–992, 1996.
35. J Feder, JC Marasa, JV Olander. The formation of capillary-like tubes by calf aortic endothelial cells grown in vitro. J Cell Physiol 116:1–6, 1983.
36. J Folkman, C Haudenschild. Angiogenesis in vitro. Nature 288:551–556, 1980.
37. T Maciag, J Kadish, L Wilkins. Organizational behavior of human umbilical vein endothelial cells. J Cell Biol 94:511–520, 1982.

20
Considerations for Angiogenic Tumor Models

William E. Fogler and Stacy M. Plum
EntreMed, Inc., Rockville, Maryland

I. INTRODUCTION

The pathogenesis of primary neoplasms and metastatic lesions is dependent on tumor-cell interactions with host homeostatic mechanisms, and there is now a considerable body of research demonstrating that the induction of a neovascular response by the malignancy is a principal determinant of growth and spread (1–5). Early experimental evidence supporting the prominent role of angiogenesis for tumor growth can be summarized as follows. First, transplantation of tumor fragments into the avascular cornea of rabbits induced new capillary growth to vascularize the expanding tumor mass. If the capillaries were physically prevented from reaching the implant or if angiogenesis was inhibited, tumor growth was impaired, restricting the tumor nodule to a diameter of approximately 0.4 mm (6). Second, tumors grown in isolated perfused organs were limited to a size of 1–2 mm^3 where the cells could receive nutrients by diffusion. Further growth of the lesions beyond 2 mm^3 was obtained following vascularization on transplantation to mice (7). These early observations demonstrated that the neovascularization of a neoplasm was absolutely required for progression of the cancer lesion. This concept, in turn, led to an explosion in research concerning tumor angiogenesis. This research has now resulted in the current paradigm, suggesting that the phenotypic changes that accompany tumor-associated angiogenesis are mediated via an angiogenic switch (8,9). This switch entails both an upregulation of endogenous angiogenic stimulators, such as the family of fibroblast and vascular endothelial growth factors (10–13), and a downregulation of angiogenic inhibitors that include endostatin and angiostatin (14–17). This paradigm further suggests that a reversal of the angiogenic phenotype in malignant cancers might be achieved through the restoration of negative regulators for vascular development (18–21).

With the increasing recognition that the progressive growth of primary neoplasms and metastatic lesions is dependent on angiogenesis, considerable effort has been focused on the potential therapeutic manipulation of the tumor vascular compartment. This, in turn, has led to the identification and subsequent development of a diverse array of agents that can target and exert an antiangiogenic, antivascular, or angiostatic effect on the neovascular processes that accompany tumor progression. Such agents have been shown to exert significant inhibitory effects on neoplastic development in a wide array of in vivo tumor models. Most often the end-point assessments in these models have included the (a) inhibition of the growth of syngeneic and xenogeneic tumors in immunocompetent or immunodeficient mice following ectopic (subcutaneous) transplantation (22–42); (b) inhibition of the number and/or growth of experimental or spontaneous metastases (14,22,43–51); (c) inhibition of growth and metastatic proclivity of syngeneic and xenogeneic tumors following orthotopic implantation (52–55); and (d) delay in onset of chemically induced tumors or malignancy in transgenic tumor models expressing oncogenes (56,57). However, what are the considerations that warrant the use of a particular tumor model to assess the therapeutic activity of an antiangiogenic, antivascular, or angiostatic agent? Given the requirement on angiogenesis for a tumor to progressively grow and disseminate, is it reasonable to only assess tumor growth delays and inhibition of metastatic disease following administration of a suspected antiangiogenic agent? What additional studies might be implemented to gain information on the phenotype of a tumor undergoing antiangiogenic therapy? These are valid questions, and it is the purpose of this article to present some of the inherent considerations required for modeling the angiogenic response in tumors. A central consideration to this review will be the suggestion that the outcome of angiogenic tumor modeling will be dependent on both tumor and host-cell properties.

II. ANGIOGENESIS ASSAYS IN THE CONTEXT OF TUMOR BIOLOGY

As stated above, neoplastic growth beyond a given volume is limited by the diffusion of nutrients and metabolic products across preexisting host vessels. While a small number of exceptions have been cited (2,58), neovascularization is required for the growth of the primary tumor and, when malignant, metastatic progression. The neovascular response induced by a progressively growing tumor is a complex process (59,60). In response to a shift in the balance of positive and negative regulators of angiogenesis at the tumor site, preexisting host capillaries and the surrounding stroma are subject to a sequence of events that include (a) retraction of pericytes from the ablumenal surface of the vessel; (b) release of proteases from the endothelial cells and degradation of the extracellular matrix surrounding the existing capillary; (c) endothelial cell migration and possibly proliferation; (d) alignment of the migrating endothelial cells into tubelike structures; (e) anastomoses of the newly formed vessels; and (f) initiation of blood flow in the newly formed capillaries.

The expression of the angiogenic phenotype in a primary malignant tumor is not sufficient to produce metastasis but it is required for progression to occur (4). The pathogenesis of cancer metastases depends on the interaction of tumor cells with the host and can be divided into several sequential steps: (a): invasion of cells from the primary tumor into the surrounding tissue, with penetration of blood and/or lymph vessels; (b) release of single or multiple tumor cell emboli into the circulation; (c) arrest of the circulating emboli in the small vascular beds of organs; (d) tumor cell extravasation through the wall of the arresting vessel and infiltration into adjacent tissue; and (e) multiplication of tumor cells and the development of vascularized stroma into a new tumor focus (61–64).

It is therefore reasonable to assume that in a given tumor model, there exists a dynamic relationship between multiple independent but interrelated processes affecting both the endothelial cell and the tumor-cell compartments. This relationship can be readily appreciated with a specific example. Recently an animal model was developed in mice in which a primary, subcutaneous Lewis lung carcinoma could almost completely suppress the growth of its pulmonary metastases (14). Following resection of the primary tumor, the previously dormant micrometastases underwent neovascularization and rapid expansion. Removal of the primary tumor was not only associated with increased neovascularization but also with a three- to-fivefold decrease in the apoptotic index of the metastatic cells (14). Hence, a primary consideration in the study of angiogenic metastatic tumor models should be the recognition that the observed outcome, whether tumor growth or inhibition, will be dependent on both host-cell influences and tumor-cell properties.

There have been a number of excellent reviews addressing methods and technical considerations for the study of cancer metastasis (65) and angiogenesis (66–68). Recently, a number of essential assay criteria have been suggested in order to relate an angiogenic/angiostatic response to a given dose of a stimulus/inhibitor (66). These criteria included (a) an understanding of the release rate and the spatial and temporal concentration distribution of angiogenic factor(s)/inhibitors(s); (b) the use of genetically well-defined neoplastic cells in terms of oncogene expression and production of growth factors (stimulators and inhibitors); (c) the use of quantitative measures of the structure of the new vasculature (vascular length, surface area, volume, number of vessels in the network, extent of basement membrane); (d) the use of quantitative measures of the functional characteristics of the new vasculature (endothelial cell migration rate, proliferation rate, canalization rate, blood flow rate, and vascular permeability); (e) the ability to distinguish between newly formed and preexisting host vessels; (f) minimization of tissue damage, since it may lead to formation of new vessels; (g) the potential to permit long-term and, if possible, noninvasive monitoring; and (h) ease and reliability. While these criteria help to focus essential model components for the study of neoplastic angiogenesis, it is often difficult to determine how these criteria may best be met in order not to compromise the overall biology of the tumor system.

The quantitation of tumor angiogenesis in vivo has been studied primarily through the use of three basic techniques that can be categorized as (a) microcirculatory preparations, (b) vascularization into polymer matrix implants, and (c) correlative studies on excised tissue. Several models have been developed that allow the direct observation of the tumor microcirculation directly. These models have included the rabbit ear chamber (69,70), the rodent dorsal skin window (71–76), the rodent cranial window (76,77), and intravital microscopy of exteriorized liver (78). The majority of these models, however, require that tumors be developed at ectopic sites, where the local host environment would be different from an orthotopic site of implantation. This fact alone can significantly impact on the biology of the tumor. Exceptions would be perhaps for the cranial window for central nervous system tumors or metastatic brain tumors and intravital microscopy of exteriorized liver for hepatic metastases. Vascularization into matrix implants uses a polymer matrix (gel or sponge), containing a known amount of angiogenic factors(s) or tumor cells, to be attached to an organ (79–85). The structure and function of the neovascular response into the sponge is then measured by a variety of techniques that include light or electron microscopy, histological examination of tissue sections stained for endothelial cells or basement membrane components, or perfusion with intravascular markers. Moreover, tumor cells and infiltrating host cells can be isolated from matrix implants for subsequent molecular and cellular biology studies (84,85). The obvious disadvantage of these methods is the inability to relate the time course of events in the same tissue or matrix implant; therefore a large number of samples need to be obtained to reach conclusions pertaining to the tumor model. More recently, advances in imaging technology (86–89)—

including magnetic resonance imaging, computed tomography, and positron emission tomography—may replace or provide adjunct studies to histology. Finally, tumor angiogenesis in vivo can be assessed qualitatively through the measurement of the plasma or urine concentration of angiogenic/antiangiogenic factors (90), although the extent of neovascularization at the local tumor site may not necessarily correlate.

Given the complex interactions involving tumor growth and metastatic spread, how can the above assays for angiogenesis be utilized to ensure that the observed responses can be adequately interpreted?

III. CONSIDERATIONS OF TUMOR MODELS IN ANGIOGENESIS

A. Host Leukocyte Involvement

It is now well established that solid tumors comprise a diverse population of host cells. Notwithstanding the subject matter of this review, in addition to vascular endothelial cells, the malignant cells are usually surrounded or infiltrated with a stromal compartment consisting of fibroblasts and leukocytes, including monocytes, lymphocytes, and neutrophils. These cells participate in the deposition of extracellular matrix proteins and elaborate a number of soluble pleotrophic cytokines that can impact on a tumor-associated neovascular response (91,92). However, although a number of studies have demonstrated the presence of infiltrating leukocytes within or at the periphery of various forms of malignant tissue, there has been considerable debate as to their biological role(s) and clinical significance (93,94). With the exception of lymphoma and certain breast carcinomas, the percentage of infiltrative leukocytes in the total cellular mass of solid tumors is low; this has suggested that this host component has little consequence for the angiogenic profile of the tumor. Based on this general observation alone, is it valid to assume that the major source of angiogenic stimulation is derived from the malignant cell population and to exclude considerations of host leukocyte involvement when attempting to study and model the neovascular response in a given tumor system? For many reasons this is not the case, and the relationship between tumor angiogenesis and the inflammatory processes that result in the recruitment and infiltration of the tumor by host leukocytes must be considered in any tumor model system.

As a specific example, the degree of host monocyte infiltration has been assessed in a number of experimental rodent tumors and initial attempts to correlate macrophage content with specific tumor characteristics—including in vivo growth rates and metastatic potential (tumor phenotypes dependent on angiogenesis)—were inconclusive (95–97). Importantly, these early studies utilized large tumors grown in the subcutis of mice for all determinations, and in each case the macrophage content was low. Subsequent to these studies, the macrophage content of spontaneous pulmonary and hepatic metastases at different stages of metastatic growth was determined. The results demonstrated that the relative content of macrophages in micrometastases was high; however, as metastases grew, the relative macrophage content fell, reaching uniformly low levels by the time the metastases were 0.5 mm in diameter (98–100). Given this observation, at the early stages of tumor growth, could the relatively high percentage of the total cellular mass attributable to host monocytes impact on tumor angiogenesis? It is now more fully understood that host cells of the monocyte/macrophage lineage are multifunctional and are capable of influencing both physiological and pathological processes, including modulation of tumor angiogenesis through production/secretion of both proangiogenic and antiangiogenic factors (101–103). Indeed even under conditions of advanced tumor growth, where the relative percentage of infiltrative macrophages is low, these secretory products can exert strong influences on the angiogenic status of experimental neoplasms.

Several lines of evidence have now also suggested direct connections between neoplastic vascular development and leukocyte/endothelial interaction. The cytokine-inducible, endothelial adhesion molecules, E-selectin and vascular cell adhesion molecule-1, mediate the rolling and firm adhesion phase, respectively, of specific leukocyte subsets (104–107). These adhesion ligands in soluble form have also recently been reported to induce chemotaxis of endothelial cells in vitro and to promote neovessel growth in the rat cornea (108). Moreover, recent in vitro studies have also suggested that the angiogenic stimulators fibroblast growth factor and vascular endothelial growth factor directly regulate VLA-4- and LFA-1/Mac-1-dependent binding of lymphocytes to endothelium (109–112). In addition, subfamilies of the C-X-C chemokines that stimulate the chemotactic transmigration of distinct sets of leukocytes can also directly regulate the process of angiogenesis (113). Studies have shown interleukin-8 to promote neovessel growth (114,115), whereas gro-β and the interferon-inducible protein 10 inhibit angiogenesis (116–119). Collectively, these studies suggest that the regulation of tumor growth by cytokine-induced alteration in the recruitment and activation of host effector cells is intimately associated with leukocyte adhesion and angiogenic modulation. With the parallel development of data implicating complementary interactions between cytokine-induced leukocyte recruitment and the direct development of the tumor neovasculature, the development of tumor models capcable of simultaneously addressing both aspects is warranted.

A number of experimental models have been developed to facilitate the in vivo study of leukocyte recruitment and to elucidate critical determinants that may impact either directly or indirectly the treatment of metastatic tumor growth (120–124). Although these models have provided important qualitative and quantitative information on leukocyte trafficking to tumors, they have not been well suited for concomitant determinants of angiogenic modulation. Similarly, models of tumor angiogenesis, such as the corneal micropocket assay (115,125) or those incorporating Matrigel implantation (82,126,127), have not been exploited to study the regulation of leukocyte recruitment. However, recently a novel tumor model was developed using recombinant human basic fibroblast growth factor—absorbed gelatin sponges containing B16F10 melanoma cells, which were affixed to the livers of syngeneic mice (85). Using this model, the relationship between the induction of hepatic leukocyte infiltration, local angiogenic modulation, and tumor growth after treatment of mice with cytokines was studied.

A number of salient features of this model deserve further comment and may be applicable for considering many models. First, it is well recognized that organ-specific endothelium can express unique characteristics and either constitutive or cytokine-inducible ligands that are important for leukocyte adhesion and recruitment (128,129). It is also known that the neovasculature of progressively growing tumors and metastatic lesions can express traits associated with the pathological process that are distinct from the surrounding tissue (110–112). The ability to implant sponges containing both tumor cells and an angiogenic stimulus to various organs within the peritoneal cavity or subcutaneously offers the opportunity to investigate site-specific influences on the development of in vivo tumor vasculature and cytokine-induced leukocyte recruitment. This latter event can be studied by in situ immunohistochemistry or by flow cytometric analysis after disaggregation of the sponge. Second, gelatin sponge implants containing endothelial cell growth factors are amenable to the study of in vivo pathological mechanisms that might control neovascularization by progressively growing tumors at a distant site. In this regard, this model would be analogous to the recently described mouse corneal pocket neovascularization assay used to discover circulating angiogenesis inhibitors generated by human tumor (125). Moreover, we believe that the ability to implant these sponges onto target organs of metastatic disease, such as the liver, in a tumor-bearing host will allow the opportunity to also identify potential organ-derived effects on neovessel growth (58). Third, in contrast to the growth factor–induced hypervascular response into gelatin sponges, the development and growth of B16-F10 melanoma was not dependent on the presence of exogenous fibroblast growth factor. This finding would suggest that a tumor cell line expressing a transfected gene product thought to be involved in angiogenic modulation and/or leukocyte recruitment could be implanted within a sponge into various normal or genetically modified mouse recipient strains for the subsequent in vivo analyses of these processes. Finally, the incorporation of exogenous fibroblast growth factor into the sponge was used to induce a rapid, acute angiogenic response independent of tumor growth and to create a hypervascular tumor nodule to facilitate studies of leukocyte infiltration and angiogenic modulation. However, the autonomous growth and vascular development of hepatic sponge implants of B16-F10 suggests that this model may also be useful to directly assess the influence of exogenous proangiogenic and antiangiogenic factors on the in situ tumor-associated vascular expression of adhesion molecules believed important for leukocyte infiltration.

B. Vascular Origin

Transplantation of a given tumor cell population into a host organ or tissue initiates a series of processes that leads to the eventual development of a neovascular network. Microscopic observations on experimental tumor models have identified arteries and arterioles, nonfenestrated capillaries, fenestrated capillaries, discontinuous capillaries, blood channels, capillary sprouts, postcapillary venules, venules and veins, and arteriovenous anastomoses as occurring in experimental tumor models (130). The tumor vasculature comprises both vessels that have been recruited from preexisting host vasculature and vessels that have resulted from a tumor-induced angiogenic response of host endothelium (59,60,131). However, although the tumor vasculature originates from the host endothelial compartment, the resulting organization within the tumor can be markedly different from that of normal tissue components, depending on the tumor type, growth rate, and location (132).

The organ site of metastatic disease also determines the source of endothelium involved in the neovascular response of the progressing tumor. For example, the angiogenic response in lung metastases is derived from both the pulmonary and broncial arteries (133,134). Using a combination of microscopy and a lanthanum perfusion technique, pulmonary B16 melanoma metastases have been analyzed with respect to location, morphological appearance, and origin and density of microvessels (135). The metastases were classified as either superficial or deep (depending on their location in lung parenchyma) and as either compact or loose in appearance. Superficial- and deep-compact metastases (5–10% of the total metastatic burden) lacked intratumoral microvessels, since no lanthanum was visualized inside these metastases. The deep-compact metastases were observed to grow around arterioles and bronchioles. The remaining metastatic burden was classified as superficial- and deep-loose metastases. Microscopic evaluation of these lesions indicated that 0.6% area of superficial-loose and 1.3% area of deep-loose metastases consisted of vessels indicating that an active angiogenesis was taking place. A majority of the capillary sprouts were derived from vessels originating from the pulmonary circulation. Deep-loose metastases were located in the lung parenchyma and were surrounded by normal alveolar tissue. Within the tumor, high numbers of nonpatent collapsed vascular channels were seen having the morphology of probable capillary sprouts. In tumor vessels, lanthanum leaked into the interstitium to a varying degree. It was also noted that, in some vascular channels, melanoma cells formed parts of the channel wall. Superficial loose metastases were, as a rule, avascular, but on very rare occasions labeled vessels could be seen in the periphery of the tumor tissue on the mesothelial side. Using this tumor model in studies of exogenously administered antiangiogenic agents, is it reasonable to expect activity against all subclasses of pulmonary metastases. The hypothesis would suggest not, since superficial- and deep-compact metastases (5–10% of the total metastatic burden) lack neovascularization.

In studies using human tumor xenografts onto immunocompromised rodents, the neovascular response of the tumor is obviously murine in origin. How might this affect studies of antiangiogenic, antivascular, or antiangiostatic agents? In an interesting model system developed to investigate the antitumor and antiangiogenic effects of human-specific anti-integrin antibodies, human foreskin was grafted onto SCID mice (136,137). When the skin graft healed, human cancer cells were injected intradermally within the human skin. The angiogenic response induced by the tumor was then shown to be human in origin. Unfortunately, within approximately 1 month, murine endothelial cells began to enter and replace the human endothelial cells in the graft (138).

C. Effect of the Organ Microenvironment

Comparison between ectopic and orthotopic tumor models has indicated that the organ microenvironment can profoundly affect the biology of cancer growth and influence metastatic potential (58,139–147). Studies have now begun to document that a contributing factor to the observed differences in tumor biology between ectopic and orthotopic transplantation involves the neovascular response. Three specific examples are provided below.

Human renal cell carcinomas are highly vascularized neoplasms, and the level of bFGF in the serum of patients has been shown to inversely correlate with survival (90). When human renal carcinoma cells are implanted subcutaneously into nude mice, they produce slow-growing, nonmetastatic tumors that are not well vascularized. In contrast, the implantation of the same cells into the kidney of mice yields rapidly growing, vascularized, highly metastatic tumors (148,149). The differences in site-dependent angiogenesis were found to be due to the expression of basic fibroblast growth factor. Kidney tumors expressed higher levels of basic fibroblast growth factor mRNA and protein compared with renal carcinoma cells growing subcutaneously (149). The authors then demostrated a reciprocal site-dependent interaction. Immunostaining of renal carcinoma tumors growing in the subcutis or kidney of nude mice revealed the presence of IFN-β in keratinocytes and skin fibroblasts but not in the kidney (149). IFN-β and IFN-α downregulate the expression of bFGF at the mRNA and protein levels in a dose-dependent manner and by a pathway that is not related to the antiproliferative activity of the IFNs (150). The fact that IFN-β is present in the subcutis but not in the kidney, coupled with its physiological role in regulating angiogenesis, suggested an active modulation of the angiogenic phenotype was occurring.

In another series of studies, various metastasis related genes (epidermal growth factor receptor, basic fibroblast growth factor, interleukin-8, collagenase type IV, carcinoembryonic antigen, and mdr-1) were upregulated during growth of human colon cancer xenografts grown in the cecum wall (orthotopic) of immunocompromised mice (151,152), whereas these genes in the same tumor lines grown in subcutaneous (ectopic) sites remained unchanged. It was then hypothesized that the hepatic microenvironment would have a different effect on the development and maintenance of the microcirculation of a colorectal tumor as compared with the subcutaneous site. To this end, a comparison was made on the microcirculation of LS174T, a human colon adenocarcinoma, that had metastasized to the liver with that of the host liver vessels and that of the same tumor grown in the subcutaneous space (153). In the liver metastasis model, cells were injected into spleens of nude mice. Four-to-eight weeks later, the liver with metastatic tumors was exteriorized, placed on a special stage, and observed under an intravital fluorescence microscope. The dorsal skin fold chamber model (154) was used to study the subcutaneous tumors. Red blood cell velocity, vessel diameter, density, permeability, and leukocyte-endothelial interactions were measured using fluorescence microscopy and image analysis. Vascular endothelial growth factor mRNA expression was determined by Northern blot analysis. LS174T tumor foci in the liver had tortuous vascular architecture, heterogeneous blood flow, significantly lower vascular density, and significantly higher vascular permeability than normal liver tissue (153). Tumors grown in the liver had significantly lower vessel density, especially in the center, coincident with central necrosis, than the subcutaneous tumors. The frequency distribution of vessel diameters of the liver tumor was slightly shifted to smaller size compared with that of the subcutaneous tumor. Leukocyte rolling in the liver tumor was twofold lower than that in the subcutaneous tumor. These physiological findings were consistent with the measurement of VEGF in that the VEF mRNA level was lower in the liver tumor than that in the subcutaneous tumor. However, macromolecular vascular permeability in the liver tumor was significantly higher than that in the subcutaneous tumor. These findings demonstrate that the liver microenvironment has different effects on some aspects of tumor angiogenesis.

In a final example, four surgically obtained human melanoma tumors were grafted intradermally and subcutaneously in immunocompromised mice (155). The tumors progressively grew and the vascular network consisted of two distinct types of microvessels: (a) normal microvessels recruited from the preexisting network in the skin during tumor growth and (b) pathological microvessels recruited by tumor-induced neovascularization. The normal microvessels included terminal arterioles invested in vascular smooth muscle, nonfenestrated capillaries, and dilated postcapillary venules. The neovasculature showed severe structural abnormalities, including incomplete endothelial lining, interrupted or absent basement membrane, and/or lack of pericytes. The extracellular matrix appeared in immunohistochemical preparations as an irregular meshwork, differing considerably in structure and density between tumor regions and individual tumors. The infiltration of host cells in the tumors was sparse, constituting less than 1% of the total number of cells. Importantly, the intradermal xenografted tumors showed organ-specific metastatic patterns, while the same tumors grown at subcutaneous sites did not develop organ-specific metastases (155). Moreover, essential biological features of the donor patient's tumors were found to be retained after the heterotransplantation. The tumor sequence from high to low values of vascular and angiogenic parameters was equal in mice and patients. Moreover, the organ-specific metastatic pattern of the xenografted tumors reflected the pattern of distant metastases in the donor patients.

IV. CONCLUSION

In conclusion, the pathogenesis of primary neoplasms and metastatic lesions is dependent on tumor-cell interactions with host homeostatic mechanisms. There is now a considerable body of research that demonstrates that the induction of a neovascular response by the malignancy is a principal determinant of growth and spread. In any given tumor model there exists a dynamic relationship between multiple interrelated processes that affect both the endothelial cell and tumor-cell compartments. The processes are dependent on host leukocyte involvement, the origin of the vascular response, and the organ's microenvironment. The study of angiogenic metastatic cancer models must therefore take into account these processes to ensure that the observed outcome, whether tumor growth or inhibition, will be reflective of the biology of the tumor.

REFERENCES

1. J Folkman. Tumor angiogenesis: therapeutic implications. N Engl J Med 285:1182–1186, 1971
2. J Folkman. What is the evidence that tumors are angiogenesis dependent? J Natl Cancer Inst 82:4–6, 1990.
3. J Folkman. Angiogenesis in cancer, vascular, rheumatoid and other diseases. Nature Med 1: 27–31, 1995.
4. IJ Fidler, LM Ellis. The implications of angiogenesis for the biology and therapy of cancer metastasis. Cell 79:185–188, 1994.
5. BR Zetter. Angiogenesis and tumor metastasis. Annu Rev Med 49:407–424, 1998.
6. MA Gimbrone Jr, RS Cotran, SB Leapman, J Folkman. Tumor growth and neovascularization: an experimental model using the rabbit cornea. J Natl Cancer Inst 52:413–427, 1974.
7. J Folkman, P Cole, S Zimmerman. Tumor behavior in isolated perfused organs: in vitro growth and metastasis of biopsy material in rabbit thyroid and canine intestinal segment. Ann Surg 164:491–502, 1966.
8. LA Liotta, PS Steeg, WG Stetler-Stevenson. Cancer metastases and angiogenesis: an imbalance of positive and negative regulation. Cell 64:327–336, 1991.
9. D Hanahan, J Folkman. Patterns and emerging mechanisms of the angiogenic switch during tumorigenesis. Cell 86:353–364, 1996.
10. N Bouck, V Stellmach, SC Hsu. How tumors become angiogenic. Adv Cancer Res 69:135–174, 1996.
11. WH Burgess WH, JA Winkles. The fibroblast growth factor family: multifunctional regulators of cell proliferation. In: L Pusztai, CE Lewis, E Yap E, eds. Regulation of the Proliferation of Neoplastic Cells. Oxford, UK: Oxford University Press, 1994, pp. 155–218.
12. M Klagsbrun, S Soker. VEGF/VPF: the angiogenesis factor found? Curr Biol 3:699–702, 1993.
13. V Joukov, A Kaipainen, M Jeltsch, K Pajusola, B Olofsson, V Kumar, U Eriksson, K Alitalo. Vascular endothelial growth factors VEGF-B and VEGF-C. J Cell Physiol 173:211–215, 1997.
14. MS O'Reilly, L Holmgren, Y Shing, C Chen, RA Rosenthal, M Moses, WS Lane, Y Cao, EH Sage, J Folkman. Angiostatin: a novel angiogenesis inhibitor that mediates the suppression of metastases by a Lewis lung carcinoma. Cell 79:315–328, 1994.
15. MS O'Reilly, T Boehm, Y Shing, N Fukai, G Vasios, WS Lane, E Flynn, JR Birkhead, BR Olsen, J Folkman. Endostatin: an endogenous inhibitor of angiogenesis and tumor growth. Cell 88:1–20, 1997.
16. BKL Sim. Angiostatin and Endostatin: endothelial cell-specific endogenous inhibitors of angiogenesis and tumor growth. Angiogenesis 2:37–48, 1998.
17. MS O'Reilly, L Holmgren, C Chen, J Folkman J. Angiostatin induces and sustains dormancy of human primary tumors in mice. Nature Med 2:689–692, 1996.
18. AL Harris. Antiangiogenesis for cancer therapy. Lancet 349:13–15, 1997.

19. JM Pluda. Tumor-associated angiogenesis: mechanisms, clinical implications, and therapeutic strategies. Semin Oncol 24:203–218, 1997.
20. J Folkman J. Antiangiogenic therapy. In: VT DeVita Jr, S Hellman, SA Rosenberg, eds. Cancer: Principles and Practice of Oncology. Philadelphia: Lippincott, 1997, pp. 3075–3085.
21. T Boehm, J Folkman, T Browder, MS O'Reilly. Antiangiogenic therapy of experimental cancer does not induce acquired drug resistance. Nature 390:404–407, 1997.
22. BKL Sim, MS O'Reilly, H Liang, AH Fortier, W He, JW Madsen, R Lapcevich, CA Nacy. A recombinant human angiostatin protein inhibits experimental primary and metastatic cancer. Cancer Res 57:1329–1334, 1997.
23. T Tanaka, Y Cao, J Folkman, HA Fine. Viral vector-targeted antiangiogenic gene therapy utilizing an angiostatin complementary DNA. Cancer Res 58:3362–3369, 1998.
24. BJ Lannutti, ST Gately, ME Quevedo, GA Soff, AS Paller. Human angiostatin inhibits murine hemangioendothelioma tumor growth in vivo. Cancer Res 57:5277–5280, 1997.
25. BT Zhu, AH Conney. Is 2-methoxyestradiol an endogenous estrogen metabolite that inhibits mammary carcinogenesis? Cancer Res 58:2269–2277, 1998.
26. N Klauber, S Parangi, E Flynn, E Hamel, RJ D'Amato. Inhibition of angiogenesis and breast cancer in mice by the microtubule inhibitors 2-methoxyestradiol and taxol. Cancer Res 57: 81–86, 1997.
27. T Fotsis, Y Zhang, MS Pepper, H Adlercreutz, R Montesano, PP Nawroth, L Schweigerer. The endogenous oestrogen metabolite 2-methoxyestradiol inhibits angiogenesis and suppresses tumor growth. Nature 386:237–239, 1994.
28. AK Sills, JI Williams, BM Tyler, DS Epstein, EP Sipos, JD Davis, MP McLane, S Pitchford, K Cheshire, FH Gannon, WA Kinney, TL Chao, M Donowitz, J Laterra, M Zasloff, H Brem. Squalamine inhibits angiogenesis and solid tumor growth in vivo and perturbs embryonic vasculature. Cancer Res 58:2784–2792, 1998.
29. F Tsuji, K Yamada. Combination therapy with fumagillin and vaccination with tumor-derived antigenic peptides in B16 melanoma transplanted mice. Int J Immunopharm 20:111–124, 1998.
30. T Fujioka, M Hasegawa, K Ogiu, Y Matsushita, M Sato, T Kubo. Antitumor effects of angiogenesis inhibitor O-(chloroacetyl-carbamoyl) fumagillol (TNP-470) against murine renal cell carcinoma. J Urol 155:1775–1778, 1996.
31. CM Coughlin, KE Salhany, MS Gee, DC LaTemle, S Kotenko, X Ma, G Gri, M Wysocka, JE Kim, L Liu, F Liao, JM Farber, S Pestka, G Trinchieri, WMF Lee. Tumor cell responses to IFN affect tumorigenicity and response to IL-12 therapy and angiogenesis. Immunity 9: 25–34, 1998.
32. K Sato, YC Yoo, M Mochizuki, I Saiki, TA Takahashi, I Azuma. Inhibition of tumor-induced angiogenesis by a synthetic lipid A analogue with low endotoxicity, DT-5461. Jpn J Cancer Res 86:374–382, 1995.
33. M Yamaoka, T Yamamoto, S Ikeyama, K Sudo, T Fujita. Angiogenesis inhibitor TNP-470 (AGM-1470) potently inhibits the tumor growth of hormone-independent human breast and prostate carcinoma cell lines. Cancer Res 53:5233–5236, 1993.
34. J Murata, I Saiki, T Makabe, Y Tsuta, S Tokura, I Azuma. Inhibition of tumor-induced angiogenesis by sulfated chitin derivatives. Cancer Res 51:22–26, 1991.
35. IJ Polakowski, MK Lewis, VR Muthukkaruppan, B Erdman, L Kubai, R Auerbach. A ribonuclease inhibitor expresses anti-angiogenic properties and leads to reduced tumor growth in mice. Am J Pathol 143:507–517, 1993.
36. F Yuan, Y Chen, Mdellian, N Safabakhsh, N Ferrara, RK Jain. Time dependent vascular regression and permeability changes in established human tumor xenografts induced by an anti-vascular endothelial growth factor/vascular permeability factor antibody. Proc Natl Acad Sci 93:14765–14770, 1996.
37. HY Min, LV Doyle, CR Vitt, CL Zandonella, JR Stratton-Thomas, MA Shuman, S Rosen-

berg. Urokinase receptor antagonists inhibit angiogenesis and primary tumor growth in syngeneic mice. Cancer Res 56:2428–2433, 1996.

38. R Pili, J Chang, RA Partis RA Mueller, FJ Chrest, A Passaniti. The alpha-glucosidase I inhibitor castanospermine alters endothelial cell glycosylation, prevents angiogenesis, and inhibits tumor growth. Cancer Res 55:2920–2926, 1995.
39. JA Low, MD Johnson, EA Bone, RB Dickson. The matrix metalloproteinase inhibitor batimastat (BB-94) retards human breast cancer solid tumor growth but not acites formation in nude mice. Clin Cancer Res 2:1207–1214, 1996.
40. G Taraboletti, A Garofalo, D Belotti, T Drudis, P Borsotti, E Scanziani, PD Brown, R Giavazzi. Inhibition of angiogenesis and murine hemangioma growth by batimastat, a synthetic inhibitor of matrix metalloproteinases. J Natl Cancer Inst 87:293–298, 1995.
41. E Pesenti, F Sola, N Mongelli, M Grandi, F Spreafico. Suramin prevents neovascularization and tumor growth through blocking of basic fibroblast growth factor activity. Br J Cancer 66:367–372, 1992.
42. JR Zhou, P Mukherjee, ET Gugger, T Tanaka, GL Blackburn, SK Clinton. Inhibition of murine bladder tumorigenesis by soy isoflavones via alterations in the cell cycle, apoptosis, and angiogenesis. Cancer Res 58:5231–5238, 1998.
43. P Borgstrom, IP Torres-Filho, B Hartley-Asp. Inhibition of angiogenesis and metastases of the Lewis-lung cell carcinoma by the quinoline-3-carboxamide, Linomide. Anticancer Res 15:719–728, 1995.
44. SA Eccles, GM Box, WJ Court, EA Bone, W Thomas, PD Brown. Control of lymphatic and hematogenous metastasis of a rat mammary carcinoma by the matrix metalloproteinase inhibitor batimastat (BB-94). Cancer Res 56:2815–2822, 1996.
45. M Yamaoka, T Yamamoto, T Masaki, S Ikeyama, K Sudo, T Fujita. Inhibition of tumor growth and metastasis of rodent tumors by the angiogenesis inhibitor O-(chloroacetyl-carbamoyl) fumagillol (TNP-470; AGM-1470). Cancer Res 53:4262–4267, 1993.
46. M Mochizuki, YC Yoo, K Matsuzawa, K Sato, I Saiki, S Tono-oka, K Samukawa, I Azuma. Inhibitory effect of tumor metastasis in mice by saponins, ginsenoside-Rb2,20- and 20(S)-gisenoside-Rg3, of red ginseng. Biol Pharm Bull 18:1197–1202, 1995.
47. K Sato, M Mochizuki, I Saiki, YC Yoo, K Samukawa, I Azuma. Inhibition of tumor angiogenesis and metastasis by a saponin of Panax ginseng, ginsenoside-Rb2. Biol Pharm Bull 17:635–639, 1994.
48. J Vukanovic, JT Isaacs. Linomide inhibits angiogenesis, growth, metastasis, and macrophage infiltration within rat prostatic cancers. Cancer Res 55:1499–1504, 1995.
49. IC Anderson, MA Shipp, AJP Docherty, BA Teicher. Combination therapy including a gelatinase inhibitor and cytotoxic agent reduces local invasion and metastasis of murine Lewis lung carcinoma. Cancer Res 56:715–718, 1996.
50. CW Crowly, RL Cohen, BK Lucas, G Liu, MA Shuman, AD Levinson. Prevention of metastasis by inhibition of the urokinase receptor. Proc Natl Acad Sci USA 90:5021–5025, 1993.
51. H Kobayashi, J Gotoh, M Fujie, H Shinohara, N Moniwa, T Terao. Inhibition of metastasis of Lewis lung carcinoma by a synthetic peptide within growth factor like domain of urokinase in the experimental and spontaneous metastasis model. Int J Cancer 57:727–733, 1994.
52. M Kirsch, J Strasser, R Allende, L Bello, J Zhang, PM Black. Angiostatin suppresses malignant glioma growth in vivo. Cancer Res 58:4654–4659, 1998.
53. M Shaw, S Ratanawong, P Chou, V Ray, Y Mirochnik, L Slobodskoy, M Rubenstein, P Guinan. Paclitaxel, bropirimine, and linomide: effect on growth inhibition in a murine prostate cancer model by different growth regulatory mechanisms. Exp Clin Pharmacol 20:111–114, 1998.
54. M Pollard. Effects of linomide on advanced prostate-seminal vesicle cancers in Lobund-Wistar rats. Prostate 35:43–49, 1998.
55. JM Wigginton, J-W Park, J-K Lee, MJ Brunda, RM Streiter, RH Wiltrout. Evaluation of mechanisms of inhibition of tumor neovascularization by IL-12/pulse IL-2: rationale for fu-

ture combination immunoangiostatic approaches to the treatment of established tumor. Proceedings Angiogenesis and Cancer, Orlando, 1988, p. B68.
56. A Takechi, T Uozumi, K Kawamoto, A Ito, K Kurisu, K Sudo. Inhibitory effect of TNP-470, a new anti-angiogenic agent, on the estrogen induced rat pituitary tumors. Anticancer Res 14:157–162, 1994.
57. S Parangi, M O'Reilly, G Christofori, L Holmgren, J Grosfeld, J Folkman, D Hanahan. Antiangiogenic therapy of transgenic mice impairs de novo tumor growth. Proc Natl Acad Sci 93:2002–2007, 1996.
58. IJ Fidler. Modulation of the organ microenvironment for treatment of cancer metastasis. J Natl Cancer Inst 87:1588–1592, 1995.
59. J Folkman. Tumor angiogenesis. In: PM Mendelsohn, MAP Howley, LA Liotta, eds. The Molecular Basis of Cancer. Philadelphia: Saunders, 1995, pp. 206–232.
60. R Bicknell. Mechanistic insights into tumour angiogenesis. In: R Bicknell, CE Lewis, N Ferrara, eds. Tumor Angiogenesis. Oxford, UK: Oxford University Press, 1997, pp. 19–28.
61. IJ Fidler. Critical factors in the biology of human cancer metastasis. Cancer Res 50:6130–6138, 1990.
62. JR MacDougall, LM Matrisian. Contributions of tumor and stromal matrix metalloproteinases to tumor progression, invasion and metastasis. Cancer Metast Rev 14:351–362, 1995.
63. GL Nicolson. Cancer progression and growth: relationship of paracrine and autocrine growth mechanisms to organ preference of metastasis. Exp Cell Res 204:171–180, 1993.
64. LA Liotta. Cancer cell invasion and metastasis. Sci Am 266:54–63, 1992.
65. DR Welch. Technical considerations for studying cancer metastasis in vivo. Clin Exp Metast 15:272–306, 1997.
66. RK Jain, K Schlenger, M Hockel, F Yuan. Quantitative angiogenesis assays: progress and problems. Nature Med 3:1203–1208, 1997.
67. R Auerbach, W Auerbach, I Polakowski. Assays for angiogenesis: a review. Pharmacol Ther 51:1–11, 1991.
68. T-P D Fan, PJ Polverini. In vivo models of angiogenesis. In: R Bicknell, CE Lewis, N Ferrara eds. Tumor Angiogenesis. Oxford, UK: Oxford University Press, 1997, pp. 5–18.
69. LE Gerlowski, RK Jain. Microvascular permeability of normal and neoplastic tissues. Microvasc Res 31:288–305, 1986.
70. TE Dudar, RK Jain. Differential response of normal and tumor microcirculation to hyperthermia. Cancer Res 44:605–612, 1984.
71. GH Algire, FY Legallais. Recent developments in the transparent-chamber technique as adapted to the mouse. J Natl Cancer Inst 10:225–253, 1949.
72. B Endrich, HS Reinhold, JF Gross, M Intaglietta. Tissue perfusion inhomogeneity during early tumor growth in rats. J Natl Cancer Inst 62:387–395, 1979.
73. K Asaishi, B Endrich, A Gotz, K Messmer. Quantitative analysis of microvascular structure and function in the amelanotic melanoma A-Mel-3. Cancer Res 41:1898–1904, 1981.
74. MW Dewhirst, CY Tso, R Oliver, CS Gustafson, TW Secomb, JF Gross. Morphologic and hemodynamic comparison of tumor and healing normal tissue microvasculature. Int J Radiat Oncol Biol Phys 17:91–99, 1989.
75. M Leunig, F Yuan, MD Menger, Y Boucher, AE Goetz, K Messmer, RK Jain. Angiogenesis, microvascular architecture, microchemodynamics, and interstitial fluid pressure during early growth of human adenocarcinoma LS174T in SCID mice. Cancer Res 52:6553–6560, 1992.
76. D Fukumura, HA Salehi, B Witmer, RF Tuma, RJ Melder, RK Jain. Tumor necrosis factor–induced leukocyte adhesion in normal and tumor vessels: effect of tumor type, transplantation site, and host strain. Cancer Res 55:4824–4829, 1995.
77. F Yuan, HA Salehi, Y Boucher, US Vasthare, RF Tuma, RK Jain. Vascular permeability and microcirculation of gliomas and mammary carcinomas transplanted in rat and mouse cranial window. Cancer Res 54:4564–4568, 1994.
78. M Dellian, BP Witwer, HA Salehi, F Yuan RK Jain. Quantitation and physiological charac-

terization of angiogenic vessels in mice: effect of basic fibroblast growth factor, vascular endothelial growth factor/vascular permeability factor, and host microenvironment. Am J Pathol 149:59–72, 1996.

79. HF Dvorak, VS Harvey, P Estrella, LF Brown, J McDonagh, AM Dvorak. Fibrin containing gels induce angiogenesis: implications for tumor stroma generation and wound healing. Lab Invest 57:673–686, 1987.
80. V Mahadevan, IR Hart, GP Lewis. Factors influencing blood supply in wound granuloma quantitated by a new in vivo technique. Cancer Res 49:415–419, 1989.
81. ML Plunkett, JA Hailey. An in vivo quantitative angiogenesis model using tumor cells entrapped in alginate. Lab Invest 62:510–517, 1990.
82. A Passaniti, RM Taylor, R Pili, Y Guo, PV Long, JA Haney, RR Pauly, DS Grant, GR Martin. A simple, quantitative method for assessing angiogenesis and antiangiogenic agents using reconstituted basement membrane, heparin, and fibroblast growth factor. Lab Invest 67:519–528, 1992.
83. J Kowalski, HH Kwan, SD Prionas, AC Allison, LF Fajardo. Characterization and applications of the disc angiogenesis system. Exp Mol Pathol 56:1–19, 1992.
84. JA Thompson, KD Anderson, JM DiPietro, JA Zwiebel, M Zametta, WF Anderson, T Maciag. Site-directed neovessel formation in vivo. Science 241:1349–1352, 1988.
85. M Watanabe, KL McCormick, K Volker, JR Ortaldo, JM Wigginton, MJ Brunda, RH Wiltrout, WE Fogler. Regulation of local host-mediated anti-tumor mechanisms by cytokines: direct and indirect effects on leukocyte recruitment and angiogenesis. Am J Pathol 150:1869–1880, 1997.
86. H Hawighorst, PG Knapstein, MV Knopp, W Weikel, G Brix, I Zuna, SO Schonberg, M Essig, P Vaupel, G van Kaick. Uterine cervival carcinoma: comparison of standard and pharmacokinetic analysis of time-intensity curves for assessment of tumor angiogenesis and patient survival. Cancer Res 58:3598–3602, 1998.
87. DA Sipkins, DA Cheresh, MR Kazemi, LM Nevin, MD Bednarski, KC Li. Detection of tumor angiogenesis in vivo by alpha Vbeta3-targeted magnetic resonance imaging. Nature Med 4:623–626, 1998.
88. JA van der Laak, JR Westphal, LJ Schalkwijk, MM Pahlplatz, DJ Ruiter, RM de Waal, PC de Wilde. An improved procedure to quantify tumour vascularity using true color image analysis: comparison with the manual hot-spot procedure in a human melanoma xenograft model. J Pathol 184:136–143, 1998.
89. H Lyng, GA Dahle, O Kaalhus, A Skretting, EK Rofstad. Measurement of perfusion rate in human melanoma xenografts by contrast-enhanced magnetic resonance imaging. Magn Reson Med 40:89–98, 1998.
90. DM Nanus, BJ Schmitz-Drager, RJ Motzer, AC Lee, V Vlamis, C Cordon-Cardo, AP Albino, VE Reuter. Expression of basic fibroblast growth factor in primary human renal tumors: correlation with poor survival. J Natl Cancer Inst 85:1597–1599, 1994.
91. RD Leek, AL Harris, CE Lewis. Cytokine networks in solid human tumors: regulation of angiogenesis. J Leukoc Biol 56:423–435, 1994.
92. CE Lewis, FR Balkwill. The complex molecular network regulating tumour angiogenesis: the importance of context. In: R Bicknell, CE Lewis, N Ferrara, eds. Tumor angiogenesis. Oxford, UK: Oxford University Press, 1997, pp. 111–113.
93. HF Dvorak, AM Dvorak. Immunohistochemical characterization of inflammatory cells that infiltrate tumors. In: S Haskill, ed. Tumor Immunity in Prognosis: The Role of Mononuclear Cell Infiltration. New York: Marcel Dekker, 1982, pp. 279–302.
94. C O'Sullivan, CE Lewis. Tumour-associated leucocytes: friends or foes in breast carcinoma. J Pathol 172:229–235, 1994.
95. JE Talmadge, M Key, IJ Fidler. Macrophage content of metastatic and nonmetastic rodent neoplasms. J Immunol 126:2245–2248, 1981.
96. M Key, JE Talmadge, IJ Fidler. Lack of correlation between the progressive growth of spon-

taneous metastases and their content of infiltrating macrophages. J Reticuloendothel Soc 32: 387–396, 1982.
97. JE Talmadge, KA Uithoven, AE Reif. Relationship of macrophage content, immunogenicity, and metastatic potential of a murine osterosarcoma of recent origin. Clin Exp Metast 3:61–72, 1985.
98. PJ Bugelski, R Kirsh, C Buscarino, SP Corwin, G Poste. Recruitment of exogenous macrophages into metastases at different stages of tumor growth. Cancer Immunol Immunother 24:93–98, 1987.
99. PJ Bugelski, RL Kirsh, JM Sowinski, G Poste. Changes in the macrophage content of lung metastases at different stages in tumor growth. Am J Pathol 118:419–424, 1985.
100. PJ Bugleski, SP Corwin, SM North, RL Kirsh, GL Nicolson, G Poste. Macrophage content of spontaneous metastases at different stage of growth. Cancer Res 47:4141–4145, 1987.
101. RD Leek, CE Lewis, AL Harris. The role of macrophages in tumour angiogenesis. In: R Bicknell, CE Lewis, N Ferrara, eds. Tumor angiogenesis. Oxford, UK: Oxford University Press, 1997, pp. 81–99.
102. Z Dong, R Kumar, X Yang, IJ Fidler. Macrophage-derived metalloelastase is responsible for the generation of angiostatin in Lewis lung carcinoma. Cell 88:801–810, 1997.
103. C Sunderkotter, K Steinbrink, M Goebeler, R Bhardwaj, and C Sorg. Macrophages and angiogenesis. J Leukoc Biol 55:410–422, 1994.
104. N Hogg, C Berlin. Structure and function of adhesion receptors in leukocyte trafficking. Immunol Today 16:327–330, 1995.
105. DN Granger, P Kubes. The microcirculation and inflammation: modulation of leukocyte-endothelial cell adhesion. J Leuko Biol 55:662–675, 1994.
106. DH Adams, S Shaw. Leucocyte-endothelial interactions and regulation of leucocyte migration. Lancet 343:831–836, 1994.
107. MP Bevilacqua. Endothelial-leukocyte adhesion molecules. Annu Rev Immunol 11:767–804, 1993.
108. AE Koch, MM Halloran, CJ Haskill, MR Shah, PJ Polverini. Angiogenesis mediated by soluble forms of E-selectin and vascular cell adhesion molecule-1. Nature 376:517–519, 1995.
109. RJ Melder, GC Koenig, BP Witwer, N Safabakhsh, LL Munn, RK Jain. During angiogenesis, vascular endothelial growth factor and basic fibroblast growth factor regulate natural killer cell adhesion to tumor endothelium. Nature Med 2:992–997, 1996.
110. AW Griffioen, CA Damen, GH Blijham, G Groenewegen. Tumor angiogenesis is accompanied by a decreased inflammatory response of tumor-associated endothelium. Blood 88:667–673, 1996.
111. AW Griffioen, CA Damen, S Martinotti, GH Blijham, G Groenewegen. Endothelial intercellular adhesion molecule-1 expression is suppressed in human malignancies: the role of angiogenic factors. Cancer Res 56:1111–1117, 1996.
112. NZ Wu, B Klitzman, R Dodge, MW Dewhirst. Diminished leukocyte-endothelium interaction in tumor microvessels. Cancer Res 52:4265–4268, 1992.
113. RM Streiter, PJ Polverini, DA Arenberg, SL Kunkel. The role of CXC chemokines as regulators of angiogenesis. Shock 4:155–160, 1995.
114. AE Koch, PJ Polverini, SL Kunkel, LA Harlow, LA DiPietro, VM Elner, SG Elner, RM Strieter. Interleukin-8 (IL-8) as a macrophage-derived mediator of angiogenesis. Science 258:1798–1801, 1992.
115. RM Streiter, SL Kunkel, VM Elner, CL Martonyl, AE Koch, PJ Polverini, SG Elner. Interleukin-8: a corneal factor that induces neovascularization. Am J Pathol 141:1279–1284, 1992.
116. Y Cao, C Chen, JA Weatherbee, M Tsang, J Folkman. Gro-b, a -C-X-C- chemokine, is an angiogenesis inhibitor that suppresses the growth of Lewis lung carcinoma in mice. J Exp Med 182:2069–2077, 1995.
117. A Angiolillo, C Sgadari, DD Taub, F Liao, JM Farber, S Maheshwari, HK Kleinman,

GH Reaman, G Tosato. Human interferon-inducible protein 10 is a potent inhibitor of angiogenesis in vivo. J Exp Med 182:155–162, 1995.
118. EE Voest, BM Kenyon, MS O'Reilly, G Truitt, RJ D'Amato, J Folkman. Inhibition of angiogenesis in vivo by interleukin 12. J Natl Cancer Inst 87:581–586, 1995.
119. C Sgadari, AL Angiolillo, G Tosato. Inhibition of angiogenesis by interleukin 12 is mediated by the interferon-inducible protein 10. Blood 87:3877–3882, 1996.
120. WE Fogler, K Volker, KL McCormick, M Watanabe, JR Ortaldo, RH Wiltrout. NK cell distribution into lung, liver and subcutaneous B16 melanoma is mediated by VCAM-1/VLA-4 interaction. J Immunol 156:4707–4714, 1996.
121. RJ Melder, HA Salehi, RK Jain. Interaction of activated natural killer cells with normal and tumor vessels in cranial windows in mice. Microvasc Res 50:35–44, 1995.
122. C Melani, SM Pupa, A Stoppacciaro, S Menard, MI Colnaghi, G Parmiani, MP Colombo. An in vivo model to compare human leukocyte infiltration in carcinoma xenografts producing different chemokines. Int J Cancer 62:572–578, 1995.
123. P Basse, RB Herberman, U Nannmark, BR Johansson, M Hokland, K Wasserman, RH Goldfarb. Accumulation of adoptively transferred adherent lymphokine-activated killer cells in murine metastases. J Exp Med 174:479–488, 1991.
124. KD Griffith, EJ Read, JA Carrasquillo, CS Carter, JC Yang, B Fisher, P Aebersold, BS Packard, MY Yu, SA Rosenberg. In vivo distribution of adoptively transferred indium-111-labeled tumor infiltration lymphocytes and peripheral blood lymphocytes in patients with metastatic melanoma. J Natl Cancer Inst 81:1709–1717, 1989.
125. C Chen, S Parangi, MJ Tolentino, J Folkman. A strategy to discover circulating angiogenesis inhibitors generated by human tumors. Cancer Res 55:4230–4233, 1995.
126. MC Kibbey, ML Corcoran, LM Wahl, HK Kleinman. Laminin SIKVAV peptide-induced angiogenesis in vivo is potentiated by neutrophils. J Cell Physiol 160:185–193, 1994.
127. R Pili, Y Guo, J Chang, H Nakanishi, GR Martin, A Passaniti. Altered angiogenesis underlying age-dependent changes in tumor growth. J Natl Cancer Inst 86:1303–1314, 1994.
128. DD Henninger, J Panes, M Eppihimer, J Russell, M Gerritsen, DC Anderson, DN Granger. Cytokine-induced VCAM-1 and ICAM-1 expression in different organs of the mouse. J Immunol 158:1825–1832, 1997.
129. U Dahmen, SD Bergese, S Qian, RP Pelletier, H Wu, DD Sedmak, JJ Fung, CG Orosz. Patterns of inflammatory vascular endothelial changes in murine liver grafts. Transplantation 60:577–584, 1995.
130. RK Jain. Determinants of tumor blood flow: a review. Cancer Res 48:2641–2658, 1988.
131. S Patan, LL Munn, RK Jain. Intussusceptive microvascular growth in solid tumors: a novel mechanism of tumor angiogenesis. Microvasc Res 51:260–272, 1996.
132. JR Less, TC Skalak, EM Sevick, RK Jain. Microvasculature architecture in a mammary carcinoma: branching patterns and vessel dimensions. Cancer Res 51:265–273, 1991.
133. AM Jonas, CB Carrington. Vascular patterns in primary and secondary pulmonary tumors in the dog. Am J Pathol 56:79–95, 1969.
134. AA Shivas, WJ Gillespie. The vascularization of Brown-Pearce carcinoma implanted in the rabbit liver. Br J Cancer 23:638–643, 1969.
135. U Nannmark, BR Johansson, JL Bryant, ML Unger, ME Hokland, RH Goldfarb, PH Basse. Microvessel origin and distribution in pulmonary metastases of B16 melanoma: implication for adoptive immunotherapy. Cancer Res 55:4627–4632, 1995.
136. PC Brooks, S Stromblad, R Klemke, D Visscher, FH Sarkar, DA Cheresh. Antiintegrin alpha v beta 3 blocks human breast cancer growth and angiogenesis in human skin. J Clin Invest 96:1815–1822, 1995.
137. M Christofidou-Solomidou, M Bridges, GF Murphy, SM Albelda, HM DeLisser. Expression and function of endothelial cell alpha v integrin receptors in wound-induced human angiogenesis in human skin/SCID chimeras. Am J Pathol 151:975–983, 1997.
138. DM Briscoe, P Ganz, SI Alexander, RJ Melder, RK Jain, RS Cotran, AH Lichtman. The

problem of chronic rejection: influence of leukocyte-endothelial interactions. Kidney Int 58: S22–S27, 1977.
139. C Manzotti, RA Audisio, G Pratesi. Importance of orthotopic implantation for human tumors as model systems: relevance to metastasis and invasion. Clin Exp Metas 11:5–14, 1993.
140. RM Mohammad, A Al-Katib, GR Pettit, VK Vaitkevicus, U Joshi, V Adsay, AP Majumdar, FH Sarkar. An orthotopic model of human pancreatic cancer in severe combined immunodeficiency mice: potential application for preclinical studies. Clin Cancer Res 4:887–894, 1998.
141. K Rembrink, JC Romijn, TH vander Kwast, H Rubben, FH Schroder. Orthotopic implantation of human prostate cancer cell lines: a clinically relevant animal model for metastatic prostate cancer. Prostate 31:168–174, 1997.
142. S Crnalic, I Hakansson, L Boquist, R Lofvenberg LA Brostrom. A novel spontaneous metastasis model of human osteosarcoma developed using orthotopic transplantation of intact tumor tissue into tibia of mice. Clin Exp Metast 15:164–172, 1997.
143. GS Oshinsky, Y Chen, T Jarret, AE Anderson, GH Weiss. A model of bladder tumor xenografts in the nude rat. J Urol 154:1925–1929, 1995.
144. T-H Kuo, T Kubota, M Watanabe, T Furukawa, S Kase, H Tanino, Y Saikawa, K Ishibiki, M Kitajima, RM Hoffman. Site-specific chemosensitivity of human small-cell lung carcinoma growing orthotopically compared to subcutaneously in SCID mice: the importance of orthotopic models to obtain relevant drug evaluation data. Anticancer Res 13:627–630, 1993.
145. T Kubota. Metastatic models of human cancer xenografted in the nude mouse: the importance of orthotopic transplantation. J Cell Biochem 56:4–8, 1994.
146. RM Hoffman. Orthotopic is orthodox: why are orthotopic-transplant metastatic models different from all other models? J Cell Biochem 56:1–3, 1994.
147. SE Cowen, MC Bibby, JA Double. Characterization of the vasculature within a murine adenocarcinoma growing in different sites to evaluate the potential of vascular therapies. Acta Oncol 34:357–360, 1995.
148. S Naito, AC von Eschenbach, R Giavazzi, IJ Fidler. Growth and metastasis of tumor cells isolated from a human renal cell carcinoma implanted into different organs of nude mice. Cancer Res 46:4109–4115, 1986.
149. RK Singh, CD Bucana, M Gutman, D Fan, MR Wilson, IJ Fidler. Organ site-dependent expression of basic fibroblast growth factor in human renal cell carcinoma cells. Am J Pathol 145:365–374, 1994.
150. RK Singh, M Gutman, CD Bucana, R Sanchez, N Llansa, IJ Fidler. Interferons a and b downregulate the expression of basic fibroblast growth factor in human carcinomas. Proc Natl Acad Sci 92:4562–4566, 1995.
151. Y Kitadai, CD Bucana, LM Ellis, H Anzai, E Tahara, IJ Fidler. In situ mRNA hybridization technique for analysis of metastasis-related genes in human colon carcinoma cells. Am J Pathol 147:1238–1247, 1995.
152. Y Takahasi, LM Ellis, MR Wilson, CD Bucana, Y Kitadai, IJ Fidler. Progressive upregulation of metastasis-related genes in human colon cancer cells implanted into the cecum of nude mice. Oncol Res 8:163–169, 1996.
153. D Fukumura, F Yuan, WL Monsky, Y Chen, RK Jain. Effect of host microenvironment on the microcirculation of human colon adenocarcinoma. Am J Pathol 151:679–688, 1997.
154. S Runkel, N Hunter, L Milas. An intradermal assay for quantification and kinetic studies of tumor angiogenesis in mice. Radiat Res 126:237–243, 1991.
155. EK Rofstad. Orthotopic human melanoma xenograft model systems for studies of tumour, angiogenesis, pathophysiology, treatment sensitivity and metastatic pattern. Br J Cancer 70: 804–812, 1994.

21

Thrombospondin-1 Modulates Angiogenesis and Tumor Progression

New Targets for Antiangiogenesis Therapy

Anitha S. John and George P. Tuszynski
MCP-Hahnemann University, Philadelphia, Pennsylvania

I. INTRODUCTION

Angiogenesis is the formation of new blood vessels from existing vasculature. Angiogenesis is a tightly regulated process occurring in a variety of physiological and pathological processes. Some of the physiological processes where transient angiogenesis is observed include would healing and reproductive and developmental processes. Pathological conditions involving aberrant angiogenic control include arthritis, atherosclerosis, and tumor growth. Tumor growth, in particular, is dependent on angiogenesis not only to obtain nutrients and oxygen requirements but also to obtain a vascular route for distant metastasis. Angiogenesis is a multistep pathway involving extracellular matrix (ECM) breakdown, endothelial cell migration and proliferation, and microvessel differentiation. The initiation of these steps has been attributed to the release of angiogenic factors, which subsequently stimulate or inhibit new vessel formation. The balance between inhibitory and stimulatory factors is a critical level of control in the initiation of angiogenesis. One of these angiogenic factors is thrombospondin (TSP-1). The role of TSP-1 in angiogenesis and tumor metastasis remains controversial. TSP-1 has been considered an angiogenesis inhibitor and therefore an inhibitor of metastasis. There have been many investigators, however, reporting data that TSP-1 promotes angiogenesis and tumor progression. Still other laboratories cite a biphasic role of TSP-1 as both an inhibitor and a promoter of angiogenesis, depending on the concentration of TSP-1. This report reviews all of the pertinent literature regarding TSP-1's involvement in angiogenesis and tumor metastasis and presents some of the data in our laboratory supporting TSP-1's role as a modulator of angiogenesis and tumor progression.

II. ANGIOGENESIS AND TUMOR METASTASIS

It has been recognized for decades that the growth and development of tumors is dependent on angiogenesis. For example, Folkman showed that tumors would not grow beyond 2

mm^3 unless vascularization occurred (1). More recently, Weidener et al. observed that the metastatic potential of several carcinomas, including breast and prostate carcinomas, depends on tumor angiogenesis as measured by intratumor microvessel density (2).

As stated in the introduction, angiogenesis is a multistep process beginning with the release of angiogenic factors. These factors can be secreted by a variety of cells, including stromal cells, inflammatory cells, and tumor cells. In addition, these factors can also be matrix-bound and subsequently be released by tumor cell–secreted proteolytic enzymes (3). In response to these angiogenic factors, endothelial cells alter their morphology by beginning to retract. In addition, endothelial cells begin subcellular protein synthesis, as evidenced by increases in the amount of endoplasmic reticulum and Golgi apparatus. The second step involved in angiogenesis includes the degradation of the capillary basement membrane and ECM by proteolytic enzymes. These enzymes are released by the endothelial cells and/or the tumor cells and include members of the matrix metalloproteinase family and plasminogen activators. After dissolution of the basement membrane and the ECM, the third step of angiogenesis involves endothelial cell migration and capillary development. Endothelial cells migrate out from the parent vessel, eventually forming a lumen. All of the cells within the new vessel are coordinately regulated; however, the exact mechanism of control remains a mystery. Some cytokines and growth factors such as interleukin-8 (IL-8), tumor necrosis factor (TNF-α), and vascular endothelial growth factor (VEGF) have been implicated in the process (3). Cell adhesion molecules such as E and P selectins and integrins have been shown to play a role in endothelial cell migration and adhesion (3). A recent study by Arap et al. showed that the arginine-glycine-aspartic acid (RGD) peptide, a peptide that selectively binds to integrins $\alpha_v\beta_3$ and $\alpha_v\beta_5$, coupled to a cytotoxic agent was capable of halting tumor growth by selectively destroying microvessels (4). The final steps of angiogenesis involve endothelial cell proliferation and microvessel maturation. Endothelial cells are normally quiescent. It has been estimated that tumor vasculature proliferates 30–40 times more quickly than the endothelial cells of normal tissues (3). By halting the migration and proliferation of these endothelial cells, tumor growth and metastasis should be inhibited.

Tumor metastasis also involves a series of steps that ultimately lead to the formation of a tumor at a distant organ site (5). The process of metastasis involes tumor-cell invasion, enabling it to migrate through the ECM; intravasation into and extravasation out of the blood or lymphatic circulation; and colonization and angiogenesis in the distant organ. the increase in the tumor microvasculature not only allows for rapid growth of tumor but also provides a means for tumor cells to enter and exit the circulation during hematogenous tumor spread. In addition, endothelial cells may also play a significant role in tumor progression by providing invading tumor cells with essential molecules necessary for matrix degradation, such as proteolytic enzymes (6). Therefore, tumor angiogenesis is critical to the growth and spread of cancer.

III. THROMBOSPONDIN-1 (TSP-1)

A. TSP-1 Structure and Functional Domains

TSP-1 is composed of three identical disulfide-linked chains each consisting of 1152 amino acids (MW 145,000), and each polypeptide chain is composed primarily of domains consisting of repeating homologous amino acid sequences (7,8). These domains are (a) the NH_2-terminal globular domain; (b) a procollagen homology domain; (c) the type 1 or properdin repeat domain, consisting of three repeating sequences homologous to sequences found in properdin; (d) the type 2 repeat domain, consisting of three repeating sequences homologous to those in epidermal growth factor; (e) the type 3 repeat domain, consisting of seven repeating Ca^{2+}-binding sequences; and (f) a COOH-terminal globular domain. These distinct domains interact with different cell-surface receptors and mediate a variety of cellular processes, including cell attachment, migration, proliferation, and differentiation (9). There are a minimum of 12 identified receptors for TSP-1 that can bind similar or different regions of TSP-1. For example, the cystein-serine-valine-threonine-cystein-glycine (CSVTCG) sequence within the type 1 repeats binds the CSVTCG receptor, recently identified and isolated from tumor-cell membranes (10). This domain has been shown to promote adhesion of tumor cells as well as endothelial cells (11–13). This domain also binds CD36, a receptor postulated to mediate the antiangiogenic activity of TSP-1 (14).

It is now known that TSP-1, originally characterized from platelet released proteins, is only one member of a family of structurally related proteins encoded by different genes, which include at least five new members, designated TSP-2, TSP-3, TSP-4, TSP-5/COMP (cartilage oligomeric matrix protein), and the most recent member, SCO-spondin (15,16). Other proteins have been identified with thrombospondin-like domains. For example, the newly discovered N-proteinase (17) and members of the ADAMTS (clone A disintegrin and metalloproteinase with thrombospondin motifs) (18) molecule family contain type I properdin repeats. Since only TSP-1 has been purified and most functional studies of TSP have been conducted with TSP-1, this review focuses on TSP-1.

B. TSP-1: Cellular and Tissue Localization

TSP-1 was originally isolated from platelets and was originally thought to be a platelet specific protein (19). Indeed, TSP-1 is stored in high concentrations in the α granules of platelets and is released into the bloodstream upon platelet activation. Since then, TSP-1 has been shown to be produced by a variety of cells including endothelial cells, fibroblasts, and immune cells (20–22). These cell types are capable of producing TSP-1 and incorporating it into the ECM. TSP-1 is usually not present under physiological conditions in most organs with the exception of a few sites within the body including the brain (23) and bone (24). TSP-1 is present at very low levels in normal plasma, and undamaged subendothelial basement membrane (25). However, TSP-1 is abundant at sites of platelet activation and in areas of remodeling tissues such as developing embryos and wounds in their early stages of repair (26,27). In particular, TSP-1 is localized in developing blood vessels (28) and vessels within and adjacent to wounds (29). Moreover, a recent immunohistochemical and in situ hybridization study has demonstrated that TSP-1 is highly expressed in the endothelium of intraplaque microvessels and of atheromatous plaques present in human atherosclerotic lesions (28). Although the functions of TSP-1 in these tissues remain unknown, these findings strongly suggest that TSP-1 may play an important role in the processes of angiogenesis.

Immunohistochemical localization studies reveal a role for TSP-1 in tumor progression. TSP-1 can be synthesized by many different types of tumor cells, although often not in as great a quantity as in their normal counterparts (15). Pratt et al. showed that the cytosolic homogenates from malignant tumors contained a much higher percentage of TSP-1 than did those fractions from benign tumors (30). In addition, high levels seemed to be associated with the center of the tumor mass, perhaps, as the authors suggest, due to the activation of endothelium within the tumor. Wong et al. reported an immunohistochemical analysis of breast cancers. In this study, 96% of the malignant tumors showed a strong expression of TSP-1 in the desmoplastic stroma adjacent to the tumor tissue (31). Benign tissues and normal tissues showed no staining of TSP-1 aside from some faint stromal staining in some fibroadenomas. No major differences were observed for other matrix proteins, suggesting that only TSP-1 was dysregulated. Our laboratory has observed similar results with breast carcinomas (32,33), squamous cell carcinomas of the head and neck (34), and pancreatic carcinoma (35). Staining was localized to the desmoplastic stroma surrounding the tumor tissue. In addition, staining of the TSP-1 CSVTCG receptor was also observed both in the tumor cells and in the endothelial cells surrounding the microvessels (33). These studies showing increased expression of TSP-1 in the tumor stroma and increased expression of a newly identified human TSP-1 specific receptor in tumor cells and the endothelium of tumor capillaries suggest a role for TSP-1 and its receptor in tumor progression and angiogenesis. The prognostic value of these staining studies is discussed in a later section.

IV. TSP-1 AND ANGIOGENESIS

A. Roles of TSP-1 as an Inhibitor of Angiogenesis

Recently, considerable evidence has implicated TSP-1 in the modulation of endothelial cell functions and angiogenesis. TSP-1 has been shown to inhibit the proliferation of endothelial cells in several studies (36,37). The ability of TSP-1 to inhibit proliferation has been suggested to be a result of inhibition of adhesion of endothelial cells to a fibronectin matrix through the amino-terminal heparin binding domains of TSP-1 (38,39). TSP-1 has also been reported to inhibit the growth and motility of endothelial cells by directly inhibiting bFGF binding (40). Dameron et al. showed that an anti-TSP-1 antibody added to early passage Li-Fraumeni fibroblast–conditioned medium promoted endothelial cell migration in a Boyden chamber assay, suggesting that TSP-1 inhibits cell motility (41). All of these potential mechanisms by which TSP-1 contributes to an antiangiogenic phenotype are controversial, as there are many studies citing the opposite observations. These results are discussed in a later section.

In vivo data first cited by Bouck and colleagues reported that TSP-1 inhibited angiogenesis as measured by a rat cornea assay of angiogenesis (42). In this report, TSP-1 (0.125 μg) was placed together with bFGF (0.05 μg) in a slow release hydron pellet, which was subsequently implanted into the rat cornea. The TSP-1 was found to inhibit bFGF-induced angiogenesis. This study was the precursor of a number of reports about the involvement of TSP-1 in in vivo angiogenesis models in a variety of disease processes (43,44). Most recently, Volpert et al. reported a study in which a human fibrosarcoma derived from the HT-1080 cell line inhibited angiogenesis and metastasis of B16/F10 melanoma cells (45). These results were attributed to the fact that HT-1080 cells secrete a large amount of TSP-1. These authors conclude that TSP-1, acting like such agents as angiostatin and endostatin, supposedly blocked the formation of micrometastases from the melanoma cells by means of antiangiogenic activity. Some important points to consider, however, are that (1) HT-1080–conditioned medium is strongly angiogenic and (2) other inhibitors, such as angiostatin, may be secreted in high amounts by these cells accounting for the inhibition of tumor growth. However, when the HT-1080 medium was tested in the rat cornea angiogenesis assays, an inhibitory effect was seen despite the presence of other known angiogenesis stimulators such as VEGF and bFGF. Another study published by Bouck and coworkers stated that the transition from a normal urothelium to an angiogenic bladder cancer is due to the downregulation of TSP-1 (43). Again, conditioned medium was used from a normal urothelium cell line and from a bladder cancer cell line. In this case, the bladder cancer medium stimulated angiogenesis in the rat aorta model and the normal urothelium inhibited both vessel formation and cell migration. The downregulation of TSP-1 while the VEGF levels remained constant was interpreted to be the switch to an angiogenic phenotype. In the Campbell study, the HT-1080 cells are capable of secreting both TSP-1 and VEGF in the same way as the normal urothelium (43). Although the HT-1080–conditioned medium was interpreted to block the formation of micrometastasis by halting angiogenesis, the fact remains that the tumor is still capable of neovascularization. The role of TSP-1 is a complex one and the arguments that TSP-1 is an angiogenesis inhibitor are countered by an increasing number of studies stating that TSP-1 promotes angiogenesis.

B. Roles of TSP-1 as a Promoter of Angiogenesis

A number of groups have also reported that TSP-1 is a stimulator of angiogenesis. All of the steps in angiogenesis discussed previously can be influenced by a number of different factors, including proteins secreted from adjacent cells. TSP-1 may be one of these proteins. ECM cells such as fibroblasts can secrete TSP-1, which, in turn, could modulate endothelial cells and other cells involved in the angiogenesis cascade (21). As stated previously, TSP-1 has been shown to inhibit the proliferation of endothelial cells. However, the neovessels of granulation tissue in early wounds express high amounts of TSP-1 (29). In addition, activated macrophages, which are angiogenic and present in the inflammatory infiltrate of wound tissue, express six times as much TSP-1 as nonactivated, nonangiogenic macrophages (46). These same authors recently reported that downregulating this activated macrophage TSP-1 production by antisense transfection caused a delay in wound healing (47). This delay was attributed to a delay in reepithelialization and dermal reorganization. In addition, proliferating endothelial cells secrete more TSP-1 than quiescent endothelial cells (48). TSP-1 also appears to be a growth requirement for primary cultures of human endothelial cells (49). These data seem to contradict the idea that TSP-1 inhibits endothelial cell proliferation.

As mentioned previously, Bouck and colleagues have a number of studies stating that TSP-1 inhibits endothelial cell migration. However, Taraboletti et al. showed that TSP-1 can modulate endothelial cell migration through the heparin-binding amino-terminal domain in a dose-dependent manner (50). High concentrations of TSP-1 (100 μg/mL) potentiated bFGF-induced cell migration, while very low concentrations (between 0.01 and 0.1 μg/mL) partially inhibited migration. The neutralization studies done to show that anti-TSP-1 antibodies stimulate endothelial cell migration may simply indicate a concentration phenomenon, depending on the amount of TSP-1 remaining after treatment with the antibody and the amount of endogenous bFGF present. Thus, endothelial cell migration is a key process in angiogenesis.

Endothelial adhesion is also an area of controversy. Our laboratory has shown previously that TSP-1 promotes endothelial cell adhesion (51). Lawler et al. showed that endothelial cell adhesion was mediated through the RGD domain in TSP-1 and integrin receptors present on endothelial cells (52). Sun et al. showed, however, that bovine aortic endothelial cells are not capable of binding TSP-1 unless the TSP-1 is disulfide-reduced

(53). The role of TSP-1 in endothelial cell proliferation, migration, and adhesion is a complex one, with factors such as concentration and environment contributing to a net effect on adhesion.

As with the in vitro data supporting TSP-1's role as an angiogenesis inhibitor, there are several studies supporting TSP-1's role as an angiogenesis promoter. Ben Ezra et al. found, in contradiction to Bouck and colleagues, that TSP-1 promotes bFGF-induced angiogenesis in the rabbit cornea model (54). These results were recently reproduced by Taraboletti et al. (55). BenEzra cites possible reasons for the discrepancy including a difference in the two models used. The rabbit cornea model is preferred over the rat cornea model due to size and technical difficulties associated with the latter model. Also, the concentration of TSP-1 and bFGF used in the study by Ben Ezra and colleagues were significantly higher. By manipulating the concentration, more TSP-1, which is capable of binding bFGF, may be available to the endothelial cells. By being able to present the bFGF as well, TSP-1 can potentiate the bFGF response. Previous studies in our laboratory demonstrated that matrix-bound TSP-1 promoted microvessel formation in a rat aorta explant angiogenesis model (56). These findings are consistent with the observation made by BenEzra et al. who reported that TSP-1 greatly potentiated the angiogenic effect of bFGF in the rabbit cornea model (54). In the rat aorta model, a rat aortic explant is embedded in a fibrin or collagen matrix that contains TSP-1. Both fibrin and collagen gel containing TSP-1 promoted the growth of microvessels. These experiments were done independently of bFGF and any effect seen was directly due to TSP-1. As the concentration of TSP-1 within the collagen matrix was increased, the number of microvessels increased. TSP-1 at a concentration of 50 μg/mL produced a 94% increase of microvessels in fibrin and a 136% increase in collagen. This effect of TSP-1 could have resulted through a series of processes including stimulating myofibroblasts to secrete angiogenic factors and stimulating endothelial cells to secrete proteases (see below), which then enables these cells to begin forming a tube morphology. The rat aorta model shows that matrix-bound TSP-1 is capable of stimulating angiogenesis. Other evidence from our laboratory includes data from a collagen gel sandwich assay in which TSP-1 stabilized the formation of microvessel network (6). Endothelial cells sandwiched between these layers form microvessel-like tubes, but in the absence of growth factors or serum, these tubelike networks dissolve. A biphasic effect was seen with TSP-1, as concentrations of 15 μg/mL inhibited this tube-forming ability, while concentrations of 5 μg/mL or less promoted stabilization. These results were correlated with the novel observation that TSP-1 is capable of stimulating MMP-9 production in endothelial cells. These findings shall be discussed in more detail in a later section. Pazouki et al. also reported a biphasic relationship between TSP-1 and vessel formation (57). Examination of TSP-1 mRNA expression in breast carcinoma samples revealed that TSP-1 expression can correlate with either a stimulatory or inhibitory effect, possibly dependent on both concentration and on other angiogenic factors. Additional evidence that TSP-1 may play a potentiating role in microvessel formation was obtained by RayChaudhury et al. (58), who showed that virus-transformed endothelial cells incapable of forming microvessels in vitro and in vivo, expressed little or no TSP-1 mRNA. However, TSP-1 was instead expressed at high levels in normal microvascular endothelial cells. When the TSP-1–deficient endothelial cells were transfected with the human TSP-1 gene, they became capable again of forming cordlike structures (59).

TSP-1 plays a complex role in angiogenesis, perhaps both stimulatory and inhibitory. The data presented from both sides of the argument appear to conflict. However, examination of some of the experimental models, such as the rat and rabbit cornea assays, may bring some light to the issue. Alternatively, the effect of TSP-1 on angiogenesis may depend on the manner by which this molecule is presented to the endothelial cells. Thus, TSP-1 may promote angiogenesis and capillary tube formation when bound to a collagen matrix, but it may reduce endothelial cell migration in response to angiogenic factors when added to the system in soluble form (6). TSP-1 binds angiogenic factors such as bFGF (60) and can either present them to the endothelium when incorporated into the matrix, or sequester them when present in solution, making them unavailable to the cells. In addition, TSP-1 in vivo may function indirectly by stimulating nonendothelial cell types which in turn promote angiogenesis (56).

V. TSP-1 AND TUMOR PROGRESSION

TSP-1 has been reported to play a role in several steps of the metastatic cascade. Again, as with angiogenesis, there is some controversy on the role of TSP-1 as a promoter or inhibitor of tumor progression. It has long been known that the hemostatic system facilitates the metastatic spread of tumors. Trousseau observed over century ago that thrombophlebitis in patients with gastrointestinal symptoms was a sign of malignancy. Other studies have shown that the propensity for malignancy decreases proportionally with a decrease in platelet counts (61). It is not surprising, then, that platelet abundant proteins, such as TSP-1, play a role in tumor progression.

As with endothelial cells, TSP-1 also promotes tumor-cell adhesion and migration. Our laboratory was the first to show that TSP-1 is a tumor cell–adhesive protein (51). This observation has been supported by a number of laboratories for a variety of different tumor types (62). There are four recognized adhesive domains within TSP-1:

1. The N-terminal heparin-binding domain, which has been shown by Roberts and colleagues to promote the attachments and spreading of human melanoma cells.
2. The CSVTCG sequence present within the type I repeats of TSP-1. An adhesion receptor, the CSVTCG receptor, was first isolated in our laboratory and has been shown to promote the adhesion of both melanoma and endothelial cells.
3. The RGD sequence present within the type III repeat, which interacts with the $\alpha v\beta 3$ integrin. Lawler and Hynes noted that this region mediates the attachment of TSP-1 to endothelial cells and platelets. These attachments are calcium dependent mechanism.
4. The carboxyl-terminal domain, which has been shown to interact with metastatic squamous cells and mediates TSP's function as a cytoadhesion molecule for progenitor cells.

These areas are the known regions which mediate TSP-1's abiltiy to act as an adhesive protein (15,63).

TSP-1 also facilitates tumor-cell migration. Taraboletti et al. first demonstrated that TSP-1 stimulated the motility of melanoma cells through its heparin-binding domain and its carboxyl-terminal domain (64). Our laboratory has shown that TSP-1 promotes the migration of oral squamous carcinoma cells, breast carcinoma cells, and pancreatic cancer cells through a collagen-gel Boyden chamber assay (65–68). These results have been attributed to the capacity of TSP-1 to regulate both urokinase plasminogen activator and MMP-9, along with their respective inhibitors.

Another study citing the importance of TSP-1 on tumor cell motility was performed by Yabkowitz et al., who found that TSP-1 stimulated motility in metastatic human squamous carcinoma cells but not in nonmetastatic cells (69). The metastatic ability of tumor cells is in part attributed to the vascularity of the tumor. Therefore, the evidence presented above in support of TSP-1's role as an angiogenesis stimulator is consistent with its role as a promoter of tumor progression. The first experiment showing that TSP-1 may play a significant role in metastasis came from our laboratory showing that TSP-1 injected prior to the injection of t241 sarcoma cells potentiated lung tumor colony formation in nude mice (70). Our laboratory has also published that patients with GI, breast, lung, and gynecological cancers have higher blood levels of TSP-1, indicating platelet involvement (71–73).

Other evidence that TSP-1 promotes tumor cell metastasis comes from a study by Varani et al. (74). They showed that the metastatic potential of a group of squamous carcinoma cells correlates with their ability to make TSP-1. The more metastatic the cell line, the more TSP-1 was secreted. Castle et al. confirmed these results by performing an antisense reduction of TSP-1 in human lung cancer cells (75). The antisense-transfected cells lost their ability to grow and metastasize in nude mice, while control transfectants retained their metastatic potential. Interestingly, a study performed by Wienstat-Saslow et al. reported that the stable transfection of human MDA-MB-435 breast cancer cells with TSP-1 cDNA inhibited their ability to grow and vascularize (76). A mutant transfectant, however, lacking the carboxyl-terminal domain, had an increased capability to metastasize and grow. The authors interpret this to mean that the carboxyl-terminal domain is responsible for TSP-1's antiangiogenic and antimetastatic ability. The results of this study could also be, once again, the issue of concentration of TSP-1. Adding an excess TSP-1, as with the transfection, may provide a form of inhibition through receptor saturation. Removing TSP-1 naturally present, as with the antisense transfection, removes TSP-1 that is needed within the system. These results could also be explained by a difference in receptor availability. These separate transfected clones may have different types and amounts of TSP-1 receptors present contributing to the results seen.

VI. MECHANISMS OF TSP-1–MEDIATED ANGIOGENESIS AND TUMOR PROGRESSION

A. TSP-1 Cell Receptors

TSP-1 is a moleucle with numerous receptors, often corresponding to different domains of the molecule. Many of these receptors have been implicated in the processes of metastasis described above. This section describes the major receptors known to date and their role in angiogenesis and tumor progression.

1. *Heparin Sulfate Proteoglycans*

A number of studies have shown that the heparin-binding region mediates cell adhesion. Roberts et al. showed that TSP-1 promotes human G361 melanoma cell and C32 melanoma cell attachment (77). This adhesion was inhibited by ligands specific for the heparin-binding domain, such as fucoidan or monoclonal antibodies. Murphy-Ulrich and Mosher examined the binding and degradation of TSP-1 by human melanoma cells and also found binding was mediated through the heparin binding domain (78). A recent report by Incordona et al. found that TSP-1's interaction with human breast cancer cells and endothelial cells is also mediated by the heparin-binding domain (79).

Sulfated glycolipids also bind to TSP-1 through the heparin-binding domain. Roberts later reported that human melanoma cells interact with TSP-1 through glycolipids and proteoglycans (80). This binding is solely through the amino-terminal domain and correlated with the cell's ability to make glucuronosyl 3-sulfate–containing glycolipids. Kaesberg et al. observed, after studying CHO K1 cells and CHO mutant cells, that cells with a reduced heparin sulfate proteoglycan content had a decreased TSP-1 adherence (81). Endothelial cell TSP-1 has been reported to possess two distinct heparin-binding sites, suggesting more than one possible receptor (82). Syndecan is one surface glycoprotein reported to be a putative receptor (83). A recent report also showed that cellular internalization and degradation of TSP-1 is mediated through the binding of the amino-terminal heparin-binding domain to the low-density-lipoprotein receptor–related protein (LRP) (84).

2. *Integrin Receptors*

TSP-1 has an RGDA sequence within its type-three repeats immediately distal to its calcium-binding domains (85). The RGD sequence is perhaps the best-known ligand of integrins—molecules involved in cell-cell and cell-matrix interactions (86). The ability of liposomes containing GPIIb-IIIa and TSP-1 to aggregate in a fibrinogen-dependent manner was first demonstrated by Rybak (87). It was later demonstrated that melanoma cells may use GPIIb-IIIa as a receptor for cellular adhesion. This adhesive ability was blocked by preincubating the cells with the RGD peptide (88,89). Lawler and Hynes also demonstrated the ability of TSP-1 to interact through integrins with a variety of cells, including platelets and endothelial cells. They identified that TSP-1 bound these cells through the integrin $\alpha_v\beta_3$ and that platelets with defective GIIb-IIIa can bind TSP-1 in a RGD-dependent manner through integrin $\alpha_v\beta_3$ (52,90).

3. *Carboxyl-Terminal Receptor*

There are two known receptors thus far for the carboxyl-terminal domain. Yabkowitz and Dixit isolated an 80,000/105,000-kDa receptor from metastatic 11B squamous carcinoma cells responsible for TSP-1 binding (91). An antibody specific to the carboxyl terminal of TSP-1 inhibited this binding. An additional 52-kDa receptor was isolated by Gao and Frazier in 1994 (92). This receptor was later identified to be integrin-associated protein (IAP) or CD47 (93). TSP-1 acts through this receptor to activate platelet integrin $\alpha IIb\beta_3$ and promote platelet spreading and platelet aggregation (94). Dorahy et al. also reported that TSP-1 binding to platelets stimulated platelet activation and aggregation (95). In addition, once TSP-1 has bound to IAP and the platelet is activated, TSP-1 can bind with two independent receptors as well (95).

4. *Type I Repeat Receptors*

The CSVTCG sequence is present in two of the three type I repeats. The CSTSCG sequence is the corresponding sequence in the remaining repeat. Many sequence homologies exist between these type I repeats and properdin, complement C8 and C9 components, and the malarial circumsporozite protein. Two well-characterized receptors for this region include the CD36 receptor and the CSVTCG receptor (10,96). The lysosomal integral membrane protein II (LIMPII) has also been reported to be a receptor binding to the type I domain (97).

CD36 was first identified as a TSP-1 receptor by Asch et al. (96). TSP-1 binding to activated platelets, C32 melanoma cells and HT1080 fibrosarcoma cells, could be inhibited by a monoclonal antibody OKM5. Using OKM5, these investigators successfully purified an 88-kDa membrane glycoprotein receptor later identified as CD36. This receptor has been identified as the means through which TSP-1 mediates platelet-monocyte and platelet-U937 cell adhesion (98). Silverstein et al. showed that antisense inhibition of CD36 in C32 melanoma cells resulted in the lack of binding of TSP-1, while sense transfection in Bowes melanoma cells, a cell line that does not express CD36 at basal levels, resulted in the ability of these cells to bind TSP-1 (99). However, some controversy exists here as well. A study examining COS cells transfected with CD36 cDNA revealed that these cells did not show an enhanced binding to human TSP-1 (100). A differential display analysis of breast carcinoma tissue revealed that treatment with tamoxifen downregulated CD36 message (101). However, a study by Clezardin et al. found that CD36 is not expressed in invasive breast carcinoma (102).

Independent of Asch et al., our laboratory identified a tumor-cell receptor specific for the CSVTCG domain of TSP-1 (10). The receptor was isolated from A549 human lung adenocarcinoma cells by CSVTCG-sepharose chromatography. Our laboratory has previously shown that the CSVTCG domain was responsible for cell adhesion in a variety of cell types, including B16-F10 melanoma cells (11,51). CSVTCG containing peptides were also capable of inhibiting platelet aggregation and tumor-cell metastasis (11). Coinjection of B16F10 melanoma cells with TSP-1 peptides was capable of reducing lung colony formation by 42–85% (see Fig. 1). Both the CSVTCG and the CSTSCG peptides were capable of significantly inhibiting metastasis. In addition, an antibody to CSTSCG was also capable of inhibiting TSP-1–dependent cell adhesion, platelet aggregation, and tumor-cell metastasis. Antibodies to our CSVTCG receptor have also been shown to inhibit breast cancer cell adhesion and intraperitoneal seeding of MDA-MB-231 human breast carcinoma in nude mice (103). Treatment of the animal with either saline, nonspecific IgG, or an anti-TSP-1 receptor antibody resulted in extensive intraperitoneal seeding in all control IgG and saline-treated animals. Only one animal out of the six treated with the anti-TSP-1 receptor antibody developed seeding. In addition, only the saline- and control IgG–treated animals exhibited micrometastases. In vitro Boyden chamber invasion assays show that human adenocarcinoma cell lines strongly expressing the CSVTCG receptor were able to invade a collagen matrix in response to TSP-1 (66). A benign mammary epithelial cell line with undetectable levels of the CSVTCG receptor showed no response to TSP-1 (66).

As mentioned previously, extensive work has been done in our laboratory to localize the expression of the CSVTCG receptor. Clinically, this receptor may be a potential chemotherapy target. In squamous cell carcinoma of the head and neck, patients with a high positive stain score for the receptor had a decreased overall survival and developed metastases faster (34). In contrast, those patients with a low score remained disease-free for at least 2 years. This receptor may be a possible test to indicate the aggressiveness of the tumor. In addition, the peptide, either alone or in combination with cytotoxic drugs, could be used to halt the formation of metastases. Other receptor antagonists such as a humanized antibody could also be a potential chemotherapuetic strategy.

B. Activation of Transforming Growth Factor Beta (TGF-β1)

TGF-β1 is a 25-kDa dimeric protein that functions as a potent cytokine, regulating growth and differentiation involved in diverse tissue remodeling processes, such as wound healing, tissue morphogenesis, angiogenesis, and tumor development. TGF-β1 is produced by

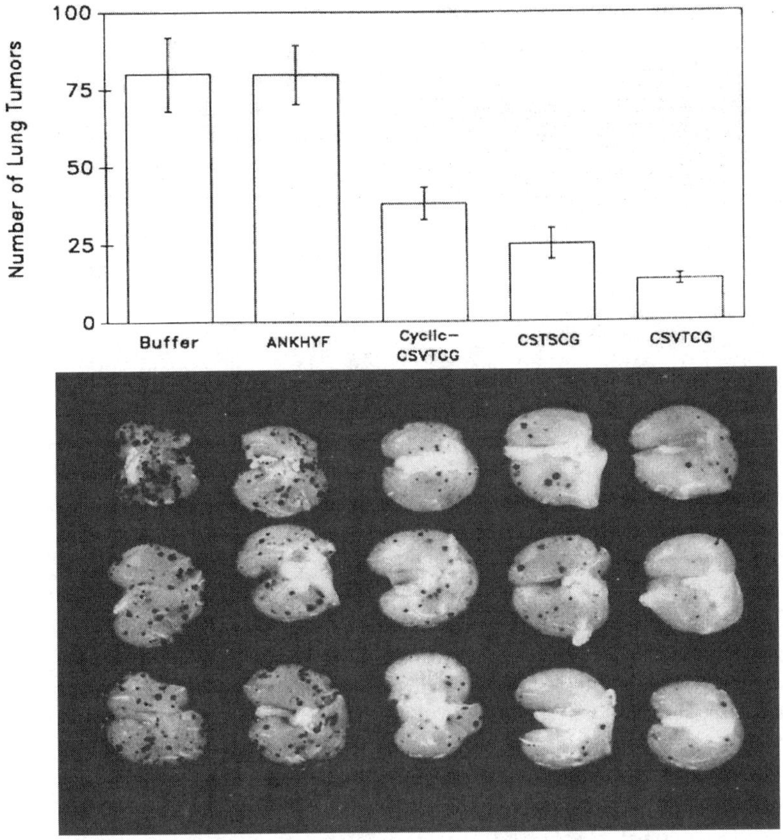

Figure 1 The effect of synthetic peptides on melanoma tumor cell metastasis after 1 mg of peptide was injected into one lateral tail vein of mice, followed immediately by injection of 100,000 B16-F10 mouse melanoma cells into the second lateral tail vein. Mice were sacrificed 2 weeks later and lung tumors were counted. Treatment groups consisted of five animals. The photograph below the bar graphs shows three lungs from each treatment group. The black spots are melanoma tumors. All metastasis experiments were repeated two to three times, showing qualitatively similar results.

virtually all cell types in a latent form, which requires activation in order to interact with its specific receptors, the transmembrane serine/threonine kinases, and elicit a biological response (104). Although TGF-β1 has a strong angiogenic effect in vivo, its in vitro effects on the angiogenic properties of endothelial cells are complex depending on the concentration of TGF-β1, matrix composition, and the presence of other cytokines (105,106). For example Iruela-Arispe and Sage showed that TGF-β1 inhibited the proliferation of endothelial cells in subconfluent monolayers but promoted the growth of cells that formed cords (107). Recently, Madri and coworkers (108) proposed a possible mechanism for these differential effects of TGF-β1 by showing that the profile of TGF-β1 receptors expressed by endothelial cells determines the response of endothelial cells to TGF-β1.

Like TSP-1, TGF-β1 is found in the platelet α granules and can be released upon platelet degranulation (7). The observation that TSP-1 and TGF-β1 colocalized in the extracellular matrix of osteoblasts suggests the possible interaction between these two molecules (109). Indeed, a number of similarities can be found between the biological effects of TGF-β1 and those of TSP-1: (a) Inhibition of endothelial cell proliferation in vitro (37,50); (b) Stimulation of angiogenesis in vivo through recruitment of inflammatory cells (54,110); (c) Promotion of effects on capillary tube formation and stabilization (59,111); (d) Upregulation of uPA and PAI-1 in cancer cells (112) and upregulation of PAI-1 in endothelial cells (37); (e) stimulation of fibroblast and smooth muscle cell proliferation (113,114); and (f) in vivo evidence showing that the TSP-1 peptide KRFK corresponding to the domain on TSP-1 reported to activate TGF-β1 reversed the lung and pancreatic abnormalities in TSP-1 mice attributed to a lack of active TGF-β1 (115,116). These findings raise the possibility that TSP-1 and TGF-β1 may mutually interact in the cellular events that contribute to the angiogenic response. Consistent with this concept is the recent work of Stacey Schultz-Cherry et al., who showed that TSP-1 can activate TGF-β1 (115).

C. TSP-1 Mediated Matrix Metalloproteinase Expression

Proteolytic degradation of the extracellular matrix (ECM) plays a major role in the growth and progression of cancer. Proteolysis is regulated by a variety of factors including cytokines and molecules of the extracellular matrix (117). We have preliminary evidence suggesting that TSP-1 is among the matrix molecules that may control pericellular proteolysis by tumor and endothelial cells. This newly discovered TSP-1 activity may provide an explanation for the apparent contradictory roles for TSP-1 in tumor progression and angiogenesis described in the literature and help to fill the gap in our understanding of the biological function of this molecule. We have discovered that TSP-1 stimulates production and activity of a 92-kDa matrix metalloproteinase, MMP-9, by breast cancer and vascular endothelial cells. We hypothesize that TSP-1 modulates cancer-cell invasion and angiogenesis by regulation of the balanced proteolytic degradation of the extracellular matrix mediated in part by the activity of TGF-β1 and the matrix metalloproteinase, MMP-9, and its natural tissue inhibitor, TIMP-1. We further hypothesize that the extent of TSP-1 stimulated cell matrix–degrading activity determines whether TSP-1 will stimulate or inhibit tumor-cell invasion and angiogenesis. Furthermore, we propose that the type I repeat peptides of TSP-1 containing the sequence CSVTCG can modulate the matrix-degrading properties of TSP-1, thus providing the basis for the development of antitumor therapies.

Matrix metalloproteinases (MMPs) represent a family of zinc-containing enzymes that degrade extracellular matrix components, allowing cells to traverse connective tissue efficiently during a number of tissue-remodeling processes, including angiogenesis (117). The 92-kDa gelatinase B (MMP-9) is one of the most recently characterized members of the MMP family, and is capable of cleaving denatured collagen, gelatin, and native collagens including the α2 chain of type I collagen, collagen types III, IV, V (118). In human coronary atherosclerotic lesions, increased expression of MMP-9 has recently been detected in atherosclerotic plaques, particularly in the regions of the plaque shoulders, core, and microvasculature (119). Recently, increased expression of MMP-9 in tumor cells as well as in stromal cells surrounding the tumor cells has been detected in several human malignant tumor tissues, such as lung and colon carcinoma (120,121). Although the mechanisms of MMP-9 upregulation and the biological functions of MMP-9 in these tumor tissues and atherosclerotic lesions have not been determined, it is evident that synthesis of MMP-9 is inducible in human microvascular endothelial cells by some inflammatory cytokines, such as tumor necrosis factor-α (TNF-α) (122), and that the expression of MMP-9 mRNA is associated with vascularization during mouse neurogenesis (123). Currently, the role of MMP-9 in angiogenesis has not yet been well established; however, it seems likely that MMP-9 may play an important role, since it has been shown that the natural inhibitor of MMP-9, TIMP-1 (tissue inhibitor of metalloproteinase-1), inhibits angiogenesis in vitro (124) and in vivo (125).

Our laboratory has studied the effect of TSP-1 mediated MMP-9 regulation on tumor cell invasion and endothelial tube formation. As mentioned previously, in a collagen sandwich gel assay, TSP-1 promoted endothelial tube formation at low concentrations (5 μg/mL) while at high concentrations (15 μg/mL), TSP-1 inhibited tube formation (see Fig. 2) (6). The conditioned media from these bovine aortic endothelial cell-tube networks were analyzed by zymography. The results showed, as the TSP-1 concentration increased, the MMP-9 secreted by these cells also increased. MMP-2, in contrast, remained constant regardless of treatment. When correlated to the change in tube length, the higher concentration of MMP-9 inhibited tube formation (see Fig. 3). These results suggest that the level of TSP-1 must be at an optimal dose for tube formation to be stabilized. An excess amount of TSP-1 may upregulate MMP-9 to a point where there is excessive proteolysis. Adding anti-MMP-9 antibodies to the collagen assay in addition to the 5 μg/mL of TSP-1 resulted in inhibition of tube formation. Treating these cells with an antibody to TSP-1 resulted in a 79% inhibition of MMP-9 production. To assess the domain of TSP-1 responsible for MMP-9 production, the BAE cells were also treated with various compounds. Heparin, at 500 μg/mL, was capable of inhibiting MMP-9 production by 50%, indicating that the heparin-binding domain plays a role in MMP-9 regulation. The CSVTCG peptide, however, from the type I repeat, blocked MMP-9 production by 86% (see Fig. 4). These results

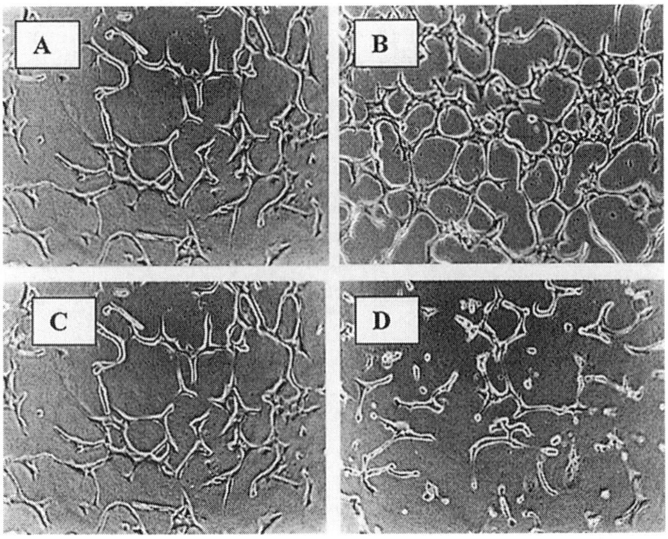

Figure 2 Effect of TSP-1 and fibronectin on endothelial tube formation. Phase-contrast micrographs showing the effects of TSP-1 and fibronectin on BAE cell microvessel-like networks after 3 days of collagen gel culture (magnification ×100). BAE cells cultured on the collagen gels under serum-free conditions were overlaid with a second layer of collagen gel to induce microvessel-like tube formation. A Buffer control. B. TSP-1 (5 μg/mL). C. Fibronectin (15 μg/mL). D. TSP-1 (15 μg/mL). Note that TSP-1 enhanced endothelial cell networks at low concentrations and inhibited network formation at high concentrations as compared with control cultures containing fibronectin.

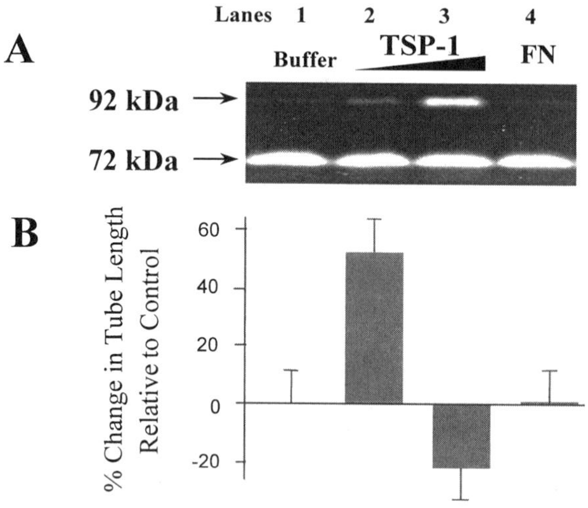

Figure 3 Effects of TSP-1 and fibronectin on BAE cell tube formation. BAE cells were cultured under serum-free conditions for 3 days between two layers of type I collagen gel containing either buffer, lane 1; TSP-1 TSP-1 (5 μg/mL) lane 2; TSP-1 (15 μg/mL) lane 3; or fibronectin (15 μg/mL) lane 4. The zymogram of the culture supernatants is shown in panel A and length of the networks in panel B. The total length of endothelial networks per field of 2×1.3 mm^2 were quantitatively determined by image analysis. Data were expressed as mean ± SD, n = 6.

are substantiated by Tolsma et al., who also showed that peptides containing the CSVTCG sequence were capable of blocking angiogenesis in a rat cornea model (126). We interpret these results to mean that the CSVTCG domain of TSP-1 is involved in mediating angiogenesis.

We have also shown in our laboratory that TSP-1 modulates the plasminogen/plasmin system. Plasmin, a known activator of many MMP family members and a proteolytic enzyme itself, is generated from plasminogen through urokinase plasminogen activator (uPa). uPa is sequestered to the cell surface by urokinase plasminogen activator receptor (uPAR) and is available on the cell surface to generate plasmin. We have shown that TSP-1 increases the amount of uPa and uPAR on a protein level in a variety of cancer

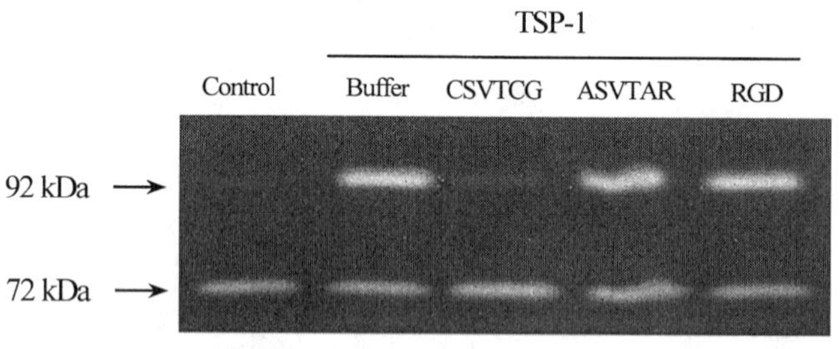

Figure 4 Zymographic analysis of the conditioned media from BAE cell monolayer cultures treated with TSP-1 and TSP-1 peptides. Cultures were treated for 24 hr with either buffer with no TSP-1–labeled control, or 15 μg/mL TSP-1 without peptides-designated buffer or 15 μg/mL TSP-1 plus 150 μg/mL of CSVTCG, ASVTAR, or RGD.

types, including breast and pancreatic carcinoma (67,68). We have also shown that TSP-1 promotes tumor cell invasion through a Boyden chamber invasion assay (68,127). This invasion is inhibited by adding anti-uPa or anti-uPAR antibodies to TSP-1–treated cells. We have also shown that TSP-1 induces MMP-9 production in tumor cells as well (unpublished data). Zymographic analysis of conditioned media of breast and pancreatic carcinoma cells revealed an upregulation of MMP-9 by TSP-1 in a dose- and time-dependent manner. This upregulation was inhibited by antibodies to TSP-1. Invasion in the in vitro Boyden chamber assay was also inhibited by adding anti-MMP9 antibodies in addition to TSP-1. Therefore, part of TSP-1's angiogenic and tumor-promoting ability results from upregulation of the MMP-9 system.

VII. CONCLUSIONS

TSP-1 is an abundant ECM molecule produced by a variety of cell types including endothelial cells, platelets, and tumor cells. TSP-1 is a complex molecule with diverse functions. Our data and the data of other investigators support the role of TSP-1 as a modulator of angiogenesis and tumor metastasis. The presence of TSP-1 in tumor stroma and the presence of the CSVTCG receptor in the tumor cells and the tumor neovasculature suggest a role for TSP-1 and its receptor in tumor progression. The mechanism we propose is summarized schematically in fig. 5. TSP-1, produced by the endothelial cell, tumor-cell, or

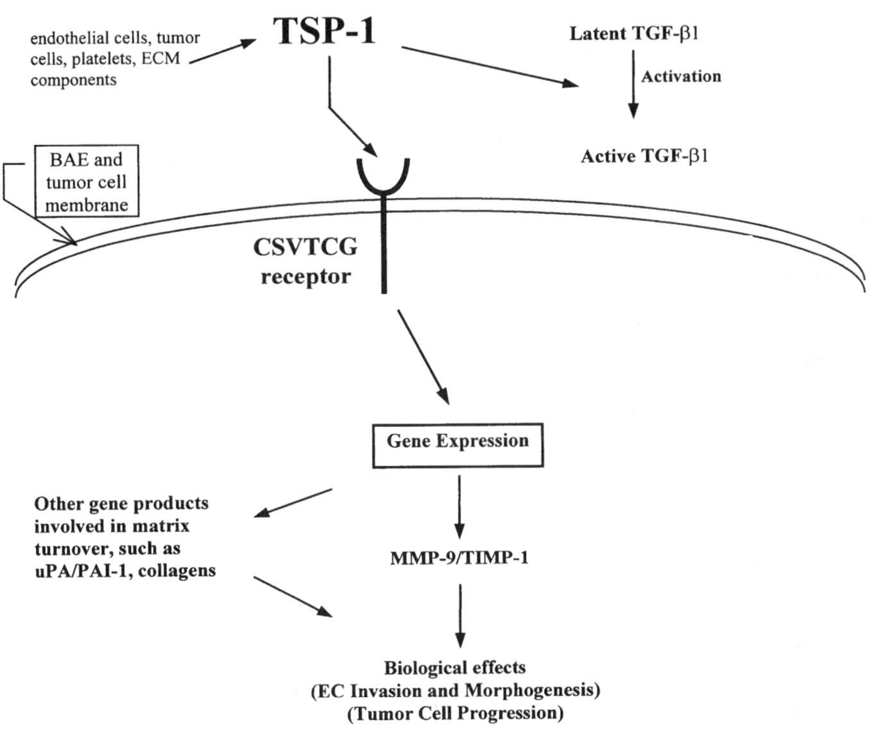

Figure 5 Hypothetical scheme for the mechanism of TSP-1-mediated angiogenesis and tumor progression.

matrix components, activates latent TGF-β. TSP-1 can act alone by binding the CSVTCG receptor, resulting in the downstream effects of the upregulation the plasmin/plasminogen system and also MMP-9 production. These enzymes and their inhibitors are important to maintain a favorable matrix environment for tumor invasion and endothelial cell morphogenesis. Ultimately, the upregulation of proteases—in addition to TSP-1's capacity to enhance endothelial cell and tumor cell proliferation, adhesion, and migration—stimulates the formation of new vessels and tumor cell migration. Once new vessels are formed within the tumor, the tumor cells have a hematogenous route through which to metastasize to different organs. Targeting TSP-1 and its receptors through receptor antagonists provides a promising avenue for reducing tumor progression.

ACKNOWLEDGMENTS

This work was supported in part by the InKine Pharmaceutical Company, Inc., and by a National Institutes of Health Grant R01 CA65675 (to G.P.T.).

REFERENCES

1. J Folkman. Tumor angiogenesis: therapeutic implications (review). N Engl J Med 285:1182–1186, 1971.
2. N Weidner. Intratumor microvessel density as a prognostic factor in cancer. Am J Pathol 147:9–19, 1995.
3. SB Fox, KC Gatter, and AL Harris. Tumour angiogenesis. J Pathol 179:232–237, 1996.
4. W Arap, R Pasqualini, and E Ruoslahti. Cancer treatment by targeted drug delivery to tumor vasculature in a mouse model (see comments). Science 279:377–380, 1998.
5. LA Liotta. Cancer cell invasion and metastasis. Sci Am 266:54–59, 1992.
6. X Qian, TN Wang, VL Rothman, RF Nicosia, and GP Tuszynski. Thrombospondin-1 modulates angiogenesis in vitro by up-regulation of matrix metalloproteinase-9 in endothelial cells. Exp Cell Res 235:403–412, 1997.
7. JW Lawler, HS Slayter, and JE Coligan. Isolation and characterization of a high molecular weight glycoprotein from human blood platelets. J Biol Chem 253:8609–8616, 1978.
8. J Lawler and RO Hynes. The structure of human thrombospondin, an adhesive glycoprotein with multiple calcium-binding sites and homologies with several different proteins. J Cell Biol 103:1635–1648, 1986.
9. WA Frazier. Thrombospondins. Curr Opin Cell Biol 3:792–799, 1991.
10. GP Tuszynski, VL Rothman, M Papale, BK Hamilton, and J Eyal. Identification and characterization of a tumor cell receptor for CSVTCG, a thrombospondin adhesive domain. J Cell Biol 120:513–521, 1993.
11. GP Tuszynski, VL Rothman, AH Deutch, BK Hamilton, and J Eyal. Biological activities of peptides and peptide analogues derived from common sequences present in thrombospondin, properdin, and malarial proteins. J Cell Biol 116:209–217, 1992.
12. CA Prater, J Plotkin, D Jaye, and WA Frazier. The properdin-like type I repeats of human thrombospondin contain a cell attachment site. J Cell Biol 112:1031–1040, 1991.
13. KA Rich, FW George IV, JL Law and WJ Martin. Cell-adhesive motif in region II of malarial circumsporozoite protein Science 249:1574–1577, 1990.
14. AS Asch, S Silbiger, D Heinegard, and RL Nachman. Thrombospondin sequence motif (CSVTCG) is responsible for CD 36 binding. Biochem Biophys Res Commun 182:1208–1217, 1992.

15. XH Qian and GP Tuszynski. Expression of thrombospondin-1 in cancer; a role in tumor progression. Proc Soc Exp Biol Med 212:199–207, 1996.
16. S Gobron, H Monnerie, R. Meiniel, I Creveaux, W Lehmann, D Lamalle, B Dastugue, and A Meiniel. SCO-spondin: a new member of the thrombospondin family secreted by the subcommissural organ is a candidate in the modulation of neuronal aggregation. J Cell Sci 109: 1053–1061, 1996.
17. DJ Prockop, AL Sieron, and SW Li. Procollagen n-proteinase and procollagen *c*-proteinase—two unusual metalloproteinases that are essential for procollagen processing probably have important roles in development and cell signaling (review). Matrix Biol 16:399–408, 1998.
18. K Kuno and K Matsushima. Adamts-1 protein anchors at the extracellular matrix through the thrombospondin type I motifs and its spacing region. J Biol Chem 273:13912–13917, 1998.
19. NL Baenziger, GN Brodie, and PW Majerus. Thrombin-sensitive protein of human platelet membranes. Proc Natl Acad Sci USA 68:240–243, 1971.
20. DF Mosher, MJ Doyle, and EA Jaffe. Synthesis and secretion of thrombospondin by cultured human endothelial cells. J Cell Biol 93:343–348, 1982.
21. EA Jaffe, JT Ruggiero, LK Leung, MJ Doyle, PJ McKeown Longo, and DF Mosher. Cultured human fibroblasts synthesize and secrete thrombospondin and incorporate it into extracellular matrix. Proc Natl Acad Sci USA 80:998–1002, 1983.
22. EA Jaffe, JT Ruggiero, and DJ Falcone. Monocytes and macrophages synthesize and secrete thrombospondin. Blood 65:79–84, 1985.
23. AS Asch, LL Leung, J Shapiro, and RL Nachman. Human brain glial cells synthesize thrombospondin. Proc Natl Acad Sci USA 83:2904–2908, 1986.
24. PG Robey, MF Young, LW Fisher, and TD McClain. Thrombospondin is an osteoblast-derived component of mineralized extracellular matrix. J Cell Biol 108:719–727, 1989.
25. BB Arbeille, FMJ Fauvel-Lafeve, MB Lemesle, D Tenza, and YJ Legrand. Thrombospondin: a component of microfibrils in various tissues. J Histochem Cytochem 39:1367–1375, 1991.
26. LA Urry, CA Whittaker, M Duquette, J Lawler, and DW Desimone. Thombospondins in early Xenopus embryos—dynamic patterns of expression suggest diverse roles in nervous system, notochord, and muscle development (review). Dev Dyn 211:390–407, 1998.
27. LA DiPietro, NN Nissen, RL Gamelli, AE Koch, JM Pyle, and PJ Polverini. Thrombospondin 1 synthesis and function in wound repair. Am J Pathol 148:1851–1860, 1996.
28. MJ Reed, ML Iruela-Arispe, ER O'Brien, T Truong, T Labell, P Bornstein, and EH Sage. Expression of thrombospondins by endothelial cells—injury is correlated with TSP-1. Am J Pathol 147:1068–1080, 1995.
29. GJ Raugi, JE Olerud, and AM Gown. Thrombospondin in early human wound tissue. J Invest Dermatol 89:551–554, 1987.
30. DA Pratt, WR Miller, and J Dawes. Thrombospondin in malignant and non-malignant breast tissue. Eur J Cancer Clin Oncol 25:343–350, 1989.
31. SY Wong, AT Purdie, and P Han. Thrombospondin and other possible related matrix proteins in malignant and benign breast disease: an immunohistochemical study. Am J Pathol 140: 1473–1482, 1992.
32. JJ Roth, DM Reiver, MS Granick, VL Rothman, RF Nicosia, and GP Tuszynski. Histopathology and clinical assessment correlate with the cysteine-serine-valine-threonine-cysteine-glycine (CSVTCG) receptor of thrombospondin-1 in breast tumors. Histol Histopathol 12:1013–1018, 1997.
33. GP Tuszynski and RF Nicosia. Localization of thrombospondin and its cysteine-serine-valine-threonine-cysteine-glycine–specific receptor in human breast carcinoma. Lab Invest 70:228–233, 1994.
34. JP Arnoletti, D Albo, N Jhala, MS Granick, MP Solomon, B Atkinson, VL Rothman, and GP Tuszynski. Computer-assisted image analysis of tumor sections for a new thrombospondin receptor. Am J Surg 168:433–436, 1994.

35. X Qian, Expression of thrombospondin-1 in human pancreatic adenocarcinoma: roles in tumor angiogenesis and matrix metalloproteinase-9 production. PhD dissertation, Allegheny University of the Health Sciences, Philadelphia, 1997.
36. P Bagavandoss and JW Wilks. Specific inhibition of endothelial cell proliferation by thrombospondin. Biochem Biophys Res Commun 170:867–872, 1990.
37. P Bagavandoss, P Kaytes, G Vogeli, PA Wells, and JW Wilks. Recombinant truncated thrombospondin-1 monomer modulatets endothelial cell plasminogen activator inhibitor 1 accumulation and proliferation in vitro. Biochem Biophys Res Commun 192:325–332, 1993.
38. JE Murphy-Ullrich and M Hook. Thrombospondin modulates focal adhesions in endothelial cells. J Cell Biol 109:1309–1319, 1989.
39. JE Murphy-Ullrich, S Gurusiddappa, WA Frazier, and M Höök. Heparin-binding peptides from thrombospondins 1 and 2 contain focal adhesion-labilizing activity. J Biol Chem 268: 26784–26789, 1993.
40. T Vogel, N Guo, HC Krutzsch, DA Blake, J Hartman, S Mendelovitz, A Panet, and DD Roberts. Modulation of endothelial cell proliferation, adhesion, and motility by recombinant heparin-binding domain and synthetic peptides from the type I repeats of thrombospondin. J Cell Biochem 53:74–84, 1993.
41. KM Dameron, OV Volpert, MA Tainsky, and N Bouck. Control of angiogenesis in fibroblasts by p53 regulation of thrombospondin-1. Science 265:1582–1584, 1994.
42. DJ Good, PJ Polverini, F Rastinejad, MM Le Beau, RS Lemons, and NP Bouck. A tumor suppressor–dependent inhibitor of angiogenesis is immunologically and functionally indistinguishable from a fragment of thrombospondin. Proc Natl Acad Sci USA 87:6624–6628, 1990.
43. SC Campbell, OV Volpert, M Ivanovich, and NP Bouck. Molecular mediators of angiogenesis in bladder cancer. Cancer Res 58:1298–1304, 1998.
44. AE Koch, Z Szekanecz, J Friedman, GK Haines, CB Langman, and NP Bouck. Effects of thrombospondin-1 on disease course and angiogenesis in rat adjuvant-induced arthritis. Clin Immunol Immunopathol 86:199–208, 1998.
45. OV Volpert, J Lawler, and NP Bouck. A human fibrosarcoma inhibits systemic angiogenesis and the growth of experimental metastases via thrombospondin-1. Proc Natl Acad Sci USA 95:6343–6348, 1998.
46. LA DiPietro and PJ Polverini. Angiogenic macrophages produce the angiogenic inhibitor thrombospondin 1. Am J Pathol 143:678–684, 1993.
47. LA DiPietro, NN Nissen, RL Gamelli, AE Koch, JM Pyle, and PJ Polverini. Thrombospondin 1 synthesis and function in wound repair. Am J Pathol 148:1851–1860, 1996.
48. AE Canfield, RP Boot-Handford, and AM Schor. Thrombospondin gene expression by endothelial cells in culture is modulated by cell proliferation, cell shape and the substratum. Biochem J 268:225–230, 1990.
49. V Morandi, F Fauvel-Lafeve, C Legrand, and YJ Legrand. Role of thrombospondin in the adhesion of human endothelial cells in primary culture. In Vitro Cell Dev Biol 29A:585–591, 1993.
50. G Taraboletti, D Roberts, LA Liotta, and R Giavazzi. Platelet thrombospondin modulates endothelial cell adhesion, motility, and growth: a potential angiogenesis regulatory factor. J Cell Biol 111:765–772, 1990.
51. GP Tuszynski, V Rothman, A Murphy, K Siegler, L Smith, S Smith, J Karczewski, and KA Knudsen. Thrombospondin promotes cell-substratum adhesion. Science 236:1570–1573, 1987.
52. J Lawler, R Weinstein, and RO Hynes. Cell attachment to thrombospondin: the role of Arg-Gly-Asp, calcium and integrin receptors. J Cell Biol 107:2351–2361, 1988.
53. X Sun, K Skorstengaard, and DF Mosher. Disulfides modulate RGD-inhibitable cell adhesive activity of thrombospondin. J Cell Biol 188:693–701, 1992.
54. D BenEzra, BW Griffin, G Maftzir, and O Aharonov. Thrombospondin and in vivo angiogen-

esis induced by basic fibroblast growth factor or lipopolysaccharide. Invest. Ophthalmol Vis Sci 34:3601–3608, 1993.
55. G Taraboletti, P Borsotti, L Morbidelli, R Giavazzi, and M Ziche. Thrombospondin-1 fragments with opposite effect on angiogenesis. Presented at the American Association for Cancer Research Annual Meeting, 1998, New Orleans, LA.
56. RF Nicosia and GP Tuszynski. Matrix-bound thrombospondin promotes angiogenesis in vitro. J Cell Biol 124:183–193, 1994.
57. S Pazouki, N Pendleton, E Heerkens, RL Smither, JV Moore, and AM Schor. Biphasic effect of thrombospondin-1 (TSP-1) in the regulation of angiogenesis in human breast carcinoma. Biochem Soc Trans 24:368S, 1996.
58. A RayChaudhury, WA Frazier, and PA D'Amore. Comparison of normal and tumorigenic endothelial cells: differences in thrombospondin production and responses to transforming growth factor-beta. J Cell Sci 107:39–46, 1994.
59. N Sheibani and WA Frazier. Thrombospondin 1 expression in transformed endothelial cells restores a normal phenotype and suppresses their tumorigenesis. Proc Natl Acad Sci USA 92:6788–6792, 1995.
60. G Taraboletti, D Belotti, P Borsotti, V Vergani, M Rusnati, M Presta, and R Giavazzi. The 140-kilodalton antiangiogenic fragment of thrombospondin-1 binds to basic fibroblast growth factor. Cell Growth Diff 8:471–479, 1997.
61. GJ Gasic, GP Tuszynski, and E Gorelik. Interactions of the hemostatic and immune systems in the metastatic spread of tumor cells. Int Rev Exp Pathol 29:173–212, 1986.
62. JC Adams and J Lawler. Diverse mechanisms for cell attachment to platelet thrombospondin. J Cell Sci 104:1061–1071, 1993.
63. GP Tuszynski and RF Nicosia. The role of thrombospondin-1 in tumor progression and angiogenesis. Bioessays. 18:71–76, 1996.
64. G Taraboletti, DD Roberts, and LA Liotta. Thrombospondin-induced tumor cell migration: haptotaxis and chemotaxis are mediated by different molecular domains. J Cell Biol 105: 2409–2415, 1987.
65. TN Wang, XH Qian, MS Granick, MP Solomon, VL Rothman, and GP Tuszynski. The effect of thrombospondin on oral squamous carcinoma cell invasion of collagen. Am J Surg 170: 502–505, 1995.
66. TN Wang, X Qian, MS Granick, MP Solomon, VL Rothman, DH Berger, and GP Tuszynski. Thrombospondin-1 (TSP-1) promotes the invasive properties of human breast cancer. J Surg Res 63:39–43, 1996.
67. D Albo, DH Berger, TN Wang, X Hu, V Rothman, and GP Tuszynski. Thrombospondin-1 and transforming growth factor-beta 1 promote breast tumor cell invasion through up-regulation of the plasminogen/plasmin system. Surgery 122:493–499; discussion, 499–500, 1997.
68. D Albo, Berger, DH Tuszynski, GP. The effect of thrombospondin-1 and TGF-beta 1 on pancreatic cancer cell invasion. J Surg Res 76:86–90, 1998.
69. R Yabkowitz, PJ Mansfield, VM Dixit, and SJ Suchard. Motility of human carcinoma cells in response to thrombospondin: relationship to metastatic potential and thrombospondin structural domains. Cancer Res 53:378–387, 1993.
70. GP Tuszynski, TB Gasic, VL Rothman, KA Knudsen, and GJ Gasic. Thrombospondin, a potentiator of tumor cell metastasis. Cancer Res 47:4130–4133, 1987.
71. GP Tuszynski, M Smith, VL Rothman, DM Capuzzi, RR Joseph, J Katz, EC Besa, J Treat, and HI Switalska. Thrombospondin levels in patients with malignancy. Thromb Haemostas 67:607–611, 1992.
72. FE Nathan, E Hernandez, CJ Dunton, J Treat, HI Switalska, RR Joseph, and GP Tuszynski. Plasma thrombospondin levels in patients with gynecologic malignancies. Cancer 73:2853–2858, 1994.
73. Y Yamashita, T Kurohiji, GP Tuszynski, T Sakai, and T Shirakusa. Plasma thrombospondin levels in patients with colorectal carcinoma cancer. 82:632–638, 1998.

74. J Varani, VM Dixit, SE Fligiel, PE McKeever, and TE Carey. Thrombospondin-induced attachment and spreading of human squamous carcinoma cells. Exp Cell Res 167:376–390, 1986.
75. V Castle, J Varani, S Fligiel, EV Prochownik, and V Dixit. Antisense-mediated reduction in thrombospondin reverses the malignant phenotype of a human squamous carcinoma. J Clin Invest 87:1883–1888, 1991.
76. DL Weinstat-Saslow, VS Zabrenetzky, K VanHoutte, WA Frazier, DD Roberts, and PS Steeg. Transfection of thrombospondin 1 complementary DNA into a human breast carcinoma cell line reduces primary tumor growth, metastatic potential, and angiogenesis. Cancer Res 54:6504–6511, 1994.
77. DD Roberts, JA Sherwood, and V Ginsburg. Platelet thrombospondin mediates attachment and spreading of human melanoma cells. J Cell Biol 104:131–139, 1987.
78. JE Murphy-Ullrich and DF Mosher. Interactions of thrombospondin with endothelial cells: receptor-mediated binding and degradation. J Cell Biology. 105:1603–1611, 1987.
79. F Incardona, J Lawler, D Cataldo, A Panet, Y Legrand, JM Foidart, and C Legrand. Heparin-binding domain, type 1 and type 2 repeats of thrombospondin mediate its interaction with human breast cancer cells. Journal of Cellular Biochemistry. 62:431–442, 1996.
80. DD Roberts. Interactions of thrombospondin with sulfated glycolipids and proteoglycans of human melanoma cells. Cancer Res 48:6785–6793, 1988.
81. PR Kaesberg, WB Ershler, JD Esko, and DF Mosher. Chinese hamster ovary cell adhesion to human platelet thrombospondin is dependent on cell surface heparan sulfate proteoglycan. J Clin Invest 83:994–1001, 1989.
82. R Dardik and J Lahav. Multiple domains are involved in the interaction of endothelial cell thrombospondin with fibronectin. Eur J Biochem 185:581–588, 1989.
83. A Rapraeger. Transforming growth factor (type β) promotes the addition of chondroitin sulfate chains to the cell surface proteoglycan (syndecan) of mouse mammary epithelia. J Cell Biol 109:2509–2518, 1989.
84. I Mikhailenko, MZ Kounnas, and DK Strickland. Low density lipoprotein receptor-related protein/alpha 2-macroglobulin receptor mediates the cellular internalization and degradation of thrombospondin: a process facilitated by cell-surface proteoglycans. J Biol Chem 270: 9543–9549, 1995.
85. J McPherson, H Sage, and P Bornstein. Isolation and characterization of a glycoprotein secreted by aortic endothelial cells in culture: apparent identity with platelet thrombospondin. J Biol Chem 256:11330–11336, 1981.
86. E Ruoslahti and MD Pierschbacher. New perspectives in cell adhesion: RGD and integrins. Science 238:491–497, 1987.
87. ME Rybak. Glycoproteins IIb and IIIa and platelet thrombospondin in a liposome model of platelet aggregation. Thromb Haemostas 55:240–245, 1986.
88. K Knudsen, L Smith, S Smith, J Karczewski, and G Tuszynski. Role of IIb-IIIa-like glycoprotein in cell-substratum adhesion of human melanoma cells. J Cell Physiol 136:471–478, 1988.
89. GP Tuszynski, J Karczewski, L Smith, A Murphy, VL Rothman, and KA Knudsen. The GPIIB-IIIa-like complex may function as a human melanoma cell adhesion receptor for thrombospondin. Exp Cell Res 182:473–481, 1989.
90. J Lawler and RO Hynes. An integrin receptor on normal and thrombasthenic platelets that binds thrombospondin. Blood 74:2022–2027, 1989.
91. R Yabkowitz and VM Dixit. Human carcinoma cells bind thrombospondin through a M_r 80,000/105,000 receptor. Cancer Res 51:3648–3656, 1991.
92. A-G Gao and WA Frazier. Identification of a receptor candidate for the carboxyl-terminal cell binding domain of thrombospondins. J Biol Chem 269:29650–29657, 1994.
93. AG Gao, FP Lindberg, MB Finn, SD Blystone, EJ Brown, and WA Frazier. Integrin-associ-

ated protein is a receptor for the C-terminal domain of thrombospondin. J Biol Chem 271: 21–24, 1996.

94. J Chung, AG Gao, and WA Frazier. Thrombspondin acts via integrin-associated protein to activate the platelet integrin alphaIIbbeta3. J Biol Chem 272:14740–14746, 1997.
95. DJ Dorahy, RF Thorne, JV Fecondo, and GF Burns. Stimulation of platelet activation and aggregation by a carboxyl-terminal peptide from thrombospondin binding to the integrin-associated protein receptor. J Biol Chem 272:1323–1330, 1997.
96. AS Asch, J Barnwell, R Silverstein, and R Nachman. Isolation of the thrombospondin membrane receptor. J Clin Invest 79:1057–1061, 1987.
97. R Crombie and R Silverstein. Lysosomal integral membrane protein Ii binds thrombospondin-1—structure-function homology with the cell adhesion molecule Cd36 defines a conserved recognition motif. J Biol Chem 273:4855–4863, 1998.
98. RL Silverstein, AS Asch, and RL Nachman. Glycoprotein IV mediates thrombospondin-dependent platelet- monocyte and platelet-U937 cell adhesion. J Clin Invest 84:546–552, 1989.
99. RL Silverstein, M Baird, SK Lo, and LM Yesner. Sense and antisense cDNA transfection of CD36 (glycoprotein IV) in melanoma cells: role of CD36 as a thrombospondin receptor. J Biol Chem 267:16607–16612, 1992.
100. P Oquendo, E Hundt, J Lawler, and B Seed. CD36 directly mediates cytoadherence of Plasmodium falciparum parasitized erythrocytes. Cell 58:95–101, 1989.
101. ID Silva, AM Salicioni, IH Russo, NA Higgy, LH Gebrim, and J Russo. Tamoxifen down-regulates CD36 messenger RNA levels in normal and neoplastic human breast tissues. Cancer Res 57:378–381, 1997.
102. P Clezardin, L Frappart, M Clerget, C Pechoux, and PD Delmas. Expression of thrombospondin (TSP1) and its receptors (CD36 and CD51) in normal, hyperplastic, and neoplastic human breast Cancer Res 53:1421–1430, 1993.
103. TN Wang, XH Qian, MS Granick, MP Solomon, VL Rothman, DH Berger, and GP Tuszynski. Inhibition of breast cancer progression by an antibody to a thrombospondin-1 receptor. Surgery 120:449–459, 1996.
104. J Massague, S Cheifetz, M Laiho, DA Ralph, FM Weis, and A Zentella. Transforming growth factor-beta. (review). Cancer Surveys 12:81–103, 1992.
105. MB Sporn and AB Roberts. Transforming growth factor-beta: recent progress and new challenges. (review). J Cell Biol 119:1017–1021, 1992.
106. JA Madri, BM Pratt, and AM Tucker. Phenotypic modulation of endothelial cells by transforming growth factor-beta depends upon the composition and organization of the extracellular matrix. J Cell Biol 106:1375–1384, 1988.
107. ML Iruela-Arispe and EH Sage. Endothelial cells exhibiting angiogenesis in vitro proliferate in response to TGF-β1. J Cell Biochem 52:414–430, 1993.
108. S Sankar, N Mahooti-Brooks, L Bensen, TL McCarthy, M Centrella, and JA Madri. Modulation of transforming growth factor β receptor levels on microvascular endothelial cells during in vitro angiogenesis. J Clin Invest 97:1436–1446, 1996.
109. M Slater, J Patava, and RS Mason. Thrombospondin co-localises with TGFβ and IGF-I in the extracellular matrix of human osteoblast-like cells and is modulated by 17β estradiol. Experientia 51:235–244, 1995.
110. PJ Mansfield and SJ Suchard. Thrombospondin promotes chemotaxis and haptotaxis of human peripheral blood monocytes. J Immunol 153:4219–4229, 1994.
111. X Qian, RF Nicosia, KM Bochenek, VL Rothman, and GP Tuszynski. The effect of thrombospondin on endothelial cell tube formation in vitro. Mol Biol Cell 5:179a, 1994.
112. D Albo, JP Arnoletti, A Castiglioni, MS Granick, MP Solomon, VL Rothman, and GP Tuszynski. Thrombospondin (TSP) and transforming growth factor beta 1 (TGF-beta) promote human A549 lung carcinoma cell plasminogen activator inhibitor type 1 (PAI-1) production

and stimulate tumor cell attachment in vitro. Biochem Biophys Res Commun 203:857–865, 1994.

113. RA Majack, LV Goodman, and VM Dixit. Cell surface thrombospondin is functionally essential for vascular smooth muscle cell proliferation. J Cell Biol 106:415–422, 1988.
114. SH Phan, RG Dillon, BM McGarry, and VM Dixit. Stimulation of fibroblast proliferation by thrombospondin. Biochem Biophys Res Commun 163:56–63, 1989.
115. S Schultz-Cherry, H Chen, DF Mosher, TM Misenheimer, HC Krutzsch, DD Roberts, and JE Murphy-Ullrich. Regulation of transforming growth factor-β activation by discrete sequences of thrombospondin 1. J Biol Chem 270:7304–7310, 1995.
116. SE Crawford, V Stellmach, JE Murphy-Ullrich, SM Ribeiro, J Lawler, RO Hynes, GP Boivin, and N Bouck. Thrombospondin-1 is a major activator of TGF-β1 in vivo. Cell 93:1159–1170, 1998.
117. LA Liotta, PS Steeg, and WG Stetler-Stevenson. Cancer metastasis and angiogenesis: an imbalance of positive and negative regulation. Cell 64:327–336, 1991.
118. Y Okada, Y Gonoji, K Naka, K Tomita, I Nakanishi, K Iwata, K Yamashita, and T Hayakawa. Matrix metalloproteinase 9 (92-kDa Gelatinase/type IV collagenase) from HT 1080 human fibrosarcoma cells. J Biol Chem 267:21712–21719, 1992.
119. ZS Galis, GK Sukhova, MW Lark, and P Libby. Increased expression of matrix metalloproteinases and matrix degrading activity in vulnerable regions of human atherosclerotic plaques. J Clin Invest 94:2493–2503, 1994.
120. BP Himelstein, R Canete-Soler, EJ Bernhard, DW Dilks, and RJ Muschel. Metalloproteinases in tumor progression: the contribution of MMP- 9 (review). Invas Metast 14:246–258, 1994.
121. R Canete-Soler, L Litzky, I Lubensky, and RJ Muschel. Localization of the 92 kd gelatinase mRNA in squamous cell and adenocarcinomas of the lung using in situ hybridization. Am J Patholo 144:518–527, 1994.
122. R Hanemaaijer, P Koolwijk, L le Clercq, WJ de Vree, and VW van Hinsbergh. Regulation of matrix metalloproteinase expression in human vein and microvascular endothelial cells: effects of tumour necrosis factor alpha, interleukin 1 and phorbol ester. Biochem J 296:(pt 3):803–809, 1993.
123. RC Soler, YH Gui, KK Linask, and RJ Muschel. MMP-9 (gelatinase B) mRNA is expressed during mouse neurogenesis and may be associated with vascularization. Dev Brain Res 88: 37–52, 1995.
124. HW Schnaper, DS Grant, WG Stetler-Stevenson, R Fridman, G D'Orazi, AN Murphy, RE Bird, M Hoythya, TR Fuerst, DL French, and et. al. Type IV collagenase(s) and TIMPs modulate endothelial cell morphogenesis in vitro. J Cell Physiol 156:235–246, 1993.
125. MD Johnson, HR Kim, L Chesler, G Tsao-Wu, N Bouck, and PJ Polverini. Inhibition of angiogenesis by tissue inhibitor of metalloproteinase. J Cell Physiol 160:194–202, 1994.
126. SS Tolsma, OV Volpert, DJ Good, WA Frazier, PJ Polverini, and N Bouck. Peptides derived from two separate domains of the matrix protein thrombospondin-1 have anti-angiogenic activity. T J Cell Biol 122:497–511, 1993.
127. D Albo, DH Berger, TN Wang, XL Hu, V Rothman, and GP Tuszynski. Thrombospondin-1 and transforming growth factor-beta1 promote breast tumor cell invasion through upregulation of the plasminogen/plasmin system. Surgery 122:493–499, 1997.

22
Inhibition of VEGF Receptor Tyrosine Kinase

Gerhard Siemeister
Research Laboratories of Schering AG, Berlin, and Institute of Molecular Medicine, Freiburg, Germany

Andreas Menrad, Michael Schirner, Karl-Heinz Thierauch, and Martin R. Schneider
Research Laboratories of Schering AG, Berlin, Germany

Georg Martiny-Baron and Dieter Marmé
Institute of Molecular Medicine, Freiburg, Germany

Guido Bold, Johannes Rösel, and Jeanette M. Wood
Novartis Pharma AG, Basel, Switzerland

I. INTRODUCTION

VEGF and its receptors on endothelial cells have been found to play a pivotal role in pathological situations that involve neovascularization and enhanced vascular permeability, such as tumor angiogenesis, diabetic retinopathy, rheumatoid arthritis, and psoriasis. Therefore, specific interference with VEGF/VEGF receptor system appears attractive for inhibition of pathological neovascularization. This chapter focuses on the validation of VEGF-receptor KDR/Flk-1 as a target for drug development and on the characterization of VEGF-receptor tyrosine kinase inhibitors that have reached early clinical trials.

II. KDR/FLK-1 AS A TARGET FOR DRUG DEVELOPMENT

Two high-affinity receptors for VEGF, Flt-1 (fms-like tyrosine kinase, VEGF-R1) and KDR (kinase insert domain-containing receptor, VEGF-R2), as well as the mouse homolog of KDR, Flk-1 (fetal-liver kinase-1), have been cloned thus far (1–3). Both receptors have seven immunoglobulin-like domains in their extracellular region, a single transmembrane-spanning domain, and an intracellular split tyrosine kinase domain. Flt-1 binds VEGF and the related placenta growth factor (4), whereas KDR binds VEGF-C and -D in addition to VEGF (5,6). VEGF-C and -D are both ligands and activators of Flt-4 (VEGF-R3), which is expressed on endothelial cells of lymphatic vessels (7). The expression level of KDR and Flt-1 is low in endothelial cells of normal tissue. In normal as well as in pathological settings with active angiogenesis or enhanced vascular permeability—such as development of the ovarian corpus luteum (8), psoriasis (9), rheumatoid arthritis (10), retinopathy (11), and tumor angiogenesis (12)—expression of Flt-1 and KDR/Flk-1 in endothelia within or near the affected areas is strongly upregulated. Recently it was shown that one prominent factor mediating VEGF-receptor upregulation is VEGF itself (13,14). These observations suggested the VEGF/VEGF receptor system as a target for suppression of pathological neovascularization.

Ferrara and coworkers were the first to show that interference with the VEGF/VEGF-receptor system by means of VEGF-neutralizing antibodies inhibits tumor growth and tumor angiogenesis in various human tumor xenograft models (15,16). Similar approaches using recombinant soluble VEGF receptor proteins to neutralize VEGF were effective for inhibition of primary tumor growth and metastasis formation in a syngeneic murine renal carcinoma model (17) as well as for inhibition of retinal neovascularization (18). Using a tetracycline-regulated VEGF expression system in a xenograft-model for glioblastoma, Benjamin and Keshet (19) recently showed that shutting off VEGF production leads to detachment of endothelial cells from the walls of preformed vessels and their subsequent death by apoptosis. This observation is consistant with the finding of VEGF acting as a survival factor for endothelial cells in vivo (20).

As enhanced expression of Flt-1 and KDR/Flk-1 was observed in tissues active in angiogenesis, the functional contribution of each receptor to pathological neovascularization was addressed by several approaches. Activation of Flt-1 was shown to mediate biological responses, such as endothelial and monocyte cell migration and tissue factor induction (21,22); however, in cells expressing only Flt-1 or in Flt-1–deficient cells transfected with *flt-1* cDNA, stimulation with VEGF induces only weak receptor phosphorylation and no significant mitogenic response (23,24). In contrast to Flt-1, KDR is strongly autophosphorylated upon VEGF-stimulation and mediates a mitogenic response (23). Furthermore, VEGF mutant proteins that bind specifically to KDR/Flk-1 induce proliferation of endothelial cells, whereas Flt-1-specific VEGF mutants fail to do so (25,26). A predominant function of KDR/Flk-1 receptor in angiogenesis was suggested by gene knockout experiments. Targeted disruption of the *flk-1* gene in mice prevented the development of mature endothelial cells (27). In contrast, inactivation of the *flt-1* gene did not affect endothelial cells, but the formation of normal vessels was impaired (28).

Millauer et al. (29) inactivated Flk-1 by retroviral transduction of a dominant-negative Flk-1 mutant to endothelial cells in vivo. This dominant-negative mutant represents a truncated Flk-1 receptor containing only the extracellular VEGF-binding domain and the transmembrane domain. Based on the observation that ligand-induced dimerization of growth factor receptors results in trans-phosphorylation and concomitant activation of their tyrosine kinase domains, it was reasoned that the truncated dominant-negative Flk-1 dimerizes with wild-type Flk-1 receptors forming inactive receptor dimers. By injection of C6 rat glioblastoma cells together with cells producing retroviruses encoding the dominant-negative Flk-1 mutant, a substantial inhibition of vascularization and tumor growth was demonstrated. Application of this dominant-negative approach to several human xenograft as well as rat and mouse tumor models showed significant inhibition of tumor growth in 6 out of 8 tumor types analyzed (30). Although these experiments suggest an important role for the KDR/Flk-1 receptor in tumor angiogenesis, the contribution of Flt-1 remains unclear. It was not convincingly ruled out that overexpression of retroviral encoded truncated Flk-1 receptors on the surface of transduced cells might neutralize tumor-derived VEGF, thereby inhibiting activation of wild-type Flk-1 as well as Flt-1. In addition, the possible heterodimerization of wild-type Flt-1 and truncated Flk-1 remained an open question. Skobe et al. (31) reported that selective blocking of VEGF binding to Flk-1 using the monoclonal antibody DC101 (32) inhibited endothelial cell proliferation, tumor cell–induced angiogenesis, and tumor cell invasion into the host tissue. Taken together, the results provide evidence that interference with KDR/Flk-1 function is sufficient to inhibit tumor angiogenesis, whereas the exact role of Flt-1 remains unclear.

III. INHIBITORS OF VEGF RECEPTOR TYROSINE KINASE

Based on these results, drug discovery programs were established to identify small molecule compounds that specifically inhibit the enzymatic activity of KDR/Flk-1 kinase and downstream signaling. SUGEN reported on the discovery of several compounds inhibiting Flk-1 kinase activity at concentrations in the low micromolar range (33–36). One of these, SU5416, has already reached clinical phase I trials. SU1433, a quinoxaline; SU1498, a tyrphostin; and SU5416, a 3-substituted indolin-2-one inhibited VEGF-stimulated Flk-1 autophosphorylation in Flk-1 overexpressing NIH3T3 cells with IC_{50} values of 9.3, 0.7, and 1 μM, respectively. They were 100-fold more selective for Flk-1 over EGF receptor and HER-2. SU1498 was 50-fold and SU5416 20-fold more selective for Flk-1 over PDGF receptor, whereas SU1433 inhibited PDGF receptor with an IC_{50} value of 5 μM. VEGF-stimulated proliferation of human umbilical vein endothelial cells (HUVEC), which express KDR receptors, was inhibited by these Flk-1 inhibitors, with IC_{50} values of 7 (SU1433), 2.5 (SU1498), and 0.04 μM (SU5416). These experiments confirmed that inhibitors of Flk-1 kinase activity also inhibited a relevant downstream response in human endothelial cells. SU1433 and SU1498 were reported to prevent the formation of new blood vessels in the chicken chorioallantoic membrane and to antagonize VEGF-induced vascular permeability in nude mice. Little information on the efficacy of these compounds in tumor models is currently available. It has been reported that SU5416, which has to be applied parenterally, inhibits growth of xenografted C6 rat glioblastoma by 92%; this was accompanied by a significant reduction in vascular density within the residual tumor (36).

A joint project of Novartis Pharma AG, Basel, Schering AG, Berlin, and Tumor Biology Center, Freiburg, succeeded in developing a potent and orally active inhibitor of VEGF receptor tyrosine kinases (37). CGP79787/ZK222584, a phtalazine derivative, represents a novel class of protein tyrosine kinase inhibitors. CGP79787/ZK222584 inhibits KDR, Flk-1, and human Flt-1 tyrosine kinase activities in vitro with IC_{50} values of 0.1, 0.3, and 0.5 μM, respectively. It also inhibits PDGF receptor tyrosine kinase but is not active against kinases from other receptor families (EGF, FGF) and nonreceptor kinases (c-Src, v-Abl, PKC). In KDR-overexpressing porcine aortic endothelial cells (23), CGP79787/ZK222584 inhibited VEGF-stimulated KDR autophosphorylation with an IC_{50} value of 0.02 μM (Fig. 1). Similar results were achieved using KDR-overexpressing Chinese hamster ovary cells, human microvascular endothelial cells, and HUVEC. Growth factor induced activation of MAP kinases was used as a measure for receptor downstream signaling. CGP79787/ZK222584 inhibited selectively VEGF-induced but not basic FGF-induced MAP kinase activation in HUVEC (Fig. 2). Proliferation (Fig. 3) and migration (Fig. 4) of HUVEC and of human microvascular endothelial cells induced by VEGF were

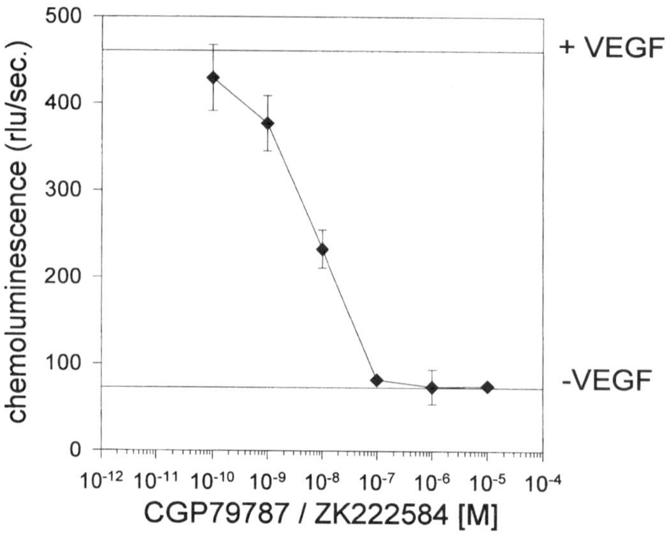

Figure 1 Inhibition of VEGF-stimulated KDR autophosphorylation by CGP79787/ZK222584. KDR overexpressing porcine aortic endothelial cells (23) were incubated on ice with 50 ng/mL VEGF for 60 min in the presence of increased concentrations of CGP79787/ZK222584. KDR tyrosine phosphorylation was determined in cell lysates by ELISA technique using anti-KDR antibody 2-10-1 (40) as catching antibody and PY-20 antiphosphotyrosine antibody for detection. Levels of KDR phosphorylation in the absence of VEGF (−VEGF) and upon VEGF-stimulation (+VEGF) are indicated by solid lines.

selectively inhibited by CGP79787/ZK222584, whereas the response to basic FGF was not affected. In concentrations up to 10 μM, CGP79787/ZK222584 does not have any antiproliferative or cytotoxic effect on cells that do not express VEGF receptors.

As a potent permeability factor, VEGF renders microvascular endothelial cells hyperpermeable. Extravasated plasma fibrinogen leads to the deposition of a fibrin gel, which could serve as a provisional matrix supporting the ingrowth of endothelial and mesenchymal cells (38). CGP79787/ZK222584 shows strong inhibitory activity at doses of 10 and 30 mg/kg PO on VEGF-induced vascular permeability as measured by Evans blue dye

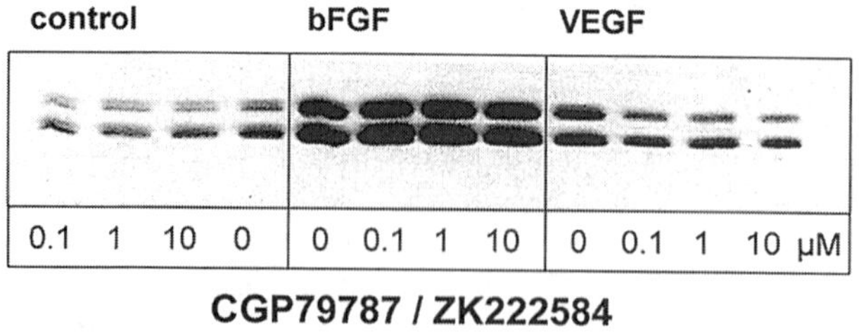

Figure 2 Inhibition of VEGF-stimulated MAP kinase activation by CGP79787/ZK222584. Quiescent HUVEC (control) were incubated for 10 min at 37°C with either 10 ng/mL basic FGF (bFGF) or 10 ng/mL VEGF (VEGF) in the presence of increased concentrations of CGP79787/ZK222584. 10 μg/lane of cell lysate was electrophoresed, blotted onto a nitrocellulose membrane, and detected with an antibody specific for activated MAP kinase.

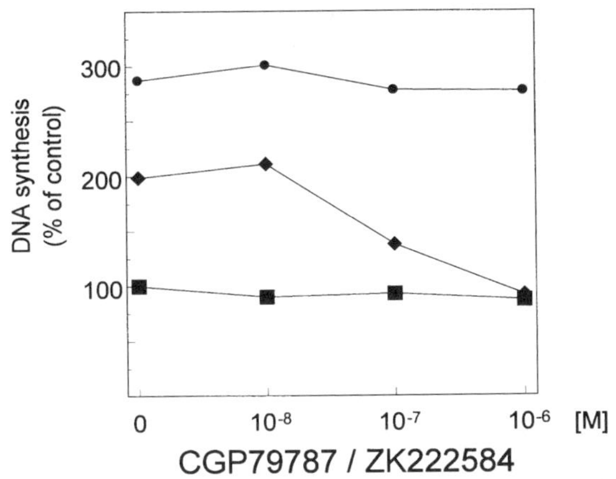

Figure 3 Inhibition of VEGF-stimulated proliferation of HUVEC by CGP79787/ZK222584. Quiescent HUVEC (squares) were stimulated with either 10 ng/mL VEGF (diamonds) or with 10 ng/mL basic FGF (circles) in the presence of increased concentrations of CGP79787/ZK222584. Growth-factor stimulated DNA synthesis was measured as [^{3}H]-thymidine incorporation.

extravasation in nude mice (Fig. 5). Interference with VEGF receptor function inhibits an important process in the formation of new blood vessels that accompanies endothelial cell proliferation and migration.

Subcutaneous implants of alginate-encapsulated tumor cells in mice represent a highly sensitive model for the quantification of in vivo tumor angiogenesis, which was recently validated by application of the antiangiogenic compound AGM-1470 (39). Angio-

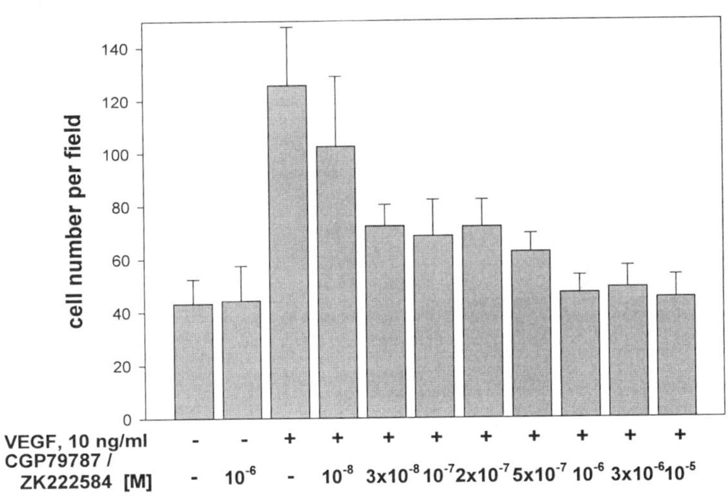

Figure 4 Inhibition of VEGF-stimulated endothelial cell migration. Human microvascular endothelial cells were seeded on collagen-coated 8-μm nucleopore filters. After 2 hr, VEGF and varying concentrations of CGP79787/ZK222584 were added. After 18 hr, the filters were stained and the upper cells were removed with a cotton swab. Migrated cells in randomized filters were counted in three visual fields using three independent chambers per concentration.

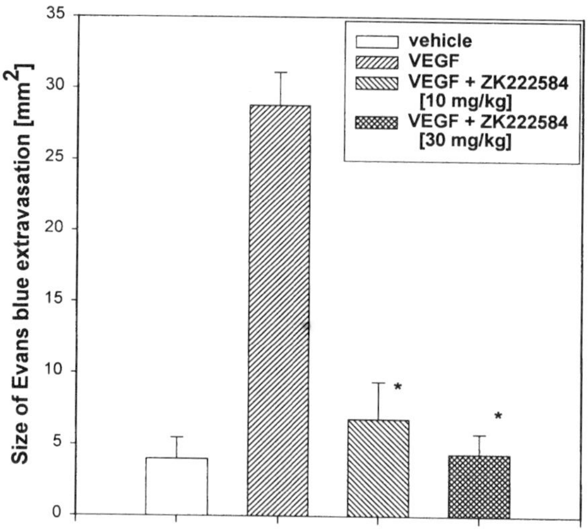

Figure 5 Effect of CGP79787/ZK222584 on VEGF-stimulated vascular permeability. Nude mice were treated with an oral dose of 10 or 30 mg/kg of CGP79787/ZK222584, respectively. After a period of 30 min, treated animals and controls received an intravenous injection of Evans blue dye. After another period of 15 min, animals were injected intradermally with 10 ng of VEGF. The area of Evans blue dye extravasation was measured 30 min after VEGF application (*$p < 0.05$ compared to control).

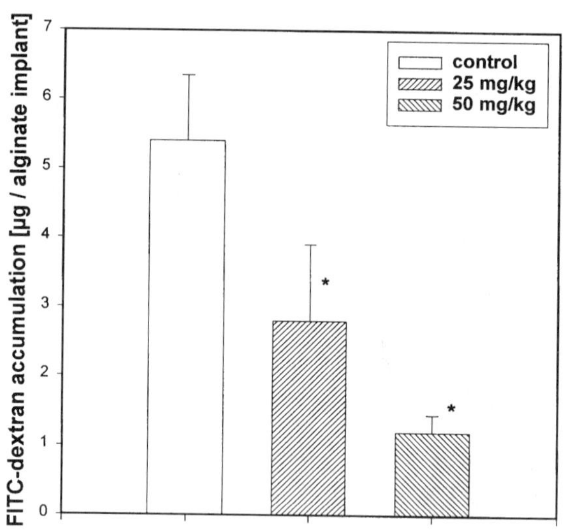

Figure 6 Effect of CGP79787/ZK222584 on vascularization induced by alginate implants containing Lewis lung carcinoma cells. Murine Lewis lung carinoma cells (5×10^4) were encapsulated in alginate beads and implanted into C57/B16 mice as described (39). Animals were treated with daily doses of 25 and 50 mg/kg of CGP79787/ZK222584, respectively, for a period of 10 days. Vascularization was determined as accumulation of FITC-dextran within the alginate implants (*$p < 0.05$ compared to control).

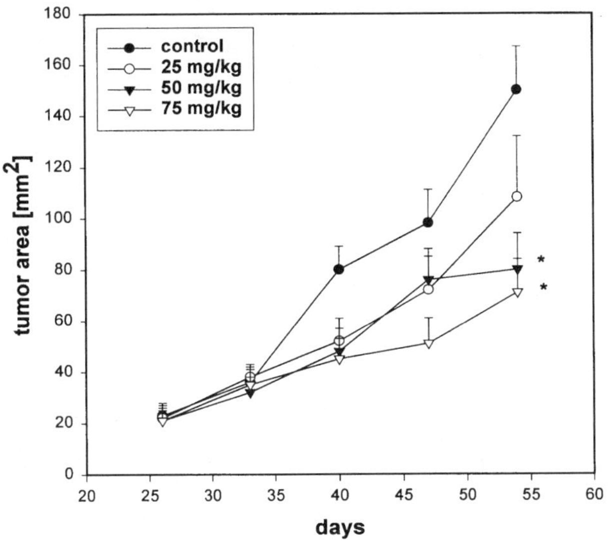

Figure 7 Effect of CGP79787/ZK222584 on tumor growth of human A431 carcinoma implanted subcutaneously into nude mice. Tumor growth was initiated by subcutaneous injection of a suspension of 1×10^6 tumor cells. When palpable primary tumors reached a size of about 25 mm^2 (day 26) animals were treated for 30 days with daily oral doses of 25, 50, or 75 mg/kg of CGP79787/ZK222584 ($*p < 0.05$ compared to control).

genesis induced by a subcutaneous alginate implant encapsulating murine Lewis lung carcinoma cells was dose-dependently inhibited by oral doses of 25 and 50 mg/kg per day of CGP79787/ZK222584 (Fig. 6). In accordance with its anti-angiogenic activity, CGP79787/ZK222584 dose-dependently inhibited the growth of A431 human epithelial carcinoma xenografts on nude mice. Tumor growth was initiated by subcutaneous injection of a tumor cell suspension, and daily oral treatment was started when palpable primary tumors reached a size of about 25 mm^2. Administration of doses of 50 and 75 mg/kg reduced tumor size by 50%, which was statistically significant (Fig. 7). In addition, CGP79787/ZK222584 showed antitumor and antimetastatic activity against a panel of human nude mouse xenografts and syngenic mouse and rat tumor models.

IV. CONCLUSION

Taken together, the results show that VEGF receptor inhibitors are agents preventing tumor growth rather than inducing reduction in tumor size. This finding was not unexpected, since VEGF receptors are predominantly expressed on proliferating endothelial cells. In contrast, VEGF receptor expression is downregulated in quiescent endothelial cells of preexisting or mature vessels, which, therefore, should not be affected by VEGF receptor inhibitors.

SU5416 as well as CGP79787/ZK222584 have currently reached clinical phase I trials. First results on SU5416 indicate that the compound appears to be well tolerated. Although both VEGF receptor inhibitors, SU5416 and CGP79787/ZK222584, looked very promising at blocking tumor angiogenesis in experimental models, ongoing clinical trials will assess how effective these agents are against cancers in humans.

ACKNOWLEDGMENT

We thank Drs. J. Waltenberger and C.H. Heldin for KDR-overexpressing porcine aortic endothelial cells.

REFERENCES

1. M Shibuya, S Yamaguchi, A Yamane, T Ikeda, A Tojo, H Matsushime, M Sato. Nucleotide sequence and expression of a novel human receptor-type tyrosine kinase gene (flt) closely related to the fms family. Oncogene 5:519–524, 1990.
2. BI Terman, ME Carrion, E Kovacs, BA Rasmussen, RL Eddy, TB Shows. Identification of a new endothelial cell growth factor receptor tyrosine kinase. Oncogene 6:1677–1683, 1991.
3. W Matthews, CT Jordan, M Gavin, NA Jenkins, NG Copeland, IR Lemischka. A receptor tyrosine kinase cDNA isolated from a population of enriched primitive hematopoietic cells and exhibiting close genetic linkage to c-kit. Proc Natl Acad Sci USA 88:9026–9030, 1991.
4. JE Park, HH Chen, J Winer, KA Houck, N Ferrara. Placenta growth factor: potentiation of vascular endothelial growth factor bioactivity, in vitro and in vivo, and high affinity binding to Flt-1 but not to Flk-1/KDR. J Biol Chem 269:25646–25654, 1994.
5. V Joukov, K Pajusola, A Kaipainen, D Chilov, I Lahtinen, E Kukk, O Saksela, N Kalkkinen, K Alitalo. A novel vascular endothelial growth vactor, VEGF-C, is a ligand for the Flt4 (VEGFR-3) and KDR (VEGFR-2) receptor tyrosine kinases. EMBO J 15:290–298, 1996.
6. MG Achen, M Jeltsch, E Kukk, T Mäkinen, A Vitali, AF Wilks, K Alitalo, SA Stacker. Vascular endothelial growth factor D (VEGF-D) is a ligand for the tyrosine kinases VEGF receptor 2 (Flk1) and VEGF receptor 3 (Flt4). Proc Natl Acad Sci USA 95:548–553, 1998.
7. L Jussila, R Valtola, TA Partanen, P Salven, P Heikkilä, MT Matikainen, R Renkonen, Kaipainen, M Detmar, E Tschachler, R Alitalo, K Alitalo. Lymphatic endothelium and Kaposi's sarcoma spindle cells detected by antibodies against the vascular endothelial growth factor receptor-3. Cancer Res 58:1599–1604, 1998.
8. N Ferrara, H Chen, T Davis-Smith, HP Gerber, TH Nguyen, D Peers, V Chisholm, KJ Hillan, RH Schwall. Vascular endothelial growth factor is essential for corpus luteum angiogenesis. Nature Med 4:336–340, 1998.
9. M Detmar, LF Brown, KP Claffey, KT Yeo, O Kocher, RW Jackman, B Berse, HF Dvorak. Overexpression of vascular permeability factor/vascular endothelial growth factor and its receptors in psoriasis. J Exp Med 180:1141–1146, 1994.
10. RA Fava, NJ Olsen, G Spencer-Green, KT Yeo, TK Yeo, B Berse, RW Jackman, DR Senger, HF Dvorak, LF Brown. Vascular permeability factor/endothelial growth factor (VPF/VEGF): accumulation and expression in human synovial fluids and rheumatoid synovial tissue. J Exp Med 180:341–346, 1994.
11. HP Hammes, J Lin, RG Bretzel, M Brownlee, G Breier. Upregulation of the vascular endothelial growth factor/vascular endothelial growth factor receptor system in experimental background diabetic retinopathy of the rat. Diabetes 47:401–406, 1998.
12. KH Plate, G Breier, B Millauer, A Ullrich, W Risau. Up-regulation of vascular endothelial growth factor and its cognate receptors in a rat glioma model of tumor angiogenesis. Cancer Res 53:5822–5827, 1993.
13. C Kremer, G Breier, W Risau, KH Plate. Up-regulation of flk-1/vascular endothelial growth

factor receptor 2 by its ligand in a cerebral slice culture system. Cancer Res 57:3852–3859, 1997.
14. B Barleon, G Siemeister, G Martiny-Baron, K Weindel, C Herzog, D Marmé. Vascular endothelial growth factor up-regulates its receptor fms-like tyrosine kinase 1 (FLT-1) and a soluble variant of FLT-1 in human vascular endothelial cells. Cancer Res 57:5421–5425, 1997.
15. KJ Kim, B Li, J Winer, M Armanini, N Gillett, HS Phillips, N Ferrara. Inhibition of vascular endothelial growth factor–induced angiogenesis suppresses tumour growth in vivo. Nature 362:841–844, 1993.
16. P Borgström, KJ Hillan, P Sriramarao, N Ferrara. Complete inhibition of angiogenesis and growth of microtumors by anti-vascular endothelial growth factor neutralizing antibody: novel concepts of angiostatic therapy from intravital videomicroscopy. Cancer Res 56:4032–4039, 1996.
17. J Drevs, I Hofmann, G Martiny-Baron, G Siemeister, A Gödde, B Barleon, C Unger, D Marmé. Effect of soluble VEGF receptor construct (sKDR-Fc) on murine renal cell carcinoma (abstr). J Cancer Res Clin Oncol 124:R43, 1998.
18. LP Aiello, EA Pierce, ED Foley, H Takagi, H Chen, L Riddle, N Ferrara, GL King, LEH Smith. Suppression of retinal neovascularization in vivo by inhibition of vascular endothelial growth factor (VEGF) using soluble VEGF-receptor chimeric proteins. Proc Natl Acad Sci USA 92:10457–10461, 1995.
19. LE Benjamin, E Keshet. Conditional switching of vascular endothelial growth factor (VEGF) expression in tumors: induction of endothelial cell shedding and regression of hemagioblastoma-like vessels by VEGF withdrawal. Proc Natl Acad Sci USA 94:8761–8766, 1997.
20. T Alon, I Hemo, A Itin, J Peér, J Stone, E Keshet. Vascular endothelial growth factor acts as a survival factor for newly formed retinal vessels and has implications for retinopathy of prematurity. Nature Med 1:1024–1028, 1995.
21. B Barleon, S Sozzani, D Zhou, HA Weich, A Mantovani, D Marmé. Migration of human monocytes in response to vascular endothelial growth factor (VEGF) is mediated via the VEGF receptor flt-1. Blood 87:3336–3343, 1996
22. M Clauss, H Weich, G Breier, U Knies, W Röckl, J Waltenberger, W Risau. The vascular endothelial growth factor receptor Flt-1 mediates biological activities. Implications for a functional role of placenta growth factor in monocyte activation and chemotaxis. J Biol Chem 271:17629–17634, 1996.
23. J Waltenberger, L Claesson-Welsh, A Siegbahn, M Shibuya, CH Heldin. Different signal transduction properties of KDR and Flt1, two receptors for vascular endothelial growth factor. J Biol Chem 269:26988–26995, 1994.
24. L Seetharam, N Gotoh, Y Maru, G Neufeld, S Yamaguchi, M Shibuya. A unique signal transduction from FLT tyrosine kinase, a receptor for vascular endothelial growth factor VEGF. Oncogene 10:135–147, 1995.
25. BA Keyt, HV Nguyen, LT Berleau, CM Duarte, J Park, H Chen, N Ferrara. Identification of vascular endothelial growth factor determinants for binding KDR and Flt-1 receptors. J Biol Chem 271:5638–5646, 1996.
26. G Siemeister, M Schirner, P Reusch, B Barleon, D Marmé, G Martiny-Baron. An antagonistic vascular endothelial growth factor (VEGF) variant inhibits VEGF-stimulated receptor autophosphorylation and proliferation of human endothelial cells. Proc Natl Acad Sci USA 95: 4625–4629, 1998.
27. F Shalaby, J Rossant, TP Yamaguchi, M Gertsenstein, XF Wu, ML Breitman, AC Schuh. Failure of blood-island formation and vasculogenesis in Flk-1-deficient mice. Nature 376:62–66, 1995.
28. GH Fong, J Rossant, M Gertsenstein, ML Breitman. Role of the Flt-1 receptor tyrosine kinase in regulating the assembly of vascular endothelium. Nature 376:66–70, 1995.
29. B Millauer, LK Shawver, KH, Plate, W Risau, A Ullrich. Glioblastoma growth inhibited in vivo by a dominant-negative Flk-1 mutant. Nature 367:576–579, 1994.

30. B Millauer, MP Longhi, KH Plate, LK Shawver, W Risau, A Ullrich, LM Strawn. Dominant-negative inhibition of flk-1 suppresses the growth of many tumor types in vivo. Cancer Res 56:1615–1620, 1996.
31. M Skobe, P Rockwell, N Goldstein, S Vosseler, NE Fusenig. Halting angiogenesis suppresses carcinoma cell invasion. Nature Med 3:1222–1227, 1997.
32. P Rockwell, G Neufeld, A Glassman, D Caron, N Goldstein. In vitro neutralization of vascular endothelial growth factor activation of flk-1 by a monoclonal antibody. Mol Cell Differ 3: 91–109, 1995.
33. LM Strawn, G McMahon, H App, R Schreck, WR Kuchler, MP Longhi, TH Hui, C Tang, A Levitzki, A Gazit, I Chen, G Keri, L Orfi, W Risau, I Flamme, A Ullrich, KP Hirth, LK Shawver. Flk-1 as a target for tumor growth inhibition. Cancer Res 56:3540–3545, 1996.
34. LK Shawver, TAT Fong, H App, C Tang, L Sun, LM Strawn, G McMahon. Flk-1 antagonists for inhibition of angiogenesis. International Business Communications Inc. Conference, ''Therapeutic Implications and Mechanisms of Angiogenesis,'' Boston, MA, 1997, pp. 3.7.1–3.7.20.
35. LM Strawn, LK Shawver. Tyrosine kinases in disease: overview of kinase inhibitors as therapeutic agents and current drugs in clinical trials. Exp Opin Invest Drugs 7:553–573, 1998.
36. P Vajkoczy, MD Menger, B Vollmar, L Schilling, P Schmiedek, A Ullrich, KP Hirth, TAT Fong. Effect of the Flk-1 antagonist SU5416 on tumor growth, angiogenesis and micro-hemodynamics (abstr). Proc Am Assoc Cancer Res 39:96, 1998.
37. J Wood, E Buchdunger, R Cozens, F Hofmann, J Mestan, H Mett, M Müller, J Roesel, C Schnell, D Stover, K Woods-Cook, G Bold, J Frei, H Towbin, B Kotyuk, M Sills, K Wass, G Martiny-Baron, G Siemeister, F Totzke, D Marmé. Pharmacological profile of a potent and orally active inhibitor of VEGF receptor tyrosine kinases (abstr). Proc Am Assoc Cancer Res 39:96, 1998.
38. HF Dvorak, LF Brown, M Detmar, AM Dvorak. Vascular permeability factor/vascular endothelial growth factor, microvascular hyperpermeability, and angiogenesis. Am J Pathol 146: 1029–1039, 1995.
39. J Hoffmann, M Schirner, A Menrad, MR Schneider. A highly sensitive model for quantification of in vivo tumor angiogenesis induced by alginate-encapsulated tumor cells. Cancer Res 57:3847–3851, 1997.
40. A Menrad, KH Thierauch, G Martiny-Baron, G Siemeister, M Schirner, MR Schneider. Novel antibodies directed against the extracellular domain of the human VEGF-receptor type II. Hybridoma 16:465–471, 1997.

23
Antiangiogenic Function of Thrombospondin-1

M. Luisa Iruela-Arispe and Maria Asunción Ortega
University of California, Los Angeles, Los Angeles, California

Francisca Vázquez
Dana-Farber Cancer Institute, Boston, Massachusetts

I. INTRODUCTION

Thrombospondin-1 (TSP1) is a large trimeric glycoprotein present in platelet α-granules and secreted by several cell types in the vessel wall, including endothelial and smooth muscle cells. A major function attributed to TSP1 relates to its ability to inhibit angiogenesis (1–4). The specific cellular and molecular mechanisms involved in this outcome are not entirely clear. TSP1 is known to affect migration, proliferation, and adhesion of endothelial cells (5–10), any of which could contribute to an antiangiogenic phenotype. The region of TSP1 responsible for the angioinhibitory effect has been mapped to two domains: (a) a procollagen-like sequence and (b) the type I, or properdin, repeats (4). This chapter reviews the current literature on TSP1 with emphasis on its ability to suppress blood vessel formation and, particularly, on its potential effect as an antiangiogenic agent in tumors.

Thrombospondin-1 is a member of a family of five related genes, TSP 1–4 and COMP (also known as TSP5) (reviewed in Refs. 11 and 12). Endothelial cells synthesize and secrete principally TSP1 (13). Transcripts for TSP1 and TSP2 have been observed in embryonic endothelial cells in vivo (days 9–1) (14). TSP2 has not been detected in either normal or pathological endothelium from adult animals, nor is TSP2 secreted by endothelial cells grown in vitro (13). Interestingly, no other TSP1 member has been detected in endothelium. Structurally, the presence of the type I or properdin repeats at the amino terminus is the main difference between TSP1/2 and all other TSP proteins (TSP3-5).

The best-characterized biological functions attributed to TSP1 have been its ability to activate TGF-β and to inhibit angiogenesis. It appears that the effects of TSP1 in the suppression of vascular growth are independent of those related to the activation of TGF-β, as will be discussed later. Most of the effects on angiogenesis have been demonstrated in vitro or by the use of specific in vivo assays, such as the chorioallantoic membrane (15) and cornea pocket assays (4). Systemically, only peptide mimetics to the protein have been used effectively in tumor assays (16), as the peptides have shown to reduce tumor growth. The relative role of TSP1 as an endogenous regulator of angiogenesis is presently unknown.

A mouse model that lacks TSP1 has been generated by homologous recombination (17). Unexpectedly, the phenotype displayed by the TSP1 knockout mouse includes pulmonary chronic inflammation and pneumonia (17). No major effect in blood vessel number was reported, however, due to the significant level of homology with TSP2, functional redundancy, and compensation for the lack of TSP1 it is likely to occur. Interestingly, the TSP2-null mouse showed enhanced vascularity with an apparent twofold increase in the number of dermal capillaries (18).

Due to the likelihood of functional compensation by TSP2 and because analysis of loss of function of a vascular inhibitor might be difficult to evaluate, we decided to generate transgenic lines that overexpress TSP1 and ask whether the protein might function as a regulator or angiogenesis in a developmental setting (19). It was our premise that increased levels of an angiogenic inhibitor might be more revealing than its complete absence from the onset of embryonic development. The organ of choice for overexpression of TSP1 was the mammary gland. Upon pregnancy, the mammary gland grows significantly. Its vascular bed also expands to compensate for the increased metabolic demands of the organ upon lactation. Finally, involution is also accompanied by reduction of capillary density and vascular remodeling. Therefore this is an excellent model system in which to ask questions related to endogenous regulation of vascular growth and regression in an adult tissue.

Transgenic animals were generated by injection of blastocysts with a construct containing the mouse mammary tumor virus (MMTV) promoter driving the hTSP1 complete cDNA. The MMTV promoter directed expression primarily to the mammary gland. Four transgenic lines with varied levels of human TSP1 protein were obtained. Vascular density in mammary glands was examined upon systemic injection of FITC-dextran (19) and by immunocytochemistry of tissue sections (Fig. 1). In all four lines, increased levels of hTSP1 was associated with significant decrease in capillary density. These transgenic

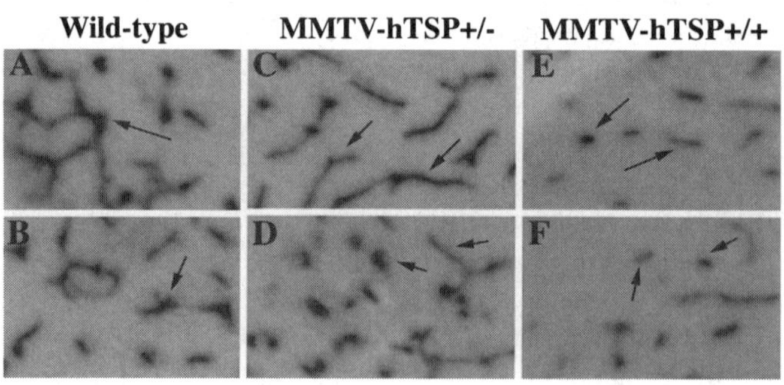

Figure 1 Overexpression of TSP1 in the mammary gland is associated with reduced capillary density. Generation of transgenic animals overexpressing TSP1 was done by injection of linearized MMTV-hTSP1 DNA into murine blastocysts. The murine mammary tumor virus (MMTV) targeted expression of TSP1 to the mammary gland. Evaluation of the mammary gland vasculature was performed by staining 30-μm sections with *Bandeira simplicifolia* lectin. (A) and (B). Mammary glands from wild-type animals, note stained vessels (arrows); (C) and (D). Mammary vessels from heterozygous animals; and (E) and (F). Mammary vessels from homozygous transgenics. Note reduction in the vascular network in E and F when compared to other panels.

lines showed between two- to sevenfold upregulation in TSP1 protein and consistently showed a 52–73% reduction in the number of capillaries (19). To our knowledge this is the first demonstration that titration of endogenous levels of TSP1 modulates angiogenesis and capillary density in vivo.

It has been our premise that potent and tissue-specific inhibitors of angiogenesis are constantly expressed in tissues providing vascular surveillance. Such molecules would function to regulate the pattern of new vessel growth and to maintain homeostasis of established tissues. Our recent results on mammary gland–targeted expression of TSP1 clearly demonstrate that this molecule has considerable angiostatic/antiangiogenic activity and that it is likely an endogenous modulator of physiological angiogenesis. This concept is also supported by the pattern of expression displayed by TSP1 in the human endometrium (20). The primate endometrium is characterized by recurrent cycles of growth during the proliferative phase (first half of the menstrual cycle) and growth inhibition in the secretory phase (second half of the menstrual cycle). We found that TSP1 is expressed in the secretory phase and is induced by progesterone (20). In fact, conditioned medium from progesterone-treated stromal cells suppresses endothelial cell proliferation in vitro. The inhibition appears to be mediated by increased levels of TSP1 in the conditioned medium, since neutralizing antibodies to TSP1 block the inhibitory effect (20).

II. EFFECTS OF TSP1 ON ENDOTHELIAL CELLS

Addition of TSP1 protein to endothelial cells in culture has been shown to inhibit proliferation (5), to induce attachment but not necessarily spreading (6,10), and to disrupt focal adhesions (6).

The domains of TSP1 that account for its antiproliferative and antiadhesive activity are located at the amino-terminal end. Based on inhibition by monoclonal antibodies, the heparin-binding domain might regulate endothelial cell proliferation (20,21); however, a 140-kDa fragment of TSP1 lacking the heparin-binding domain also suppresses endothelial cell growth (1).

The effects of TSP on endothelial cells are further complicated by its interaction with a variety of other extracellular macromolecules and growth factors (11,12). In the presence of type I collagen, TSP1 appears to facilitate rather than inhibit tube formation in vitro (22), and TSP1 binds and activates TGF-β (23). Therefore, the nature of the extracellular environment or the state of TSP1 (i.e., bound or in solution) might modulate the responses of TSP1 on capillary morphogenesis through the activation of different receptor subsets.

Currently three domains of TSP1 have been implicated in interactions with endothelial cells: (a) $\alpha v\beta 3$, binds to an RGD sequence located in the carboxyl-terminus of TSP1 (last type III repeat) (7), (b) CD-36, binds to the second and third type I repeats (24–26), and (c) a partially characterized receptor that binds to the type I repeats (27). Four additional receptor activities have been characterized in other cells (Table 1); however their relevance to cells of the vascular wall has not been established.

The antiproliferative effect mediated by TSP1 is specific to endothelial cells; in fact, smooth muscle cells spread and proliferate in response to TSP1. The role of TSP1 in smooth muscle cell proliferation was initially demonstrated by Majack (31). In these studies, monoclonal antibodies directed against several regions of TSP1 inhibited the proliferation of aortic smooth muscle cells. We have been able to confirm Majack's experiments

Table 1 Binding Sites Present in TSP1, Receptors, and Cell Types in Which These Receptors Have Been Shown to Bind to TSP1

Binding sequence	Receptor	Cell type	Reference
Type I repeats and heparin-binding	Glycoconjugates (includes heparin, fucoidin, and sulfatide)	Melanoma cells	34
TSP1 (site not defined)	LRP	Fibroblasts, smooth muscle cells	28
CSVTCG	CD-36	Endothelial cells, platelets, monocytes adipocytes, tumor cells, epithelial cells	24–26
GRDS	$\alpha_4\beta_1$ and $\alpha_5\beta_1$	Lymphocytes,	29
	$\alpha_v\beta_1$	Endothelial, smooth muscle	7
RFYVVMWK and FIRVVMYEGKK	Carboxyl-terminal receptor (52-kDa)	K569 cell line	30

with purified TSP1 in the absence of TGF-β1 as well as with recombinant truncated fragments corresponding to the TSP1 amino-terminal region.

It is likely that cell-specific effects of TSP1 reflect differences in the type and/or number of cell-surface receptors utilized by endothelial and smooth muscle cells to bind to TSP1. To date, seven TSP1 receptors have been identified (Table 1), but little of their signal transduction pathways or functions has been elucidated. One might anticipate that occupancy of different TSP1 receptors would result in different metabolic changes, such as protein phosphorylation, with alternate consequences on gene expression in different cell types. The search for signaling pathways and also for the identification of the minimal region responsible for TSP1 effect in the vasculature has engendered a great deal of interest and led to the identification of the region(s) that effectively inhibit angiogenesis.

III. ANTIANGIOGENIC DOMAINS IN THROMBOSPONDIN

Two regions of TSP1—the procollagen-homology sequence and the type I repeats—were initially identified as domains that could independently suppress angiogenesis (4). Subsequently, a more detailed series of structure-function analyses indicated that the amino acid sequence associated with the CD36 binding domain, located in the second and third type I repeats, was the specific region that conferred angiosuppressive features to TSP1 (32). In addition, the same study implicated CD36 as the signaling receptor for the antiangiogenic effects. We and others have found that the procollagen region is not effective in suppressing angiogenesis in the chorioallantoic membrane or in endothelial cell proliferation (15).

The relative contribution of the type I repeats to the inhibitory signals mediated by TSP1 has been studied in the cornea pocket (4) and chorioallantoic membrane (15); both assays concur that this region of the molecule is effective in blocking angiogenic signals by growth factors. TSP1 contains three tandem type I or properdin repeats; nevertheless, only the second and third repeats appear relevant to the suppressive vascular effect of TSP1 in the cornea pocket assay (4). We have examined the role of the second and third type I repeats in the inhibition of tumor growth and vascularization using tumor xenographs. Briefly, HT1080 human tumor cell lines were transfected with a truncated construct containing either all three or only the first type I domains. The resulting stable lines were tested for secretion of recombinant protein and injected into the dorsal flanks of nude mice. The results of these experiments showed that the second and third type I repeats but not the first repeat are essential to the antiangiogenic response mediated by TSP1 (Fig. 2). In addition, peptide mimetics of the type I repeats as well as truncated recombinant proteins containing these regions were able to antagonize growth factor–mediated angiogenesis in the chorioallantoic membrane (15,33) and suppress growth of tumors in nude mice (16).

The type I repeats of TSP1 have been characterized extensively (4,23,34). The second type I repeat of TSP1 and shown to (a) bind heparin in three regions (KRFK, WSPW, CSVTCG), (b) activate latent forms of TGF-β via a nonproteolytic pathway (KRFKR), (c) bind sulfatide, (d) promote cell adhesion, and (e) inhibit angiogenesis through the CSVTCG sequence.

Recently, much interest has focused on the role of TSP1 as an activator of latent TGF-β. In fact, the pulmonary phenotype displayed by the TSP1 knockout mouse can be compensated and reverted by systemic injection of peptides containing the TGF-β activating sequence (KRFK) (35). However, it is important to note that peptides that activate TGF-β do not appear to have a bearing on the antiangiogenic functions of TSP1. Peptides mimetics to the amino-terminal region of the second type I repeats containing mutations in the KRFK sequence are as efficient as wild-type peptides in suppressing tumor growth in nude mice (16) and inhibiting angiogenesis in the chorioallantoic membrane (33).

Additional dissection of the second and third type I domains has been performed by several laboratories using nonoverlapping assays. These studies have generated more specific information as to which amino acids, within the type I repeats, are responsible for the antiangiogenic effect of TSP1. Using the cornea pocket assay, Bouck and col-

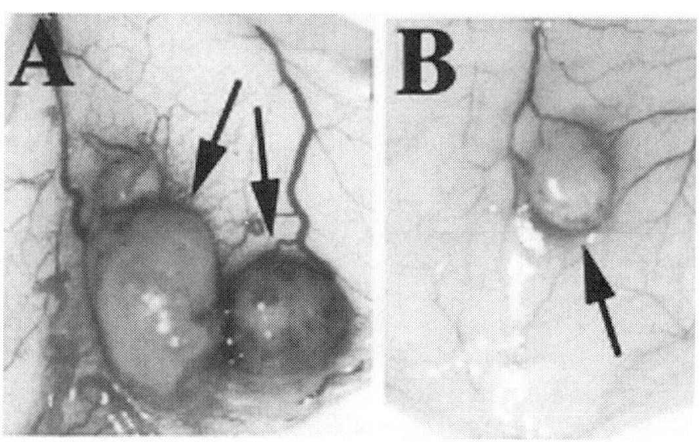

Figure 2 Effect of TSP1 overexpression on tumor growth. Truncated TSP1 mammalian expression constructs expanding from the amino-terminus until the first type I repeat (A) or containing all three type I repeats (B) were used to transfect HT1080 fibrosarcoma cell lines. The resulting stable lines were used to generate tumors in immunocompromised mice. Dorsal injection of 10^5 cells was allowed to grow as a tumor for 5 weeks. Examination of the resulting tumors revealed that tumors from cells transfected with all three type I repeats were significantly smaller (B) than the ones resulting from cells containing the first type I repeat alone (A).

leagues found that the CD-36 binding sequence (located in middle region of the second and third type I repeats) is the most active angioinhibitory region (32). Roberts and colleagues have generated a series of peptide mimetics to the amino-terminal region of the second and third type I repeats. By injecting these peptides into nude mice bearing tumors, they demonstrated that this region is also effective at suppressing vascularization and tumor growth (16). Since the assays are quite different, it is difficult to compare the results directly and evaluate if these regions are equally effective at inhibiting angiogenesis. In collaboration with Roberts, we set out to perform structure-function analysis of the type I repeats utilizing recombinant protein and peptides in the chorioallantoic membrane (33). The objective was to evaluate side by side and in the same assay the efficacy of these subregions within the type I repeats. Our results were clear and revealed that both regions are equal contributors to the antiangiogenic effect mediated by TSP1. The findings also revealed growth-factor selectivity. While the amino-terminal end of the type I repeats suppressed angiogenesis driven by bFGF, the sequence carboxyl-terminal to that (the region that binds to CD36) was able to suppress angiogenesis driven by either bFGF or VEGF (33).

IV. TSP1 AS AN INHIBITOR OF ANGIOGENESIS IN TUMORS

The dependency of tumor growth on neovascularization was demonstrated over three decades ago by Folkman and colleagues (reviewed in Refs. 36 and 37). The initial studies demonstrated that tumors maintained in organ culture grew to approximately 1–2 mm in diameter, after which no further increase in total mass was observed regardless of time in culture (38). In contrast, tumors introduced into the rabbit cornea within 6 mm of the iris promoted neovascularization, grew exponentially, and metastasized (39). These findings implicated an absolute requirement for angiogenesis in the progression of tumor growth. Thirty-five years later, this original hypothesis has gained solid acceptance, and the basic observation has been validated in a variety of tumors including cutaneous malignant melanomas (40), breast carcinomas (41), and brain tumors (42), among others. Indeed, tumor-mediated angiogenesis has been identified as a significant and independent prognostic indicator of early-stage breast carcinoma, among several other tumor types, as well as a prerequisite for the expansion and metastatic progression of malignant disease (43,44).

The possibility that tumor progression could be regulated by pharmacological and/or genetic suppression of blood vessel growth has engendered a long-standing interest in the identification of molecules or synthetic compounds that block angiogenesis.

It has been our premise that endogenous physiological regulators of vascular growth are also likely to be effective at suppressing the vascularization of tumors, offer no resistance, and perhaps add tissue-selectivity to the process. TSP1 and related molecules/genes could likely be among such candidates. Injection of TSP1, truncated proteins, and peptide mimetics has proven effective in the suppression of vascular growth of tumors and the inhibition of tumor growth itself (16). Questions that still remain relate to tissue-specificity of the effect mediated by TSP1 and its relative potency as compared with other angiogenic inhibitors. Finally, the unraveling of TSP1-mediated signal transduction pathways in endothelial cells will be fundamental to our understanding of the effects mediated by this protein in the suppression of vascular growth.

ACKNOWLEDGMENTS

We would like to thank the members of the laboratory for helpful discussions and suggestions. This work was performed in part with support provided by the following grants: American Heart Association, grant-in-aid 96-1218, Department of Defense—Army DAMD 17-94-J-4346 and National Institutes of Health—National Cancer Institute R29 CA63356-01.

REFERENCES

1. DJ Good, PJ Polverini, F Rastinejad, M Le Beau, RS Lemons, WA Frazier, NP Bouck. A tumor suppressor-dependent inhibitor of angiogenesis is immunologically and functionally indistinguishable from a fragment of thrombospondin. Proc Natl Acad Sci USA 87:6624–6628, 1990.
2. ML Iruela-Arispe, P Bornstein, H Sage. Thrombospondin exerts and antiangiogenic effect on tube formation by endothelial cells in vitro. Proc Natl Acad Sci USA 878:5026–5030, 1991.
3. LA DiPietro, DR Nebgen, PJ Polverini. Downregulation of endothelial cell thrombospondin 1 enhances in vitro angiogenesis. J Vasc Res 31:178–185, 1994.
4. SS Tolsma, OV Volpert, DJ Good, WA Frazier, PJ Polverini, N Bouck. Peptides derived from two separate domains of the matrix protein thrombospondin 1 have anti-angiogenic activity. J Cell Biol 122:497–511, 1993.
5. P Bagavandoss, JW Wilks. Specific inhibition of endothelial cell proliferation by thrombospondin. Biochem Biophys Res Commun 170:867–872, 1990.
6. JE Murphy-Ullrich, M Höök. Thrombospondin modulates focal adhesions in endothelial cells. J Cell Biol 109:1309–1319, 1989.
7. J Lawler, R Weinstein, RO Hynes. Cell attachment to thrombospondin: the role of ARG-GLY-ASP, calcium, and integrin receptors. J Cell Biol 107:2351–2361, 1989.
8. G Taraboletti, DD Roberts, LA Liotta. Thrombospondin-induced tumor cell migration: haptotaxis and chemotaxis are mediated by different molecular domains. J Cell Biol 105:2409–2415, 1987.
9. T Vogel, N-h Guo, HC Krutzsch, DA Blake, J Hartman, S Mendlovitz, A Panet, DD Roberts. Modulation of endothelial cell proliferation, adhesion, and motility by recombinant heparin-binding domain and synthetic peptides from the type I repeats if thrombospondin. J Cell Biochem 53:74–84, 1993.
10. G Taraboletti, DD Roberts, LA Liotta, J Giavazzi. Platelet thrombospondin modulates endothelial cell adhesion, motility, and growth: a potential angiogenesis regulatory factor. J Cell Biol 111:765–772, 1990.
11. P Bornstein, EH Sage. The thrombospondins: structure and regulation of expression. Methods Enzymol 245:3–85, 1994.
12. JC Adams, J Lawler. The thrombospondin family. Curr Biol 3:188–190, 1993.
13. MJ Reed, ML Iruela-Arispe, ER O'Brien, T Truong, T LaBell, P Bornstein, H Sage. Expression of thrombospondins by endothelial cells: injury is correlated with TSP1. Am J Pathol 147:1068–1080, 1995.
14. ML Iruela-Arispe, DA Liska, EH Sage, P Bornstein. Differential expression of thrombospondin 1, 2, and 3 during murine development. Dev Dynam 197:40–56, 1993.
15. ML Iruela-Arispe, H Dvorak. Angiogenesis: a dynamic balance of stimulators and inhibitors. Thromb Haemostas 78:672–677, 1997.
16. Guo NH, Krutzsch HC, Inman JK, Shannon CS, Roberts DD. Antiproliferative and antitumor activities of D-reverse peptides derived from the second type-1 repeat of thrombospondin-1. J Pept Res 50:210–221, 1997.

17. JM Lawler, V Sunday, V Thibert, M Duquette, EL George, H Rayburn, RO Hynes. Thrombospondin-1 is required for normal murine pulmonary homeostasis and its absence causes pneumonia. J Clin Invest 101:982–992, 1998.
18. TR Kyriakides, YH Zhu, LT Smith, SD Bain, Z Yang, MT Lin, KG Danielson, RV Iozzo, M LaMarca, CE McKinney, EI Ginns, P Bornstein. 1998. Mice that lack thrombospondin 2 display connective tissue abnormalities that are associated with disordered collagen fibrillogenesis, an increased vascular density, and a bleeding diathesis. J Cell Biol 140:419–430, 1998.
19. MA Ortega, ML Iruela-Arispe. Thrombospondin-1 is an endogenous regulator of angiogenesis in the mammary gland. J Cell Biol, 1999. In press.
20. ML Iruela-Arispe, P Porter, P Bornstein, EH Sage. Thrombospondin-1, an inhibitor of angiogenesis, is regulated by progesterone in human stromal cells. J Clin Invest 97:403–412, 1996.
21. G Taraboletti, DD Roberts, LA Liotta, R Giavazzi. Platelet thrombospondin modulates endothelial cell adhesion, motility, and growth: a potential angiogenesis regulatory factor. J Cell Biol 111:765–772, 1990.
22. RF Nicosia, GP Tuszynski. Matrix-bound thrombospondin promotes angiogenesis in vitro. J Cell Biol 124:183–193, 1994.
23. S Schultz-Cherry, S Ribeiro, L Gentry, JE Murphy-Ullrich. Thrombospondin binds and activates the small and large forms of latent transforming growth factor-β in a chemically defined system. J Biol Chem 269:6775–26782, 1994.
24. HY Huh, SK Lo, LM Yesner, RL Silverstein. CD36 induction on human monocytes upon adhesion to tumor necrosis factor-activated endothelial cells. J Biol Chem 270:6267–6271, 1995.
25. GP Tuszynski, VL Rothman, M Papale, BK Hamilton, E. Jacob. Identification and characterization of a tumor cell receptor for CSVTCG, a thrombospondin adhesive domain. J Cell Biol 120:513–521, 1993.
26. W-X Li, RJ Howard, LLK Leung. Identification of SVTCG in thrombospondin as the conformation-dependent, high affinity site for its receptor, CD36. J Biol Chem 268:16179–16184, 1993.
27. VM Dixit, GA Grant, SA Santoro, WA Frazier. Isolation and characterization of a heparin-binding domain from the amino terminus of platelet thrombospondin. J Biol Chem 259:10100–10105, 1984.
28. S Godyna, G Liau, I Popa, S Stefansson, WS Argraves. Identification of the low density lipoprotein receptor-related protein (LRP) as an endocytic receptor for thrombospondin-1. J Cell Biol 129:1403–1410, 1995.
29. R Yabkiwitz, VM Dixit, N Guo, DD Roberts, Y Shimizu. Activated T-Cell adhesion to thrombospondins mediated by the $\alpha 4\beta 1$ (VLA-4) and $\alpha 5\beta 1$ (VLA-5) integrins. J Immunol 151: 149–158, 1993.
30. AG Gao, WA Frazier. Identification of a receptor candidate for the carboxyl-terminal cell binding domain of thrombospondins. J Biol Chem 269:29650–29657, 1994.
31. RA Majack, LV Goodman, VM Dixit. Cell surface thrombospondin is functionally essential for vascular smooth muscle cell proliferation. J Cell Biol 106:415–422, 1988.
32. DW Dawson, SF Pearce, Zhong R, Silverstein RL, Frazier WA, Bouck NP. CD36 mediates the in vitro inhibitory effects of thrombospondin-1 on endothelial cells. J Cell Biol 138:707–717, 1997.
33. ML Iruela-Arispe, M Lombardo, HC Krutzsch, J Lawler, DD Roberts. Inhibition of angiogenesis by thrombospondin-1 is mediated by two independent regions within the type 1 repeats. Circulation, 1999. In press.
34. NH Guo, HC Krutzsch, E Negre, T Vogel, DA Blake, DD Roberts. Heparin- and sulfide-binding peptides from the type I repeats of human thrombospondin promote melanoma cell adhesion. Proc Natl Acad Sci USA 89:3040–3044, 1992.

35. SE Crawford, V Stellmach, JE Murphy-Ullrich, SM Ribeiro, J Lawler, RO Hynes, GP Boivin, N Bouck. Thrombospondin-1 is a major activor of TGF-β1 in vivo. Cell 26:1159–1170, 1998.
36. J Folkman. What is there evidence that tumors are angiogenic-dependent? J Natl Cancer Inst 82:4–6, 1990.
37. J Folkman. New perspectives in clinical oncology from angiogenesis research. Eur J Cancer 32:2534–2539, 1996.
38. J Folkman, DM Long, FF Becker. Growth and metastasis of tumor in organ culture. Cancer 16:453–467, 1963
39. MA Gimbrone, S Leapman, RS Cotran, J Folkman. Tumor angiogenesis: iris vascularization at a distance from experimental intraocular tumors. J Natl Cancer Inst 50:219–228, 1973.
40. RL Barnhill, MA Levy. Regressing thin cutaneous malignant melanomas (<1.0 mm) are associated with angiogenesis. Am J Pathol 143:99–104, 1993.
41. S Bosari, AK Lee, RA DeLellis, BD Wiley, GJ Heatley, ML Silverman. Microvesssel quantitation and prognosis in invasive breast carcinoma. Human Pathol 23:755–761, 1992.
42. SS Brem, D Zagzag, AMC Tsanaclis, S Gately, MP Elkouby, SE Brien, Inhibition of angiogenesis and tumor growth in the brain. Am J Pathol 3:1121–1142, 1990.
43. N Weidner, JP Semple, WR Welch, J Folkman. Tumor angiogenesis and metastasis-correlation in invasive breast carcinoma. N Engl J Med 324:1–8, 1991.
44. N Weidner, J Folkman, F Possa, B Pierantonio, EN Allred, DH Moore, S Meli, and G Gasparini. Tumor angiogenesis: a new significant and independent prognostic indicator in early-stage breast carcinoma. J Natl Cancer Inst 84:1875–1887, 1992.

24
Thalidomide

Erwin A. Kruger, Eddie Reed, and William D. Figg
National Cancer Institute, National Institutes of Health, Bethesda, Maryland

I. HISTORICAL BACKGROUND

Thalidomide (*N*-phthalidoglutarimide) was developed in the 1950s by Chemic Grünenthal of Stolberg, Germany, as an effective, nontoxic antiemetic and sedative. While no reported toxicities were observed throughout its development, thalidomide usage was tragically linked to embryotoxicity (particularly dysmelia) in humans (1,2). Thalidomide's teratogenicity was discovered in 1961, when McBride and Lenz independently documented the association of thalidomide usage during pregnancy with fetal abnormalities, particularly stunted limb growth (1,2). Their discovery eventually led to the withdrawal of the drug from general circulation. However, despite the well-established finding of thalidomide as a potent human teratogen in pregnancy, the drug is relatively well tolerated and has remained in continuous clinical use for a variety of human diseases. It has been studied extensively, yet its mechanisms of action in causing teratogenicity and other biological effects remain to be described in detail.

Thalidomide treatment has been or is being evaluated in a wide spectrum of human diseases (Table 1). Currently, the relevance of thalidomide therapy in these diseases appears to stem from the compound's immunosuppressive as well as antiangiogenic properties. The exact mechanisms of action of thalidomide in autoimmune disease, inflammatory disease, and cancer require further investigation. However, several studies outlined below suggest that the mechanism of action of thalidomide may correlate with many of its observed in vivo and in vitro biological effects. The following discussion aims to outline the current available data on thalidomide's clinical profile as an anticancer agent, and discussion of thalidomide in other clinical contexts will be described as necessary.

II. IMMUNOMODULATORY PROPERTIES

In 1965, Sheskin made the fortuitous discovery that patients with erythema nodosum leprosum associated with lepromatous leprosy improved while concurrently receiving thalidomide, which was given as a sedative (3,4). This observation led to the hypothesis that thalidomide had immunosuppressive properties. Consequently, in addition to its use in

Table 1 Main Clinical Applications of Thalidomide

Disease by type	Reference(s)
Skin disorders	
Prurigo nodularis	6–8
Leprosy and erythema nodosum leprosum	3,4,57
Behcet's disease	5,9–16
Pyoderma gangrenosum	17–19
Immunologic and rheumatological disorders	
Lupus erythematosus	20–24
Arthritis	25,26
Crohn's disease and ulcerative colitis	27,28
Hematological or oncological disorders	
Graft vs. host disease (GVHD)	31–36
Glioma	29,30
Prostatic cancer	37,38
AIDS-associated Kaposi's sarcoma	39,40
Breast cancer	41–43
Infectious diseases	
Tuberculosis	44–46
HIV/AIDS and wasting syndrome	47–51
Apthous ulcers	52–56

reactional lepromatous leprosy (3,4,57), thalidomide was or is being currently evaluated in a host of immunological disorders: lupus erythematosus (20–24), Behcet's syndrome (5,9–16), prurigo nodularis (6–8), ulcerative colitis (27,28), rheumatoid arthritis (25,26), AIDS (47–51), Kaposi's sarcoma (39,40), graft-versus-host disease (31–36) and others. However, the mechanism of thalidomide's immunosuppressive properties still remains under investigation.

Progress made by several groups in elucidating thalidomide's effect on the immune system has shown that thalidomide modulates important immunological cytokines. Studies have shown that thalidomide accelerates the degradation of tumor necrosis factor-alpha (TNF-α) mRNA by twofold in human monocytes, corresponding to a decrease in both in vitro and in vivo TNF-α levels (58,59). This effect has also been demonstrated in the clinic, where serum levels of TNF-α have decreased in patients with leprosy who were receiving thalidomide (73). Interestingly, cachexia, fever, and other symptoms characteristic of AIDS-associated wasting and advanced malignancies have been associated with excessive production of TNF-α by monocytes/macrophages or elevated serum levels of TNF-α (84–89). Thus, while thalidomide treatment in AIDS is primarily given for its anticachectic properties (47–51), it has been suggested that thalidomide may have a palliative benefit in suppressing TNF-α production in other clinical contexts, such as symptomatic advanced malignant disease (70,84–87,90).

Other cytokines have also demosntrated thalidomide modulation. Thalidomide enhances production of interleukin-4 (IL-4) and IL-5 and inhibits interferon-γ in peripheral blood mononuclear cell (PBMC) culture (74). Moller et al. suggested that thalidomide potently suppressed the production of IL-12 from human PBMC and primary human monocytes in a concentration-dependent manner (60).

Neubert and colleagues have evaluated the effect of thalidomide on surface markers of circulating leukocytes of marmosets and humans (61–64). Interestingly, they have suggested that thalidomide downregulates the expression of specific α and β subunits of various adhesion receptors on white blood cells. In particular, the $\beta 2$ integrin subunit (CD18) was significantly diminished on granulocytes, monocytes, and lymphocytes, while the $\beta 1$ integrin subunit (CD29) was reduced slightly (63). The $\alpha 4$ chain (VLA$\alpha 4$) was consistently downregulated, while some adhesion receptors were observed to vary in expression profile, such as the expression of ICAM-1 (CD54), which increased on granulocytes, was eliminated in monocytes, and varied with cell type on lymphocytes (63). Since the $\beta 2$ integrins, expressed solely on leukocytes, are crucial in the adhesion of vascular endothelium and migration of activated leukocytes to sites of inflammation (i.e., diapedesis), it has been postulated that thalidomide's anti-inflammatory activity is mediated by antagonism of the diapedesis cascade (65,67).

The integrin modulatory activity of thalidomide prompted Neubert et al. to advance the hypothesis that thalidomide's teratogenicity was mediated by suppressed expression of adhesion factors in the developing fetus (66). To test this hypothesis, they administered a teratogenic analogue of thalidomide (EM-12) to pregnant marmosets and measured the expression of adhesion receptors in growing fetuses at various stages in development. A significant finding was the consistent suppression of the $\beta 3$ integrin subunit (CD61) in all parts of the fetus, and a suppression of $\beta 1$ and $\alpha 4$ subunits in the limb buds and heart primordia but not in the head and body (66). Both the $\beta 1$ and $\beta 3$ integrin subunits have been primarily associated with cell binding to extracellular matrix proteins (68,69). Neubert et al. offer their observation of suppressed β-integrin subunits as support of their hypothesis that thalidomide, given at a particularly susceptible stage in development, may disrupt migration of cells from the somatic mesoderm to the limb bud primordium (66).

II. ANTIANGIOGENIC PROPERTIES

Previous to the hypothesis of Neubert et al, support for a distinct mechanism of thalidomide's teratogenicity came in 1994, when D'Amato et al. demonstrated that orally administered thalidomide and a teratogenic analog, EM-12, significantly inhibited bFGF-induced angiogenesis in the in vivo rabbit cornea micropocket assay (70). Administration of thalidomide was performed at teratogenic doses (200 mg/kg), resulting in inhibition of vascularized cornea from 30–51% (median 36%) compared to controls. The teratogenic analog EM-12 was inhibitory as well, with a median inhibition of 42% compared to controls. While both the parent compound and the EM-12 analogue failed to have angioinhibitory activity in endothelial cell culture or in the chick chorioallantoic membrane (CAM) assay, the authors hypothesized that hepatic metabolism of thalidomide is required to form an antiangiogenic metabolite (70,71). Electron microscopic analysis of the new blood vessels in the cornea and the deformed limbs showed similar specific ultrastructural changes after treatment with thalidomide (70). These observations led the investigators to postulate that the teratogenic properties of thalidomide may be directly correlated to its antiangiogenic effect in the developing limb bud and that thalidomide could potentially have therapeutic applications in treating angiogenesis-dependent diseases, such as proliferative retinopathy and solid tumors (70–72). Recently, Kruse et al. demonstrated that oral thalidomide treatment at 200 mg/kg has also shown inhibition of VEGF-induced angiogenesis (750- or 500-ng pellet of VEGF) in the rabbit corneal micropocket assay, suggesting the antiangiogenic properties observed by D'Amato were not specific to the angiogenic stimulus used (74).

In this study, significant corneal vascularity inhibition ranged from 55–62% of controls starting after day 5 of treatment (74).

Kenyon et al. further characterized thalidomide and related metabolites in a mouse corneal model of neovascularization (91). In this recent study, the authors evaluated 200 mg/kg doses of thalidomide, phthaloylglutamic acid (PGA; a breakdown metabolite of thalidomide), EM-12, and EM-138 (a stable breakdown metabolite of EM-12) in both bFGF- and VEGF-induced models of neovascularization (91). They tested both oral administration and intraperitoneal injections, since rodents seemingly have no susceptibility to the teratogenicity of orally administered thalidomide, yet they do demonstrate vascular abnormalities when given intraperitoneal treatment (2,92–95). Isomers of thalidomide were also evaluated to detect differences among S(−)-enantiomers, R(+)-enantiomers, and the racemic parent compound (96). The study concluded that while oral administration of thalidomide and EM-12 failed to demonstrate significant inhibition of angiogenesis, intraperitoneal administration of all the metabolites tested inhibited bFGF- and VEGF-induced angiogenesis (91). Moreover, since the metabolites PG acid and EM-138 retained antiangiogenic activity, it was suggested that the open piperidine ring pattern of metabolism was important for antiangiogenic activity (91) (see Sec. IV below). Furthermore, the authors noted greater inhibition of neovascularization with the S(−)-isomer of thalidomide and ruled out inhibition of TNF-α as a mechanism of thalidomide's antiangiogenic activity (91).

While the in vivo studies mentioned above have supported thalidomide's role as an antiangiogenic molecule, other studies have elucidated details of the angiogenic cascade at a molecular level. Interestingly, two angiogenic pathways have been described that are mediated by distinct α_v integrins in the CAM assay and the rabbit corneal assay: a bFGF-induced angiogenic pathway dependent on $\alpha_v\beta_3$ and a distinct VEGF-induced pathway dependent on $\alpha_v\beta_5$ (75,76). In the CAM studies, topical applications of LM609, a monoclonal antibody antagonist of $\alpha_v\beta_3$, inhibited angiogenesis without an effect on preexisting vessels (76). In other studies, LM609 shows inhibition of blood vessel formation in early embryonic development. When microinjected into quail embryos, the antibody inhibited the formation of the dorsal aorta (78). Moreover, Brooks et al. further described how the LM609 antagonist of integrin $\alpha_v\beta_3$ promoted tumor regression and induced apoptosis of angiogenic blood vessels (77). These studies suggest a correlation between antagonizing integrin receptors with the β_3 integrin subunit and antiangiogenic properties in both artificially induced angiogenesis and embryogenesis. Referring back to studies by Neubert et al. of thalidomide treatment in marmoset fetuses, where they noted the down-regulation of the β_3 integrin subunit, it has been postulated that thalidomide will likely suppress the expression of $\alpha_v\beta_3$ by endothelial cells stimulated by an angiogenic agent (65). If further investigation supports this hypothesis, it would lend support to the notion that thalidomide's antiangiogenic mechanism may be related to its pharmacological modulation of angiogenic vascular integrins, which is in accord with its known modulation of surface markers on mammalian leukocytes (65). However, Neubert et al. have expressed skepticism that thalidomide's teratogenicity is limited to its antiangiogenic properties (66).

III. THALIDOMIDE AS AN ANTICANCER AGENT: PRECLINICAL DATA

The dependence on angiogenesis for the growth and metastasis of solid malignancies is well recognized and has amassed scientific support for over thirty years (79–81). The data supporting the notion that progressive tumor growth and metastasis are angiogenesis-dependent has led to a growing interest in antiangiogenic therapy as a potential therapeutic strategy against cancer. Thalidomide, after being characterized as an antiangiogenic compound by D'Amato et al. in 1994, was advanced as a candidate therapeutic against cancer and other angiogenesis-dependent pathologies (70). However, despite the therapeutic promise of thalidomide as an antiangiogenic agent for cancer, the data supporting the proposed antitumor activity of this compound in preclinical animal studies is discrepant.

Thalidomide treatment against rodent tumor cell lines was studied without success in two animal models (82,83). One study by Pollard suggested that thalidomide, administered by gavage at 50 mg/kg body weight per day, promoted the lung metastasis of prostate adenocarcinoma cells (PA-III) implanted subcutaneously in L-W male rats (82). Though the author did not offer a hypothesis explaining the seeming lack of antitumor activity in this study, it could be explained by the dose regimen used. Historically, rodents have demonstrated a lack of susceptibility to oral doses of thalidomide (2,92–95). The previous work in rodent models outlined above by D'Amato et al., Kruse et al., and Kenyon et al. has demonstrated antiangiogenic activity in rodent models at teratogenic thalidomide doses of 200 mg/kg per day (70,74,91). It could be postulated that the relatively low dose (50 mg/kg per day) of thalidomide used in the PA-III tumor model was insufficient to mount an antiangiogenic/antitumor activity but may have had immunomodulatory activity. A previous animal study has shown immunomodulatory activity of thalidomide at comparable levels (up to 100 mg/kg) to the Pollard study (97). Paradoxically, immunosuppression by thalidomide could potentiate metastasis in the absence of an appreciable antiangiogenic effect, which could explain the promotion of lung metastasis in this model (82).

In discordance with the negative Pollard study, Minchinton et al. have documented thalidomide's inhibitory effect on the metastatic potential and radiosensitivity of SCCVII or Lewis Lung (LL) tumors in a murine model (100). Intraperitoneal administration of thalidomide given daily at 200 mg/kg (0.77 mmol/kg) was started on the day of tumor implantation. While thalidomide treated demonstrated little effect on the growth of primary LL or SCCVII tumors followed by local irradiation of 20-Gy x-rays, thalidomide-treated animals were shown to have significantly fewer metastases than controls (100). Interestingly, the authors also observed an enhanced antitumor activity of thalidomide and combined radiation/tirapazamine treatment, suggesting that thalidomide decreases tumor oxygenation in the tumor mass (100). To explain the paradoxical suppression of metastatic potential but lack of activity against the primary tumor in this study the authors postulated that thalidomide may inhibit an early stage of neovascularization (100). In accordance with this study's synergistic antitumor activity of thalidomide and radiation/tirapazamine, enhancement of antitumor activity with combination therapy of thalidomide and 5,6-dimethylxanthenone-4-acetic acid has been shown in a murine setting, suggesting thalidomide in combination with potent antitumor agents may demonstrate significant activity but thalidomide alone may not (102).

Thalidomide's lack of activity against primary tumors has been seen in another in vivo study, where thalidomide administered by daily gavage ranging from 300 μg per mouse to 1 mg per mouse failed to show antiangiogenic or antitumor properties in mice inoculated with either B16-F10 melanoma or CT-26 colon carcinoma cells (83). Thalidomide treatment in this study started at either 2 or 10 days following tumor injection. While subcutaneous B16-F10 tumors demonstrated some growth retardation, the effect was not statistically significant compared to controls (83). Furthermore, morphological evaluation of microvessels showed that both thalidomide treatment groups and controls exhibited an intact network of new blood vessels (83). The negative results were explained as a function of host and tumor factors. The authors suggested that the extensive involvement of the immune system in solid tumor formation and progression in a whole animal system adds complexity to the assay in comparison to the rabbit or mouse cornea studies, which are immunoprivileged (83). Based on previous observations of tumor adaptation to new microenvironments, another hypothesis that the authors offered to account for the negative results was tumor heterogeneity and adaptation (83,98–99).

Another possible explanation for the lack of tumor activity in the in vivo rodent models described above was recently advanced by Bauer et al. (101). This group evaluated cross-species metabolic activation of thalidomide in the ex vivo angiogenesis rat aortic ring assay and endothelial cell culture (101). Their findings suggest that while thalidomide coincubated with human or rabbit liver microsomes inhibited the microvessel formation from rat aortas and slowed human aortic endothelial cell proliferation, neither activity was observed with thalidomide alone or thalidomide coincubated with rat microsomes (101). The group advanced the hypothesis that the antiangiogenic effects of thalidomide are mediated by a metabolite that can be formed in both humans and rabbits but not in rodents (101). Interestingly, this hypothesis lends a biological explanation to the lack of antitumor activity in the previous in vivo rodent models (82,83,100), and it also suggests, assuming a correlation between thalidomide's antiangiogenic properties and teratogenicity, why rodents have historically been tolerant of relatively high doses of thalidomide (1,2,92–95).

The preclinical antitumor activity of thalidomide has thus proven complex to assess. Nevertheless, insofar as thalidomide has proven antiangiogenic activity (70,74,91), and with the historical scientific support of the hypothesis that tumorigenesis, progression, and metastasis are angiogenesis-dependent (79–81), thalidomide has moved into the clinic for the treatment of brain glioblastoma, androgen-independent prostate cancer, AIDS-associated Kaposi's sarcoma, breast cancer, and multiple myeloma (29–30,37–43) (Table 1).

IV. PHARMACOLOGY/PHARMACOKINETICS: PRECLINICAL AND CLINICAL DATA

Thalidomide (*N*-phthalidoglutarimide; $C_{13}H_9O_4N_2$; MW = 258.23 D) is a racemic compound whose S(−)/L and R(+)/D forms represent derivatives of the L- and D-glutamic acid, respectively. The maximal solubility of racemic thalidomide in water is approximately 2×10^4 mol/L (45 to 60 mg/L) (104). All four amide bonds present in the molecule are susceptible to hydrolytic cleavage in vitro at pH values higher than 6 to form three primary products [4-phthalimidoglutaramic acid, 2-phthalimidoglutaramic acid, and α-(o-carboxybenzamido) glutarimide] and eight minor products (104,133–138). To date, there are five known metabolites of thalidomide (4-OH-thalidomide, 3-OH-thalidomide, 3′-OH-thalidomide, 4′-OH-thalidomide, and 5′-OH-thalidomide) (104,133–138). However, considering the possible combinations of hydrolysis, hydroxylation, and optical activity, there may be more than fifty metabolites of thalidomide in vivo.

Nonenzymatic cleavage of one or more of the amide bonds in the thalidomide molecule produces hydrolysis products which contain at least one carboxyl group (104). They are thus polar and can be expected to cross biological membranes less efficiently than the parent compound. Thalidomide constitutes a transport form for its hydrolysis products; the nonpolar parent compound enters cells or tissues and is converted to polar derivatives which have been shown to accumaulate in erythrocytes and in the embryo (105).

Studies in experimental animals showed high concentrations of the drug in the gastrointestinal tract, liver, and kidney and lower concentrations in muscle, brain and adipose tissue. In pregnant animals, thalidomide is able to pass across the placental barrier (104–106). Other animal studies suggest the main pathway of degradation is nonenzymatic hydrolytic cleavage (104,105). Minor amounts of hydroxylated products have been detected in the urine of some species (71,104). Studies suggest only the parent compound is enzymatically modified and that hepatic metabolism of thalidomide probably involves enzymes of the cytochrome P-450 family (104,107,108). While studies show the parent compound does not cause enzyme induction, thalidomide may possibly interfere with the enzyme induction caused by other compounds (109).

Oral administration of thalidomide at 100 – 200 mg in humans results in maximal blood concentrations of 0.9–1.5 μg/L after 4 to 6 h (102). Absorption and elimination half-lives calculated from data of eight healthy subjects were 1.7 ± 1.05 and 8.7 ± 4.11 hr, respectively; a lag time of 0.41 ± 0.17 hr was observed in six individuals (102). Using a one-compartment model, the authors estimated a volume of distribution of 120.64 ± 45.46 L, a total body clearance of 10.41 ± 2.04 L/hr, and a renal clearance 0.08 ± 0.03 L/hr. Only 0.6 ± 0.22% of the administered dose was excreted as unchanged compound in urine. The hydrolytic cleavage in serum is much slower than that in vitro at pH 7.4 (102,104). The difference may be explained by the fact that thalidomide has shown to be highly bound to plasma proteins (110).

Drug interactions with thatlidomide have not been systematically studied. While thatlidomide enhances the sedative activity of barbiturates, alcohol, chlorpromazine, and reserpine, the drug is antagonized by methyl amphetamine and methylphenidate. Thalidomide was found to antagonize the action of histamine, serotonin, acetylcholine, and prostaglandins in organ bath experiments but had no influence on the uterine reaction to oxytocin, vasopressin, and histamine (111).

Based on the clinical history of thalidomide in a wide variety of clinical settings, the toxicities of the drug in humans are well known (Table 2). The main adverse effects include somnolence, constipation, peripheral sensory neuropathy, swelling of the limbs and face, erythema of the limbs, hair loss, fever, rash, and amenorrhea, yet others have

Table 2 Adverse Effects of Thalidomide Treatment

Peripheral sensory neuropathy
Somnolence
Constipation and xerostomia
Confusion
Edema of the face and limbs
Erythema of the limbs
Hair loss
Fever
Rash
Increased appetite
Loss of libido
Nausea
Pruritus
Amenorrhea
Menstrual abnormalities

been noted (Table 2). However, there is evidence to suggest thalidomide may not be responsible for all side effects observed during treatment. In a crossover study in patients with aphthous stomatitis, some side effects such as headache, dryness of the skin, gastric pain and impairment of vision occurred in placebo and thalidomide groups with similar frequency (112).

The most serious and worrisome of thalidomide's toxicities is peripheral sensory neuropathy. Peripheral neuropathy, clinically characterized by a symmetrical distal paresthesia with sensory loss in distal limbs (predominantly the toes), has been estimated at 1% incidence in lepra-reaction patients, 12% in rheumatoid arthritis, 22% in prurigo nodularis, 25% in patients with discoid lupus erythematosus, and 21% in electrophysiological studies in a mixed group of patients (25,113–116). Some studies noted a dose-dependent pattern, with neuropathies occurring with cumulative thalidomide dosages of greater than 40 g (117,118). Paresthesias are often noted early in the course of neuropathy, before other neurological signs develop (116). A higher incidence of neuropathy has been reported in females and the elderly (119). The neuropathy does not appear to progress after discontinuation of the drug, and many cases of neuropathy seem to be reversible if treatment is discontinued once it is detected (120–122). Some patients with preexisting neuropathies may be particularly sensitive to the development of thalidomide polyneuritis (123,124). However, in conditions such as erythema nodosum leprosum (ENL), which present neuritis as a consequence of disease, thalidomide treatment has improved symptoms (3,4,113). Thus, while thalidomide neuropathy was not observed after repeated ENL episodes treated with thalidomide, it may be difficult to distinguish neuropathy as an adverse drug effect or neuropathological changes caused by the underlying disease (113).

V. THALIDOMIDE CLINICAL TRIALS IN CANCER

The preclinical data supporting thalidomide's antiangiogenic activity prompted the initiation of four phase II clinical trials in solid tumors: AIDS-associated Kaposi's sarcoma (KS), refractory glioblastoma, breast cancer, and prostate cancer. While these trials are ongoing, preliminary data has been presented recently regarding the therapeutic efficacy of thalidomide in these four clinical settings. Furthermore, recent findings in a pilot study of thalidomide treatment in multiple myeloma have been reported.

A. AIDS-Associated Kaposi's Sarcoma

Oral thalidomide in AIDS patients has been used to treat HIV wasting and AIDS-associated apthous ulcers (see Table 1). A phase II trial of thalidomide in AIDS-associated Kaposi's sarcoma was initiated by Yarchoan et al. for 13 patients, median [range] age 31 [31–49] (127,128). Eight patients were poor risks and five were good risks as assessed using the TIS staging criteria for Kaposi's sarcoma (127,128). Preliminary results of this study were available for 11 of 13 patients on a dose-escalation regimen from 200 mg/day up to 1000 mg/day. The authors have recently reported an overall partial response rate of 36% (4 of 11 patients; 3 of the 4 responding patients were poor risks by TIS staging) that was not dose-dependent (127,128). Of the responding patients, 3 were on stable anti-HIV therapy, 2 had previous cytotoxic chemotherapy. Responses were observed at 4 weeks (3 of 11 patients) and 8 weeks (1 of 11 patients) at daily doses of 300 mg, 400 mg (2 patients), and 800 mg. The authors note that two partial responders have not progressed at 50 and 30 weeks, while two other partial responders progressed at 16 and 49 weeks (127,128). Toxicities requiring cessation of treatment were reported for 3 patients: grade 3 rash with fever recurring after drug rechallenge (1 patient); myositis (1 patient); and depression (1 patient). Three patients had grade 3 neutropenia. Dose reduction was required to resolve sedation in 5 patients. Finally, serial assays for basic fibroblast growth factor (bFGF) and TNF-α demonstrated no clear trends with thalidomide treatment.

The authors have concluded thalidomide is well tolerated and appears to have activity in AIDS-related Kaposi's sarcoma. The trial is ongoing and further study will establish thalidomide's therapeutic potential in Kaposi's saracoma.

B. Glioblastoma

The evidence of thalidomide's antiangiogenic activity made it a promising drug for long-term therapy in patients with vascular cerebral tumors. Fine et al. are currently conducting a phase II trial of thalidomide in 38 patients with recurrent high-grade astrocytomas (35% of patients) and mixed gliomas (65% of patients) (124,125). Of the 38 patients, 32 have been evaluable for response. Thalidomide was relatively well tolerated. The most common side effect observed was somnolence, which did not seem aggravated by increasing thalidomide concentrations. Four patients reported seizures during thalidomide treatment. However, three of these patients had prior history of regular episodes of seizures, and there was no obvious increase in activity with thalidomide. The one patient who reported a seizure without prior history had significant tumor progression at the time of the episode. Thus, the investigators speculate that there is little conclusive evidence that thalidomide treatment alone can cause or increase seizure incidence independent of disease progression, but the cases were noted. Interestingly, the authors noted that peripheral neuropathy was not a major side effect observed in this study.

Of the 32 patients evaluable for response, 2 had a partial response, 2 a minimal response, 12 had stable disease (38% overall response rate), and 16 (50%) progressed (124,125). Of the 32 evaluable patients, 8 were alive 1 year after thalidomide treatment. Serial assays for bFGF were conducted but no clear trends were observed.

C. Breast Cancer

A multicenter (Georgetown University, University of Chicago, Duke University, and the Dana-Farber Cancer Institute in Boston) phase II trial of thalidomide for metastatic breast cancer has been initiated (131). Two dosing regimens are being evaluated: a low dose of 200 mg/day and a high dose–escalating regimen from 800 mg/day up to 1200 mg/day dependent on patient tolerance. The multicenter study has had 28 evaluable patients (14 patients at low dose; 14 patients on high dose: 4 patients on 1200 mg/day, 1 on 1000 mg/day, 3 on 800 mg/day; dose decrease due to somnolence resulted in 4 on 600 mg/day and 2 on 400 mg/day). The main on-study toxicity in the low-dose arm was one patient with grade 3 peripheral neuropathy at week 8 of treatment. Off-study toxicities included somnolence (4 low dose and 8 high dose), peripheral neuropathy (5 low and 4 high), dizziness/instability (2 low and 4 high), arrhythmia (1 low and 0 high), and neutropenia (1 low and 1 high). All patients starting at 800 mg/day failed to remain on study beyond 8 weeks due to progression. Two patients on 200 mg/day had stable disease at 8 weeks, but both progressed, one at 10 weeks, the other at 16 weeks (131).

D. Prostate Cancer

Figg et al are currently evaluating two dosing regimens of thalidomide (200 or 1200 mg/day) in patients with androgen-independent prostate cancer (AIPC) (129,130). Forty-three patients have been enrolled, 13 being evaluated at the high dose, and the rest (30 patients) at the low dose. All patients have failed androgen deprivation and several had failed secondary hormonal manipulations. Two patients had grade 3 complications (neuromotor and constipation), with the remaining complications limited to grade 1 or 2 (dizziness, drowsiness, xerostoma, neurosensory, and rash) (129,130). Pharmacokinetic analysis revealed a half-life for thalidomide of 13.1 $\pm$ 14.5 hr, clearance (CL/F) of 7.1 $\pm$ 3.7 L/hr, and a volume distribution (VD/F) of 88.9 $\pm$ 44.5 L. Steady-state concentrations for the high-dose group have average $>$ 10 μg/mL. The authors report that 4 patients (30%) have had PSA declines (from 20–37%, 2 in each group). Currently, the authors have reported 2 patients with partial responses and 5 patients with stable disease independent of dose regimen (personal communication) (129).

E. Multiple Myeloma

Mehta and colleagues at the Arkansas Cancer Research Center are conducting a clinical study of thalidomide for the treatment of multiple myeloma (132). In 26 patients with refractory multiple myeloma (previous therapy including bone marrow transplantation/chemotherapy), thalidomide treatment stopped progression of disease in nearly 60% of patients. According to the authors, an unprecedented observation was that the results of three patients approached complete response. The study has been expanded to include over 120 evaluable patients.

VI. SUMMARY AND CONCLUSIONS

Thalidomide was removed from general circulation in the early 1960s after evidence of its teratogenicity in humans was discovered. However, thalidomide has remained in clinical use for a number of diseases ranging from leprosy-related ENL to malignant disease. The therapeutic efficacy of thalidomide in the diverse clinical settings for which it has been or is being evaluated seem associated with its known immunomodulatory and antiangiogenic biological activities.

Preclinical data suggest that thalidomide has an antiangiogenic effect in rabbit and mouse cornea models. Since angiogenesis has been implicated as a requirement in the growth, progression, and ultimately metastasis of primary solid tumors, thalidomide has moved into clinical trials for a number of malignancies. The teratogenic action of thalidomide necessitates contraceptive measures in all women of childbearing potential receiving the drug. The most frequently observed side effects are related to the sedative action of thalidomide, namely somnolence, dizziness, and mood changes. Severe adverse effects relate to thalidomide's neuropathological reactivity, and though motor function is seemingly not effected, the main toxicity is loss of nerve fiber associated with peripheral sensory neuropathy. Previous experience suggest the peripheral neuropathy is more prevalent in the legs than in the arms.

The description of thalidomide as an antiangiogenic agent has stimulated interest in the potential for its use in anticancer therapy. While the in vitro and in vivo preclinical studies of thalidomide have differed in results, the complexity of studying the biological effects of thalidomide in various experimental designs may be related to its need for metabolic activation, its species-dependent activity, its dosage, or its mode of administration. The cumulative data from preclinical studies did, however, justify the initiation of clinical trials of thalidomide in human cancer.

Clinical trials evaluating thalidomide treatment in AIDS-associated Kaposi's sarcoma, glioblastoma, breast cancer, prostatic cancer, and, most recently, multiple myeloma are ongoing. Preliminary results suggest that thalidomide is well tolerated in these clinical settings and treatment has demonstrated various degrees of activity. Further investigation is required to define the role thalidomide treatment could potentially play as an anticancer therapeutic.

The exact mechanisms by which thalidomide modulates the immune response and the angiogenic cascade are currently under investigation. Thalidomide has demonstrated modulation of cell-surface markers on human and marmoset leukocytes. Based on these experiments, it has been suggested that thalidomide's teratogenic, immunomodulatory, and antiangiogenic activity may be mediated by regulation of specific cell-surface integrin subunits, particularly the β integrin subunits. The description of at least two distinct pathways of angiogenesis mediated by the $\alpha_v\beta_3$ and the $\alpha_v\beta_5$ integrins lends credibility to the hypothesis that perhaps thalidomide's antiangiogenic activity is associated with vascular integrin expression or antagonism. However, this hypothesis remains speculative. If thalidomide does prove to mediate its antiangiogenic activity by modulating vascular integrins, it would have direct implications for effective anticancer therapy. As the mechanisms of tumor angiogenesis are elucidated, it could be anticipated that tumor heterogeneity in the angiogenic profile of primary tumor versus micrometastases will define which clinical settings could benefit most from thalidomide therapy.

In conclusion, thalidomide is a promising drug for anticancer therapy based on its oral bioavailability, minimal and treatable side effects, and preliminary clinical results. Ongoing studies in defining the mechanisms of thalidomide's immunomodulatory and antiangiogenic activity may provide further insight in its clinical development and maximizing its potential in treating a host of diseases, particualrly refractory human cancers.

REFERENCES

1. McBride WG (1961). Thalidomide and congenital abnormalities. Lancet 2:1358.
2. Lenz W (1962). Thalidomide and congenital abnormalities. Lancet 1:45.
3. Sheskin J (1965). Thalidomide in treatment of lepra reactions. Clin Pharmacol Ther 6:303–306.
4. Sheskin J (1965). Further observations with thalidomide in lepra reactions. Leprosy Rev 36: 183–187.
5. Gardner-Medwin JM, Smith NJ, Powel RJ (1994). Clinical experience with thalidomide in the management of severe oral and genital ulceration in conditions such as Behçet's disease: use of neurophysiological studies to detect thalidomide neuropathy. Ann Rheumatol Dis 53: 828–832.
6. Berger TG, Hoffman C, Thieberg MD (1995). Prurigo nodularis and photosensitivity in AIDS: treatment with thalidomide. J Am Acad Dermatol 33 (5 pt 1):837–838.
7. Grabczyska SA, Hawk JL (1997). Managin PLE and actinic prurigo. Practitioner 241: 74,77,79.

8. Maurer TA, Poncelet A, Badger J, Berger TG (1996). Thalidomide for the treatment of prurigo nodularis of HIV-infected patients (abstr.) Int Conf AIDS 11:312.
9. Waters MF (1971). An internally-controlled double blind trial of thalidomide in severe erythema nodosum leprosum. Lepr Rev 42:26–42.
10. Sheskin J (1980). The treatment of lepra reaction in lepromatous leprosy: fifteen years' experience with thalidomide. Int J Dermatol 19:318–322.
11. Barnhill RL, McDougall AC (1982). Thalidomide: use and possible mode of action in reactional lepromatous leprosy and in various other conditions. J Am Acad Dermatol 7:317–323.
12. Jew LJ, Middleton RK (1990). Thalidomide in erythema nodosum leprosum. DICP 24:482–483.
13. Kaplan G (1994). Cytokine regulation of disease progression in leprosy and tuberculosis. Immunobiology 19:564–568.
14. Meyerson MS (1996). Erythema nodosum leprosum. Int J Dermatol 35:389–392.
15. Sampaio EP, Kaplan G, Miranda A, Nery JA, Miguel CP, Viana SM, Sarno EN (1993). The influence of thalidomide on the clinical and immunologic manifestation of erythema nodosum leprosum. J Infect Dis 168:408–414.
16. Waters MFR, Jacobs JM (1996). Leprous neuropathies. Baillieres Clin Neurol 5:171–197.
17. Chow RKP, Ho VC (1996). Treatment of pyoderma gangrenosum. J Am Acad Dermatol 34:1047–1060.
18. Rustin MH, Gilkes JJ, Robinson TW (1990). Pyoderma gangrenosum associated with Behcet's disease: treatment with thalidomide. J Am Acad Dermatol 23(5 pt 1):941–944.
19. Farrell AM, Black MM, Bracka A, Bunker CB (1998). Pyoderma gangrenosum of the penis. Br J Dermatol 138:337–340.
20. Alarcon-Segovia D (1994). The treatment of systemic lupus erythematosus. Clin Exp Rheumatol 12(Suppl 11):S49–S53.
21. Bessis D, Guillot B, Monpoint S, Danduran M, Guilhou JJ (1992). Thalidomide for systemic lupus erythematosus (letter). Lancet 339:549–550.
22. Burrows NP, Walport MJ, Jones RR (1991). Treatment of chronic widespread cutaneous lupus erythematosus with thalidomide. Br J Dermatol Suppl 125:65–66.
23. Stevens RJ, Andujar C, Edwards CJ, Ames PR, Barwick AR, Khamashta MA, Hughes GR (1997). Thalidomide in the treatment of the cutaneous manifestations of lupus erythematosus: experience in sixteen consecutive patients. Br J Rheumatol 36:353–359.
24. Warren KJ, Nopper AJ, Crosby DL (1998). Thalidomide for recalcitrant discoid lesions in a patient with systemic lupus erythematosus. J Am Acad Dermatol 39(2 pt 1):293–295.
25. Gutierrez-Rodriguez O, Starusta-Bacal P, Gutierrez-Montes O (1989). Treatment of refractory rheumatoid arthritis—the thalidomide experience. J Rheumatol 16:158–163.
26. Huzinga TW, Dijkmans BA, van der Velde EA, van de Pouw Kraan TC, Verweij CL, Breedveld FC (1996). An open study of pentoxyfylline and thalidomide as adjuvant therapy in the treatment of rheumatoid arthritis. Ann Rheum Dis 55:833–836.
27. Stevens CA, Shah SA, Bousvaros A (1995). Immunosuppressive agents in gastrointestinal disease. Curr Opin Gastroenterol 11:554–561.
28. Waters MF, Laing AB, Ambikapathy A, Lennard-Jones JE (1979). Treatment of ulcerative colitis with thalidomide. Br Med J 1:792.
29. Fine HA (1995). Novel biologic therapies for malignant gliomas: antiangiogenesis, immunotherapy, and gene therapy. Neurol Clin 13:827–846.
30. Kaba SE, Kyritsis AP (1997). Recognition and management of gliomas. Drugs 53:235–244.
31. Chen F, Tsang K, Liang R (1997). Thalidomide-responsive chronic pulmonary GVHD (letter). Bone Marrow Transplant 19(1):95–96.
32. Cole CH, Rogers PC, Pritchard S, Phillips G, Chan KW (1994). Thalidomide in the management of chronic graft-versus-host disease in children following bone marrow transplantation. Bone Marrow Transplant 14:937–942.

33. Ehninger G, Eger K, Stuhler A, Schuler U (1993). Thalidomide—the need for a new clinical evaluation of an old drug. Bone Marrow Transplant 12(suppl 3):S26–S28.
34. Field EO, Gibbs JE, Tucker DF, Hellmann K (1966). Effect of thalidomide on the graft versus host reaction. Nature 211:1308–1310.
35. Vogelsang GB (1990). Thalidomide treatment for graft-vs-host disease. Hematology 12:293–300.
36. Vogelsang GB, Farmer ER, Hess AD, Altamonte V, Beschorner WE, Jabs DA, Corio RL, Levin LS, Colvin OM, Wingard JR, et al (1992). Thalidomide for the treatment of chronic graft-versus-host disease. N Engl J Med 326:1055–1058.
37. Carducci MA, DeWeese TL, Nelson WG, Simons JW, Sinibaldi V, Eisenberger MA (1996). Prostate cancer treatment strategies based on tumor-specific biological principles: future directions. Semin Oncol 23(6 suppl 14):56–62.
38. Karp JE, Chiarodo A, Brawley O, Kelloff GJ (1996). Prostate cancer prevention: investigational approaches and opportunities. Cancer Res 56:5547–5556.
39. Carlesimo M, Giustini S, Rossi A, Bonaccorsi P, Calvieri S (1995). Treatment of cutaneous and pulmonary sarcoidosis with thalidomide. J Am Acad Dermatol 32(5 pt 2):866–869.
40. Soler RA, Howard M, Brink NS, Gibb D, Tedder RS, Nadal D (1996). Regression of AIDS-related Kaposi's sarcoma during therapy with thalidomide. Clin Infect Dis 23:501–505.
41. Gasparini G, Harris AL (1995). Clinical importance of the determination of tumor angiogenesis in breast carcinoma: much more than a new prognostic tool. J Clin Oncol 13:765–782.
42. Marshall JI, Hawkins MJ (1995). The clinical experience with antiangiogenic agents. Breast Cancer Res Treat 36:253–261.
43. Nguyen M, Tran C, Barsky S, Sun J-R, McBride W, Pegram M, Pietras R, Love S, Glaspy J (1997). Thalidomide and chemotherapy combination: preliminary results of preclinical and clinical studies. Int J Oncol 10:965–969.
44. Cornea P, Vladutiu A, Marcovici G (1967). The immunosuppressive effect of thalidomide in experimental tuberculosis in guinea-pigs. Tubercle 48:51–53.
45. Tramontana JM, Utaipat U, Molloy A, Akarasewi P, Burroughs M, Makonkawkeyoon S, Johnson B, Klausner JD, Rom W, Kaplan G (1995). Thalidomide treatment reduces tumor necrosis factor alpha production and enhances weight gain in patients with pulmonary tuberculosis. Mol Med 1:384–397.
46. Veloso VG, Grinsztejn B, Liporage J, Sampaio E, Kaplan G (1996). Thalidomide for AIDS patients with tuberculosis (abstr) Int Conf AIDS 11:303.
47. Gunzler V (1992). Thalidomide in human immunodeficiency virus (HIV) patients: a review of safety considerations. Drug Safety 7:116–134.
48. James JS (1992). Thalidomide for wasting syndrome: progress toward compromise. AIDS Treat News 234:1–5.
49. Makonkawkeyoon S, Limson-Pobre RN, Moreira AL, Schauf V, Kaplan G (1993). Thalidomide inhibits the replication of human immunodeficiency virus type 1. Proc Natl Acad Sci USA 90:5974–5978.
50. Marriot JB, Cookson S, Carlin E, Youle M, Hawkins D, Nelson M, Pearson M, Vaughan A, Gazzard B, Dalgleish A (1996). A placebo controlled phase II trial of thalidomide in asymptomatic HIV-positive patients; clinical tolerance and effect on activation markers and cytokines (abstr). Int Conf AIDS 11:221.
51. Minor JR, Piscitelli SC (1996). Thalidomide in diseases associated with human immunodeficiency virus infection. Am J Health Sys Pharm 53:429–431.
52. Alexander LN, Wilcox CM (1997). A prospective trial of thalidomide for the treatment of HIV-associated idiopathic esophageal ulcers. AIDS Res Hum Retroviruses 13(4):301–304.
53. Grinspan D (1985). Significant response of oral aphthosis to thalidomide treatment. J Am Acad Dermatol 12 (1 pt 1):85–90.
54. Hoy JF, Stewart K, Mijch AM, Lucas CR (1991). Placebo-controlled trial of thalidomide in the treatment of severe aphthous ulceration in AIDS (abstr). Int Conf AIDS 7:266.

55. Randall P (1995). Thalidomide effective treatment for AIDS-related mouth ulcers. NIAID AIDS Agenda Dec. 2.
56. Weidle PJ (1996). Thalidomide for aphthous ulcers in patients infected with the human immunodeficiency virus. Am J Health Syst Pharm 53:368, 371–372, 378.
57. Iyer CG, Languillon J, Ramanujam K, et al (1971). WHO coordinated short-term double-blind trial with thalidomide in the treatment of acute lepra reactions in male lepromatous patients. Bull WHO 45:719–732.
58. Sampaio E, Sarno E, Galilly V, Cohn ZA, Kaplan G (1991). Thalidomide selectively inhibits tumor necrosis factor alpha production by stimulated human monocytes. J Exp Med 173: 699–703.
59. Moreira A, Sampaio E, Zmuidzinas A, Frindt A, Smith K, Kaplan G (1993). Thalidomide exerts its inhibitory activity on tumor necrosis factor alpha (TNF-alpha) by enhancing mRNA degradation. J Exp Med 177:1675–1680.
60. Moller DR, Wysocka M, Greenlee BM, Ma X, Wahl L, Flockhart DA, Trinchieri G, Karp CL (1997). Inhibition of IL-12 production by thalidomide. J Immunol 159:5157–5161.
61. Neubert R, Nogueira AC, Neubert D (1992). Thalidomide and the immune system. 2: Changes in receptors on blood cells of a healthy volunteer. Life Sci 51:2107–2116.
62. Neubert R, Nogueira AC, Neubert D (1993). Thalidomide derivatives and the immune system. I: Changes in the pattern of integrin receptors and other surface markers on T lymphocyte subpopulations of marmoset blood. Arch Toxicol 67:1–17.
63. Nogueira AC, Neubert R, Helge H, Neubert D (1994). Thalidomide and the immune system. 3: Simultaneous up- and down-regulation of different integrin receptors on human white blood cells. Life Sci 55:77–92.
64. Neubert R, Helge H, Neubert D (1995). Thalidomide and the immune system. 4: Down-regulation of the CD26 receptor, probably involved in the binding of HIV components to T cells in primates. Life Sci 56:407–420.
65. McCarty MF (1997). Thalidomide may impede cell migration in primates by down-regulating integrin beta-chains: potential therapeutic utility in solid malignancies, proliferative retinopathy, inflammatory disorders, neointimal hyperplasia, and osteoporosis. Med Hypoth 49:123–131.
66. Neubert R, Hinz N, Thiel R, Neubert D (1996). Down-regulation of regulation of adhesion receptors on cells of primate embryos as a probable mechanism of action of thalidomide. Life Sci 58:295–316.
67. Zimmerman GA, Prescott SM, McIntyre TM (1992). Endothelial cell interactions with granulocytes: tethering and signaling molecules. Immunol Today 13:93–99.
68. Hynes RO, Lander AD (1992). Contact and adhesive specificities in the associations, migrations, and targeting of cells and axons. Cell 69:11–25.
69. Brooks, PC (1996). Role of integrins in angiogenesis. European J Cancer 32A(14):2423–2429.
70. D'Amato RJ, Loughnan MS, Flynn E, Folkman J (1994). Thalidomide is an inhibitor of angiogenesis. Proc Natl Acad Sci USA 91:4082–4085.
71. Gordon GB, Spielberg SP, Blake DA, Balasubramanian V (1981). Thalidomide teratogenesis: evidence for a toxic arene oxide metabolite. Proc Natl Acad Sci USA 78:2545–2548.
72. Folkman J. (1990). What is the evidence that tumors are angiogenesis-dependent? J Natl Cancer Inst 82:4–6.
73. Sarno EN, Grau GE, Vieira LLM, Nery JA (1991). Serum levels of tumour necrosis factor-alpha and interleukin-1β during leprosy reactional states. Clin Exp Immunol 84:103–108.
74. McHugh SM, Rifkin IR, Dieghton J, Wilson AB, Lachmann PJ, Lockwood CM, Ewan PW (1995). The immunosuppressive drug thalidomide induces T helper cells in mitogen- and antigen-stimulated human peripheral blood mononuclear cell cultures. Clin Exp Immunol 99:160–167.

75. Brooks PC, Clark RAF, Cheresh DA (1994). Requirement of vascular integrin alpha-v beta-3 for angiogenesis. Science 264:569–571.
76. Friedlander, M, Brooks PC, Shaffer RW, Kincaid CM, Varner JA, Cheresh DA (1995). Definition of two angiogenic pathways by distinct alpha-v integrins. Science 270:1500–1502.
77. Brooks PC, Montgomery AMP, Rosenfeld M, Reisfeld RA, Hu T, Klier G, Cheresh DA (1994). Integrin alpha-v beta-3 antagonists promote tumor regression by inducing apoptosis of angiogenic blood vessels. Cell 79:1157–1164.
78. Drake CJ, Cheresh DA, Little CD (1995). An antagonist of integrin alpha-v beta-3 prevents maturation of blood vessels during embryonic neovascularization. J Cell Sci 108:2655–2661.
79. Folkman J (1998). Tumor angiogenesis. In: Holland JR, Frei E, Bast R, Kufe D, Morton D, Weichselbaum R, eds. Cancer Medicine, 4th ed. Baltimore: Williams & Wilkins, 1997.
80. Folkman J (1971). Tumor angiogenesis: therapeutic implications. N Engl J Med 285:1182–1186.
81. Folkman J (1972). Anti-angiogenesis: new concept for therapy of solid tumors. Ann Surg 175:409–416.
82. Pollard M (1996). Thalidomide promotes metastasis of prostate adenocarcinoma cells (PA-III) in L-W rats. Cancer Lett 101:21–24.
83. Gutman M, Szold A, Ravid A, Lazzauskas T, Merimsky O, Klausner JM (1996) Failure of thalidomide to inhibit tumor growth and angiogenesis in vivo. Anticancer Res 16:3673–3676.
84. Beutler B, Cerami A. (1987). Cachectin: more than a tumor necrosis factor. N Engl J Med 316:379–385.
85. Nakashima J, Tachibana M, Ueno M, Miyajima A, Baba S, Murai M (1998). Association between tumor necrosis factor in serum and cachexia in patients with prostate cancer. Clin Cancer Res 4:1743–1748.
86. Ling PR, Schwartz JH, Bistrian BR (1997). Mechanisms of host wasting induced by administration of cytokines in rats. Am J Physiol 272(3 pt 1):E333–E339.
87. Tisdale MJ (1997). Biology of cachexia. J Natl Cancer Inst 89:1763–1773.
88. Lahdevirta J, Maury CP, Teppo AM, Repo H (1998). Elevated levels of circulating cachectin/tumor necrosis factor in patients with acquired immunodeficiency syndrome. Am J Med 85: 289–291.
89. Belec L, Meillet D, Hernvann A, Gresenguet G, Gherardi R (1994). Differential elevation of circulating interleukin-1 beta, tumor necrosis factor alpha, and interleukin-6 in AIDS-associated cachectic states. Clin Diagn Lab Immunol 1:117–120.
90. Deaner P (1998). Thalidomide for distressing night sweats in advanced malignant disease. Palliat Med 12:208–209.
91. Kenyon BM, Browne F, D'Amato RJ (1997). Effects of thalidomide and related metabolites in a mouse corneal model of neovascularization. Exp Eye Res 64:971–978.
92. Scott WJ, Fradkin R, Wilson JG (1970). Non-confirmation of thalidomide induced teratogenesis in rats and mice. Teratology 5:33–36.
93. Cahen R (1964). Evaluation of the teratogenicity of drugs. Clin. Pharmacol Ther 5:480–514.
94. Szabo KT, Steelman RL (1967). Effects of maternal thalidomide treatment on pregnancy, fetal development, and mortality of the offspring in random-bred mice. Am J Vet Res 28: 1823–1828.
95. Köhler F, Ockenfels H (1970). Teratogene Wirkung der N-Phthalyl-DL-glutaminsaüre nach intraperitonealer: Applikation bei der Maus. Experientia 26:1157–1158.
96. Kean WF, Lock C, Howard-Lock HE (1991). Chirality in antirheumatic drugs. Lancet 338: 1565–1568.
97. Ching LM, Xu ZF, Gummer BH, Palmer BD, Joseph WR, Baguley BC (1995). Effect of thalidomide on tumor necrosis factor production and antitumor activity induced by 5,6 dimethylanthenone-4-acetic acid. Br J Cancer 72:339–343.

98. Fidler IJ (1990). Critical factors in the biology of human cancer metastasis: twenty-eighth G.H.A. Clowes Memorial Award Lecture. Cancer Res 50:6130–6138.
99. Gutman M, Singh RK, Xie K, Bucana CD, Fidler IJ (1995). Regulation of interleukin-8 expression in human melanoma cells by the organ microenvironment. Cancer Res 55:2470–2475.
100. Minchiton AI, Fryer KH, Wendt KR, Clow KA, Hayes MMM (1996). The effect of thalidomide on experimental tumors and metastases. Anticancer Drugs 7:339–343.
101. Bauer KS, Dixon SC, Figg WD (1998) Inhibition of angiogenesis by thalidomide requires metabolic activation, which is species-dependent. Biochem Pharmacol 55:1827–1834.
102. Ching LM, Browne WL, Tchernegovski R, Gregory T, Baguley BC, Palmer BD (1998). Interaction of thalidomide, phthalimide analogues of thalidomide and pentoxifylline with the antitumour agent 5,6-dimethylxanthenone-4-acetic acid: concomitant reduction of serum tumour necrosis factor-alpha and enhancement of anti-tumour activity. Br J Cancer 78:336–343.
103. Chen TL, Vogelsang GB, Petty BG, Brundrett RB, Noe DA, et al. (1989). Plasma pharmacokinetics and urinary excretion of thalidomide after oral dosing in healthy male volunteers. Drug Metab Dispos 17:402–405.
104. Schumacher H, Smith RL, Williams RT (1965). The metabolism of thalidomide and some of its hydrolysis products in various species. Br J Pharmacol 25:338–351.
105. Fabro S, Smith RL, Williams RT (1967). Toxicity and teratogenicity of optical isomers of thalidomide. Nature 215:296.
106. Nicholls PJ (1966). A note on the absorption and excretion of 14C-labeled thalidomide in pregnant mice. J Pharm Pharmacol 18:46–48.
107. Braun AG, Harding FA, Wienreb SL (1986). Teratogen metabolism: thalidomide activation is mediated by cytochrome p450. Toxicol Appl Pharmacol 82:175–179.
108. Braun AG, Wienreb SL (1985). Teratogen metabolism: spontaneous decay products of thalidomide and thalidomide analogues are not bioactivated by liver microsomes. Teratol Carcinog Mutagen 5:149–158.
109. Brode E (1968). Experimental studies with cyclic imides: modification of hypohyseal-adrenal axis by thalidomide. Arzneimittelforschung 18:661–665.
110. Bakay B, Nyhan WL (1968). Binding of thalidomide by macromolecules in the fetal and maternal rat. J Pharmacol Exp Ther 161:348–360.
111. Locker D, Superstine E, Sulman FG (1971). The metabolism of the push and pull principle: endocrine effects of thalidomide and its analogues. Arch Int Pharmacodyn Ther 194:39–55.
112. Revuz J, Guillaume JC, Janier M, Hans P, Marchand C, et al (1990). Crossover study of thalidomide vs placebo in severe recurrent apththous stomatitis. Arch Dermatol 126:923–927.
113. Sheskin J, Yaar I (1979). Motor-conduction velocity of cubital nerves in patients with leprosy reactions: summary of a 13-year observation series during thalidomide therapy. Hautarzt 30: 376–379.
114. Johnke H, Zachariae H (1993). Thalidomide treatment of prurigo nodularis. Ugeskr Laeger 155:3028–3030.
115. Knop J, Bonsmann G, Happle R, Ludolph A, Matz DR, et al. (1983). Thalidomide in the treatment of sixty cases of chronic discoid lupus erythematosus. Br J Dermatol 108:461–466.
116. Ochonisky S, Verroust J, Bastuju-Garin S, Cherardi R, Revus J (1994). Thalidomide neuropathy incidence and clinico-electrophysiologic findings in 42 patients. Arch Dermatol 130:66–69.
117. Fullerton PM, O'Sullivan DJ (1968). Thalidomide neuropathy: a clinical, electrophysiological, and histological follow-up study. J Neurol Neurosurg Psychiatry 31:543–551.
118. Wulff CH, Hoyer H, Asboe-Hansen G, Brodthagen H (1985). Development of polyneuropathy during thalidomide therapy. Br J Dermatol 112:475–480.

119. Hafstrom T (1967). Polyneuropathy after neurosedyn (thalidomide and its prognosis). Acta Neurol Scand 43 (suppl 32):1–41.
120. Hess CW, Hunziker T, Kupfer A, Ludin HP (1986). Thalidomide-induced peripheral neuropathy: a prospective clinical, neurophysiological and pharmacogenetic evaluation. J Neurol 233:83–89.
121. Chapon F, Lechevalier B, Da Silva CD, Rivrain Y, Dupuy B, et al (1985). Neuropathies a la thalidomide. Rev Neurol 141:719–728.
122. Clemmensen OJ, Olsen PZ, Andersen KE (1984). Thalidomide neurotoxicity. Arch Dermatol 120:338–341.
123. Gutierrez-Rodriguez O, Starusta-Bacal P, Gutierrez-Montes O (1988). Thalidomide-induced neuropathy in rheumatoid arthritis patients. Arthritis Rheum 31:145–146.
124. Hoyer H, Wulff CH, Asboe-Hansen G, Brodthagen H (1983). Behandling of prurigo nodulans Hyde med talidomid. Ugeskrift Laeger 145:3243–3245.
125. Fine, HA, Loeffler JS, Kyritsis A, Wen P, Black PM, Levin VA, Graham S, Yung WKA (1997). A phase II trial of the antiangiogenic agent, thalidomide, in patients with recurrent high-grade gliomas (abstr). Proc ASCO 16:385a.
126. Fine, HA (1998). Phase II study of thalidomide in patients with refractory glioblastoma multiforme. National Cancer Institute: Fall 1998 NCI Phase I Meeting, Sept. 15–16, 1998.
127. Yarchoan R (1998). Phase II study of oral thalidomide in patients with AIDS-associated Kaposi's sarcoma. National Cancer Institute: Fall 1998 NCI Phase I Meeting Sept. 15–16, 1998.
128. Yarchoan R, Little RF, Wyvill K, Welles L, Pluda JM, Figg WD, Tosato G (1998). A phase II study of oral thalidomide in patients with AIDS-related Kaposi's sarcoma (KS). 12th World AIDS Conference Geneva: 318.
129. Figg WD, Bergan R, Brawley O, Tompkins A, Linehan M, Duray P, Bauer KS, Pluda J, Reed E (1998). Randomized phase II study of thalidomide in androgen-independent prostate cancer (AIPC). Proc ASCO 16:333a.
130. Figg WD (1998). Randomized phase II trial of thalidomide in androgen-independent prostate cancer. National Cancer Institute: Fall 1998 NCI Phase I Meeting Sept 15–16, 1998.
131. Baidas S (1998). Phase II study of thalidomide in patients with breast cancer [lecture presentation]. National Cancer Institute: Fall 1998 NCI Phase I Meeting Sept. 15–16, 1998.
132. Reuters (1998). Thalidomide stops myeloma progression in U.S. study (press release). Little Rock, Arkansas, September 28, 1998.
133. Schmahl HJ, Nau H, Neubert D (1988). The enantiomers of the teratogenic thalidomide analogue EM 12: 1. Chiral inversion and plasma pharmacokinetics in the marmoset monkey. Arch Toxicol 62:200–204.
134. Winckler K, Klinkmüller KD, Schmahl HJ (1989). Determination of the thalidomide analogues 2-(2,6-dioxopiperidine-3-y10 phthalimidine (EM 12), 2-(2,6-dioxopiperidine-4-y1)phthalimidine (EM 16) and their metabolities in biological samples. J Chromatogr 488: 417–425.
135. Czejka MJ, Koch HP (1987). Determination of thalidomide and its major metabolities by high-performance liquid chromatography. J Chromatogr 413:181–187.
136. Fabro S, Schumacher H, Smith RL, Stagg RBL, William RT (1965). The metabolism of thalidomide: some biological effects of thalidomide and its metabolites. Br J Pharmacol 25: 352–362.
137. Günzler V, Hanauske-Abel HM, Tschank G, Schum-Wissermann H (1986). Immunological effects of thalidomide: inactivity of the drug and several of its hydrolysis products in mononucleocyte proliferation tests. Arzneimittelforschung 36:1138–1141.
138. Jönsson NA (1972). Chemical structure and teratogenic properties: III. A review of available data on structure-activity relationships and mechanism of action of thalidomide analogues. Acta Pharm Suec 9:521–542.

25
CM101

Russell F. DeVore III
Vanderbilt University Medical Center, Nashville, Tennessee

I. INTRODUCTION

CM101 is a complex bacterial polysaccharide isolated from group B β-hemolytic *Streptococcus* (GBS) cultures recovered from newborn infants who died from early-onset septicemia (1). CM101 is a biologically more potent subfraction of ''GBS toxin,'' which was first described by Hellerqvist and colleagues to be instrumental in the pathophysiology of early-onset disease of the newborn characterized by septic shock, pneumonia, and respiratory failure (2–4). More recent clinical studies have demonstrated that CM101 is common to all GBS serotypes and is found in the serum, spinal fluid, and urine of infants diagnosed with GBS pneumonia or meningitis (5). When administered intravenously to sheep, CM101 binds to the endothelium of capillaries and arterioles and induces extensive endothelial damage, dense inflammatory infiltrates, and hemorrhage in the lungs—findings virtually identical to those found in the lungs of infants with early-onset disease (2,3,6).

GBS toxin is highly species-, tissue-, and age-specific. For instance, the lungs of mice and humans beyond the age of 1 week are unaffected by GBS toxin, whereas newborn mice and humans and young and adult sheep are highly susceptible. Premature infants are significantly more susceptible to the severe respiratory sequelae of GBS infection than full-term infants (4). The specificity of GBS toxin for newborns is hypothesized to be related to the expression of embryonal receptors on the pulmonary vasculature prior to and shortly following birth. The indication that there might be specificity for certain types of immature vasculature led Hellerqvist to speculate that malignant neovasculature might exhibit susceptibility to GBS toxin/CM101 and thereby present a selective target and opportunity to exploit CM101 as a therapeutic agent.

II. PRECLINICAL STUDIES

Non-tumor-bearing mice are not adversely affected by intravenous administration of CM101. Up to 40 mg/kg has been administered without causing lethal toxicity, and the LD_{10} is between 4 and 16 mg/kg. In immunocompetent BALB/c mice bearing established Madison lung carcinoma, CM101 binds to the endothelium of the tumor neovasculature and induces a severe inflammatory response evidenced by vasodilation, endothelial and tumor-cell necrosis, invasion by inflammatory cells, and capillary thrombosis (7,8). Short-term treatment improves median survival for treated mice, and protracted therapy renders some animals disease-free. In these animal models, no organ toxicity or evidence of toxicity to nontumor vasculature is observed. These observations prompted the clinical development of CM101 as an anticancer treatment.

III. PHASE I CLINICAL TRIAL DESIGN

Based on the promising preclinical data, a phase I trial of CM101 was undertaken in patients with advanced refractory malignancies (9). At the time of study initiation a protracted treatment schedule was considered desirable, but owing to safety concerns, the first 15 patients were treated three times in one week on a Monday, Wednesday, Friday (MWF) schedule. Patients who had no evidence of disease progression and favorable tolerance received repeated treatment cycles every 3–4 weeks. The primary endpoint was toxicity. Patients were also observed for evidence of antitumor activity, development of anti-CM101 antibodies, and development of a systemic inflammatory cytokine response. After safety was established in the first 15 patients and a maximally tolerated dose was estimated, a protracted schedule was evaluated. Patients were treated weekly for 10 consecutive weeks, given 2 weeks of rest, and then evaluated for subsequent treatment cycles if toxicity was acceptable and there was no evidence of tumor progression.

To date 24 patients have been treated with CM101. All patients had malignancies considered refractory or resistant to standard therapy. Most patients were heavily pretreated with chemotherapy. The initial dose was 1 U/kg given every other day for a total of three doses on a MWF schedule. One unit is defined as the amount of drug, when administered intravenously to sheep, required to induce an elevation in the pulmonary artery blood pressure by 100–150% and to reduce the circulating white blood cell count by 40–60% of baseline (10). CM101 was administered as a 15-min intravenous infusion and patients were monitored closely for 12 hr afterward.

IV. TOXICITIES

Four dose levels of CM101 were studied using the intermittent MWF schedule: 1, 2, 3.3, and 5 units/kg. Fifteen patients were treated by this schedule, five of whom had at least stable disease after cycle 1 of therapy and therefore received additional treatment cycle(s). Mild to moderate fever and chills lasting 1–2 hr were universal and accompanied by mild tachypnea, tachycardia, and an asymptomatic decline in blood pressure in most patients (Table 1). Two patients experienced moderate or severe hypotension, one at the 2 U/kg and one at 5 U/kg dose levels. The latter patient had a history of intermittent supraventricular tachycardia and experienced hypotension as a result of this, which occurred following a treatment with CM101. A specific dose-limiting toxicity was not well characterized in the MWF patients; however, two of the three patients receiving the 5 U/kg dose experienced severe dose-limiting toxicities (Table 2). In addition to the patient who experienced severe hypotension from a supraventricular tachycardia, a patient with extensive pulmonary metastases experienced severe dyspnea following the first dose of CM101. These were the only grade IV toxicities observed.

Grade III dyspnea occurred in two patients, one at the dose level of 1 U/kg and one

Table 1 Common Grade I/II Toxicities on MWF Schedule in 15 Patients

Fever/chills	15
Nausea	8
Cough	5
Dyspnea	5
Headache	9
Myalgias	6
Facial flushing	8
Acute pain	4[a]

[a] Pain at tumor site.

at 2 U/kg dose levels. All three patients who experienced grade III/IV dyspnea had lung metastases and compromised pulmonary function, with a carbon monoxide diffusion capacity of $\leq 76\%$ of predicted value. These were the only patients with that degree of pulmonary compromise. Dyspnea was always short-lived lasting 1–3 hr. Some patients had a component of bronchospasm. It was unclear if the administration of bronchodilators was beneficial. The mechanism of dyspnea is not defined, but we believe this is related to the generalized cardiopulmonary compromise that occurs with the sepsis-like toxicities of CM101, and patients with significant underlying pulmonary insufficiency were naturally

Table 2 Number of Patients Experiencing Grades III–IV Toxicities on the MWF vs. Weekly CM101 Schedules[a]

	MWF toxicity grade		Weekly[c] toxicity grade	
	III	IV	III	IV
Arrhythmia	0	1 (5)	0	0
Pain[b]	3 (3.3)	0	4	0
Dyspnea	2 (1,2)	1 (5)	0	0
Headache	1 (5)	0	2	0
Fever	0	0	4	0
Fatigue	0	0	3	0
Anemia	0	0	3	0
Edema	0	0	1	0
Confusion	0	0	1	0
Constipation	0	0	1	0
Hypokalemia	0	0	0	1

[a] Numbers in parentheses represent dose levels at which grade III/IV toxicities occurred on the MWF schedule, expressed in units per kilogram. All grade III–IV toxicities on the weekly schedule occurred at the 3.3 U/kg dose level.
[b] Acute tumor-related pain.
[c] Two patients developed deep venous thrombosis on the weekly schedule.

more susceptible. Although unexpected, we cannot rule out the possibility that CM101 induces inflammation within the pulmonary vasculature either by direct vascular injury or perhaps indirectly by a cytokine-driven mechanism.

A dose-dependent drop in blood pressure was observed, with a maximum decrease in systolic blood pressure of 42 mmHg at the dose level of 3.3 U/kg. There were no significant renal, cardiac (other than the single episode of supraventricular tachycardia), hepatic, neurological, or hematological toxicities. Interestingly, acute pain at the site of known tumor occurred in 7/15 patients. This occurred in 1/9 patients treated at the first two dose levels and 6/6 patients treated at the two highest dose levels. Pain was severe in 3 patients. There was not a clear relationship between acute tumor-related pain and antitumor response. However, two of the responding patients did not have significant tumor bulk at sites where pain would be expected. Another responding patient with a large liver tumor did experience severe pain requiring intravenous narcotics. In summary, CM101 toxicities were compatible with a cytokine-driven mechanism, were short-lived, and outside of dyspnea in patients with compromised pulmonary function, were not associated with any direct organ damage.

V. BIOLOGICAL CORRELATIVE STUDIES

Patients with early-onset septicemia and sheep treated with CM-101 experienced rapid and substantial drops in leukocyte counts due to sequestration of leukocytes in the pulmonary vasculature (3). Likewise, patients treated with CM-101 experienced a rapid decline in granulocytes, lymphocytes, and monocytes (9). The total leukocyte decline is profound, with a decrease of almost 6000 cells per cubic millimeter at the dose level of 5 U/kg (Fig. 1). The nadir occurs at about 1 hr following CM101 administration. The granulocyte de-

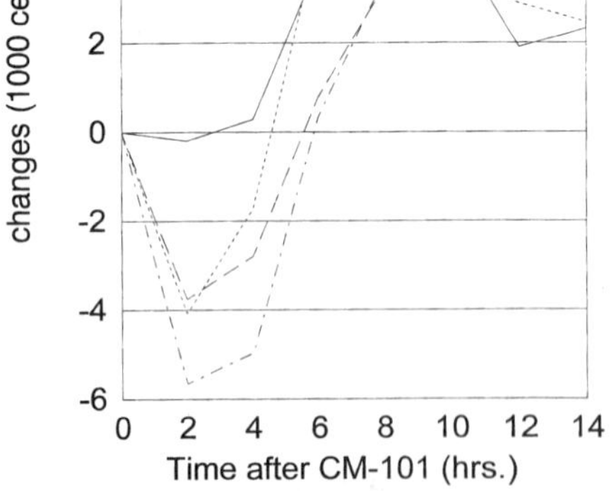

Figure 1 Leukocyte response to CM101.

Table 3 Peak Cytokine Levels Following Treatment with CM101

	TNFa (pg/mL)	IL-6 (pg/mL)	IL-8 (pg/mL)	IL-10 (pg/mL)	sE-sel (ng/mL)	MIP-1a (pg/mL)
MWF schedule						
Baseline	9.7	13.2	122	5.9	85.4	35.9
1 μ/kg	253	3501	3241	129	165	627
2 μ/kg	1305	1313	5603	562	241	1915
3.3 μ/kg	4194	7055	9447	445	353	5208
5 μ/kg	1090	1347	6746	1422	681	2079
Weekly schedule						
Baseline	11.8		199		93.5	
2 μ/kg	2070		4589		324	
3.3 μ/kg	2000		5654		574	

cline is followed by a rapid rise above baseline occurring 2–4 hr following treatment. Extrapolating from the sheep model, we postulate that at least some of the observed leukocyte decline is associated with recruitment to the tumor rather than a generalized margination throughout the vasculature. To address this possibility, indium-labeled leukocyte scans are planned in future clinical trials.

All patients treated with CM101 experienced a time- and dose-dependent elevation of systemic cytokines (Table 3). TNF-α, MIP-1α, IL-8, IL-6, and IL-10 peaked in a se-

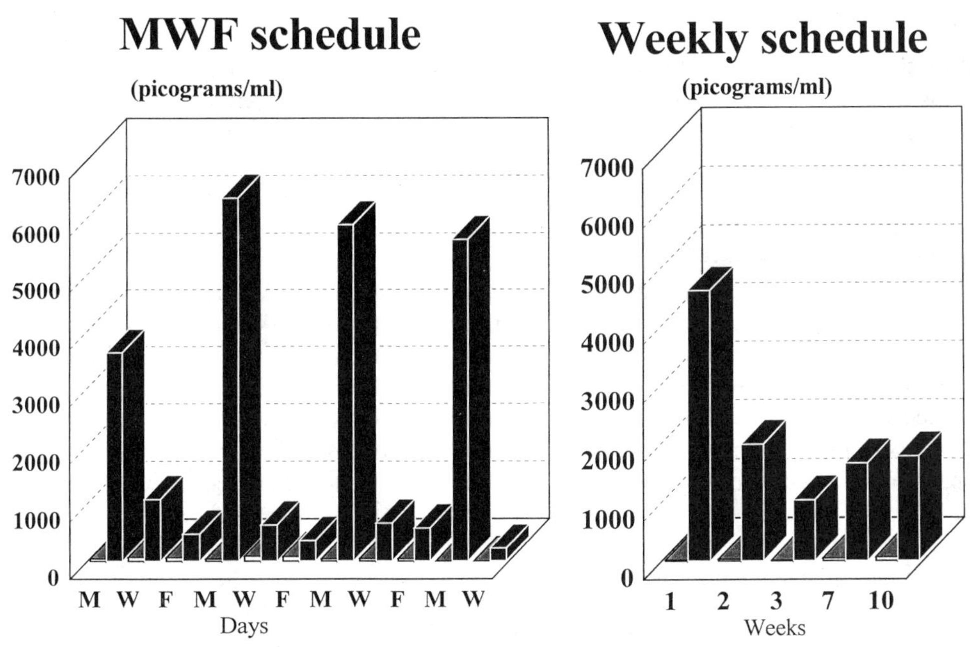

Figure 2 Representative cytokine (IL-8) response patterns in an individual patient treated on both the MWF and weekly schedules. Gray bars represent baseline levels. Black bars represent peak levels. M, W, F represent days of treatment within a given cycle (MWF cycles were repeated every 4 weeks). Weekly schedule values represent peak levels from weeks 1, 2, 3, 7, 10.

quential manner between 60 and 240 min after treatment, suggesting an interdependent cascade of inflammatory events. Cytokine levels generally returned to baseline within 4–8 hr. The cytokine responses corroborate the proposed inflammatory mechanism of action of CM101 and their timing paralleled that of the observed toxicities. Arguably the cytokine response does little to substantiate the supposition that CM101 is specifically binding to the tumor endothelium and inducing inflammation. However, soluble E-selectin is considered a specific marker for endothelial engagement, and levels were elevated in a dose-dependent manner in patients receiving CM101 (11,12). Soluble E-selectin levels did not return to baseline following therapy but remained elevated and increased further with the Wednesday and Friday treatments. The persistent elevation in soluble E-selectin suggests CM101 may induce a sustained process within the neovasculature even following its clearance from the system.

Excepting soluble E-selectin, the inflammatory cytokine levels were markedly attenuated on subsequent treatment days—Wednesday and Friday (Fig. 2). Two predominant hypotheses have been considered. First, the lack of cytokine response may indicate that a significantly lower number of CM101 receptors are present during subsequent treatments. This could be related to downregulation or destruction of a significant proportion of the neovasculature-expressing receptors. Second, perhaps the degree of cytokine elevation is in proportion to the immune system's capacity to respond to CM101, and there may be a "fatiguing" that occurs, thus rendering the drug less effective.

VI. ANTITUMOR ACTIVITY

Three patients exhibited objective evidence of antitumor activity. A patient with classic Kaposi's sarcoma and multiple lower extremity lesions was developing multiple new lesions on the left lower extremity each month. One week after receiving his first dose of 2 U/kg CM101, the largest marker lesion (raised approximately 1 cm from the skin surface and 1 cm in diameter) became flattened. It was no longer detectable at 3 weeks. Three additional marker lesions, each measuring approximately 5 mm in diameter, were unchanged in size. Several smaller lesions became pale, decreased in size, and desquamated by 3 weeks. The disease-related discoloration and severe edema of the extremity improved dramatically and almost completely resolved during therapy. This patient received a total of five treatment cycles before a new lesion appeared and CM101 was discontinued.

A 43-year-old female with metastatic adenocarcinoma of the small bowel involving a supraclavicular lymph node had previously failed treatment with 5-fluorouracil and was treated with 2 U/kg of CM101. Reduction in the size of the lymph node occurred within 2 weeks of the first dose of CM101. An approximately 90% reduction in tumor size was apparent after four monthly cycles. However, following the second course of therapy, a new supraclavicular lymph node appeared and enlarged after the subsequent treatment cycle. Both lymph nodes were excised approximately 5 hr following CM101 administration on day 1, treatment cycle 4. The biopsy demonstrated leukocytic infiltration, something not observed in a previous biopsy. The patient was taken off study at that point and eventually developed recurrent disease in the supraclavicular area 7 months after receiving the last dose of CM101.

A 45-year-old male with hepatocellular carcinoma previously treated with etoposide and cisplatin was treated with 3.3 U/kg of CM101. He had three hepatic lesions, one measuring 13 cm in greatest diameter. After two monthly cycles, there was a 25% decrease in tumor volume; this was most remarkable in the smaller lesions. Approximately 1 hr after CM101 treatment, he experienced acute onset of severe pain in the areas of previous tumor-related pain, necessitating premedication with intravenous narcotics in subsequent treatments. He developed progressive disease after five monthly cycles of CM101 and study treatment was withdrawn.

VII. PHASE I EVALUATION OF WEEKLY CM101

We originally intended to change to a more protracted administration schedule once safety and feasibility were established with the MWF regimen. A continuous MWF schedule without the 2–3 week treatment-free interval was planned. The cytokine data, however, generated concern that the WF treatments were of little effect and that perhaps a weekly schedule would not only give the system more time to recover from the previous treatment but also be more practical. Thus, we chose to work on the hypothesis that the WF attenuation in cytokine response represented an undesirable ''fatiguing'' of the immune system, possibly limiting the drug's therapeutic potency if used on that schedule. Therefore, patients were treated once weekly for 10 consecutive weeks followed by a 2-week rest. There was concern that repeating restaging scans on a conventional monthly or 6-weekly schedule was potentially misleading due to the possibility that inflammation within the tumor could alter the radiographic appearance and possibly even mimic progression. Therefore restaging studies were not done until 12 weeks following the first day of a given treatment cycle and 2–3 weeks following the last dose of CM101.

The first group of 6 patients treated on the weekly schedule received 3.3 U/kg of CM101. Three of the patients who previously experienced antitumor responses to CM101 on the MWF schedule were treated again, but with the weekly regimen. This decision was largely based on the hypothesis that ''resistance'' to CM101 was related to schedule rather than some intrinsic resistance of the tumor or its vasculature. Because a protracted schedule was considerably more active in animals, we considered it reasonable to re-treat previously responding patients with the new schedule. Although the previously treated patients with small bowel adenocarcinoma and Kaposi's sarcoma both exhibited early signs of antitumor activity, no patient in this cohort experienced an objective tumor shrinkage when restaged at 12 weeks.

The weekly schedule of 3.3 U/kg was poorly tolerated, with three patients experiencing grade III fatigue and two patients developing deep venous thrombosis (DVT) (Table 2). The latter two both had conditions associated with DVT. One patient had prostate cancer with progressive pelvic lymphadenopathy and the second had a previous history of DVT thought secondary to a hypercoagulable state related to his underlying small bowel adenocarcinoma. Based on CM101's mechanism of action, it is conceivable that coagulation abnormalities could occur; however, no other patients experienced this toxicity, and currently these events are considered possibly but not probably related to CM101.

Because of the poor tolerance to the dose level of 3.3 U/kg, six additional patients received the weekly $\times$ 10 schedule at a dose of 2 U/kg. This regimen was very well tolerated, with no patient experiencing $\geq$ grade 3 toxicity. Again, no objective antitumor activity was observed in this patient group, although one patient with metastatic colon cancer experiencing disease progression at the time of study entry had stable disease at the end of cycle 1 (12 weeks) and progressed midway through cycle 2. A patient with breast cancer developed erythema, pruritus, and edema of skin metastases following each

CM101 treatment but experienced progressive disease by the end of cycle 1. Other patients exhibited early subjective or minor objective improvements but had progressive disease when restaged at 12 weeks. The toxicity profile for the MWF and weekly regimens was similar, although no patients on the weekly regimen developed significant dyspnea or hypotension.

The inflammatory cytokine response was not as expected with the weekly schedule but exhibited the same attenuation in peak levels with repetitive dosing as observed with the MWF schedule (Fig. 2). Therefore, attenuation of the inflammatory cytokine response occurs when treatments are separated by 2–7 days but does not occur when MWF cycles are repeated at 3–4 weeks.

VIII. SUMMARY AND FUTURE DIRECTIONS

Phase I testing has shown that CM101 is safe and exhibits a toxicity profile consistent with cytokine-mediated phenomena. Several findings corroborate the drug's proposed mechanism of action. The marked elevations in inflammatory cytokines and the elevation in soluble E-selectin indicate an inflammatory mechanism of action consistent with endothelial engagement. As seen in early-onset disease and in the sheep model, CM101 induces profound leukopenia, suggesting recruitment of leukocytes, possibly to the tumor. A patient biopsy supports this supposition. Acute tumor-related pain was observed in several patients, suggesting acute onset of tumor inflammation. Finally and most significantly, evidence of antitumor activity was observed.

Future clinical studies will investigate the optimal dose and schedule and attempt to define the drug's mechanism of action in vivo. Current protocols incorporate labeled leukocyte scans, repeat tumor biopsies for immunolocalization of CM101, and blood flow studies. The drug's interaction with chemotherapeutic agents and radiation will be explored. There may be certain tumors for which CM101 will exhibit single-agent activity with associated tumor shrinkage, something not expected for most agents targeting the tumor vasculature. Therefore, developmental strategies may allow for a more traditional phase I, II, and III sequence of studies. Smaller tumors may be more susceptible, making adjuvant and neoadjuvant approaches attractive. Laboratory efforts are focused on the identification of the endothelial cell receptor for CM101 and further definition of the drug's mechanism of action. Future clinical trials will attempt to confirm these laboratory observations in humans and improve our understanding of this interesting new agent.

REFERENCES

1. CG Hellerqvist, J Rojas, RS Green, S Sell, H Sundell, MT Stahlman. Studies on group B β-hemolytic Streptococcus: I. Isolation and partial characterization of an extracellular toxin. Pediatr Res 15:892–898, 1981.
2. J Rojas, RS Green, CG Hellerqvist, R Olegard, KL Brigham, MT Stahlman. Studies on group B β-hemolytic Streptococcus: II. Effects on pulmonary hemodynamics and vascular permeability in unanesthetized sheep. Pediatr Res 15:899–904, 1981.
3. J Rojas, LE Larrson, CG Hellerqvist, KL Brigham, ME Gray, MT Stahlman. Pulmonary hemodynamic and ultrastructural changes associated with group B β-hemolytic streptococcal toxemia in adult sheep and newborn lambs. Pediatr Res 17:1002–1008, 1983.

4. LE Weisman, B Stoll, DF Cruess, RT Hall, GB Merenstein, VG Hemming, GW Fischer. Early onset group B streptococcal sepsis: a current assessment. J Pediatr 121:428–433, 1992.
5. HW Sundell, H-P Yan, K Wu, B Wamil, R Gaddipati, CE Carter, M Stahlman, CG Hellerqvist. Isolation and identification of group B beta-hemolytic streptococcal (GBS) toxin from septic newborn infants (abstr). Pediatr Res 39:302A, 1997.
6. BA Russell, RS Pappas, JM Brandt, HW Sundell, CG Helerqvist. Immunological and histochemical studies on the pathophysiology of group B streptococcal toxin manifested in early onset disease (abstr). Glycoconj J 6:399, 1988.
7. CG Hellerqvist, GB Thurman, DL Page, Y-F Wang, BA Russell, CA Montgomery, HW Sundell. Anti-tumor effects of GBS toxin: a polysaccharide exotoxin from group B β-hemolytic Streptococcus. Cancer Res Clin Oncol 120:63–70, 1993.
8. GB Thurman, BA Russell, GE York, Y-F Wang, DL Page, HW Sundell, CG Hellerqvist. Effects of GBS toxin on long term survival of mice bearing transplanted Madison lung tumors. Cancer Res Clin Oncol 120:479–484, 1994.
9. RF DeVore, CG Hellerqvist, GB Wakefield, BD Wamil, GB Thurman, PA Minton, HW Sundell H-P Yan, CE Carter, Y-F Wang, GE York, M-H Zhang, DH Johnson. Phase I study of the antineovascularization drug CM101. Clin Cancer Res 3:365–372, 1997.
10. K Sandberg, KE Edberg, W Fish, RA Parker, CG Hellerqvist, H Sundell. Thromboxane receptor blockage (SQ 29548) in group B streptococcal (GBS) toxin challenge in young lambs. Pediatr Res 35:571–579, 1994.
11. AJH Gearing, W Newman. Circulating adhesion molecules in disease. Immunol Today 14: 506–512, 1993.
12. BD Wamil, GB Thurman, HW Sundell, RF DeVore, G Wakefield, DH Johnson, Y-F Wang, CG Hellerqvist. Soluble E-selectin in cancer patients as a marker of the therapeutic efficacy of CM101, a tumor-inhibiting anti-neovascularization agent, evaluated in a phase I clinical trial. J Cancer Res Clin Oncol 123:173–179, 1997.

26
TNP-470

Pankaj Bhargava
Lombardi Cancer Center, Georgetown University Medical Center, Washington, D.C.

Margaret S. Dordal
TAP Holdings Inc., Deerfield, Illinois

Michael J. Hawkins
Washington Cancer Institute, Washington Hospital Center, Washington, D.C.

I. DISCOVERY OF TNP-470

The therapeutic potential of angiogenesis inhibitors was realized when Judah Folkman suggested in 1971 that tumor growth was angiogenesis-dependent (1). Ten years later, steroids such as medroxyprogesterone acetate, dexamethasone, and cortisone were identified as the first selective antiangiogenic compounds (2). The combination of heparin and steroids was subsequently shown to cause objective tumor regressions in mice through synergistic antiangiogenic activity (3). These observations, coupled with an increased understanding of the process of angiogenesis, led many investigators to search for drugs acting on specific steps in the angiogenic cascade. In this environment of rational drug design, the discovery of TNP-470 stands out as an example in serendipity in drug discovery.

In 1985, Donald Ingber, while working as a postdoctoral fellow in Folkman's lab, observed a unique phenomenon in endothelial cell cultures contaminated with fungus (4). While most fungal contaminants induced complete cell detachment and death in endothelial cells, one variety of fungus appeared to induce a gradient of cell retraction and rounding in the adjacent endothelial monolayer (5). Endothelial cells along the edge of the fungal colony appeared completely spherical, while cells only a few diameters away exhibited a normal morphology. The therapeutic potential of this discovery lay in previous observations that endothelial cells induced to round failed to grow when stimulated by any angiogenic factor (6–8). The fungus was identified as *Aspergillus fumigatus fresenius*, which secreted a soluble factor that induced endothelial cell rounding in culture. A collaboration between Takeda Chemical Industries Limited, Japan, and Folkman's lab led to isolation of the active compound secreted by the fungus, which was identified as fumagillin, a known antibiotic that had previously been used to treat amebiasis in humans (9–11).

Purified fumagillin was shown to completely inhibit endothelial cell proliferation in the presence of saturating levels of basic fibroblast growth factor (half maximal inhibition of human umbilical vein endothelial cells at 0.5 ng/mL) (5). Fumagillin also suppressed tumor-induced neovascularization in the mouse dorsal air sac model, but its effectiveness as an antitumor agent was limited by severe weight loss in mice (>15% of their starting weight). This led to the search for fumagillin analogues that would retain the potent antiangiogenic activity of fumagillin without its side effects. Alkaline hydrolysis of fumagillin yielded fumagillol, from which over a hundred derivatives were synthesized and tested. Among these analogues, a new subset of angiostatic antibiotics called angioinhibins was identified; one of the most potent of these was O-(chloroacetylcarbamoyl) fumagillol, or AGM-1470 (later termed TNP-470) (Fig. 1).

TNP-470 was found to be a potent inhibitor of endothelial cell proliferation in vitro, with a half maximal inhibition of endothelial cell proliferation at ~10 pg/mL (50 times greater potency than the fumagillin parent) (5). It also inhibited angiogenesis in vivo in the chorioallantoic membrane (CAM) assay, rat corneal micropocket assay, and murine sponge implantation assay (12). At doses required to inhibit endothelial cell proliferation (10 pg–1 μg/mL), TNP-470 caused cytostatic rather than cytotoxic inhibition of endothelial cells (4). Cytotoxicity for endothelial cells was observed at doses >1 μg/mL, while inhibition of tumor-cell proliferation was not observed until doses at least 10 times higher (10 μg/mL) of TNP-470 were reached (5). In a rat blood vessel organ culture assay, TNP-470 inhibited tube formation at 1 ng/mL, while inhibition of endothelial cell growth was not observed until 1 μg/mL (12). When administered systemically to mice, TNP-470 inhibited the growth of several solid tumors that were otherwise refractory to TNP-470 in vitro (5). These studies demonstrated that TNP-470 exerted its anti-tumor effect by acting on the tumor vasculature. A major advantage over the parent fumagillin was that TNP-470

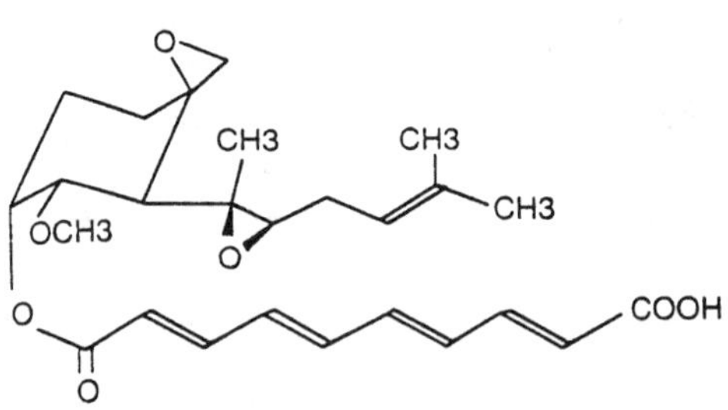

Fumagillin

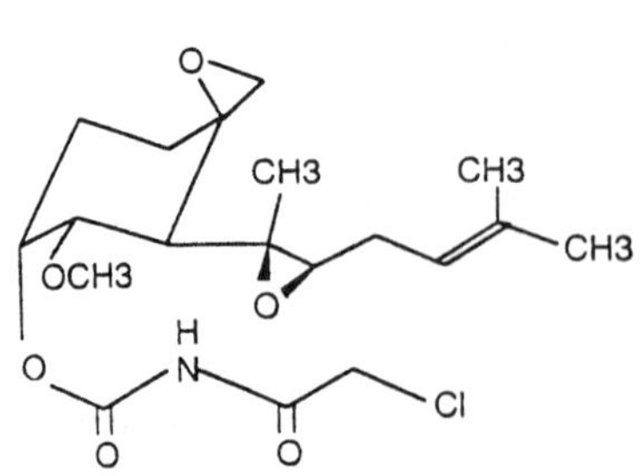

TNP-470

Figure 1 Structure of fumagillin and TNP-470.

produced much less weight loss or other side effects like hair loss, intestinal disturbance, or infection in mice treated at 30 mg/kg subcutaneously every other day for 100 days (5).

Because of evidence of antitumor activity coupled with a favorable toxicity profile in animal experiments, TNP-470 was chosen for clinical development as an anticancer agent.

II. MECHANISM OF ACTION

While the effects of TNP-470 on endothelial cells and angiogenesis have been well described, the mechanisms responsible for these effects remain unclear. Kusaka et al. (13) showed that TNP-470 inhibited the growth of human umbilical vein endothelial cells (HUVEC) in a biphasic manner, with cytostatic inhibition in the first phase (concentrations from 300 pg/mL–3 μg/mL) and cytotoxic inhibition in the second phase (≥30 μg/mL). Cytostatic inhibition of endothelial cell growth by TNP-470 continued for several days (up to 6 days after a 2-hr incubation at 100 ng/mL) even after TNP-470 was removed from the culture medium. This sustained inhibition was felt to be responsible for antitumor effects seen with intermittent administration of TNP-470 in animal experiments. In the same study, TNP-470 was found to suppress DNA synthesis and caused an increased proportion of endothelial cells to go into the G0/G1 phase. Antoine et al. (14) showed that TNP-470 prevented the entry of normal endothelial cells (HUVEC and bovine aortic endothelial cells) into the G1 phase of the cell cycle, but this effect could be bypassed in transformed endothelial cells (Ea.hy926 and cEnd.1 cells). This suggested that TNP-470 specifically inhibited a cell-cycle control pathway active in normal cells but which could be bypassed in altered or transformed cells. In vitro studies have demonstrated up to a 100-fold resistance to TNP-470 in immortalized HUVECs transformed by human pappillomavirus E6 and E7 (15). Interestingly, the resistance to TNP-470 was not observed in immortalized HUVECs transformed by Ki-ras, indicating a close interaction between the cell proliferation pathway stimulated by E6 and E7 and that inhibited by TNP-470.

Recent studies indicate that TNP-470 acts in a complex manner and may affect several key points in the cell-cycle pathway. TNP-470 suppressed the mRNA expression and activation of specific cyclins and cyclin-dependent kinases (cdc2 and cdk2) and inhibited phosphorylation of RB protein, a tumor suppressor retinoblastoma gene product (16). TNP-470 did not inhibit cellular protein tyrosyl phosphorylation or the expression of immediate early genes c-*fos* and c-*myc*, thus suggesting that its angioinhibitory action was located late in the G1 phase. Other studies confirmed the suppression of cyclin D1 mRNA in mid-G1 phase in HUVEC (17) and decreased levels of cdk2 mRNA in bovine smooth muscles after treatment with TNP-470 (18). A brief treatment with TNP-470 (15 min duration) was shown to cause sustained inhibition of G1 phase– associated protein kinase C (isoform PKCδ) in experiments with HUVECs (19). TNP-470 was shown to selectively increase levels of E-selectin mRNA and its expression in subconfluent growing bovine endothelial cells, while confluent cells appeared relatively resistant to this effect (20). This indicated relative specificity of TNP-470 for growing endothelial cells, without apparent effects on quiescent endothelium.

A mammalian protein, methionine aminopeptidase-2 (MetAP-2) was recently identified as a target for fumagillin, the parent compound for TNP-470 (21). Fumagillin was found to be highly selective for MetAP-2, which is conserved between human and *Saccharomyces cerevisiae*. It was further shown that TNP-470 and ovalicin, a structurally similar compound with anti-angiogenic activity, potently inhibit the activity of MetAP-2 without affecting its inhibition of eukaryotic initiation factor 2α (eIF-2α) phosphorylation

(22). A significant correlation was found between the inhibition of endothelial cell proliferation and inhibition of MetAP-2 activity for these drugs. These findings suggest that MetAP-2 is a common target for TNP-470 and ovalicin.

III. PRECLINICAL STUDIES

In vitro studies revealed that TNP-470 inhibited tumor cell proliferation at concentrations many times higher than those required to inhibit endothelial cell proliferation. Human breast and prostate carcinoma cell lines were inhibited at doses ranging from 0.05–4.9 μg/mL (in soft agar cultures and monolayer cultures, respectively), while endothelial cell proliferation was inhibited at doses in the picogram range (23). These data suggest that the in vivo antitumor activity of TNP-470 is mediated via its angioinhibitory effects rather than direct effects on tumor cell proliferation. Human glioblastoma cells were found to be the only cancer cell type that were inhibited by TNP-470 at doses similar to those required for angioinhibitory activity (picogram range) (24).

The antitumor activity of TNP-470 was evaluated in several experimental rodent and human tumors (summarized in Ref. 25). TNP-470 was found to inhibit tumor growth in a variety of murine and human tumor models, and the antitumor effects were considered to be due mainly to its angioinhibitory activity. TNP-470 also inhibited the development of metastasis in several tumor models in a dose-dependent manner. Several of these studies demonstrated improved survival duration in rodents treated with TNP-470. In models designed to study the development and progression of tumors, TNP-470 inhibited tumorigenesis and the growth of de novo tumors by inhibiting neovascularization (26–28).

Most rodent studies of human tumor models found an effective dose in the range of 30 mg/kg injected subcutaneously every other day (Table 1). No major side effects

Table 1 Antitumor Activity of TNP-470 in Animal Experiments

Effect of TNP-470 on the growth of human tumors				
Tumor type	Treatment schedule	Dose (mg/kg)	Tumor growth inhibition (%)	Ref.
Glioblastoma	Three times/week for 3 weeks SC	30	94.5	24
Neurofibrosarcoma	Three times/week for 6 weeks SC or IP	30	91.5	77
Breast carcinoma	Once a week for 6 weeks SC	50–200	88 maximal inhibition	23
Prostate carcinoma	Once a week for 6 weeks SC	50–200	96 maximal inhibition	23
Choriocarcinoma	Once every other day for 5 weeks SC	30	60	78
Ovarian cancer	Once every other day for 5 weeks SC	30	No effect	78
Esophageal and gastric cancer	Three times a week for 2 weeks SC	30	27	53
Glioma	Three times a week for 6 weeks SC	30	Suppression of growth but no effect on neovascularization	79

Table 1 Continued

Effect of TNP-470 on the growth of human tumors				
Tumor type	Treatment schedule	Dose (mg/kg)	Tumor growth inhibition (%)	Ref.
Breast carcinoma	Every other day for 3–7 weeks SC	30–50	4–49	80
Anaplastic thyroid carcinoma	Twice a week intratumoral, peritumoral, SC and IP	50	Complete inhibition by intratumoral route	81
Gastric cancer	Every other day for 5–6 weeks SC	30	Significant inhibition of growth	82
Neuroblastoma	Every other day for 12 days SC	10	66	83
Hepatocellular carcinoma	Every other day SC for 2 weeks	30	52	84
Medulloblastoma	?? for 4 weeks	1–100	66 (in 100-mg/kg group)	85
Meningioma	Three times a week for 4–6 weeks SC	30	63–76	86
Head and neck cancer	Every other day SC for 30 days	30	No effect	87
Glioma	Every other day IP. for ??	30	Significant growth inhibition	88

Effect of TNP-470 on metastasis formation from human tumors				
Tumor type	Treatment schedule	Dose (mg/kg)	Effect on metastasis	Ref.
Breast carcinoma	Three times/week for 6 weeks	30	No effect	89
Choriocarcinoma	Once every other day for 5 weeks	30	Reduction in the number and rate of metastases	78
Colon cancer	Every other day for 6 weeks SC	20–30	70–100% inhibition of the number of metastases	90
Colon cancer	On alternate days for 5 weeks SC	30	Decreased the number of mice with metastases and the number of metastatic foci	91
Pancreatic cancer	Every other day for 6 or 10 weeks SC	30	Decrease in size and number of metastases in mice treated for 10 weeks	92
Breast cancer	Every other day for 3–7 weeks SC	30–50	50 mg/kg dose significantly reduced lymph node metastases	80
Gastric cancer	Every other day for 5–6 weeks SC	30	Inhibited liver and peritoneal metastases	82
Hepatocellular carcinoma	Every other day for 2 weeks SC	30	Decreased number of mice with lung metastases	84
Gastric cancer	Every other day for 5 weeks SC	15–30	30 mg/kg completely inhibited liver metastases	93
Colorectal cancer	Every other day for 5 weeks SC	30	Decrease number of liver metastases	94

Source: Adapted from Ref. 25.

other than weight loss (up to 10% loss in body weight) were found in treated animals at these doses. In these studies, TNP-470 was found to be slightly less active when injected intravenously compared with subcutaneous administration (25). Studies are currently evaluating route and schedule dependency of TNP-470 in inhibition of primary tumor growth and experimental pulmonary metastases in rodent models of Lewis lung carcinoma and B16 melanoma (29). It has been shown that twice-a-day subcutaneous treatment is most efficacious against subcutaneously implanted tumors, while subcutaneous schedules (both once a day and twice a day at doses up to 15 mg/kg per day) and daily oral treatments (50–100 mg/kg per day) are equally efficacious against tumors implanted intravenously. The use of embolic substances (microspheres and medium-chain triglyceride solutions), in which TNP-470 is very stable, have been shown to prolong the retention of the drug at the tumor site and augmented the efficacy of antitumor therapy (30,31).

Intravenous toxicology studies in rats and dogs examined the effects of dosing for 90 days (1 hr every day) followed by 13 weeks of recovery. The maximum tolerated dose (that dose resulting in minimal residual changes) was 3 mg/kg (18 mg/m^2) in rats and 0.5 mg/kg (10 mg/m^2) in dogs. At higher doses, the main toxicological manifestations in rats were loss of weight and injection site changes (intimal necrosis, fibrosis, inflammation, thrombosis, and vasculitis). In dogs, toxicity included decreases in food consumption and body weight, low incidence of convulsions, aggressive behavior, ataxia, decreased activity, tremors, cataracts, leukocytopenia, anemia, and decrease in phosphorous. At the end of recovery, all findings resolved except cataract, the decline in phosphorous, and injection site thrombosis in dogs; however, recovery was complete in rats (32).

IV. PHARMACOKINETICS AND METABOLISM

The metabolic pathway of TNP-470 was studied in cultured human hepatocytes and microsomal fractions from various human tissues (33). The drug was rapidly and extensively metabolized to at least six metabolites within 30 min in human hepatocytes, and TNP-470 was undetectable after 60 min. The predominant extracellular metabolites were M-II and M-IV, while M-III and M-V represented the major intracellular metabolites detected. The formation of M-IV was associated with esterase-like enzymatic clevage of TNP-470, and this metabolite was further metabolized by microsomal epoxide hydrolase to M-II (Fig. 2). Large interindividual variations were observed in the metabolism of TNP-470 in liver microsomes obtained from different donors, with a variability factor as high as 10 between the lowest and the highest metabolic rates. Evaluation of pharmacokinetics and metabolism in the rhesus monkey confirmed the rapid and extensive metabolism of TNP-470 into six metabolites, with a substantial variation in plasma M-IV levels between the monkeys (34,35). M-I, M-III, and M-IV exhibited short plasma half-lives of 30 min or less, and TNP-470 was undetectable in plasma and urine as early as 6 min following intravenous administration. Urinary recovery within 36 hr represented only ~20% of the administered dose, leading the authors to conclude that most of the administered dose was probably excreted in feces.

Human pharmacokinetics were studied in two phase I trials using validated assays for TNP-470 and M-IV (later characterized as AGM-1883, an active metabolite) (36,37). A study in patients with HIV-associated Kaposi's sarcoma (38) reported rapid clearance of TNP-470 and AGM-1883 from the circulation, with a short terminal half-lives (0.88 $\pm$ 2.5 hr for TNP-470 and 0.276 $\pm$ 0.193 hr for AGM-1883). Pharmacokinetic modeling

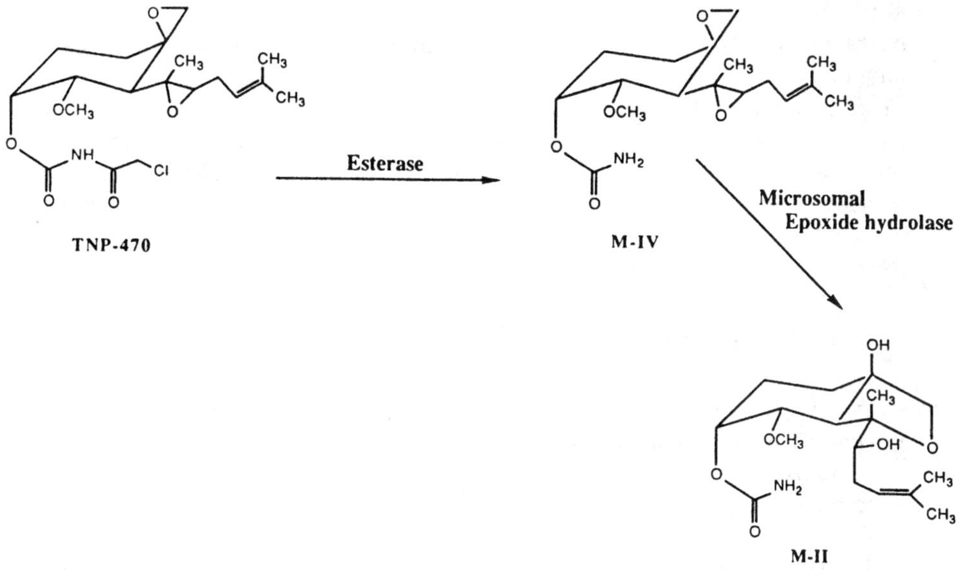

Figure 2 Proposed metabolic pathway for TNP-470. (From Ref. 33.)

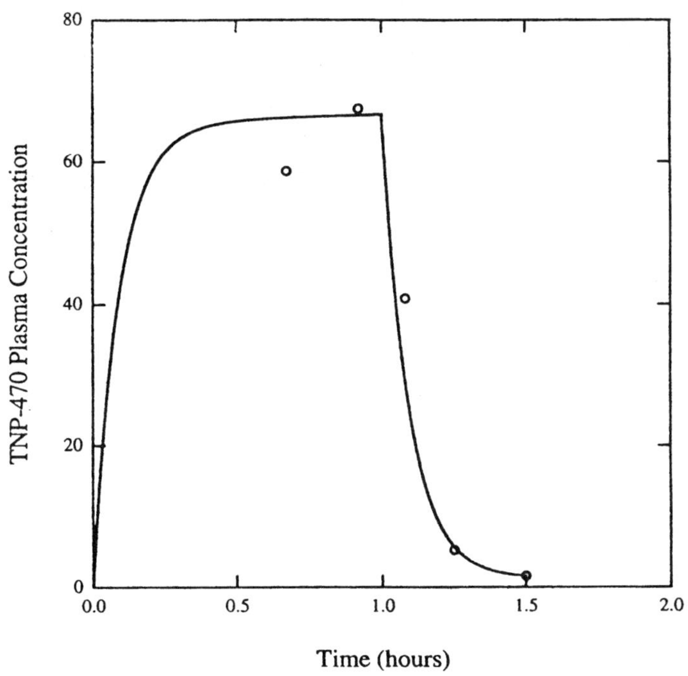

Figure 3 A TNP-470 plasma concentration-versus-time curve for a patient who received TNP-470 15.4 mg/m^2. The observed concentrations (○) can be visualized in relation to the fitted concentrations (plotted line). The terminal half life was estimated from a two-compartment model to be 1.07 hr in this patient. (From Ref. 38.)

showed that the agent best fit to a two-compartment model, with a linear relationship between the dose of TNP-470 and both area under the curve ($AUC_{[inf]}$) and time to maximal concentration (C_{max}) (Fig. 3). The C_{max} for TNP-470 ranged between 6.6 and 597.6 ng/mL and for AGM-1883 between 0.4 and 158.1 ng/mL. Another study giving TNP-470 to patients with advanced cancers (39) reported similarly short half-life for TNP-470 (3 $\pm$ 3 min) and AGM-1883 (9 $\pm$ 4 min) with C_{max} of TNP-470 $>$ 200 ng/mL at doses $\geq$177 mg/m^2. The short half-life of the drug has prompted exploration of frequent dosing and protracted infusion schedules in ongoing trials.

V. TNP-470 COMBINATION THERAPY

Tumors are composed of a malignant cell compartment and a recognized normal tissue compartment (such as vascular components, extracellular matrix, stromal cells, etc.). Conventional cytotoxic therapies are targeted at the malignant cell compartment, while antiangiogenic drugs target the normal tissue (vascular) compartment. A combined attack at these two compartments would therefore be expected to have a greater antitumor effect. Studies showed that antiangiogenic agents could potentiate cytotoxic therapies against primary and metastatic disease (40). Combinations of TNP-470 with cytotoxic chemotherapies, radiotherapy, hormonal therapy and cytokines were evaluated in various tumor models (Table 2). Although TNP-470 did not increase the cytotoxicity of anticancer therapies in cell culture, TNP-470 and minocycline individually and especially in combination increased the tumor growth delay produced by several standard cytotoxic therapies in a Lewis lung carcinoma model (41). Biodistribution studies showed that TNP-470 and minocycline increased the tissue and tumor levels of cytotoxic drugs in this murine model, which may explain the increased tumor growth delay seen with this combination (42). TNP-470/minocycline was shown to enhance tumor growth delay by cyclophosphamide, cisplatin, melphalan, BCNU, mitomycin C, Adriamycin, 5-fluorouracil, thiotepa, and temozolomide in different murine tumor models (see Table 2).

In a recent study, increased antitumor activity of paclitaxel/carboplatin (a combination frequently used for the clinical treatment of human non-small-cell lung carcinoma) was demonstrated by the addition of TNP-470/minocycline (43). Fewer studies of TNP-

Table 2 Effect of Combination of TNP-470 with Other Treatment Modalities in Experimental Animal Models

Tumor type	Combined treatment	Additional effect	Ref.
	Animal tumors		
Lewis lung carcinoma	Interleukin-12	Increase in tumor growth inhibition	95
Lewis lung carcinoma	Minocycline, cyclophosphamide	Increase in tumor growth delay, 40–50% animals were long term survivors	41
FSaIIC fibrosarcoma	Minocycline, cyclophosphamide	Eightfold increase in tumor cell killing	41
Rat pituitary tumores	Bromocriptine	Suppression of vascular formation	96,97

Table 2 Continued

Tumor type	Combined treatment	Additional effect	Ref.
	Animal tumors		
Murine islet cell cancer	Interferon alfa/beta, minocycline	Reduction of tumor volume and capillary density	45
Rabbit VX-2 carcinoma	Doxorubicin	Antitumor effect	31
Murine squamous carcinoma	Hyperthermia	tumor growth delay and reduction of vascularized area	52
Murine EMT-6 mammary carcinoma	Minocycline, cyclophosphamide, cisplatin	Increase in tumor growth delay	47
Murine melanoma	Mitomycin, doxorubicin, cisplain, 5-fluorouracil	Decrease in tumor growth and number of metastases	98
Murine mammary carcinoma	Fractionated radiotherapy	Increase probability of tumor control	51
Murine endometrial cancer cell lines	Estrogen, progesterone	Decreased secretion of bFGF in undifferentiated endometrial cancer cell line	48
Murine glioma model	Temozolomide	Reduction in interstitial levels of temozolomide in tumors, but growth inhibition still occurred	99
Murine liver cancer	Cisplatin	Increased antitumor effect	100
Lewis lung carcinoma	Micnocycline, suramin, genistein, interferon delta-4, THC, cisplatin, BCNU	Increased tumor growth delay	49
Lewis lung carcinoma	Paclitaxel, carboplatin, minocycline	Increased antitumor activity and efficacy	43
	Human tumors		
PC-3 prostate cancer	Cisplatin	Increase in antitumor activity	23
MDA-MB-231 breast cancer	5-Fluorouracil, doxorubicin	No additional effect	23
MDA-MB-231 breast cancer	Tamoxifen	Inhibition of tumor growth	47
Esophageal and gastric cancer	Hyperthermia	Increased inhibition of tumor growth	53
Non-small-cell lung cancer	Paclitaxel	Sequential exposure to taxol followed by TNP-470 caused synergistic cytotoxicity	44
Non-small-cell lung cancer	SN-38	Sequential exposure to SN-38 followed by TNP-470 caused synergistic cytotoxicity	45

Source: Adapted from Ref. 25.

470 combinations have been done in human cancer models, but recent studies have shown enhanced antitumor efficacy of TNP-470 administered in combination with paclitaxel (44) and SN-38 (45) in human non-small-cell lung cancer cell lines. Both studies showed a sequence-dependent effect, with exposure to cytotoxic drugs followed by TNP-470 showing greater synergy. The combination studies of TNP-470 and paclitaxel or TNP-470 and gemcitabine for 28 days in dogs with 28 days of recovery demonstrated no unexpected types of toxicity as compared with dogs treated with any of the drugs alone (32).

Combination therapy with TNP-470 and α/β-interferon (also an angiogenesis inhibitor) suppressed tumor growth by 80% and inhibited pulmonary metastatic tumor growth by sevenfold compared with controls in a murine Lewis lung carcinoma model (46). In a transgenic mouse islet-cell carcinoma model, combination of TNP-470, minocycline, and interferon α/β markedly restricted neovascularization and inhibited de novo tumor growth (28). Pretreatment with tamoxifen was found to increase the efficacy of TNP-470 in inhibiting the growth of MCF-7 breast cancer cells inoculated in mice (47). Based on the activity on differentiated and undifferentiated cells, a combination of TNP-470 and progestin therapy may be effective in endometrial cancer (48). Concurrent administration of TNP-470 was found to increase the effect of radiation in the Lewis lung carcinoma (41,49) and rat 9L gliosarcoma models (50). However, a negative effect was seen in a mouse mammary carcinoma model, in which administration of TNP-470 during fractionated radiotherapy decreased the effect of radiation (51). Inhibition of reoxygenation by TNP-470 during fractionated radiotherapy was thought to be responsible for this effect, and an additive effect was seen when TNP-470 was administered after the end of radiotherapy in this model. TNP-470 enhanced the antitumor effect of hyperthermia in murine squamous cell carcinoma and human esophageal and gastric cancer models, thought to be attributable to inhibition of angiogenesis following heat-induced vascular damage (52,53).

VI. EFFECTS OF TNP-470 IN OTHER DISEASES CHARACTERIZED BY NEOVASCULARIZATION

The effects of TNP-470 were studied in rat collagen-induced arthritis and rat adjuvant arthritis, animal models of rheumatoid arthritis in which neovascularization of the synovium is associated with pannus formation and destruction of joints (54,55). TNP-470 when used alone (27.5 mg/kg SC on alternate days) significantly reduced the incidence and severity of arthritis (by clinical, radiographic, and histological criteria) without affecting T-cell function in vitro or phenotype in vivo. Moreover, the combination of TNP-470 with paclitaxel (56) or cyclosporine (57) was more effective than single-agent therapy.

VII. OTHER EFFECTS OF TNP-470

A. Bone Formation

TNP-470 inhibited bone morphogenetic protein (BMP)–induced ectopic bone formation in a dose-dependent manner in mice (58). Histological examination revealed that TNP-470 prevented proliferation of mesenchymal cells and chondrogenesis at the initial step of endochondral bone formation, suggesting that angiogenesis may play an essential role in the recruitment of BMP receptor–positive cells that differentiate into chondrocytes and osteoblasts. In a nude mouse model of human breast cancer (MDA-MB-231 cell line) with bone metastasis, TNP-470 (30 mg/kg SC three times a week) reduced the number and area of osteolytic bone metastases (approximately 60 and 70%, respectively) in addition to reducing their radiolucency (59). In a murine bone marrow culture under 1,25-dihydroxy-vitamin D3 in which mature functional osteoclasts formed in vitro, TNP-470 significantly inhibited the formation of multinucleated osteoclast-like cells. TNP-470 also suppressed in vivo bone resorption in calvaria treated with interleukin-1β, an osteoclast stimulator. Together, these data suggested that TNP-470 inhibited bone metastases not only through its antiangiogenic action but also by the inhibition of osteoclastic bone resorption, making it a potentially beneficial drug in the treatment of osteolytic bone metastasis (59).

B. Kidney

TNP-470 potently inhibited PDGF- and bFGF-stimulated proliferation of rat mesangial cells in vitro (IC50 = 50 pg/mL) and mesangial cell proliferation and extracellular matrix expansion in anti-Thy 1.1 glomerulonephritis, a rat model of mesangial proliferative glomerulonephritis (60). High doses of TNP-470 (20 mg/kg per day) markedly suppressed mesangial cell proliferation and mesangial matrix expansion on day 6; however, mesangiolysis remained. Low doses (10 mg/kg per day) moderately inhibited mesangial cell proliferation and mesangial matrix synthesis and induced appropriate mesangial healing by day 14 in this model.

C. Immune System

TNP-470 enhanced the proliferation of normal human B lymphocytes in the presence of T cells in vitro, suggesting that this activity was mediated through action on normal T cells (61). However, TNP-470 was ineffective on the proliferation of transformed T and B cells derived from lymphoid tumors in vitro (62). Moreover, TNP-470 failed to stimulate the proliferation of normal B lymphocytes cocultured with tumor T cells or tumor B lymphocytes cocultured with normal T cells, thus demonstrating that its effect was restricted to normal lymphocytes. In vivo studies demonstrated that TNP-470, when injected subcutaneously or intraperitoneally into mice at pharmacological doses, induced a significant increase in size of axillary and mesenteric lymph nodes (63). Histological analysis revealed significant increase in size of germinal centers in the lymph nodes of TNP-470–treated mice. No such effect was seen in T cell–deficient nude mice. These data suggest that TNP-470 may act as an immune modulator in immunocompetent mice.

D. Hematopoietic Precursors

The effects of TNP-470 on human hematopoietic progenitors and isolated CD34–positive cells were reported in a study using bone marrow cells from healthy volunteers (64). TNP-470 inhibited both erythroid and myeloid colony formation from bone marrow progenitors in a dose-dependent manner, with IC_{50} of 0.5 pg/mL and 1.0 pg/mL respectively. TNP-470 was also a potent inhibitor of colony formation from CD34–positive cells, with an IC_{50} of 0.05 pg/mL. Although hematopoietic precursors were more sensitive to the inhibitory effects of TNP-470 than endothelial cells, the effects of TNP-470 on hematopoietic proliferative diseases remain to be investigated.

E. Female Reproductive System

Since angiogenesis occurs normally in the female reproductive organs, the effects of antiangiogenic agents on fertility were investigated (65). Administration of TNP-470 (30 mg/kg SC) to pregnant mice between days 1 and 7 of gestation resulted in complete failure of embryonic growth due to interference with decidualization, placental yolk sac formation, and embryonic vascular development. There was functional recovery of the reproductive system within 6–8 weeks of TNP-470 treatment, resulting in the successful conception of normal litters in all treated mice. Additionally, when nonpregnant cycling female mice were chronically treated with TNP-470, inhibition of endometrial maturation and corpora lutea formation were observed. However, a study investigating the effects of TNP-470 on the primate ovulatory cycle, a model more relevant to humans, failed to demonstrate any effect on reproductive function (66). In this study, short-term treatment of macaque and marmoset with TNP-470 (6 mg/kg IV and 18 mg/kg per day SC respectively) had no major effects on the primate corpus luteum. The difference between results in mouse and primate may reflect interspecies variance or differential susceptibility to the antiangiogenic effects of TNP-470 and suggests that the use of TNP-470 may be without deleterious effects on the ovulatory cycle in women.

F. Antimicrosporidial Activity

Fumagillin, the parent compound of TNP-470, is used as topical treatment for ocular infections and keratitis caused by microsporiosis, an infection seen in patients with AIDS and other immune deficiency syndromes. Systemic treatment with fumagillin was, however, limited by severe toxicity seen in mice. In vitro studies revealed that the MICs of TNP-470 at which 50% of isolates were killed ($MIC_{50} = 0.35 \pm 0.21$ and 0.38 ± 0.11 ng/mL for *Encephalitozoon intestinalis* and *Vittaforma cornea* respectively) were similar to the MIC_{50} of fumagillin and significantly superior to MIC_{50} of albendazole, a drug used for the treatment of systemic microsporidiosis (67). Another in vitro study with RK13 cells infected with microsporidia of the family Encephalitozoonidae found TNP-470 to be highly effective, with an ID_{50} (50% inhibitory dose compared with controls) of 1 ng/mL (68). Moreover, TNP-470 demonstrated in vivo activity against *Encephalitozoon cuniculi*, with prolonged survival and prevention of the development of ascites in infected athymic mice (68). These data suggest that TNP-470 may be a promising new agent for the systemic treatment of microsporidiosis.

VIII. CLINICAL STUDIES

TNP-470 entered clinical trials in 1992, with the first phase I trial initiated in patients with HIV-associated Kaposi's sarcoma (HIV-KS). To date, seven phase I trials of TNP-470 as a single agent in a variety of schedules and tumor types have been conducted or are ongoing. Completed studies are summarized in Table 3. These studies were designed to determine the maximum tolerated dose (MTD) of a 1-hr infusion either weekly, three times a week (MWF), or every other day (qod); a 4-hr infusion once a week; or a 5-day (120-hr) infusion every other week. Phase II studies using a 1-hr infusion MWF have been initiated in renal cell, breast, cervical, and pancreatic carcinomas and glioblastoma.

Dose-limiting toxicities were observed with the 1-hr infusion of TNP-470 at a dose

Table 3 Summary of Clinical Studies Reported with TNP-470 in Patients with Cancer

Phase I trials							
					Responses		
Investigators	Patient population	No. of patients	Schedule	Maximum tolerated dose	CR	PR	SD
Pluda et al. (74)	HIV-associated Kaposi's sarcoma	15	IV every other day	N/A	0	0	4
Kudelka et al. (75)	Squamous cell carcinoma of cervix	18	IV every other day	60 mg/m^2	1	0	3
Zukiwski et al. (69)	Androgen-independent prostate cancer	20	IV every other day	N/A	0	1	0
Dezube et al. (72)	HIV-associated Kaposi's sarcoma	38	IV once a week	N/A (70 mg/m^2)	0	7	5
Bhargava et al. (39)	Advanced cancers	33	IV once a week	177 mg/m^2	0	0	3

of 71 mg/m^2 qod in studies of prostate cancer (69) and cervical cancer patients (70) and at a dose of 76.5 mg/m^2 MWF in solid tumor patients (71). In contrast, Dezube et al. did not observe dose-limiting toxicities with the 1-hr infusion administered weekly at doses up to 70 mg/m^2 (72). The primary toxicities observed were reversible neurological changes, consisting of cognitive and cerebellar effects such as memory impairment, incoordination and gait disturbance, dizziness, visual disturbances, and heightened anxiety and emotional lability that developed 3–7 weeks after initiation of dosing. The resulting maximal tolerated doses of TNP-470 are shown in Table 3. The dose selected for initial phase II studies was 60 mg/m^2 in the carcinomas and 50 mg/m^2 in glioblastoma.

We have evaluated TNP-470 given as a weekly intravenous infusion (dose ranging from 25–235 mg/m^2) to patients with advanced cancers (39). The dose-limiting toxicity was neurotoxicity at a dose of 235 mg/m^2—characterized by cerebellar symptoms consisting of ataxia, dizziness and nystagmus—which were dose-related and resolved completely within 2 weeks of stopping treatment. This is similar to the pattern reported by other investigators using a 1-hr qod or MWF schedule (71). In this study, TNP-470 and its principal active metabolite, AGM-1883, were found to have an extremely short half-life (3 $\pm$ 3 min for TNP-470 and 9 $\pm$ 4 min for AGM-1883), while M-II, an inactive metabolite, had a longer half-life. Based on the results of this study, we are currently testing TNP-470 in a protracted (120-hr continuous IV infusion) dosing schedule in a phase I trial in patients with advanced cancers.

In preclinical studies, efficacy was generally defined as a decrease in tumor growth rate rather than as tumor shrinkage. Clinical end points that reflect a decrease in tumor growth rate include increased time to progression and decreased frequency of metastasis. Efficacy was not an end point of the phase I studies, and these studies were not designed to detect decreases in tumor growth rate or increases in time to tumor progression. Nevertheless, individual patients described in reports of the phase I trials have experienced prolonged stabilization of disease, including three patients with cervical cancer (5, 8, and 20 months) (70). In our study, one patient who had undergone surgical resection and radiation for a high-grade sarcoma of the kidney—with positive margins, which had a high likelihood of recurrence—has been treated with TNP-470 (50 mg/m^2 IV once a week) for over 3 years without any evidence of disease recurrence or long-term toxicity (39). In the phase II study of renal cell carcinoma, Stadler, et al. report that 18% of patients remained in study for 6 or more months without toxicity or disease progression (73).

Although in preclinical studies TNP-470 delayed tumor growth rather than causing regressions, and objective tumor regressions have been uncommon in phase II studies, in patients with HIV-associated Kaposi's sarcoma, Dezube, et al. observed an objective tumor response (defined as flattening of ≥50% of all baseline raised lesions) in 18% of patients treated with weekly 1-hr infusions. Responses occurred at all doses (10–70 mg/m^2); the median duration of response was 11 weeks (range 3–26+ weeks) (72). In contrast, Pluda et al. reported no objective tumor responses when doses of TNP-470 up to 40 mg/m^2 qod were administered (38,74). Nine additional patients in these two studies experienced periods of disease stabilization.

Perhaps the most dramatic tumor response with TNP-470 was that of a patient with advanced squamous cell carcinoma of the cervix and biopsy-proven metastates to the lungs in whom a complete response was seen after 18 weeks of treatment. The patient received TNP-470 for 22 months and remained free of disease for 8+ months after the drug had been discontinued (70,75). Partial responses have also been reported in a patient with renal cell carcinoma (73) and a patient with androgen-independent prostate cancer (69). In the prostate cancer study, a partial response was defined as decline of prostate-specific antigen (PSA) of >50% of baseline. Dixon et al. have reported that in vitro, TNP-470 increases the expression of PSA (1 to 1.5-fold) even when the total cell number is decreased (76). This complicates the use of PSA as a surrogate marker for tumor response and raises the possibility that other responses were obscured by this phenomenon.

As previously mentioned, combination of TNP-470 with cytotoxic therapy has been shown to produce durable tumor remission in mice. Recently initiated studies will evaluate the combination of TNP-470 with Taxol and with BCNU.

IX. CONCLUSIONS AND FUTURE DIRECTIONS

Antiangiogenic agents may represent a significant advance in the treatment of cancer; however, it is likely that these agents by themselves may not be sufficient to cure cancer. The redundancy of the vascular system and the slow turnover of endothelial cells may require prolonged treatment with antiangiogenic agents to cause an antitumor effect. TNP-470 was one of the first antiangiogenic agents to enter clinical trials in cancer patients. Phase I trials completed thus far have provided important information regarding toxicities and pharmacokinetics with different dosing schedules of TNP-470. These studies were not designed to detect clinical benefit; however, objective tumor regression and disease stabilization have been reported in individual patients. Some of the challenges to ongoing clinical development of TNP-470 are the following:

1. Since TNP-470 has an extremely short half-life, it may require prolonged administration for clinical effect. A phase I trial is currently evaluating a protracted intravenous infusion schedule for TNP-470 in patients with advanced cancers. If this approach proves beneficial, future trials may evaluate subcutaneous or oral dosing for ease of administration during prolonged treatment.
2. While it is apparent that angiogenesis is a multistep process, the factors regulat-

ing each step of this pathway are not completely understood. Inhibition of more than one step in this cascade may be required to maximally inhibit angiogenesis. TNP-470 alone may not be causing complete inhibition of angiogenesis in tumors. Therefore, the drug may need to be administered in combination with other antiangiogenic agents or chemotherapeutic drugs to see an optimal clinical effect. Phase I trials have been initiated to investigate this approach.

3. Since antiangiogenic agents may induce dormancy of tumors without actually causing tumor regression, new paradigms in clinical trial design and end points (such as increase in time to progression, decreased frequency of metastasis) may be required to evaluate the clinical utility of these agents.

Finally, the effects of TNP-470 on other diseases characterized by neovascularization remain to be explored. Ongoing preclinical and clinical studies are expected to provide important information to guide future clinical development of this novel drug.

REFERENCES

1. Folkman, J. Tumor angiogenesis: therapeutic implications. N Engl J Med 285:1182–1186, 1971.
2. Folkman, J, and Ingber, DE. Angiostatic steroids, method of discovery and mechanism of action. Ann Surg 206:374–383, 1987.
3. Crum, R, Szabo, S, and Folkman, J. A new class of steroids inhibits angiogenesis in the presence of heparin or a heparin fragment. Science 230:1375–1378, 1985.
4. Ingber, DE Discovery of TNP-470 and other angiogenesis inhibitors. In: B Teicher (ed.), Cancer Therapeutics: Experimental and Clinical Agents. Totowa, NJ: Humana Press, 1996, pp. 283–298.
5. Ingber, D, Fujita, T, Kishimoto, S, Sudo, K, Kanamaru, T, Brem, H, and Folkman, J. Synthetic analogues of fumagillin that inhibit angiogenesis and suppress tumour growth. Nature 348:555–557, 1990.
6. Ingber, DE, Madri, JA, and Folkman, J. A possible mechanism for inhibition of angiogenesis by angiostatic steroids: induction of capillary basement membrane dissolution. Endocrinology 119:1768–1775, 1986.
7. Ingber, DE Fibronectin controls capillary endothelial cell growth by modulating cell shape. Proc Natl Acad Sci USA 87:3579–3583, 1990.
8. Ingber, DE, Madri, JA, and Folkman, J. Endothelial growth factors and extracellular matrix regulate and synthesis through modulation of cell and nuclear expansion. In Vitro Cell Dev Biol 23:387–394, 1987.
9. Killough, J, Magill, G, and Smith, R. The treatment of amebiasis with fumagillin. Science 115:71–72, 1952.
10. McCowen, M, Callender, M, and Lawalis Jr, J. Fumagillin (H-3), a new antibiotic with amebicidal properties. Science 113:202–203, 1951.
11. Hanson, F, and Eble, T. An antiphage agent isolated from Aspergillus sp. J Bacteriol 58: 527–529, 1949.
12. Kusaka, M, Sudo, K, Fujita, T, Marui, S, Itoh, F, Ingber, D, and Folkman, J. Potent antiangiogenic action of AGM-1470: comparison to the fumagillin parent. Biochem Biophys Res Commun 174:1070–1076, 1991.
13. Kusaka, M, Sudo, K, Matsutani, E, Kozai, Y, Marui, S, Fujita, T, Ingber, D, and Folkman, J. Cytostatic inhibition of endothelial cell growth by the angiogenesis inhibitor TNP-470 (AGM-1470). Br J Cancer 69:212–216, 1994.
14. Antoine, N, Greimers, R, De Roanne, C, Kusaka, M, Heinen, E, Simar, LJ, and Castronovo

V. AGM-1470, a potent angiogenesis inhibitor, prevents the entry of normal but not transformed endothelial cells into the G1 phase of the cell cycle. Cancer Res 54:2073–2076, 994.

15. Toi, M, Seki, M, Umezawa, K, and Rhim, J. Resistance to TNP-470 in immortalized human umbilical vein endothelial cells transformed by human papilloma virus E6 and E7 (Abstr. 306). Proc Am Assoc Cancer Res 39:45, 1998.
16. Abe, J, Zhou, W, Takuwa, N, Taguchi, J, Kurokawa, K, Kumada, M, and Takuwa, Y. A fumagillin derivative angiogenesis inhibitor, AGM-1470, inhibits activation of cyclin-dependent kinases and phosphorylation of retinoblastoma gene product but not protein tyrosyl phosphorylation or protooncogene expression in vascular endothelial cells. Cancer Res 54: 3407–3412, 1994.
17. Hori, A, Ikeyama, S, and Sudo, K. Suppression of cyclin d1 MRNA expression by the angiogenesis inhibitor TNP-470 (AGM-1470) in vascular endothelial cells. Biochem Biophys Res Commun 204:1067–1073, 1994.
18. Koyama, H, Nishizawa, Y, Hosoi, M, Fukumoto, S, Kogawa, K, Shioi, A, and Morii, H. The fumagillin analogue TNP-470 inhibits DNA synthesis of vascular smooth muscle cells stimulated by platelet-derived growth factor and insulin-like growth factor-1: possible involvement of cyclin-dependent kinase 2. Circ Res 79:757–764, 1996.
19. Tudan, C, Pelech, S, and Attardo, G Inhibition of G1 phase-associated protein kinase C activity by AGM-1470 (Abstr. 299). Ann Oncol 9: 1998.
20. Budson, AE, Ko, L, Brasel, C, and Bischoff, J. The angiogenesis inhibitor AGM-1470 selectively increases E-selectin. Biochem Biophys Res Commun 225:141–145, 1996.
21. Sin, N, Meng, L, Wang, MQ, Wen, JJ, Bornmann, WG, and Crews, CM. The anti-angiogenic agent fumagillin covalently binds and inhibits the methionine aminopeptidase, MetAP-2. Proc Natl Acad Sci USA 94:6099–6103, 1997.
22. Griffith, EC, Su, Z, Turk, BE, Chen, S, Chang, YH, Wu, Z, Biemann, K, and Liu, JO. Methionine aminopeptidase (type 2) is the common target for angiogenesis inhibitors AGM-1470 and ovalicin. Chem Biol 4:461–471, 1997.
23. Yamaoka, M, Yamamoto, T, Ikeyama, S, Sudo, K, and Fujita, T. Angiogenesis inhibitor TNP-470 (AGM-1470) potently inhibits the tumor growth of hormone-independent human breast and prostate carcinoma cell lines. Cancer Res 53:5233–5236, 1993.
24. Takamiya, Y, Brem, H, Ojeifo, J, Mineta, T, and Martuza, RL. AGM-1470 inhibits the growth of human glioblastoma cells in vitro and in vivo. Neurosurgery 34:869–875; discussion 875, 1994.
25. Castronovo, V and Belotti, D. TNP-470 (AGM-1470): mechanisms of action and early clinical development. Eur J Cancer 32A:2520–2527, 1996.
26. Ikebe, T, Yamamoto, T, Kubo, S, Hirohashi, K, Kinoshita, H, Kaneda, K, and Sakurai, M. Suppressive effect of the angiogenesis inhibitor TNP-470 on the development of carcinogen-induced hepatic nodules in rats. Jpn J Cancer Res 89:143–149, 1998.
27. Tanaka, Y, Kawamata, H, Fujimoto, K, and Oyasu, R. Angiogenesis inhibitor TNP-470 suppresses tumorigenesis in rat urinary bladder. J Urol 157:683–688, 1997.
28. Parangi, S, O'Reilly, M, Christofori, G, Holmgren, L, Grosfeld, J, Folkman, J, and Hanahan, D. Antiangiogenic therapy of transgenic mice impairs de novo tumor growth. Proc Natl Acad Sci USA 93:2002–2007, 1996.
29. Carter, C, Gilbert, K, Bopp, B, and Milkowski, D. TNP-470 inhibition of primary tumor growth and experimental pulmonary metastases of Lewis lung carcinoma and B16 melanoma is route and schedule dependent (Abstr 309). Proc Am Soc Cancer Res 39:45, 1998.
30. Yanai, S, Okada, H, Misaki, M, Saito, K, Kuge, Y, Ogawa, Y, and Toguchi, H. Antitumor activity of a medium-chain triglyceride solution of the angiogenesis inhibitor TNP-470 (AGM-1470) when administered via the hepatic artery to rats bearing Walker 256 carcinosarcoma in the liver. J Pharmacol Exp Ther 271:1267–1273, 1994.
31. Yanai, S, Okada, H, Saito, K, Kuge, Y, Misaki, M, Ogawa, Y, and Toguchi, H. Antitumor

effect of arterial administration of a medium-chain triglyceride solution of an angiogenesis inhibitor, TNP-470, in rabbits bearing vx-2 carcinoma. Pharm Res 12:653–657, 1995.
32. TAP Holdings Inc., Deerfield, IL. Data on file (N-5-261), 1994.
33. Placidi, L, Cretton-Scott, E, de Sousa, G, Rahmani, R, Placidi, M, and Sommadossi, JP. Disposition and metabolism of the angiogenic moderator o-(chloroacetyl-carbamoyl) fumagillol (TNP-470; AGM-1470) in human hepatocytes and tissue microsomes. Cancer Res 55: 3036–3042, 1995.
34. Cretton-Scott, E, Placidi, L, McClure, H, Anderson, DC, and Sommadossi, JP. Pharmacokinetics and metabolism of o-(chloroacetyl-carbamoyl) fumagillol (TNP-470, AGM-1470) in rhesus monkeys. Cancer Chemother Pharmacol 38:117–122, 1996.
35. Placidi, L, Scott, EC, de Sousa, G, Rahmani, R, Placidi, M, and Sommadossi, JP. Interspecies variability of TNP-470 metabolism, using primary monkey, rat, and dog cultured hepatocytes. Drug Metab Dispos 25:94–99, 1997.
36. Figg, WD, Yeh, HJ, Thibault, A, Pluda, JM, Itoh, F, Yarchoan, R, and Cooper, MR. Assay of the antiangiogenic compound TNP-470, and one of its metabolites, AGM-1883, by reversed-phase high-performance liquid chromatography in plasma. J Chromatogr 652:187–194, 1994.
37. Ong, VS, Stamm, GE, Menacherry, S, and Chu, S. Quantitation of TNP-470 and its metabolites in human plasma: sample handling, assay performance and stability. J Chromatogr B Biomed Sci Appl 710:173–182, 1998.
38. Figg, WD, Pluda, JM, Lush, RM, Saville, MW, Wyvill, K, Reed, E, and Yarchoan, R. The pharmacokinetics of TNP-470, a new angiogenesis inhibitor. Pharmacotherapy 17:91–97, 1997.
39. Bhargava, P, Marshall, J, Rizvi, N, Dahut, W, Yoe, J, Figuiera, M, Phipps, K, Hawkins, M, Ong, V, and Cato, A. A study of TNP-470 in patients with advanced cancer (abstr. 1489). Proc Am Assoc Cancer Res 38:221, 1997.
40. Teicher, BA, Sotomayor, EA, and Huang, ZD. Antiangiogenic agents potentiate cytotoxic cancer therapies against primary and metastatic disease. Cancer Res 52:6702–6704, 1992.
41. Teicher, BA, Holden, SA, Ara, G, Sotomayor, EA, Huang, ZD, Chen, YN, and Brem, H. Potentiation of cytotoxic cancer therapies by TNP-470 alone and with other anti-angiogenic agents. Int J Cancer 57:920–925, 1994.
42. Teicher, BA, Dupuis, NP, Robinson, MF, Emi, Y, and Goff, DA. Antiangiogenic treatment (TNP-470/minocycline) increases tissue levels of anticancer drugs in mice bearing Lewis lung carcinoma. Oncol Res 7:237–243, 1995.
43. Herbst, RS, Takeuchi, H, and Teicher, BA. Paclitaxel/carboplatin administration along with antiangiogenic therapy in non-small-cell lung and breast carcinoma models. Cancer Chemother Pharmacol 41:497–504, 1998.
44. Satoh, H, Ishikawa, H, Fujimoto, M, Fujiwara, M, Yamashita, YT, Yazawa, T, Ohtsuka, M, Hasegawa, S, and Kamma, H. Combined effects of TNP-470 and Taxol in human non-small cell lung cancer cell lines. Anticancer Res 18:1027–1030, 1998.
45. Satoh, H, Ishikawa, H, Fujimoto, M, Fujiwara, M, Yamashita, YT, Yazawa, T, Ohtsuka, M, Hasegawa, S, and Kamma, H. Angiocytotoxic therapy in human non-small cell lung cancer cell lines—advantage of combined effects of TNP-470 and SN-38. Acta Oncol 37:85–90, 1998.
46. Brem, H, Gresser, I, Grosfeld, J, and Folkman, J. The combination of antiangiogenic agents to inhibit primary tumor growth and metastasis. J Pediatr Surg 28:1253–1257, 1993.
47. Teicher, BA, Holden, SA, Dupuis, NP, Kakeji, Y, Ikebe, M, Emi, Y, and Goff, D. Potentiation of cytotoxic therapies by TNP-470 and minocycline in mice bearing EMT-6 mammary carcinoma. Breast Cancer Res Treat 36:227–236, 1995.
48. Fujimoto, J, Hori, M, Ichigo, S, Hirose, R, Sakaguchi, H, and Tamaya, T. Plausible novel therapeutic strategy of uterine endometrial cancer with reduction of basic fibroblast growth

factor secretion by progestin and o-(chloroacetyl-carbamoyl) fumagillol (TNP-470; AGM-1470). Cancer Lett 113:187–194, 1997.
49. Teicher, BA, Holden, SA, Ara, G, Korbut, T, and Menon, K. Comparison of several antiangiogenic regimens alone and with cytotoxic therapies in the Lewis lung carcinoma. Cancer Chemother Pharmacol 38:169–177, 1996.
50. Teicher, BA, Dupuis, NP, Emi, Y, Ikebe, M, Kakeji, Y, and Menon, K. Increased efficacy of chemo- and radio-therapy by a hemoglobin solution in the 9L gliosarcoma. In Vivo 9: 11–18, 1995.
51. Murata, R, Nishimura, Y, and Hiraoka, M. An antiangiogenic agent (TNP-470) inhibited reoxygenation during fractionated radiotherapy of murine mammary carcinoma. Int J Radiat Oncol Biol Phys 37:1107–1113, 1997.
52. Nishimura, Y, Murata, R, and Hiraoka, M. Combined effects of an angiogenesis inhibitor (TNP-470) and hyperthermia. Br J Cancer 73:270–274, 1996.
53. Yano, T, Tanase, M, Watanabe, A, Sawada, H, Yamada, Y, Shino, Y, Nakano, H, and Ohnishi, T. Enhancement effect of an anti-angiogenic agent, TNP-470, on hyperthermia-induced growth suppression of human esophageal and gastric cancers transplantable to nude mice. Anticancer Res 15:1355–1358, 1995.
54. Peacock, DJ, Banquerigo, ML, and Brahn, E. A novel angiogenesis inhibitor suppresses rat adjuvant arthritis. Cell Immunol 160:178–184, 1995.
55. Peacock, DJ, Banquerigo, ML, and Brahn, E. Angiogenesis inhibition suppresses collagen arthritis. J Exp Med 175:1135–1138, 1992.
56. Oliver, SJ, Banquerigo, ML, and Brahn, E. Suppression of collagen-induced arthritis using an angiogenesis inhibitor, AGM-1470, and a microtubule stabilizer. Taxol Cell Immunol 157:291–299, 1994.
57. Oliver, SJ, Cheng, TP, Banquerigo, ML, and Brahn, E. Suppression of collagen-induced arthritis by an angiogenesis inhibitor, AGM-1470, in combination with cyclosporin: reduction of vascular endothelial growth factor (VEGF). Cell Immunol 166:196–206, 1995.
58. Mori, S, Yoshikawa, H, Hashimoto, J, Ueda, T, Funai, H, Kato, M, and Takaoka, K. Antiangiogenic agent (TNP-470) inhibition of ectopic bone formation induced by bone morphogenetic protein-2. Bone 22:99–105, 1998.
59. Sasaki, A, Alcalde, RE, Nishiyama, A, Lim, DD, Mese, H, Akedo, H, and Matsumura, T. Angiogenesis inhibitor TNP-470 inhibits human breast cancer osteolytic bone metastasis in nude mice through the reduction of bone resorption. Cancer Res 58:462–467, 1998.
60. Haraguchi, M, Okamura, M, Konishi, M, Konishi, Y, Negoro, N, Inoue, T, Kanayama, Y, and Yoshikawa, J. Anti-angiogenic compound (TNP-470) inhibits mesangial cell proliferation in vitro and in vivo. Kidney Int 51:1838–1846, 1997.
61. Antoine, N, Bours, V, Heinen, E, Simar, LJ, and Castronovo, V. Simulation of human B-lymphocyte proliferation by AGM-1470, a potent inhibitor of angiogenesis. J Natl Cancer Inst 87:136–139, 1995.
62. Antoine, N, Daukandt, M, Locigno, R, Heinen, E, Simar, LJ, and Castronovo, V. The potent angioinhibin AGM-1470 stimulates normal but not human tumoral lymphocytes. Tumori 82: 27–30, 1996.
63. Antoine, N, Daukandt, M, Heinen, E, Simar, LJ, and Castronovo, V. In vitro and in vivo stimulation of the murine immune system by AGM-1470, a potent angiogenesis inhibitor. Am J Pathol 148:393–398, 1996.
64. Hasuike, T, Hino, M, Yamane, T, Nishizawa, Y, Morii, H, and Tatsumi, N. Effects of TNP-470, a potent angiogenesis inhibitor, on growth of hematopoietic progenitors (letter). Eur J Haematol 58:293–294, 1997.
65. Klauber, N, Rohan, RM, Flynn, E, and D'Amato, RJ. Critical components of the female reproductive pathway are suppressed by the angiogenesis inhibitor AGM-1470. Nature Med 3:443–446, 1997.
66. Fraser, H, Dickson, S, and Lunn, S. The effect of the anti-angiogenic agent TNP-470 on

luteal establishment and function in the macaque and marmoset (abstr.). IBC Angiogenesis Conference, 1998.
67. Didier, ES. Effects of albendazole, fumagillin, and TNP-470 on microsporidial replication in vitro. Antimicrob Agents Chemother 41:1541–1546, 1997.
68. Coyle, C, Kent, M, Tanowitz, HB, Wittner, M, and Weiss, LM. TNP-470 is an effective antimicrosporidial agent. J Infect Dis 177:515–518, 1998.
69. Zukiwski, A, Gutterman, J, Bui, C, Sella, A, Ellerhorst, J, Tu, S, Amato, R, Figg, W, Kilbourn, R, and Logothetis, C. Phase I trial of the angiogenesis inhibitor TNP-470 (AGM-1470) in patients with androgen-independent prostate cancer. Proc Am Soc Clin Oncol 13: A795, 1994.
70. Kudelka, AP, Verschraegen, CF, and Loyer, E. Complete remission of metastatic cervical cancer with the angiogenesis inhibitor TNP-470 (letter). N Engl J Med 338:991–992, 1998.
71. Milkowski, D and Weiss, R TNP-470. In: B Teicher, ed. Antiangiogenic Agents in Cancer Therapy. Totowa, NJ: Humana Press, 1999, pp. 385–398.
72. Dezube, BJ, Von Roenn, JH, Holden-Wiltse, J, Cheung, TW, Remick, SC, Cooley, TP, Moore, J, Sommadossi, JP, Shriver, SL, Suckow, CW, and Gill, PS. Fumagillin analog in the treatment of Kaposi's sarcoma: a phase I AIDS clinical trial group study. AIDS clinical trial group no. 215 team. J Clin Oncol 16:1444–1449, 1998.
73. Stadler, W, Shapiro, C, Sossman, J, Clark, J, Vogelzang, N, and Kuzel, TA multi-institutional study of the angiogenesis inhibitor TNP-470 in metastatic renal cell carcinoma (Abst. 1192). Proc Am Soc Clin Oncol 17:310A, 1998.
74. Pluda, J, Wyvill, K, Figg, WD, Whitcup, S, Lietzau, S, Saville, M, Cohen, R, Feigal, E, Parks, D, Foli, A, et al. A phase I study of angiogenesis inhibitor, TNP-470 (AGM-1470), administered to patients with HIV-associated Kaposi's sarcoma. Proc Am Soc Clin Oncol 13:A8, 1994.
75. Kudelka, A, Levy, T, Verschraegen, C, Edwards, C, Piamsomboon, S, Termrungruanglert, W, Freedman, R, Kaplan, A, Kieback, D, Meyers, C, Jaeckle, K, Loyer, E, Steger, M, Mante, R, Mavligit, G, Killian, A, Tang, R, Gutterman, JU, and Kavanagh, J. A phase I trial of TNP-470 administered to patients with advanced squamous cell carcinoma of the cervix. Clin Cancer Res 3:1502–1505, 1997.
76. Dixon, S, Horti, J, Logothetis, C, Guo, Y, Bauer, K, Reed, D, and Figg, W. TNP-470 up regulates the expression of prostate specific antigen (abstr. 298). Proc Am Soc Cancer Res 39:44, 1998.
77. Takamiya, Y, Friedlander, RM, Brem, H, Malick, A, and Martuza, RL. Inhibition of angiogenesis and growth of human nerve-sheath tumors by AGM-1470. J Neurosurg 78:470–476, 1993.
78. Yanase, T, Tamura, M, Fujita, K, Kodama, S, and Tanaka, K. Inhibitory effect of angiogenesis inhibitor TNP-470 on tumor growth and metastasis of human cell lines in vitro and in vivo. Cancer Res 53:2566–2570, 1993.
79. Bernsen, HJ, Rijken, PF, Peters, H, Bakker, H, and van der Kogel, AJ. The effect of the anti-angiogenic agent TNP-470 on tumor growth and vascularity in low passaged xenografts of human gliomas in nude mice. J Neurooncol 38:51–57, 1998.
80. Singh, Y, Shikata, N, Kiyozuka, Y, Nambu, H, Morimoto, J, Kurebayashi, J, Hioki, K, and Tsubura, A. Inhibition of tumor growth and metastasis by angiogenesis inhibitor TNP-470 on breast cancer cell lines in vitro and in vivo. Breast Cancer Res Treat 45:15–27, 1997.
81. Hama, Y, Shimizu, T, Hosaka, S, Sugenoya, A, and Usuda, N. Therapeutic efficacy of the angiogenesis inhibitor o-(chloroacetyl-carbamoyl) fumagillol (TNP-470; AGM-1470) for human anaplastic thyroid carcinoma in nude mice. Exp Toxicol Pathol 49:239–247, 1997.
82. Kanai, T, Konno, H, Tanaka, T, Matsumoto, K, Baba, M, Nakamura, S, and Baba, S. Effect of angiogenesis inhibitor TNP-470 on the progression of human gastric cancer xenotransplanted into nude mice. Int J Cancer 71:838–841, 1997.
83. Wassberg, E, Pahlman, S, Westlin, JE, and Christofferson, R. The angiogenesis inhibitor

TNP-470 reduces the growth rate of human neuroblastoma in nude rats. Pediatr Res 41:327–333, 1997.

84. Xia, JL, Yang, BH, Tang, ZY, Sun, FX, Xue, Q, and Gao, DM. Inhibitory effect of the angiogenesis inhibitor TNP-470 on tumor growth and metastasis in nude mice bearing human hepatocellular carcinoma. J Cancer Res Clin Oncol 123:383–387, 1997.
85. Isobe, N, Uozumi, T, Kurisu, K, and Kawamoto, K. Experimental studies of the antitumor effect of TNP-470 on malignant brain tumors. Antitumor effect of TNP-470 on a human medulloblastoma xenograft line. Neuropediatrics 27:136–142, 1996.
86. Yazaki, T, Takamiya, Y, Costello, PC, Mineta, T, Menon, AG, Rabkin, SD, and Martuza, RL. Inhibition of angiogenesis and growth of human non-malignant and malignant meningiomas by TNP-470. J Neurooncol 23:23–29, 1995.
87. Gleich, LL, Zimmerman, N, Wang, YO, and Gluckman, JL. Angiogenic inhibition for the treatment of head and neck cancer. Anticancer Res 18:2607–2609, 1998.
88. Taki, T, Ohnishi, T, Arita, N, Hiraga, S, Saitoh, Y, Izumoto, S, Mori, K, and Hayakawa, T. Anti-proliferative effects of TNP-470 on human malignant glioma in vivo: potent inhibition of tumor angiogenesis. J Neurooncol 19:251–258, 1994.
89. McLeskey, SW, Zhang, L, Kharbanda, S, Kurebayashi, J, Lippman, ME, Dickson, RB, and Kern, FG. Fibroblast growth factor overexpressing breast carcinoma cells as models of angiogenesis and metastasis. Breast Cancer Res Treat 39:103–117, 1996.
90. Tanaka, T, Konno, H, Matsuda, I, Nakamura, S, and Baba, S. Prevention of hepatic metastasis of human colon cancer by angiogenesis inhibitor TNP-470. Cancer Res 55:836–839, 1995.
91. Konno, H, Tanaka, T, Matsuda, I, Kanai, T, Maruo, Y, Nishino, N, Nakamura, S, and Baba, S. Comparison of the inhibitory effect of the angiogenesis inhibitor, TNP-470, and mitomycin c on the growth and liver metastasis of human colon cancer. Int J Cancer 61:268–271, 1995.
92. Kawarada, Y, Ishikura, H, Kishimoto, T, Saito, K, Takahashi, T, Kato, H, and Yoshiki, T. Inhibitory effects of the antiangiogenic agent TNP-470 on establishment and growth of hematogenous metastasis of human pancreatic carcinoma in scid beige mice in vivo. Pancreas 15:251–257, 1997.
93. Shishido, T, Yasoshima, T, Denno, R, Sato, N, and Hirata, K. Inhibition of liver metastasis of human gastric carcinoma by angiogenesis inhibitor TNP-470. Jpn J Cancer Res 87:958–962, 1996.
94. Konno, H, Tanaka, T, Kanai, T, Maruyama, K, Nakamura, S, and Baba, S. Efficacy of an angiogenesis inhibitor, TNP-470, in xenotransplanted human colorectal cancer with high metastatic potential. Cancer 77:1736–1740, 1996.
95. Voest, EE, Kenyon, BM, O'Reilly, MS, Truitt, G, D'Amato, RJ, and Folkman, J. Inhibition of angiogenesis in vivo by interleukin 12 (see comments). J Natl Cancer Inst 87:581–586, 1995.
96. Takechi, A. Effect of angiogenesis inhibitor TNP-470 on vascular formation in pituitary tumors induced by estrogen in rats. Neurol Med Chir (Tokyo) 34:729–733, 1994.
97. Takechi, A, Uozumi, T, Kawamoto, K, Ito, A, Kurisu, K, and Sudo, K. Inhibitory effect of TNP-470, a new anti-angiogenic agent, on the estrogen induced rat pituitary tumors. Anticancer Res 14:157–162, 1994.
98. Kato, T, Sato, K, Kakinuma, H, and Matsuda, Y. Enhanced suppression of tumor growth by combination of angiogenesis inhibitor o-(chloroacetyl-carbamoyl)fumagillol (TNP-470) and cytotoxic agents in mice. Cancer Res 54:5143–5147, 1994.
99. Devineni, D, Klein-Szanto, A, and Gallo, JM. Uptake of temozolomide in a rat glioma model in the presence and absence of the angiogenesis inhibitor TNP-470. Cancer Res 56:1983–1987, 1996.
100. Matsuoka, S. Anti-tumor effects of an angiogenesis inhibitor, TNP-470 (AGM-1470) alone and in combination with cisplatinum, and changes in serum copper levels in liver cancer bearing rats. Hokkaido. Igaku Zasshi 71:69–80, 1996.

27
Chronic Inflammation and Angiogenesis

James D. Winkler and Jeffrey R. Jackson
SmithKline Beecham Pharmaceuticals, King of Prussia, Pennsylvania

I. RELATIONSHIPS BETWEEN INFLAMMATION AND ANGIOGENESIS

Angiogenesis is an important aspect of new tissue development, growth, and tissue repair. It is also a component of many diseases including cancer, blindness, and chronic inflammation such as rheumatoid arthritis (RA) and psoriasis (1–3). There is considerable evidence to suggest that chronic inflammation plays a key role in the process and progession of angiogenesis in certain settings. The codependence of chronic inflammation and angiogenesis, a survey of the mediators involved, the models used, and the potential benefits of targeting chronic inflammation to affect angiogenesis are the subjects of this chapter.

There is growing evidence that there is a link between the processes of angiogenesis and inflammation. This link has recently been reviewed (4) and is discussed only briefly here. Chronic inflammatory disease is by its nature a proliferative process, similar in many ways to cancer. In chronic inflammation there is an influx of inflammatory cells as well as an in situ proliferation of both inflammatory and other resident cells. As this proliferation occurs, there is an increased need for blood supply to both provide nutrients and remove waste. In addition, the neovasculature facilitates the access of invading inflammatory cells to the inflammatory site. The increased vascular surface area resulting from the angiogenesis can enhance the activation of inflammatory cells via the activation of adhesion molecules and increase their ability to produce proinflammatory mediators. Thus, angiogenesis supports the chronic inflammatory process.

The migration, proliferation, and activation of tissue and inflammatory cells in chronic inflammation can be extremely damaging to normal tissue. Two examples of this phenomenon are pannus formation in RA and skin proliferation in psoriatic diseases (5,6). In both cases, the proliferating tissue contains an abundance of inflammatory cells as well as angiogenic blood vessels and a variety of inflammatory/angiogenic mediators. There are also zones of relative hypoxia where the tissue proliferation has outstripped blood vessel growth, and this can induce further capillary development via the release of large quantities of angiogenic factors by hypoxic macrophages. Thus, a chronic inflammatory state supports the angiogenic process.

II. INFLAMMATORY MEDIATORS IMPLICATED IN ANGIOGENESIS

The signals controlling angiogenesis, while directed at the endothelial cells, come from cells in the nearby tissues, and many cells are capable of producing angiogenic factors when they sense the correct stimuli. Cells shown to produce angiogenic signaling molecules include tumor cells, keratinocytes, corneal fibroblasts, synovial fibroblasts, monocytes, and macrophages (2). Macrophages appear at most sites where angiogenesis is occurring in an abnormal environment, including wound healing and disease states, and the presence of these cells is sufficient to induce angiogenesis in the avascular cornea (7,8). In fact, almost every growth factor and cytokine known to regulate angiogenesis can be produced by macrophages (9,10).

Factors capable of inducing angiogenesis are categorized as either indirect or direct. Indirect factors act via intermediary mechanisms and often induce angiogenesis only in vivo. Directly acting mediators are usually able to induce proliferation, migration, and/or differentiation of endothelial cells directly in vitro. Good examples of direct factors are basic fibroblast growth factor (bFGF) (11,12) and vascular endothelial growth factor (VEGF) (13,14). In vitro, VEGF causes endothelial cell proliferation and migration; in vivo, it is potently angiogenic and causes vascular permeability and ocular angiogenesis (15–17). Other growth factors with some direct effects on endothelial cells include transforming growth factors α and β (TGF-α and TGF-β), and platelet-derived growth factor (PDGF) (18–20). Another example of a directly acting angiogenic factor is angiopoietin. This protein, a ligand for the endothelial cell Tie2 receptor, has been shown to stimulate endothelial cell migration in vitro and blood vessel growth in vivo (21,22).

Many inflammatory mediators have both direct and indirect angiogenic activities. Prostaglandins E_1 and E_2 (PGE_1 and PGE_2), tumor necrosis factor α (TNF-α), interleukins 1, 6, and 8 (IL-1, IL-6, and IL-8) have all been shown to induce angiogenesis in vivo (23). The E prostaglandins induce angiogenesis in both the cornea and chicken chorioallantoic membrane assays and may mediate the angiogenic actions of bFGF (4,24). Leukotrienes (LT) have also been shown to promote angiogenesis. LTC_4 can promote tube formation in vitro, and LTC_4 and LTD_4 increase vessel growth in the chorioallantoic membrane assay (25,26). IL-8 is both chemotactic and mitogenic for endothelial cells in addition to causing angiogenesis in vivo in the cornea assay (27). Aberrant increases in IL-8 produced by psoriatic keratinocytes have been said to be an important aspect of their ability to produce an angiogenic response (28). IL-1β has been reported to be significantly more potent than VEGF in inducing corneal angiogenesis, requiring as little as 1 ng (29). TNF-α appears to have dose-dependent effects on angiogenesis: in vivo, it is a potent inducer at low doses in corneal implants and an inhibitor at higher doses (30). Finally, platelet-activating factor, a potent proinflammatory lipid, has been shown to directly stimulate endothelial cell proliferation and to have angiogenic effects in vivo, some of which may be mediated by the production of nitric oxide (31–33).

The agents described above can induce many facets of inflammation, and their angiogenic effects could be linked to the recruitment of inflammatory cells or increased vascular permeability. However, recent data suggest that many of these inflammatory mediators can induce the expression of more directly acting angiogenic factors, such as VEGF. PGE_2, IL6, and IL-1 have been shown to induce VEGF mRNA and protein expression (34–37). VEGF is also regulated by the cellular oxygen concentration, with hypoxia inducing its expression (17,37,38). Thus, during growth of inflammatory tissue, the hypoxic conditions, coupled with high concentrations of inflammatory mediators, can cause an increase of VEGF production. Such increases have been seen in rheumatoid arthritis patients, showing that the results seen in culture have relevance to the disease state (39).

III. MODELS OF INFLAMMATORY ANGIOGENESIS

There are numerous models that mimic aspects of angiogenesis (5,40,41). These range from in vitro models (endothelial cells in various culture systems), to vessel culture models, to in vivo models, such as the chicken chorioallantioic membrane assay, ocular models, and tumor metastasis and angiogenesis in response to inflammatory insults. To study the effects of anti-inflammatory compounds on angiogenesis, models are often used in which inflammation has been shown to be the driving force for the angiogenic response.

In order to study the interactions between inflammation and angiogenesis, we have employed a murine air-pouch granuloma model, first described by Kimura et al. (42) and modified and characterized by Willoughby's laboratory (43). This model involves the creation of a subcutaneous air pouch. The cells lining the pouch resemble a normal synovium in many ways. The inflammatory response is started by the injection of Freund's complete adjuvant and croton oil. A key feature of this model is that the granuloma growth is not dependent on the angiogenesis; excluding the croton oil results in the growth of an avascular granuloma (42). Thus, treatments that selectively block angiogenesis can be observed to have different effects from those that block inflammation.

The granuloma within the air pouch grows over time, typically peaking around 7 days. Over that time, a number of changes occur at both the structural and molecular levels (41,43). One of the most striking changes is in the histological organization of the granuloma tissue. The early granuloma is extremely cellular, with little organization observed. By 3 days, clear structure is emerging, with developed blood vessels easily seen. This structural aspect is heightened by 12–14 days, when there emerges a clear fibrotic nature to the granuloma. As in many chronic inflammatory situations, there are waves of different inflammatory cells that come to the site, first polymorphonuclear cells, followed by mononuclear cells (44). Finally there appear more ''residential'' inflammatory cells, such as tissue macrophages, mast cells, and fibroblasts, completing the transition from acute to chronic inflammation. Because these different cells respond to inflammatory stimuli in different ways, it is not surprising that different inflammatory mediators are produced at different times (Table 1).

There is a rapid yet transient expression of VEGF in the granuloma model (45). The peak of VEGF expression appears around 1 day after insult and may account for the earliest vascularization of the granuloma. Several inflammatory mediators appear to peak at later times. For example, there are dramatic increases in the content of IL-1β and TNF-α in the granuloma at 5–7 days after inflammatory insult. This peak of mediator production coincides with the influx of mononuclear cells into the granuloma and the change from acute to chronic inflammatory character. Increases in other mediators or growth factors may occur at later times, as during the fibrotic phase of the granuloma (days 10–21). However, these later times are much less studied. There are indications that bFGF and TGFβ are increased at later times (44). Certainly, as the complexity of the tissue increases, it is possible that numerous mediators are interacting with the vasculature to support angiogenesis, and it may become correspondingly more difficult to disrupt angiogenesis at later times.

Table 1 Change in Granuloma Characteristics Over Time

	Time period		
	Early (0–3 days)	Middle (3–7 days)	Late (7–21 days)
Cells	Neutrophils	Monocytes, lymphocytes	Macrophages, fibroblasts
Structure	Diffuse	Mixed cellular	Fibrotic nature, matrix deposition
Mediators	PDGF, EGF	IL1-β	BFGF
	VEGF		TGF-β
		TNF-α	
Angiogenesis	Large branches	Fine network from large branches	Reorganization and regression

Another model for the assessment of angiogenesis in a setting of inflammation that is an alternative to the crotin oil granuloma model is the sponge implant model (46,47). This model has a granuloma induced by a forgein body, which is usually a sponge or a piece of cartilage. Although not as robust as the crotin oil granuloma, these implant granulomas are more dependent on angiogenesis for growth. Modulation of IL-1, vasoactive intestinal peptide, and angiotensin II has been detected in this model (48,49). Another useful model is the Matrigel implant, in which a proangiogenic substance is mixed with Matrigel and implanted subcutaneously. The Matrigel forms a plug that will be infiltrated by blood vessels if an angiogenic signal is present. After removal, the extent of angiogenesis can be assessed by morphometric analysis and the results expressed as a percentage of the Matrigel that is vascularized, or assessment of cellular infiltration can be done using image analysis (33,50,51). Inflammatory stimuli such as platelet-activating factor, interleukins, and TNF-α have been shown to promote angiogenesis in this model. In addition, authors have shown that more complex additions can be made, such as synovial fluid (32) and whole cells (52). Matrigel is a complex mixture of matrix proteins. This is a limitation for some researchers, who may want to study specific matrix molecules and their effects on angiogenesis. In these cases, researchers can use implants of inert material, such as alginate beads (53,54).

A notable model for the study of angiogenesis driven by inflammatory stimuli is the corneal implant model. The cornea is normally devoid of vasculature, and inflammatory angiogenesis is most commonly induced by implantation of proangiogenic stimuli in pellets (40). Implants with VEGF or bFGF produce a strong vascular invasion of the cornea over 3–7 days (55–57). However, equally strong angiogenic effects are observed with pellets containing either IL-1β or TNF-α (29,58,59), both cytokines being profoundly proinflammatory in vivo. This occurs without the massive inflammation inherent in the granuloma models. Nevertheless, implantation of inflammatory cells, such as macrophages or fibroblasts, in the cornea produces angiogenesis (58).

IV. EFFECTS OF ANTI-INFLAMMATORY DRUGS ON ANGIOGENESIS

While an association between angiogenesis and inflammation appears obvious on observational grounds, the difficulties of assessing angiogenesis in vivo have prevented the proof

Table 2 Summary of Various Inhibitors of Angiogenesis in the Mouse Air Pouch Granuloma Model

Compound	Molecular target	Reference
Hyaluronan + cortisone	Unknown	43
Medroxyprogesterone	Unknown	77
Tetrahydrocortisol	Unknown	43
SB 203347	PLA_2	77
SB 220025	p38 MAP kinase	69
Ro 24-4736	PAF receptor antagonist	68
Ibuprofen	COX	1
Methotrexate	DNA synthesis	1

that angiostasis would have potential to restrict chronic inflammation. The following section reviews several classes of anti-inflammatory drugs, and these studies support the hypothesis that inflammatory mediators can drive the angiogenesis associated with chronic inflammation (Table 2) (60,61).

A. Steroids

Steroids provide an example of the problems in interpretation of anti-inflammatory versus antiangiogenesis effects. It has been argued for some time whether anti-inflammatory glucocorticoids are angiostatic. Some steroids, such as tetrahydrocortisol and medroxyprogesterone, are effective in reducing granulomatous inflammatory angiogenesis (43,60). In contrast, dexamethasone appears to increase vascular density. This is an artifactual finding, as dexamethasone has such a potent anti-inflammatory effect that granuloma formation is severely inhibited, preventing comparative assessment of angiogenesis (1,62).

Using the cotton pellet/cartilage coimplant model, Colville-Nash and coworkers (63) inhibited angiogenesis with angiostatic steroid therapy; low-dose cortisone with heparin was also effective, neither ingredient having an effect alone (64). As a result of this, granuloma development was reduced, inflammatory cell recruitment reduced, and cartilage degradation abolished. Angiostatic steroid therapy is also effective in inhibiting delayed-type hypersensitivity granuloma development (65).

B. Drugs Affecting Lipid Mediators

Although nonsteroidal anti-inflammatory drugs (NSAIDs) are the mainstay of anti-inflammatory therapy, they have been little studied in angiogenesis, their effects being assumed to be purely symptomatic. The NSAIDs indomethacin and ibuprofen significantly reduced the vascularity of the granuloma in the air-pouch model, suggesting that the inflammatory angiogenesis may be partially induced by PGE_2 (1,43). A combination of another NSAID, diclofenac, and hyaluronan as a delivery agent can decrease vascularity (66) as well as induce vascular regression. Clinically, NSAIDs have been shown to decrease angiogenesis in patients during gastric ulcer healing (67). The advent of selective cyclooxygenase-2 inhibitors means that these effects need to be examined in greater detail.

The effects of other compounds that block lipid mediator production has recently been reported (68). Inhibition of 14-kDa PLA_2 blocked angiogenesis in the granuloma model and reduced the content of both LT and platelet-activating factor. However, inhibition of 5-lipoxygenase had no effect on angiogenesis, whereas inhibition of platelet-activating factor receptors reduced angiogenesis. Together, these data indicate that platelet-activating factor plays a key role in angiogenesis in this model.

C. Drugs that Block Inflammatory Cytokines

We have recently observed that inhibition of cytokine production, using inhibitors of p38 MAP kinase, reduced angiogenesis in the air-pouch granuloma model (69). This reduction of angiogenesis occurred during days 5–7, the time of peak cytokine production, and was accompanied by dramatic reductions in TNF-α and IL-1β in the granuloma. These results suggest that, during the vascularization of the granuloma, many inflammatory cytokines contribute to the proangiogenic environment. Drugs that can decrease these mediators can affect this environment, shifting the balance away from neovascularization.

D. Antirheumatic Drugs

Drugs such as gold compounds, D-penicillamine, and methotrexate have been used extensively to treat rheumatoid arthritis, demonstrating some efficacy in slowing the progress of the disease, yet their mechanisms of action are not completely understood. Several antirheumatic compounds block angiogenesis in the air-pouch model, including D-penicillamine, chloroquine, auranofin, cyclophosphamide, azathioprine, methotrexate, and levamisole, confirming the antiangiogenic activities seen for many of these in vitro (5,62). Aurothiomalate and auranofin both inhibit angiogenesis in the cotton pellet/cartilage model, whereas only aurothiomalate is effective in inhibiting cartilage erosion (70). Koch et al. (71) demonstrated potent inhibition of angiogenic factor secretion by macrophages with gold compounds. Methotrexate, on the other hand, also inhibits endothelial cell DNA synthesis in vitro at physiologically relevant concentrations (72) and inhibits endothelial cell growth factor-induced corneal angiogenesis. A similar study suggested that D-penicillamine also inhibits endothelial cell proliferation in vitro and corneal angiogenesis in vivo (73). Thus, these drugs may exert their effect at least partially through inhibition of angiogenesis.

E. Inhibitors of Metalloproteinase Enzymes

Matrix metalloproteinase–induced breakdown of extracellular matrix is important for the initiation and progression of angiogenesis (5). In 1994, Galardy et al. (74) showed that the collagenase inhibitor BB-94 could inhibit angiogenesis induced by carcinoma cells implanted in the cornea. This compound has also shown some efficacy against haemangioma growth (75), which has a profound angiogenic component. Of relevance, BB16, a compound from the same series, is very effective in inhibiting adjuvant arthritis (76). These actions could be due in part to the inhibition of angiogenesis.

V. CONCLUSIONS

There is a clear association of angiogenesis with chronic inflammation: an inflammatory state can promote angiogenesis and angiogenesis can facilitate chronic inflammation. Many inflammatory mediators have been shown to be directly or indirectly angiogenic.

Taking advantage of this interaction, many models have been developed that measure angiogenesis in a setting of chronic inflammation. Because of the interconnection between inflammation and angiogenesis, many compounds that block chronic inflammatory responses—such as antirheumatic drugs, steroids, and nonsteroidal anti-inflammatory compounds and cytokine inhibitors—have also been shown to inhibit angiogenesis. As more specific inhibitors of angiogenesis become available, this intriguing connection between angiogenesis and inflammation can be further explored, and pharmacological control of angiogenesis may hold great promise for the alleviation of chronic inflammatory states.

REFERENCES

1. Colville-Nash, PR, Scott, DL. Angiogenesis and rheumatoid arthritis: pathogenic and therapeutic implications. Ann Rheum Dis 51:919–925, 1992.
2. Folkman, J, Brem, H. Angiogenesis and inflammation. In: JI Gallin, IM Goldstein, and R Snyderman, eds. Inflammation: Basic Principles and Clinical Correlates, 2d ed. New York: Raven Press, 1992, pp. 821–839.
3. Folkman, J. Angiogenesis in cancer, vascular, rheumatoid and other disease. Nature Med 1: 27–31, 1995.
4. Jackson, JR, Seed, MP, Kircher, CH, Willoughby, DA, Winkler, JD. The co-dependence of angiogenesis and chronic inflammation. FASEB J 11:457–465, 1997.
5. Colville-Nash, PR, Seed, MP. The current state of angiostatic therapy, with special reference to rheumatoid arthritis. Curr Opin Invest Drugs 2:763–813, 1993.
6. Detmar, M, Brown, LF, Claffey, KP, Yeo, KT, Kocher, O, Jackman, RW, Berse, B, Dvorak, HF. Overexpression of vascular permeability factor/vascular endothelial growth factor and its receptors in psoriasis. J Exp Med 180:1141–1146, 1994.
7. Koch, AE, Polverini, PJ, Leibovich, JL. Induction of neovascularization by activated human monocytes. J Leuk Biol 39:233–238, 1986.
8. Knighton, DR, Hunt, TK, Scheuenstuhl, H, Halliday, BJ, Werb, Z, Banda, MJ. Oxygen tension regulates the expression of angiogenesis factor by macrophages. Science 221:1283–1285, 1983.
9. Sunderkotter, C, Steinbrink, K, Goebeler, M, Bhardwaj, R, Sorg, C. Macrophages and angiogenesis. J Leukoc Biol 55:410–422, 1994.
10. Polverini, PJ. Role of the macrophage in angiogenesis-dependent diseases. In: ID Goldberg and EM Rosen, eds. Regulation of Angiogenesis. Basel, Switzerland: Birkhauser, 1997, pp. 11–28.
11. Baird, A, Walicke, PA. Fibroblast growth factors. Br Med Bull 45:438–452, 1989.
12. Ernestein, J, Waleh, NS, Kramer, RH. Basic FGF and TGF-beta differentially modulate integrin expression of human microvascular endothelial cells. Exp Cell Res 203:499–503, 1992.
13. Thomas, KA. Vascular endothelial growth factor, a potent and selective angiogenic agent. J Biol Chem 271:603–606, 1996.
14. Neufeld, G, Tessler, S, Gitay-Goren, H, Cohen, T, Levi, B-Z. Vascular endothelial growth factor and its receptors. Prog Growth Factor Res 5:89–97, 1994.
15. Leung, DW, Cachianes, G, Kuang W-J, Goeddel, DV, Ferrara, N. Vascular endothelial growth factor is a secreted angiogenic mitogen. Science 246:1306–1309, 1989.
16. Miller, JW, Adamis, AP, Shima, DT, D'Amore, PA, Moulton, RS, O'Reilly, MS, Folkman, J, Dvorak, HF, Brown, LF, Berse, B, et al. Vascular endothelial growth factor/vascular permeability factor is temporally and spatially correlated with ocular angiogenesis in a primate model. Am J Pathol 145:574–584, 1994.
17. Shima, DT, Adamis, AP, Ferrara, N, Yeo, KT, Yeo, TK, Allende, R, Folkman, J, Damore, PA. Hypoxic induction of endothelial-cell growth-factors in retinal cells—identification and

characterization of vascular endothelial growth-factor (VEGF) as the mitogen. Mol Med 1: 182–193, 1995.

18. Schreiber, AB, Winkler, ME, Derynck, R. Transforming growth factor-α: a more potent angiogenic mediator than epidermal growth factor. Science 232:1250–1253, 1986.
19. Bell, L, Madri, JA. Effect of platelet factors on migration of cultured bovine aortic endothelial and smooth muscle cells. Circ Res 65:1057–1065, 1989.
20. Mustonen, T, Alitalo, K. Endothelial receptor tyrosine kinases involved in angiogenesis. J Cell Biol 129:895–898, 1995.
21. Koblizek, TI, Weiss, C, Yancopoulos, GD, Deutsch, U, Risau, W. Angiopoietin-1 induces sprouting angiogenesis in vitro. Curr Biol 8:529–532, 1998.
22. Yancopoulos, GD, Klagsbrun, M, Folkman, J. Vasculogenesis, angiogenesis, and growth factors: ephrins enter the fray at the border (comment) (review). Cell 93:661–664, 1998.
23. Strieter, RM, Polverini, PJ, Arenberg, DA, Kunkel, SL. The role of CXC chemokines as regulators of angiogenesis. Shock 4:155–160, 1995.
24. Spisni, E, Manica, F, Tomasi, V. Involvement of prostanoids in the regulation of angiogenesis by peptide growth factors. Prostag Leuko Essen Fatty Acids 47:111–115, 1992.
25. Kanayasu, T, Nakao Hayashi, J, Asuwa, N, Morita, I, Ishii, T, Ito, H, Murota, S. Leukotriene C-4 stimulates angiogenesis in bovine carotid artery endothelial cells in vitro. Biochem Biophys Res Comm 159:572–578, 1989.
26. Tsopanoglou, NE, Pipilo Synetos, E, Maragoudakis, ME. Leukotrienes C-4 and D-4 promote angiogenesis via a receptor-mediated interaction. Eur J Pharmacol 258:151–154, 1994.
27. Moore, BB, Keane, MP, Addison, CL, Arenberg, DA, Strieter, RM. CXC chemokine modulation of angiogenesis: the importance of balance between angiogenic and angiostatic members of the family (review). J Invest Med 46:113–120, 1998.
28. Nickoloff, BJ, Mitra, RS, Varani, J, Dixit, VM, Polverini, PJ. Aberrant production of interleukin-8 and thrombospondin-1 by psoriatic keratinocytes mediates angiogenesis. Am J Pathol 144:820–828, 1994.
29. BenEzra, D, Hemo, I, Maftzir, G. In vivo angiogeneic activity of interleukins. Arch Ophthalmol 108:573–576, 1990.
30. Fajardo, LF, Kwan, HH, Kowalski, J, Prionas, SD, Allison, AC. Dual role of tumor necrosis factor-α in angiogenesis. Am J Pathol 140:539–544, 1992.
31. Camussi, G, Montrucchio, G, Lupia, E, De Martino, A, Perona, L, Arese, M, Vercellone, A, Toniolo, A, Bussolino, F. Platelet-activating factor directly stimulates in vitro migration of endothelial cells and promotes in vivo angiogenesis by a heparin-dependent mechanism. J Immunol 154:6492–6501, 1995.
32. Lupia, E, Montrucchio, G, Battaglia, E, Modena, V, Camussi, G. Role of tumor necrosis factor-alpha and platelet-activating factor in neoangiogenesis induced by synovial fluids of patients with rheumatoid arthritis. Eur J Immunol 26:1690–1694, 1996.
33. Montrucchio, G, Lupia, E, De Martino, A, Battaglia, E, Arese, M, Tizzani, A, Bussolino, F, Camussi, G. Nitric oxide mediates angiogenesis induced in vivo by platelet-activating factor and tumor necrosis factor-alpha. Am J Pathol 151:557–563, 1997.
34. Cohen, T, Nahari, D, Cerem LW, Neufeld, G, Levi, BZ. Interleukin-6 induces the expression of vascular endothelial growth-factor. J Biol Chem 271:736–741, 1996.
35. Ben-Av, P, Crofford, LJ, Wilder, RL, Hla, T. Induction of vascular endothelial growth factor expression in synovial fibroblasts by prostaglandin E and interleukin-1: a potential mechanism for inflammatory angiogenesis. FEBS Lett 372:83–87, 1995.
36. Li, J, Perrella, MA, Tsai, J-C, Yet, S-F, Hsieh, C-M, Yoshizumi, M, Patterson, C, Endege, WO, Zhou, F, Lee, M-E. Induction of vascular endothelial growth factor gene expression by interleukin-1β in rat aortic smooth muscle cells. J Biol Chem 270:308–312, 1995.
37. Jackson, JR, Minton, JAL, Ho, ML, Wei, N, Winkler, JD. Expression of VEGF in synovial fibroblasts is induced by hypoxia and interleukin-1β. J Rheumetol 24:1253–1259, 1997.
38. Shweiki, D, Neeman, M, Itin, A, Keshet, E. Induction of vascular endothelial growth factor

expression by hypoxia and by glucose deficiency in multicell spheroids: implications for tumor angiogenesis. Proc Natl Acad Sci USA 92:768–772, 1995.
39. Nagashima, M, Yoshino, S, Ishiwata, T, Asano, G. Role of vascular endothelial growth factor in angiogenesis of rheumatoid arthritis. J Rheumatol 22:1624–1630, 1995.
40. Cockerill, GW, Gamble, JR, Vadas, MA. Angiogenesis: models and modulators. Int Rev Cytol 159:113–160, 1995.
41. JD Winkler, MP Seed, JR Jackson Inflammatory angiogenic factors in a model of chronic inflammation. In: M. E. Maragoudakis, ed. Angiogenesis: Models, Modulators and Clinical Applications. New York: Plenum Press, 1998, pp. 113–120.
42. Kimura, M, Suzuki, J, Amemiya, K. Mouse granuloma pouch induced by Freund's complete adjuvant with croton oil. J Pharmacobiol Dyn 8:393–400, 1985.
43. Colville-Nash, PR, Alam, CAS, Appleton, I, Browne, JR, Seed, MP, Willoughby, DA. The pharmacological modulation of angiogenesis in chronic granulomatous inflammation. J Pharmacol Exp Ther 274:1463–1472, 1995.
44. Appleton, I, Tomlinson, A, Colville-Nash, PR, Willoughby, DA. Temporal and spatial immunolocalization of cytokines in murine chronic granulomatous tissue. Lab Invest 69:405–414, 1993.
45. Appleton, I, Brown, NJ, Willis, D, Colville-Nash, PR, Alam, C, Brown, JR, Willoughby, DA. The role of vascular endothelial growth factor in a murine chronic granulomatous tissue air pouch model of angiogenesis. J Pathol 180:90–94, 1996.
46. Andrade, SP, Fan, TP, Lewis, GP. Quantitative in-vivo studies on angiogenesis in a rat sponge model. Br J Exp Pathol 68:755–766, 1987.
47. Hu, DE, Hiley, CR, Smither, RL, Gresham, GA, Fan, TP. Correlation of 133Xe clearance, blood flow and histology in the rat sponge model for angiogenesis: further studies with angiogenic modifiers. Lab Invest 72:601–610, 1995.
48. Hu, DE, Hori, Y, Presta, M, Gresham, GA, Fan, TP. Inhibition of angiogenesis in rats by IL-1 receptor antagonist and selected cytokine antibodies. Inflammation 18:45–58, 1994.
49. Hu, DE, Hiley, CR, Fan, TP. Comparative studies of the angiogenic activity of vasoactive intestinal peptide, endothelins-1 and -3 and angiotensin II in a rat sponge model. Br J Pharmacol 117:545–551, 1996.
50. Haralabopoulos, GC, Grant, DS, Kleinman, HK, Maragoudakis, ME. Thrombin promotes endothelial cell alignment in Matrigel in vitro and angiogenesis in vivo. Am J Physiol 273: C239–C245, 1997.
51. Angiolillo, AL, Kanegane, H, Sgadari, C, Reaman, GH, Tosato, G. Interleukin-15 promotes angiogenesis in vivo. Biochem Biophys Res Commun 233:231–237, 1997.
52. Ito, Y, Iwamoto, Y, Tanaka, K, Okuyama, K, Sugioka, Y. A quantitative assay using basement membrane extracts to study tumor angiogenesis in vivo. Int J Cancer 67:148–152, 1996.
53. Plunkett, ML, Hailey, JA. An in vivo quantitative angiogenesis model using tumor cells entrapped in alginate. Lab Invest 62:510–517, 1990.
54. Hoffman, J, Schirner, M, Menrad, A, Schneider, MR. A highly sensitive model for quantification of in vivo tumor angiogenesis induced by alginate-encapsulated tumor cells. Cancer Res 57:3847–3851, 1997.
55. BenEzra, D. Neovasculogenic ability of prostaglandins, growth factors, and synthetic chemoattractants. Am J Ophthalmol 86:455–461, 1978.
56. BenEzra, D, Griffin, BW, Maftzir, G, Aharonov, O. Thrombospondin and in vivo angiogenesis induced by basic fibroblast growth factor or lipopolysaccharide. Invest Ophthalmol Vis. Sci 34:3601–3608, 1993.
57. Gaudric, A, N'guyen, T, Moenner, M, Glacet-Bernard, A, Barritault, D. Quantification of angiogenesis due to basic fibroblast growth factor in a modified rabbit corneal model. Ophthalm Res 24:181–188, 1992.
58. Wiedemann, P. Growth factors in retinal diseases: proliferative vitreoretinopathy, proliferative diabetic retinopathy and retinal degeneration. Surv Ophthalmol 36:373–384, 1992.

59. BenEzra, D, Maftzir, G. Antibodies to IL-1 and TNF-alpha but not to bFGF or VEGF inhibit angiogenesis. Invest Ophthalmol Vis Sci 37:4664, 1996.
60. Jackson, JR. Angiogenesis and inflammation. Pharmacol Rev Commun 8:81–86, 1996.
61. MP Seed, PR Colville-Nash, JR Jackson, JD Winkler. Angiogenesis and inflammation. In: T-P. Fan and R. Auerbach, eds. The New Angiotherapy. Humana Press. In press, 1999.
62. Colville-Nash, PR, Seed, MP, Willoughby, DA. Anti-rheumatic drugs and the development of vasculature in murine chronic granulomatous air pouches. Br J Pharmacol 107:421P–4210, 1992.
63. Colville-Nash, PR, El-Ghazaly, M, Willoughby, DA. The use of angiostatic steroids to inhibit cartilage destruction in an in vivo model of granuloma mediated cartilidge destruction. Agents Actions 38:126–134, 1993.
64. Crum, R, Szabo, S, Folkman, J. A new class of steroids inhibits angiogenesis in the presense of heparin or heparin fragments. Science 230:1375–1378, 1985.
65. Nugent, RA, Schlachter, ST, Murphy, M, Dunn, CJ, Staite, ND, Galinet, LA, Shields, SK, Wu, H, Aspar, DG, Richard, KA. Carbonyl-containing bisphosphonate esters as novel anti-inflammatory and antiarthritic agents. J Med Chem 37:4449–4454, 1994.
66. Alam, CAS, Seed, MP, Willoughby, DA. Angiostasis and vascular regression in chronic granulomatous inflammation induced by diclofenac in combination with hyaluronan in mice. J Pharm Pharmacol 47:407–411, 1995.
67. Hudson, N, Balsitis, M, Everitt, S, Hawkey, CJ. Angiogenesis in gastric ulcers: impaired in patients taking non-steroidal anti-inflammatory drugs. Gut 37:191–194, 1995.
68. Jackson, JR, Bolognese, B, Hubbard, WC, Marshall, LA, Winkler, JD. The role of platelet-activating factor and other lipid mediators in inflammatory angiogenesis. Biochim Biophys Acta 1392:145–152, 1998.
69. Jackson, JR, Bolognese, B, Hillegass, L, Kassis, S, Adams, J, Griswold, DE, Winkler, JD. Pharmacological effects of SB 220025, a selective inhibitor of p38 mitogen-activated protein kinase, in angiogenesis and chronic inflammatory disease models. J Pharmacol Exp Ther 284: 687–692, 1998.
70. Bottomley, KM, Griffiths, RJ, Rising, TJ, Steward, A. A modified mouse air pouch model for evaluating the effects of compounds on granuloma induced cartilage degradation. Br J Pharmacol 93:627–635, 1988.
71. Koch, AE, Cho, M, Burrows, J, Leibovich, SJ, Polverini, PJ. Inhibition of production of macrophage-derived angiogeneic activity by the anti-rheumatic agents gold sodium thiomalate and auranofin. Biochem Biophys Res Comm 154:205–212, 1988.
72. Hirata, S, Matsubara, T, Saura, R, Tateishi, H, Hirohata, K. Inhibition of in vitro vascular endothelial cell proliferation and in vivo neovascularization by low-dose methotrexate. Arthritis Rheum 32:1065–1073, 1989.
73. Matsubara, T, Saura, R, Hirohata, K, Ziff, M. Inhibition of human endothelial cell proliferation in vitro and neovascularization in vivo by D-penicillamine. J Clin Invest 83:158–167, 1989.
74. Galardy, RE, Grobelny, D, Foellmer, HG, Fernandez, LA. Inhibition of angiogenesis by the matrix metalloprotease inhibitor N-[2R-2-(hydroxamidocarbonymethyl)-4-methylpentanoyl)]-L-tryptophan methylamide. Cancer Res 54:4715–4718, 1994.
75. Taraboletti, G, Garofalo, A, Belotti, D, Drudis, T, Borsotti, P, Scanziani, E, Brown, PD, Giavazzi, R. Inhibition of angiogenesis and murine hemangioma growth by batimastat, a synthetic inhibitor of matrix metalloproteinases. J Natl Cancer Inst 87:293–298, 1995.
76. DiMartino, MJ, High, W, Galloway, WA, Crimmin, MJ. Preclinical antiarthritic activity of matrix metalloproteinase inhibitors. Ann NY Acad Sci 732:411–413, 1994.
77. Jackson, JR, Bolognese, B, Kircher, CH, Marshall, LA, Winkler, JD. Modulation of angiogenesis in a model of chronic inflammation. Inflamm Res 46:S129–S130, 1997.

28

Angiogenic Mediators in Healing Wounds

Nicholas N. Nissen
University of California, Los Angeles, Los Angeles, California

Luisa A. DiPietro
Loyola University Medical Center, Maywood, Illinois

I. INTRODUCTION

Angiogenesis during wound repair serves the dual function of providing nutrients to healing tissues and contributing to the structural repair through the formation of granulation tissue. Wound angiogenesis follows an orderly and carefully regulated pattern. During the proliferative phase of wound healing, capillary growth occurs until the capillary density reaches nearly twice that of uninjured normal tissue. As wound healing resolves, these capillaries regress, leaving behind a residual vascularity that is similar to that of uninjured tissue. This pattern of carefully regulated growth and regression of vessels occurs in adult mammals in only a few physiological circumstances, including healing wounds, follicular development in the ovary, and lactation in the female breast.

Angiogenesis in wounds provides an ideal model for the study of the regulation of physiological angiogenesis. Neovascularization within wounds, as in all tissues, depends upon many factors, including cell-to-cell interaction, cell-to–extracellular matrix interaction, the activity of proteases, and the activity of soluble mediators. This chapter centers on studies that have tried to determine the role of soluble angiogenic mediators in healing wounds.

Numerous models have been utilized to study the angiogenic process during wound repair, including tissue culture systems, animal models, and human surgical wounds. This chapter focuses on the three most widely reported models: the subcutaneous wound chamber, rodent excisional skin wounds, and human wound fluid analysis. Each of these models suffers from an inability to reproduce the human wound environment or to sample it over the entire course of healing, but jointly they provide a great deal of information about the proangiogenic mediators during the wound repair process.

II. ANGIOGENIC MEDIATORS IN WOUND CHAMBERS

Schilling first reported the technique of sampling the fluid from subcutaneous chambers in an effort to define the environment of healing wounds (1). Since this original report,

Table 1 Angiogenic Mediators Produced by Wound Macrophages

Tumor necrosis factor alpha (TNF-α)
Platelet-derived growth factor (PDGF)
Transforming growth factor alpha (TGF-α)
Transforming growth factor beta (TGF-β)
Insulinlike growth factor I (IGF-I)
Vascular endothelial cell growth factor (VEGF)

multiple chamber types have been utilized in various species. These chambers allow sampling of the soluble and cellular components of wound repair at various time points. Limitations of the wound chamber include the potential to stimulate a mild foreign-body reaction and to prolong the proliferative phase of wound repair by creating a large dead space. In addition, the subcutaneous space may not be reflective of many clinically relevant wounds.

The greatest usefulness of the wound chamber technique has been in elucidating the role of the macrophage in wound angiogenesis. Following the classic report by Leibovich and Ross that identified macrophages as critical to wound repair (2), Hunt and coworkers undertook the study of the role the macrophage using a subcutaneous wire mesh cylinder in a rabbit model. In a series of reports, they demonstrated that cellular isolates from wound chambers stimulated angiogenesis in vivo and that this activity peaks 21 days after implantation, at a time when macrophages are maximal (3–5). They also demonstrated that cell-free supernatant from wound chambers, also obtained at 21 days, stimulates endothelial cell migration and proliferation as well as in vivo angiogenesis (5–7). These results imply that after an initial delay, macrophages become the predominate cellular mediators of angiogenesis in this wound model. Macrophages, in turn, function by producing soluble factors capable of mediating all phases of angiogenesis. Numerous macrophage-derived angiogenesis factors have been isolated, and many of these are produced by wound macrophages (Table 1) (8–10). The functional significance of these many potential mediators for wound chamber angiogenesis has not been reported (11).

III. ANGIOGENIC MEDIATORS IN THE RODENT SKIN MODEL

The mouse skin wound model is an extremely popular model for studing wound repair because of its ease of reproduction and manipulation. The typical wound in this model is a full-thickness excisional wound that is left open to heal by a combination of contraction, granulation tissue formation, and reepithelialization. Each of these components of the repair process has been studied in attempts to identify the primary active mediators. The study of the angiogenic response in murine wounds has provided considerable evidence regarding the role of two important angiogenic mediators, VEGF and FGF-2.

VEGF, known originally as vascular permeability factor, is an endothelial cell–specific mitogen and chemoattractant with potent in vivo angiogenic activity (12,13). A role for VEGF in wound repair was first demonstrated by Brown and others, who observed that in rat and guinea pig skin wounds, vascular permeability peaked at 2–3 days after injury and lasted as long as 7 days. This activity coincided with VEGF production by keratinocytes (14). A study by Frank et al. similarly demonstrated a pattern of VEGF mRNA expression in mouse skin wounds that peaked between 3 and 7 days after injury, at the time of maximal granulation tissue formation. Interestingly, this study also reported that genetically diabetic mice showed much lower levels of VEGF mRNA expression than control mice, suggesting that deficient production of VEGF may lead to impaired healing in these animals (15).

Further evidence that VEGF mediates angiogenesis in murine wounds was provided by a report examining the temporal regulation of one of the VEGF receptors known as VEGFR-1, or Flt-1. Peters et al. demonstrated in a mouse punch wound model that VEGFR-1 is initially upregulated in ectatic vessels at the wound margin and later in the vessels coursing through newly formed granulation tissue (16). The localization and temporal regulation of both VEGF and VEGF-receptor production in mouse wounds supports a role for VEGF as a mediator of wound angiogenesis.

Fibroblast growth factor 2 (FGF-2, or basic fibroblast growth factor, bFGF) is a heparin-binding protein that, like VEGF, stimulates endothelial cell activation and induces angiogenesis in vivo (17). Unlike VEGF, the receptors for FGF-2 are not specific for endothelial cells but are distributed on numerous cell types involved in wound healing (18). For unknown reasons, FGF-2 is sequestered in most normal tissues (19,20). The broad distribution of both FGF-2 and the FGF receptors has led to speculation that FGF-2 is important to tissue repair.

The temporal pattern of FGF-2 production in mouse wounds was addressed in a study by Werner et al. Using a mouse excisional punch wound model, these authors reported that FGF-2 mRNA increases by only fourfold during the first week of injury. This was compared to a 150-fold increase in keratinocyte growth factor (21). This relatively small increase in FGF-2 production is unlikely to affect the soluble wound environment, as FGF-2 lacks a secretory signal sequence and is not known to be actively secreted by cells (22). Werner et al. went on to show that the production of FGF-R1, a receptor for FGF-2, was not increased during wound healing over levels found in normal skin. These findings provide evidence against a prolonged FGF-2–mediated angiogenic stimulus in wounds and suggest that the active production or secretion of FGF-2 after injury is unlikely to contribute to wound angiogenesis. Increased FGF-2 protein has, however, been demonstrated in mouse wounds by immunohistochemical staining (10,23). The source of this FGF-2 has been hypothesized to be intra- or extracellular tissue stores released at the time of injury. In support of this, our laboratory has examined soluble FGF-2 levels in murine excisional wound exudates. Although high levels of soluble FGF-2 are present immediately after skin injury, the soluble FGF-2 level rapidly decreases to the point of being nearly undetectable within 48 hr. (Nissen and DiPietro, unpublished). Taken together, these studies suggest that FGF-2 may serve to initiate angiogenesis in murine wounds but is unlikely to have a regulatory role beyond the first few days.

IV. ANGIOGENIC MEDIATORS IN HUMAN WOUNDS

The study of human wounds is more difficult than that of animal models because of the obvious limitations in long-term sampling, obtaining tissue biopsies, ensuring wound uniformity, and manipulating the wound environment. Most reports on the angiogenic environment of human wounds are derived from studies on wound exudate. For example, several studies have demonstrated that surgical wound and skin-graft donor-site exudates stimulate endothelial cell proliferation and migration (5,24) Another study reported the presence of several cytokines in skin-graft donor fluid, suggesting that some of these might contribute to wound repair (25).

Recent studies in our laboratory have sought to characterize the angiogenic profile of human surgical wounds and to identify the soluble proangiogenic mediators that are critical to wound healing. These studies examined wound exudate collected from mastectomies, neck dissections, and myocutaneous flap procedures. These wound types each involve dissection of skin, fascial, and muscle components, and are routinely drained postoperatively. Wound drain samples were obtained from the immediate postoperative period (within 6 hr of surgery) to as much as 7 days postoperatively (26,27).

In order to determine the timing of angiogenic activity of human wound fluid, samples collected from surgical drains were tested for the ability to (a) stimulate endothelial cell proliferation, (b) stimulate endothelial cell chemotaxis, and (c) induce angiogenesis in vivo in the rat corneal pocket assay (Table 2). The endothelial cell proliferative activity of surgical wound fluids peaked immediately after surgery and then fell to near serum levels within 2 days. In contrast, endothelial cell chemotactic activity was present in wound fluid at all time points sampled from immediately after surgery through postoperative day 6. In vivo angiogenic activity correlated well with chemotactic activity, as samples from immediately postoperative through postoperative day 6 stimulated noninflammatory neovascularization in the rat cornea assay. Endothelial cell proliferative activity therefore appears to peak early in human surgical wounds, while endothelial cell chemotactic activity and a net positive angiogenic stimulus are maintained for at least 6 days.

Two of the potential mediators of the angiogenic activity in surgical wound fluid are VEGF and FGF-2. As discussed previously, these factors are potent angiogenic agonists and have been identified in animal wound models. When measured in surgical wound fluids, FGF-2 and VEGF demonstrated distinct patterns. FGF-2 levels were highest immediately after surgery and then fell rapidly to near-serum levels by postoperative day 2. In contrast, VEGF levels increased slowly over the first few days and eventually peaked on the seventh postoperative day. Neutralizing antibody experiments confirmed that endothelial cell proliferation, chemotaxis, and in vivo angiogenesis induced by early wound samples was primarily due to FGF-2 (26). In contrast, antibody neutralization studies demonstrated that VEGF is the primary mediator of endothelial cell chemotaxis and in vivo angiogenic activity during postoperative days 3 through 6 (27). These studies suggest that FGF-2 provides an immediate angiogenic stimulus in surgical wounds that is subsequently sustained by VEGF. Recent studies in a pig model of omental repair support

Table 2 Effect of Human Wound Fluids on Endothelial Cell Proliferation, Endothelial Cell Chemotaxis, and In Vivo Angiogenesis[a]

Postoperative day	Endothelial cell proliferation	Endothelial cell chemotaxis	Angiogenesis
0	+++	++	++
1	++	+++	+++
3	−	+++	+++
6	−	+++	+++

[a] The effect of human wound fluid on endothelial cell proliferation, chemotaxis, and in vivo angiogenesis in the rat cornea is shown. +++ strongly promotes, ++ moderately promotes, + weakly promotes, − no activity.

this concept, as they demonstrate that VEGF is a critical proangiogenic mediator in this system as well (28).

The identification of the primary mediators of angiogenesis during skin wound healing unlike that of the surgical wound, has not been reported. Information derived from rodent skin wound models is difficult to interpret because of the structural and functional differences between human and rodent skin. Porcine skin is a closer approximation to human skin and is being increasingly utilized as a model of both human burn and skin-graft donor-site wounds.

Despite the relative lack of information about angiogenesis in human skin wounds, it seems likely that VEGF assumes an important role during skin repair. Two lines of evidence support this reasoning. First, VEGF has been identified in conditions such as psoriasis and bullous diseases (29,30), demonstrating the potential for VEGF-mediated activity in the dermis. Second, cutaneous elements such as keratinocytes, dermal microvessels, dermal fibroblasts, and skin explants produce VEGF in response to injury, hypoxia, or cytokines such as TGF-α (31,32). This paracrine regulation of VEGF by cutaneous elements remains to be substantiated in a porcine or human model.

V. A MODEL OF WOUND ANGIOGENESIS

A large body of evidence has accumulated implicating FGF-2 and VEGF as important mediators of wound angiogenesis. The temporal patterns of FGF-2 and VEGF-mediated angiogenic responses in wounds suggest a model in which these two mediators are key components of an angiogenic cascade (Fig. 1). FGF-2 is well suited to initiate angiogenesis because it is immediately available after injury and affects numerous cell types. Preformed

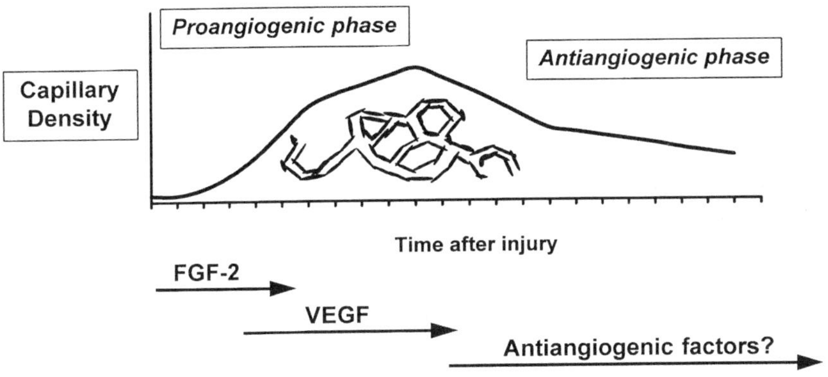

Figure 1 A model of the regulation of wound angiogenesis. Tissue injury causes the immediate release of FGF-2, providing a nearly instantaneous angiogenic stimulus. This stimulus can initiate cell proliferation and collagen synthesis and may serve as an antiapoptotic signal for endothelial cells. Within a day or two of injury, cells within the wound, such as macrophages, begin to synthesize proangiogenic factors such as VEGF. This second wave of positive angiogenic factors induces endothelial cell chemotaxis into the wound bed and promotes capillary formation. As wound healing proceeds, the production of these proangiogenic mediators abates, granulation tissue formation is completed, and tissue hypoxia diminishes. The resolution of wound repair, with its attendant regression of capillaries, probably depends upon the production of specific but as yet unidentified antiangiogenic signals.

FGF-2 that is sequestered in uninjured tissue is likely released at the time of injury either by cell damage, enzymatic degradation of the extracellular matrix, or platelet degranulation (33–37). This release, in turn, accounts for the rapid rise in soluble FGF-2 demonstrated in both human and murine wounds. Because multiple cell types carry FGF receptors, soluble FGF-2 stimulates multiple early wound events, including the proliferation of both endothelial cells and fibroblasts. Other potential effects of early wound FGF-2 include stimulating the production of VEGF and acting synergistically with VEGF to maximize new vessel growth (38–40). An interesting study by Broadley et al. demonstrated that loss of the early FGF-2–mediated stimulus diminishes early granulation tissue formation in a rat model (41). More recent studies in mice rendered genetically deficient for FGF-2 confirm the importance of FGF-2 in wound healing. FGF-2 −/− mice exhibit delayed wound healing, including impaired collagen synthesis and reepithelialization (42). The release of preformed FGF-2 therefore appears to initiate angiogenesis and collagen deposition prior to the synthesis of additional mediators such as VEGF.

By the time the FGF-2-mediated stimulus in wounds declines, VEGF-mediated angiogenic activity is well under way. Both the human and animal studies cited above demonstrate a pattern in which VEGF-mediated activity appears in the wound after several days and persists for at least 1–2 weeks. Unlike the broadly distributed FGF receptors, VEGF-receptors are with a few exceptions exclusive to endothelium (43). Transition to a predominantly VEGF-mediated angiogenic stimulus might therefore benefit wound repair by allowing more specific regulation of vessel growth and regression during the proliferative and maturation phases of wound repair (Fig. 1). Whether or not VEGF persists as the primary mediator of angiogenesis throughout the entire proangiogenic phase of tissue repair remains to be determined.

The primary source of VEGF in a given wound probably varies according to wound type. In the mouse skin wound, the primary source of VEGF is the keratinocyte (14). Several dermal cell types such as endothelial cells and fibroblasts may also contribute to VEGF production in human skin (31). Because of the relatively small degree of skin injury in most surgical wounds, keratinocytes are unlikely to produce much VEGF in these wounds. Instead, VEGF production in these wounds appears to be largely mediated by fibroblasts and macrophages (27).

Several mechanisms regulating the production of VEGF in wounds have been proposed. Cytokines such as TGF-α, TGF-β, and PDGF stimulate the production of VEGF in vitro and may function similarly in wounds (32,44,45). Wound hypoxia has also been proposed as a potential stimulant for VEGF production. Extensive work by Knighton, Hunt, and others has demonstrated that wounds are characteristically hypoxic during the early phases of wound repair and that angiogenic activity in wounds varies inversely with wound oxygen levels (46,47) Hypoxia stimulates upregulation of VEGF and VEGF-receptor production in vitro and may function similarly in the early wound (48). Recent reports have also implicated the redox state of the wound in modulating VEGF production and bioactivity (49). Finally, there is increasing evidence that nitric oxide metabolites are involved in VEGF-mediated angiogenic activity (50). To date, none of these factors has been firmly established in regulating VEGF-mediated angiogenesis in wounds in vivo.

VI. CONCLUSIONS AND UNANSWERED QUESTIONS

We and others have proposed that angiogenesis in acute wounds during the inflammatory and proliferative phases of healing is mediated largely by FGF-2 and VEGF. There remain, however, many unanswered questions regarding the generation and activity of soluble positive angiogenic factors in wounds. First, the source of FGF-2 in the early wound is not yet defined. Certainly FGF-2 may be released from damaged and dying cells within the wound bed itself. However, the total FGF-2 content of normal tissue is substantially less than that found in a similar amount of injured tissue (Nissen and DiPietro, unpublished). This suggests that FGF-2 is delivered to the wound shortly after injury by serum, platelets, or regional tissue pools. A serum source of FGF-2 seems unlikely in light of the low levels of FGF-2 measured in normal or surgical patient serum (26). Platelets have been shown to contain small amounts of FGF-2 and thus might contribute to the FGF-2 found in the early wound (37). Mast cells have also been reported to contain FGF-2, and may add to the initial pool as they degranulate after injury (51). Perhaps the most likely scenario is that FGF-2 is mobilized from intracellular and extracellular matrix and endothelial cell surface pools within the surrounding tissues.

Another question that arises from these studies is the relative importance of soluble versus bound FGF-2 in regulating wound angiogenesis. While the amount of FGF-2 in the fluid phase of the wound drops off precipitously after the first 24 hr, the level of FGF-2 that remains bound in the extracellular matrix or to cell surfaces within the wound bed is not yet known. Healing wounds that are subjected to immunostaining for FGF-2 show that substantial FGF-2 remains present in the wound bed long after the observed drop in soluble FGF-2 (23,52). The function of extracellular FGF-2 in the ongoing regulation of wound angiogenesis is not clear. In vitro studies demonstrate that FGF-2 bound to the extracellular matrix can regulate endothelial cell chemotaxis, proliferation, and differentiation (53,54). This raises the possibility that FGF-2 bound in the wound matrix may continue to influence vessel growth long after soluble FGF-2 levels fall.

A third set of questions that is largely unstudied relates to the mechanisms of downregulation of the angiogenic stimulus in wounds. As mentioned earlier, the healing wound provides an extraordinary model of physiological vascular regression. To date, the mechanism that regulates the controlled regression of capillaries in wounds is unknown. Recent studies of resolving granulation tissue in a sponge model suggest that physiological capillary regression begins with apoptosis of an individual cell (55). Adjacent endothelial cells, apparently recognizing the apoptotic cell, undergo contraction, which leads to capillary closure. Once the capillary is closed, the endothelial cells in the closed vascular branch die by a combination of apoptosis and necrosis.

The specific signals in healing wounds that mediate capillary regression are not known. Do levels of proangiogenic mediators simply decline once a cytokine stimulus fades or wound hypoxia resolves? Are antiangiogenic molecules actively synthesized in wounds? What role does restoration of the extracellular matrix play in downregulating angiogenesis? Answering these questions in the scenario of wound healing might provide insight into the regulation of pathological angiogenesis, such as that associated with chronic inflammatory conditions. Recent studies provide hints that the regulation of wound vascularization may also involve the modulation of other regulatory pathways, such as vascular integrins or the angiopoietins (56–59). Additional studies will no doubt define the mechanism of physiological vascular regression in resolving wounds.

Perhaps the most difficult dilemma in the study of angiogenesis in human wounds is determining how the angiogenic process goes awry in conditions of deficient wound healing. The best-known clinical example of this is diabetes mellitus, which predisposes patients to delayed wound healing and wound complications. Studies in genetically diabetic mice demonstrate decreased granulation tissue formation and delayed closure of skin wounds (23,60). Production of both FGF-2 and VEGF is decreased in these animals compared to controls (15,60). Whether similar deficiencies are present in diabetic patients remains to be determined. Our understanding of the angiogenic response in venous ulcers is also poor. Alterations in protease environment or the sequestration of growth factors by macromolecules have both been suggested to be operative in these wounds (61). The challenge for the future is to more completely understand the mechanisms that regulate angiogenesis in healing wounds. A better understanding of the regulation of the growth-factor response in normal wounds should lead to more suitable scientific and clinical approaches to the problem of impaired wound healing.

ACKNOWLEDGMENTS

This work was supported by the Dr. Ralph and Marion C. Falk Foundation and the National Institutes of Health (GM50875).

REFERENCES

1. JA Schilling, BV Favata, M Radakovich. Studies of fibroplasia in wound healing. Surg Gynecol Obstet 96:143–149, 1953.
2. SJ Leibovich, R Ross. The role of the macrophage in wound repair: a study with hydrocortisone and antimacrophage serum. Am J Pathol 78:71–91, 1975.
3. RA Clark, RD Stone, DYK Leung, I Silver, DC Hohn, TK Hunt. Role of macrophages in wound healing. Surg Forum 27:16–18, 1976.
4. TK Hunt, DR Knighton, KK Thakral, WH Goodson III, WS Andrews. Studies on inflammation and wound healing: angiogenesis and collegen synthesis stimulated in vivo by resident and activated wound macrophages. Surgery 96:48–54, 1984.
5. G Greenburg, TK Hunt. The proliferative response in vitro of vascular endothelial and smooth muscle cells exposed to wound fluids and macrophages. J Cell Physiol 97:353–360, 1978.
6. SU Orredson, DR Knighton, H Scheuenstuhl, TK Hunt. A quantitative in vitro study of fibroblast and endothelial cell migration in response to serum and wound fluid. J Surg Res 35: 249–258, 1983.
7. MJ Banda, DR Knighton, TK Hunt, Z Werb. Isolation of a nonmitogenic angiogenesis factor from wound fluid. Proc Natl Acad Sci USA 79:7773–7777, 1982.
8. LA DiPietro. Wound healing: the role of macrophages and other immune cells. Shock 4:233–240, 1995.
9. DA Rappolee, D Mark, MJ Banda, Z Werb. Wound macrophages express TGF-α and other growth factors in vivo: analysis by mRNA phenotyping. Science 241:708–712, 1988.
10. I Appleton, A Tomlinson, PR Colville-Nash, DA Willoughby. Temporal and spatial immunolocalization of cytokines in murine chronic granulomatous tissue. Lab Invest 69:405–414, 1993.
11. C Sunderkotter, M Goebeler, K Schulze-Osthoff, R Bhardwaj, C Sorg. Macrophage-derived angiogenesis factors. Pharmacol Ther 51:195–216, 1991.
12. HF Dvorak, LF Brown, M Detmar, AM Dvorak. Vascular permeability factor/vascular endothelial growth factor, microvascular hyperpermeability, and angiogenesis. Am J Pathol 146: 1029–1039, 1995.
13. N Ferrara, K Houck, L Jakeman, D Leung. Molecular and biological properties of the vascular endothelial growth factor family of proteins. Endocr Rev 13:18–32, 1992.
14. LF Brown, Y Kiang-Teck, B Berse, T Yeo, DR Senger, HF Dvorak, L Van De Water. Expres-

sion of vascular permeability factor (vascular endothelial growth factor) by epidermal keratinocytes during wound healing. J Exp Med 176:1375–1379, 1992.
15. S Frank, G Hubner, G Breie, MR Longaker, DG Greenhalgh, S Werner. Regulation of vascular endothelial growth factor expression in cultured keratinocytes. J Biol Chem 270:12607–12613, 1995.
16. KG Peters, C DeVries, LT Williams. Vascular endothelial growth factor receptor expression during embryogenesis and tissue repair suggests a role in endothelial differentiation and blood vessel growth. Proc Natl Acad Sci USA 90:8915–8919, 1993.
17. J Folkman, M Klagsbrun. Angiogenesis factors. Science 235:442–447, 1987.
18. JA Abraham, M Klagsbrun. Modulation of wound repair by members of the fibroblast growth factor family. In: RAF Clark, ed. Molecular and Cellular Biology of Wound Repair. New York: Plenum Press, 1996.
19. SE Hughes, PA Hall. Immunolocalization of fibroblast growth factor receptor 1 and its ligands in human tissue. Lab Invest 69:173–182, 1993.
20. C Cordon-Cardo, I Vlodavsky, A Haimovitz-Friedman, AD Hicklin, Z Fuks. Expression of basic fibroblast growth factors in normal human tissues. Lab Invest 63:832–840, 1990.
21. S Werner, KG Peters, MT Longaker, F Fuller-Pace, MJ Banda, LT Williams. Large induction of keratinocyte growth factor expression in the dermis during wound healing. Proc Natl Acad Sci USA 89:6896–6900, 1992.
22. JA Abraham, A Mergia, JL Whang, K Hjerrild, D Gospodarowicz, JC Fiddes. Nucleotide sequence of a bovine clone encoding the angiogenic protein basic fibroblast growth factor. Science 233:545–548, 1986.
23. Y Kurita, R Tsuboi, R Ucki, DB Rifkin, H Ogawa. Immunohistochemical localization of basic fibroblast growth factor in wound healing sites of mouse skin. Arch Dermatol Res 284:193–197, 1992.
24. MH Katz, AF Alvarez, RS Kirsner, WH Eaglstein, V Falanga. Human wound fluid from acute wounds stimulates fibroblast and endothelial cell growth. J Am Acad Dermatol 25:1054–1058, 1991.
25. LS Grayson, JF Hansborough, RL Zapata-Sirvent, CA Dore, JL Morgan, MA Nicolson. Quantitation of cytokine levels in skin graft donor site wound fluid. Burns 19:401–405, 1993.
26. NN Nissen, RL Gamelli, PJ Polverini, LA DiPietro. Basic fibroblast growth factor mediates angiogenic activity in early surgical wounds. Surgery 119:457–465, 1996.
27. NN Nissen, PJ Polverini, AE Koch, MV Volin, RL Gamelli, LA DiPietro. Vascular endothelial growth factor mediates angiogenic activity during the proliferative phase of wound healing. Am J Pathol 152:1445–1452, 1998.
28. VK Gupta, PL McNeil, C Riegner, TR Howdieshell. Vascular endothelial growth factor is a key mediator of omental angiogenesis. Surg Forum 47:746–749, 1996.
29. M Detmar, LF Brown, YP Claffey, KT Yeo, O Kocher, RW Jackman, B Berse, HF Dvorak. Overexpression of vascular permeability factor/vascular endothelial growth factor and its receptors in psoriasis. J Exp Med 180:1141–1146, 1994.
30. LF Brown, TJ Harrist, KT Yeo, M Stahle-Backdahl, RW Jackman, B Berse, K Tognazzi, HF Dvorak, M Detmar. Increased expression of vascular permeability factor (vascular endothelial growth factor) in bullous pemphigoid, dermatitis herpetiformis, and erythema multiforme. J Invest Dermatol 104:744–749, 1995.
31. M Detmar, LF Brown, B Berse, RW Jackman, BM Elicker, HF Dvorak, KP Claffey. Hypoxia regulates the expression of vascular permeability factor/vascular endothelial growth factor (VPF/VEGF) and its receptors in human skin. J Invest Dermatol 108:263–268, 1997.
32. M Detmar, KT Yeo, JA Nagy, L Van de Water, LF Brown, B Berse, BM Elicker, S Ledbetter, HF Dvorak. Keratinocyte-derived vascular permeability factor (vascular endothelial growth factor) is a potent mitogen for dermal microvascular endothelial cells. J Invest Dermatol 105: 44–50, 1995.

33. CM Gajdusek, S Carbon. Injury-induced release of basic fibroblast growth factor from bovine aortic endothelium. J Cell Physiol 139:570–579, 1989.
34. L Muthukrishnan, E Warder, P McNeil. Basic fibroblast growth factor is efficiently released from a cytosolic storage site through plasma membrane disruptions of endothelial cells. J Cell Physiol 148:1–16, 1991.
35. M BenEzra, I Vlodavsky, R Ishai-Michaeli, G Neufeld, R Bar-Shavit. Thrombin-induced release of active basic fibroblast growth factor-heparan sulfate complexes from subendothelial extracellular matrix. Blood 81:3324–3331, 1993.
36. P Bashkin, S Doctrow, M Klagsbrun, CM Svahn, J Folkman, I Vlodavsky. Basic fibroblast growth factor binds to subendothelial extracellular matrix and is released by heparitinase and heparin-like molecules. Biochemistry 28:1737–1743, 1989.
37. G Brunner, H. Nguyen, J Gabrilove, DB Rifkin, EL Wilson. Basic fibroblast growth factor expression in human bone marrow and peripheral blood cells. Blood 81:631–638, 1993.
38. GT Stavri, IC Zachary, PA Baskerville, JF Martin, JD Erusalimsky. Basic fibroblast growth factor upregulates the expression of vascular endothelial growth factor in vascular smooth muscle cells: synergistic interaction with hypoxia. Circulation 92:11–14, 1995.
39. P Koolwijk, MG van Erck, WJ de Vree, MA Vermeer, HA Weich, R Hanemaaijer, V van Hinsbergh. Cooperative effects of TNF-α, bFGF, and VEGF on the formation of tubular structure of human microvascular endothelial cells in a fibrin matrix: role of urokinase activity. J Cell Biol 132:1177–1188, 1996.
40. T Asahara, C Bauters, LP Zheng, S Takeshita, S Bunting, N Ferrara, JF Symes, JM Isner. Synergistic effects of vascular endothelial growth factor and basic fibroblast growth factor on angiogenesis in vivo. Circulation 92:11365–11371, 1995.
41. K Broadley, A Aquino, S Woodward, A Buckley-Sturrock, Y Sato, DB Rifkin, JM Davidson. Monospecific antibodies implicate basic fibroblast growth factor in normal wound repair. Lab Invest 61:571–575, 1989.
42. S Ortega, M Ittmann, SH Tsang, M Erlich, C Basilico. Neuronal defects and delayed wound healing in mice lacking fibroblast growth factor 2. Proc Natl Acad Sci USA 95:5672–5677, 1998.
43. T Mustonen, K Alitalo. Endothelial receptor tyrosine kinases involved in angiogenesis. J Cell Biol 129:895–898, 1995.
44. L Pertovaara, A Kaipainen, T Mustonen, A Orpana, N Ferrara, O Saksela, K Alitalo. Vascular endothelial growth factor is induced in response to transforming growth factor-beta in fibroblastic and endothelial cells. J Biol Chem 269:6271–6274, 1994.
45. GJ Dolecki, DT Connolly. Effects of a variety of cytokines and inducing agents on vascular permeability factor mRNA levels in U937 cells. Biochem Biophys Res Commun 180:572–578, 1991.
46. DR Knighton, IA Silver, TK Hunt. Regulation of wound-healing angiogenesis—Effect of oxygen gradients and inspired oxygen concentration. Surgery 90:262–270, 1981.
47. DR Knighton, TK Hunt, H Scheuenstuhl, BJ Halliday, Z Werb, MJ Banda. Oxygen tension regulates the expression of angiogenesis factor by macrophages. Science 221:1283–1285, 1983.
48. D Schweiki, A Itin, D Soffer, E Keshet. Vascular endothelial growth factor induced by hypoxia may mediate hypoxia-initiated angiogenesis. Nature 359:843–845, 1992.
49. DD Zabel, JJ Feng, J Scheuenstuhl, TK Hunt, MZ Hussain. Lactate stimulation of macrophage-derived angiogenic activity is associated with inhibition of poly(ADP-ribose) synthesis. Lab Invest 74:644–649, 1996.
50. A. Papapetropoulos, G Garcia-Cardena, JA Madri, WC Sessa. Nitric oxide contributes to the angiogenic properties of vascular endothelial growth factor in human endothelial cells. J Clin Invest 100:3131–3139, 1997.
51. Z Qu, JM Liebler, MR Powers, T Galey, P Ahmadi, XN Huang, JC Ansel, JH Butterfield, SR Planck, JT Rosenbaum. Mast cells are a major source of basic fibroblast growth

factor in chronic inflammation and cutaneous hemangioma. Am J Pathol 147:564–573, 1995.

52. NS Gibran, FF Isik, DM Heimbach, D Gordon. bFGF in the early human burn wound. J Surg Res 56:226–234, 1994.
53. S Rogelj, M Klagsbrun, R Atzmon, M Kurokawa, A Haimowitz, Z Fuks, I Vlodavsky. Basic fibroblast growth factor is an extracellular matrix component required for supporting the proliferation of vascular endothelial cells and the differentiation of PC12 cells. J Cell Biol 109: 823–831, 1989.
54. S Biro, ZX Yu, YM Fu, G Smale, J Sasse, J Sanchez, VJ Ferrans, W Casscells. Expression and subcellular distribution of basic fibroblast growth factor are regulated during migration of endothelial cells. Circ Res 74:485–494, 1994.
55. Honma, T., and Hamasaki, T. 1998. Ultrastructure of blood vessel regression in involution of foreign-body granuloma. J Submicrosc Cytol Pathol 30:31–44, 1998.
56. RAF Clark, MG Tonnesen, J Gailit, DA Cheresh. Transient functional expression of $\alpha v\beta 3$ on vascular cells during wound repair. Am J Pathol 148:1407–1421, 1996.
57. M Christofidou-Solomidou, M Bridges, GF Murphy, SM Albelda, HM DeLisser. Expression and function of endothelial cell αv integrin receptors in wound-induced human angiogenesis in human/SCID mice chimeras. Am J Pathol 151:975–983, 1997.
58. PC Maisonpierre, C Suri, PF Jones, S Bartunkova, SJ Weigand, C Radziejewski, DL Compton, J McClain, TH Aldrich, N Papdopoulos, TJ Daly, S Davis, TN Sato, GD Yancopoulos. Angiopoietin-2, a natural antagonist for Tie2 that disrupts in vivo angiogenesis. Science 277:55–60, 1997.
59. AL Wong, ZA Haroon, S Werner, MW Dewhirst, CS Greenberg, KG Peters. Tie2 expression and phosphorylation in angiogenic and quiescent adult tissues. Circ Res 81:567–574, 1997.
60. S Werner, M Breeden, G Hübner, DG Greenhalgh, MT Longaker. Induction of keratinocyte growth factor expression is reduced and delayed during wound healing in the genetically diabetic mouse. J Invest Dermatol 103:469–473, 1994.
61. V Falanga, WH Eaglstein. The ''trap'' hypothesis of venous ulceration. Lancet 341:1006–1008, 1993.

29
Angiogenesis in Rheumatoid Arthritis

Zoltan Szekanecz
University Medical School of Debrecen, Debrecen, Hungary

Alisa E. Koch
Northwestern University Medical School and Department of Veterans' Affairs, Chicago Healthcare System, Chicago, Illinois

I. INTRODUCTION

Angiogenesis is crucial in a number of physiological processes—such as reproduction, development, and tissue repair—as well as in pathological conditions including, among others, rheumatoid arthritis (RA) and other inflammatory diseases. The angiogenic process, its mediators and inhibitors, as well as the basic mechanisms, cellular and molecular interactions underlying neovascularization, are extensively discussed in a number of other chapters in this volume (also reviewed in Refs. 1–7).

In RA, leukocytes emigrate into the synovium through the vascular endothelium, resulting in synovial inflammation and, eventually, joint destruction. The RA synovial tissue is rich in newly formed vessels. The high turnover of capillary formation perpetuates the extravasation of inflammatory leukocytes and thus the progression of RA. Therefore, RA is considered an important candidate for an ''angiogenic disease'' (1–3, 8–10).

The neovascularization process and the outcome of ''angiogenic diseases,'' such as RA, are dependent on the balance or imbalance between angiogenic mediators and inhibitors. The suppression of neovascularization by blocking the action of angiogenic mediators, or by the administration of angiostatic compounds, may be useful in controlling the progression of RA (reviewed in Refs. 1–7). This chapter reviews those angiogenic mediators and inhibitors that have been shown to be involved in arthritis. In addition, a number of angiostatic compounds—such as angiostatin, endostatin, angiopoietin-2—that have been investigated for their therapeutic potential in various tumors have not yet been studied in context with antirheumatic therapy. The most relevant angiogenic and angiostatic factors that may play a role in RA are listed in Tables 1 and 2, respectively. Angiogenesis and its mediators and inhibitors are summarized in Fig. 1). The discussion of these factors is followed by a review on the most important interactions between endothelial cells, angiogenic mediators, and inhibitors. Present and future angiogenesis-targeting strategies that may be included in the therapy of RA are also discussed.

Table 1 Some Angiogenic Mediators in Rheumatoid Arthritis[a]

		Refs.		Refs.
Growth factors	bFGF	12–14	PD-ECGF/gliostatin	25
	aFGF	11	EGF	26
	VEGF	15,17	IGF-1	27,28
	HGF	19–20	TGF-β[b]	29–38
	PDGF	21–24		
Cytokines	TNF-α[b]	41,42,45,46	IL-15	39–56
	IL-1α, β[b]	47,48	G-CSF	57
	IL-6[b]	49–53	GM-CSF	57
	IL-13	54,55	Oncostatin M	58,59
Chemokines	IL-8	42,43,61–63	groβ	62,66
	ENA-78	61,64	CTAP-III	61,67
	groα	61,63,65		
ECM macromolecules	Type I collagen	1,68	Heparan sulfate	1,68
	Fibronectin	1,69	Laminin	1,68
	Heparin	1,68	Tenascin	70
Proteolytic enzymes	Matrix metalloproteinases	72,73	Plasminogen activators	75,76
Cell adhesion molecules	β1 integrins	77,86	P-selectin	92,93
	β3 integrins	77,87	CD31	83,94
	VCAM-1	79,90	endoglin	36,84,85
	E-selectin	79,82,89		
Other mediators	Angiogenin	14	Substance P	6,97
	PAF	46,95,96		

[a] See text for abbreviations. Selected references are included in table.
[b] Exhibits variable stimulatory and inhibitory activity (see text).

II. ANGIOGENIC MEDIATORS IN RHEUMATOID ARTHRITIS

Angiogenic mediators shown to play a role in the pathogenesis of arthritis include growth factors, cytokines, chemokines, extracellular matrix macromolecules, adhesion receptors, proteolytic enzymes, and other factors (Table 1, Fig. 1). Most of these mediators are released by endothelial cells and macrophages; cells also present in high quantities in the arthritic synovium (reviewed in Refs. 1–7). Angiogenesis and thus the perpetuation of synovial inflammation in RA may be blocked by suppressing the production of these mediators (1–7).

A. Heparin-Binding Growth Factors

Basic (bFGF) and acidic fibroblast growth factors (bFGF) are bound to heparin and heparan sulfate in the extracellular matrix. During the angiogenic response, these mediators are mobilized by endothelial-derived heparanase and plasmin (4–6). Both FGFs are expressed in the RA synovium in situ (11,12). Synovial fibroblasts in culture also express bFGF (13). We have detected bFGF in macrophages, lining cells, and fibroblasts in the RA synovial tissue (14).

Vascular endothelial growth factor (VEGF) is also bound to heparin (4–6). RA

Table 2 Angiostatic Compounds with Potential Relevance for Rheumatoid Arthritis[a]

		Refs.		Refs.
Growth factors	TGF-β[b]	29,38		
Cytokines	IL-1α, β[b]	47,48	IL-12	102,104
	IL-4	100,101	IFN-α, γ	29,37,47
	IL-6[b]	49,53	LIF	105–106
Chemokines	IP-10	60,62,107,108	PF4	60,62,109
Heparin-binding factors	Thrombospondin-1	110,111	PF4	60,62,109
Antirheumatic drugs	Gold salts	1,114,115	Dexamethasone	1,29
	Chloroquine	1,29	Methotrexate	1,113
	NSAIDs	1,112	Cyclophosphamide	1,29
	Azathioprine	1,29	Cyclosporine A	116
	D-penicillamine	1,29	Sulfasalazine	1,29
Angiostatic steroids	2-methoxyestradiol	117,118		
Protease inhibitors	TIMPs	73,104	PAIs	75,76
Antibiotics	Fumagillin-derivatives	116,119	Deoxy spergualin	125
	Minocycline	120–122	Rapamycin	123,124
	Clarithromycin	126,127		
Tissue-derived inhibitors	Cartilage-derived inhibitors	128		
Cytoskeleton disrupting agents	Taxol	119		
Other compounds	SPARC	129	Opioids	130,131

[a] See text for abbreviations. Selected references are included in table.
[b] Exhibits variable stimulatory and inhibitory activity (see text).

synovial fluids contain large amounts of antigenic VEGF (15,16). There is strong VEGF protein and mRNA expression in RA synovial lining cells, macrophages, fibroblasts, and smooth muscle cells (16). We and others have published data on the role of VEGF in RA-associated endothelial cell migration, proliferation, and chemotaxis (1,3,15). In addition, synovial vessels have fenestrated endothelia (17). It has recently been shown that VEGF induces endothelial fenestration in vitro (18).

Hepatocyte growth factor (HGF)/scatter factor is also a potent heparin-binding angiogenic mediator (19). We detected large amounts of HGF in RA synovial fluids. RA synovial tissue macrophages and lining cells express HGF. HGF expression on RA synovial fibroblasts could be induced by interleukin (IL)-1 (20).

B. Other Growth Factors

Other angiogenic growth factors in the RA synovium, which do not bind heparin, include platelet-derived (PDGF), platelet-derived endothelial cell (PD-ECGF), epidermal (EGF), insulin-like (IGF-I) and transforming growth factor (TGF)-β (1–3,15,21).

PDGF, although it is structurally related to VEGF, differs from VEGF in a number of ways. PDGF, unlike FGFs and VEGF, is likely to act indirectly on angiogenesis by inducing the production of other angiogenic factors including bFGF, VEGF, and type I collagen (21,22). PDGF has been found in RA synovial fluids and is expressed by RA

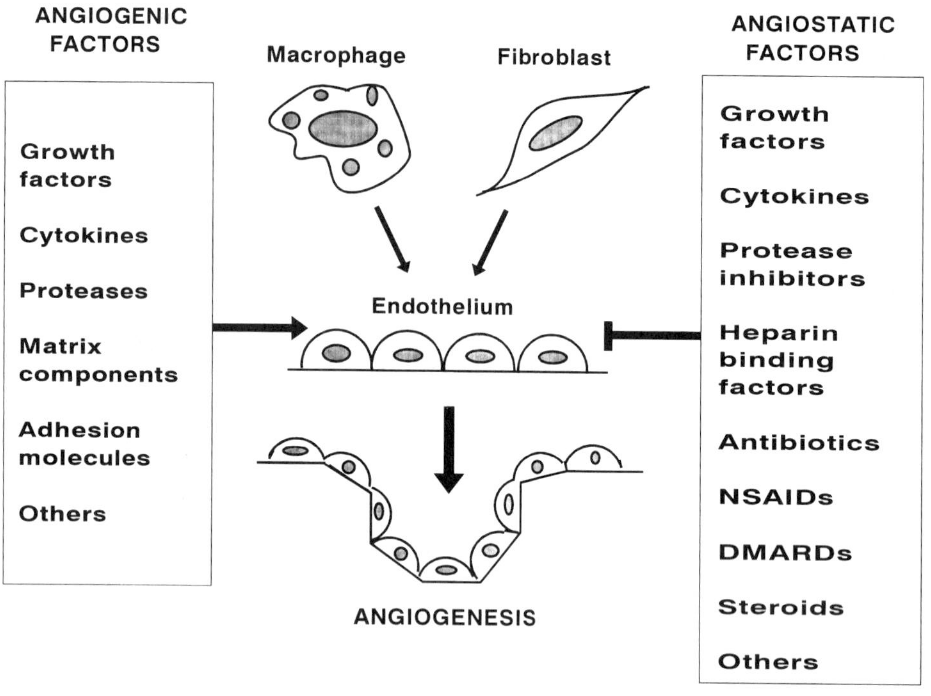

Figure 1 Angiogenesis, angiogenic mediators, and inhibitors in the rheumatoid synovium. Capillary formation is triggered by a series of macrophage-, fibroblast-, or endothelium-derived angiogenic mediators (left). Neovascularization can be blocked by angiostatic compounds (right). The outcome of angiogenesis depends highly on the balance between mediators and inhibitors.

synovial tissue macrophages (23). PDGF is also a mitogen for synovial fibroblasts and thus enhances synovial proliferation (24).

PD-ECGF/gliostatin is structurally similar to FGFs (4–6). Large quantities of PD-ECGF have been detected in the sera and synovial fluids of RA patients (25). Synovial PD-ECGF levels correlate with the acute-phase response in RA (25).

EGF stimulates most steps of angiogenesis; however, it is less potent than heparin-binding factors (4–6). EGF has been detected in RA synovial fluids (26).

The amount of IGF-I is increased in RA compared to normal synovial fluids (27). IGF-I mRNA has also been detected in RA synovial tissues (28).

TGF-β has a bifunctional effect on angiogenesis, which appears to be dose-dependent (1–6,29). Similar observations have been made in the context of arthritis. It is likely that TGF-β stimulates angiogenesis indirectly by recruiting angiogenic macrophages (30). TGF-β stimulates the secretion of both the angiogenic urokinase-type plasminogen activator, and the angiostatic metalloproteinase inhibitors by synovial fibroblasts, which is one possible mechanism by which this growth factor could both stimulate and inhibit neovascularization (31,32). TGF-β has disparate effects on synovial inflammation as well: it induces neutrophil recruitment into rat joints (33), while it suppresses rat adjuvant-induced arthritis (34). In the RA synovial tissue, we and others detected TGF-β in macrophages, lining cells, and fibroblasts (35,36). TGF-β mRNA has been detected in RA synovial fluid leukocytes (37). TGF-β partially antagonizes the effects of IL-1 in RA synovial fluids (38).

C. Cytokines

Among proinflammatory cytokines, which also play a role in RA, tumor necrosis factor (TNF)-α, IL-1, IL-8, and possibly IL-6 are involved in angiogenesis (1–6). A recently described IL-2-like cytokine, IL-15, also stimulates neovascularization (39). Interestingly, among the anti-inflammatory cytokines, IL-13 may also have proangiogenic effects in arthritis (40). Most of these cytokines are produced by macrophages, and thus they, especially TNF-α, account for the majority of macrophage-derived angiogenic activity in RA (1,2,4,6,41–44).

TNF-α is a potent angiogenic cytokine with efficacy comparable to that of FGFs (41,42). Large amounts of TNF-α have been detected in the sera and synovial fluids of RA patients (45). TNF-α accounts for a portion of RA synovial fluid–mediated neovascularization (46). In the RA synovial tissue, macrophages and endothelial cells express antigenic TNF-α (45). TNF-α levels correlate with clinical and laboratory markers of RA activity (45).

The effects of IL-1α and IL-1β on angiogenesis are in many ways similar to those of TGF-β, as some groups showed the angiogenic effects of these cytokines in two different animal models for angiogenesis, while others reported their angiostatic activities in the same systems (1,5,6,29). In RA, synovial tissue macrophages and fibroblasts as well as peripheral blood monocytes release significant amounts of IL-1 (47,48).

The role of IL-6 in angiogenesis is also controversial. It stimulated endothelial cell migration (49) but inhibited endothelial proliferation in vitro (50). Therefore, the role of IL-6 in angiogenesis is not fully clear. A significant amount of IL-6 is found in RA synovial fluids (51). In RA, IL-6 is produced by synovial tissue macrophages and fibroblasts as well as by lipopolysaccharide-stimulated synovial fluid monocytes (52,53).

IL-13 is an anti-inflammatory cytokine, as it inhibits the production of several other inflammatory mediators (54,55). Recently, we found that IL-13 may have proangiogenic effects, as it stimulates endothelial chemotaxis (40). We found little IL-13 in the sera, synovial fluids, and synovial tissues of RA patients (55), while another group reported the presence of high amounts of IL-13 in RA synovia (54). The reason for the differences between these two studies remains to be determined.

IL-15 induces neovascularization in nude mice (39). High amounts of IL-15 are present in RA synovial fluids and in the RA synovial tissue lining layer (56).

Granulocyte- and granulocyte-macrophage colony stimulating factors (G-CSF and GM-CSF) exhibit modest angiogenic activities compared to heparin-binding angiogenic mediators (4–6). RA synovial tissue macrophages constitutively express GM-CSF, while IL-1 or TNF-α stimulate fibroblasts to secrete this cytokine (57).

Oncostatin M has recently been found to stimulate angiogenesis via bFGF-dependent mechanisms (58). Macrophages are the major source of this cytokine in the RA synovial tissue. Elevated levels of oncostatin M were found in the synovial fluids of patients with RA in comparison to those with osteoarthritis (59).

D. Chemokines

A number of C-X-C chemokines have been implicated in neovascularization (60). These chemotactic mediators are also involved in the inflammatory mechanisms and angiogenesis underlying RA (60,61). It has been suggested that the ELR (glutamyl-leucyl-arginyl-) motif found in a number of C-X-C chemokines—such as IL-8, epithelial neutrophil activating protein (ENA)-78, growth-related oncogene α (groα), groβ, and connective tissue activating protein (CTAP)-III—is responsible for the angiogenic activity of these chemotactic cytokines (60,62). In contrast, other C-X-C chemokines lacking the ELR motif, such as platelet factor-4 (PF4) and interferon-γ–inducible protein (IP-10), are not only unable to stimulate neovascularization but, on the contrary, are potent angiostatic factors (60,62). Most of these chemokines play an important role in the recruitment of inflammatory leukocytes into the RA synovium (61).

Regarding ELR^+ C-X-C chemokines, we and other investigators have shown that IL-8 induces neovascularization both in vivo and in vitro (6,42,43,62). Synovial fibroblasts produce IL-8 mRNA in response to IL-1α (63). In RA, IL-8 is produced by synovial macrophages, chondrocytes, and by IL-1– or TNF-α–activated fibroblasts (43,61). We have detected IL-8 in RA synovial fluids (43).

We have identified significantly increased levels of ENA-78 in RA in comparison to osteoarthritic synovial fluids (61,64). This chemokine is present in higher quantities in the sera of RA patients than in normal subjects (61,64). TNF-α and IL-1 stimulate RA synovial fibroblasts to produce ENA-78 (64).

Synovial tissue fibroblasts and synovial fluid mononuclear cells produce groα upon activation by proinflammatory cytokines and mitogens, respectively (61,65). The expression of groα in isolated synovial fibroblasts is induced by IL-1 or TNF-α (63,65). We found that groα is present in RA synovial lining cells and subsynovial macrophages (65).

The chemokine groβ is also angiogenic (62). A relative increase in the constitutive expression of groβ on synovial fibroblasts has been found in osteoarthritic in comparison to rheumatoid synovial tissues (66), suggesting a differential role of groα and groβ in RA.

RA sera contain high levels of the angiogenic CTAP-III (61,67). CTAP-III stimulates the synthesis of various extracellular matrix molecules by synovial fibroblasts (67).

E. Extracellular Matrix Components and Matrix-Degrading Proteolytic Enzymes

During synovial angiogenesis, endothelial cells migrate into the extracellular matrix of the synovial tissue, which contains type I collagen, fibronectin, laminin, vitronectin, tenascin, thrombospondin, proteoglycans, and other matrix components (8,68). Endothelial migration is stimulated by type I collagen but inhibited by type IV collagen (1,6,68). Fibronectin is chemotactic for endothelial cells and promotes microvessel elongation in tissue explant cultures (1,69). Laminin and proteoglycans are abundantly produced during angiogenesis, and they are involved in endothelial cell attachment to the basement membrane and the underlying matrix (1,68). Heparin and heparan-sulfate proteoglycans are crucial for the action of heparin-binding angiogenic growth factors (1–6). Tenascin is also an important component in endothelial sprouting-permissive matrices (70). RA synovial fibroblasts show increased binding to collagen, fibronectin, laminin, and tenascin in comparison to normal synoviocytes (71). Thus, all these matrix macromolecules are important in RA-associated neovascularization (1–3,8).

Angiogenesis in RA requires proteolytic enzymes—such as matrix metalloproteinases, including collagenase, gelatinase, stromelysin, and others—as well as urokinase- and tissue-type plasminogen activators. A number of metalloproteinases were found in high quantities in RA synovial fluids and tissues (1,72,73). Synovial fluid collagenase concentration is a marker of synovial inflammation (72). In the RA synovial tissue, lining cells are the major producers of collagenase (72). We have also reported increased expression of the urokinase receptor (CD87) in RA compared to normal synovial tissues (74). Others have found abundant plasminogen activator concentrations in RA synovial fluids and synovial tissue extracts (75,76). RA synovial lining cells express both urokinase- and tissue-type plasminogen activators (76).

F. Cellular Adhesion Molecules

Adhesion molecules play a crucial role in leukocyte emigration into the RA synovium (1–3,7,77,78). It has been recently shown by us and others that a number of adhesion receptors, especially endothelial adhesion molecules, mediate neovascularization (46,77,79–85; reviewed in Ref. 7) (Fig. 2).

β1 integrins, as well as the αVβ3 integrin, which are present on endothelia and mediate endothelial cell-extracellular matrix interactions, have been implicated in angiogenesis (7,80,81,86). Most β1 integrins are highly expressed on RA synovial fluid lymphocytes as well as synovial tissue lining cells, leukocytes, fibroblasts, and endothelia

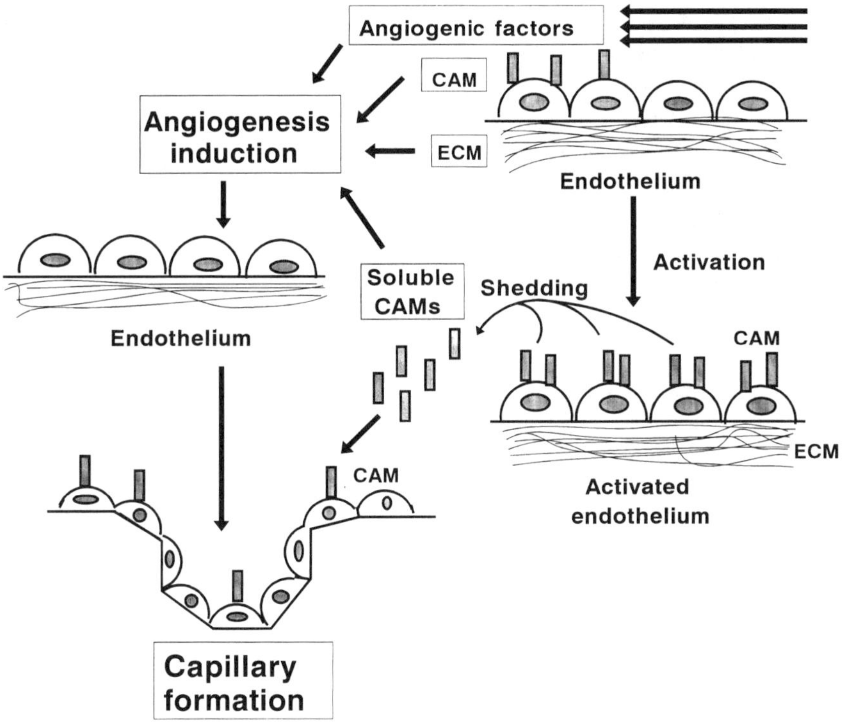

Figure 2 The role of adhesion molecules in arthritis-associated angiogenesis. Soluble angiogenic mediators, such as cytokines present in the arthritic synovium, stimulate the expression and shedding of endothelial adhesion receptors. In turn, soluble and surface-bound adhesion molecules, especially VCAM-1, E-selectin, and β3 integrins, may trigger neovascularization. CAM = cellular adhesion molecule; ECM = extracellular matrix.

(71,77,86). The $\alpha V\beta 3$ integrin has been detected by us and other investigators on RA synovial macrophages, lining cells, and fibroblasts (71,77,87). It has recently been suggested that VEGF, at least in part, acts via $\alpha 1\beta 1$ and $\alpha 2\beta 1$ integrin-dependent mechanisms (88).

Regarding other endothelial adhesion molecules, VCAM-1, E-selectin, P-selectin, CD31, and endoglin (CD105) may be involved in the neovascularization process. We have shown, that recombinant soluble VCAM-1 and E-selectin may act as proangiogenic factors based on in vitro and in vivo angiogenesis studies (79). Others have reported that antibodies to E-selectin inhibited tube formation of bovine capillary endothelial cells (82). High levels of soluble VCAM-1 and E-selectin have been found in the synovial fluids of RA patients (89,90). We and others have detected VCAM-1 and E-selectin on RA synovial endothelia. VCAM-1 is also present in the synovial lining layer (10,91). In our hands, antibodies to VCAM-1 and E-selectin partially neutralized RA synovial fluid-mediated endothelial chemotaxis (79). Thus, VCAM-1 and E-selectin may be important in RA-associated neovascularization.

Soluble P-selectin was found to stimulate endothelial cell migration in vitro (92). We found elevated levels of soluble P-selectin in the synovial fluids of patients with RA compared to those with osteoarthritis.

The angiogenic effects of CD31 and the TGF-β receptor endoglin have been reported (83–85). We have detected both CD31 and endoglin on most RA synovial endothelial cells as well as on lining cells and macrophages (36,94).

Thus, as reviewed in Fig. 2, a number of soluble and possibly also surface-bound adhesion receptors may be involved in the angiogenic process. The synthesis and expression of these molecules are regulated by soluble factors, such as proinflammatory cytokines, which are also abundantly produced in the RA synovium (reviewed in Ref. 7).

G. Other Angiogenic Mediators

Angiogenin stimulates angiogenesis more effectively than several growth factors; however, its mode of action is not fully clear (3–6). We have detected angiogenin in the synovial fluid and tissue of RA patients (14).

Platelet-activating factor (PAF) stimulates angiogenesis via a heparin-dependent mechanism (95). The role of PAF in RA synovial fluid-mediated neovascularization has been reported (46). A PAF antagonist, BN5730, has undergone a clinical trial in RA, resulting in clinical improvement in these patients (96).

Substance P is also a macrophage-derived angiogenic factor (4–6). Increased amounts of substance P have been detected in RA compared to osteoarthritic synovial fluids and tissues (97).

Using the techniques of molecular biology, the role of receptor tyrosine kinases in angiogenesis has recently been evaluated. The VEGF receptor kinase flk-1 has been implicated in neovascularization by the discovery of putative flk-1–expressing endothelial cell progenitors with angiogenic capacity (98). Recently, the structure of angiopoietin-1, the ligand for the tie-1 tyrosine kinase, has been established. Angiopoietin-1 exerts no proangiogenic activity in vitro. Instead, it regulates the assembly of nonendothelial vessel wall components such as smooth muscle cells (99). The role of these tyrosine kinases has not as yet been explored in RA. (The angiopoietin-1 inhibitor, termed angiopoietin-2, is described below, in Sec. III.J, among the angiostatic factors.)

III. ANGIOGENESIS INHIBITORS IN THE RHEUMATOID SYNOVIUM

Angiogenesis inhibitors include a few cytokines and growth factors—some of which may also stimulate neovascularization under different circumstances—as well as ELR C-X-C chemokines, heparin-binding factors, antirheumatic drugs, steroids, protease inhibitors, antibiotics, tissue-derived inhibitors, anticytoskeletal agents, and other compounds (Table 2, Fig. 1). Some of these factors influence the progression of RA and thus may be useful for the management of this disease.

A. Cytokines and Growth Factors

Some of the growth factors and cytokines—such as TGF-β, IL-1α, IL-1β and IL-6—were already discussed above, as these mediators can both stimulate or inhibit neovascularization under various conditions (1–6,29).

Interferon (IFN)-α and IFN-γ have been shown to block FGF- and VEGF-dependent angiogenesis (4–6,29). The IFN-γ protein, however, is hardly detectable in RA (37,47). Its mRNA has been found in RA synovial tissues and synovial fluid cells (37,47).

IL-4, an anti-inflammatory cytokine, antagonizes several effects of IL-1 and TNF-α in RA (100). We have recently shown that IL-4 inhibited angiogenesis both in vivo and in vitro (101).

IL-12 inhibits angiogenesis by inducing the production of two angiostatic mediators, IFN-γ and IP-10 (102,103). IL-12 has been detected in RA synovial fluids (104), while IL-12 mRNA is expressed in synovial fluid leukocytes of RA patients (37).

RA synovial fluids contain high levels of the angiostatic leukemia inhibitory factor (LIF) (105). Synovial fibroblasts produce LIF in response to IL-1 or TNF-α stimulation (106).

B. Chemokines

As described above, a number of C-X-C chemokines lacking the ELR motif are potent angiogenesis inhibitors. IP-10 and PF4 block bFGF-, IL-8–, and ENA-78–induced neovascularization (60,62,107). Very little IP-10 has been detected in the sera of RA patients (108). A PF4 peptide sequence inhibited collagen-induced murine arthritis, a representative model for RA (109).

C. Protease Inhibitors

Tissue inhibitors of metalloproteinases (TIMP) -1, -2 and -3 inhibit angiogenesis by preventing the breakdown of matrix components by metalloproteinases (1–6,29). TIMPs have been found in large quantities in RA synovial fluids and tissues (73).

Plasminogen activator inhibitors (PAI) have also been detected in RA synovia. PAI-1 concentrations were increased in RA in comparison to osteoarthritic synovial fluids (75,76). PAI-1 is expressed on RA synovial lining cells (76).

D. Heparin-Binding Factors

Heparin-binding molecules inhibit the binding of angiogenic growth factors such as FGFs to heparin. Among these compounds, thrombospondin-1 is an important extracellular matrix component, which also serves as an adhesion molecule, while PF4, a C-X-C chemokine, is described above. We have shown thrombospondin-1 expression on RA synovial endothelial cells and macrophages (110). In addition, in the rat adjuvant-induced arthritis model for RA, the implantation of thrombospondin-1-containing pellets into the rat ankles after the induction of arthritis resulted in a biphasic modulation of synovial blood vessel counts (111).

E. Antirheumatic Drugs

A number of antirheumatic compounds—such as nonsteroidal anti-inflammatory drugs (NSAIDs), gold salts, D-penicillamine, chloroquine, sulfasalazine, methotrexate, azathioprine, cyclophosphamide, corticosteroids, and possibly cyclosporine A—may inhibit angiogenesis by various mechanisms (1–6,29). Among NSAIDs, indomethacin blocks neovascularization by suppressing the production of the angiogenic prostaglandin E2 (1,3). A selective cyclooxygenase-2 inhibitor, NS-398, inhibited angiogenesis in the rat sponge implant model (112). Methotrexate blocks blood vessel formation, at least in part by suppressing synovial metalloproteinase gene expression (113). We and others have shown that gold compounds inhibit angiogenesis in vitro and in vivo (114,115). Cyclosporine A acts synergistically with fumagillin in suppressing murine collagen-induced arthritis by inhibiting VEGF production (116). Sulfasalazine, azathioprine, cyclophosphamide, chloroquine, and dexamethasone are also able to inhibit endothelial proliferation and migration (reviewed in Refs. 1, 3, 29).

F. Angiostatic Steroids

These steroid act by inhibiting basement membrane degradation (5,29). Among the several known compounds, 2-methoxyestradiol suppressed type II collagen–induced arthritis in mice (117). New-generation angiostatic steroids have been developed by the combination of heparin and cortisone. This complex prevents cartilage degradation in an animal arthritis model (118).

G. Antibiotics

The angiostatic analogues of fumagillin, such as AGM-1470/angioinhibin, suppressed collagen-induced arthritis in rats (116,119). In addition, AGM-1470 ameliorated rat adjuvant-induced arthritis (84). AGM-1470 acts at least in part by decreasing serum VEGF levels in arthritis (116).

Minocycline inhibits metalloproteinases and thus angiogenesis (29). Minocycline also suppressed both adjuvant- and collagen-induced arthritis in rats (120). There have been a few trials using this antibiotic drug in RA, with some clinical improvement in mild or moderate disease (121,122).

Rapamycin inhibits endothelial proliferation (29). There are a few clinical studies on the efficacy of rapamycin in animal models of arthritis (123,124).

Deoxyspergualin has angiostatic activity similar to other antibiotics (29). Deoxyspergualin successfully abrogated type II collagen induced murine arthritis (125).

Clarithromycin, which is widely used in several types of infections, may also inhibit angiogenesis (126). This antibiotic drug was shown to suppress IL-1, IL-6, IL-8, and GM-CSF production by cultured synovial fibroblasts (127).

H. Tissue-Derived Inhibitors

Such inhibitors have been isolated from a number of tissues. Cartilage-derived angiogenesis inhibitors have been successfully tried in animal models of arthritis (128).

I. Cytoskeleton-Disrupting Agents

Among these compounds, paclitaxel (Taxol) was found to suppress type II collagen-induced arthritis in rats (119).

J. Other Angiogenesis Inhibitors

Other angiostatic compounds that have been studied in context with arthritis include secreted-protein-acidic and rich-in-cysteine (SPARC)/osteonectin (29,129). Increased SPARC levels were found in synovial fluids of RA patients. In addition, SPARC secretion in chondrocyte cultures is regulated by angiogenic growth factors and cytokines suggesting the existence of a regulatory feedback in arthritis (129).

Opioids are also known angiogenesis inhibitors (29). both kappa- and mu opioids attenuated the progression of rat adjuvant-induced arthritis (130,131).

Recently, a naturally occuring inhibitor of angiopoietin-1, termed angiopoietin-2 has been identified. This antagonist of the tie-2 receptor tyrosine kinase inhibits neovascularization (132).

There is growing amount of data on the inhibitory effects of angiostatin and endostatin on tumor-associated angiogenesis. Angiostatin, a fragment of plasminogen, was purified from the urine of lung tumor–bearing mice (133), while endostatin, a fragment of collagen type XVIII, is produced by murine hemangioendothelioma cells (134). The possible relevance of these potent angiostatic compounds for the management of RA has not as yet been elucidated.

IV. REGULATION OF ANGIOGENESIS IN THE RHEUMATOID SYNOVIUM

RA is an inflammatory disease characterized by leukocyte ingress into and extensive neovascularization in the synovial tissue leading to pannus formation and eventually cartilage destruction (1–3,8,9). While the normal synovial tissue is relatively poor in microvasculature, the inflamed synovial tissue contains large numbers of vessels with activated endothelium resembling the morphology of high endothelial venules (9,135). The intense angiogenic activity was found to correlate with clinical and inflammatory scores as well as synovial hyperplasia (1,9,87).

The outcome of angiogenesis, and thus the extent of leukocytic invasion through the newly formed vessels into the synovium, greatly depends on the imbalance between angiogenic and angiostatic mediators mostly produced by synovial macrophages, endothelial cells, and fibroblasts (reviewed in Refs. 1–3). Several interactive and feedback mechanisms existing in the RA synovial tissue regulate the angiogenic process. Angiogenic mediators, both soluble and membrane-bound, may stimulate each other, resulting in the enhancement of neovascularization. Some of these factors themselves exert biphasic effects on angiogenesis. At the same time, as a negative feedback, these factors may induce the production of angiostatic factors, thus downregulating angiogenesis. In addition to these naturally occurring processes, externally administered therapeutic agents may interfere with both sides of this balance. It is important to understand these interactions so that more effective antiangiogenic and thus anti-inflammatory therapy can be established. Therefore, although these mechanisms cannot be discussed in their complexity, some of these interactions are summarized here by showing a number of examples (Figs. 1 and 2).

1. Possibly the most important regulatory mechanism is the balance between antagonistic angiogenic and angiostatic couples, such as matrix metalloproteinases versus TIMPs, plasminogen activators versus PAIs, heparin-dependent growth factors versus heparin-binding antagonists, or ELR^+ versus ELR-C-X-C chemokines (5,29,60,62). For example, an excess of matrix metalloproteinases over TIMPs has been described in RA, which may play a role in cartilage degradation as well as angiogenesis (136).

2. As described above, TGF-β, IL-1, and IL-6 may stimulate or inhibit angiogenesis depending on their concentration. Thus, the production of these mediators may be a key regulator of neovascularization (33–38,47,48,51–53).

3. There are several examples of the additive interactions of angiogenic mediators resulting in the perpetuation of neovascularization. Among soluble factors, proinflammatory cytokines, especially TNF-α and IL-1, stimulate RA synovial fibroblasts to produce IL-8, ENA-78, groα, GM-CSF, HGF, and other angiogenic factors (20,43,57,61,63–65). There are "indirect" mediators of angiogenesis, such as PDGF, TGF-β, or oncostatin M, which induce bFGF and VEGF release (1,6,21,22,58). The interactions between angiogenic matrix molecules and integrins (e.g., β1 integrin–laminin, fibronectin, or β3 integrin–vitronectin binding) are also essential in RA-associated neovascularization (reviewed in Refs. 7,77,78). Proinflammatory cytokines can upregulate the expression of endothelial angiogenic adhesion molecules, such as E-selectin or VCAM-1 (77). As one of the possible ligands for E-selectin is closely related to bFGF, soluble adhesion molecules may also participate in bFGF-triggered signaling in the angiogenic process (79). Basic FGF also contains the DGR (inverse RGD) sequence, which suggests that bFGF may also interact with integrins (137). As described above, endoglin is a known TGF-β receptor. Both endoglin and TGF-β have been implicated in angiogenesis (1,29,84,85). In fact, we found correlated expression of endoglin and TGF-β in the RA synovium (36).

4. Angiogenic mediators may stimulate the production of angiostatic factors, resulting in a negative feedback downregulating the angiogenic process. For example, the angiogenic TNF-α, as well as IL-1 induce synovial LIF release (106).

5. A number of angiostatic compounds act by downregulating the production of angiogenic mediators. For example, thrombospondin-1 interferes with bFGF on endothelial cells by inhibiting the chemoattractant activity of this growth factor (138).

6. Angiostatic agents may also interact with each other or induce the release of other angiogenesis inhibitors. Cyclosporin A and the fumagillin-derivative AGM-1470 act synergistically in blocking the angiogenic activity of VEGF and abrogating murine arthri-

tis (116). IL-12 is an ''indirect'' inhibitor of neovascularization, as it stimulates the release of two other angiostatic mediators, IFN-γ and IP-10 (102,103).

7. Drugs currently used in the therapy of RA and other inflammatory diseases may suppress the production of angiogenic factors and thus control the progression of this disease. These clinical applications are discussed in Sec. V. Regarding the mechanism of action of these drugs, methotrexate and minocyclin downregulate matrix metalloproteinase synthesis (29,113), fumagillin and its derivatives antagonize VEGF (116), while clarithromycin inhibits proinflammatory and proangiogenic cytokine production (127).

In conclusion, a regulatory network of inflammatory and endothelial cells as well as soluble angiogenic and angiostatic factors, extracellular matrix components, and adhesion molecules exists in RA and likely in other ''angiogenic diseases.'' Angiogenesis inhibitors can interact with these mediators at any steps of angiogenesis (Fig. 1).

V. CLINICAL RELEVANCE OF ANGIOGENESIS IN RHEUMATOID ARTHRITIS

Angiogenesis research has important practical diagnostic and therapeutic relevance. Regarding its diagnostic, prognostic value, the number of newly formed blood vessels in biopsy specimens may reflect the progression of the disease (1,3). We and others have demonstrated significantly higher degrees of vascularity in RA than in osteoarthritic or normal synovial tissues (9,87). The elevated concentration of the angiogenic-soluble E-selectin and VCAM-1 in the synovial fluid of RA patients may also be a useful marker of increased neovascularization as well as inflammation (139,140).

Regarding applications for therapy, as discussed above, a number of antirheumatic drugs currently used in RA—including NSAIDs, gold salts, sulfasalazine, methotrexate, D-penicillamine, chloroquine, azathioprine, cyclophosphamide, corticosteroids, and possibly cyclosporine A—inhibit the production of macrophage-derived angiogenic mediators (1,114–116).

Future antiangiogenic and anti-inflammatory targeting in RA may include the inhibition of cytokine, growth factor or chemokine production. At this moment TNF-α seems to be the main target for these trials. Both engineered monoclonal anti-TNF-α antibodies and soluble TNF-α receptors have undergone clinical trials (141–143). Other cytokines, such as IL-1 or IL-6, may also be targeted using antibodies or gene therapy (142,144). Among angiostatic chemokines, a peptide derived from PF4 was able to abrogate murine arthritis (109).

The synthesis and expression of angiogenic cellular adhesion molecules could also be inhibited (7). For example, fibronectin peptides containing the RGD sequence, which is recognized by several β1 and β3 integrins, suppress angiogenesis as well as arthritis in rats (145). Gold compounds given to RA patients suppressed the expression of synovial E-selectin, an angiogenic adhesion molecule, in the synovium (146).

Other angiostatic compounds could also be used to target neovascularization. Type II collagen-induced arthritis in rats was suppressed by the fumagillin-derivative AGM-1470 as well as by the cytoskeleton-disrupting agent paclitaxel (119). Cartilage-derived tissue inhibitors of angiogenesis have also been tried in arthritis (6,29,128). Opioids also attenuated rat adjuvant–induced arthritis (130,131). Theoretically, most angiogenesis inhibitors described above, especially protease inhibitors and antibiotics, may undergo further trials in arthritis models.

VI. SUMMARY

In summary, angiogenesis is important in leukocyte extravasation and thus the pathogenesis of RA. The outcome of neovascularization in the rheumatoid synovium depends on the imbalance between angiogenic and angiostatic mediators. Thus, it is important to understand the role of capillary formation in the pathogenesis of inflammatory arthritis. In addition, existing and potential angiostatic drugs may be useful for future therapy of RA.

ACKNOWLEDGMENTS

We would like to apologize to those investigators whose work could not be included due to space considerations. Supported by NIH grants AR-30692, AR-41492 and AI-40987 (A.E.K.); funds from the Veterans' Administration Research Service (A.E.K.); the Arthritis Foundation, Illinois Chapter Wolkonsky Award for Rheumatoid Arthritis research grant (A.E.K.); funds from Northwestern Memorial Hospital (A.E.K.); the Gallagher Professorship for Arthritis Research (A.E.K.); the Hungarian National Scientific Research Fund No. F 025813 (Z.S.); and the Medical Research Council Fund No. T-05 314/96 (Z.S.).

REFERENCES

1. Z Szekanecz, G Szegedi, AE Koch. Angiogenesis in rheumatoid arthritis. J Invest Med 46: 27–41, 1998.
2. AE Koch. Angiogenesis: implications for rheumatoid arthritis. Arthritis Rheum 41:951–962, 1998.
3. PR Colville-Nash, DL Scott. Angiogenesis in rheumatoid arthritis: pathogenic and therapeutic implications. Ann Rheum Dis 51:919–925, 1992.
4. J Folkman, M Klagsbrun. Angiogenic factors. Science 235:442–447, 1987.
5. M Klagsbrun, PA D'Amore. Regulators of angiogenesis. Annu Rev Physiol 53:217–239, 1991.
6. Z Szekanecz, MM Halloran, CJ Haskell, MR Shah, PJ Polverini, AE Koch. Mediators of angiogenesis: the role of cellular adhesion molecules. Trends Glycosci Glycotechnol 11:73–93, 1999.
7. Z Szekanecz, AE Koch. Adhesion molecules: potent inducers of endothelial cell chemotaxis. In: PP Zilla, HP Greisler, eds. Tissue Engineering of Prosthetic Vascular Grafts. New York: Academic Press 271–277, 1999.
8. ED Harris Jr. Rheumatoid arthritis: pathophysiology and implications for therapy. N Engl J Med 332:1277–1287, 1990.
9. Z Szekanecz, GK Haines, TR Lin, LA Harlow, S Goerdt, G Rayan, AE Koch. Differential distribution of intercellular adhesion molecules (ICAM-1, ICAM-2, and ICAM-3) and the MS-1 antigen in normal and diseased human synovia: their possible pathogenetic and clinical significance in rheumatoid arthritis. Arthritis Rheum 37:221–231, 1994.
10. AE Koch, JC Burrows, GK Haines, TM Carlos, J Harlan, SJ Leibovich. Immunolocalization

of leukocyte and endothelial adhesion molecules in human rheumatoid and osteoarthritic synovial tissue. Lab Invest 64:313–320, 1991.
11. H Sano, R Forough, JAM Maier, JP Case, A Jackson, K Engleka. Detection of high levels of heparin binding growth factor-1 (acidic fibroblast growth factor) in inflammatory arthritic joints. J Cell Biol 110:1417–1426, 1990.
12. Z Qu, XN Huang, P Ahmadi, J Andresevic, SR Planck, CE Hart, JT Rosenbaum. Expression of basic fibroblast growth factor in synovial tissue from patients with rheumatoid arthritis and degenerative joint disease. Lab Invest 73:339–346, 1995.
13. DH Goddard, SL Grossman, WV Williams, DB Weiner, JL Gross, K Eidsvoog. Regulation of synovial cell growth: coexpression of transforming growth factor β and basic fibroblast growth factor by cultured synovial cells. Arthritis Rheum 35:1296–1303, 1992.
14. S Hosaka, MR Shah, N Barquin, GK Haines, AE Koch. Expression of basic fibroblast growth factor and angiogenin in arthritis. Pathobiology 63:249–256, 1995.
15. AE Koch, LA Harlow, GK Haines, EP Amento, EN Unemori, WL Wong, RM Pope, N Ferrara. Vascular endothelial growth factor: a cytokine modulating endothelial function in rheumatoid arthritis. J Immunol 152:4149–4156, 1994.
16. RA Fava, NJ Olsen, G Spencer-Green, KT Yeo, TK Yeo, B Berse, RW Jackman, DR Senger, HF Dvorak, LF Brown. Vascular permeability factor/endothelial growth factor (VPF/VEGF): accumulation and expression in human synovial fluids and rheumatoid synovial tissue. J Exp Med 180:341–346, 1994.
17. RS Cotran. Endothelial cells. In: WN Kelley, ED Harris, S Ruddy, C Sledge, eds. Textbook of Rheumatology, 3d ed. Philadelphia: Saunders, 1989, pp. 389–411.
18. S Esser, K Wolburg, H Wolburg, G Brier, T Kurzchalia, W Risau. Vascular endothelial growth factor induces endothelial fenestrations in vitro. J Cell Biol 140:947–959, 1998.
19. DS Grant, HK Kleinman, ID Goldberg, MM Bhargava, BJ Nickoloff, JL Kinsella, PJ Polverini, EM Rosen. Scatter factor induces blood vessel formation in vivo. Proc Natl Acad Sci USA 90:1937–1941, 1993.
20. AE Koch, MM Halloran, S Hosaka, MR Shah, CJ Haskell, SK Baker, RJ Panos, GK Haines, GL Bennett, RM Pope, N Ferrara. Hepatocyte growth factor: a cytokine mediating endothelial migration in inflammatory arthritis. Arthritis Rheum 39:1566–1575, 1996.
21. N Sato, JG Beitz, J Kato, M Yamamoto, JW Clark, P Calabresi, AR Frackelton Jr. Platelet-derived growth factor indirectly stimulates angiogenesis in vitro. Am J Pathol 142:1119–1130, 1993.
22. E Brogi, T Wu, A Namiki, JM Isner. Indirect angiogenic cytokines upregulate VEGF and bFGF gene expression in vascular smooth muscle cells, whereas hypoxia upregulates VEGF expression only. Circulation 90:649–652, 1994.
23. EF Remmers, H Sano, RL Wilder. Platelet-derived growth factors and heparin-binding (fibroblast) growth factors in the synovial tissue pathology of rheumatoid arthritis. Semin. Arthritis Rheum 21:191–199, 1991.
24. R Lafyatis, EF Remmers, AB Roberts, DE Yocum, MB Sporn, RL Wilder. Anchorage-independent growth of synoviocytes from arthritic and normal joints. J Clin Invest 83:1267–1276, 1989.
25. M Takeuchi, T Otsuka, N Matsui, K Asai, T Hirano, A Moriyama. Aberrant production of gliostatin/platelet-derived endothelial cell growth factor in rheumatoid synovium. Arthritis Rheum 37:662–672, 1994.
26. J Kusada, T Otsuka, N Matsui, T Hirano, K Asai, T Kato. Immuno-reactive human epidermal growth factor (h-EGF) in rheumatoid synovial fluids. Nippon Seikeigeka Gakkai Zasshi 67: 859–865, 1993.
27. T Matsumoto, SE Gargosky, K Iwasaki, RG Rosenfeld. Identification and characterization of insulin-like growth factors (IGFs), IGF-binding proteins (IGFBPs), and IGFBP proteases in human synovial fluid. J Clin Endocrinol Metab 81:150–155, 1996.
28. GM Keyszer, AH Heer, J Kriegsmann, T Geiler, C Keysser, RE Gay, S Gay. Detection of

insulin-like growth factor I and II in synovial tissue specimens of patients with rheumatoid arthritis and osteoarthritis by in situ hybridization. J Rheumatol 22:275–281, 1995.
29. W Auerbach, R Auerbach. Angiogenesis inhibition: a review. Pharmac Ther 63:265–311, 1994.
30. DM Wiseman, PJ Polverini, DW Kamp, SJ Leibovich. Transforming growth factor-beta (TGFβ) is chemotactic for human monocytes and induces their expression of angiogenic activity. Biochem Biophys Res Commun 157:793–800, 1988.
31. JK Wright, TE Cawston, BL Hazleman. Transforming growth factor beta stimulates the production of the tissue inhibitor of metalloproteinases (TIMP) by human synovial and skin fibroblasts. Biochim Biophys Acta 1094:207–210, 1991.
32. JA Hamilton, DS Piccoli, T Leizer, DM Butler, M Croatto, AK Royston. Transforming growth factor beta stimulates urokinase-type plasminogen activator and DNA synthesis, but not prostaglandin E2 production, in human synovial fibroblasts. Proc Natl Acad Sci USA 88:7180–7184, 1991.
33. RA Fava, NJ Olsen, AE Postlethwaite, KN Broadley, JM Davidson, LB Nanney. Transforming growth factor β1 (TGF-β1) induced neutrophil recruitment to synovial tissues: implications for TGF-β-driven synovial inflammation and hyperplasia. J Exp Med 173:1121–1132, 1991.
34. ME Brandes, JB Allen, Y Ogawa, SM Wahl. Transforming growth factor β1 suppresses acute and chronic arthritis in experimental animals. J Clin Invest 87:1108–1113, 1991.
35. CQ Chu, M Field, E Abney, RQH Zheng, S Allard, M Feldmann, RN Maini. Transforming growth factor-β1 in rheumatoid synovial membrane and cartilage/pannus junction. Clin Exp Immunol 86:380–386, 1991.
36. Z Szekanecz, GK Haines, LA Harlow, MR Sha, TW Fong, R Fu, SJ-W Lin, G Rayan, AE Koch. Increased synovial expression of transforming growth factor (TGF)-β receptor endoglin and TGF-β1 in rheumatoid arthritis: possible interactions in the pathogenesis of the disease. Clin Immunol Immunopathol 76:187–194, 1995.
37. A Bucht, P Larsson, L Weisbrot, C Thorne, P Pisa, G Smedegard, EC Keystone, A Gronberg. Expression of interferon-gamma (IFN-gamma), IL-10, IL-12 and transforming growth factor-beta (TGF-beta) mRNA in synovial fluid cells from patients in the early and late phases of rheumatoid arthritis (RA). Clin Exp Immunol 103:357–367, 1996.
38. SM Wahl, JB Allen, HL Wong, SF Dougherty, LR Ellingsworth. Antagonistic and agonistic effects of transforming growth factor-β and IL-1 in rheumatoid synovium. J Immunol 145: 2514–2519, 1990.
39. AL Angiolillo, H Kanegane, C Sgadari, GH Reaman, G Tosato. Interleukin-15 promotes angiogenesis in vivo. Biochem Biophys Res Commun 233:231–237, 1997.
40. MM Halloran, CJ Haskell, JM Woods, S Hosaka, AE Koch. Interleukin-13 is an endothelial chemotaxin. Pathobiology 65:287–292, 1997.
41. SJ Leibovich, PJ Polverini, HM Shepard, DM Wiseman, V Shively, N Nuseir. Macrophage-induced angiogenesis is mediated by tumour necrosis factor-α. Nature 329:630–632, 1987.
42. AE Koch, PJ Polverini, SL Kunkel, LA Harlow, LA DiPietro, VM Elner, SG Elner, RM Strieter. Interleukin-8 as a macrophage-derived mediator of angiogenesis. Science 258:1798–1801, 1992.
43. AE Koch, SL Kunkel, JC Burrows, HL Evanoff, GK Haines, RM Pope, RM Strieter. The synovial tissue macrophage as a source of the chemotactic cytokine interleukin-8. J Immunol 147:2187–2195, 1991.
44. AE Koch, PJ Polverini, SJ Leibovich. Stimulation of neovascularization by human rheumatoid synovial tissue macrophages. Arthritis Rheum 29:471–479, 1986.
45. T Saxne, MA Palladino Jr, D Heinegard, N Talal, FA Wollheim. Detection of tumor necrosis factor α but not tumor necrosis factor β in rheumatoid arthritis synovial fluid and serum. Arthritis Rheum 31:1041–1045, 1988.
46. E Lupia, G Montrucchio, E Battaglia, V Modena, G Camussi. Role of tumor necrosis factor-

alpha and platelet-activating factor in neoangiogenesis induced by synovial fluids of patients with rheumatoid arthritis. Eur J Immunol 26:1690–1694, 1996.
47. FM Brennan, M Field, CQ Chu, M Feldmann, RN Maini. Cytokine expression in rheumatoid arthritis. Br J Rheumatol 30:76–80, 1991.
48. A Shore, S Jaglal, EC Keystone. Enhanced interleukin 1 generation by monocytes in vitro is temporally linked to an early event in the onset or exacerbation of rheumatoid arthritis. Clin Exp Immunol 65:293–302, 1986.
49. EM Rosen, D Liu, E Setter, M Bhargava, ID Goldberg. Interleukin-6 stimulates motility of vascular endothelium. EXS 59:194–205, 1991.
50. L Sachs. Angiogenesis—cytokines as part of a network. EXS 59:20–22, 1991.
51. FA Houssiau, J-P Devogelaer, J Van Damme, CN De Deuxchaisnes, J Van Snick. Interleukin-6 in synovial fluid and serum of patients with rheumatoid arthritis and other inflammatory arthritides. Arthritis Rheum 31:784–788, 1988.
52. P-A Guerne, BL Zuraw, JH Vaughan, DA Carson, M Lotz. Synovium as a source of interleukin 6 in vitro. J Clin Invest 83:585–592, 1989.
53. M Field, C Chu, M Feldmann, RN Maini. Interleukin-6 localisation in the synovial membrane in rheumatoid arthritis. Rheumatol Int 11:45–50, 1991.
54. P Isomaki, R Luukkainen, P Toivanen, J Punnonen. The presence of interleukin-13 in rheumatoid synovium and its antiinflammatory effects on synovial fluid macrophages from patients with rheumatoid arthritis. Arthritis Rheum 39:1693–1702, 1996.
55. JM Woods, GK Haines, MR Shah, G Rayan, AE Koch. Interleukin-13 in synovial fluid and tissue of patients with rheumatoid arthritis. Clin Immunol Immunopathol 85:210–220, 1997.
56. IB McInnes, J Al-Mughales, M Field, BP Leung, F Huang, R Dixon, RD Sturrock, PC Wilkinson, FY Liew. The role of interleukin-15 in T cell migration and activation in rheumatoid arthritis. Nature Med 2:175–182, 1996.
57. JM Alvaro-Garcia, NJ Zvaifler, GS Firestein. Granulocyte/macrophage colony-stimulating factor-mediated induction of class II MHC antigen on human monocytes: a possible role in rheumatoid arthritis. J Exp Med 170:865–875, 1989.
58. ES Wijelath, B Carlsen, T Cole, J Chen, S Kothari, WP Hammond. Oncostatin M induces basic fibroblast growth factor expression in endothelial cells and promotes endothelial cell proliferation, migration and spindle morphology. J Cell Sci 110:871–879, 1997.
59. H Okamoto, M Yamamura, Y Morita, S Harada, H Makino, Z Ota. The synovial expression and serum levels of interleukin-6, interleukin-11, leukemia inhibitory factor, and oncostatin M in rheumatoid arthritis. Arthritis Rheum 40:1096–1105, 1997.
60. RM Strieter, SL Kunkel, AB Shanafelt, DA Arenberg, AE Koch, PJ Polverini. The role of C-X-C chemokines in regulation of angiogenesis. In: AE Koch, RM Strieter, eds. Chemokines in Disease. Austin, TX: R.G. Landes Company, 1996, pp. 195–209.
61. AE Koch, SL Kunkel, RM Strieter. Chemokines in arthritis. In: AE Koch, RM Strieter, eds. Chemokines in Disease. Austin, TX: R.G. Landes Company, 1996, pp. 103–116.
62. RM Strieter, PJ Polverini, SL Kunkel, DA Arenberg, MD Burdick, J Kasper, J Dzuiba, J Van Damme, A Walz, D Marriott, S-Y Chan, S Roczniak, AB Shanafelt. The functional role of the ELR motif in CXC chemokine-mediated angiogenesis. J Biol Chem 270:27348–27357, 1995.
63. S Hosaka, T Akahoshi, C Wada, H Kondo. Expression of the chemokine superfamily in rheumatoid arthritis. Clin Exp Immunol 97:451–457, 1994.
64. AE Koch, SL Kunkel, LA Harlow, DD Mazarakis, GK Haines, MD Burdick, RM Pope, A Walz, RM Strieter. Epithelial neutrophil activating peptide-78: a novel chemotactic cytokine for neutrophils in arthritis. J Clin Invest 94:1012–1018, 1994.
65. AE Koch, SL Kunkel, MR Shah, S Hosaka, MM Halloran, GK Haines, MD Burdick, RM Pope, RM Strieter. Growth-related gene product alpha: A chemotactic cytokine for neutrophils in rheumatoid arthritis. J Immunol 155:3660–3666, 1995.

66. M Hogan, B Sherry, C Ritchlin, M Fabre, R Winchester, A Cerami, R Bucala. Differential expression of the small inducible cytokines GRO alpha and GRO beta by synovial fibroblasts in chronic arthritis: possible role in growth regulation. Cytokine 6:61–69, 1994.
67. CW Castor, PC Andrews, RD Swartz, SC Ellis, PA Hossler, MR Clark, EL Matteson, EF Sachter. The origin, variety, distribution and biological fate of connective tissue activating peptide III isoforms: characteristics in patients with rheumatoid, renal, and arterial disease. Arthritis Rheum 36:1142–1153, 1993.
68. JA Madri, KS Williams. Capillary endothelial cell cultures: phenotypic modulation by matrix components. J Cell Biol 97:153–165, 1983.
69. RF Nicosia, E Bonanno, M Smith. Fibronectin promotes the elongation of microvessels during angiogenesis in vitro. J Cell Physiol 154:654–661, 1993.
70. AE Canfield, AM Schor. Evidence that tenascin and thrombospondin-1 modulate sprouting of endothelial cells. J Cell Sci 108:797–809, 1995.
71. N Rinaldi, M Schwarz-Eywill, D Weis, P Leppelmann-Jansen, M Lukoschek, U Keilholz, TF Barth. Increased expression of integrins on fibroblast-like synoviocytes from rheumatoid arthritis in vitro correlates with enhanced binding to extracellular matrix proteins. Ann Rheum Dis 56:45–51, 1991.
72. S Maeda, T Sawai, M Uzuki, Y Takahashi, H Omoto, M Seki, M Sakurai. Determination of interstitial collagenase (MMP-1) in patients with rheumatoid arthritis. Ann Rheum Dis 54:970–975, 1995.
73. RM Hembry, MR Bagga, JJ Reynolds, DL Hamblen. Immunolocalisation studies on six matrix metalloproteinases and their inhibitors, TIMP-1 and TIMP-2, in synovia from patients with osteo- and rheumatoid arthritis. Ann Rheum Dis 54:25–32, 1995.
74. Z Szekanecz, GK Haines, AE Koch. Differential expression of the urokinase receptor (CD87) in arthritic and normal synovial tissues. J Clin Pathol 50:314–319, 1997.
75. C Belcher, F Fawthrop, R Bunning, M Doherty. Plasminogen activators and their inhibitors in synovial fluids from normal, osteoarthritis, and rheumatoid arthritis knees. Ann Rheum Dis 55:230–236, 1996.
76. HK Ronday, HH Smits, GN Van Muijen, MS Pruszczynski, RJ Dolhain, EJ Van Langelaan, FC Breedveld, JH Verheijen. Difference in expression of the plasminogen activation system in synovial tissue of patients with rheumatoid arthritis and osteoarthritis. Br J Rheumatol 35:416–423, 1996.
77. Z Szekanecz, G Szegedi, AE Koch. Cellular adhesion molecules in rheumatoid arthritis. Regulation by cytokines and possible clinical importance. J Investig Med 44:124–135, 1996.
78. DO Haskard. Cell adhesion molecules in rheumatoid arthritis. Curr Opin Rheumatol 7:229–234, 1995.
79. AE Koch, MM Halloran, CJ Haskell, MR Shah, PJ Polverini. Angiogenesis mediated by soluble forms of E-selectin and vascular cell adhesion molecule-1. Nature 376:517–519, 1995.
80. SM Albelda. Differential expression of integrin cell-substratum adhesion receptors on endothelium. EXS 59:188–192, 1991.
81. PC Brooks, RA Clark, DA Cheresh. Requirement of vascular integrin alpha v beta 3 for angiogenesis. Science 264:569–571, 1994.
82. N Nguyen, NA Strubel, J Bischoff. A role for sialyl Lewis-X/A glycoconjugates in capillary morphogenesis. Nature 365:267–269, 1993.
83. ER Horak, R Leek, N Klenk, S LeJeune, K Smith, N Stuart, M Greenall, K Stepniewska, AL Harris. Angiogenesis, assessed by platelet/endothelial cell adhesion molecule antibodies, as indicator of node metastases and survival in breast cancer. Lancet 340:1120–1124, 1992.
84. JA Maier, D Delia, PE Thorpe, G Gasparini. In vitro inhibition of endothelial cell growth by the antiangiogenic drug AGM-1470 (TNP-470) and the anti-endoglin antibody TEC-11. Anticancer Drugs 8:238–244, 1997.

85. P Kumar, JM Wang, C Bernabeu. CD 105 and angiogenesis (editorial). J Pathol 178:363–366, 1996.
86. H El Gabalawy, J Wilkins. Beta 1 (CD29) integrin expression in rheumatoid synovial membranes: an immunohistologic study of distribution patterns. J Rheumatol 20:231–237, 1993.
87. B Johnson, GK Haines, LA Harlow, AE Koch. Adhesion molecule expression in human synovial tissues. Arthritis Rheum 36:137–146, 1993.
88. DR Senger, KP Claffey, JE Benes, CA Perruzzi, AP Sergiou, M Detmar. Angiogenesis promoted by vascular endothelial growth factor: regulation through $\alpha1\beta1$ and $\alpha2\beta1$ integrins. Proc Natl Acad Sci USA 94:13612–13617, 1997.
89. AE Koch, W Turkiewicz, LA Harlow, RM Pope. Soluble E-selectin in arthritis. Clin Immunol Immunopathol 69:29–35, 1993.
90. SM Wellicome, P Kapahi, JC Mason, Y Lebranchu, H Yarwood, DO Haskard. Detection of a circulating form of vascular cell adhesion molecule-1: raised levels in rheumatoid arthritis and systemic lupus erythematosus. Clin Exp Immunol 92:412–418, 1993.
91. LS Wilkinson, JCW Edwards, RN Poston, DO Haskard. Expression of vascular cell adhesion molecule-1 in normal and inflamed synovium. Lab Invest 68:82–88, 1993.
92. L Morbidelli, L Brogelli, HJ Granger, M Ziche. Endothelial cell migration is induced by soluble P-selectin. Life Sci 62:7–11, 1998.
93. S Hosaka, MR Shah, RM Pope, AE Koch. Soluble forms of P-selectin and intercellular adhesion molecule-3 in synovial fluids. Clin Immunol Immunopathol 78:276–282, 1996.
94. Z Szekanecz, GK Haines, LA Harlow, MR Shah, TW Fong, R Fu, SJ-W Lin, AE Koch. Increased synovial expression of the adhesion molecules CD66a, CD66b and CD31 in rheumatoid and osteoarthritis. Clin Immunol Immunopathol 76:180–186, 1995.
95. G Camussi, G Montrucchio, E Lupia, A De Martino, L Perona, M Arese, A Vercellone, A Toniolo, F Bussolino. Platelet-activating factor directly stimulates in vitro migration of endothelial cells and promotes in vivo angiogenesis by a heparin-dependent mechanism. J Immunol 154:6492–6501, 1995.
96. P Halliquin, P Guinot, V Chermat-Izard, X Puechal, CJ Menkes. Treatment of rheumatoid arthritis with platelet activating factor antagonist BN 50730. J Rheumatol 22:1651–1654, 1995.
97. CJ Menkes, M Renoux, S Laoussadi, A Mauborgne, J Bruxelle, F Cesselin. Substance P levels in the synovium and synovial fluid from patients with rheumatoid arthritis and osteoarthritis. J Rheumatol 20:714–717, 1993.
98. T Asahara, T Murohara, A Sullivan, M Silver, R van der Zee, T Li, B Witzenbichler, G Schatteman, JM Isner. Isolation of putative progenitor endothelial cells for angiogenesis. Science 275:964–967, 1997.
99. S Davis, TH Aldrich, PF Jones, A Acheson, DL Compton, V Jain, TE Ryan, J Bruno, C Radziejewski, PC Maisonpierre, GD Yancopoulos. Isolation of angiopoietin-1, a ligand for the TIE2 receptor, by secretion-trap expression cloning. Cell 87:1161–1169, 1996.
100. E Sugiyama, A Kuroda, H Taki, M Ikemoto, T Hori, N Yamashita, M Maruyama, M Kobayashi. Interleukin 10 cooperates with interleukin 4 to suppress inflammatory cytokine production by freshly prepared adherent rheumatoid synovial cells. J Rheumatol 22:2020–2026, 1995.
101. OV Volpert, T Fong, AE Koch, JD Peterson, C Waltenbaugh, RI Tepper, N Bouck. Inhibition of angiogenesis by interleukin 4. J Exp Med 188:1039–1046, 1998.
102. EE Voest, BM Kenyon, MS O'Reilly, G Truitt, RJ D'Amato, J Folkman. Inhibition of angiogenesis in vivo by interleukin 12. J Natl Cancer Inst 87:581–586, 1995.
103. A Angiolillo, C Sgadari, G Tosato. A role for the interferon-inducible protein 10 in inhibition of angiogenesis by interleukin-12. Ann NY Acad Sci 795:158–167, 1996.
104. JF Schlaak, I Pfers, KH Meyer, Zum Buschenfelde, E Marker Hermann. Different cytokine profiles in the synovial fluid of patients with osteoarthritis, rheumatoid arthritis and seronegative spondylarthropathies. Clin Exp Rheumatol 14:155–162, 1996.

105. PM Waring, GJ Carroll, DA Kandiah, G Buirski, D Metcalf. Increased levels of leukemia inhibitory factor in synovial fluid from patients with rheumatoid arthritis and other inflammatory arthritides. Arthritis Rheum 36:911–915, 1993.
106. JA Hamilton, PM Waring, EL Filonzi. Induction of leukemia inhibitory factor in human synovial fibroblasts by IL-1 and tumor necrosis factor-α. J Immunol 150:1496–1502, 1993.
107. RM Strieter, SL Kunkel, DA Arenberg, MD Burdick, PJ Polverini. Interferon gamma-inducible protein 10 (IP-10), a member of the C-X-C chemokine family, is an inhibitor of angiogenesis. Biochem Biophys Res Commun 210:51–57, 1995.
108. S Narumi, Y Tominaga, M Tamaru, S Shimai, H Okumura, K Nishioji, Y Itoh, T Okanoue. Expression of IFN-inducible protein-10 in chronic hepatitis. J Immunol 158:5536–5544, 1997.
109. PH Wooley, C Schaefer, JD Whalen, JA Dutcher, DF Counts. A peptide sequence from platelet factor 4 (CT-112) is effective in the treatment of type II collagen induced arthritis in mice. J Rheumatol 24:890–898, 1997.
110. AE Koch, J Friedman, JC Burrows, GK Haines, NP Bouck. Immunolocalization of the angiogenesis inhibitor thrombospondin in human arthritic synovial tissues. Pathobiology 161:1–6, 1993.
111. AE Koch, Z Szekanecz, J Friedman, GK Haines, CB Langman, NP Bouck. Effects of thrombospondin-1 on disease course and angiogenesis in rat adjuvant-induced arthritis. Clin Immunol Immunopathol 86:199–208, 1998.
112. M Majima, M Isono, Y Ikeda, I Hayashi, K Hatanaka, Y Harada, O Katsumata, S Yamashina, M Katori, S Yamamoto. Significant roles of inducible cyclooxygenase (COX)-2 in angiogenesis in rat sponge implants. Jpn J Pharmacol 75:105–114, 1997.
113. GS Firestein, MM Paine, DL Boyle. Mechanisms of methotrexate action in rheumatoid arthritis: selective decrease in synovial collagenase gene expression. Arthritis Rheum 37:193–200, 1994.
114. AE Koch, M Cho, J Burrows, SJ Leibovich, PJ Polverini. Inhibition of production of macrophage-derived angiogenic activity by the anti-rheumatic agents gold sodium thiomalate and auranofin. Biochem Biophys Res Commun 154:205–212, 1988.
115. R Saura, T Matsubara, K Mizuno. Inhibition of neovascularization in vivo by gold compounds. Rheumatol Int 14:1–7, 1994.
116. SJ Oliver, TP Cheng, ML Banquerigo, E Brahn. Suppression of collagen-induced arthritis by an angiogenesis inhibitor, AGM-1470, in combination with cyclosporin: reduction of vascular endothelial growth factor (VEGF). Cell Immunol 166:196–206, 1995.
117. E Josefsson, A Tarkowski. Suppression of type II collagen-induced arthritis by the endogenous estrogen metabolite 2-methoxyestradiol. Arthritis Rheum 40:154–163, 1997.
118. PR Colville-Nash, M El-Ghazaly, DA Willoughby. The use of angiostatic steroids to inhibit cartilage destruction in an in vivo model of granuloma-mediated cartilage degradation. Agents Actions 38:126–134, 1993.
119. SJ Oliver, ML Banquerigo, E Brahn. Suppression of collagen-induced arthritis using an angiogenesis inhibitor, AGM-1470, and a microtubule stabilizer, Taxol Cell Immunol 157:291–299, 1994.
120. KL Sewell, F Breedveld, E Furrie, J O'Brien, C Brinckerhoff, R Dynesius-Trentham, Y Nosaka, DE Trentham. The effect of minocycline in rat models of inflammatory arthritis: correlation of arthritis suppression with enhanced T cell calcium flux. Cell Immunol 167: 195–204, 1996.
121. DE Trentham, RA Dynesius-Trentham. Antibiotic therapy for rheumatoid arthritis: scientific and anecdotal appraisals. Rheum Dis Clin North Am 21:817–834, 1995.
122. M Kloppenburg, FC Breedveld, JP Terwiel, C Mallee, BA Dijkmans. Minocycline in active rheumatoid arthritis: a double-blind, placebo-controlled trial. Arthritis Rheum 37:629–636, 1994.
123. RP Carlson, DA Hartman, LA Tomchek, TL Walter, JR Lugay, W Calhoun, SN Sehgal, JY

Chang. Rapamycin, a potential disease-modifying antiarthritic drug. J Pharmacol Exp Ther 266:1125–1138, 1993.
124. RP Carlson, WL Baeder, RG Caccese, LM Warner, SN Sehgal. Effects of orally administered rapamycin in animal models of arthritis and other autoimmune diseases. Ann NY Acad Sci 685P:86–113, 1993.
125. K Nemoto, T Mae, F Abe, T Takeuchi. Successful treatment with a novel immunosuppressive agent, deoxyspergualin, in type II collagen-induced arthritis in mice. Ann NY Acad Sci 685P: 148–154, 1993.
126. J Yatsunami, N Turuta, K Wakamatsu, N Hara, S Hayashi. Clarithromycin is a potent inhibitor of tumor-induced angiogenesis. Res Exp Med (Berl) 197:189–197, 1997.
127. N Matsuoka, K Eguchi, A Kawakami, M Tsuboi, Y Kawabe, T Aoyagi, S Nagataki. Inhibitory effect of clarithromycin on costimulatory molecule expression and cytokine production by synovial fibroblast-like cells. Clin Exp Immunol 104:501–508, 1996.
128. MA Moses, D Wiederschain, I Wu, CA Fernandez, V Ghazizadeh, WS Lane, E Flynn, A Sytkowski, T Tao, R Langer. Troponin I is present in human cartilage and inhibits angiogenesis. Proc Natl Acad Sci (USA) 96:2645–2650, 1999.
129. S Nakamura, K Kamihagi, H Satakeda, M Katayama. Enhancement of SPARC (osteonectin) synthesis in arthritic cartilage: increased levels in synovial fluids from patients with rheumatoid arthritis and regulation by growth factors and cytokines in chondrocyte cultures. Arthritis Rheum 39:539, 1996.
130. JL Wilson, V Nayanar, JS Walker. The site of anti-arthritic action of the kappa-opioid, U-50, 488H, in adjuvant arthritis: importance of local administration. Br J Pharmacol 118: 1754–1760, 1996.
131. JS Walker, AK Chandler, JL Wilson, W Binder, RO Day. Effect of mu-opioids morphine and buprenorphine on the development of adjuvant arthritis in rats. Inflamm Res 45:557–563, 1996.
132. PC Maisonpierre, C Suri, PF Jones, S Bartunkova, SJ Wiegand, C Radziejewski, D Compton, J McClain, TH Aldrich, N Papadopoulos, TJ Daly, S Davis, TN Sato, GD Yancopoulos. Angiopoietin-2, a natural antagonist for Tie2 that disrupts in vivo angiogenesis. Science 277: 55–60, 1997.
133. MS O'Reilly, L Holmgren, Y Shing, C Chen, RA Rosenthal, M Moses, WS Lane, Y Cao, EH Sage, J Folkman. Angiostatin: a novel angiogenesis inhibitor that mediates the suppression of metastases by a Lewis lung carcinoma. Cell 79:315–328, 1994.
134. MS O'Reilly, T Boehm, Y Shing, N Fukai, G Vasios, WS Lane, E Flynn, JR Birkhead, BR Olsen, J Folkman. Endostatin: an endogenous inhibitor of angiogenesis and tumor growth. Cell 88:277–285, 1997.
135. ACHM van Dinther-Janssen, E Horst, G Koopman, W Newmann, RJ Scheper, CJ Meijer, ST Pals. The VLA-4/VCAM-1 pathway is involved in lymphocyte adhesion to endothelium in rheumatoid synovium. J Immunol 147:4207–4210, 1991.
136. J Martel-Pelletier, R McCollum, N Fujimoto, K Obata, JM Cloutier, JP Pelletier. Excess of metalloproteases over tissue inhibitor of metalloprotease may contribute to cartilage degradation in osteoarthritis and rheumatoid arthritis. Lab Invest 70:807–815, 1994.
137. M Presta, M Rusnati, C Urbinati, E Tanghetti, M Statuto, A Pozzi, A Gualandris, G Ragnotti. Basic fibroblast growth factor bound to cell substrate promotes cell adhesion, proliferation, and protease production in cultured endothelial cells. EXS 59:205–209, 1991.
138. G Taraboletti, D Belotti, R Giavazzi. Thrombospondin modulates basic fibroblast growth factor activities on endothelial cells. EXS 59:210–213, 1991.
139. AE Koch, W Turkiewicz, LA Harlow, RM Pope. Soluble E-selection in arthritis. Clin Immunol Immunopathol 69:29–35, 1993.
140. SM Wellicome, P Kapahi, JC Mason, Y Lebranchu, H Yarwood, DO Haskard. Detection of a circulating form of vascular cell adhesion molecule-1: raised levels in rheumatoid arthritis and systemic lupus erythematosus. Clin Exp Immunol 92:412–418, 1993.

141. RN Maini, MJ Elliott, FM Brennan, RO Williams, CQ Chu, E Paleolog, PJ Charles, PC Taylor, M Feldmann. Monoclonal anti-TNF-α antibody as a probe of pathogenesis and therapy of rheumatoid disease. Immunol Rev 144:195–221, 1995.
142. G Kingsley, G Panayi, J Lanchbury. Immunotherapy of rheumatic diseases—practice and prospects. Immunol Today 12:177–179, 1991.
143. LW Moreland, SW Baumgartner, MH Schiff, EA Tindall, RM Fleischmann, AL Weaver, RE Ettlinger, S Cohen, WJ Koopman, K Mohler, MB Widmer, CM Blosch. Treatment of rheumatoid arthritis with a recombinant human tumor necrosis factor receptor (p75) Fc fusion protein. N Engl J Med 337:141–147, 1997.
144. I Saiki, J Murata, T Makabe, N Nishi, S Tokura, I Azuma. Inhibition of tumor angiogenesis by a synthetic cell-adhesive polypeptide containing the Arg-Gly-Asp (RGD) sequence of fibronectin, poly(RGD). Jpn J Cancer Res 81:668–675, 1990.
145. CH Evans, PD Robbins. Pathways to gene therapy in rheumatoid arthritis. Curr Opin Rheumatol 8:230–234, 1996.
146. MM Corkill, BW Kirkham, DO Haskard, C Barbatis, T Gibson, GS Panayi. Gold treatment of rheumatoid arthritis decreases synovial expression of the endothelial leukocyte adhesion receptor ELAM-1. J Rheumatol 18:1453–1460, 1991.

30

Growth Factor Therapy in Ischemic Heart Disease

Roger J. Laham and Michael Simons
Beth Israel Deaconess Medical Center and Harvard Medical School, Boston, Massachusetts

I. INTRODUCTION

The rapid development of growth-factor therapy in patients with advanced ischemic heart disease over the last 5 years offers hope of a new treatment strategy that is based on generation of new blood supply in the diseased heart. A number of growth factors including the fibroblast growth-factor (FGF) family, vascular endothelial growth-factor (VEGF) family, platelet-derived growth (PDGF) family, angiopoietins, as well as cytokines such as IL-6 and IL-8, ''master switch'' genes such as hypoxia-inducible factor (HIF)-1α, and molecules such as nitric oxide (NO), platelet-activating factor (PAF), and substance P have all been shown to stimulate blood vessel growth under different circumstances. This chapter reviews recent results of animal and human growth-factor therapy studies primarily concentrating on FGF and VEGF families, the most widely studied among angiogenic growth factors.

In evaluating studies of therapeutic angiogenesis, it is important to consider outcome efficacy measures. These may include assessment of myocardial perfusion, microvascular reactivity, global or regional ventricular function, exercise performance, and freedom from symptoms and cardiac events. While each of these end points conveys important information regarding the functional state of the myocardium, all are subject to considerable variability and may change under the influence of various medications or changes in hemodynamic state. Therefore, a comprehensive assessment of any therapeutic angiogenic intervention should be a key feature of any clinical study.

II. MECHANISM OF GROWTH-FACTOR ACTION

All growth factors are presumed to induce neovascularization by stimulation of endothelial cell proliferation and migration. In addition, such effects as stimulation of extracellular matrix breakdown, attraction of pericytes and macrophages, stimulation of smooth muscle-cell proliferation and migration, formation and ''sealing'' of new vascular structures, and deposition of new matrix all play a role in the angiogenic process.

The ability of growth factors such as FGF-2 and VEGF to increases the rate endothelial cell proliferation has been demonstrated in several animal studies (1–5). Interestingly, in the setting of regional myocardial ischemia, endothelial cell proliferation is largely limited to the ischemic zone even during systemic (or nonselective coronary) growth-factor administration. Similarly, in the setting of arterial injury, endothelial cell growth is limited to the immediate surrounding area and not the entire vascular tree. It is apparent, therefore, that potent local factors, including perhaps both angiogenesis promoters as well as inhibitors, control the local angiogenic response. A number of angiogenic inhibitors have recently been identified, including angiostatin (6–9), endostatin (10), the endothelial proliferation inhibitor *btg* 1 (11), thrombospondin 1 (12–14), and thrombospondin 2 (12,13,15–17). Furthermore, limited expression of high-affinity growth-factor receptors such as FGF R1, flk-1, and flt-1 may limit the angiogenic potency of these cytokines in normal tissues. Finally, changes in the composition of the extracellular matrix—including expression of matrix-digesting enzymes (such as matrix metalloproteinases) as well as their inhibitors (such as TIMP 1) (18,19) in addition to changes in endothelial cell expression of integral matrix proteins (such as $\alpha_v\beta_3$ and $\alpha_v\beta_5$ integrins) (20–22) and heparan sulfate–carrying proteoglycans (such as syndecans) (23–25)—may regulate the ability of tissues to respond to angiogenic signaling.

A. The Role of Nitric Oxide

In addition to direct stimulation of endothelial cell growth, all growth factors possess a wide spectrum of biological activities that may be involved in angiogenic response. Particularly interesting and important may be the ability to stimulate production of NO. NO has a wide spectrum of activity, including the ability to induce significant vasodilation, stimulate or suppress proliferation and migration of a variety of cell types including smooth muscle and endothelial cells, affect myocardial contractility and platelet aggregation, and activate expression of various inflammatory cytokines. NO generation in tissue is controlled by the availability of the substrate L-arginine and the activity of NO synthases (NOS). Three isoforms of NO synthase are present in cardiac tissues, including calcium-dependent isoforms (eNOS and nNOS) and one calcium-independent isoform (iNOS). The nNOS isoform is found almost exclusively in the atria, in the specialized conduction system, and along the epicardial coronary arteries. In contrast, iNOS is present in the endocardium, microvascular endothelial cells, vascular smooth muscle cells, infiltrating macrophages, fibroblasts, and cardiac myocytes (26–28). iNOS expression in microvascular endothelial cells and cardiac myocytes is induced in response to lipopolysaccharide (LPS) or inflammatory cytokines including IL-1β, IFN-γ, and TNF-α (28–32). Expression of the third NO synthase, eNOS, is observed in macrophages, endothelial cells throughout the myocardium from epicardial coronary arteries to capillaries and venules, and cardiac myocytes (28). While eNOS is constitutively expressed, its expression is subject to important degrees of regulation. Factors such as shear stress (27,33), state of endothelial cell growth, hypoxia (34–36), and exposure to cytokines (37,38) and LPS (31) can affect eNOS expression. While the activity of eNOS (and nNOS) is thought to be controlled allosterically by Ca^{2+}/calmodulin, recent investigations have demonstrated the existence of another control mechanism in vivo involving direct interaction of eNOS in endothelial cells with specific residues of the scaffolding domain of caveolin (39). A number of investigations have linked generation of NO with regulation of blood vessel growth. However, this relationship is not well defined and sometimes contradictory. Thus, VEGF, FGF, and substance P all stimulate NO release and are associated with development of neovascularization in vivo (35,40–50). Furthermore, proliferative effects of VEGF in cell culture and the ability of TGF-β to induce vessel formation in 3D gels is suppressed in presence of NO synthase inhibitors (51,52). However, in a rabbit corneal model of angiogenesis, NO was found to be involved in VEGF- but not bFGF-induced angiogenesis (40,47), while tumor development and vascularity following Lewis lung carcinoma transplant in mice were inhibited by NO donor treatment (53). In addition, studies in cell culture and the chick embryo chorioallantoic membrane demonstrated that sodium nitroprusside and L-arginine inhibit angiogenesis while inhibitors of NO synthase stimulate it (44,45,54–57).

Since virtually all of heparin-binding growth factors have the ability to release NO and NOS inhibitors are able to severely limit angiogenic effect of these growth factors, NO generation may be a key event in many angiogenic states. In this regard, it is particularly interesting to note defective angiogenesis in a peripheral limb ischemia model in the eNOS knockout mouse (35). The administration of L-arginine, previously shown to augment endogenous production of NO, to a rabbit model of hindlimb ischemia resulted in improvement in limb perfusion, angiographic collaterals, and capillary density compared to controls. Angiogenesis in the ischemic hindlimb of eNOS knockout mice was significantly impaired compared to wild-type controls when evaluated by laser Doppler flow analysis and capillary density measurement. Impaired angiogenesis in eNOS knockout mice was not improved by administration of VEGF (35).

Both VEGFs and FGFs provoke NO release in vitro as well as in vivo (35,50,58–64), and long-term stimulation with VEGF increases expression of eNOS (38,65). Furthermore, inhibitors of NO release apparently have the ability to block VEGF, FGF-1 (but perhaps not FGF-2) and TNF-α induced angiogenesis (35,40,66,67). Very little is known about the mechanisms of interaction between NO and growth-factor stimulation or the molecular events involved in NO-dependent stimulation of blood vessel growth.

B. Growth-Factor Amplification Cascade

Another intriguing and poorly understood feature of angiogenic growth-factor therapy is the occurrence of an extensive response (in terms of neovascularization) 3–4 weeks following a single-bolus growth-factor administration (68,69). Since no known pharmacokinetic profile can explain this effect and the half life of most growth factors is far shorter than the time span before their biological activity becomes observable, one is left with speculations that either there is a prolonged retention of the growth factors in ischemic tissues secondary, perhaps, to heparan sulfate binding or that the single bolus administration sets off a positive feedback loop that function over the ensuing weeks.

While absorption of the growth factors (at least in the case of VEGF and FGF-2) is indeed increased in ischemic compared to normal areas of the myocardium, there is little evidence to show prolonged retention of the growth factor. The initial differential in absorption in favor of the ischemic myocardium can be explained by alterations in heparan sulfate matrix (higher expression of heparan sulfate carrying core proteins such as syndecans and glypican) and increased expression of high-affinity receptors. However, there are no data to suggest that these or any other known mechanisms, can account for prolonged retention of the growth factor in tissues.

Thus we are left with the possibility of a positive feedback loop as the only plausible explanation for prolonged growth-factor activity. One potential mechanism for such a feedback loop is the ability of growth factors, particularly VEGF-A, to induce a local inflammatory reaction. The nature of this ''inflammatory'' stimulus varies but includes the ability to induce vascular permeability, thus leading to an influx of plasma proteins into the extracellular space that can then act as a nidus for a cascade of inflammatory events as well as macrophage chemoattractant properties of VEGF and FGFs. The combination of the two will lead to rapid accumulation of macrophages at the site of transient growth factor deposition (thus preferring ischemic regions, where growth factor deposition is transiently increased several fold compared to normal tissues). The presence of leaky blood vessels with accompanying accumulation of plasma proteins such as fibrin in extracellular space will serve as additional factors in promoting macrophage accumulation in these areas. In addition, FGF-2 itself promotes macrophage adhesion by acting as a bridge binding heparan sulfates on the cell surface to high-affinity receptors on the endothelial vessel lining (70–77). The attachment of macrophages will, in turn, lead to their activation, followed by secretion of a number of growth factors and cytokines including FGFs, VEGF, Il-6, and IL-8. This sequence of events, then, sets up a positive feedback loop, leading to further local accumulation of the growth factors, increased macrophage adhesion, and further growth-factor secretion (75,77–79).

A number of experimental observations point to the plausibility of such a chain of events. The ability of VEGF to induce blood vessel leakiness, perhaps mediated by NO and PAF, is well documented (80). Likewise, both VEGF and FGF serve as macrophage chemoattractants (81–85). Furthermore, stimulation of macrophage adhesion per se can lead to a profound angiogenic response even in the absence of external growth-factor stimulation (78,79,86–88), while the absence of macrophages leads to severe reduction in angiogenesis even in the presence of growth-factor therapy (85,89). In addition to these considerations, the macrophage/inflammation feedback-loop model explains several other observations in the field of angiogenic growth-factor therapy. First, there appears to be little difference in response to therapy with a number of different growth factors, including VEGF, HGF, FGF-1, FGF-2, FGF-4, and FGF-5. This suggests that either all of these factors are equipotent or they all act through the same mechanism. Since most of these agents interact with distinct high-affinity receptors, the latter explanation appears untenable. However, if they all induce local inflammation/macrophage adhesion, even if by different routes, then the end result would appear to be the same no matter the nature of the initial stimulus. It should also be noted that there is considerable cross-talk between different growth factor families, with FGF-2 inducing prompt expression of VEGF in endothelial cells and VEGF also capable of stimulating FGF-2 expression (90–95). These synergies are further confirmed by studies observing exponential responsiveness of endothelial cell proliferation following administration of a combination of various growth factors (92,93).

In addition to promoting neovascularization, therapy with growth factors is associated with improved microvascular vasodilation in response to endothelial receptor-dependent agonists such as ADP and serotonin (96–98), which may itself lead to significant improvement in myocardial performance. The relative contribution of this effect, especially in light of the far greater sensitivity of ischemic tissue to VEGF- and FGF-2–mediated vasomotion, to the overall improvement in myocardial blood flow and function is not well understood (58).

III. ANGIOGENESIS VERSUS ARTERIOGENESIS

Therapeutic administration of either FGFs or VEGF results in both intramyocardial as well as epicardial neovascularization. The relative importance of those events is not clear, however. The process of intramyocardial collateral development (angiogenesis) is characterized by (a) the appearance of thin-walled vessels under 200 μm in diameter with a poorly developed tunica media and (b) an increase in the number of true capillaries (<20 μm in diameter, containing only a single endothelial layer). Neoarteriogenesis, on the other hand, is characterized by the development of larger vessels (>200 μm in diameter), with well-developed media and adventitia, that usually form close to the site of the occlusion of a major epicardial coronary artery (bridging collaterals) or extend from one coronary artery to another (99,100). The distinction between these two groups of newly formed vessels is important not only because of their location but also because stimuli for their development appear to be quite different and because they exhibit different physiological properties. While increased intramyocardial vessel growth may promote better distribution of flow throughout the ischemic myocardium, the development of epicardial collaterals may well be necessary to provide the source of blood flow to intramyocardial vessels, which is impaired by obstruction of native epicardial vessels. It is not known whether both of these processes respond to the same stimuli, although it seems clear that ischemia is far more important in stimulation of intramyocardial than epicardial collateral development. Furthermore, while angiographic and morphometric studies show that both FGFs and VEGF augment both the presence of epicardial collaterals and intramyocardial neovascularization, it is not clear if these modes of new vessel growth are in any way growth factor–specific. While, on theoretical grounds, FGFs may be more likely to induce neoarteriogenesis and VEGF angiogenesis, the cross-talk between various components of the angiogenic response and the ability of bFGF to induce VEGF expression (92) suggest that there may be little difference in the final result.

IV. END POINTS AND OUTCOME MEASURES FOR ANGIOGENESIS STUDIES

A. Coronary Angiography

Coronary angiography is usually performed in multiple views with standard manual contrast injections. Evaluation of angiographic collateral density is performed by either assessing the presence of new collaterals and counting first-, second-, and third-generation collaterals or by measuring the collateral index, usually done by two independent angiographers blinded to treatment assignment (101). The ‘‘collateral index’’ is assessed for left-to-left and right-to-left collaterals using a four-point scale: 0 = no visible collateral vessels, 1 = faint filling of side branches of the main epicardial vessel without filling the main vessel, 2 = partial filling of the main epicardial vessel, and 3 = complete filling of the main vessel (101). In addition, TIMI grade flow can be measured if the artery is subtotally occluded. TIMI grades are as follows: 0 = no flow within the native vessel; 1 = faint, slow filling of the native vessel, without opacification of the distal vessel; 2 = slow filling of the entire vessel length; and 3 = brisk, normal flow within the entire vessel (102,103). Finally, spectral Doppler flow velocity distal to the occlusion or in the contralateral circulation may be used to assess average peak collateral flow velocity (104,105). Most of the angiographic methods to assess collateral perfusion are confounded by catheter engagement of the coronary artery, speed and force of manual injection, and vasodilators used; they are limited by the spatial resolution of digital angiography. Furthermore, angiography does not provide any myocardial functional data.

B. Myocardial Perfusion

Radioactive or colored microspheres can be used in preclinical studies and have provided the mainstay of myocardial perfusion assessment in animal models of angiogenesis (106,107). Typically, microspheres (10–15 μm in diameter) are injected into a proximal "mixing chamber," such as the left atrium, at different time points, and absolute regional myocardial blood flow is determined at the time of sacrifice. Microsphere flow, however, does not provide any functional left ventricular information. Contrast echocardiography may be used in animal and clinical studies to assess myocardial perfusion and left ventricular function but is limited by frequent poor echocardiographic windows, reactions to contrast agents, operator variability, and poor edge detection. Nuclear perfusion imaging has been used clinically to evaluate myocardial perfusion. This modality has several limitations, including poor spatial resolution, relative rather than absolute flow, difficult quantitative analysis, motion artifact, and nonstandardized acquisition windows and protocols. Magnetic resonance imaging has emerged as an attractive tool for the serial and noninvasive assessment of myocardial function and perfusion (108). This modality, however, needs to be validated before use in multicenter trials; it requires the availability of magnetic resonance imaging facilities; and it cannot be performed in a minority of patients with claustrophobia and pacemakers.

C. Morphometric Analysis

Although histological and morphometric analysis can detect neovascularization, it provides no functional information. Another important factor to consider in immunocytochemical or in situ hybridization analysis of new vessel formation is the time course and spatial heterogeneity of expression of various endothelial cell markers such as flk-1, flt-1, vWF, and PDGF-α receptor (109–111). Additional considerations pertain to the use of other endothelial cell markers. For example, CD-31 (PECAM) is expressed not only on endothelial cells but also on mononuclear cells, platelets, and granulocytes and thus would overestimate the extent of angiogenesis in infarct and other inflammation-dependent models. Thus while a variety of physiological end points can show beneficial effects of therapeutic growth-factor administration, attributing these effects to angiogenesis or neoarteriogenesis per se is much more difficult.

V. PRECLINICAL STUDIES OF GROWTH FACTOR–INDUCED ANGIOGENESIS

A. Acidic Fibroblast Growth Factor (FGF-1)

FGF-1 is a potent endothelial cell mitogen and chemotactic factor that is active in animal models of angiogenesis (112), dermal repair and capillary growth, and large vessel reendothelialization (113,114). Like all FGFs, FGF-1 is a pleiotropic molecule capable of stimulating growth of a variety of cell types, including vascular smooth muscle and endothelial cells. It can signal efficiently through all seven ligand-selective FGF receptor subtypes generated by alternative splicing of the four known FGF receptor genes.

Ischemic myocardium contains abundant transcripts of FGF-1 message that are predominantly localized to the growing collateral vessels (99,100,115). In contrast, normal porcine myocardium has very low to undetectable levels of FGF-1 message, although it does possess mature protein (115,116). Acidic FGF's role in cardiac angiogenesis is further supported by observation of expression of aFGF protein as well as its receptor (*flg*) in the developing rodent and avian myocytes (117,118) that persists in the immediate neonatal period, at which time both transcripts localize predominantly to nonmuscle tissues and microvasculature, suggesting that they have a role in the capillary growth known to occur during that period (117,118). Hypoxic capillary endothelial cells show a threefold increase in expression of high-affinity FGF receptors in culture along with an increased mitogenic and chemotactic responsiveness to FGF-2 (119).

Despite all these attributes, initial studies of FGF-1–induced angiogenesis were disappointing. Application of polytetrafluoroethylene (PTFE) fiber or collagen I sponges soaked with up to 800 μg of FGF-1 to an internal mammary artery pedicle positioned over the left anterior descending coronary artery distal to ameroid occlusion of the coronary artery failed to stimulate formation of angiographically visible collaterals (120). Similarly, infusion of 2.4 mg of FGF-1 with heparin over an 8-week period into an internal mammary artery (IMA) implanted into the myocardium subtended by an ameroid-occluded coronary artery failed to improve IMA flow above that seen with infusion of heparin alone (121). Finally, a 4-week infusion of FGF-1 (30 μg/hr, total dose ~20 mg) with heparin failed to improve collateral zone perfusion to a greater extent than did infusion of heparin alone (122). On the other hand, implantation of 1 μg of *E. coli*–derived FGF-1 in a PTFE sponge between the aorta and the left ventricle in rats resulted in the appearance of new vessels between the aorta and the myocardium (123). Gene therapy using replication-deficient recombinant adenovirus vectors coding for the nonsecreted form of the peptide or the secreted form resulted in histological evidence of neovascularization in the Matrigel mouse angiogenesis model (3). However, direct gene transfer into porcine arteries of a eukaryotic expression vector encoding a secreted form of FGF-1 induced intimal thickening in porcine arteries 21 days after gene transfer, with substantial intimal hyperplasia and neocapillary formation (124). Intraarterial injection of plasmids encoding for secreted and nonsecreted FGF-1 resulted in an increase in angiographic collaterals, a reduction in vascular resistance, and an increase in capillary density in a rabbit hindlimb ischemia model (125). The nonsecreted FGF-1 was less effective than the secreted form (125).

One possible explanation for the failures of FGF-1 to induce vessel formation in chronic myocardial ischemia is that wild-type protein is rapidly inactivated in vivo. The instability of wild-type FGF-1 results from formation of inactivating disulfide bonds promoted by air oxidation (126), probably catalyzed by trace metals (127). The protein's half-life may be greatly increased by replacement of one of the three buried cysteines (128) with a serine residue (S^{117}) (127). This single atomic substitution does not alter the mitogenic potency of FGF-1 and, because of its lack of exposure, does not present a distinct immunological surface epitope (126). Thus altered, S^{117}-FGF-1 has an activity half-life of 1.4 hr (126), compared to a half-life of only 15 min for the wild-type protein at 37°C, apparently reflecting a diminished rate of disulfide bond formation. Binding to heparin increases the thermal stability of wild-type FGF-1 to approximately 60°C (129) and its mitogenic half-life at 37°C by nearly 100-fold, to 24 hr (126). However, even a 1-day half-life can lead to a nearly complete loss of activity well before the end of the long period of time needed for slow-release dosing. On the other hand, in the presence of heparin, the S^{117} mutant exhibits a 10-fold greater 240-hr activity half-life (126).

With these considerations in mind, the therapeutic efficacy of the S^{117}-FGF-1 preparation was tested by incorporating 10 μg of the mutant protein into ethylene vinyl acetate copolymer (EVAc) matrices that were then applied to the ameroid-compromised myocardium of a pig (130). Comparison of FGF-1-treated and control animals demonstrated significant improvement of myocardial flow in the treated territory, accompanied by improvement in global and regional left ventricular function. Analysis of microvascular responses demonstrated normalization of abnormal responses to β-adrenergic and endothelium-dependent vasodilation by FGF-1 treatment in the ischemic myocardium (96). Thus, the S^{117}-FGF-1 mutant form of FGF-1 demonstrates significant therapeutic potential in a pig chronic ischemia model.

B. Basic Fibroblast Growth Factor (FGF-2)

FGF-2 is indeed the most extensively studied growth factor in a variety of angiogenesis models. FGF-2 is detectable in both atrial and ventricular adult myocardium as well as in intramyocardial vessels (131). Unlike FGF-1, the pattern of FGF-2 expression during embryogenesis does not reveal any spatial association between FGF-2 expression and the developing vasculature. Furthermore, a homozygous deletion of the FGF-2 gene is not associated with any significant phenotype with decreased vascular smooth muscle contractility, low blood pressure, and thrombocytosis (132). However, like FGF-1, FGF-2 can induce neovascularization in a variety of in vitro and in vivo systems (133,134). Owing to its unique properties, FGF-2 can be stored in the basement membrane and then released by administration of heparin or by enzymatic matrix degradation.

The ability of FGF-2 to induce significant angiogenesis in mature tissues was suggested by studies in the canine and porcine infarction models. In these experiments, intracoronary injections of FGF-2 in the setting of acute coronary thrombosis in dogs or intracoronary injections of 75–150 μm of FGF-2 containing Affigel beads in pigs led to a significantly higher vessel count compared to controls (135,136).

Studies of the therapeutic efficacy of FGF-2 in chronic myocardial ischemia have explored a variety of routes of administration in both canine and porcine models of myocardial ischemia. Daily injections of 110 μg of FGF-2 for 18 days directly into the circumflex coronary artery distal to an ameroid occluder hastened restoration of flow in the compromised territory compared to normal saline controls (2). Morphometric analysis of LCX myocardium demonstrated equal capillary densities in both FGF-2–treated and control animals but a significant (twofold) increase in the number of larger (>20 μm) vessels (2). To examine the effect of systemic FGF-2 administration, the same group of investigators carried out daily left atrial injections of 1.74 mg of FGF-2 for 18 days. This mode of FGF-2 administration resulted in early augmentation of coronary flow in the growth factor–treated animals that was comparable to that seen with direct intracoronary injections but was lost by day 38 (137). Prolonged (up to 6 months) infusions of FGF-2 did not result in any further increase in collateral blood flow, which was identical in both FGF-2- and saline-treated animals. This underscored the priming role of ischemia for bFGF-induced collateral development (138). The myocardial angiogenic effect of FGF-2 was not associated with structural or vasoproliferative effect on the nonischemic retina of dogs with myocardial ischemia (139). In the same canine model, 7-day systemic arterial administration of FGF-2 enhanced collateral development without increasing neointimal accumulation at sites of vascular injury (140). In summary, these studies have clearly documented increased coronary perfusion following FGF-2 treatment, which is probably predominately mediated by neoarteriogenesis, as suggested by the lack of increase in capillary density. At the same time, in addition to employing pharmacological doses of the growth factor, these investigations did not evaluate the functional significant of FGF-2–induced angiogenesis.

Local perivascular delivery of FGF-2 was evaluated in a porcine model of chronic myocardial ischemia (LCX ameroid occlusion). Heparin-alginate microcapsules were used for sustained delivery of FGF-2. This form of delivery is characterized by zero-order release kinetics of the growth factor from the polymer over a 4–5 week period, ease of implantation, and the absence of any inflammatory reaction associated with polymer placement (141–145). Furthermore, perivascular delivery has a potential advantage of bypassing the endothelial barrier and avoiding rapid washout of growth factor due to rapid arterial blood flow (145). Animals implanted with heparin-alginate pellets containing 8 μg of FGF-2 at the time of ameroid placement (~5 μg released over the course of the experiment—approximately 328-fold less than in previously discussed studies) demonstrated significantly better preservation of perfusion of the ischemic zone during pacing compared to control animals (143). In addition, ventricular function studies demonstrated better preservation of regional left ventricular function in the ameroid-compromised territory at rest and faster recovery following pacing in FGF-2–treated animals (143). However, rapid pacing induced equal and significant deterioration in both FGF-2–treated and control groups. Examination of the effect of progressively larger amounts of FGF-2 (10 and 100 μg) delivered in a similar manner in a pig model (142) demonstrated substantial improvement in resting coronary blood flow in the chronically ischemic myocardium in both FGF-2 groups compared to controls and an increase in angiographic collaterals (142,143).

Analysis of left ventricular function demonstrated a higher ejection fraction at rest and during pacing in both 10- and 100-μg FGF-2 groups compared to controls. Similarly, regional wall motion in the ischemic territory was better preserved at rest in both FGF-2 groups, although only the 100 μg FGF-2 group maintained normal wall thickening during rapid pacing (142). Thus, perivascular sustained-release FGF-2 resulted in a dose-dependent improvement of coronary perfusion accompanied by an improvement in left ventricular function (142,143).

Given this increase in the number of epicardial collaterals, the observed improvement in coronary blood flow and left ventricular function in these studies may have been equally due to FGF-2-induced neoarteriogenesis as well as true myocardial angiogenesis. In addition to improving angiographic collaterals, myocardial perfusion, and left ventricular function, perivascular delivery of FGF-2 resulted in normalization of ischemia-induced impairment of endothelium-dependent vasodilation in the LCX ischemic area (97,146). Additional studies of FGF-2 delivery modalities demonstrated enhanced new epicardial small-vessel growth following intrapericardial infusion using an osmotic pump in a rabbit model of angiotensin-II induced left ventricular hypertrophy (147).

Most attempts to stimulate myocardial angiogenesis have employed methods of prolonged growth-factor delivery. However, some of these options, such as repeated intracardiac injections or infusions, are either unfeasible or impractical in patients with ischemic heart disease. Therefore, single-dose delivery—either intracoronary, intravenous, or intrapericardial—may be more practical clinically. Intrapericardial delivery of FGF-2 (from 30 μg up to 2 mg) in a porcine LCX ameroid occlusion model resulted in significant increases in left-to-left angiographic collaterals and blood flow, which was accompanied by improvements in myocardial perfusion and function in the ischemic territory as measured by magnetic resonance imaging as well as by histological evidence of increased myocardial vascularity (69). None of these benefits were seen in saline- or heparin-treated ischemic animals.

Local intramyocardial injection of FGF-2 given 4–5 weeks after creation of an infarct in pigs resulted in an increase in the number of arterioles but not capillaries compared to control injections. FGF-2's effect was potentiated by coadministration with heparin (148).

In addition to stimulation of neovascularization in the setting of chronic ischemia, FGF-2 administration also results in beneficial effects in acute myocardial ischemia. Intracoronary administration of 20 μg of FGF-2 in a canine experimental myocardial infarct model improved cardiac systolic function, reduced infarct size, and increased the number of arterioles and capillaries in the infarct zone (135). Similar findings were observed with intrapericardial injection of 30 μg of FGF-2 in dogs with acute myocardial infarction; this resulted in significant reduction of the size of myocardial infarction as well as improvement in left ventricular function, both of which were augmented by a simultaneous administration of heparan sulfate (149). These findings were duplicated in an acute ischemia/reperfusion study in which 10 μg of FGF-2 was given 10 min after initiation of coronary occlusion and again immediately before reperfusion, resulting in significant salvage of myocardium at risk compared to control animals (150). These beneficial effects may have resulted from the vasodilatory action of FGF-2 (151,152), a myocardial protective effect, possibly via a NO-dependent pathway (153,154), as well as an antiapoptotic effect (155).

C. Fibroblast Growth Factor-4 (FGF-4)

Disruption of the fibroblast growth-factor receptor 2 gene results in a recessive embryonic lethal mutation, suggesting an important role of FGF-4 in embryonic development. Preimplantation development is normal until the blastocyst stage (156). FGF-4 appears to play an important role in tumor growth and tumor angiogenesis (157). Conditioned media from mammary cells transfected with FGF-4 cDNA induced an angiogenic phenotype in human umbilical vein endothelial cells cultured on type I collagen gel, with formation of tubelike structures on invasion of the gel, an effect completely suppressed by adding a neutralizing antibody directed against VEGF (90). Recombinant human FGF-4 did not induce an angiogenic phenotype in HUVECs by itself; however, FGF-4–transfected cells had an increased expression of VEGF, which was further increased in a dose-dependent fashion by increasing concentration of recombinant human FGF-4 (90). Therefore, FGF-4's angiogenic effect may be mediated through VEGF induction.

D. Fibroblast Growth Factor-5 (FGF-5)

FGF-5 is expressed in choroidal neovascular membranes removed from patients with age-related macular degeneration and is found primarily in blood vessels and the surrounding extracellular matrix (158). A single study has examined the efficacy of an adenovirus mediated transfer of a human FGF-5 construct into ischemic myocardium in a pig ameroid model (159). Two intracoronary injections of 2×10^{11} viral particles of recombinant adenovirus carrying full-length cDNA encoding human FGF-5 gene into the left main and right coronary arteries were performed 3 weeks following ameroid implantation; this led to improved perfusion in FGF-5–treated but not control lacZ-treated pigs compared to the preinjection baseline and it was associated with an improvement in regional left ventricular function. Interestingly, increased capillary numbers were seen both in ischemic and nonischemic territories. Microscopic examination did not demonstrate any significant inflammation associated with adenoviral-FGF-5 injections (159).

E. Vascular Endothelial Growth Factor

Four major forms of VEGF are generated as a result of alternative splicing of a single gene: $VEGF_{121}$, $VEGF_{165}$, $VEGF_{189}$, and $VEGF_{206}$. VEGF transcripts are detected in all cardiac tissues. VEGF signaling is mediated by two tyrosine kinase receptors, *flk-1* and *flt-1*, which are expressed in a variety of endothelial cells. Like FGFs, VEGF and its receptors' expression in the heart is induced sevenfold by hypoxia/ischemia (160–164).

While the work of Isner and colleagues has demonstrated considerable therapeutic efficacy of VEGF administration either in protein or gene therapy form in animal models of limb ischemia and in patients with ischemic limb disease (165–169), experience with VEGF administration in the setting of myocardial ischemia has been more controversial. Two studies carried out by Unger and colleagues in a dog ameroid model produced opposite results. In one study daily intracoronary injections of 45 μg of VEGF delivered distal to the occluder over a 28-day period (total dose 900 μg) resulted in faster restoration of collateral zone flow than similar injections of normal saline (170). However, a 7-day course of VEGF infusions (total dose 720 μg) (140) failed to produce any effect. It is not clear why there was such a dramatic difference between the two studies; the reduction in total dose and a shorter duration of infusion in the second study may have played a role.

The therapeutic efficacy of VEGF delivered via an implantable pump periadventitially to the circumflex coronary artery distal to the ameroid occluder was tested in a porcine ameroid model (107). In these studies, VEGF treatment was associated with better preservation of coronary flow in the ameroid zone during pacing. Assessment of myocardial perfusion using magnetic resonance imaging demonstrated not only significantly better perfusion of the compromised territory in VEGF-treated animals but also a reduction in size of this territory (108). Morphometric analysis found nearly a fourfold increase in the number of collateral vessels in VEGF-treated animals, limited to the ischemic zone (107).

Analysis of microvascular function demonstrated significantly better restoration of endothelium-mediated, receptor-dependent relaxation in VEGF-treated animals (98,107). These improvements in coronary flow and microvascular function were reflected in enhanced left ventricular performance in VEGF-treated animals, as demonstrated by higher ejection fraction and better preservation of regional wall shortening during pacing stress (107). Thus, VEGF administration in this porcine ameroid model of myocardial ischemia increased capillary density, improved coronary blood flow and myocardial perfusion, restored microvascular function, and potentiated recovery of global and regional left ventricular function.

Despite these effects of perivascular delivery, bolus intracoronary delivery is much more appealing for clinical use. The feasibility of intracoronary delivery was tested by Hariawala et al., who injected 2 mg of rh $VEGF_{165}$ into the left coronary artery of 8 pigs. Four of the 8 animals survived the injection (with 4 animals dying acutely of refractory hypotension); however, 30 days later, the remaining animals demonstrated improved coronary flow compared to the control group (171). Given 50% mortality of high-dose single-bolus VEGF therapy, the efficacy of a lower dose (20 μg) single-bolus intracoronary LCX infusion was compared to the same amount of VEGF delivered either by perivascular, intracoronary, or local injection in the coronary artery using an InfusaSleeve catheter (68,98). The studies conducted in a porcine ameroid constrictor model demonstrated that both intracoronary bolus injection and local delivery resulted in a significant increase in angiographically visible left-to-left but not right-to-left collaterals and improvement in myocardial blood flow, regional left ventricular function, and microvascular function. Although there was a trend toward better results with local compared to intracoronary bolus delivery, both methods appeared equally efficacious. There was no significant hemodynamic compromise associated with any of these delivery approaches (68,98). Thus, VEGF, delivered by either intracoronary or periadventitial methods, appears to be an effective angiogenic agent.

Although most of the angiogenic studies were performed using $VEGF_{165}$, other VEGF isoforms appear to have angiogenic properties. Direct administration of an adenoviral vector expressing VEGF121 cDNA into areas of ischemic myocardium in a porcine ameroid constrictor model resulted in increased angiographic collaterals, a significant reduction in the ischemic area at stress (using nuclear imaging), and improved segmental wall thickening (by echocardiography), compared to controls (172).

The hemodynamic effects of VEGF administration are mediated by NO release and are secondary to microvasular dilation (40,58–60,173). Intracoronary VEGF results in a dose-dependent increase in Doppler-measured coronary flow and dose-dependent hypotension, both effects prevented by pretreatment with L-NAME (59). In addition, VEGF's angiogenic effects may be mediated at least in part via a NO-dependent pathway (35,38,67). L-arginine, which augments endogenous production of NO, resulted in enhanced angiogenesis in a rabbit ischemic hindlimb model (35). Angiogenesis measured by laser Doppler flow analysis and capillary density in a mouse ischemic hindlimb model was significantly impaired in eNOS knockout mice and not augmented by VEGF treatment (35).

VI. CLINICAL STUDIES OF GROWTH FACTOR–INDUCED ANGIOGENESIS

With the presence of a large population of patients with ischemic heart disease who are suboptimal candidates for coronary artery bypass grafting (CABG) or angioplasty and with the relatively high percentage (up to 37% in some series) of patients undergoing CABG with incomplete revascularization (174,175), the above described promising preclinical results of growth factor–induced angiogenesis propelled the use of these growth factors in clinical studies of therapeutic angiogenesis. To date few studies have been published. However, several ongoing studies are testing the therapeutic potential of growth factors in ischemic heart disease (176–179).

A double-blind placebo-controlled trial sponsored by the National Institutes of Health is examining the effect of local perivascular bFGF (10 or 100 μg) delivered using heparin-alginate beads implanted at the time of CABG surgery in patients who have a viable but underperfused myocardial territory not amenable to bypass grafting. Implantation of growth factor–containing beads added on average 2.8 ± 1.1 min to operative time. Interim analysis of the first 10 patients demonstrated no acute or chronic hemodynamic effects following growth-factor implantations and no significant changes in hematological or chemistry profiles during follow-up; no other adverse events were noted (177,178). Plasma bFGF levels did not increase above baseline. Nuclear perfusion scans and magnetic resonance imaging demonstrated improved results with administration of 100 μg of bFGF (177,178). FGF-1 (0.01 mg/kg body weight) was injected close to the bypassed vessels after the completion of internal mammary artery → left anterior descending coronary artery anastomosis in 20 patients with three-vessel coronary disease. LIMA angiography 12 weeks later demonstrated somewhat increased vessel density at the site of injections as assessed by digital angiography (179). A phase I study of intracoronary (15 patients) and intravenous (28 patients) VEGF in patients with severe coronary disease has been completed and showed a dose-dependent hypotensive effect of VEGF, which was the dose-limiting toxicity. Preliminary efficacy data were promising, with improved symptoms (176). However, the phase II double-blind placebo-controlled trial of intracoronary/intravenous recombinant vascular endothelial growth factor trial (VIVA) did not show any benefit of VEGF as compared to placebo (179A). Unger and colleagues are examining the safety of intracoronary infusion of progressively larger doses of FGF-2 in patients with documented coronary disease (E. Unger, personal communication). A phase I dose-escalation study of intracoronary (52 patients) and intravenous (14 patients) bFGF in patients with severe coronary disease has been completed and demonstrated the safety and tolerability of bFGF. The doses tested ranged from 0.33 μg/kg (ideal body weight) to 48 μg/kg. Hypotension occurred at 48 μg/kg in 2/10 patients in that group and was successfully treated with intravenous fluids. The maximum tolerated dose was determined at 36 μg/kg (defined by hypotension). There were no changes in ECG or cardiac enzymes and no peripheral or pulmonary edema. Transient leukocytosis was noted on Day 2 at doses > 24 μg/kg and transient proteinuria was observed in 5 of 66 patients and was not dose-related. No retinal toxicity was observed at 60 days (179B). The preliminary efficacy data is promising, with improved exercise time, quality of life using the Seattle Angina Questionnaire, and improved target wall motion and perfusion using magnetic resonance imaging (179B). The phase II study is currently underway. A phase I study with a gene-therapy strategy used an E1(-)E3(-) adenovirus (Ad) gene-transfer vector expressing human vascular endothelial growth factor (VEGF) 121 cDNA to induce therapeutic angiogenesis in the myocardium of individuals with clinically significant coronary artery disease. Ad(GV)VEGF121 was administered to 21 individuals by direct myocardial injection into an area of reversible ischemia either as an adjunct to conventional coronary artery bypass grafting ($n = 15$) or as sole therapy via a minithoracotomy ($n = 6$). There was no evidence of systemic or cardiac-related adverse events related to vector administration. In both groups, coronary angiography and stress sestamibi scan assessment of wall motion 30 days after therapy suggested improvement in the area of vector administration. All patients reported improvement in angina class after therapy (179C). In addition, a gene therapy trial using surgical intramyocardial injection of adenoviral constructs with $VEGF_{121}$ and plasmid $VEGF_{165}$ and intracoronary injection of an adenoviral construct with FGF-4 are currently under way.

VII. SIDE EFFECTS OF GROWTH FACTOR–INDUCED ANGIOGENESIS

The administration of pluripotent agents such as FGFs and VEGF is likely to produce untoward side effects. These growth factors have several effects besides their angiogenic potential, which, of course, may induce neovascularization in other organs and promote tumor angiogenesis. Growth factors have been implicated in the development of diabetic retinopathy and various ocular neovascularization processes (158,180–186). Intraocular expression of VEGF is closely correlated with diabetic retinopathy, retinal vein occlusion, and retinopathy of prematurity as well as ischemia-induced retinal neovascularization in mice (181,182,184–186). VEGF-neutralizing chimeric proteins reduced retinal neovascularization in an animal model of ischemic retinopathy (185). Tumor angiogenesis and progression is another theoretical side effect, particularly with mounting evidence implicating growth factors in tumor angiogenesis and investigating the antitumor effects of angiogenesis inhibitors (187–193). However, no clinical data are yet available on VEGF- or FGF-induced retinopathy or tumorogenicity.

Furthermore, growth factors, particularly VEGF, have significant hemodynamic effects. Both VEGF and FGF-2 induce microvascular vasodilation in vitro and in vivo (58–60,171). Intracoronary VEGF results in a dose-dependent hypotension that was dose-limiting in both animals and human studies (59,60,176,194). The hypotensive effect of VEGF appears to be NO-dependent (inhibited by L-NAME) and is caused by microvascular dilation (59,60,176,194). In addition, VEGF, also known as vascular permeability factor, results in increased vascular permeability that appears to be mediated via release of platelet activating factor and nitric oxide (62,63,80). In limited clinical experience, local plasmid-mediated VEGF gene transfer to patients with limb ischemia has led to development of extremity edema (165,166). Administration of basic FGF in high dosages is also associated with a number of side effects, including anemia, thrombocytopenia, membranous nephropathy with morphological evidence of glomerular damage, and hyperostosis (64,141,195).

Of particular concern is the potential for various growth factors to promote atherosclerosis and increase plaque angiogenesis with the propensity for plaque rupture and acute coronary syndromes as well as to induce neoinitimal hyperplasia following percutaneous or surgical revascularization. Plaque instability may also result from increased permeability (VEGF). Dogs underwent balloon denudation injury of the iliofemoral artery followed by a 7-day systemic treatment with FGF-2 and VEGF (140). FGF-2 did not exacerbate neointimal accumulation, while VEGF exacerbated neointimal thickening after vascular injury (140). However other studies did not confirm this finding. When VEGF was administered for 4 weeks (120 μg IV twice weekly) following balloon angioplasty–induced endothelial denudation of rabbit carotid artery, there was a significant increase in the in vivo regeneration of endothelium compared to controls (196). Plasmid VEGF gene transfer to a rabbit femoral balloon injury model resulted in more rapid reendothelialization, diminished neointimal thickening, reduction in thrombogenicity, and improved endothelium-dependent vasomotor reactivity compared to controls. Finally, plasmid-VEGF delivered locally to rabbit stented iliac artery resulted in accelerated endothelialization and reduced intimal thickening (197). Similar results were observed for FGF-1 and FGF-2 (114,198–200).

REFERENCES

1. J Muhlhauser, MJ Merrill, R Pili, H Maeda, M Bacic, B Bewig, A Passaniti, NA Edwards, RG Crystal, MC Capogrossi. VEGF165 expressed by a replication-deficient recombinant adenovirus vector induces angiogenesis in vivo. Circ Res 77:1077–1086, 1995.
2. EF Unger, S Banai, M Shou, DF Lazarous, MT Jaklitsch, M Scheinowitz, R Correa, C Kling-

beil, SE Epstein. Basic fibroblast growth factor enhances myocardial collateral flow in a canine model. Am J Physiol 266:H1588–H1595, 1994.
3. J Muhlhauser, R Pili, MJ Merrill, H Maeda, A Passaniti, RG Crystal, MC Capogrossi. In vivo angiogenesis induced by recombinant adenovirus vectors coding either for secreted or nonsecreted forms of acidic fibroblast growth factor. Hum Gene Ther 6:1457–1465, 1995.
4. ER Edelman, MA Nugent, LT Smith, MJ Karnovsky. Basic fibroblast growth factor enhances the coupling of intimal hyperplasia and proliferation of vasa vasorum in injured rat arteries. J Clin Invest 89:465–473, 1992.
5. PA Burke, K Lehmann-Bruinsma, JS Powell. Vascular endothelial growth factor causes endothelial proliferation after vascular injury. Biochem. Biophys Res Commun 207:348–354, 1995.
6. MS O'Reilly, Y Shing, Y Cao, J Folkman. Endogenous inhibitors of angiogenesis (abstr). Proc Annu Meet Am Assoc Cancer Res 37:669, 1996.
7. MS O'Reilly, L Holmgren, Y Shing, C Chen, RA Rosenthal, M Moses, WS Lane, Y Cao, EH Sage, J Folkman. Angiostatin: a novel angiogenesis inhibitor that mediates the suppression of metastases by a Lewis lung carcinoma. Cell 79:315–328, 1994.
8. MS O'Reilly, L Holmgren, Y Shing, C Chen, RA Rosenthal, Y Cao, M Moses, WS Lane, EH Sage, J Folkman. Angiostatin: a circulating endothelial cell inhibitor that suppresses angiogenesis and tumor growth. Cold Spring Harb Symp Quant Biol 59:471–482, 1994.
9. Z Wu, MS O'Reilly, J Folkman, Y Shing. Suppression of tumor growth with recombinant murine angiostatin. Biochem Biophys Res Commun 236:651–654, 1997.
10. MS O'Reilly, T Boehm, Y Shing, N Fukai, G Vasios, WS Lane, E Flynn, JR Birkhead, BR Olsen, J Folkman. Endostatin: an endogenous inhibitor of angiogenesis and tumor growth. Cell 88:277–285, 1997.
11. D Westernacher, W Schaper. A novel heart derived inhibitor of vascular cell proliferation: purification and biological activity. J Mol Cell Cardiol 27:1535–1543, 1995.
12. N Bertin, P Clezardin, R Kubiak, L Frappart. Thrombospondin-1 and -2 messenger RNA expression in normal, benign, and neoplastic human breast tissues: correlation with prognostic factors, tumor angiogenesis, and fibroblastic desmoplasia. Cancer Res 57:396–399, 1997.
13. TS Panetti, H Chen, TM Misenheimer, SB Getzler, DF Mosher. Endothelial cell mitogenesis induced by LPA: inhibition by thrombospondin-1 and thrombospondin-2. J Lab Clin Med 129:208–216, 1997.
14. VA Patel, DJ Hill, MC Eggo, MC Sheppard, GP Becks, A Logan. Changes in the immunohistochemical localisation of fibroblast growth factor-2, transforming growth factor-beta 1 and thrombospondin-1 are associated with early angiogenic events in the hyperplastic rat thyroid. J Endocrinol 148:485–499, 1996.
15. OV Volpert, V Stellmach, N Bouck. The modulation of thrombospondin and other naturally occurring inhibitors of angiogenesis during tumor progression. Breast Cancer Res Treat 36: 119–126, 1995.
16. OV Volpert, SS Tolsma, S Pellerin, JJ Feige, H Chen, DF Mosher, N Bouck. Inhibition of angiogenesis by thrombospondin-2. Biochem Biophys Res Commun 217:326–332, 1995.
17. TR Kyriakides, YH Zhu, LT Smith, et al. Mice that lack thrombospondin 2 display connective tissue abnormalities that are associated with disordered collagen fibrillogenesis, an increased vascular density, and a bleeding diathesis. J Cell Biol 140:419–430, 1998.
18. M Nguyen, J Arkell, CJ Jackson. Active and tissue inhibitor of matrix metalloproteinase-free gelatinase B accumulates within human microvascular endothelial vesicles. J Biol Chem 273:5400–5404, 1998.
19. CJ Jackson, J Arkell, M Nguyen. Rheumatoid synovial endothelial cells secrete decreased levels of tissue inhibitor of MMP (TIMP1). Ann Rheum Dis 57:158–161, 1998.
20. DA Sipkins, DA Cheresh, MR Kazemi, LM Nevin, MD Bednarski, KC Li. Detection of tumor angiogenesis in vivo by alphaVbeta3-targeted magnetic resonance imaging. Nature Med 4:623–626, 1998.

21. PC Brooks, S Silletti, TL von Schalscha, M Friedlander, DA Cheresh. Disruption of angiogenesis by PEX, a noncatalytic metalloproteinase fragment with integrin binding activity. Cell 92:391–400, 1998.
22. BP Eliceiri, R Klemke, S Stromblad, DA Cheresh. Integrin alphavbeta3 requirement for sustained mitogen-activated protein kinase activity during angiogenesis. J Cell Biol 140:1255–1263, 1998.
23. FY Zhou, RT Owens, J Hermonen, M Jalkanen, M Hook. Is the sensitivity of cells for FGF-1 and FGF-2 regulated by cell surface heparan sulfate proteoglycans? Eur J Cell Biol 73: 166–174, 1997.
24. MS Filla, P Dam, AC Rapraeger. The cell surface proteoglycan syndecan-1 mediates fibroblast growth factor-2 binding and activity. J Cell Physiol 174:310–321, 1998.
25. M Kato, H Wang, V Kainulainen, ML Fitzgerald, S Ledbetter, DM Ornitz, M Bernfield. Physiological degradation converts the soluble syndecan-1 ectodomain from an inhibitor to a potent activator of FGF-2. Nature Med. 4:691–697, 1998.
26. S Moncada, A Higgs. The L-arginine-nitric oxide pathway. N Engl J Med 329:2002–2012, 1993.
27. M Uematsu, Y Ohara, JP Navas, K Nishida, TJ Murphy, RW Alexander, RM Nerem, DG Harrison. Regulation of endothelial cell nitric oxide synthase mRNA expression by shear stress. Am J Physiol 269:C1371–C1378, 1995.
28. RA Kelly, JL Balligand, TW Smith. Nitric oxide and cardiac function. Circ Res 79:363–380, 1996.
29. A Orucevic, S Hearn, P Lala. The role of active inducible nitric oxide synthase expression in the pathogenesis of capillary leak syndrome resulting from interleukin 2 therapy in mice. Lab Invest 76:53–65, 1997.
30. K Singh, JL Balligand, TA Fischer, TW Smith, RA Kelly. Regulation of cytokine-inducible nitric oxide synthase in cardiac myocytes and microvascular endothelial cells. Role of extracellular signal-regulated kinases 1 and 2 (ERK1/ERK2) and STAT1 alpha. J Biol Chem 271: 1111–1117, 1996.
31. JL Lu, LM Schmiege, 3rd, L Kuo, J Liao. Downregulation of endothelial constitutive nitric oxide synthase expression by lipopolysaccharide. Biochem Biophys Res Commun 225:1–5, 1996.
32. R Bryk, DJ Wolff. Mechanism of inducible nitric oxide synthase inactivation by aminoguanidine and L-N6-(1-iminoethyl)lysine. Biochemistry 37:4844–4852, 1998.
33. V Ranjan, Z Xiao, SL Diamond. Constitutive NOS expression in cultured endothelial cells is elevated by fluid shear stress. Am J Physiol 269:H550–H555, 1995.
34. RM Tuder, BE Flook, NF Voelkel. Increased gene expression for VEGF and the VEGF receptors KDR/Flk and Flt in lungs exposed to acute or to chronic hypoxia: modulation of gene expression by nitric oxide. J Clin Invest 95:1798–1807, 1995.
35. T Murohara, T Asahara, M Silver, et al. Nitric oxide synthase modulates angiogenesis in response to tissue ischemia. J Clin Invest 101:2567–2578, 1998.
36. LP McQuillan, GK Leung, PA Marsden, SK Kostyk, S Kourembanas. Hypoxia inhibits expression of eNOS via transcriptional and posttranscriptional mechanisms. Am J Physiol 267: H1921–H1927, 1994.
37. SK Kostyk, S Kourembanas, EL Wheeler, D Medeiros, LP McQuillan, PA D'Amore, SJ Braunhut. Basic fibroblast growth factor increases nitric oxide sythase production in bovine endothelial cells. Am J Physiol 269:H1583–H1589, 1995.
38. JD Hood, CJ Meininger, M Ziche, HJ Granger. VEGF upregulates ecNOS message, protein, and NO production in human endothelial cells. Am J Physiol 274:H1054–H1058, 1998.
39. G Garcia-Cardena, P Martasek, B Masters, P Skidd, J Couet, S Li, M Lisanti, W Sessa. Dissecting interaction between nitric oxide synthase (NOS) and caveolin. J Biol Chem 272: 25437–25440, 1997.
40. M Ziche, L Morbidelli, R Choudhuri, HT Zhang, S Donnini, HJ Granger, R Bicknell. Nitric

oxide synthase lies downstream from vascular endothelial growth factor-induced but not basic fibroblast growth factor-induced angiogenesis. J Clin Invest 99:2625–2634, 1997.
41. M Ziche, L Morbidelli, E Masini, S Amerini, H Granger, C Maggi, P Geppetti, F Ledda. Nitric oxide mediates angiogenesis in vivo and endothelial cell growth and migration in vitro promoted by substance P. Am J Physiol 94:2036–2044, 1994.
42. W Yang, J Ando, R Korenaga, T Toyo-oka, A Kamiya. Exogenous nitric oxide inhibits proliferation of cultured vascular endothelial cells. Biochem. Biophys. Res. Commun. 203: 1160–1167, 1994.
43. R Sarkar, R Webb, J Stanley. Nitric oxide inhibition of endothelial cell mitogenesis and proliferation. Surgery 118:274–279, 1995.
44. E Pipili-Synetos, A Papageorgiou, E Sakkoula, G Sotiropoulou, T Fotsis, G Karakiulakis, ME Maragoudakis. Inhibition of angiogenesis, tumor growth and metastasis by the NO-releasing vasodilators, isosorbide mononitrate and dinitrate. Br J Pharmacol 116:1829–1834, 1995.
45. E Pipili-Synetos, E Sakkoula, G Haralabopoulos, P Andriopoulou, P Peristeris, ME Maragoudakis. Evidence that nitric oxide is an endogenous antiangiogenic mediator. Br J Pharmacol 111:894–902, 1994.
46. G Montrucchio, E Lupia, A de Martino, E Battaglia, M Arese, A Tizzani, F Bussolino, G Camussi. Nitric oxide mediates angiogenesis induced in vivo by platelet-activating factor and tumor necrosis factor-alpha. Am J Pathol 151:557–563, 1997.
47. L Morbidelli, CH Chang, JG Douglas, HJ Granger, F Ledda, M Ziche. Nitric oxide mediates mitogenic effect of VEGF on coronary venular endothelium. Am. J. Physiol. 270:H411–H415, 1996.
48. G Garcia-Cardena, J Folkman. Is there a role for nitric oxide in tumor angiogenesis? (editorial; comment). J Natl Cancer Inst 90:560–561, 1998.
49. J Chambers, G Voss, J Snider, S Meyer, J Cartland, R Wilson. The direct in vivo effects of nitric oxide on the coronary circulation. Am J Physiol 271:H1584–H1593, 1996.
50. S Babaei, K Teichert-Kuliszewska, JC Monge, F Mohamed, MP Bendeck, DJ Stewart. Role of nitric oxide in the angiogenic response in vitro to basic fibroblast growth factor. Circ Res 82:1007–1015, 1998.
51. A Papapetropoulos, G Garcia-Cardena, JA Madri, WC Sessa. Nitric oxide production contributes to the angiogenic properties of vascular endothelial growth factor in human endothelial cells. J Clin Invest 100:3131–3139, 1997.
52. P Pintavorn, BJ Ballermann. TGF-beta and the endothelium during immune injury. Kidney Int 51:1401–1412, 1997.
53. T Yamamoto, N Terada, Y Nishizawa, H Tanaka, H Akedo, A Seiyama, T Shiga, H Kosaka. Effects of NG-nitro-L-arginine and/or L-arginine on experimental pulmonary metastasis in mice. Cancer Lett 87:115–120, 1994.
54. JH Guh, TL Hwang, FN Ko, SC Chueh, MK Lai, CM Teng. Antiproliferative effect in human prostatic smooth muscle cells by nitric oxide donor. Mol Pharmacol 53:467–474, 1998.
55. K Sumitani, R Kamijo, M Nagumo. Cytotoxic effect of sodium nitroprusside on cancer cells: involvement of apoptosis and suppression of c-myc and c-myb proto-oncogene expression. Anticancer Res 17:865–871, 1997.
56. F Blachier, V Robert, M Selamnia, C Mayeur, PH Duee. Sodium nitroprusside inhibits proliferation and putrescine synthesis in human colon carcinoma cells. FEBS Lett 396:315–318, 1996.
57. QP Lee, HW Park, J Thayer, PE Mirkes, MR Juchau. Apoptosis induced in cultured rat embryos by intra-amniotically microinjected sodium nitroprusside. Teratology 53:21–30, 1996.
58. FW Sellke, SY Wang, A Stamler, JJ Lopez, J Li, M Simons. Enhanced microvascular relaxations to VEGF and bFGF in chronically ischemic porcine myocardium. Am J Physiol 271: H713–H720, 1996.

59. JJ Lopez, RJ Laham, JP Carrozza, M Tofukuji, FW Sellke, S Bunting, M Simons. Hemodynamic effects of intracoronary VEGF delivery: evidence of tachyphylaxis and NO dependence of response. Am J Physiol 273:H1317–H1323, 1997.
60. R Laham, M Tofukuji, F Sellke, M Simons. Vascular endothelial growth factor (VEGF) affects microvessels but not epicardial arteries and veins. Circulation 96:I-551, 1997.
61. B Malavaud, I Tack, F Jonca, F Praddaude, F Moro, JL Ader, J Plouet. Activation of Flk-1/KDR mediates angiogenesis but not hypotension. Cardiovasc Res 36:276–281, 1997.
62. H Wu, Q Huang, Y Yuan, H Granger. VEGF induces NO-dependent hyperpermeability in coronary venules. Am J Physiol 271:H2735–H2739, 1996.
63. T Murohara, J Horowitz, B Silver, Y Tsurumi, D Chen, A Sullivan, J Isner. Vascular endothelial growth factor/vascular permeability factor enhances vascular permeability via nitric oxide and prostacyclin. Circulation 97:99–107, 1998.
64. J Floege, M Burg, C Hugo, KL Gordon, H Van Goor, M Reidy, WG Couser, KM Koch, RJ Johnson. Endogenous fibroblast growth factor-2 mediates cytotoxicity in experimental mesangioproliferative glomerulonephritis. J Am Soc Nephrol 9:792–801, 1998.
65. A Dembinska-Kiec, J Dulak, L Partyka, R Krzesz, D Dudek, S Bartus, M Polus, I Guevara, I Wybranska, T Krzeminski. Induction of nitric oxide synthase (NOS) and vascular endothelial growth factor (VEGF) in experimental model of angioplasty and heart ischemia. Adv Exp Med Biol 433:163–167, 1997.
66. Z Huang, K Chen, PL Huang, SP Finklestein, MA Moskowitz. bFGF ameliorates focal ischemic injury by blood flow-independent mechanisms in eNOS mutant mice.
67. A Parenti, L Morbidelli, XL Cui, JG Douglas, JD Hood, HJ Granger, F Ledda, M Ziche. Nitric oxide is an upstream signal of vascular endothelial growth factor–induced extracellular signal-regulated kinase 1/2 activation in postcapillary endothelium. J Biol Chem 273:4220–4226, 1998.
68. JJ Lopez, RJ Laham, A Stamler, JD Pearlman, S Bunting, A Kaplan, JP Carrozza, FW Sellke, M Simons. VEGF administration in chronic myocardial ischemia in pigs. Cardiovasc Res 40:272–281, 1998.
69. R Laham, M Rezaee, M Post, M Novicki, F Sellke, J Pearlman, D Hung, M Simons. A single intrapericardial dose of basic fibroblast growth factor induces functional angiogenesis in a porcine model of chronic myocardial ischemia (abstr). Circulation 98:I–794, 1998.
70. M Lansink, P Koolwijk, V van Hinsbergh, T Kooistra. Effect of steroid hormones and retinoids on the formation of capillary-like tubular structures of human microvascular endothelial cells in fibrin matrices is related to urokinase expression. Blood 92:927–938, 1998.
71. Z Qu, X Huang, P Ahmadi, P Stenberg, JM Liebler, AC Le, SR Planck, JT Rosenbaum. Synthesis of basic fibroblast growth factor by murine mast cells: regulation by transforming growth factor beta, tumor necrosis factor alpha, and stem cell factor. Int Arch Allergy Immunol 115:47–54, 1998.
72. ML Iruela-Arispe, HF Dvorak. Angiogenesis: a dynamic balance of stimulators and inhibitors. Thromb Haemost 78:672–677, 1997.
73. I Appleton, NJ Brown, D Willis, PR Colville-Nash, C Alam, JR Brown, DA Willoughby. The role of vascular endothelial growth factor in a murine chronic granulomatous tissue air pouch model of angiogenesis. J Pathol 180:90–94, 1996.
74. AW Griffioen, CA Damen, GH Blijham, G Groenewegen. Tumor angiogenesis is accompanied by a decreased inflammatory response of tumor-associated endothelium. Blood 88:667–673, 1996.
75. DR Senger, L Van de Water, LF Brown, JA Nagy, KT Yeo, TK Yeo, B Berse, RW Jackman, AM Dvorak, HF Dvorak. Vascular permeability factor (VPF, VEGF) in tumor biology. Cancer Metast Rev 12:303–324, 1993.
76. HF Dvorak, M Detmar, KP Claffey, JA Nagy, L van de Water, DR Senger. Vascular permeability factor/vascular endothelial growth factor: an important mediator of angiogenesis in malignancy and inflammation. Int Arch Allergy Immunol 107:233–235, 1995.

77. JA Varner, DA Cheresh. Tumor angiogenesis and the role of vascular cell integrin alphavbeta3. Important Adv Oncol 69–87, 1996.
78. PJ Polverini. Role of the macrophage in angiogenesis-dependent diseases. EXS 79:11–28, 1997.
79. V Kodelja, C Muller, S Tenorio, C Schebesch, CE Orfanos, S Goerdt. Differences in angiogenic potential of classically vs alternatively activated macrophages. Immunobiology 197: 478–493, 1997.
80. MG Sirois, ER Edelman. VEGF effect on vascular permeability is mediated by synthesis of platelet-activating factor. Am J Physiol 272:H2746–H2756, 1997.
81. NN Nissen, PJ Polverini, AE Koch, MV Volin, RL Gamelli, LA DiPietro. Vascular endothelial growth factor mediates angiogenic activity during the proliferative phase of wound healing. Am J Pathol 152:1445–1452, 1998.
82. I Ure, B Partsch, K Wolff, P Petzelbauer. Granulocyte/macrophage colony-stimulating factor increases wound-fluid interleukin 8 in normal subjects but does not accelerate wound healing. Br J Dermatol 138:277–282, 1998.
83. M Ichinose, M Sawada, K Sasaki, Y Oomura. Effect of acidic fibroblast growth factor (aFGF) on phagocytosis in mouse peritoneal macrophages. Microbiol Immunol 42:139–142, 1998.
84. M Arras, WD Ito, D Scholz, B Winkler, J Schaper, W Schaper. Monocyte activation in angiogenesis and collateral growth in the rabbit hindlimb. J Clin Invest 101:40–50, 1998.
85. S Hiratsuka, O Minowa, J Kuno, T Noda, M Shibuya. Flt-1 lacking the tyrosine kinase domain is sufficient for normal development and angiogenesis in mice (in process citation). Proc Natl Acad Sci USA 95:9349–9354, 1998.
86. J Satta, Y Soini, M Mosorin, T Juvonen. Angiogenesis is associated with mononuclear inflammatory cells in abdominal aortic aneurysms. Ann Chir Gynaecol 87:40–42, 1998.
87. SK Smith. Angiogenesis. Semin Reprod Endocrinol 15:221–227, 1997.
88. W Fiedler, U Graeven, S Ergun, S Verago, N Kilic, M Stockschlader, DK Hossfeld. Vascular endothelial growth factor, a possible paracrine growth factor in human acute myeloid leukemia. Blood 89:1870–1875, 1997.
89. S Aharinejad, SC Marks, Jr., P Bock, A Mason-Savas, CA MacKay, EK Larson, ME Jackson, M Luftensteiner, E Wiesbauer. CSF-1 treatment promotes angiogenesis in the metaphysis of osteopetrotic (toothless, tl) rats. Bone 16:315–324, 1995.
90. CF Deroanne, A Hajitou, CM Calberg-Bacq, BV Nusgens, CM Lapiere. Angiogenesis by fibroblast growth factor 4 is mediated through an autocrine up-regulation of vascular endothelial growth factor expression. Cancer Res 57:5590–5597, 1997.
91. Y Nakamura, R Morishita, J Higaki, et al. Hepatocyte growth factor is a novel member of the endothelium-specific growth factors: additive stimulatory effect of hepatocyte growth factor with basic fibroblast growth factor but not with vascular endothelial growth factor. J Hypertens 14:1067–1072, 1996.
92. GT Stavri, IC Zachary, PA Baskerville, JF Martin, JD Erusalimsky. Basic fibroblast growth factor upregulates the expression of vascular endothelial growth factor in vascular smooth muscle cells: synergistic interaction with hypoxia. Circulation 92:11–14, 1995.
93. MS Pepper, N Ferrara, L Orci, R Montesano. Potent synergism between vascular endothelial growth factor and basic fibroblast growth factor in the induction of angiogenesis in vitro. Biochem Biophys Res Commun 189:824–831, 1992.
94. F Jonca, N Ortega, PE Gleizes, N Bertrand, J Plouet. Cell release of bioactive fibroblast growth factor 2 by exon 6-encoded sequence of vascular endothelial growth factor. J Biol Chem 272:24203–24209, 1997.
95. G Seghezzi, S Patel, CJ Ren, A Gualandris, G Pintucci, ES Robbins, RL Shapiro, AC Galloway, DB Rifkin, P Mignatti. Fibroblast growth factor-2 (FGF-2) induces vascular endothelial growth factor (VEGF) expression in the endothelial cells of forming capillaries: an autocrine mechanism contributing to angiogenesis. J Cell Biol 141:1659–1673, 1998.
96. FW Sellke, J Li, A Stamler, JJ Lopez, KA Thomas, M Simons. Angiogenesis induced by

acidic fibroblast growth factor as an alternative method of revascularization for chronic myocardial ischemia. Surgery 120:182–188, 1996.
97. FW Sellke, SY Wang, M Friedman, K Harada, ER Edelman, W Grossman, M Simons. Basic FGF enhances endothelium-dependent relaxation of the collateral-perfused coronary microcirculation. Am J Physiol 267:H1303–H1311, 1994.
98. FW Sellke, M Tofukuji, RJ Laham, J Li, MD Hariawala, S Bunting, M Simons. Comparison of VEGF delivery techniques on collateral-dependent microvascular reactivity. Microvasc Res 55:175–178, 1998.
99. W Schaper, W Ito. Molecular mechanisms of collateral vessel growth. Circ Res 79:911–919, 1996.
100. W Schaper. Collateral vessel growth in the human heart: role of fibroblast growth factor-2 (editorial; comment). Circulation 94:600–601, 1996.
101. KP Rentrop, M Cohen, H Blanke, RA Phillips. Changes in collateral channel filling immediately after controlled coronary artery occlusion by an angioplasty balloon in human subjects. J Am Coll Cardiol 5:587–592, 1985.
102. DS Baim, E Braunwald, F Feit, GL Knatterud, ER Passamani, TL Robertson, WJ Rogers, RE Solomon, DO Williams. The Thrombolysis in Myocardial Infarction (TIMI) Trial phase II: additional information and perspectives. J Am Coll Cardiol 15:1188–1192, 1990.
103. JH Chesebro, G Knatterud, R Roberts, J Borer, LS Cohen, J Dalen, HT Dodge, CK Francis, D Hillis, P Ludbrook, et al. Thrombolysis in Myocardial Infarction (TIMI) Trial, Phase I: a comparison between intravenous tissue plasminogen activator and intravenous streptokinase. Clinical findings through hospital discharge. Circulation 76:142–154, 1987.
104. JJ Piek, JJ Koolen, AC Metting van Rijn, H Bot, G Hoedemaker, GK David, AJ Dunning, JA Spaan, CA Visser. Spectral analysis of flow velocity in the contralateral artery during coronary angioplasty: a new method for assessing collateral flow. J Am Coll Cardiol 21:1574–1582, 1993.
105. MJ Kern, TJ Donohue, RG Bach, FV Aguirre, EA Caracciolo, EO Ofili. Quantitating coronary collateral flow velocity in patients during coronary angioplasty using a Doppler guidewire. Am J Cardiol 71:34D–40D, 1993.
106. JT Willerson, JT Watson, I Hutton, GH Templeton, DE Fixler. Reduced myocardial reflow and increased coronary vascular resistance following prolonged myocardial ischemia in the dog. Circ Res 36:771–781, 1975.
107. K Harada, M Friedman, JJ Lopez, SY Wang, J Li, PV Prasad, JD Pearlman, ER Edelman, FW Sellke, M Simons. Vascular endothelial growth factor administration in chronic myocardial ischemia. Am J Physiol 270:H1791–1802, 1996.
108. JD Pearlman, MG Hibberd, ML Chuang, K Harada, JJ Lopez, SR Gladstone, M Friedman, FW Sellke, M Simons. Magnetic resonance mapping demonstrates benefits of VEGF-induced myocardial angiogenesis. Nature Med 1:1085–1089, 1995.
109. J Li, LF Brown, MG Hibberd, JD Grossman, JP Morgan, M Simons. VEGF, flk-1, and flt-1 expression in a rat myocardial infarction model of angiogenesis. Am J Physiol 270:H1803–H1811, 1996.
110. J Edelberg, W Aird, R Rosenberg. Cardiac myocyte regulation of cardiac microvascular endothelial cell gene expression: a critical role for PDGF. Circulation 96:I-414, 1997.
111. WC Aird, N Jahroudi, H Weiler-Guettler, HB Rayburn, RD Rosenberg. Human von Willebrand factor gene sequences target expression to a subpopulation of endothelial cells in transgenic mice. Proc Natl Acad Sci USA 92:4567–4571, 1995.
112. KA Thomas. Biochemistry and molecular biology of fibroblast growth factors. In: Fallon JH, Loughlin SE, eds. Neurotropic Factors. Orlando, FL: Academic Press, 1992, pp 285–312.
113. TN Mellin, RJ Mennie, DE Cashen, JJ Ronan, J Capparella, ML James, J Disalvo, J Frank, D Linemeyer, G Gimenez-Gallego, et al. Acidic fibroblast growth factor accelerates dermal wound healing. Growth Factors 7:1–14, 1992.

114. TD Bjornsson, M Dryjski, J Tluczek, R Mennie, J Ronan, TN Mellin, KA Thomas. Acidic fibroblast growth factor promotes vascular repair. Proc Natl Acad Sci USA 88:8651–8655, 1991.
115. H Sasaki, H Hoshi, YM Hong, T Suzuki, T Kato, H Sasaki, M Saito, H Youki, K Karube, S Konno, et al. Purification of acidic fibroblast growth factor from bovine heart and its localization in the cardiac myocytes. J Biol Chem 264:17606–17612, 1989.
116. E Speir, V Tanner, AM Gonzalez, J Farris, A Baird, W Casscells. Acidic and basic fibroblast growth factors in adult rat heart myocytes: localization, regulation in culture, and effects on DNA synthesis. Circ Res 71:251–259, 1992.
117. SE Hughes, D Crossman, PA Hall. Expression of basic and acidic fibroblast growth factors and their receptor in normal and atherosclerotic human arteries. Cardiovasc Res 27:1214–1219, 1993.
118. HL Weiner, JL Swain. Acidic fibroblast growth factor mRNA is expressed by cardiac myocytes in culture and the protein is localized to the extracellular matrix. Proc Natl Acad Sci USA 86:2683–2687, 1989.
119. R Shreeniwas, S Ogawa, F Cozzolino, G Torcia, N Braunstein, C Butura, J Brett, HB Lieberman, MB Furie, J Joseph-Silverstein, et al. Macrovascular and microvascular endothelium during long-term hypoxia: alterations in cell growth, monolayer permeability, and cell surface coagulant properties. J Cell Physiol 146:8–17, 1991.
120. S Banai, MT Jaklitsch, W Casscells, M Shou, S Shrivastav, R Correa, SE Epstein, EF Unger. Effects of acidic fibroblast growth factor on normal and ischemic myocardium. Circ Res 69: 76–85, 1991.
121. EF Unger, M Shou, CD Sheffield, E Hodge, M Jaye, SE Epstein. Extracardiac to coronary anastomoses support regional left ventricular function in dogs. Am J Physiol 264:H1567–H1574, 1993.
122. EF Unger, S Banai, M Shou, M Jaklitsch, E Hodge, R Correa, M Jaye, SE Epstein. A model to assess interventions to improve collateral blood flow: continuous administration of agents into the left coronary artery in dogs. Cardiovasc Res 27:785–791, 1993.
123. K Schlaudraff, B Schumacher, BU Specht, R Seitelberger, V Schlosser, R Fasol. Growth of new coronary vascular structures by angiogenic growth factors. Eur J Cardiothorac Surg 7: 637–641, 1993.
124. EG Nabel, ZY Yang, G Plautz, R Forough, X Zhan, CC Haudenschild, T Maciag, GJ Nabel. Recombinant fibroblast growth factor-1 promotes intimal hyperplasia and angiogenesis in arteries in vivo. Nature 362:844–846, 1993.
125. H Tabata, M Silver, JM Isner. Arterial gene transfer of acidic fibroblast growth factor for therapeutic angiogenesis in vivo: critical role of secretion signal in use of naked DNA. Cardiovasc Res 35:470–479, 1997.
126. S Ortega, MT Schaeffer, D Soderman, J DiSalvo, DL Linemeyer, G Gimenez-Gallego, KA Thomas. Conversion of cysteine to serine residues alters the activity, stability, and heparin dependence of acidic fibroblast growth factor. J Biol Chem 266:5842–5846, 1991.
127. DL Linemeyer, JG Menke, LJ Kelly, J DiSalvo, D Soderman, MT Schaeffer, S Ortega, G Gimenez-Gallego, KA Thomas. Disulfide bonds are neither required, present, nor compatible with full activity of human recombinant acidic fibroblast growth factor. Growth Factors 3:287–298, 1990.
128. M Blaber, J DiSalvo, KA Thomas. X-ray crystal structure of human acidic fibroblast growth factor. Biochemistry 35:2086–2094, 1996.
129. DB Volkin, PK Tsai, JM Dabora, JO Gress, CJ Burke, RJ Linhardt, CR Middaugh. Physical stabilization of acidic fibroblast growth factor by polyanions. Arch Biochem Biophys 300: 30–41, 1993.
130. JJ Lopez, ER Edelman, A Stamler, et al. Angiogenic potential of perivascularly delivered aFGF in a porcine model of chronic myocardial ischemia. Am J Physiol 274:H930–H936, 1998.

131. E Kardami, RR Fandrich. Basic fibroblast growth factor in atria and ventricles of the vertebrate heart. J Cell Biol 109:1865–1875, 1989.
132. M Zhou, RL Sutliff, RJ Paul, et al. Fibroblast growth factor 2 control of vascular tone. Nature Med 4:201–207, 1998.
133. C Bauters. Growth factors as a potential new treatment for ischemic heart disease. Clin. Cardiol 20:II-52–II57, 1997.
134. JA Ware, M Simons. Angiogenesis in ischemic heart disease. Nature Med 3:158–164, 1997.
135. A Yanagisawa-Miwa, Y Uchida, F Nakamura, T Tomaru, H Kido, T Kamijo, T Sugimoto, K Kaji, M Utsuyama, C Kurashima, et al. Salvage of infarcted myocardium by angiogenic action of basic fibroblast growth factor. Science 257:1401–1403, 1992.
136. A Battler, M Scheinowitz, A Bor, D Hasdai, Z Vered, E Di Segni, N Varda-Bloom, D Nass, S Engelberg, M Eldar, et al. Intracoronary injection of basic fibroblast growth factor enhances angiogenesis in infarcted swine myocardium. J Am Coll Cardiol 22:2001–2006, 1993.
137. DF Lazarous, M Scheinowitz, M Shou, E Hodge, S Rajanayagam, S Hunsberger, WG Robison Jr, JA Stiber, R Correa, SE Epstein, et al. Effects of chronic systemic administration of basic fibroblast growth factor on collateral development in the canine heart. Circulation 91:145–153, 1995.
138. M Shou, V Thirumurti, S Rajanayagam, DF Lazarous, E Hodge, JA Stiber, M Pettiford, E Elliott, SM Shah, EF Unger. Effect of basic fibroblast growth factor on myocardial angiogenesis in dogs with mature collateral vessels. J Am Coll Cardiol 29:1102–1106, 1997.
139. JL Jacot, NM Laver, JP Glover, DF Lazarous, EF Unger, WG Robison Jr. Histological evaluation of the canine retinal vasculature following chronic systemic administration of basic fibroblast growth factor. J Anat 188:349–354, 1996.
140. DF Lazarous, M Shou, M Scheinowitz, et al. Comparative effects of basic fibroblast growth factor and vascular endothelial growth factor on coronary collateral development and the arterial response to injury. Circulation 94:1074–1082, 1996.
141. J Lopez, E Edelman, A Stamler, J Morgan, F Sellke, M Simons. Local perivascular administration of basic fibroblast growth factor: drug delivery and toxicological evaluation. Drug Metab Dispos 24:922–924, 1996.
142. JJ Lopez, ER Edelman, A Stamler, MG Hibberd, P Prasad, RP Caputo, JC Carrozza, PS Douglas, FW Sellke, M Simons. Basic fibroblast growth factor in a porcine model of chronic myocardial ischemia: a comparison of angiographic, echocardiographic and coronary flow parameters. J Pharmacol Exp Ther 282:385–390, 1997.
143. K Harada, W Grossman, M Friedman, ER Edelman, PV Prasad, CS Keighley, WJ Manning, FW Sellke, M Simons. Basic fibroblast growth factor improves myocardial function in chronically ischemic porcine hearts. J Clin Invest 94:623–630, 1994.
144. ER Edelman, E Mathiowitz, R Langer, M Klagsbrun. Controlled and modulated release of basic fibroblast growth factor. Biomaterials 12:619–626, 1991.
145. ER Edelman, MA Nugent, MJ Karnovsky. Perivascular and intravenous administration of basic fibroblast growth factor: vascular and solid organ deposition. Proc Natl Acad Sci USA 90:1513–1517, 1993.
146. FW Sellke, JE Quillen, LA Brooks, DG Harrison. Endothelial modulation of the coronary vasculature in vessels perfused via mature collaterals. Circulation 81:1938–1947, 1990.
147. C Landau, AK Jacobs, CC Haudenschild. Intrapericardial basic fibroblast growth factor induces myocardial angiogenesis in a rabbit model of chronic ischemia. Am Heart J 129:924–931, 1995.
148. E Watanabe, DM Smith, J Sun, FW Smart, JB Delcarpio, TB Roberts, CH Van Meter Jr, WC Claycomb. Effect of basic fibroblast growth factor on angiogenesis in the infarcted porcine heart. Basic Res Cardiol 93:30–37, 1998.
149. Y Uchida, A Yanagisawa-Miwa, F Nakamura, K Yamada, T Tomaru, K Kimura, T Morita. Angiogenic therapy of acute myocardial infarction by intrapericardial injection of basic fi-

broblast growth factor and heparin sulfate: an experimental study. Am Heart J 130:1182–1188, 1995.

150. M Horrigan, A MacIsaac, F Nicolini, D Vince, P Lee, S Ellis, E Topol. Reduction in myocardial infarct size by basic fibroblast growth factor after temporary coronary occlusion in a canine model. Circulation 94:1927–1933, 1996.
151. P Cuevas, F Carceller, S Ortega, M Zazo, I Nieto, G Gimenez-Gallego. Hypotensive activity of fibroblast growth factor. Science 254:1208–1210, 1991.
152. P Cuevas, AM Gonzalez, F Carceller, A Baird. Vascular response to basic fibroblast growth factor when infused onto the normal adventitia or into the injured media of the rat carotid artery. Circ Res 69:360–369, 1991.
153. Z Huang, K Chen, P Huang, S Finkelstein, M Moskowitz. bFGF ameliorates focal ischemic injury by blood flow-independent mechanism in eNOS mutant mice. Am J Physiol 272: H1401–1405, 1996.
154. RR Padua, R Sethi, NS Dhalla, E Kardami. Basic fibroblast growth factor is cardioprotective in ischemia-reperfusion injury. Mol Cell Biochem 143:129–135, 1995.
155. P Cuevas, D Reimers, F Carceller, V Martinez-Coso, M Redondo-Horcajo, I Saenz de Tejada, G Gimenez-Gallego. Fibroblast growth factor-1 prevents myocardial apoptosis triggered by ischemia reperfusion injury. Eur J Med Res 2:465–468, 1997.
156. E Arman, R Haffner-Krausz, Y Chen, JK Heath, P Lonai. Targeted disruption of fibroblast growth factor (FGF) receptor 2 suggests a role for FGF signaling in pregastrulation mammalian development. Proc Natl Acad Sci USA 95:5082–5087, 1998.
157. R Bagheri-Yarmand, Y Kourbali, C Mabilat, JF Morere, A Martin, H Lu, C Soria, J Jozefonvicz, M Crepin. The suppression of fibroblast growth factor 2/fibroblast growth factor 4-dependent tumour angiogenesis and growth by the anti-growth factor activity of dextran derivative (CMDB7). Br J Cancer 78:111–118, 1998.
158. T Kitaoka, LS Morse, S Schneeberger, H Ishigooka, LM Hjelmeland. Expression of FGF5 in choroidal neovascular membranes associated with ARMD. Curr Eye Res 16:396–399, 1997.
159. FJ Giordano, P Ping, MD McKirnan, S Nozaki, AN DeMaria, WH Dillmann, O Mathieu-Costello, HK Hammond. Intracoronary gene transfer of fibroblast growth factor-5 increases blood flow and contractile function in an ischemic region of the heart. Nature Med 2:534–539, 1996.
160. A Ladoux, C Frelin. Hypoxia is a strong inducer of vascular endothelial growth factor mRNA expression in the heart. Biochem Biophys Res Commun 195:1005–1010, 1993.
161. A Ladoux, C Frelin. Cobalt stimulates the expression of vascular endothelial growth factor mRNA in rat cardiac cells. Biochem Biophys Res Commun 204:794–798, 1994.
162. D Shweiki, A Itin, D Soffer, E Keshet. Vascular endothelial growth factor induced by hypoxia may mediate hypoxia-initiated angiogenesis. Nature 359:843–845, 1992.
163. S Banai, D Shweiki, A Pinson, M Chandra, G Lazarovici, E Keshet. Upregulation of vascular endothelial growth factor expression induced by myocardial ischaemia: implications for coronary angiogenesis. Cardiovasc Res 28:1176–1179, 1994.
164. J Waltenberger, U Mayr, S Pentz, V Hombach. Functional upregulation of the vascular endothelial growth factor receptor KDR by hypoxia. Circulation 94:1647–1654, 1996.
165. JM Isner, A Pieczek, R Schainfeld, R Blair, L Haley, T Asahara, K Rosenfield, S Razvi, K Walsh, JF Symes. Clinical evidence of angiogenesis after arterial gene transfer of phVEGF165 in patient with ischaemic limb. Lancet 348:370–374, 1996.
166. JM Isner, K Walsh, J Symes, A Pieczek, S Takeshita, J Lowry, K Rosenfield, L Weir, E Brogi, D Jurayj. Arterial gene transfer for therapeutic angiogenesis in patients with peripheral artery disease. Hum Gene Ther 7:959–988, 1996.
167. JM Isner. Angiogenesis for revascularization of ischaemic tissues (editorial). Eur Heart J 18: 1–2, 1997.
168. S Takeshita, LP Zheng, E Brogi, M Kearney, LQ Pu, S Bunting, N Ferrara, JF Symes, JM Isner. Therapeutic angiogenesis: a single intraarterial bolus of vascular endothelial growth

factor augments revascularization in a rabbit ischemic hind limb model. J Clin Invest 93: 662–670, 1994.
169. S Takeshita, L Weir, D Chen, LP Zheng, R Riessen, C Bauters, JF Symes, N Ferrara, JM Isner. Therapeutic angiogenesis following arterial gene transfer of vascular endothelial growth factor in a rabbit model of hindlimb ischemia. Biochem Biophys Res Commun 227: 628–635, 1996.
170. S Banai, MT Jaklitsch, M Shou, DF Lazarous, M Scheinowitz, S Biro, SE Epstein, EF Unger. Angiogenic-induced enhancement of collateral blood flow to ischemic myocardium by vascular endothelial growth factor in dogs. Circulation 89:2183–2189, 1994.
171. MD Hariawala, JJ Horowitz, D Esakof, DD Sheriff, DH Walter, B Keyt, JM Isner, JF Symes. VEGF improves myocardial blood flow but produces EDRF-mediated hypotension in porcine hearts. J Surg Res 63:77–82, 1996.
172. CA Mack, SR Patel, EA Schwarz, et al. Biologic bypass with the use of adenovirus-mediated gene transfer of the complementary deoxyribonucleic acid for vascular endothelial growth factor 121 improves myocardial perfusion and function in the ischemic porcine heart. J Thorac Cardiovasc Surg 115:168–176; discussion 176–167, 1998.
173. R Yang, GR Thomas, S Bunting, A Ko, N Ferrara, B Keyt, J Ross, H Jin. Effects of vascular endothelial growth factor on hemodynamics and cardiac performance. J Cardiovasc Pharmacol 27:838–844, 1996.
174. ELp Jones, JMp Craver, RAp Guyton, et al. Importance of complete revascularization in performance of the coronary bypass operation. Am J Cardiol 51:7–12, 1983.
175. JF McNeer, MJ Conley, CF Starmer, VS Behar, Y Kong, RH Peter, AG Bartel, HN Oldham, Jr., WG Young Jr, DC Sabiston Jr, RA Rosati. Complete and incomplete revascularization at aortocoronary bypass surgery: experience with 392 consecutive patients. Am Heart J 88: 176–182, 1974.
176. T Henry, K Rocha-Singh, J Isner, et al. Results of intracoronary recombinant human vascular endothelial growth factor (rh VEGF) administration trial. J Am Coll Cardiol 31:65A, 1998.
177. FW Sellke, RJ Laham, ER Edelman, JD Pearlman, M Simons. Therapeutic angiogenesis with basic fibroblast growth factor: technique and early results. Ann Thorac Surg 65:1540–1544, 1998.
178. R Laham, F Sellke, J Edelman, M Simons. Local perivascular basic fibroblast growth factor (bFGF) treatment in patients with ischemic heart disease. J Am Coll Cardiol 31:394A, 1998.
179. B Schumacher, P Pecher, BU von Specht, T Stegmann. Induction of neoangiogenesis in ischemic myocardium by human growth factors: first clinical results of a new treatment of coronary heart disease. Circulation 97:645–650, 1998.
179A. T Henry, B Annex, M Azrin, G McKendall, J Willerson, R Hendel, F Giordano, R Klein, C Gibson, D Berman, C Luce, T McCluskey. Double blind placebo controlled trial of recombinant human vascular endothelial growth factor—the VIVA trial. J Am Coll Cardiol 33: 384A, 1999.
179B. R Laham, M Leimbach, N Chronos, J Vansant, J Pearlman, R Pettigrew, H Guler, M Whitehouse, D Hung, D Baim, S King, M Simons. Intracoronary administration of recombinant fibroblast growth factor-2 (rFGF-2) in patients with severe coronary artery disease: results of phase I. J Am Coll Cardiol 33:383A, 1999.
179C. TK Rosengart, LY Lee, SR Patel, TA Sanborn, M Parikh, GW Bergman, R Hachamovitch, M Szulc, PD Kligfield, PM Okin, RT Hahn, RB Devereux, MR Post, NR Hackett, T Foster, TM Grasso, MI Lesser, OW Isom, RG Crystal. Angiogenesis gene therapy: phase I assessment of direct intramyocardial administration of an adenovirus vector expressing VEGF121 cDNA to individuals with clinically significant severe coronary artery disease. Circulation 100:468–474, 1999.
180. A Yoshida, S Yoshida, AK Khalil, T Ishibashi, H Inomata. Role of NF-kappaB–mediated interleukin-8 expression in intraocular neovascularization. Invest Ophthalmol. Vis Sci 39: 1097–1106, 1998.
181. FM Grignolo, L Emanuelli, CM Eandi, B Brogliatti. Ocular angiogenesis: a new in vivo experimental model. Acta Ophthalmol Scand Suppl 224:10–11, 1997.

182. LP Aiello. Vascular endothelial growth factor and the eye: biochemical mechanisms of action and implications for novel therapies. Ophthalmic Res 29:354–362, 1997.
183. RN Frank, RH Amin, D Eliott, JE Puklin, GW Abrams. Basic fibroblast growth factor and vascular endothelial growth factor are present in epiretinal and choroidal neovascular membranes. Am J Ophthalmol 122:393–403, 1996.
184. GA Lutty, DS McLeod, C Merges, A Diggs, J Plouet. Localization of vascular endothelial growth factor in human retina and choroid. Arch Ophthalmol 114:971–977, 1996.
185. LP Aiello, EA Pierce, ED Foley, H Takagi, H Chen, L Riddle, N Ferrara, GL King, LE Smith. Suppression of retinal neovascularization in vivo by inhibition of vascular endothelial growth factor (VEGF) using soluble VEGF-receptor chimeric proteins. Proc Natl Acad Sci USA 92: 10457–10461, 1995.
186. RF Gariano, RE Kalina, AE Hendrickson. Normal and pathological mechanisms in retinal vascular development. Surv Ophthalmol 40:481–490, 1996.
187. RA Payson, MA Chotani, IM Chiu. Regulation of a promoter of the fibroblast growth factor 1 gene in prostate and breast cancer cells. J Steroid Biochem Mol Biol 66:93–103, 1998.
188. H Anzai, Y Kitadai, CD Bucana, R Sanchez, R Omoto, IJ Fidler. Expression of metastasis-related genes in surgical specimens of human gastric cancer can predict disease recurrence. Eur J Cancer 34:558–565, 1998.
189. K Murakami-Mori, S Mori, S Nakamura. Endogenous basic fibroblast growth factor is essential for cyclin E-CDK2 activity in multiple external cytokine-induced proliferation of AIDS-associated Kaposi's sarcoma cells: dual control of AIDS-associated Kaposi's sarcoma cell growth and cyclin E-CDK2 activity by endogenous and external signals. J Immunol 161: 1694–1704, 1998.
190. JM Connolly, DP Rose. Angiogenesis in two human prostate cancer cell lines with differing metastatic potential when growing as solid tumors in nude mice. J Urol 160:932–936, 1998.
191. H Konno, T Tanaka, M Baba, T Kanai, K Matsumoto, K Kamiya, S Nakamura, S Baba. Quantitative analysis of vascular endothelial growth factor in colon cancer: clinical and experimental. Eur Surg Res 30:273–278, 1998.
192. F Larcher, R Murillas, M Bolontrade, CJ Conti, JL Jorcano. VEGF/VPF overexpression in skin of transgenic mice induces angiogenesis, vascular hyperpermeability and accelerated tumor development. Oncogene 17:303–311, 1998.
193. C Linder, S Linder, E Munck-Wikland, H Strander. Independent expression of serum vascular endothelial growth factor (VEGF) and basic fibroblast growth factor (bFGF) in patients with carcinoma and sarcoma. Anticancer Res 18:2063–2068, 1998.
194. DD Ku, JK Zaleski, S Liu, TA Brock. Vascular endothelial growth factor induces EDRF-dependent relaxation in coronary arteries. Am J Physiol 265:H586–H592, 1993.
195. G Mazue, F Bertolero, C Jacob, P Sarmientos, R Roncucci. Preclinical and clinical studies with recombinant human basic fibroblast growth factor. Ann NY Acad Sci 638:329–340, 1991.
196. AD Callow, ET Choi, JD Trachtenberg, SL Stevens, DT Connolly, C Rodi, US Ryan. Vascular permeability factor accelerates endothelial regrowth following balloon angioplasty. Growth Factors 10:223–228, 1994.
197. E Van Belle, FO Tio, D Chen, L Maillard, M Kearney, JM Isner. Passivation of metallic stents after arterial gene transfer of phVEGF165 inhibits thrombus formation and intimal thickening. J Am Coll Cardiol 29:1371–1379, 1997.
198. V Lindner, RA Majack, MA Reidy. Basic fibroblast growth factor stimulates endothelial regrowth and proliferation in denuded arteries. J Clin Invest 85:2004–2008, 1990.
199. MA Corson, BC Berk. Growth factors and the vessel wall. Heart Dis Stroke 2:166–170, 1993.
200. AK Hanna, JC Fox, DG Neschis, SD Safford, JL Swain, MA Golden. Antisense basic fibroblast growth factor gene transfer reduces neointimal thickening after arterial injury. J Vasc Surg 25:320–325, 1997.

31
Cerebral Ischemia

Alan Z. Segal and Seth P. Finklestein
Massachusetts General Hospital and Harvard Medical School, Boston, Massachusetts

I. INTRODUCTION

Growth-factor therapy presents an exciting new possibility for the treatment of stroke. Potential strategies include the prevention of cerebral ischemia through pharmacological "revascularization," cytoprotection of neurons in acute stroke, and promotion of functional recovery in established stroke. This chapter briefly discusses the last two of these strategies, but focuses primarily on revascularization—the induction of new blood vessels using angiogenic factors in the brain. Several growth factors with pluripotential functions posses angiogenic capability, including acidic and basic fibroblast growth factors (FGFs), vascular endothelial growth factor (VEGF), insulin-like growth factor (IGF), and platelet-derived growth factor (PDGF). Among these, the FGFs and VEGF have been the most extensively investigated as cerebral growth factors in animals. More importantly, these agents have begun to enter the clinical arena for use in patients with vascular disease.

II. FGFs AS GROWTH FACTORS IN THE BRAIN

Basic fibroblast growth factor (bFGF) is a potent nerve growth factor, promoting the survival and proliferation of a wide variety of embryonic CNS neurons in vitro (1). In addition, bFGF is a mitogen for glial cells, oligodendrocytes, and capillary endothelial cells (2). Because of this latter function, bFGF is a potentially potent angiogenic factor in the brain (3).

Production of bFGF has been shown to occur in astrocytes (4) and neurons (5), whereas at least one of the high-affinity bFGF receptors (flg) is widely localized on neurons and nonneuronal cells. bFGF is found throughout the brain as well as the circumventricular organs and choroid plexus, where bFGF mRNA has been localized to capillary endothelial cells as well as ependymal and meningeal cells (6).

III. ROLE OF bFGF IN ACUTE NEUROPROTECTION

bFGF has been well studied as a promoter of neuronal cell survival and is a promising agent to spare vulnerable neurons in stroke. Following acute cerebral ischemia, there is a massive release of excitatory amino acids, primarily glutatmate, from presynaptic nerve terminals (7). Excessive glutamate-mediated excitation of neurons results in inward currents of calcium and ultimately irreversible cellular injury (8). Sustained waves of depolarization are generated that propagate from the ischemic zone into normal tissue. Oxygen free radicals are formed, which lead to the peroxidation of lipid membranes and the destruction of key enzymes. In vitro, bFGF and other growth factors have been shown to protect cultured neurons against these insults. Particularly susceptible are vulnerable nearby neurons in the ''ischemic penumbra'' at the border of an infarct (9). In addition, distant neurons projecting to infarcted tissue are prone to delayed retrograde cell death, a process that may be attenuated by administration of bFGF (10).

Endogenous expression of bFGF in the brain is markedly upregulated in the setting of ischemia. Our own studies have shown that following MCA ligation in rats, bFGF mRNA measured by in situ hybridization peaks at 1 day, while bFGF protein levels, studied by immunohistochemistry, peak at 3 days after infarction. bFGF immunoreactivity is increased throughout the cortex surrounding the infraction as well as in deep nuclei such as the caudate, putamen, and nucleus acumbens. bFGF expression was localized to both astrocytes and neurons (11).

Exogenous administration of bFGF has been extensively studied as a neuroprotective factor in ischemia models in rodents. Our group initially studied the effects of bFGF in neonatal rats. bFGF administered systemically (100 μg/kg intraperitoneally), 30 min before common carotid artery ligation and hypoxia, decreased infarct volume by 40% (12). In subsequent studies, we examined the infarct-sparing effects of bFGF in mature rats. bFGF 1.2 μg/day was delivered by continuous infusion into the lateral ventricle of rats. Following MCA ligation, infarct size was 25–30% smaller in bFGF-treated than in vehicle-treated animals (13). Intravenous systemic bFGF in adult rats, 45 μg/kg per hour for 3 hr, started 30 min after onset of ischemia, reduced infarct volume by 40–50% in a MCA occlusion model when assessed 24 hr after onset of ischemia. This reduction in infarct volume was accompanied by improved neurological outcome as assessed by a behavioral rating scale. Autoradiography showed that labeled intravenously administered bFGF crossed the damaged blood-brain barrier to enter the ischemic but not the nonischemic hemisphere (14).

The use of bFGF (FIBLAST) in humans has been studied in a randomized, double-blind safety and tolerance study. Sixty-six patients with acute ischemic stroke were randomized to placebo or different doses of FIBLAST up to 150 μg/kg or placebo within 12 hr of symptom onset. Side effects were limited to a dose-dependent, self-limited, reversible leukocytosis, peaking at 2 days and returning to baseline at 5 days. This leukocytosis was not associated with fever, infection, or other safety findings (15). Two large-scale clinical trials of intravenous FIBLAST have been initiated. The North American trial is currently on hold owing to safety concerns; however, a European trial using a different dosing schedule is ongoing.

IV. bFGF AS A VASODILATOR

bFGF may also play a role because of its action as a vasodilator, an effect mediated via nitric oxide (NO) production by endothelial nitric oxide synthase (eNOS). When bFGF is applied topically on cerebral pial arterioles in rats, a significant arteriolar vasodilation is observed (16,17). Infarct size, however, does not appear to be affected by NO function.

While NO ''knockout'' mice deficient in the eNOS gene undergoing MCA occlusion showed no bFGF-mediated increase in rCBF in the ischemic penumbra compared to wild-type mice, this did not affect the size of the subsequent stroke (18). The degree of infarct reduction was equivalent in eNOS knockout and in wild-type animals. The potency of bFGF related vasodilation may thus be insufficient to be functionally relevant in the brain. Indeed, in a focal ischemia model in rats, perfusion-weighted MRI did not detect significant improvement in perfusion following bFGF infusion (19).

V. ROLE OF bFGF IN RECOVERY

Following stroke, partial spontaneous recovery of function occurs. The mechanism of this recovery might involve reversal of metabolic ''shock,'' unmasking of existing but functionally inactive brain pathways, or new neuronal sprouting from surviving neurons and formation of new synapses. Upregulation of polypeptide growth factors, such as bFGF, may be measured in tissues surrounding focal infarcts (20).

Treatment with exogenous neurotrophic growth factors may accelerate functional recovery after stroke. In cell cultures, bFGF increases axonal outgrowth and upregulates the expression of growth-associated protein 43 (GAP-43), a molecular marker of axonal sprouting and synapse formation. Chronic intracisternal administration of bFGF enhances sensorimotor recovery of the impaired limbs following stroke in rats (21).

VI. ROLE OF bFGF IN ANGIOGENESIS

As noted previously, bFGF may be a potent angiogenic factor in the brain. In moyamoya disease—a progressive narrowing of the major cerebral arterial trunks—often diagnosed in children, there is evidence that bFGF may act as an ''endogenous angiogenic factor.'' Levels of bFGF are elevated in the spinal fluid of these patients (22), especially those in whom neovascularization is well developed (23). While bFGF may promote neovascularization in moyamoya, it has also been proposed that bFGF-induced endothelial cell proliferation may contribute to the occlusive pathogenesis of the disease. Interestingly, other factors expected to be upregulated in hypoxia (e.g., VEGF) do not appear to be increased in moyamoya patients compared to controls (23).

The angiogenic effects of bFGF in the brain have been studied experimentally. Puumala et al. performed chronic intraventricular infusion of bFGF in the rat. They showed increases in capillary count and in vascular density (including capillaries, penetrating arterioles, venules, and draining veins). The effect was seen in animals in the high-dose group, which received 10 μg × 6 doses = 60 μg total, compared to 2 μg × 6 doses = 12 μg total in the ''low'' dose group. Increased vasculature was restricted to a region bordering the left lateral ventricle (the site of bFGF infusion). Neovascularization did not occur at the border of the right lateral ventricle or any of the other brain regions sampled (periphery of the third ventricle, interhemispheric fissure, or cortex). This work raises important issues regarding optimal delivery of drug into the brain. Effects were probably limited in this study by dilution of drug by CSF produced at the level of the choroid plexus and limitation of penetration of drug across ependyma and diffusion into tissue (24). An intravascular route of administration might more directly promote endothelial cell proliferation within blood vessels.

Optimal effectiveness of bFGF-induced angiogenesis may depend on a tropism for ischemic tissues. In a model of chronic ischemia, Lyons et al. performed carotid artery ligation bilaterally in rats, with intraventricular infusions of bFGF every 4 days for 28 days. Dosing varied from 0.5–50 ng. A dose dependent increase in capillary density was observed in all brain regions examined, most prominently at the cortical level. The cortical effect was presumed to be due to circulation of drug through the ventricles into the cisternal system and out over the cortex. In vehicle-treated animals, chronic ischemia itself produced a smaller increase in capillaries, possibly due to recruitment of existing capillaries and endogenous release of bFGF (25). It is therefore likely that endogenous upregulation of bFGF can be pharmacologically augmented to effect a more robust angiogenic response.

In a model of acute focal infarction, Tenjin et al. performed MCA occlusion in 50 rats. Animals received either vehicle, serum, or 50 ng bFGF biweekly through an indwelling intraventricular catheter, commencing 2 days after MCA occlusion. No significant change in capillary bed density or viable neurons in the peri-infarction cortex was observed (26). It is likely that neuronal cell death was complete by the time FGF was administered at 2 days. While bFGF-induced angiogenesis might prevent ischemia in a chronic occlusion, its response is unlikely to be rapid enough to affect acute reperfusion.

VII. VEGF AS AN ANGIOGENIC FACTOR

VEGF is a potent mitogen for endothelial cells. The VEGF gene has been assigned to chromosome 6p21.3, which codes for at least four subforms varying in length from 121–206 amino acids. The most common form, $VEGF_{165}$, a 45-kDa protein, includes a signal sequence that permits its secretion and diffusion into the extracellular matrix, where it binds to heparin and proteoglycans with heparin-like moieties. Longer isoforms (e.g., $VEGF_{189}$) may require activation by plasmin-mediated proteolysis. VEGF receptors on endothelial cells (Flt-1 and Flk-KDR) activate tyrosine kinases and intracellular signal transduction cascades. VEGF is also a promoter of vascular permeability, which may augment these processes (27).

VIII. VEGF AS A GROWTH FACTOR IN THE BRAIN

VEGF is normally expressed in the brain and has been isolated in ependyma, choroid plexus, and granule cells in cerebellum (28). During brain development, VEGF transcript levels are abundant in the ventricular neuroectoderm and postnatal brain when endothelial cells proliferate rapidly, but they are reduced in adults when endothelial cell proliferation ceases (29). The role of VEGF in the normal adult brain is not well understood. Increased expression of VEGF, however, has been well characterized in pathological situations; for instance, in glioma, flt VEGF receptor transcripts have been localized in neovascular endothelial cells (30).

Hypoxia is known to increase production of VEGF in cultured glia (31) and cerebral endothelial cells (32,33). It is well known that ischemic stroke causes active angiogenesis, particularly in the ischemic penumbra, which correlates with longer survival in humans (34). In animals, VEGF upregulation has been documented in both chronic and acute ischemia models. Kovacs et al. documented increased VEGF levels in an MCA occlusion model in the rat. Western blot showed banding at 38 and 45 kDa, corresponding to the

$VEGF_{121}$ and $VEGF_{165}$ isoforms. Immunohistochemistry showed that macrophages, neurons, and glial cells expressed VEGF. This peaked at 1 day in the ischemic core and 7 days in the penumbra. VEGF and its flt receptor were also detected on endothelial cells, along with an increase in endothelial cell thickness (35).

VEGF is also upregulated in transient cerebral ischemia. Hayashi et al. performed a MCA occlusion for 2 hr in the rat. Both VEGF mRNA and protein were increased. Northern blot analysis showed no detectable VEGF mRNA in control brains but showed increased gene product at 1 hr, peaking at 3 hr after reperfusion. These effects were short-lived, as no VEGF mRNA was found at 3 or 7 days. The length of ischemia from 1–3 hr made no difference in the degree and temporal profile of VEGF induction. While Western blot showed no immunoreactivity in controls, VEGF isoforms 121 and 165 were isolated as in the ischemic brains. Immunohistochemical analysis showed VEGF reactivity restricted to the ependymal cells of control brains. Under conditions of ischemia, VEGF staining was found in cortical neurons at 1 hr, with a peak at 3 hr. This diminished at 1 day and was absent at 3 or 7 days. Pial cells, however, showed sustained VEGF staining up to 7 days. While Kovacs' fixed occlusion model showed increased VEGF expression in neurons, endothelial cells and glia, this study of transient occlusion showed changes restricted to neurons. Interestingly, cortical neuronal staining was limited to layers III and V, areas known to be vulnerable to hypoxia (36).

IX. OTHER EFFECTS OF VEGF

Unlike FGF, VEGF is not known to possess neuroprotective effects. However, a protective role for VEGF in ischemia might relate to prevention of apoptotic cell death of capillary endothelial cells (37). Alternatively, VEGF may be an unfavorable factor in acute ischemia. Through its vascular permeability effects, VEGF has been shown to increase cerebral edema in brain tumors (38). Edema also occurs in stroke. However, the mechanism of this edema (''cytotoxic''—related to cell death) differs from that seen in tumors (''vasogenic''—related to leaky blood vessels). Thus, it is unlikely that VEGF would be a cause of cytotoxic edema in stroke.

Like FGF, VEGF is also known to be a smooth muscle relaxant, with potent vasodilatory effects. These effects have been well documented in the coronaries (39) and leg (40), but to our knowledge have not yet been investigated in the CNS.

X. CLINICAL POTENTIAL FOR ANGIOGENIC THERAPY IN CEREBRAL ISCHEMIA

Angiogenesis may be of potential therapeutic benefit for patients suffering from both ''large'' and ''small'' vessel cerebrovascular disease. Large vessel atherosclerotic disease affects both vessels branching from the aorta [such as the common carotid artery (CCA) or vertebral artery (VA)] and the major cerebral arteries intracranially [such as the internal carotid artery (ICA), middle cerebral artery (MCA), anterior cerebral artery (ACA), posterior cerebral artery (PCA), or basilar artery (BA)]. Stenoses at the carotid artery siphon (just prior to its bifurcation into the ACA/MCA) or in the basilar artery present commonly

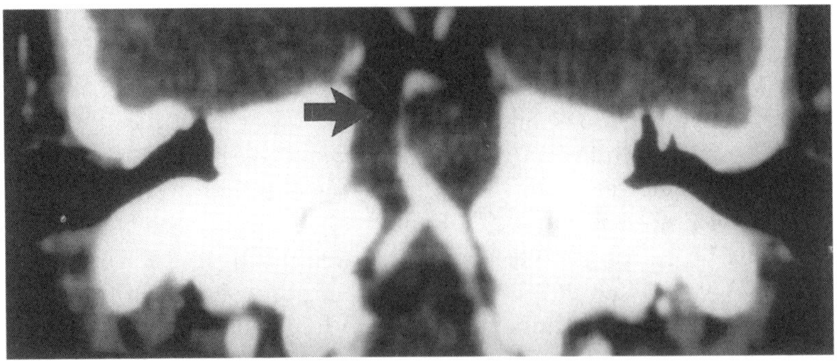

Figure 1 Computed tomography with angiographic contrast enhancement. The vertebral and basilar arteries are seen at the level of the skull base. A severe narrowing is present in the middle basilar (arrow). (Photo courtesy of Michael H. Lev.)

and pose few therapeutic options (see Figs. 1 and 2). In these patients, compromised flow results in brain hypoperfusion analogous to that of arterial insufficiency in the lower extremity or severe coronary artery disease. These patients present with chronic, repeated, stereotyped transient ischemic attacks (TIAs). These TIAs may affect any aspect of neurological function,—for example, causing motor or sensory impairment, dysarthria, dizzi-

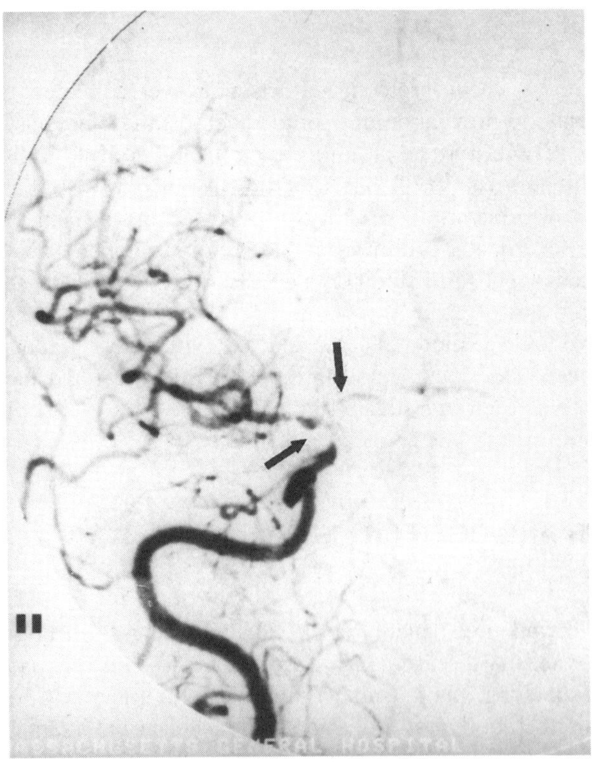

Figure 2 Angiogram at the level of the intracranial carotid. Severe atherosclerotic narrowing is seen involving the top of the carotid and proximal middle cerebral artery (left arrow), with near total occlusion of the anterior cerebral artery (right arrow). (Photo courtesy of Walter Koroshetz.)

ness, or ataxia. Such ischemia may be reversible but often leads to fixed infarction if episodes of insufficiency are severe or prolonged enough. Low flow may also predispose to in situ platelet fibrin thrombi, which can cause stroke due to local thrombosis, distal propagation of clot, or remote infarction due to an artery-to-artery embolic phenomenon.

Patients with long-standing hypertension may suffer from extensive small vessel ''microangiopathic'' disease. Depending on the clinical setting, this process has been referred to as subcortical arteriosclerotic encephalopathy (Binswanger's disease), cerebral leukoaraiosis (with preferential white-matter degeneration), or multi-infarct dementia (MID). MID may be distinguished from the former entities if a natural history of multiple discrete small stroke events has occurred in a stepwise fashion. Pathologically, these processes are related to a process of lipohyalinosis of small vessels, which is compounded by risk factors of hypertension, diabetes, tobacco use, and elevated cholesterol levels. Treatment of these diseases—currently limited to control of risk factors and, in some cases, antiplatelet therapy—does little to arrest these progressive processes.

XI. SURGICAL TREATMENT OF CEREBROVASCULAR DISEASE

Stenosis affecting the internal carotid artery (see Fig. 3) and occurring at the bifurcation of the common carotid artery (CCA) into its internal and external branches is easily amenable to surgical treatment, such as carotid endarterectomy (CEA) or new therapies such as stenting. Well-established randomized studies (41) have confirmed the benefit of CEA in preventing stroke. Other surgically correctable disease might occur at the level of the aortic arch, at the origins of the vertebral or carotid arteries, which are amenable to bypass. For intracranial disease, however, surgical treatment is not recommended. Extracranial-intracranial (EC-IC) bypass surgery in which the superficial temporal artery (STA) is anastomosed to the middle cerebral artery has been done successfully in children with moyamoya. However, in adult patients with atherosclerotic disease, surgery (as opposed to medical therapy) does not reduce the risk of subsequent ischemic stroke (42). EC-IC bypass procedures have also been performed in the posterior circulation—for instance, STA to superior cerebellar artery to bypass a basilar stenosis. These procedures have shown increased perfusion in small numbers of patients (43).

Interventional neuroradiology may offer other possible solutions, such as percutaneous transluminal angioplasty (PTCA) and stenting. PTCA has been performed in the vertebral-basilar system as well as in the distal ICA or proximal MCA, with significant improvement in symptoms and low restenosis rates (44,45). No long-term data are available, however, with respect to long-term incidence of subsequent stroke. Endovascular stenting, in conjunction with angioplasty, is another promising technique. Stenting of the cervical carotid artery in high-risk patients in lieu of endarterectomy has shown complication rates comparable to those of CEA (46). These may also be utilized intracranially. Polymer-coated stents may also offer the possibility of drug delivery in high concentrations to local arterial wall (47).

XII. MEDICAL THERAPIES

Current medical treatment for these patients is extremely limited. Blood pressure is often maintained at a higher level to maximize the pressure head proximal to the stenosis and

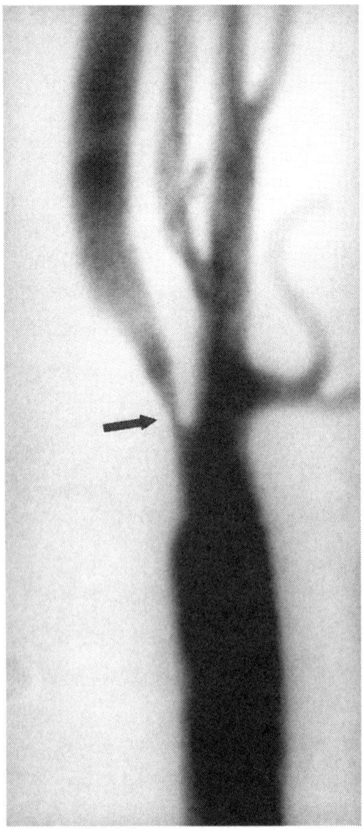

Figure 3 Angiogram at the level of carotid artery bifurcation. Stenosis is present in the proximal internal carotid artery (arrow). This lesion would be amenable to surgical correction or stenting. (Photo courtesy of Michael H. Lev.)

optimize flow to tissue at risk. Such blood pressure manipulation, however, poses the risk of detriment, as chronic hypertension is a risk factor for progressive coronary disease, ischemic stroke, and intracerebral hemorrhage. Blood pressure management can be particularly difficult in patients with postural hypotension (e.g., diabetics), who may be at risk for supine hypertension but for orthostatic hypotension with TIA symptoms upon standing.

Pharmacological therapy is also limited. Antiplatelet agents such as aspirin and ticlopidine are the current standard of care for prevention of recurrent TIA (48). In our institution, antithrombotic therapy, such as warfarin, is frequently used in patients with severe stenoses, particularly those who fail aspirin. In the acute setting, among patients who are threatening to have major strokes, we use intravenous heparin, which may stabilize the endothelium and prevent in situ thrombosis. Randomized trials of heparin (subcutaneous rather than intravenous) in stroke, however, have shown no benefit from this therapy (49).

XIII. FUTURE DIRECTIONS

Given the limitations of medical and surgical therapy for TIA, ''revascularization'' with angiogenic factors has exciting potential. As with coronary arteries and peripheral vascular disease, any therapeutic algorithm for cerebral ischemia would require decisions regarding the choice of an angiogenic agent as well as the mode and timing of drug delivery. Current experience with exogenous induction of cerebral angiogenesis is limited to the animal experiments with intraventricular FGF, as outlined earlier in this chapter. More effective drug delivery might be achieved with VEGF gene therapy—either via naked plasmid DNA or adenoviral vectors. Gene product might be injected intraarterially proximal to a stenosis, be coated on angioplasty balloons, or potentially be seeded on endovascular stents (Fig. 4). Intravenous bFGF, currently under investigation for acute stroke, might also be employed on a more chronic basis. Timing of drug delivery would also need to be determined, including the frequency and duration of therapy to achieve angiogenesis. Further animal modeling would be helpful to clarify these many variables.

Any human study of angiogenic therapy would also depend on definitions of reliable end points. Clinical markers such as frequency or severity of symptoms and incidence of stroke could be measured. Radiological end points would include angiography; however, noninvasive methods such as transcranial Doppler (TCD), perfusion magnetic resonance imaging (MRI), MRI angiography, or positron emission tomography (PET) might provide more easily obtained functional information.

The future use of growth factors in the brain will thus reflect a multifaceted approach. This would include protection of neurons in acute stroke, enhancement of recovery after stroke, and, ideally, prevention of stroke using angiogenic revascularization.

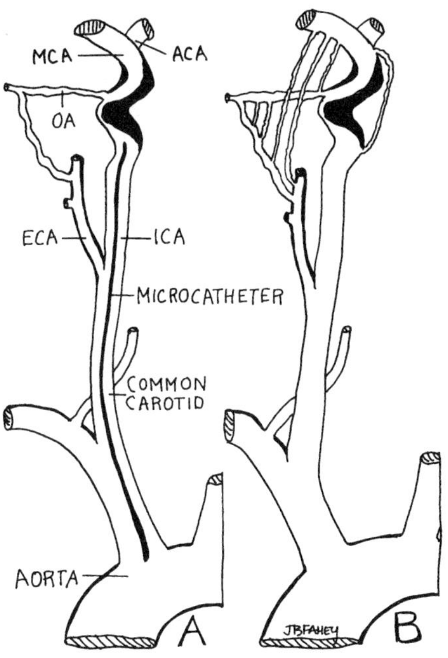

Figure 4 A. Schematic drawing of a severe stenosis involving the carotid artery siphon and middle cerebral artery (MCA). B. Following injection of an angiogenic agent through an intraarterial microcatheter (e.g., VEGF), new collateral vascular channels are formed. (Graphic by Jean Fahey.)

REFERENCES

1. Walicke PA. Basic and acidic fibroblast growth factors have trophic effects on neurons from multiple CNS regions. J Neurosci 8:2618–2627, 1988.
2. Folkman J, Klagsbrun M. Angiogenic Factors. Science 235:442–447, 1987.
3. Gospodarowicz D, Massoglia S, Cheng J, Fujii DK. Effect of fibroblast growth factor and lipoproteins on the proliferation of endothelial cells derived from bovine adrenal cortex, brain cortex, and corpus luteum capillaries. J Cell Physiol 127:121–136, 1986.
4. Ferrara N, Ousley F, Gospodarowicz D. Bovine brain astrocytes express basic fibroblast growth factor, a neurotropic and angiogenic mitogen. Brain Res 462:223–232, 1988.
5. Pettmann B, Labordette G, Weibel M, Sensenbrenner M. The brain fibroblast growth factor (FGF) is localized in neurons. Neurosci Lett 68:175–179, 1986.
6. Wanaka A, Johnson EM Jr, Milbrandt J. Localization of FGF receptor mRNA in the adult rat central nervous system by in situ hybridization. Neuron 5:267–281, 1990.
7. Shimada N, Graf R, Rosner G, et al. Ischemic flow threshold for extracellular glutamate increase in cat cortex. J Cereb Blood Flow Metab 9:603–606, 1989.
8. Siesjo BK, Bengtsson F. Calcium fluxes, calcium antagonists and calcium related pathology in brain ischemia, hypoglycemia and spreading depression: a unifying hypothesis. J Cereb Blood Flow Metab 9:127–140, 1989.
9. Finklestein SP. The potential use of neurotrophic growth factors in the treatment of cerebral ischemia. Adv Neurol 71:413–418, 1996.
10. Yamada K, Kinoshita A, Kohmura E, et al. Basic fibroblast growth factor prevents thalamic degeneration after cortical infarction. J Cereb Blood Flow Metab 11:472–478, 1991.
11. Kawamata T, Speliotes EK, Finklestein SP. The role of polypeptide growth factors in recovery from stroke: brain plasticity. Adv Neurol 73:377–382, 1997.
12. Nozaki K, Finklestein SP, Beal MF. Basic fibroblast growth factor protects against hypoxia-ischemia and NMDA neurotoxicity in neonatal rats. J Cereb Blood Flow Metab 13:221–228, 1993.
13. Koketsu N, Berlove D, Moskowitz MA, et al. Pretreatment with intraventricular basic fibroblast growth factor (bFGF) decreases infarct size following focal cerebral ischemia in rats. Ann Neurol 35:451–457, 1994.
14. Fisher M, Meadows ME, Do T, et al. Delayed treatment with intravenous basic fibroblast growth factor reduces infarct size following permanent focal cerebral ischemia in rats. J Cereb Blood Flow Metab 15:953–959, 1995.
15. The FIBLAST Study Group. Clinical safety trial of IV bFGF in acute stroke. Stroke 29:287, 1998.
16. Regli L, Anderson RE, Meyer FB. Basic fibroblast growth factor increases cortical blood flow in vivo. Brain Res 665:155–157, 1994.
17. Rosenblatt S, Irikura K, Caday CG, et al. Basic fibroblast growth factor (bFGF) dilates rat pial arterioles. J Cereb Blood Flow Metab 14:70–74, 1994.
18. Huang Z, Chen K, Huang PL, et al. bFGF ameliorates focal ischemic injury by blood flow-independent mechanisms in eNOS mutant mice. Am J Physiol 272:H1401–H1405, 1996.
19. Tatlisumak T, Takano K, Carano RA, et al. Effect of basic fibroblast growth factor on experimental focal ischemia studied by diffusion-weighted and perfusion imaging. Stroke 27:2292–2297, 1996.
20. Finklestein SP, Apostolides PJ, Caday CG, et al. Increased basic fibroblast growth factor (bFGF) immunoreactivity at the site of focal brain wounds. Brain Res 460:253–259, 1988.
21. Kawamata T, Alexis NE, Dietrich WD, et al. Intracisternal basic fibroblast growth factor (bFGF) enhances behavioral recovery following focal cerebral infarction in the rat. J Cereb Blood Flow Metab 11:542–547, 1996.
22. Takahasi A, Sawamura Y, Houkin K, et al. The cerebrospinal fluid in patients with moyamoya

disease (spontaneous occlusion of the circle of Willis) contains high levels of basic fibroblast growth factor. Neurosci Lett 160:214–216, 1994.
23. Yoshimoto T, Houkin K, Takahashi A, et al. Angiogenic factors in moya-moya disease. Stroke 27:2160–2165, 1996.
24. Puumala M, Anderson RE, Meyer FB. Intraventricular infusion of HBGF-2 promotes cerebral angiogenesis in Wistar rat. Brain Res 534:283–286, 1990.
25. Lyons MK, Anderson RE, Meyer FB. Basic fibroblast growth factor promotes in vivo cerebral angiogenesis in chronic forebrain ischemia. Brain Res 558:315–320, 1991.
26. Tenjin H, Anderson RE, Meyer FB. Treatment with basic fibroblast growth factor following focal cerebral ischemia does not prevent neuronal injury. J Neurol Sci 128:66–70, 1995.
27. Ferrara N, Davis-Smyth T. The biology of vascular endothelial growth factor. Endocr Rev 18:4–25, 1997.
28. Monacci WT, Merrill MJ, Oldfield EH. Expression of vascular permeability factor/vascular endothelial growth factor in normal rat tissues. Am J Physiol 268:F240–F250, 1993.
29. Breier G, Sterrer S, Risau W. Expression of vascular endothelial growth factor during embryonic angiogenesis and endothelial cell differentiation. Development 114:521–532, 1992.
30. Plate KH, Breier G, Millauer B, Ulrich A, Risau W. Up-regulation of vascular endothelial growth factor and its cognate receptors in a rat glioma model of tumor angiogenesis. Cancer Res 53:5822–5827, 1993.
31. Stein I, Neeman M, Shweiki D, Itin A, Keshet E. Stabilization of vascular endothelial growth factor mRNA by hypoxia and hypoglycemia and coregulation with other ischemia-induced genes. Mol Cell Biol 15:5363–5368, 1995.
32. Fischer S, Sharma HS, Karliczek GF, Schaper W. Expression of vascular permeability factor/vascular endothelial growth factor in pig cerebral microvascular endothelial cells and its upregulation by adenosine. Mol Brain Res 28:141–148, 1995.
33. Ladoux A, Frelin C. Expression of vascular endothelial growth factors by cultured endothelial cells from brain vessels. Biochem Biophys Res Comun 1993:799–803, 1993.
34. Krupinski J, Kaluza J, Kumar P, Kumar S, Wang JM. Role of angiogenesis in patients with cerebral ischemic stroke. Stroke 25:1794–1798, 1994.
35. Kovacs Z, Kiyonobu I, Ken S, Takanori I, Masashi F. VEGF and flt: expression time kinetics in rat brain infarct. Stroke 27:1865–1872, 1996.
36. Hayashi I, Abe K, Suzuki H, Itoyama Y. Rapid induction of vascular endothelial growth factor gene expression after transient middle cerebral artery occlusion in rats. Stroke 28:2039–2044, 1997.
37. Alon T, Hemo I, Itin A, Stone J, Keshet E. Vascular endothelial growth factor acts as a survival factor for newly formed retinal vessels and has implications for retinopathy of prematurity. Nature Med 1:1024–1028, 1995.
38. Strugar J, Rothbart D, Harrington W, Criscuolo R. Vascular permeability factor in brain metastasis: correlation with vasogenic brain edema and tumor angiogenesis. J Neurosurg 81:560–566, 1994.
39. Ku DD, Zaleski JK, Liu S, Brock TA. Vascular endothelial growth factor induces EDRF-dependent relaxation in coronary arteries. Am J Physiol 265:H586–H592, 1993.
40. Bauters C, Asahara T, Sheng LP, Takeshita S, Bunting S, Ferrara N, Symes JF, Isner JM. Recovery of disturbed endothelium-dependent flow in the collateral-perfused rabbit ischemic hindlimb after administration of vascular endothelial growth factor. Circulation 91:2802–2809, 1995.
41. North American Symptomatic Carotid Endarterectomy Trial Collaborators. Beneficial effect of carotid endarterectomy in symptomatic patients with high-grade carotid stenosis. N Engl J Med 325:445–453, 1991.
42. Holohan TV. Extracranial-intracranial bypass to reduce the risk of ischemic stroke. Can Med Assoc J 144:1457–1465, 1991.
43. Ogawa A, Kameyama M, Muraishi K, Yoshimoto T, Ito M, Sakurai. Cerebral blood flow

and metabolism following superficial temporal artery to superior cerebellar artery bypass for vertebrobasilar occlusive disease. J Neurosurg 76:955–960, 1992.
44. Higashida RT, Tsai FY, Halbach VV, Dowd CF, Smith T, Fraser K, Hieshima GB. Transluminal angioplasty for atherosclerotic disease of the vertebral and basilar arteries. J Neurosurg 78:192–198, 1993.
45. Callahan AS III, Berger BL. Balloon angioplasty of intracranial arteries for stroke prevention. J Neuroimaging 7:232–235, 1997.
46. Roubin GS, Yadav S, Iver SS, Vitek J. Carotid-stent-supported angioplasty: a neurovascular intervention to prevent stroke. Am J Cardiol 78:8–12, 1996.
47. Lambert TL, Dev V, Rechavia E, Forrester JS, Litvack F, Eigler NL. Localized arterial wall drug delivery from a polymer-coated removable metallic stent: kinetics, distribution and bioactivity of forskolin. Circulation 90:1003–1011, 1994.
48. Feinberg WM, Albers GW, Barnett HJM et al. Guidelines for the management of transient ischemic attacks. Stroke 25:1320–1335, 1994.
49. The International Stroke Trial (IST). A randomised trial of aspirin, subcutaneous heparin, or both or neither among 19,435 patients with acute ischaemic stroke. Lancet 349:1569–1581, 1997.

32
Therapeutic Angiogenesis

Jeffrey M. Isner and Takayuki Asahara
Tufts University School of Medicine, Boston, Massachusetts

I. INTRODUCTION

The development of blood vessels may be considered in several contexts. Vasculogenesis and angiogenesis are the processes responsible for the development of the circulatory system, the first functional unit in the developing embryo (1). Pathological angiogenesis includes the role of postnatal neovascularization in the pathogenesis of arthritis, diabetic retinopathy, and, most notably, tumor growth and metastasis (2). Therapeutic angiogenesis involves the development of collateral blood vessels supplying ischemic tissues, either endogenously or in response to administered growth factors. The purpose of this review is to consider the mechanisms responsible for therapeutic angiogenesis, which develops endogenously, as well as novel strategies that have been devised to augment this response. Because recapitulation of the embryonic paradigm forms the conceptual basis for therapeutic as well as pathological angiogenesis, selected aspects of embryonic blood vessel development are included. While pathological angiogenesis is beyond the scope of the current paper, certain principles that have emerged from studies of pathological neovascularization are considered for the implications they may have for cardiovascular disease.

II. DISCUSSION

A. Vasculogenesis and Angiogenesis

Vasculogenesis refers to the in situ formation of blood vessels from progenitor endothelial cells (ECs), or angioblasts (3). It is necessary to distinguish between the vascular development that takes place in the yolk sac of the embryo from that which occurs in the embryo proper. Extraembryonic vasculogenesis begins as a cluster formation, or blood island. Growth and fusion of multiple blood islands in the yolk sac of the embryo ultimately gives rise to the yolk sac capillary network (4); after the onset of blood circulation, this network differentiates into an arteriovenous vascular system (5). The integral relationship between the elements that circulate in the vascular system—the blood cells—and the cells that are principally responsible for the vessels themselves—ECs—is implied by the composition of the embryonic blood islands. The cells destined to generate hematopoietic cells are situated in the center of the blood island and are termed *hematopoietic stem cells* (HSCs). Endothelial progenitor cells, or angioblasts, are located at the periphery of the blood islands. In addition to this spatial association, HSCs and angioblasts share certain antigenic determinants, including Flk-1, Tie-2, and CD-34. These progenitor cells have consequently been considered to derive from a common precursor, putatively termed a *hemangioblast* (6–8).

Vasculogenesis that occurs within the embryo proper is currently considered to involve differentiation of ''solitary'' angioblasts, i.e., angioblasts that are not intimately associated with concomitantly differentiating HSCs (3). These angioblasts may migrate and fuse with other angioblasts and capillaries or form vessels in situ. In contrast to the in situ differentiation of progenitor cells required to establish the primordial vascular network, extension of the primitive vasculature involves angiogenesis, i.e., sprouting of new capillaries from the pre-existing network. By definition, this implicates differentiated ECs as the responsible cellular element. Full development of the circulatory system involves recurrent remodeling as some vessels regress, presumably by apoptosis, and others branch and/or are invested with a multilayer architecture characteristic of medium to large arteries and veins. The extensive EC proliferative activity that constitutes the basis for angiogenesis in the embryo contrasts with extraordinary EC quiescence in the adult, where the interval for EC turnover is estimated to be >1000 days (9).

1. *Classic Paradigm for Angiogenesis*

Until recently (see below), vasculogenesis was considered retricted to the embryo, while new blood vessel formation in adult species was inferred to be the exclusive consequence of angiogenesis. The full paradigm for angiogenesis has been suggested to begin with ''activation'' of ECs within a parent vessel, followed by disruption of the basement membrane and subsequent migration of ECs into the interstitial space, possibly in the direction of an ischemic stimulus (10) (Fig. 1). Concomitant and/or subsequent EC proliferation, intracellular-vacuolar lumen formation, pericyte ''capping,'' and production of a basement membrane complete the development sequence.

During angiogenesis, migration always precedes proliferation by approximately 24 hr (11). sholley et al. (12) documented the critical if not exclusive roles of migration and redistribution of preexisting ECs in the commencement of neovascularization. Subsequent studies have established the critical role played by plasmin and other proteases in promoting migration through preexisting matrix (13,14).

In contrast to these in vivo inflammatory and in vitro organ culture models, angiogenesis that develops in response to experimental vascular obstruction—i.e., collateral vessel development—has been shown by several previous investigators to involve proliferation not only of ECs but also of SMCs. Peak EC proliferation, which contributes to naturally occurring collateral development in the setting of vascular occlusion varies from 2.6–3.5% in the canine coronary circulation (15,16); from 5–6% in the rodent renal vasculature (17); and is less than 1% in swine coronaries (18). Proliferation of SMCs, the additional requisite cell type for the formation of larger blood vessels, is an implicit component of angiogenesis regardless of animal species or circulatory site. Schaper et al., in fact, speculated nearly 25 years ago that ''it is tempting to assume that EC proliferation not only serves the purpose of forming the endothelium of a finally larger artery but rather actively participates in the development of the tunica media'' (19). Proliferative activity—for SMCs as well as ECs—is highest at the level of the smallest-diameter collateral vessels, the so-called midzone collateral segments (15,16,20,21). Fifth, while evidence of EC

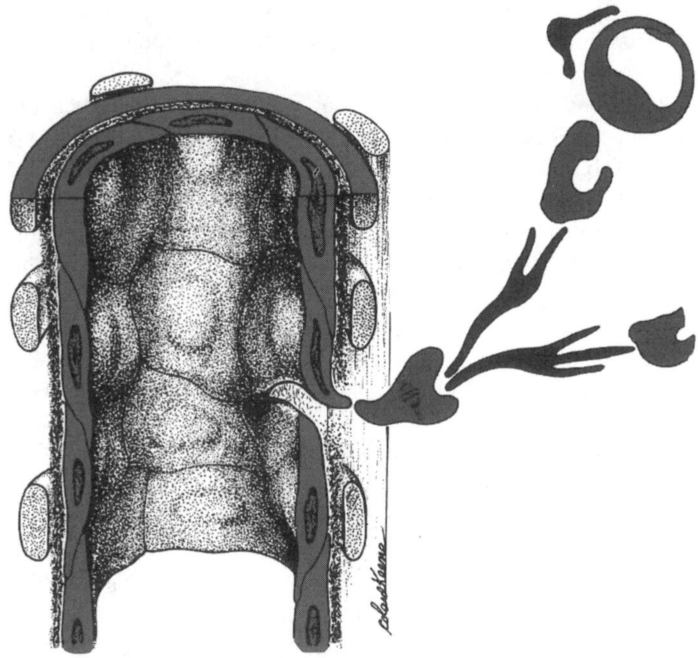

Figure 1 Classic paradigm for angiogenesis. ECs break free from their basement membrane and surrounding extracellular matrix, migrate, proliferate, and remodel (i.e., form a lumen), thus generating new blood vessels or "sprouts" from the parent vessel. (From Ref. 10.)

and SMC proliferation alone does not necessarily distinguish new vessel development from an increase in the size of preexisting vessels, adjunctive data regarding increased capillary density (22,23) support the notion that proliferative activity does in fact reflect true angiogenesis.

2. *Ligand-Receptor Systems Modulate Vasculogenesis and Angiogenesis*

A series of gene-targeting studies have elucidated the role of certain ligands and/or their receptors in vasculogenesis and angiogenesis. The phenotypic characteristics of these knockout mice are relevant to adult cardiovascular disease because of the implications they may have for the role played by these same ligand-receptor systems in promoting postnatal angiogenesis.

As indicated above, KDR (the murine equivalent is known as flk-1 or VEGFR-2), the principal receptor for vascular endothelial growth factor (VEGF), is expressed by both angioblasts and HSCs. It is perhaps not surprising, therefore, that mice deficient in this gene die in utero between 8.5 and 9.5 days postcoitum due to an early defect in the development of hematopoietic and ECs. Yolk-sac blood islands were absent at 7.5 days, organized blood vessels could not be observed in the embryo and yolk sac at any stage, and hematopoietic progenitors were severely reduced (24). Markers of early endothelial precursors such as flt-1, flt-4, and tie-2 were expressed, but a marker of later endothelial development, tie-2, could not be detected, indicating a deficiency of mature ECs (24). Expression of CD34, a marker of HSCs, was greatly reduced as well. The absence of blood islands and blood vessels in these mice established that the flk-1 signaling pathway is required very early in the development of endothelial lineage and may be important for blood cell development as well.

Findings in the flk-1 knockout mouse predicted what was to be found when the ligand, VEGF, was deficient. Mice deficient in even one of two VEGF alleles die in utero between days 10.5 and 12. Both blood island formation (vasculogenesis) and vascular sprouting from preexisting vessels (angiogenesis) were again impaired (25,26). The failure of blood-vessel ingrowth was accompanied by apoptosis and disorganization of neuroepithelial cells. The heterozygous lethal phenotype was interpreted as evidence for tight, dose-dependent regulation of embryonic vessel development by VEGF. Parenthetically, the aortas of VEGF-deficient mice have been noted to be hypoplastic, similar to those observed in mice deficient in endothelial nitric oxide (NO) synthase (P. Huang, personal communication). It is interesting to speculate that this may reflect the role of NO in VEGF-modulation of EC function (see below).

The tyrosine kinase flt-1 receptor (VEGF-R1) constitutes a second high-affinity binding receptor for VEGF. Mouse embryos homozygous for a targeted mutation in the flt-1 locus formed fully differentiated ECs in both embryonic and extraembryonic regions but assembled these cells into abnormal vascular channels and died in utero (27). Blood islands, for example, were disorganized, consisting of intermixed angioblasts and HSCs. In the head mesenchyme, instead of progressive development of individual small vessels, large, fused vessels were seen that contained internally localized groups of ECs. These findings were interpreted as evidence that the flt-1 signaling pathway may regulate normal EC cell-cell or cell-matrix interactions during vascular development.

Tie-1 and tie-2 (28) comprise a second family of receptor tyrosine kinases, other than the VEGF family, in which expression is nearly specific for ECs. Mice embryos deficient in tie-1 fail to establish structural integrity of otherwise differentiated ECs (29); consequently, erythrocytes extravasate through the blood-vessel EC (but not between ECs), leading to death immediately after birth with widespread hemorrhage. Embryos homozygous mutant for tie-2 die earlier (day 10.5 in utero) with dilated vessels lacking distinction between small and large vessels; absence of ordered branching has been inferred as evidence of disordered angiogenesis. Vasculogenesis per se was not disrupted.

Successful disruption of the ligand for tie-2, angiopoietin-1 (30,31), resulted in embryonic lethality by day 12.5 with defects similar to those seen in the tie-2 receptor knockout. These included defects in organized branching, so that vessels remained dilated and almost syncytial. There was no change, however, in the total number of ECs.

The critical roles played by VEGF and its receptors in governing vasculogenesis and angiogenesis, and the tie receptor/ligand family in maturation of the vascular network, have implications for the roles of these EC mitogens in promoting angiogenesis under circumstances of tissue ischemia in adults. In contrast to the lethal consequences of VEGF and tie deficiencies, it is interesting to note that mice in which gene targeting has been used to disrupt the gene for basic fibroblast growth factor survive to maturity with no apparent phenotypic abnormalities of either the vascular or hematopoietic systems (Gerald Dorn and Tom Deutschman, personal communication).

3. Remodeling

Remodeling in angiogenesis refers to the formation of a vascular lumen. Whereas multiple cell types, including ECs, grown in vitro on a collagen-matrix gel will form cords, the presence of a lumen distinguishes vessels, or tubes, from solid cords. The presence of a lumen is clearly fundamental to the function of the circulatory system; yet the mechanisms responsible for tube formation are perhaps the least well understood aspect of angiogenesis (32). As Risau has pointed out, since isolated ECs may combine to form a lumen in vitro (33), lumen formation must represent an intrinsic feature or differentiation program of these cells (3). Mechanisms that have been discussed include the joining of polarized ends of capillary ECs in a ringlike fashion, or, alternatively, simple deletion of a portion of the cell (vacuole formation) (34). Adding further to the complexity is the requirement to form qualitatively differing luminal and abluminal surfaces. Finally, it is inferred that remodeling comprises a coordinated process of lumen formation and vessel extension, with fusion of individual cells and their lumina via cell-cell adhesion molecules. The specific molecules responsible for vessel extension remain ambiguous. It is presumed that this aspect of vessel formation is subject to the regulatory factors responsible for vascular development in the embryo, as discussed above. In this respect, it is interesting to note that VEGF, for example, upregulates gap junction expression (specifically connexin 43) in ECs (35).

4. *Nascent Vessels*

Capillary growth rates (i.e., the velocity of neovascularization) range from 0.23–0.8 mm/day, depending on the experimental system used and/or the type of tumor (36). The light microscopic features of newly formed vessels have been distinguished from those of native vessels (37). Whereas a histological section of a capillary blood vessel in the normal brain reveals one or two ECs per lumen, in a brain tumor such as a glioblastoma, 5–10 ECs may occupy one lumen. Tumor-induced vessels often appear dilated and saccular. Moreover, tumors may contain giant capillaries and arteriovenous shunts without intervening capillaries, so that blood may even flow from one venule to another.

Ultrastructural analysis of newly formed vessels has focused on features potentially responsible for augmented permeability. Dvorak et al. (38,39), for example, found that vascular leakage could not be attributed to passage of molecules through interendothelial cell junctions or injured tumor endothelium but instead involved transendothelial transport via a novel cytoplasmic organelle which they termed the vesicular-vacuolar organelle (VVO). Others (40) have reported VEGF-induced ultrastructural features consistent with endothelial fenestration.

5. *Matrix-Integrin Interactions*

Activated or proliferative ECs have been shown to express high levels of $\alpha v\beta 3$ (41,42). In nonhuman primates subjected to focal cerebral ischemia, for example, microvascular expression of $\alpha v\beta 3$ was noted in ischemic but not nonischemic, tissues (43). Ligation of $\alpha v\beta 3$ on proliferating ECs promotes a critical adhesion-dependent cell survival signal leading to inhibition of p53 activity, decreased expression of $p21^{WAF1/CIP1}$ and suppression of the bax cell-death pathway (44). The intracellular molecular conflict that results from blocking $\alpha v\beta 3$ thus leads to unscheduled apoptosis and the abrogation of angiogenesis (42,44,45). Failure to ligate $\alpha v\beta 3$ may therefore inhibit the ability of actively cycling cells to ligate extracellular matrix proteins—including fibronectin, vitronectin, fibrinogen, and osteopontin—thereby influencing adhesion, migration, and ultimately survival of these cells (46).

Consistent with this notion, the integrin α_v β_3 has been shown to be required for angiogenesis in vivo (41) and antagonists of this integrin have been shown to inhibit angiogenesis by inducing apoptosis (42). Basic FGF, which has been shown to protect ECs from apoptosis (47), is known to modulate integrin expression by ECs (48–50). VEGF, by upregulation of both the β3 integrin and fibronectin, may similarly inhibit apoptosis by enhancing EC adhesion to matrix proteins (51). This notion is thus consistent with the concept that VEGF may exert a survival effect on ECs (41,42,52–54). These findings suggest that the net increase in EC viability following VEGF administration is not limited to the mitogenic effects of VEGF on ECs but is supplemented by the potential for VEGF to inhibit apoptosis.

6. *Vasculogenesis in the Adult*

Postnatal neovascularization has been previously considered to result exclusively from the proliferation, migration, and remodeling of fully differentiated ECs derived from preexisting blood vessels—i.e., angiogenesis (3,9,15). The formation of blood vessels from EC progenitors, or angioblasts—i.e., vasculogenesis—has been considered to be restricted to embryogenesis (5,55).

We reasoned, however, that the use of HSCs derived from peripheral blood in lieu of bone marrow to provide sustained hematopoietic recovery constituted inferential evidence for circulating stem cells (56). Given the common ancestry of HSCs and angioblasts, we investigated the hypothesis that stem cells circulating in peripheral blood might, under selected circumstances, differentiate into ECs (57,58). Flk-1 and a second antigen, CD-34, shared by angioblasts and HSCs (24,59–69), were used to isolate putative angioblasts from the leukocyte fraction of peripheral blood. In vitro, these cells differentiated into ECs. In animal models of ischemia, heterologous, homologous, and autologous EC progenitors incorporated into sites of active angiogenesis (Fig. 2). These findings thus suggest that circulating EC progenitors may contribute to neoangiogenesis in adult individuals, consistent with vasculogenesis.

Parenthetically, these findings may have implications for augmenting collateral vessel growth to ischemic tissues (therapeutic angiogenesis, see below) and for delivery of anti- or proangiogenic agents, respectively, to sites of pathological or utilitarian angiogenesis. A potentially limiting factor in strategies designed to promote neovascularization of ischemic tissues (70) is the resident population of ECs that is competent to respond to administered angiogenic cytokines (71). This issue may be successfully addressed with autologous EC transplants. The fact that progenitor ECs home to foci of angiogenesis suggests potential utility as autologous vectors for gene therapy. For antineoplastic therapies, CD34-positive cells could be isolated and transfected with or coupled to antitumor drugs or angiogenesis inhibitors. For treatment of regional ischemia, angiogenesis could be amplified by transfection of MB^{CD34+} to achieve constitutive expression of angiogenic cytokines and/or provisional matrix proteins (72).

B. Angiogenic Cytokines

Beginning a little over a decade ago (73), a series of polypeptide growth factors (Table 1) were purified, sequenced, and demonstrated to be responsible for natural as well as pathological angiogenesis. These angiogenic cytokines all share in common the ability to act as mitogens for ECs.

Among the various growth factors that have been shown to promote angiogenesis, VEGF (74), also known as vascular permeability factor (VPF) (75) and vasculotropin (VAS) (76), is an EC-specific mitogen. Moreover, a plethora of studies have documented upregulation of VEGF in various cell types following exposure to other angiogenic cyto-

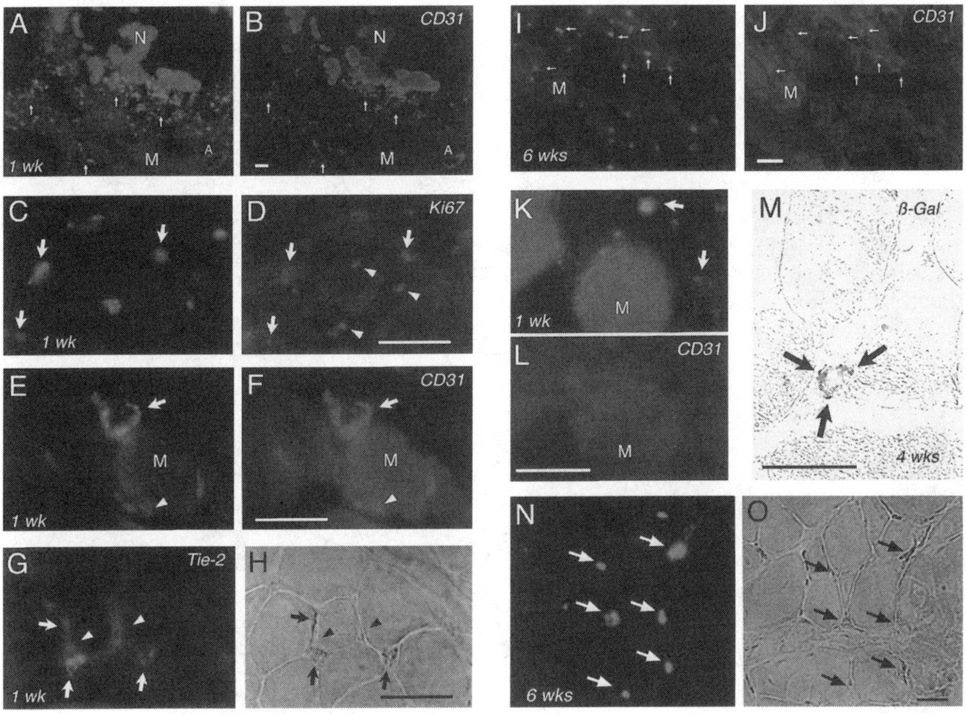

Figure 2 Heterologous (A to L), homologous (M), and autologous (N and O) EC progenitors incorporate into sites of angiogenesis in vivo. A and B. CD34+ mononuclear peripheral blood cells (MB^{CD34+}) (arrows), labeled with the fluorescent dye DiI, between skeletal myocytes (M), including necrotic (N) myocytes 1 week after injection; most are co-labeled with CD31 (arrows). Note preexisting artery (small A), identified as CD31-positive, but DiI-negative. C and D. Evidence of proliferative activity among several DiI-labeled MB^{CD34+} derived cells (arrows), indicated by coimmunostaining for Ki67 antibody. Proliferative activity is also seen among DiI negative, Ki67-positive capillary ECs (arrowhead); both cell types comprise neovasculature. E. DiI and CD31 in capillary ECs (arrows) between skeletal myocytes, photographed through double filter 1 week after DiI-labeled MB^{CD34+} injection. F. Single filter shows CD31 expression in DiI-labeled capillary ECs, integrated into capillary with native (DiI-negative, CD31-positive) ECs (arrowheads). G. Immunostaining 1 week after MB^{CD34+} injection showing capillaries comprising DiI-labeled MB^{CD34+}-derived cells expressing Tie-2 receptor. Several MB^{CD34+}-derived cells (arrow) are Tie-2 positive and are integrated with some Tie-2 positive host capillary cells (arrowhead) identified by the absence of fluorescence. H. Phase-contrast photomicrograph of same tissue section shown in G indicates corresponding DiI-labeled (arrows) and -unlabeled (arrowheads) capillary ECs. I and J. Six weeks after administration, MB^{CD34+} derived cells colabel for CD31 in capillaries between preserved skeletal myocytes. K and L. One week after injection of MB^{CD34+}, isolated MB^{CD34+}-derived cells (arrows) are observed between myocytes, but do not express CD31. M. Immunostaining of β-galactosidase in tissue section harvested from ischemic muscle of B6,129 mice 4 weeks after administration of MB^{Flk-1+} isolated from transgenic mice constitutively expressing β-gal. (Flk-1 cell isolation was used for selection of EC progenitors due to lack of a suitable antimouse CD34 antibody.) Cells overexpressing β-gal (arrows) have been incorporated into capillaries and small arteries; these cells were identified as ECs by anti-CD31 and anti-BS-1 lectin. N and O. Sections of muscles harvested from rabbit ischemic hindlimb 4 weeks after administration of autologous MB^{CD34+}. MB^{CD34+}-derived cells in capillaries, all indicated by (arrows) in a phase-contrast photomicrograph (O). Each scale bar indicates 50 μm. (From Ref. 56.)

Table 1 Angiogenic Cytokines

Acidic fibroblast growth factor (aFGF) (124,159,164,167,168)
Angiopoietin[a] (30,31)
Basic fibroblast growth factor (bFGF) (97,125,126,135,156,160,162,169–172)
Heparin-binding epidermal growth factor (HB-EGF) (101)
Insulin-like growth factor (IGF) (173)
Placental growth factor[a] (PIGF) (88)
Platelet derived growth factor (PDGF) (173)
Scatter factor (hepatocyte growth factor, HGF) (117)
Transforming growth factor-beta (TGF-beta) (101)
Vascular endothelial growth factor (VEGF)[a] (23,25,26,70,127,128,157,158,163,169,172–175)

[a] Specific for endothelial cells.

kines. VEGF may thus be considered a prototypical angiogenic cytokine and for this reason is discussed here in further detail. Specific aspects of the remaining angiogenic cytokines in Table 1 may be found in the accompanying lists of citations.

Four homodimeric species of VEGF have been identified, each monomer having 121, 165, 189, or 206 amino acids respectively (77). The secretion pattern of the four isoforms differs markedly. $VEGF_{121}$ is a weakly acidic polypeptide that does not bind to heparin and is freely soluble in the conditioned medium of transfected cells. The heparin-binding capabilities of the remaining three isoforms are progressively augmented as the result of a stepwise enrichment in basic residues. Thus $VEGF_{165}$, the predominant form secreted by a variety of normal and transformed cells (38), is a basic heparin-binding glycoprotein with an isoelectric point of 8.5; while it is secreted, a significant portion remains bound to the cell surface or extracellular matrix. The $VEGF_{189}$ isoform includes 24 additional amino acids and has been shown not to be freely secreted but instead remains nearly completely bound to the cell surface and/or extracellular matrix (78). $VEGF_{206}$ is a rare isoform so far identified only in a human fetal liver cDNA library.

The possibility of hierarchial efficacy among these three isoforms by performing arterial gene transfer of $phVEGF_{121}$, $phVEGF_{165}$, and $phVEGF_{189}$ was investigated in the rabbit ischemic hindlimb model (79). Remarkably, no differences with regard to anatomic or physiological evidence of angiogenesis could be demonstrated—although all three isoforms yielded statistically significant improvement in every parameter measured compared to the LacZ controls. Moreover, separate experiments performed using 100, 200, and 400 μg of each plasmid failed to disclose a differential dose-response curve among the three isoforms with respect to angiographic score, calf blood pressure ratio, resting/maximum flow, or capillary/myocyte ratio (Y. Tsurumi, unpublished data).

These findings may be interpreted to support the observation made previously by Houck et al. (78) regarding the proteolytically clipped VEGF species that result from the action of plasmin on the 165 and 189 isoforms. The size of the resulting monomers, which in each case are mitogenic for ECs and enhance vascular permeability in a Miles assay (80), is similar to the size of the intact 121 isoform. It is therefore possible that the proteolytic cascade of plasminogen activation, a key step during angiogenesis (13), cleaves the longer forms of VEGF, releasing a soluble $VEGF_{121}$-like species that is the final common mediator of angiogenesis in vivo.

Besides VEGF, also referred to as VEGF-1 or VEGF-A, two additional proteins with particular abundance in heart and skeletal muscle, VEGF-2 or VEGF-C (81) and

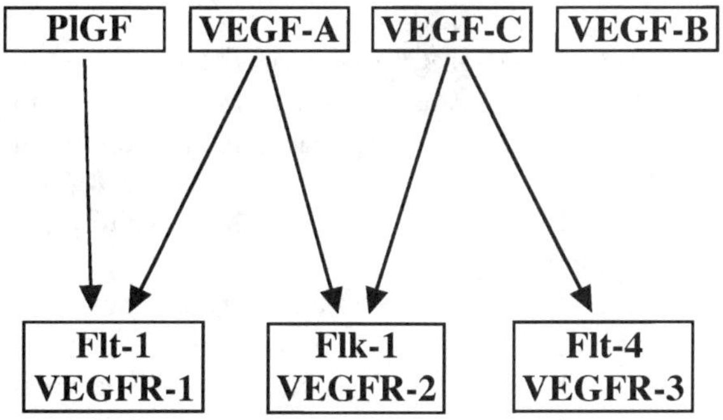

Figure 3 Ligand-receptor relationships for placental growth factor (PIGF) and proteins encoded for by the three vascular endothelial growth factor (VEGF) genes. (VEGFR = vascular endothelial growth factor receptor; alternative receptor nomenclature is shown directly above.) (From Ref. 90.)

VEGF-3 or VEGF-B (82), have been isolated. In contrast to the widespread distribution of these VEGFs, the fourth member of the VEGF-family, placenta growth factor (PlGF) (83) appears to be restricted in vivo to placenta and certain tumors (84,85). All four proteins share structural homology among themselves as well as with the platelet-derived growth factor A and B polypeptides (PDGF-A and -B), in particular a conserved motif, including eight cysteine residues in the putative receptor-binding domain (86). Similar to the a and b chains of PDGF, VEGF-1/VEGF-A and PlGF can form heterodimers with biological activity (87–89). VEGF-3/VEGF-B is also able to heterodimerize with VEGF-1/VEGF-A in cells expressing both factors (82). Although not shown thus far, it seems reasonable to assume that VEGF-2/VEGF-C, by virtue of its shared homology to the other members of the VEGF family, can form heterodimers in a similar manner, therefore adding another level of complexity to processes related to angiogenesis.

Whereas VEGF-1/VEGF-A is a high-affinity ligand for the receptors flt-1 and KDR/flk-1, VEGF-2/VEGF-C was shown to bind to another recently identified endothelial-specific receptor tyrosine kinase, flt-4, as well as to flk-1/KDR, but not to flt-1 (81,90) (Fig. 3). PlGF was shown to bind with high affinity to flt-1 but not to flk-1/KDR (88) or flt-4 (91). The receptor(s) for VEGF-3/VEGF-B (82) have not yet been characterized.

The constitutive coexpression of all three VEGF types in many adult tissues (81,82,92) suggests an interactive or at least redundant capacity of the VEGF members to regulate angiogenesis and modulate EC function. Data regarding the bioactivity of VEGF-2/VEGF-C and VEGF-3/VEGF-B, however, are currently limited. In vitro, both factors exhibit mitogenic activity (81,82,92), and VEGF-2/VEGF-C stimulated EC outgrowth in a collagen gel assay (81). In contrast, the fact that expression of PlGF is restricted to placental tissue and certain tumors (84,85) suggests that this protein plays a minor role in the maintenance of vascular integrity.

1. Autocrine Loop as a Feature of Angiogenic Cytokines

Certain angiogenic cytokines, notably aFGF and bFGF, lack a secretory signal sequence (93,94). This prompted previous investigators to study the possibility that these cytokines act via an alternative pathway. Indeed, several groups established evidence to support the notion that aFGF and bFGF can indeed modulate EC behavior via an autocrine pathway (95–98).

Evidence for a similar autocrine pathway in the case of VEGF was established by Namiki et al. (99). Under quiescent conditions, VEGF mRNA was not detected in either HUVECs or human microvascular ECs (HMECs). Because ECs also express the high-affinity VEGF receptors flt-1 and KDR, these cells in humans appear to include the requisite elements for an autocrine pathway. The finding that hypoxia induces activation of KDR in HUVECs is consistent with an external autocrine pathway (100), given that all VEGF isoforms include a secretory signal sequence.

The extent to which such an autocrine pathway may complement endogenous production of VEGF from SMCs, macrophages, and tumor cells or facilitate the response to exogenous administration of VEGF remains to be determined. With regard to the latter, it has been recognized that single-bolus administration of VEGF may stimulate development of new collateral vessels over a period of several days, despite the fact that the circulating half-life of VEGF is less than 3 min. One explanation for this observation is that as VEGF is rapidly cleared from the circulation, it binds avidly to heparan sulfate proteoglycans present on the luminal surface of the vascular endothelium. Experimental evidence to support this concept in the case of bFGF has been previously reported by Fuks et al. (47). Alternatively, an autocrine loop, activated under hypoxic conditions known to stimulate angiogenesis, might also serve to amplify and thereby protract the response in ECs stimulated by exogenously administered VEGF (Fig. 4). Such autocrine amplification may, in fact, represent a common motif shared by angiogenic cytokines in general.

2. *Direct Versus Indirect Cytokines*

Certain stimuli capable of inducing the development of neovessels in vivo, specifically certain cytokines (101–104) and hypoxia, fail to stimulate EC proliferation in vitro, suggesting a role for additional mediators and/or cell types. We considered whether these so-called indirect angiogenic growth factors might stimulate vascular SMCs to express genes encoding direct EC mitogens. Such a sequential cascade would provide a mechanism by which growth factors that are otherwise nonmitogenic or frankly inhibitory for EC proliferation in vitro could instead stimulate angiogenesis in vivo.

Although PDGF BB, for example, can directly stimulate proliferation of selected populations of microvascular ECs in vitro (105–107), ECs from most vascular districts do not respond to treatment with PDGF BB (108). Nonetheless, in vivo PDGF BB has been clearly demonstrated to induce supportive angiogenesis in both a wound-healing model (103) and chorioallantoic membrane assay (104). Furthermore, microvascular ECs isolated from rat epididymal fat pads and not directly responsive to PDGF BB stimulation, when cocultured with myofibroblasts from the same tissue source, form capillary chords in the presence of PDGF BB (102). The angiogenic action of PDGF BB documented in this way was attributed to production of soluble EC mitogen(s) by myofibroblasts in response to stimulation with PDGF BB.

We demonstrated that PDGF BB treatment of human vascular SMCs concurrently induces VEGF and bFGF mRNA species (109). This finding indicates that PDGF BB, a known mitogen for SMCs, could also stimulate SMCs to produce direct angiogenic factors and thereby promote angiogenesis.

Likewise, the plurality of mechanisms that contribute to the angiogenic activity of transforming growth factor-β1 (TGF-β1) (110) has been the subject of controversy. Although clearly inhibiting growth and migration of subconfluent ECs in vitro (111), TGF-β1 facilitates capillary formation in vivo (110). Furthermore, TGF-β1 modulates the composition of the extracellular matrix (112) and may therefore act in part by locally creating

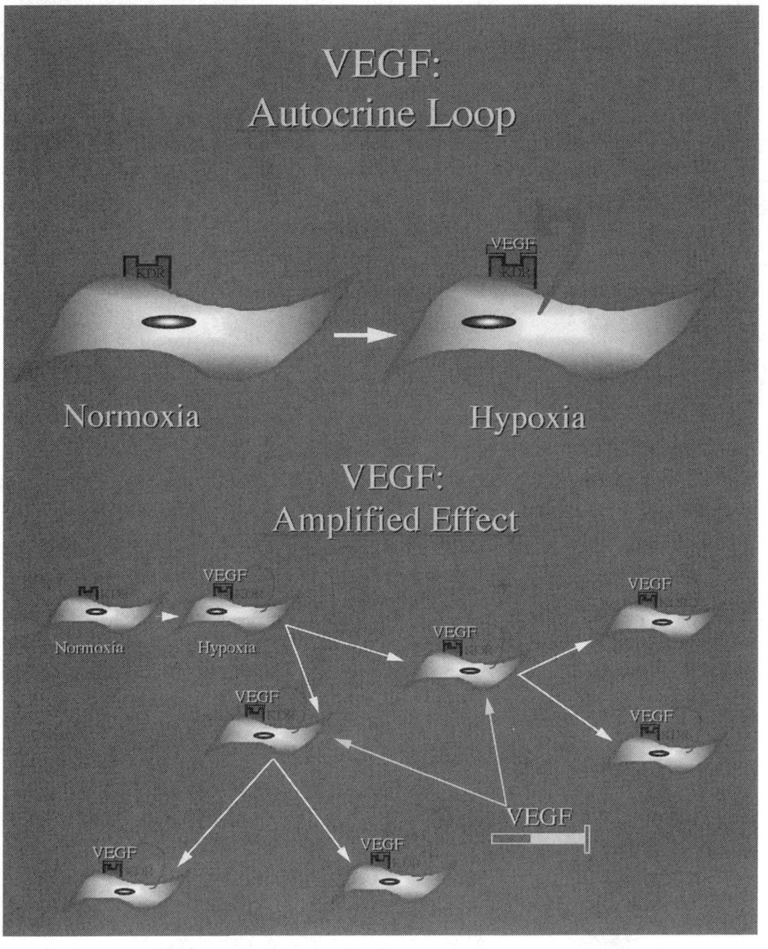

Figure 4 Under conditions of hypoxia, ECs upregulate VEGF, which, once secreted, may then interact with its receptor. Such an autocrine loop provides the basis for amplification of any given VEGF secreted or administered into an ischemic territory. ECs stimulated to proliferate in response to VEGF may then serve as additional sources of VEGF synthesis, thus amplifying the effect of the initial dose of VEGF.

a proangiogenic environment. While TGF-β1 directly inhibits EC proliferation (111), it may nevertheless direct organization of ECs into tubelike structures, depending on the conditions of culture tested (113). TGF-β1 in high concentrations was found to inhibit the formation of capillary-like structures in an in vitro model of angiogenesis, whereas low doses were observed to potentiate VEGF- or bFGF-induced neovessel formation in the same assay system (111). This phenomenon was interpreted by Pepper et al., as an example of ''contextual'' angiogenesis to indicate that the angiogenic response to a given cytokine is dependent upon the presence and concentration of other mediators in the pericellular environment of the target ECs. Winkles et al. have also observed that TGF-β1 stimulates bFGF expression by SMCs, thus supporting the notion that direct angiogenic mediators (114) might contribute in part to TGF-β1–induced angiogenesis. Our finding

(109) that TGF-β1 stimulation induces VEGF as well as bFGF gene expression, in concert with the above-described mechanisms, suggests that both EC mitogens are likely instrumental for mediating the indirect angiogenic effects of TGF-β1 in vivo.

Subsequently, this same motif has been demonstrated for insulin-like growth factor [IGF(115), bFGF (116), and scatter factor (hepatocyte growth factor (117)]. Basic FGF and scatter factor, however, are further distinguished by the fact that they exert a direct effect on EC proliferation and migration, in addition to upregulating VEGF synthesis in vascular SMCs.

In the context of factors that upregulate VEGF expression, it is important to emphasize the critical role played by hypoxia, since this explains, at least in part, the clinical observation that tissue ischemia is a fundamental stimulus for angiogenesis. Increased expression of VEGF has been documented at the periphery of necrotic foci of certain neoplasms, as well as hypoxia-stimulated induction of VEGF in cultured cells from glial tumor and rat skeletal muscle (118). We have shown that low O_2 tension selectively modulates VEGF but not bFGF in vascular SMCs (109). This observation is intriguing in light of homology between a sequence in the VEGF promoter (77) and a nucleotide sequence in the erythropoietin promoter identified as a binding site for an hypoxia-specific transcription factor (HIF-1) (119). It is now clear that while transcriptional upregulation of VEGF does occur in response to hypoxia, posttranslational mechanisms constitute the dominant basis for hypoxia-induced synthesis of VEGF protein (120–123).

3. *Site-Specific Effects of Angiogenic Cytokines*

From a teleologic perspective, it would appear critical that EC mitogens not promote angiogenesis in an indiscriminate fashion; it is clearly preferable for the survival of the organism that angiogenesis be limited to sites of wound healing and tissue ischemia, where it may have facilitatory effects. There is, in fact, evidence to indicate that this is the case. Studies in a variety of animal models using aFGF (124), bFGF (125,126), VEGF (23,127,128), and scatter factor (117) have shown that systemic administration of angiogenic cytokines selectively produces neovascularization in the ischemic limb (129); neither in the contralateral normal limb nor in any other organs were foci of neovascularization observed. In fact, when rhVEGF was injected into the normal or ischemic limb of rabbits with unilateral hindlimb ischemia, angiogenesis was observed only in the ischemic limb.

Furthermore, patients with peripheral vascular disease have been shown to have detectable levels of circulating bFGF (130), yet evidence of angiogenesis appeared limited in these patients to collateral vessel development in the ischemic lower extremities. More recently, following intramuscular injection of the gene encoding VEGF, we have documented circulating levels of VEGF (I. Baumgartner and J. Isner, unpublished data, see below), yet, again, evidence of angiogenesis was limited to the ischemic limb.

There is evidence to suggest that regional ischemia is the principal factor responsible for localization of angiogenesis. For example, in rabbits treated with rhVEGF, the magnitude of augmented collateral development is inversely related to the extent of hindlimb collateral arteries seen prior to rhVEGF therapy (23). The same is true for blood flow in the ischemic limb (131). Thus, if a paucity of angiographically visible collaterals and reduced blood flow are indicative of limb ischemia, then it would appear that the extent to which angiogenesis is enhanced is related to the severity of ischemia.

Experiments performed on ECs in vitro suggest that the basis for localized bioactivity of angiogenic cytokines such as VEGF is due to upregulation of the VEGF receptor in response to hypoxia. In certain types of ECs, notably bovine retinal ECs, this may occur

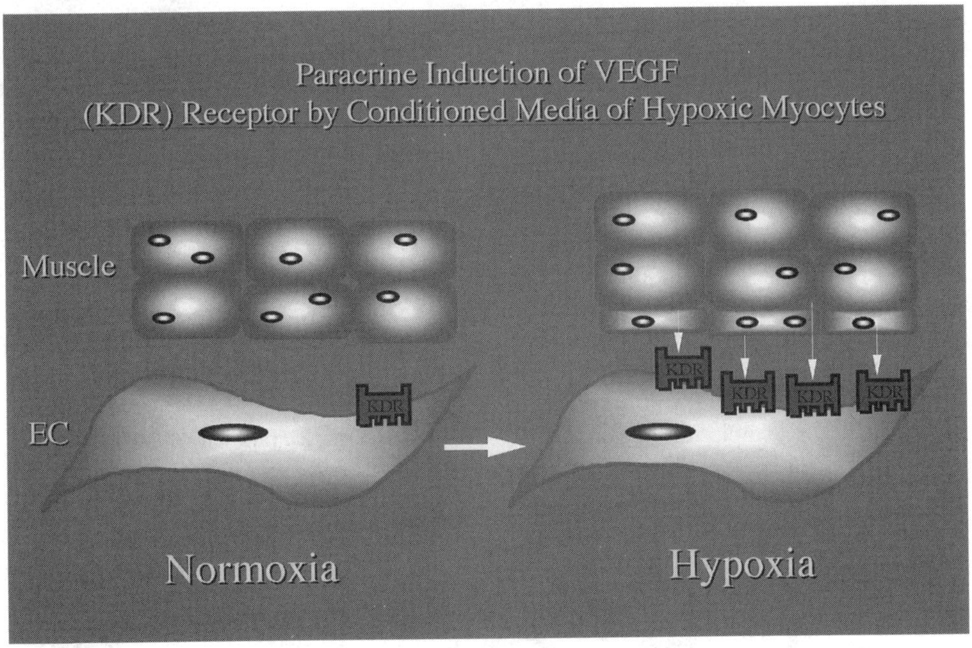

Figure 5 Factors secreted by hypoxic myocytes upregulate VEGF receptor expression on ECs within the hypoxic milieu. Such localized receptor expression may explain the finding that angiogenesis does not occur indiscriminately but rather at sites of tissue ischemia. (From Ref. 133.)

as a direct response to hypoxia (132); in our experience, direct upregulation of receptor expression does not occur in human umbilical vein ECs (HUVECs) or microvascular ECs. The difference in these observations may be due to the fact that bovine retinal ECs possess threefold more VEGF receptors (132).

In the case of HUVECs and microvascular ECs, receptor upregulation appears to occur indirectly in response to hypoxia (Fig. 5). Specifically, we observed an increase in KDR mRNA levels in HUVECs treated for 3 hr with conditioned medium from hypoxic myoblasts, as compared with KDR mRNA levels of HUVECs treated for the same time with conditioned medium from normoxic myoblasts (133). Such an increase suggests synthesis of new receptors in conjunction with mobilization of preformed receptors from the cellular pool. Scatchard analysis of ^{125}I-VEGF binding indicated no substantial change in affinity of the KDR receptor in HUVECs treated with myoblast medium conditioned either in normoxia or hypoxia. Thus, increased binding observed using hypoxia-conditioned medium was due to a 13-fold increase in the number of KDR receptors. These data suggest that a ''factor'' secreted from hypoxic myocytes in ischemic tissues upregulates VEGF receptor expression on adjacent ECs; consequently the ECs in these ischemic tissues act as a magnet for circulating VEGF.

C. Therapeutic Angiogenesis

The therapeutic implications of angiogenic growth factors were identified by the pioneering work of Folkman and colleagues over two decades ago (134). Their work documented the extent to which tumor development was dependent upon neovascularization and suggested that this relationship might involve angiogenic growth factors that were specific for neoplasms.

More recent investigations have established the feasibility of using recombinant formulations of such angiogenic growth factors to expedite and/or augment collateral artery development in animal models of myocardial and hindlimb ischemia. This novel strategy for the treatment of vascular insufficiency has been termed therapeutic angiogenesis (23). The angiogenic growth factors first employed for this purpose comprised members of the FGF family. Baffour et al. administered bFGF in daily intramuscular (IM) doses of 1 or 3 μg to rabbits with acute hindlimb ischemia; at the completion of 14 days of treatment, angiography and necropsy measurement of capillary density showed, as compared with controls, evidence of augmented collateral vessels in the lower limb (125). Pu et al. used aFGF to treat rabbits in which the acute effects of surgically induced hindlimb ischemia were allowed to subside for 10 days before beginning a 10-day course of daily 4-mg IM injections; at the completion of 30 days' follow-up, both angiographic and hemodynamic evidence of collateral development was superior to that of ischemic controls treated with IM saline (124). Yanagisawa-Miwa et al. likewise demonstrated the feasibility of bFGF for salvage of infarcted myocardium, but in this case growth factor was administered intraarterially at the time of coronary occlusion, followed 6 hr later by a second intraarterial bolus (135).

Evidence that VEGF stimulates angiogenesis in vivo had been developed in experiments performed on rat and rabbit cornea (136,137), the chorioallantoic membrane (74), and the rabbit bone graft model (137). The finding that VEGF could be employed to achieve angiogenesis that was therapeutic was first demonstrated by Takeshita et al. (23). The 165–amino acid isoform of VEGF ($VEGF_{165}$) was administered as a single intraarterial bolus to the internal iliac artery of rabbits in which the ipsilateral femoral artery was excised to induce unilateral hindlimb ischemia. The severity of hindlimb ischemia in this animal model has been shown in previous studies to include reduced tissue oxygen saturation (125), increased femoral venous lactate (138), and skeletal muscle necrosis (125). Doses of 500–1000 μg of VEGF produced statistically significant augmentation of angiographically visible collateral vessels (Fig. 6) and histologically identifiable capillaries (Fig. 7); consequent amelioration of the hemodynamic deficit in the ischemic limb was significantly greater in animals receiving VEGF than in untreated controls (calf blood pressure ratio $= 0.75 \pm 0.14$ versus 0.48 ± 0.19, $p < 0.05$). Serial (baseline, as well as 10 and 30 days post-VEGF) angiograms disclosed progressive linear extension of the collateral artery of origin (stem artery) to the distal point of parent-vessel (reentry artery) reconstitution in 7 of 9 VEGF-treated animals. Similar results were achieved in a separate series of experiments in which VEGF was administered by an IM route daily for 10 days (127). By 30 days post-$VEGF_{165}$, flow at rest as well as maximum flow velocity and maximum blood flow provoked by 2 mg papaverine were all significantly higher in the VEGF-treated group (131). These findings thus established proof of principle for the concept that the angiogenic activity of VEGF is sufficiently potent to achieve therapeutic benefit.

1. *Therapeutic Angiogenesis Preserves Vasomotor Reactivity of Collateral Vessels*

Abnormal vascular reactivity may limit the facilitatory effects of collateral vessels on tissue perfusion (139–143). Previous studies have established that chronic perfusion through native coronary collateral vessels produces endothelial dysfunction in the recipient, downstream reconstituted vasculature (139,142). Therapeutic angiogenesis may pro-

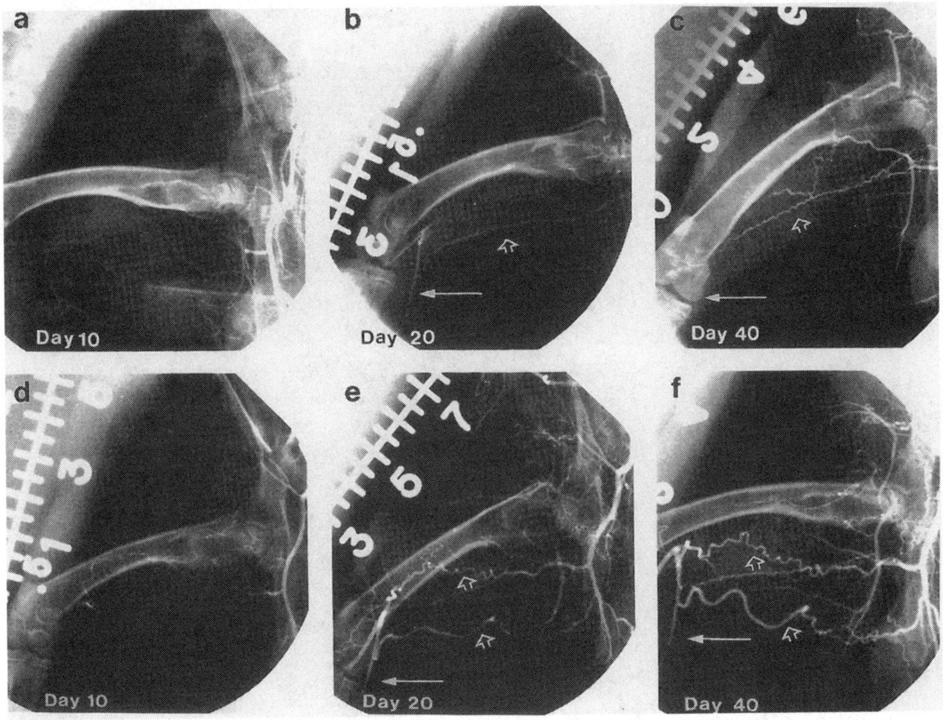

Figure 6 Angiogenesis at the level of medium-sized arteries. Selective internal iliac angiography of control rabbit with unilateral hindlimb ischemia performed at (a) day 10 (baseline), (b) day 20, and (c) day 40, and of VEGF-treated rabbit at (d) day 10 (baseline), (e) day 20, and (f) day 40. The angiograms shown here yielded angiographic scores of (a) 0.10, (b) 0.11, (c) 0.17, (d) 0.12, (e) 0.36, and (f) 0.41. Distal reconstitution, barely apparent in the control group (b, c, arrows), was evident in the VEGF-treated group (e, f, arrows). Direct and linear extension of internal iliac artery to popliteal and/or saphenous arteries was also more evident in VEGF-treated group (b, c, e, f, open arrows). (From Ref. 157.)

mote recovery of endothelium-dependent flow. In the rabbit model of hindlimb ischemia, endothelium-independent and endothelium dependent hindlimb blood flow were essentially restored 30 days following administration of a single intraarterial bolus of VEGF (144).

At least three mechanisms could explain an improvement in endothelium-dependent flow responses of the collateral-dependent limb after VEGF therapy. The first possibility relates to the characteristics of flow and perfusion pressure in arterioles distal to collaterals. We have previously demonstrated that VEGF therapy produces a significant increase in the calf blood pressure of the ischemic limb (23); it is entirely possible that such improved perfusion pressure may lead to repair of dysfunctional endothelium in the collateral-perfused distal vasculature. A second and intriguing possibility relates to a direct improvement of endothelial function by VEGF. In the case of bFGF, for example, in vitro studies have recently demonstrated that endothelial function in the coronary microcirculation perfused via collateral vessels is preserved by chronic administration of this EC mitogen (145). The fact that VEGF may also modulate qualitative aspects of EC function (146) suggests that it too may directly repair ECs presumed damaged by protracted ischemia

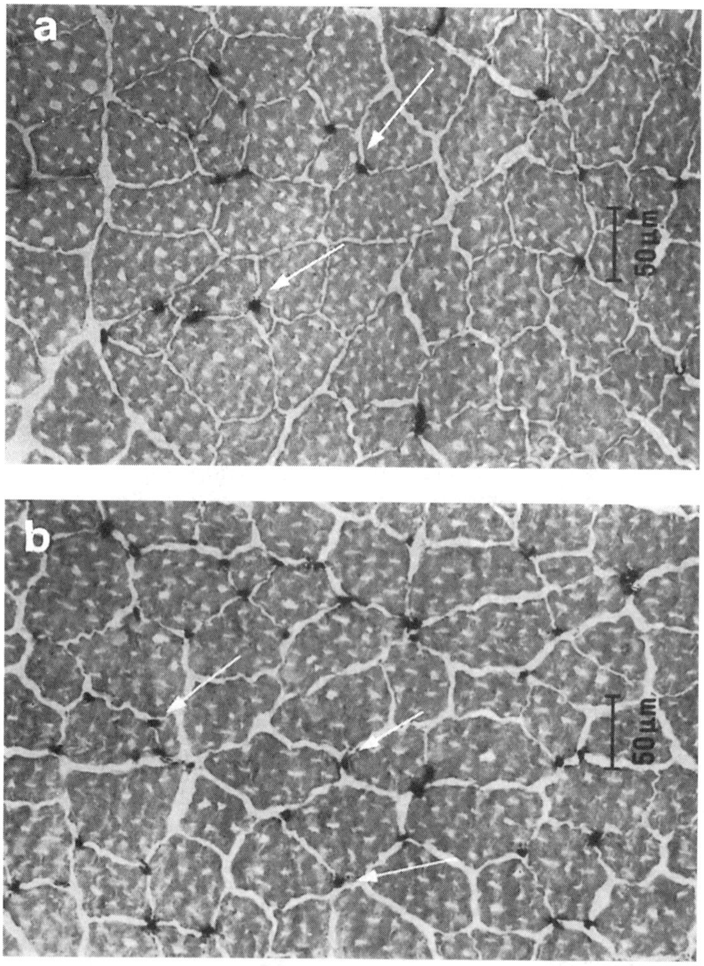

Figure 7 Angiogenesis at the capillary level. Alkaline phosphatase staining of rabbit ischemic hindlimb muscle harvested at day 40: (a) control group and (b) VEGF-treated group. Black dots indicate capillaries (arrows). (Counterstained with eosin.) (From Ref. 157.)

in the collateral-dependent limb and thereby restore normal endothelium-dependent flow. Third, the possibility that the documented improvement in endothelium-dependent flow is the result of the newly formed, VEGF-induced collateral vessels, cannot be discounted.

The finding that angiogenic growth factors restore the responses of the ischemic hindlimb to endothelium-dependent vasodilators may have important clinical implications. The hypersensitivity to serotonin of the collateral circulation is not limited to animal models. Platelet activation releases vasoactive quantities of serotonin in vitro (147), and the S_2-receptor antagonist ketanserin dilates limb collaterals in over 50% of human patients with advanced atherosclerosis (148). Ketanserin has also been reported to improve limb perfusion in selected patients with peripheral artery disease (149), suggesting that abnormal reactivity may largely limit the beneficial consequences of collaterals in humans.

2. Therapeutic Angiogenesis Achieved by Arterial Gene Transfer

No recombinant protein formulation of any of the three principal VEGF isoforms or any other angiogenic cytokine is currently approved or available for human clinical application. Arterial gene transfer constitutes an alternative strategy for accomplishing therapeutic angiogenesis in patients with limb ischemia. In the case of VEGF, this is a particularly appealing strategy because, as indicated previously, the VEGF gene encodes a signal sequence that permits the protein to be naturally secreted from intact cells (77). Previous studies from our laboratory (150,151) indicated that arterial gene transfer of cDNA encoding for a secreted protein could potentially yield meaningful biological outcomes in spite of a low transfection efficiency. We therefore performed preclinical animal studies to establish the feasibility of site-specific gene transfer of $phVEGF_{121}$, $phVEGF_{165}$, and $phVEGF_{189}$ applied to the hydrogel polymer coating of an angioplasty balloon (152) and delivered percutaneously to the iliac artery of rabbits in which the femoral artery had been excised to cause unilateral hindlimb ischemia (138).

Site-specific transfection of $phVEGF_{165}$ was confirmed by analysis of the transfected internal iliac arteries using reverse transcriptase–polymerase chain reaction (RT-PCR) (153) and then sequencing the RT-PCR product. Augmented development of collateral vessels was documented by serial angiograms in vivo and increased capillary density at necropsy. Consequent amelioration of the hemodynamic deficit in the ischemic limb was documented by improvement in the calf blood pressure ratio (ischemic/normal limb) to 0.70 ± 0.08 in the VEGF-transfected group versus 0.50 ± 0.18 in controls ($p < 0.05$). Similar findings were achieved with the 121 and 189 VEGF isoforms (79). These findings thus established that site-specific arterial gene transfer can be used to achieve physiologically meaningful therapeutic modulation of vascular disorders, including therapeutic angiogenesis.

The relevance of these findings in the rabbit ischemic hindlimb model is supported by recent clinical application of arterial gene therapy (154). Using a dose-escalating design, treatment was initiated with 100 μg of phVEGF. Three patients presenting with rest pain (but no gangrene) and treated with 1000 μg were subsequently shown at 1-year follow-up to have improved blood flow to the ischemic limb and to remain free of rest pain. With the increase in dose of $phVEGF_{165}$ to 2000 μg, angiographic (Fig. 8) and histological evidence of new blood vessel formation became apparent (70) (Fig. 9). Subsequently, the use of intramuscular gene transfer, employed initially as a means of treating patients in whom vascular disease in the ischemic limb was too extensive to permit an intraarterial approach, resulted in circulating levels of VEGF detectable by enzyme-linked immunosorbent assay (ELISA) as well as reproducible hemodynamic and angiographic improvement (I. Baumgartner and J. Isner, unpublished observations). In at least one case, therapeutic angiogenesis administered in this fashion accomplished genuine limb salvage (Fig. 10). The clinical reproducibility of these findings may ultimately be influenced by certain features of host diversity, including the extent of native VEGF and VEGF receptor expression among patients with peripheral and/or myocardial ischemia.

3. Recombinant Protein Versus Gene Therapy

There are now abundant data in animal models to support the potential utility of both recombinant protein therapy and arterial gene transfer for myocardial angiogenesis (135,145,155–161). The same is true for lower extremity vascular insufficiency

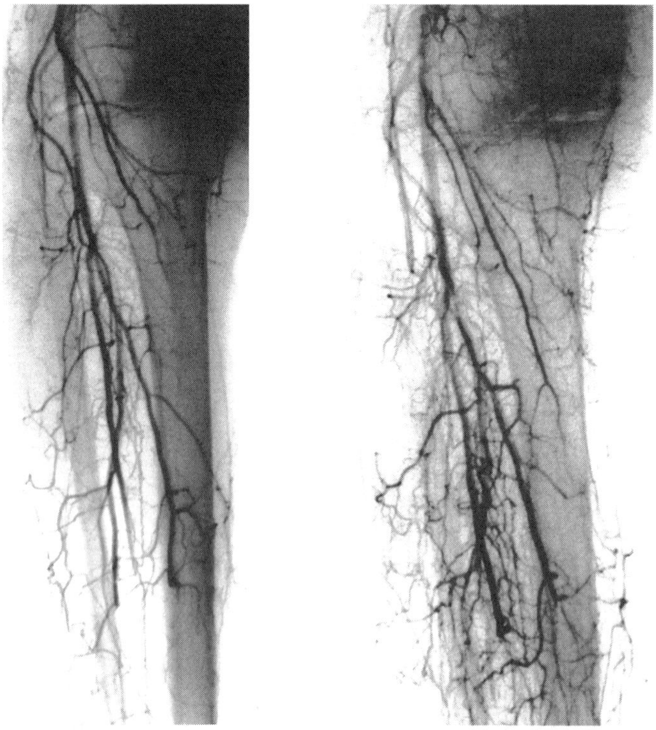

Figure 8 Selective digital subtraction angiograms performed in patient with critical limb ischemia due to occlusion of all three infrapopliteal vessels at the level of the midcalf: (Left) immediately prior to and (right) 1 month after gene therapy with 2000 μg of naked DNA encoding VEGF. The latter angiogram disclosed a plethora of new collateral vessels in ischemic limb. (From Ref. 177.)

(23,117,124–126,126–128,162–164). It is therefore reasonable to ask why, if the recombinant protein has been or can be manufactured, it has to consider transferring the gene encoding that protein (i.e., gene therapy)? There are at least four issues to consider in this regard. First, and perhaps most critical in the case of arterial gene therapy, is the potential requirement to maintain an optimally high and local concentration over time. In the case of therapeutic angiogenesis, for example, it may be preferable to deliver a lower dose over a period of several days or more from an actively expressing transgene in the ischemic limb, rather than a single- or multiple-bolus doses of recombinant protein. Second, there is the matter of economics: namely, which therapy would ultimately cost more to develop, implement, and reimburse, particularly for those indications requiring multiple or even protracted treatment? Third, in certain species—namely rats, rabbits, and swine—both rhbFGF and rhVEGF have been shown to produce varying degrees of systemic hypotension, even in the setting of hypercholesterolemia (20,157,165,166) (Fig. 11). And fourth, the route of administration may be a factor, depending upon the target site for neovascularization. For example, the ready access to ischemic skeletal muscle together with the promising results of IM gene therapy make this approach eminently suitable for the treatment of lower extremity vascular insufficiency. In contrast, IM administration for myocardial angiogenesis cannot be accomplished as readily and may therefore be more amenable to recombinant protein therapy.

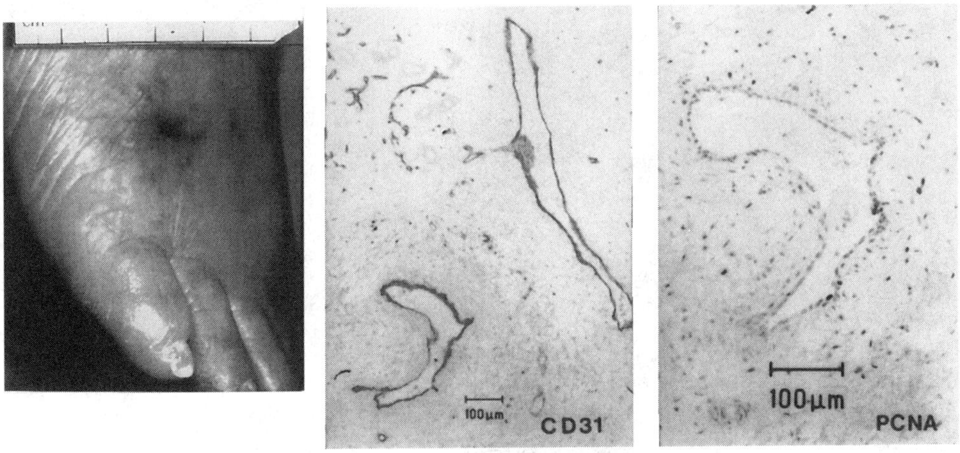

Figure 9 One of three spider angiomata that developed approximately 1 week after gene therapy in the distal portion of the ischemic limb of a patient. Photomicrographs of tissue sections immunostained with antibody to endothelial antigen CD31 indicate the vascularity of the lesion, while immunostain of adjacent section for proliferating cell nuclear antigen (PCNA) indicates the extent of proliferative activity among ECs in the lesion. (From Ref. 177.)

Resolution of these issues will almost certainly require the empirical experience of human clinical trials. Recombinant bFGF and VEGF are now being investigated in human clinical trials for patients with lower extremity vascular disease and myocardial ischemia respectively, so that the magnitude of bioactivity that can be achieved with either of these drugs should be ultimately clarified.

V. PERSPECTIVES

The identification of angiogenic growth factors has generated the opportunity for novel therapies in the treatment of a variety of diseases. This includes pathological angiogenesis, such as diabetic retinopathy, rheumatoid arthritis, and cancer. In these cases, antibodies and/or naturally occurring angiogenesis inhibitors are being investigated clinically to antagonize key angiogenic factors. The strategy is designed to eliminate the vascular infrastructure and thereby minimize pathological consequences to the patient.

A complementary strategy is likely to emerge for the treatment of cardiovascular diseases. Clinical trials of therapeutic angiogenesis have already been initiated in patients with myocardial ischemia and peripheral vascular disease. These include trials of recombinant protein therapy as well as gene transfer. While preliminary applications of gene therapy have established proof of the concept that angiogenic growth factors can augment collateral artery development in human subjects, many questions remain to be answered. What are the relative advantages of protein versus gene therapy, both in terms of bioactivity as well as cost and safety? What is the maximum extent of clinical improvement that can be expected? How is the therapeutic outcome affected by certain features of the host?

It is also likely that future investigations may clarify the impact on established risk factors for vascular disease, such as lipid dyscrasias and diabetes, on native angiogenesis; thus risk factor modifications may be useful for indirectly aiding natural angiogenesis in

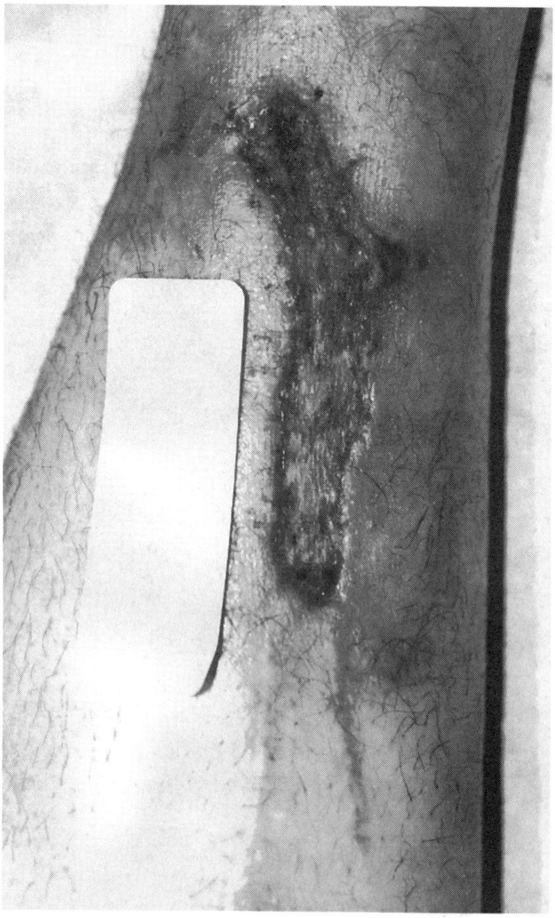

Figure 10 Successful placement of a split-thickness skin graft in the first patient treated with intramuscular (IM) administration of naked DNA encoding VEGF ($phVEGF_{165}$). Following two IM injections of 2000 μg of $phVEGF_{165}$, the patient's ankle-brachial index increased from 0.28 to 0.55, angiography disclosed new collateral vessels in ischemic limb, and a wound of 9 × 3cm at the site of vein harvest in medial calf healed sufficiently to permit successful grafting.

ischemic territories. It is intriguing to consider the possibility that one may be able to identify certain genotypic characteristics indicating why some patients form robust collateral networks while others fail to do so.

The role of endothelial cell (EC) progenitors in both natural as well as therapeutic angiogenesis will be clarified by future studies. Recent studies have forced us to modify the classic concept that collateral development occurs solely by migration of fully differentiated ECs. It is now clear that circulating stem cells contribute to such neovascularization. What remains to be determined is the proportion of new vessel growth that results from circulating stem cells versus parent vessel ECs. Finally, it is possible that EC progenitors may be employed in strategies of ''supply side'' therapeutic angiogenesis; specifically can administration of EC progenitors, including cells engineered to secrete pro-angiogenic agents, complement the impact of angiogenic cytokines on collateral vessel growth?

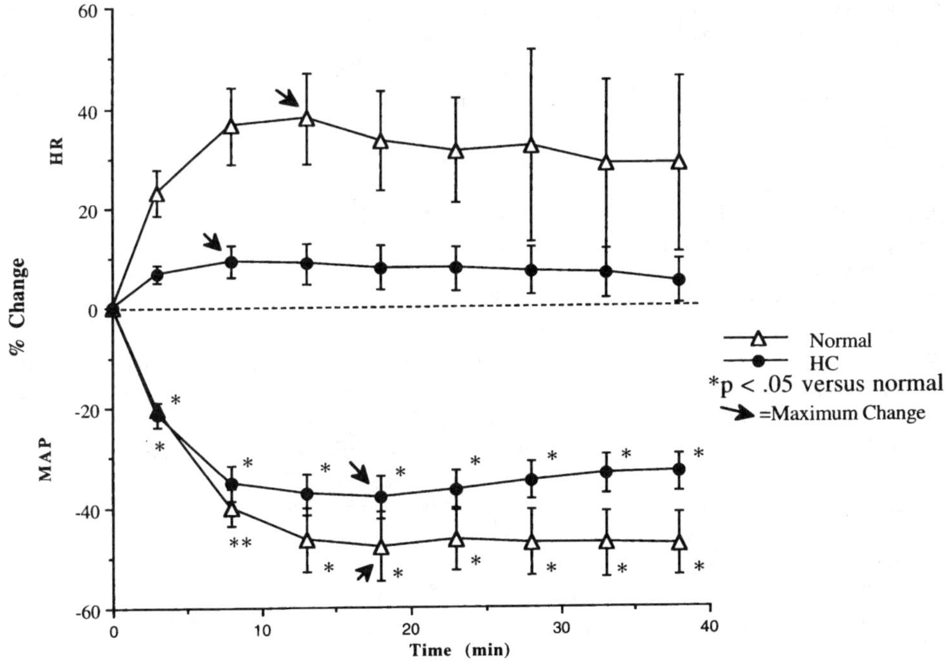

Figure 11 (Top) Administration of rhVEGF to normal and hypercholesterolemic (HC) rabbits leads to prompt reduction in mean arterial pressure (MAP), with a compensatory increase in heart rate (HR). The magnitude and duration (≥ 40 min) of hypotension are not significantly different in HC versus normal rabbits. (Bottom) MAP does not drop below baseline when administration of VEGF is preceded by administration of an inhibitor of nitric oxide synthase (L-NNA).

ACKNOWLEDGMENTS

Supported in part by an Academic Award in Vascular Medicine (HL02824) and grants HL53354 and HL57516 from the National Institutes of Health, Bethesda, Maryland, and the E.L. Wiegand Foundation, Reno, Nevada.

REFERENCES

1. SF Gilbert. Developmental Biology, 4th ed. Sunderland, MA: Sinauer, 1997. p. 342.
2. J Folkman. Angiogenesis in cancer, vascular, rheumatoid and other disease. Nature Med 1: 27–30, 1995.
3. W Risau. Differentiation of endothelium. FASEB J 9:926–933, 1995.
4. W Risau, I Flamme. Vasculogenesis. Annu Rev Cell Dev Biol 11:73–91, 1995.
5. W Risau, H Sariola, H-G Zerwes, J Sasse, P Ekblom, R Kemler, T Doetschman. Vasculogenesis and angiogenesis in embryonic stem cell-derived embryoid bodies. Development 102: 471–478, 1988.
6. I Flamme, W Risau. Induction of vasculogenesis and hematopoiesis in vitro. Development 116:435–439, 1992.
7. W His. Leoithoblast und Angioblast der Wirbelthiere. Abhandl K S Ges Wiss Math Phys 22:171–328, 1900.

8. M Weiss, SH Orkin. In vitro differentiation of murine embryonic stem cells: new approaches to old problems. J Clin Invest 97:591–595, 1996.
9. J Folkman, Y Shing. Angiogenesis. J Biol Chem 267:10931–10934, 1992.
10. PA D'Amore, RW Thompson. Mechanisms of angiogenesis. Annu Rev Physiol 49:453–464, 1987.
11. DH Ausprunk, J Folkman. Migration and proliferation of endothelial cells in preformed and newly formed blood vessels during tumor angiogenesis. Microvasc Res 14:53–65, 1977.
12. MM Sholley, GP Ferguson, HR Seibel, JL Montour, JD Wilson. Mechanisms of neovascularization: vascular sprouting can occur without proliferation of endothelial cells. Lab Invest 51:624–634, 1984.
13. MS Pepper, R Montesano. Proteolytic balance and capillary morphogenesis. Cell Differ Dev 32:319–328, 1990.
14. MS Pepper, N Ferrara, L Orci, R Montesano. Vascular endothelial growth factor (VEGF) induces plasminogen activators and plasminogen activator inhibitor-1 in microvascular endothelial cells. Biochem Biophys Res Comm 181:902–906, 1991.
15. W Schaper, MD Brahander, P Lewi. DNA synthesis and mitoses in coronary collateral vessels of the dog. Circ Res 28:671–679, 1971.
16. S Pasyk, W Schaper, J Schaper, K Pasyk, G Miskiewicz, B Steinseifer. DNA synthesis in coronary collaterals after coronary artery occlusion in conscious dog. Am J Physiol 242: H1031–H1037, 1982.
17. N Ilich, NK Hollenberg, DH Williams, HL Abrams. Time course of increased collateral arterial and venous endothelial cell turnover after renal artery stenosis in the rat. Circ Res 45:579–582, 1979.
18. FC White, SM Carroll, A Magnet, CM Bloor. Coronary collateral development in swine after coronary artery occlusion. Circ Res 71:1490–1500, 1992.
19. W Schaper, J Schaper, R Xhonneux, R Vandesteene. The morphology of intercoronary anastomoses in chronic coronary artery occlusion. Cardiovasc Res 3:315–323, 1969.
20. P Cuevas, F Carceller, S Ortega, M Zazo, I Nieto, G Gimenez-Gallego. Hypotensive activity of fibroblast growth factor. Science 254:1208–1210, 1991.
21. M Bucay, JH Nguy, R Barrios, SA Kerns, PD Henry. Impaired macro- and microvascular growth in hypercholesterolemic rabbits (abstr). J Am Coll Cardiol 19:151A, 1992.
22. AM Graham, R Baffour, T Burdon, B DeVarennes, MA Ricci, A Common, R Lisbona, AD Sniderman, JF Symes. A demonstration of vascular proliferation in response to arteriovenous reversal in the ischemic canine hind limb. J Surg Res 47:341–347, 1989.
23. S Takeshita, LP Zheng, E Brogi, M Kearney, LQ Pu, S Bunting, N Ferrara, JF Symes, JM Isner. Therapeutic angiogenesis: a single intra-arterial bolus of vascular endothelial growth factor augments revascularization in a rabbit ischemic hindlimb model. J Clin Invest 93: 662–670, 1994.
24. F Shalaby, J Rossant, TP Yamaguchi, M Gertsenstein, X-F Wu, ML Breitman, AC Schuh. Failure of blood-island formation and vasculogenesis in Flk-1 deficient mice. Nature 376: 62–66, 1995.
25. P Carmeliet, V Ferreira, G Breier, S Pollefeyt, L Kieckens, M Gertsenstein, M Fahrig, A Vandenhoeck, H Kendraprasad, C Eberhardt, C Declercq, J Pawling, L Moons, D Collen, W Risau, A Nagy. Abnormal blood vessel development and lethality in embryos lacking a single VEGF allele. Nature 380:435–439, 1996.
26. N Ferrara, K Carver-Moore, H Chen, M Dowd, L Lu, KS O'Shea, L Powell-Braxton, KJ Hilan, MW Moore. Heterozygous embryonic lethality induced by targeted inactivation of the VEGF gene. Nature 380:439–442, 1996.
27. G-H Fong, J Rossant, M Gersenstein, ML Breitman. Role of the Flt-1 receptor tyrosine kinase in regulating the assembly of vascular endothelium. Nature 376:66–70, 1995.
28. TN Sato, Y Qin, CA Kozak, KL Audus. Tie-1 and tie-2 define another class of putative

receptor tyrosine kinase genes expressed in early embryonic vascular system. Proc Natl Acad Sci USA 90:9355–9358, 1993.
29. TN Sato, Y Tozawa, U Deutsch, K Wolburg-Buchholz, Y Fujiwara, M Gentron-Maguire, T Gridley, H Wolburg, W Risau, Y Qin. Distinct roles of the receptor tyrosine kinases Tie-1 and Tie-2 in blood vessel formation. Nature 376:70–74, 1995.
30. S Davis, TH Aldrich, PF Jones, A Acheson, DL Compton, V Jian, TE Ryan, J Bruno, C Radziejewdki, PC Maisonpierre, GD Yancopoulos. Isolation of angiopoietin-1, a ligand for the TIE2 receptor by secretion-trap expression cloning. Cell 87:1161–1169, 1996.
31. C Suri, PF Jones, S Patan, S Bartunkova, PC Maisonpierre, S Davis, T Sato, GD Yancopoulos. Requisite role of angiopoietin-1, a ligand for the TIE2 receptor, during embryonic angiogenesis. Cell 87:1171–1180, 1996.
32. DE Ingber, J Folkman. How does extracellular-matrix control capillary morphogenesis? Cell 58:803–805, 1989.
33. J Folkman, C Haudenschild. Angiogenesis in vitro. Nature 288:551–556, 1980.
34. T Bar, F-H Guldner, JR Wolff. ''Seamless'' endothelial cells of blood capillaries. Cell Tissue Res 235:99–106, 1984.
35. JE Saffitz, A Sullivan, JM Isner. Regulation of endothelial cell connexin expression by vascular endothelial growth factor (abstr). Circulation 94:I-238, 1996.
36. J Folkman. Tumor angiogenesis. In: FF Becker, ed. Cancer Biology. New York: Plenum Press, 1975, pp. 355–388.
37. RK Jain. Determinants of tumor blood flow. Cancer Res 48:2641–2647, 1988.
38. HF Dvorak, LF Brown, M Detmar, AM Dvorak. Vascular permeability factor/vascular endothelial growth factor, microvascular hyperpermeability, and angiogenesis. Am J Pathol 146:1029–1039, 1995.
39. S Kohn, JA Nagy, HF Dvorak, AM Dvorak. Pathways of macromolecular tracer transport across venules and small veins: structural basis for the hyperpermeability of tumor blood vessels. Lab Invest 67:596–607, 1992.
40. WG Roberts, GE Palade. Increased microvascular permeability and endothelial fenestration induced by vascular endothelial growth factor. J Cell Sci 108:2369–2379, 1995.
41. PC Brooks, RAF Clark, DA Cheresh. Requirement of vascular integrin alpha-v-beta-3 for angiogenesis. Science 264:569–571, 1994.
42. PC Brooks, AMP Montgomery, M Rossenfeld, RA Reisfeld, T Hu, G Killer, DA Cheresh. Integrin alpha-v-beta-3 antagonists promote tumor regression by inducing apoptosis of angiogenic blood vessels. Cell 79:1157–1164, 1994.
43. Y Okada, BR Copeland, GF Hamann, JA Koziol, DA Cheresh, GJ del Zoppo. Integrin alpha-v-beta-3 is expressed in selected microvessels after focal cerebral ischemia. Am J Pathol 149:37–44, 1996.
44. S Stromblad, JC Becker, M Yebra, PC Brooks, DA Cheresh. Suppression of p53 activity and $p21^{WAF1CIP1}$ expression by vascular cell integrin alpha-v-beta-3 during angiogenesis. J Clin Invest 98:426–433, 1996.
45. G Breier, U Albrecht, S Sterrer, W Risau. Expression of vascular endothelial growth factor during embryonic angiogenesis and endothelial cell differentiation. Development 114:521–532, 1992.
46. E Ruoslahti, JC Reed. Anchorage dependence, integrins, and apoptosis. Cell 77:477–478, 1994.
47. Z Fuks, RS Persaud, A Alfieri, M McLoughlin, D Ehleiter, JL Schwartz, AP Seddon, C Cordon-Cardo, A Haimovitz-Friedman. Basic fibroblast growth factor protects endothelial cells against radiation-induced programmed cell death in vitro and in vivo. Cancer Res 54: 2582–2590, 1994.
48. P Defilippi, L Silengo, G Tarone. Regulation of adhesion receptors expression in endothelial cells. Microbiol Immunol 184:87–98, 1993.

49. J Enenstein, NS Walsh, RH Kramer. Basic FGF and TGF-b differentially modulate integrin expression of human microvascular endothelial cells. Exp Cell Res 203:499–503, 1992.
50. S Klein, FG Giancotti, M Presta, SM Albelda, CA Buck, DB Rifkin. Basic fibroblast growth factor modulates integrin expression in microvascular endothelial cells. Mol Biol Cell 4: 973–982, 1993.
51. I Spyridopoulos, E Brogi, M Kearney, AB Sullivan, C Cetrulo, JM Isner, DW Losordo. Vascular endothelial growth factor inhibits endothelial cell apoptosis induced by tumor necrosis factor-alpha: balance between growth and death signals. J Mol Cell Cardiol 29:1321–1330, 1997.
52. M Friedlander, PC Brooks, RW Shaffer, CM Kincaid, JA Varner, DA Cheresh. Definition of two angiogenic pathways by distinct $alpha_v$ integrins. Science 270:1500–1502, 1995.
53. T Alon, I Hemo, A Itin, J Pe'er, J Stone, E Keshet. Vascular endothelial growth factor acts as a survival factor for newly formed retinal vessels and has implications for retinopathy of prematurity. Nature Med 1:1024–1028, 1995.
54. O Katoh, H Tauchi, K Kawaishi, A Kimura, Y Satow. Expression of the vascular endothelial growth factor (VEGF) receptor gene KDR, in hematopoietic cells and inhibitory effect of VEGF on apoptotic cell death caused by ionizing radiation. Cancer Res 55:5687–5692, 1995.
55. L Pardanaud, C Altman, P Kitos, F Dieterien-Lievre. Relationship between vasculogenesis, angiogenesis and haemopoiesis during avian ontogeny. Development 105:473–485, 1989.
56. T Asahara, T Murohara, A Sullivan, M Silver, R van der Zee, T Li, B Witzenbichler, G Schatteman, JM Isner. Isolation of putative progenitor endothelial cells for angiogenesis. Science 275:965–967, 1997.
57. MH-D Wu, QWechezak, AR Shi, AW Clowes, IL Gordon, LR Sauvage. Definitive proof of endothelialization of a Dacron arterial prosthesis in a human being. J Vasc Surg 21:862–867, 1995.
58. Q Shi, MH Wu, N Hayashida, AR Wechezak, AW Clowes, LR Sauvage. Proof of fallout endothelialization of impervious dacron grafts in the aorta and inferior vena cava of the dog. J Vasc Surg 20:546–556, 1994.
59. CI Civin, ML Banquerigo, LC Strauss, MR Loken. Antigen analysis of hematopoiesis. VI. Flow cytometric characterization of My-10-positive progenitor cells in normal human bone marrow. Exp Hematol 15:10–17, 1987.
60. F Katz, RW Tindle, DR Sutherland, MD Greaves. Identification of a membrane glycoprotein associated with hemopoietic progenitor cells. Leuk Res 9:191–198, 1985.
61. RG Andrews, JW Singer, ID Bernstein. Monoclonal antibody 12-8 recognizes a 115-kd molecular present on both unipotent and multipotent hematopoietic colony-forming cells and their precursors. Blood 67:842–845, 1986.
62. J Fina, HV Molgard, D Robertson, NJ Bradley, PD Managhan, DR Sutherland, MA Baker, MF Graeves. Expression of the CD34 gene in vascular endothelial cells. Blood 75:2417–2426, 1990.
63. D Soligo, D Delia, A Oriani, G Cattoretti, A Orazi, V Bertolli, N Quirici, G Lambertenghi Deliliers. Identification of $CD34^+$ cells in normal and pathological bone marrow biopsies by OBEND10 monoclonal antibody. Leukemia 5:1026–1030, 1991.
64. A Ito, S Nomura, S Hirota, T Suda, Y Kitamura. Enhanced expression of CD34 messenger RNA by developing endothelial cells of mice. Lab Invest 72:532–538, 1995.
65. C deVries, JA Escobedo, H Ueno, K Houck, N Ferrara, LT Williams. The fms-like tyrosine kinase, a receptor for vascular endothelial growth factor. Science 255:989–991, 1992.
66. BI Terman, ME Carrion, E Kovacs, BA Rasmussen, RL Eddy, TB Shows. Identification of a new endothelial cell growth factor receptor tyrosine kinase. Oncogene 6:1677–1683, 1991.
67. W Matthews, CT Jordan, M Gavin, NA Jenkins, NG Copeland, IR Lemischka. A receptor tyrosine kinase cDNA isolated from a population of enriched primitive hematopoietic cells and exhibiting close genetic linkage to c-kit. Proc Natl Acad Sci USA 88:9026–9030, 1991.
68. B Millauer, S Wizigmann-Voos, H Schnurch, R Martinez, NPH Moller, W Risau, A Ulrich.

High affinity VEGF binding and developmental expression suggest Flk-1 as a major regulator of vasculogenesis and angiogenesis. Cell 72:835–846, 1993.

69. TP Yamaguchi, DJ Dumont, RA Conlon, ML Breitman, J Rossant. flk-1, and flt-related receptor tyrosine kinase is an early marker for endothelial cell precursors. Development 118:489–498, 1993.
70. JM Isner, A Pieczek, R Schainfeld, R Blair, L Haley, T Asahara, K Rosenfield, S Razvi, K Walsh, J Symes. Clinical evidence of angiogenesis following arterial gene transfer of phVEGF$_{165}$. Lancet 348:370–374, 1996.
71. MR Tschudi, M Barton, NA Bersinger, P Moreau, F Cosentino, G Noll, T Malinski, TF Luscher. Effect of age on kinetics of nitric oxide release in rat aorta and pulmonary artery. J Clin Invest 98:899–905, 1996.
72. DR Senger, SR Ledbetter, KP Claffey, A Papadopoulos-Sergiou, CA Perruzzi, M Detmar. Stimulation of endothelial cell migration by vascular permeability factor/vascular endothelial growth factor through cooperative mechanisms involving the avb3 integrin, osteopontin, and thrombin. Am J Pathol 149:293–305, 1996.
73. Y Shing, J Folkman, J Sullivan, R Butterfield, J Murray, M Klagsbrun. Heparin-affinity purification of a tumor-derived capillary endothelial cell growth factor. Science 223:1296–1299, 1984.
74. N Ferrara, WJ Henzel. Pituitary follicular cells secrete a novel heparin-binding growth factor specific for vascular endothelial cells. Biochem Biophys Res Commun 161:851–855, 1989.
75. PJ Keck, SD Hauser, G Krivi, K Sanzo, T Warren, J Feder, DT Connolly. Vascular permeability factor, an endothelial cell mitogen related to PDGF. Science 246:1309–1312, 1989.
76. J Plouet, J Schilling, D Gospodarowicz. Isolation and characterization of a newly identified endothelial cell mitogen produced by AtT-20 cells. EMBO J 8:3801–3806, 1989.
77. E Tischer, R Mitchell, T Hartmann, M Silva, D Gospodarowicz, J Fiddes, J Abraham. The human gene for vascular endothelial growth factor: multiple protein forms are encoded through alternative exon splicing. J Biol Chem 266:11947–11954, 1991.
78. KA Houck, DW Leung, AM Rowland, J Winer, N Ferrara. Dual regulation of vascular endothelial growth factor bioavailability by genetic and proteolytic mechanisms. J Biol Chem 267:26031–26037, 1992.
79. S Takeshita, Y Tsurumi, T Couffinhal, T Asahara, C Bauters, JF Symes, N Ferrara, JM Isner. Gene transfer of naked DNA encoding for three isoforms of vascular endothelial growth factor stimulates collateral development in vivo. Lab Invest 75:487–502, 1996.
80. AA Miles, EM Miles. Vascular reactions to histamine, histamine liberators or leukotoxins in the skin of the guinea pig. J Physiol 118:228–257, 1952.
81. V Joukov, K Pajusola, A Kaipainen, D Chilov, I Lahtinen, E Kukk, O Saksela, N Kalkkinen, K Alitalo. A novel vascular endothelial growth factor, VEGF-C, is a ligand for the Flt4 (VEGFR-3) and KDR (VEGFR-2) receptor tyrosine kinases. EMBO J 15:290–298, 1996.
82. B Olofsson, K Pajusola, A Kaipainen, G vonEuler, V Joukov, O Saksela, A Orpana, RF Pettersson, K Alitalo, U Eriksson. Vascular endothelial growth factor B, a novel growth factor for endothelial cells. Proc Natl Acad Sci USA 93:2576–2581, 1996.
83. D Maglione, V Guerriero, G Viglietto, P Delli-Bovi, MG Persico. Isolation of a human placenta cDNA coding for a protein related to the vascular permeability factor. Proc Natl Acad Sci USA 88:9267–9271, 1991.
84. D Maglione, V Guerriero, G Viglietto, MG Ferraro, O Aprelikova, K Alitalo, S Del Vecchio, KJ Lei, JY Chou, MG Persico. Two alternative mRNAs coding for the angiogenic factor, placenta growth factor (PIGF), are transcribed from a single gene of chromosome 14. Oncogene 8:925–931, 1993.
85. S Hauser, HA Weich. A heparin-binding form of placenta growth factor (PIGF-2) is expressed in human umbilical vein endothelial cells and in placenta. Growth Factors 9:259–268, 1993.
86. JE Park, G-A Keller, N Ferrara. The vascular endothelial growth factor (VEGF) isoforms:

differential deposition into the subepithelial ECM and bioactivity of ECM-bound VEGF. Mol Biol Cell 4:1317–1326, 1993.

87. J DiSalvo, ML Bayne, G Conn, PW Kwok, PG Trivedi, DD Soderman, TM Palisi, KA Sullivan, KA Thomas. Purification and characterization of a naturally occuring vascular endothelial growth factor. Placenta growth factor heterodimer. J Biol Chem 270:7717–7723, 1995.
88. JE Park, HH Chen, J Winer, KA Houck, N Ferrara. Placenta growth factor: potentiation of vascular endothelial growth factor bioactivity, in vitro and in vivo, and high affinity binding to Flt-1 but not to Flk-1/KDR. J Biol Chem 269:25646–25654, 1994.
89. Y Cao, P Linden, D Shima, F Browne, J Folkman. In vivo angiogenic activity and hypoxia induction of heterodimer of placenta growth factor/vascular endothelial growth factor. J Clin Invest 98:2507–2511, 1996.
90. P Carmeliet, D Collen. Molecular analysis of angiogenesis. Am J Physiol. In press, 1999.
91. K Pajusola, O Aprelikova, G Pelicci, H Weich, Welsh Claesson, L., K Alitalo. Signaling properties of FLT4, a proteolytically processed receptor tyrosine kinase related to two VEGF receptors. Oncogene 9:3545–3555, 1994.
92. J Lee, A Gray, J Yuan, S Luoh, H Avraham, WI Wood. Vascular endothelial growth factor-related protein: a ligand and specific activator of the tyrosine kinase receptor Flt4. Proc Natl Acad Sci USA 93:1988–1992, 1996.
93. P Mignatti, T Morimoto, DB Rifkin. Basic fibroblast growth factor released by single, isolated cells stimulates their migration in an autocrine manner. Proc Natl Acad Sci USA 88: 11007–11011, 1991.
94. M Jaye, RM Lyall, R Mudd, J Schlessinger, N Sarver. Expression of acidic fibroblast growth factor cDNA confers growth advantage and tumorigenesis to Swiss 3T3 cells. EMBO J 7: 963–969, 1988.
95. Y Myoken, T Okamoto, M Kan, D Sato, K Takada. Release of fibroblast growth factor-1 by human squamous cell carcinoma correlates with autocrine cell growth. In Vitro Cell Dev Biol 30A:790–795, 1994.
96. L Schweigerer, G Neufeld, J Friedman, JA Abraham, JC Fiddes, D Gospodarowicz. Capillary endothelial cells express basic fibroblast growth factor, a mitogen that promotes their own growth. Nature 325:257–259, 1987.
97. S Villaschi, RF Nicosia. Angiogenic role of endogenous basic fibroblast growth factor released by rat aorta after injury. Am J Pathol 143:181–190, 1993.
98. M Bensaid, F Malecaze, H Prats, F Bayard, JP Tauber. Autocrine regulation of bovine retinal capillary endothelial cell (BREC) proliferation by BREC-derived basic fibroblast growth factor. Exp Eye Res 48:801–813, 1989.
99. A Namiki, E Brogi, M Kearney, T Wu, T Couffinhal, L Varticovski, JM Isner. Hypoxia induces vascular endothelial growth factor in cultured human endothelial cells. J Biol Chem 270:31189–31195, 1995.
100. MB Sporn, AB Roberts. Autocrine secretion—10 years later. Ann Intern Med 117:408–414, 1992.
101. M Klagsbrun, PA D'Amore. Regulators of angiogenesis. Annu Rev Physiol 53:217–239, 1991.
102. N Sato, JG Beitz, J Kato, M Yamamoto, JW Clark, P Calabresi Jr, RA Frackelton. Platelet-derived growth factor indirectly stimulates angiogenesis in vitro. Am J Pathol 142:1119–1130, 1993.
103. GF Pierce, JE Tarpley, D Yanagihara, TE Mustoe, GM Fox, A Thomason. Platelet derived growth factor (BB homodimer), transforming growth factor-beta, and basic fibroblast growth factor in dermal wound healing. Am J Pathol 140:1375–1388, 1992.
104. W Risau, H Drexler, V Mironov, A Smits, A Siegbahn, K Funa, C-H Heldin. Platelet-derived growth factor is angiogenic in vivo. Growth Factors 7:261–266, 1992.
105. JG Beitz, IS Kim, P Calabresi, AR Frackelton. Human microvascular endothelial cells ex-

press receptors for platelet-derived growth factor. Proc Natl Acad Sci USA 88:2021–2025, 1991.
106. A Smits, M Hermansson, M Nister, I Karnushina, C-H Heldin, K Oberg. Rat brain capillary endothelial cells express functional PDGF B-type receptors. Growth Factors 2:1–8, 1989.
107. RS Bar, M Boes, BA Booth, BL Dake, S Henley, MN Hart. The effects of platelet-derived growth factor in cultured microvessel endothelial cells. Endocrinology 124:2021–2025, 1989.
108. P D'Amore, SR Smith. Growth factor effects on cells of the vascular wall: a survey. Growth Factors 8:61–75, 1993.
109. E Brogi, T Wu, A Namiki, JM Isner. Indirect angiogenic cytokines upregulate VEGF and bFGF gene expression in vascular smooth muscle cells, while hypoxia upregulates VEGF expression only. Circulation 90:649–652, 1994.
110. AB Roberts, MB Sporn, RK Assoian, JM Smith, NS Roche, LM Wakefield, UI Heine, LA Liotta, V Falanga, JH Kehrl, AS Fauci. Transforming growth factor type-beta:rapid induction of fibrosis and angiogenesis in vivo and stimulation of collagen formation in vitro. Proc Natl Acad Sci USA 83:4167–4171, 1986.
111. MS Pepper, JD Vassalli, L Orci, R Montesano. Biphasic effect of transforming growth factor-b1 on in vitro angiogenesis. Exp Cell Res 204:356–363, 1993.
112. JK Chen, H Hoshi, WL McKeehan. Transforming growth factor type b specifically stimulates synthesis of proteoglycans in human adult arterial smooth muscle cells. Proc Natl Acad Sci USA 84:5287–5291, 1987.
113. J Madri, B Pratt, A Tucker. Phenotypic modulation of endothelial cells by transforming growth factor-beta depends upon the composition and organization of the extracellular matrix. J Cell Biol 106:1375–1382, 1988.
114. JA Winkles, CG Gay. Serum, phorbol ester, and polypeptide mitogens increase class 1 and 2 heparin binding (acidic and basic fibroblast growth factor) gene expression in human vascular smooth muscle cells. Cell Growth Diff 2:531–540, 1991.
115. RS Warren, H Yuan, MR Matli, N Ferrara, DB Donner. Induction of vascular endothelial growth factor by insulin-like growth factor 1 in colorectal carcinoma. J Biol Chem 271: 29483–29488, 1996.
116. GT Stavri, IC Zachary, PA Baskerville, JF Martin, JD Erusalimsky. Basic fibroblast growth factor upregulates the expression of vascular endothelial growth factor in vascular smooth muscle cells: synergistic interaction with hypoxia. Circulation 92:11–14, 1995.
117. B Witzenbichler, E Van Belle, L Chang, R Schwall. Scatter factor (SF) induces vascular endothelial growth factor (VEGF) expression in vascular smooth muscle cells (VSMC) and acts synergistic to VEGF on endothelial cell (EC) migration in vitro (abstr). Circulation 94:I-593–I-594, 1996.
118. D Shweiki, A Itin, D Soffer, E Keshet. Vascular endothelial growth factor induced by hypoxia may mediate hypoxia-initiated angiogenesis. Nature 359:843–845, 1992.
119. GL Wang, GL Semenza. Characterization of hypoxia-inducible factor 1 and regulation of DNA binding activity by hypoxia. Blood 268:21513–21518, 1993.
120. DT Shima, U Deutsch, PA D'Amore. Hypoxic induction of vascular endothelial growth factor (VEGF) in human epithelial cells is mediated by increases in mRNA stability. FEBS Lett 370:203–208, 1995.
121. AP Levy, NS Levy, S Wegner, MA Goldberg. Transcriptional regulation of the rat vascular endothelial growth factor gene by hypoxia. J Biol Chem 270:13333–13340, 1995.
122. MA Goldberg, TJ Schneider. Similarities between the oxygen-sensing mechanisms regulating the expression of vascular endothelial growth factor and erythropoietin. J Biol Chem 269:4355–4359, 1994.
123. NS Levy, AP Levy, MA Goldberg, G Koren. Regulation of vascular endothelial growth factor isoforms by hypoxia and identification of a novel truncated vascular endothelial growth factor isoform in cardiac myocytes (abstr). Circulation 90:I-521, 1994.
124. LQ Pu, AD Sniderman, R Brassard, KJ Lachapelle, AM Graham, R Lisbona, JF Symes.

Enhanced revascularization of the ischemic limb by means of angiogenic therapy. Circulation 88:208–215, 1993.
125. R Baffour, J Berman, JL Garb, SW Rhee, J Kaufman, P Friedmann. Enhanced angiogenesis and growth of collaterals by in vivo administration of recombinant basic fibroblast growth factor in a rabbit model of acute lower limb ischemia: dose-response effect of basic fibroblast growth factor. J Vasc Surg 16:181–191, 1992.
126. T Asahara, C Bauters, LP Zheng, S Takeshita, S Bunting, N Ferrara, JF Symes, JM Isner. Synergistic effect of vascular endothelial growth factor and basic fibroblast growth factor on angiogenesis in vivo. Circulation 92:II-365–II-371, 1995.
127. S Takeshita, L-Q Pu, L Zheng, N Ferrara, LA Stein, AD Sniderman, JM Isner, JF Symes. Vascular endothelial growth factor induces dose-dependent revascularization in a rabbit model of persistent limb ischemia. Circulation 90:II-228–II-234, 1994.
128. C Bauters, T Asahara, LP Zheng, S Takeshita, S Bunting, N Ferrara, JF Symes, JM Isner. Site-specific therapeutic angiogenesis following systemic administration of vascular endothelial growth factor. J Vasc Surg 21:314–325, 1995.
129. RA Berkman, MJ Merrill, WC Reinhold, WT Monacci, A Saxena, WC Clark, JT Robertson, IU Ali, EH Oldfield. Expression of the vascular permeability factor/vascular endothelial growth factor gene in central nervous system neoplasms. J Clin Invest 91:153–159, 1993.
130. S Rohovsky, M Kearney, A Pieczek, K Rosenfield, R Schainfeld, PA D'Amore, JM Isner. Elevated levels of basic fibroblast growth factor in patients with limb ischemia. Am Heart J 132:1015–1019, 1996.
131. C Bauters, T Asahara, LP Zheng, S Takeshita, S Bunting, N Ferrara, JF Symes, JM Isner. Physiologic assessment of augmented vascularity induced by VEGF in ischemic rabbit hindlimb. Am J Physiol 267:H1263–H1271, 1994.
132. H Thieme, LP Aiello, H Takagi, N Ferrara, GL King. Comparative analysis of vascular endothelial growth factor receptors on retinal and aortic vascular endothelial cells. Diabetes 44:98–103, 1995.
133. E Brogi, G Schatteman, T Wu, EA Kim, L Varticovski, B Keyt, JM Isner. Hypoxia-induced paracrine regulation of VEGF receptor expression. J Clin Invest 97:469–476, 1996.
134. J Folkman. Tumor angiogenesis: therapeutic implications. N Engl J Med 285:1182–1186, 1971.
135. A Yanagisawa-Miwa, Y Uchida, F Nakamura, T Tomaru, H Kido, T Kamijo, T Sugimoto, K Kaji, M Utsuyama, C Kurashima, H Ito. Salvage of infarcted myocardium by angiogenic action of basic fibroblast growth factor. Science 257:1401–1403, 1992.
136. AP Levy, R Tamargo, H Brem, D Nathans. An endothelial cell growth factor from the mouse neuroblastoma cell line NB41. Growth Factors 2:9–19, 1989.
137. DT Connolly, DM Hewelman, R Nelson, JV Olander, BL Eppley, JJ Delfino, RN Siegel, RS Leimgruber, J Feder. Tumor vascular permeability factor stimulates endothelial cell growth and angiogenesis. J Clin Invest 84:1470–1478, 1989.
138. LQ Pu, S Jackson, KJ Lachapelle, Z Arekat, AM Graham, R Lisbona, R Brassard, S Carpenter, JF Symes. A persistent hindlimb ischemia model in the rabbit. J Invest Surg 7:49–60, 1994.
139. C Orlandi, JL Blackshear, NK Hollenberg. Specific increase in sensitivity to serotonin of the canine hindlimb collateral arterial tree via the 5-hydroxytryptamine-2 receptor. Microvasc Res 32:121–130, 1986.
140. FW Sellke, Y Kagaya, RG Johnson, T Shafique, FJ Schoen, W Grossman, RM Weintraum. Endothelial modulation of porcine coronary microcirculation perfused via immature collaterals. Am J Physiol 262:H1669–H1675, 1992.
141. NK Hollenberg. Collateral arterial growth and reactivity: lessons from the limb and renal blood supply. In: W Schaper, J Schaper, eds. Collateral Circulation. Boston: Kluwer, 1992, pp. 1–15.
142. DG Harrison. Neurohumoral and pharmacologic regulation of collateral perfusion. In: W Schaper, J Schaper, eds. Collateral Circulation. Boston: Kluwer, 1992, pp. 329–343.
143. EP McFadden, JC Clarke, GJ Davis, JC Kaski, AW Haider, A Maseri. Effect of intracoronary

serotonin on coronary vessels in patients with stable and patients with variant angina. N Engl J Med 324:648–654, 1991.

144. C Bauters, T Asahara, LP Zheng, S Takeshita, S Bunting, N Ferrara, JF Symes, JM Isner. Recovery of disturbed endothelium-dependent flow in the collateral-perfused rabbit ischemic hindlimb after administration of vascular endothelial growth factor. Circulation 91:2802–2809, 1995.
145. FW Sellke, SY Wang, M Friedman, K Harada, ER Edelman, W Grossman, M Simons. Basic FGF enhances endothelium-dependent relaxation of the collateral-perfused coronary microcirculation. Am J Physiol 267:H1303–H1311, 1994.
146. KG Peters, C deVries, LT Williams. Vascular endothelial growth factor receptor expression during embryogenesis and tissue repair suggests a role in endothelial differentiation and blood vessel growth. Proc Natl Acad Sci USA 90:8915–8919, 1993.
147. RA Cohen, JT Shepherd, PM Vanhoutte. Inhibitory role of the endothelium in the response of isolated coronary arteries to platelets. Science 221:273–274, 1983.
148. MJ Janicek, M Meyerovitz, DP Harrington, NK Hollenberg. Arterial response to ketanserin and aspirin in patients with advanced peripheral atherosclerosis. Invest Radiol 27:415–421, 1992.
149. J DeCree, J Leempoels, H Geukens, H Verhaegen. Placebo-controlled double blind trial of ketanserin in intermittent claudication. Lancet 2:775–779, 1984.
150. DW Losordo, JG Pickering, S Takeshita, G Leclerc, D Gal, L Weir, M Kearney, J Jekanowski, JM Isner. Use of the rabbit ear artery to serially assess foreign protein secretion after site specific arterial gene transfer in vivo: evidence that anatomic identification of successful gene transfer may underestimate the potential magnitude of transgene expression. Circulation 89:785–792, 1994.
151. S Takeshita, DW Losordo, M Kearney, JM Isner. Time course of recombinant protein secretion following liposome-mediated gene transfer in a rabbit arterial organ culture model. Lab Invest 71:387–391, 1994.
152. R Riessen, H Rahimizadeh, E Blessing, S Takeshita, JJ Barry, JM Isner. Arterial gene transfer using pure DNA applied directly to a hydrogel-coated angioplasty balloon. Hum Gene Ther 4:749–758, 1993.
153. S Takeshita, L Weir, D Chen, LP Zheng, R Riessen, C Bauters, JF Symes, N Ferrara, JM Isner. Therapeutic angiogenesis following arterial gene transfer of vascular endothelial growth factor in a rabbit model of hindlimb ischemia. Biochem Biophys Res Commun 227: 628–635, 1996.
154. JM Isner, K Walsh, J Symes, A Pieczek, S Takeshita, J Lowry, K Rosenfield, L Weir, E Brogi, D Jurayj. Arterial gene transfer for therapeutic angiogenesis in patients with peripheral artery disease. Human Gene Ther 7:959–988, 1996.
155. S Banai, MT Jaklitsch, M Shou, DF Lazarous, M Scheinowitz, S Biro, SE Epstein, EF Unger. Angiogenic-induced enhancement of collateral blood flow to ischemic myocardium by vascular endothelial growth factor in dogs. Circulation 89:2183–2189, 1994.
156. EF Unger, S Banai, M Shou, DF Lazarous, MT Jaklitsch, M Scheinowitz, C Klingbeil, SE Epstein. Basic fibroblast growth factor enhances myocardial collateral flow in a canine model. Am J Physiol 266:H1588–H1595, 1994.
157. M Hariawala, JR Horowitz, D Esakof, DD Sheriff, DH Walter, GM Chaudhary, V Desai, B Keyt, JM Isner, JF Symes. VEGF improves myocardial blood flow but produces EDRF-mediated hypotension in porcine hearts. J Surg Res 63:77–82, 1996.
158. JD Pearlman, MG Hibberd, ML Chuang, K Harada, JJ Lopez, SR Gladston, M Friedman, FW Sellke, M Simons. Magnetic resonance mapping demonstrates benefits of VEGF-induced myocardial angiogenesis. Nature Med 1:1085–1089, 1995.
159. FW Selke, L Jianyi, A Stamler, JJ Lopez, KA Thomas, M Simons. Angiogenesis induced by acidic fibroblast growth factor as an alternative method of revascularization for chronic myocardial ischemia. Surg 120:182–188, 1996.

160. K Harada, W Grossman, M Friedman, ER Edelman, PV Prasad, CS Keighlcy, WJ Manning, FW Selke, M Simons. Basic fibroblast growth factor improves myocardial function in chronically ischemic porcine hearts. J Clin Invest 94:623–630, 1994.
161. FJ Giordano, P Ping, D McKirnan, S Nozaki, AN DeMaria, WH Dillmann, O Mathieu-Costello, HK Hammond. Intracoronary gene transfer of fibroblast growth factor 5 increases blood flow and contractile function in an ischemic region of the heart. Nature Med 2:534–539, 1996.
162. HT Yang, MR Deschenes, RW Ogilvie, RL Terjung. Basic fibroblast growth factor increases collateral blood flow in rats with femoral arterial ligation. Circ Res 79:62–69, 1996.
163. CE Walder, CJ Errett, J Ogez, H Heinshon, S Bunting, P Lindquist, N Ferrara, GR Thomas. Vascular endothelial growth factor (VEGF) improves blood flow and function in a chronic ischemic hind limb model. J Cardiovasc Pharmacol 27:91–98, 1996.
164. H Tabata, M Silver, JM Isner. Arterial gene transfer of acidic fibroblast growth factor for therapeutic angiogenesis in vivo: critical role of secretion signal in use of naked DNA. Cardiovasc Res 35:470–479, 1997.
165. R Yang, S Bunting, A Ko, B Keyt, H Jin. Hemodynamic effects of vascular endothelial growth factor in conscious rats (abstr). Circulation 92:I-659, 1995.
166. J Horowitz, M Hariawala, DD Sheriff, B Keyt, JF Symes. In vivo administration of vascular endothelial growth factor is associated with EDRF-dependent systemic hypotension in procine and rabbit animal models (abstr). Circulation 92:I-630–I-631, 1995.
167. J Muhlhauser, R Pili, M Merrill, H Maeda, A Passaniti, R Crystal, MC Capogrossi. In vivo angiogenesis induced by recombinant adenovirus vectors coding either for secreted or nonsecreted forms of acidic fibroblast growth factor. Human Gene Ther 6:1457–1465, 1995.
168. EG Nabel, ZY Yang, G Plautz, R Forough, X Zhan, CC Haudenschild, T Maciag, GJ Nabel. Recombinant fibroblast growth factor-1 promotes intimal hyperplasia and angiogenesis in arteries in vivo. Nature 362:844–846, 1993.
169. F Goto, K Goto, K Weindel, J Folkman. Synergistic effects of vascular endothelial growth factor and basic fibroblast growth factor on the proliferation and cord formation of bovine capillary endothelial cells within collagen gels. Lab Invest 69:508–517, 1993.
170. MK Lyons, RE Anderson, FB Meyer. Basic fibroblast growth factor promotes in vivo cerebral angiogenesis in chronic forebrain ischemia. Brain Res 558:315–320, 1991.
171. P Mignatti, R Tsuboi, E Robbins, DB Rifkin. In vitro angiogenesis on the human amniotic membrane: requirement for basic fibroblast growth factor-induced proteinases. J Cell Biol 108:671–682, 1989.
172. MS Pepper, N Ferrara, L Orci, R Montesano. Potent synergism between vascular endothelial growth factor and basic fibroblast growth factor in the induction of angiogenesis in vitro. Biochem Biophys Res Commun 189:824–831, 1992.
173. RF Nicosia, SV Nicosia, M Smith. Vascular endothelial growth factor, platelet-derived growth factor, and insulin-like growth factor-1 promote rat aortic angiogenesis in vitro. Am J Pathol 145:1023–1029, 1994.
174. DH Walter, U Hink, T Asahara, E Van Belle, J Horowitz, Y Tsurumi, R Vandlen, H Heinsohn, B Keyt, N Ferrara, JF Symes, JM Isner. The in vivo bioactivity of vascular endothelial growth factor/vascular permeability factor is independent of N-linked glycosylation. Lab Invest 74:546–556, 1996.
175. DW Leung, G Cachianes, WJ Kuang, DV Goeddel, N Ferrara. Vascular endothelial growth factor is a secreted angiogenic mitogen. Science 246:1306–1309, 1989.
176. JR Horowitz, A Rivard, R van der Zee, MD Hariawala, DD Sheriff, DD Esakof, M Chaudhry, JF Symes, JM Isner. Vascular endothelial growth factor/vascular permeability factor produces nitric oxide-dependent hypotension. Arterioscler Thromb Vasc Biol 17:2793–2800, 1997.
177. JM Isner, L Feldman. Gene therapy for arterial disease. Lancet 344:1653–1654, 1994.

33

Lessons to Be Learned from Clinical Trials of Angiogenesis Modulators in Ischemic Diseases

William W. Li, Michael Jaffe, and Vincent W. Li
The Angiogenesis Foundation, Cambridge, Massachusetts

Dimitris Tsakayannis
The Angiogenesis Foundation and University of Athens School of Medicine, Athens, Greece

I. INTRODUCTION

The field of angiogenesis, begun in the 1960s as an inquiry into the role of neovascularization in solid tumor growth, has richly yielded both fundamental and functional insights into the growth control of the human vascular system. From the standpoint of basic research, major cellular and molecular cascades of events in angiogenesis have now been elucidated. (For a comprehensive review, see Ref. 1.) These events are being correlated to the neovascularization observed in diseases ranging from cancer to coronary artery disease. It is now understood that the process of angiogenesis plays at least two critical roles in human pathology (2). On the one hand, the occurrence of new blood vessel growth can drive the progression of conditions such as cancer, diabetic retinopathy, psoriasis, rheumatoid arthritis, and AIDS-related Kaposi's sarcoma. On the other hand, angiogenesis plays a crucial role in maintaining healthy tissues and repairing damaged tissues by increasing circulatory perfusion and providing paracrine survival factors to perivascular cells. A massive pharmaceutical industry effort has been undertaken to develop new drugs that can inhibit undesirable angiogenesis (antiangiogenic therapy) as well as drugs that can stimulate angiogenesis (therapeutic angiogenesis, or angiogenic therapy). This discussion focuses primarily on the strategies and lessons being learned for therapeutic angiogenesis in medical conditions characterized by ischemia.

Therapeutic angiogenesis, a concept proposed in 1993 by Hockel and colleagues (3), can broadly be defined as the use of biological agents, bioactive materials, or environmental conditions to stimulate the growth of new blood vessels, to restore or augment circulatory perfusion of tissues, to reverse ischemia, or to accelerate healing. This approach is now being developed for the treatment of ischemic heart disease, cerebrovascular disease, critical limb ischemia, and delayed wound healing—conditions that together account for more than 10 million disease cases annually in the United States alone (4). Although peptic ulcer disease, a condition related to *Helicobacter pylori* infection and hypersecretion of gastric acid, is a nonischemic disease, it is characterized by poor angiogenesis in the ulcer and can be treated through therapeutic angiogenesis (5).

At the time of this writing, at least seven biopharmaceutical companies are sponsoring various angiogenic cytokines in clinical trials of angiogenic therapy for these diseases, most in the early phase I/II stage. One cytokine, rhPDGF-BB (Ortho-McNeil Pharmaceuticals), has completed clinical trials and been approved by the U.S. Food and Drug Administration (FDA) for human use. A growing number of tissue engineering firms are developing bioactive living skin equivalents that possess angiogenic cytokine gene expression activity. One product, Apligraf (Organogenesis, Inc.), has been FDA-approved. The hyperbaric oxygen chamber, an FDA-approved device that delivers oxygen at high tension levels to patients suffering tissue hypoxia, has been used to increase angiogenesis in wound healing granulation tissue in patients.

The following discussion reviews the development of new angiogenesis-stimulating drugs for treating ischemic diseases and presents critical lessons learned from new clinical trials in this area. It is intended to introduce readers to the current state of therapeutic angiogenesis drug development and to present information that may be used to increase the efficiency and effectiveness of future clinical development in this area.

Portions of this report are based on an Angiogenesis Foundation–sponsored study of clinical trials that was presented, in parts, at the 1998 Keystone Symposium on Angiogenesis and Microcirculation, held in Steamboat Springs, Colorado; at the 1998 Annual Wound Healing Society Meeting, held in Salt Lake City, Utah; at the 1998 CHI Tissue Engineering and Tissue Regeneration Conference, held in Pittsburgh, Pennsylvania; and the 4th Samsung International Symposium on Molecular Medicine, held in Seoul, Korea.

II. ANGIOGENESIS AS A RESPONSE TO ISCHEMIA

Multiple angiogenic pathways are activated under conditions of ischemia. Early studies on tumor angiogenesis speculated, correctly, that hypoxia is a stimulus for the release of angiogenic factors such as VEGF by tumors (6). Hypoxia-inducible factor 1 alpha (HIF-1α) is linked to increased VEGF expression and angiogenesis (7). VEGF and its receptors are upregulated in ischemic nonneoplastic tissues, such as the heart and lung parenchyma (8,9). Ischemia-induced cell injury contributes to the angiogenic response by the release of angiogenic growth factors such as basic fibroblast growth factor (bFGF) from the cytosol (10,11). Thrombosis is another mechanism for ischemic-related angiogenesis. Tissue ischemia and thrombosis are interrelated clinical phenomena, where poor circulation is a predisposing state for blood clot formation and thrombosis exacerbates sluggish circulatory flow dynamics. Thrombin, a primary element of the blood clot, is an angiogenic mediator (12,13). Platelets, the first cells to accumulate in thrombotic regions, contain several angiogenic factors (bFGF, VEGF, PDGF), and these are known to be released in ischemic wound conditions (14–16). Inflammation also accompanies ischemic tissue damage, and inflammatory mediators, such as leukotrienes and interleukins, and macrophages, monocytes, and neutrophils, which themselves elaborate angiogenic factors, and potentiate the angiogenic response to ischemia (17–20).

Despite this plethora of reactive angiogenic mechanisms, most ischemic tissues are generally unable to mount an adequate physiological response to reverse disease processes. Patients with advanced coronary artery disease often have insufficient collateralization to

Table 1 Major Ischemic Conditions Characterized by an Inadequate Angiogenic Response

Disease	U.S. prevalence (1998)	Characteristic
Ischemic		
Stroke	4 million (survivors)	Inadequate brain angiogenesis
Ischemic heart disease	1.45 million	Inadequate collateralization
Peripheral vascular disease	2.3 million	Inadequate limb angiogenesis
Delayed wound healing	3.9 million	Inadequate granulation
Nonischemic		
Peptic ulcer disease	4.6 million	Inadequate granulation

Source: 1998 Market Study, The Angiogenesis Foundation.

prevent ischemic damage and, without medical or surgical intervention, eventual infarction. Acute stroke patients are unable to generate sufficient neovascularization to salvage ischemic brain and prevent loss of neurological function. Patients with end-stage peripheral vascular disease suffer from critical limb ischemia and are unable to induce sufficient angiogenesis to reverse skin ulceration and gangrene. Patients with chronic wound ischemia lack the appropriate angiogenic response to generate granulation tissue for complete healing. Table 1 lists highly prevalent diseases characterized by an inadequate angiogenesis response to ischemia. These clinical situations provide the basis for biopharmaceutical industry efforts to develop therapeutic technologies that augment or amplify angiogenesis in ischemic tissues. Currently, clinical trials of recombinant angiogenic growth factors and other angiogenic modalities are under way in patients suffering from ischemia of the myocardium, brain, skin, and limb (21).

III. ANGIOGENESIS-STIMULATING TECHNOLOGIES

Therapeutic angiogenesis developers face the challenge of designing angiogenic modulators that can (a) initiate and promote the entire cascade of angiogenesis; (b) overcome the action of endogenous antiangiogenic molecules in the ischemic environment; (c) drive the formation of a stabilized mature blood vessel capable of supporting blood flow; and (d) achieve the desired clinical result of improving tissue perfusion, enhancing quality of life, and prolonging survival. The four primary modes of therapeutic angiogenesis now under study are outlined in Table 2.

Table 2 Current Therapeutic Modalities for Angiogenesis

Modality	FDA-approved product	Indication
Growth factors	rhPDGF-BB (becaplermin, REGRANEX)	Diabetic ulcers
Bioactive materials	Apligraf	Venous ulcers
	Graftcyte	Hair implants
Laser therapy	The Heart System	Ischemic heart
Hyperbaric oxygen	Hyperbaric oxygen chamber	Ischemic wounds

Table 3 Known Angiogenic Factors

Angiogenin
del-1
Fibroblast growth factors: acidic (aFGF)[a] and basic (bFGF)[a]
Follistatin
Granulocyte colony-stimulating factor (G-CSF)[a]
Hepatocyte growth factor (HGF)/scatter factor (SF)
HIV Tat
Interleukin-8 (IL-8)
Leptin
Midkine
Placental growth factor
Platelet-derived endothelial cell growth factor (PD-ECGF)
Platelet-derived growth factor (PDGF)
Pleiotrophin (PTN)
Proliferin
Transforming growth factor-alpha (TGF-alpha)
Transforming growth factor-beta (TGF-beta)[a]
Tumor necrosis factor-alpha (TNF-alpha)
Vascular endothelial growth factor (VEGF)/vascular permeability factor (VPF)[a]

[a] In clinical trial for therapeutic angiogenesis, 1998.

A. Peptide Angiogenic Growth Factors and Other Pharmacological Therapy

Of the 19 known polypeptide angiogenic factors (see Table 3), 6 are being tested in clinical trials for ischemic diseases: acidic fibroblast growth factor (aFGF); basic fibroblast growth factor (bFGF); vascular endothelial growth factor/vascular permeability factor (VEGF/VPF); platelet-derived growth factor (PDGF); transforming growth factor beta (TGF-β); and granulocyte colony-stimulating factor (G-CSF). Some clinical trials are delivering recombinant growth factors while others are transfecting endothelial cells with adenoviral gene vectors encoding specific growth factors. Recombinant human PDGF (becaplermin) received FDA-approval for the treatment of diabetic neuropathic ulcers in December 1997.

B. Bioactive Materials

Biocompatible polymers and matrices containing angiogenesis modulators have been manufactured for placement on ischemic tissue beds such as wounds. One such modulator, copper peptide, has a binding site on FGF molecules and can serve as a local FGF concentrating device (22–24). Graftcyte (Procyte Corporation) is a copper-containing wound dressing designed for enhancing wound healing following hair-restoration surgery. Another type of bioactive material is the living skin equivalent, a tissue-engineered biological dressing whose cellular constituents exhibit synthetic activity for angiogenic factors, such as bFGF and VEGF. Apligraf (Organogenesis, Inc.) is the first living skin equivalent to be FDA-approved for use in the treatment of venous stasis ulcers (25).

C. Laser Therapy

Laser therapy, termed *transmyocardial revascularization* (TMR), has been designed to treat the ischemic heart and stimulates angiogenesis as a secondary event to photoacoustic injury (26,27). The laser-created channel become thrombosed and clot elements, such as thrombin and platelets, serve as local angiogenic stimuli. Myocyte lysis and the local inflammation response to laser therapy may also release angiogenic factors in the treated site. The first TMR device, the Heart Laser System (PLC Systems, Inc.) employs a high-powered CO_2 laser system and was FDA-approved in August 1998 for use in the United States for the treatment of stable angina that is not amenable to conventional surgical bypass or angioplasty. The Heart Laser System has already been in use in the People's Republic of China. A percutaneous version of TMR, called PTMR (Eclipse Surgical Technologies), utilizes a holmium:YAG laser and is commercially available in Europe.

D. Hyperbaric Oxygen

Hyperbaric oxygenation (HBO) chambers are designed to deliver oxygen at high pressures to the entire body surface of patients suffering from derangements in tissue gas exchange. The patient breathes 100% oxygen while the environment within the chamber is progressively increased to greater-than-sea level pressures [2.0 atmosphere absolute (ATA) or 33 ft of seawater (fsw)]. At 2.0 ATA, as compared with normal atmospheric pressure, blood plasma begins to carry increased amounts of dissolved oxygen and enhances the total oxygen-carrying capacity of blood. While HBO is primarily used for the treatment of severe burns and for decompression sickness, this modality is also used to treat ischemic wounds, such as diabetic, venous stasis and decubitus ulcers, and frostbite (28). High tension oxygen (P_{O_2}) stimulates endothelial cell proliferation and angiogenesis in ischemic wounds (29–31). Available in many major university medical centers, HBO chambers are costly and require specially trained personnel and facilities.

IV. OVERVIEW OF THERAPEUTIC ANGIOGENESIS IN CLINICAL TRIAL

A. Ischemic Heart Disease

Coronary artery disease, which can lead to myocardial ischemia, afflicts 13.9 million patients in the United States. While available interventions (anti-ischemic medications, angioplasty, stents, surgical bypass) are effective, an estimated 80,000 patients cannot be treated with conventional revascularization techniques because of comorbidities. Cardiologists are therefore developing molecular techniques to create ''natural'' bypasses via angiogenesis stimulation in the heart. Amelioration of symptoms of myocardial ischemia (angina pectoris), angiographic demonstration of increased vascularity, and improvement of physical function is the current goal of clinical trials of therapeutic angiogenesis for the heart.

Preclinical studies have validated therapeutic angiogenesis as a novel approach to inducing neovascularization in the heart (32–35). Cardiologists and cardiothoracic surgeons are now administering recombinant human acidic FGF, basic FGF, and rhVEGF to stimulate myocardial neovascularization in ischemic territories. Three primary drug delivery modes are being employed: (a) direct injection into ischemic myocardium following thoractomy (performed by cardiothoracic surgeons with or without surgical bypass grafting) (36); (b) growth factor incorporation into alginate beads, which are implanted in the epicardium of ischemic myocardial territories during surgery (37); and (c) intraarterial injection of protein or adenoviral gene vector into proximal regions of occluded coronary vessels. All studies conducted to date have focused on the administration of a single angiogenic factor. Arterial gene transfer of naked plasmid DNA encoding $phVEGF_{165}$ into the human heart is another approach under study. This method has effectively induced neovascularization in ischemic skeletal muscle (38).

A photoinjury-based approach employs a laser to drill multiple channels directly into the ischemic myocardium employing transmyocardial revascularization (TMR). To date, more than 4000 patients have been treated with TMR at over 200 centers worldwide. Laser-created channels experience transient perfusion via coronary collaterals but then become thrombosed. Subsequently, it is thought that platelets and inflammatory cells initiate secondary angiogenesis in the photoacoustically injured tissue.

The primary tool for the evaluation of cardiac patients receiving therapeutic angiogenesis has relied upon coronary angiography, thallium imaging, and observation of symptomatic relief from unstable angina. Patients enrolled in these studies are primarily those who have advanced coronary artery disease and have failed available interventions as well as pharmacological therapy.

B. Peripheral Vascular Disease

The amelioration of the symptoms of limb ischemia (claudication or rest pain) and prevention of limb amputation are the primary goals of therapeutic angiogenesis for peripheral vascular disease. Preclinical studies have validated therapeutic angiogenesis as a novel approach to inducing neovascularization in the ischemic limb (39–41). Vascular surgeons and cardiologists are administering arterial gene therapy of plasmid DNA encoding $phVEGF_{165}$ (Vascular Genetics Inc.) in patients with critical limb ischemia caused by severe atherosclerotic narrowing of their distal arteries, known as peripheral vascular disease (38). While the initial technique employed a hydrogel-coated balloon angioplasty catheter, clinical investigators have discovered that direct intramuscular injection of the plasma DNA in patients results in $phVEGF_{165}$ gene expression and angiogenesis localized to the ischemic limb. The mechanism for this tissue- and expression-selectivity is not known.

The primary tool for the evaluation of patients receiving therapeutic angiogenesis for the ischemic limb has relied upon angiography, functional exercise testing, observation of ischemic ulcer healing, and documentation of relief from pain (claudication and rest pain). Patients enrolled in these studies are primarily those who face limb amputation and are not suitable candidates for surgical bypass of major limb vessels.

C. Delayed Wound Healing

The complete closure of a delayed-healing wound, reduction in risk for infection, and prevention of gangrene, osteomyelitis, and the need for limb amputation are the primary goals of therapeutic angiogenesis for wounds. Numerous preclinical studies have demonstrated the beneficial effects of topical growth factor application on experimental wounds (42–45). Vascular surgeons, plastic surgeons, and dermatologists are testing a variety of recombinant angiogenic growth factors on patients with various types of delayed-closure wounds, including neuropathic diabetic foot ulcers, venous stasis ulcers, and decubitus ulcers. The molecules under investigation include platelet-derived growth factor (rhPDGF-BB, Ortho-McNeil Pharmaceuticals), which has been FDA-approved for the treatment of diabetic ulcer; basic fibroblast growth factor (bFGF, Scios); granulocyte colony-stimulating factor (G-CSF, Neupogen, Amgen, Inc.), and transforming growth factor beta (TGF-β, Novartis). Recombinant human PDGF-BB received FDA approval for the topical treatment of neuropathic diabetic foot ulcers in December 1997 and is now in use throughout the United States (46).

Living skin equivalents are being applied as a biological wound dressing to diabetic, venous stasis, and decubitus ulcers. These products are capable of gene expression of angiogenic factors and other cytokines and may be considered a type of solid-state growth-factor delivery system. Some clinicians are testing the value of combining rhPDGF-BB with living skin equivalents (Apligraf) for chronic skin ulcer therapy.

Dermatologists and plastic surgeons are applying copper-containing skin dressings (Graftcyte™, Procyte Corporation) to cutaneous wounds. This technology has been used for the treatment of severe cold injury (frostbite).

Surgeons are utilizing hyperbaric chambers to deliver oxygen therapy to patients suffering from a variety of delayed-closure wounds, including diabetic and venous stasis ulcers. Hyperbaric therapy has also been used to treat frostbite.

The primary tool for the evaluation of patients receiving angiogenic therapy for delayed-healing wounds has been clinical observation of granulation, epithelialization, and wound closure. Active wound infection has been a contraindication for treatment with any angiogenic growth factor.

D. Stroke

More than 500,000 patients suffer stroke each year in the United States, and there are estimated to be up to 4 million stroke survivors who are at risk for another event. The amelioration of neurological disability and the preservation of neurological function are goals of clinical trials of angiogenic factors in acute stroke (47). Preclinical studies have correlated the onset of angiogenesis with neuronal salvage and sprouting and functional recovery in experimental stroke (48–53). Neurologists are testing recombinant basic fibroblast growth factor (Fiblast, Wyeth-Ayerst Laboratories) in 900 European patients presenting to hospital emergency rooms with the symptoms of acute stroke (54). These phase II/III clinical trials are designed to assess the neuroprotective benefits of bFGF, but the angiogenic property of bFGF is also a feature of the therapy. A U.S. arm of the Fiblast trial was terminated in July 1998 after an excessive number of unfavorable outcomes, including mortality, were observed in the first 450 patients treated. As of this writing, the European arm is being continued (55). One major difference between the U.S. and the European trials was the duration of drug infusion, with the U.S. arm employing a more aggressive and shorter infusion period.

V. EMERGING LESSONS FROM THERAPEUTIC ANGIOGENESIS

The Angiogenesis Foundation is conducting analytical studies of clinical trials of both therapeutic angiogenesis and antiangiogenic drugs in order to achieve three major goals: (a) to correlate new basic science discoveries and molecular mechanisms of angiogenesis modulation with actual clinical results; (b) to identify and classify clinically successful strategies for controlling angiogenesis in human disease; and (c) to identify major challenges and areas of high risk to drug developers in order to promote efficiency in therapeutic development. Further basic and applied research can then be swiftly undertaken to answer key questions and develop practical solutions. Based on the early experiences from clinical trials of therapeutic angiogenesis technologies, five critical lessons are emerging from the clinical use of angiogenic peptides to help guide current and future efforts in the development of therapeutic angiogenesis.

Lesson 1: Administration of single exogenous cytokines can trigger the formation of functional neovascular networks

Current clinical trials have shown that the application of several discrete angiogenic factors—acidic FGF, basic FGF, VEGF, and PDGF—can successfully generate functional neovascular channels in ischemic tissue of patients. This insight is important for two reasons: (a) successful angiogenesis requires a coordinated cascade of molecular steps, and there is no a priori reason to believe that a single cytokine can successfully initiate and propagate all the steps needed to generate vascular functionality in the clinical setting, and (b) multiple endogenous angiogenesis inhibitors are present in human tissues, and a single cytokine must exert an effect that can offset the antiangiogenic activity of these molecules. The cytokines that have been tested, however, appear to pull the ''angiogenic trigger'' as well as overcome local inhibitory forces.

In a clinical trial of angiogenic therapy for ischemic heart disease, Schumacher and colleagues (Fulda Medical Center, Fulda, Germany) showed angiographic evidence of a neovascular ''blush'' in 20 patients with severe LAD (left anterior descending) coronary artery stenosis following intramyocardial injection of acidic FGF (FGF-1). Intraarterial digital subtraction angiography showed accumulation of contrast medium extending peripherally up to 4 cm beyond the LAD artery originating at the site of FGF injection. This effect was not seen in control patients who received heat-denatured FGF-1 (56). While these were the first data to be published in a peer-reviewed scientific journal, other cardiologists have achieved similar success. Simon and colleagues (Beth Israel Medical Center, Boston, Massachusetts, USA) have treated more than 21 patients in a phase I clinical trial examining cardiac angiogenesis following intravenous delivery of basic FGF (FGF-2) and noted anatomic neovascularization (57). Henry and colleagues (Hennepin County Medical Center, Minneapolis, Minnesota, USA) have presented phase I data on a study of 15 patients receiving recombinant human VEGF administered by intracoronary infusion (dose range 0.005–0.167 μg/kg/min $\times$ 10 min $\times$ 2) in which 7/15 patients exhibited significantly increased numbers of collateral vessels (58).

In clinical trials of angiogenic therapy for critical limb ischemia, Baumgartner and colleagues (St. Elizabeth's Medical Center, Boston, Massachusetts, USA) reported results of a phase I study of intramuscular gene transfer therapy of VEGF (total dose of 4000 μg naked plasmid DNA over 20 treatments) (59). Seven of ten limbs showed newly visible collateral vessels (200 to $>$800 μm by luminal diameter) by contrast angiography, and 8/10 limbs exhibited improved distal circulatory flow by magnetic resonance angiography. This pivotal trial is the first demonstrating clinical success using angiogenic gene therapy.

In the field of wound healing, numerous individual angiogenic factors, including TGF-β, GM-CSF, PDGF, and basic FGF, have been topically applied to delayed-healing wounds with a resultant increase in granulation tissue (wound neovascularization) (60–62). Steed and colleagues (Presbyterian University Hospital, Pittsburgh, Pennsylvania, USA) reported increased granulation in patients receiving topical recombinant human PDGF-BB (2.2 μg/cm^2 $\times$ 20 weeks or until complete wound closure) for chronic diabetic ulcers. A total of 29/61 (48%) patients receiving rhPDGF-BB healed completely, as compared with 14/57 (25%) patients receiving a placebo vehicle (46). As previously mentioned, this drug (Regranex, Ortho-McNeil Pharmaceuticals) has been FDA-approved. The efficacy of rhPDGF-BB for chronic wound therapy is greatly enhanced by proper wound care techniques, specifically using sharp debridement to remove necrotic and fibrinous tissue (63). Such debridement likely increases local cytokine concentrations in the wound, including angiogenic factors such as bFGF, VEGF, and PDGF.

The similar anatomic results caused by different angiogenic cytokines suggest that the angiogenic potential of ischemic tissues possesses functional plasticity that can be triggered by multiple initiators and is not limited by a single ''key and lock'' mechanism. At this time, the choice of angiogenic factor does not appear to be the rate-limiting step for myocardial neovascularization of ischemic heart disease. Insufficient data have been released about the results of ongoing phase I/II clinical trials of basic FGF for peripheral vascular disease to permit a comparison with VEGF studies. It remains unknown whether one cytokine is more efficient in inducing therapeutic angiogenesis versus another, nor is it yet known whether the combination of several cytokines in a therapeutic setting would result in an enhanced clinical response.

Lesson No. 2: Continuous administration of angiogenic stimulators is not necessary

Clinical trials of therapeutic angiogenesis are demonstrating that the angiogenic effect appears to be achievable in at least some ischemic tissues without continuous infusion of the angiogenesis stimulator. This result has been somewhat surprising given the pharmacokinetic profile of most angiogenic peptides. Early research workers speculated that the short half-life in serum of basic FGF (30 min) and VEGF (2–3 min) would pose a therapeutic delemma because, in theory, a sustained continuous infusion of growth factor might be required to induce angiogenesis in human tissues.

Remarkably, limited or intermittent applications of therapeutic angiogenic stimulators appear sufficient to initiate sustained neovascularization and collateral blood vessel formation for prolonged periods of time. Patients in Fulda, Germany, treated with a single intramyocardial injection of FGF-1 (0.01 mg/kg body weight) still exhibited evidence of neovascularization by transfemoral, intraarterial, and digital subtraction angiography at 12 weeks following therapy (56,64). Patients in Boston, Massachusetts, treated with monthly intramuscular ph$VEGF_{165}$ gene injections for critical limb ischemia were assessed by hemodynamic [resting ankle-brachial index (ABI)] and angiographic studies 4 weeks following each monthly gene injection, and 3 months after the last of a total of 20 treatments. Digital subtraction angiography showed persistent new collateral vessels in 7 of 10 VEGF-treated limbs at 2 years after the initiation of therapy, and no evidence of vascular regression during follow-up films in this study. Immunohistochemistry of a limb muscle specimen obtained from one patient 10 weeks after treatment initiation showed foci of proliferating endothelial cells. PCR analyses showed persistence of DNA fragments unique to ph$VEGF_{165}$ (38).

In the case of chronic diabetic neuropathic ulcers, Weiman and colleagues (Louisville, Kentucky) demonstrated in 123 patients that a once-daily application of the recombinant angiogenic cytokine PDGF-BB would accelerate wound closure by as much as 8 weeks compared with controls, with clinical evidence of new granulation (wound angiogenesis) within 2 weeks of initiating therapy (65). Thus, once-daily administration of PDGF appears sufficient to promote neovascularization in the wound setting.

The lack of the need for continuous drug administration is most likely explained by the multiplicity of endogenous angiogenic stimuli already present in the ischemic tissues. In the atherosclerotic heart and limb, for example, it is likely that chronic hypoxia leads to reactive expression of aFGF, bFGF, VEGF, and PDGF (6,9,66). Posttreatment angiogenesis following exogenous stimulation is most likely due to an amplification of existing physiological forces. Experimental studies have shown, for example, that a combination of different angiogenic growth factors may have additive, synergistic, or interdependent effects (67–71). Therefore, therapeutic angiogenesis with exogenous cytokines may be analogous to the pouring of gasoline onto a lighted match—a provocatory external stimulus that amplifies existing forces. Additionally, many angiogenic factors have pluripotent effects on cell types besides the endothelial cells. Basic FGF and PDGF stimulate smooth muscle cell and fibroblast proliferation as well as collagen synthesis. Because new blood vessel formation involves stabilization and maturation of the vascular architecture, these exogenous cytokines likely facilitate other steps beyond endothelial cell activation in the angiogenesis cascade (72,73).

Theoretically, as long as a critical degree of hypoxia is present, these endogenous stimuli will continue to be elaborated. It is suggestive then, from the persistence of therapeutic angiogenesis, that current cytokine therapy ameliorates but does not completely eliminate tissue hypoxia. Alternatively, it may be speculated that once a mature blood vessel is created, it is destined to persist anatomically. Long-term studies have not yet been conducted to determine the true nature of the durability of neovascularization induced by therapeutic angiogenesis.

The lack of a continuous infusion requirement in cardiovascular diseases confers additional theoretical advantages because several of the angiogenic factors have potentially negative effects on target tissues, unrelated to new blood vessel growth. Basic FGF, for example, is known to have a negative inotopic effect on the myocardium (74). In animal models, intracoronary rhVEGF infusion can cause severe and lethal hypotension (75,76), and hypotension has indeed been observed in some patients receiving VEGF therapy. Repeated VEGF infusion of coronary vessels can also lead to tachyphylaxis to VEGF (77). Thus, the overall safety observed to date in clinical trials employing bFGF and VEGF for ischemic diseases suggest that single or intermittent drug delivery of these agents might enable patients to avoid these undesirable and adverse effects.

Lesson 3: A variety of drug delivery routes are feasible

Therapeutic angiogenesis has been achieved in ischemic hearts, limbs, and wounds using a variety of drug delivery modalities. Particularly for angiogenic growth factors, this is significant because of both the theoretical and practical challenges of delivering peptide drugs.

For the ischemic heart, angiogenic peptides have been delivered to patients by (a) direct intraarterial coronary infusion; (b) epicardial implantation of growth factor–impregnated heparin-alginate microspheres (78,79); and (c) adenoviral-based angiogenic factor gene vector (implanted at the time of surgery and via an outpatient catheter delivery system) (80). Transmyocardial revascularization is an approach that does not employ a drug entity but stimulates secondary angiogenesis following laser-initiated photoacoustic tissue injury (81). These methods have all resulted in angiographically documented increased collateralization in ischemic myocardium.

For the ischemic limb, naked $phVEGF_{165}$-DNA has been delivered to endothelial cells of the compromised vasculature via a hydrogel polymer coating on an angioplasty balloon (38,82). Surprisingly, direct intramuscular injection of naked plasmid DNA has also led to constitutive expression of $phVEGF_{165}$ and clinical angiogenesis in the critically ischemic limb (59).

Topical application of rhPDGF-BB (0.01%) successfully stimulates wound angiogenesis in the neuropathic diabetic ulcer (65). The drug is delivered in a sodium carboxymethylcellulose-based vehicle only once per day. Present formulations utilize a nonsterile low-bioburden vehicle that is appropriate for wounds healing by secondary intention but not for wounds healed by primary intention. Unpublished studies conducted by the drug sponsor indicated that more frequent application does not improve wound healing. It is not known if systemically delivered rhPDGF-BB would speed wound healing or initiate therapeutic angiogenesis in other target tissues.

Lesson No. 4: Local angiogenic therapy does not result in systemic angiogenesis

The current limited experience from clinical trials of therapeutic angiogenesis *suggests* that local angiogenic therapy does not result in systemic neovascularization. This is reassuring, given the concern for stimulating diffuse angiogenesis in normally avascular tissues such as the cornea, retina, and joints or initiating angiogenesis in dormant malignancies.

When acid FGF or basic FGF is delivered by perivascular or intraarterial injection to the ischemic heart, neovascularization appears to be delimited to the heart only, specifically around the tissue territories threatened by ischemia (56). Phase I studies of intracoronary rhVEGF in which ophthalmic and dermatological exams were conducted did not reveal evidence for angiogenesis in these end organs. Similarly, studies of $phVEGF_{165}$ gene therapy for ischemic limbs have shown evidence for limb neovascularization only, despite the fact that treatment was delivered by intramuscular injection and that expressed $phVEGF_{165}$ was detectable in the serum by immunoassay for up to 12 weeks (59). Interestingly, an early report of rhVEGF gene therapy observed the appearance of three ''spider angioma'' skin lesions on the dorsal forefoot and ankle of a treated 71-year-old patient 1 week following a single gene transfer by hydrogel-balloon transfection. Excisional biopsy of one lesion revealed proliferating endothelial cells by CD31 and proliferating cell nuclear antigen (PCNA) immunostaining. The two other lesions apparently regressed spontaneously by 8 weeks. These angiomatous lesions appeared only downstream from the site of gene transfer and only in the ischemic limb (38).

Based on available studies, therapeutic angiogenesis appears to preferentially manifest in ischemic tissues, despite the presence of augmented growth factor present in the systemic circulation. Virtually all clinical trials to date, however, have employed angiographic techniques that are unable to visualize arteries <200 μm in diameter. Microangiography, a technique capable of illuminating the true microvasculature, is still an experimental technology (83). Therefore, conventional angiographic evidence of increased collateralization actually demonstrates venules, venous channels, arterioles, and small arteries—not angiogenesis per se. True neovascularization is difficult to visualize in situ in most ischemic tissues; biopsy of the ischemic site before and after treatment may be required to substantiate angiogenesis and distinguish between pharmacologically induced neovascularization and mere vasodilatation of existing collateral channels, as has been observed following VEGF infusion. Furthermore, microscopic neovascular changes may be taking place in many tissue beds following growth-factor treatment but are undetectable using conventional technologies. Further animal studies are warranted on this topic because generalized subclinical angiogenesis might still place patients at risk for potential provocation of tumor angiogenesis from dormant neoplastic colonies or advancement of neovascularization in the diabetic retina (2,21,84).

Lesson No. 5: Therapeutic angiogenesis improves quality of life (QOL)

The most encouraging insight to date from clinical trials of therapeutic angiogenesis is that some patients treated with investigational drugs and devices have clearly experienced an improved quality of life following treatment.

Therapeutic angiogenesis has relieved angina caused by ischemic heart disease and improved the ability of patients to carry on with their activities of daily living. Of 15 patients, 13 experienced improvement in their anginal symptoms following intracoronary administration of rhVEGF by phase I study at Hennepin County Medical Center (85). An open-label, placebo-controlled study of adenosine and heparin (Ado/H) as angiogenic stimulators for refractory angina showed a decrease in the amount exercise-induced myocardial ischemia from 30–23% ($p = 0.02$) by thallium scintigraphy in Ado/H-treated patients (86).

Gene transfer of phVEGF$_{165}$ has also led to dramatic symptomatic improvement in treated patients with critical limb ischemia (59). Three of nine patients treated with angiogenic gene therapy were able to avoid below-the-knee amputation, and 57% of ischemic ulcers were healed or markedly improved. All three patients who suffered from ischemic rest pain of many months' duration had complete resolution of their pain after angiogenic therapy. For all nine patients, the frequency of ischemic pain episodes (the number of afflicted nights per week) declined by 75% at 8 weeks after treatment. These patients each experienced a significant improvement in QOL as compared with their condition prior to treatment. Recent unpublished observations also suggest that phVEGF$_{165}$ gene transfer may lead to improvement of sensory neuropathies as well.

Angiogenic therapy employing topical rhPDGF-BB has improved the lives of diabetic patients by accelerating the healing of chronic foot ulcers. Phase III clinical trials conducted by Wieman and colleagues showed that rhPDGF-BB treatment (0.01%) resulted in complete closure at 20 weeks of 50% of chronic ulcers as compared with a rate of 35% among placebo controls. Furthermore, this drug accelerated healing over that of controls by up to 8 weeks. For patients who are at risk for gangrene and sepsis, who are unable to walk normally or to wear shoes, the use of rhPDGF-BB in conjunction with good wound care confers considerable therapeutic advantage (65).

VI. FUTURE CONSIDERATIONS

While the insights outlined in this discussion foster optimism for the prospects of therapeutic angiogenesis in the coming years, some important theoretical concerns have not been adequately addressed by the current short-term clinical experience. For example, will neovascularization stimulated by angiogenic therapy in an atherosclerotic plaque located in a coronary artery precipitate its rupture, leading to thrombosis and myocardial infarction? Indeed, will therapeutic angiogenesis itself stimulate atherosclerosis or restenosis, particularly when pluripotent growth factors such as basic FGF are employed (87,88)? How stable is the neovascularization induced by exogenously applied angiogenic factors? Will therapeutically induced new blood vessels possess the same architectural integrity as those that are mediated physiologically? Or will fragile vascular malformations result (21, 89,90)? Will VEGF-induced microvasculature possess excessive permeability? Fragile blood vessels that leak or bleed can cause devastating clinical consequences, as witnessed in proliferative diabetic retinopathy or cerebral arteriovenous malformations (90–93).

Finally, the universal concern about therapeutic angiogenesis is that it should restore perfusion to ischemic target tissues without provoking neovascularization in premalignant or dormant neoplastic tissues or in tissue beds such as the diabetic eye or arthritic joint (21,84,91). In addition to enhancing oxygen delivery and metabolic waste removal, the endothelial cells of newly generated blood vessels are known to elaborate numerous bioactive products that may affect the behavior of adjacent pathological cell types (95). So far, clinical experience has not uncovered such undesirable effects. As the therapeutic field of angiogenesis advances, however, it will be prudent for basic and clinical scientists alike to study the balancing point of angiogenic therapy—the point where molecular revascularization has been achieved and beyond which pathological neovascularization may result. At this juncture, the availability of angiogenesis stimulators as well as angiogenesis inhibitors for clinical use may be not only helpful but perhaps mandatory for reasons of patient safety.

VII. CONCLUSION

In summary, therapeutic angiogenesis has now moved from theoretical concept to a clinical reality. Although pharmaceutical industry efforts are still nascent, significant insights have already been gleaned from the current experience of clinical trials aimed at stimulating angiogenesis in the heart, brain, limb, and wound. In ischemic tissues, therapeutic angiogenesis appears to be relatively easily initiated and maintained for periods of weeks or more. Its effects appear to localize easily and not to provoke harmful systemic side effects. Important ''proof of concept'' has been achieved by demonstrating that therapeutic angiogenesis can improve the quality of life of patients.

As observed by Nobel Laureate Albert Szent-Gyorgi: ''Discovery is seeing what everyone has seen, and thinking what no one has thought.'' The emerging lessons from therapeutic angiogenesis challenges basic and clinical scientists to take the current observations gleaned from clinical trials and further inquire into the mechanistic, pharmacological, and clinical consequences of new blood vessel growth in the human body in the dawn of the coming era of molecular revascularization.

ACKNOWLEDGMENTS

We wish to acknowledge the following expert individuals for their helpful discussions and insights: Dr. Karol Watson, Dr. David Steed, Professor Roy Bicknell, Professor Michael Maragoudakis, Dr. Michael Simon, Dr. Kevin Malloy, Dr. Richard Kawamoto, Dr. Richard Casey, and Mr. Jay Lee.

REFERENCES

1. Risau W. Mechanisms of angiogenesis. Nature 1997; 386:671–674.
2. Casey R, Li WW. Factors controlling ocular angiogenesis. Am J Opthalmol 1997; 124:521–529.
3. Hockel M, Schlenger K, Doctrow S, Kissel T, Vaupel P. Therapeutic angiogenesis. Arch Surg 1993; 128:423–429.
4. Casey R, Jaffe M, Li WW, Li VW, Tsakayannis D. Market Study of Angiogenesis-Dependent Diseases: Classification and Sizing of the Potential Market for the Pharmaceutical Industry in the United States and in the European Union. Cambridge, MA: The Angiogenesis Foundation, 1998, pp. 7–20.
5. Folkman J, Szabo S, Stovroff M, McNeil P, Li W, Shing Y. Duodenal ulcer: discovery of a new mechanism and development of angiogenic therapy which accelerates healing. Ann Surg 1991; 214:414–427.
6. Shweiki D, Itin A, Soffer D, Keshet E. Vascular endothelial growth factor induced by hypoxia may mediate hypoxia-induced angiogenesis. Nature 1992; 359:843–845.
7. Carmeliet P, Dor Y, Herbert J-M, Fukumura D, Brusselmans K, Dewerchin M, Neeman M, Bono F, Abramovitch R, Maxwell P, Koch CJ, Ratcliffe P, Moons L, Jain RK, Collen D, Keshet E. Role of HIF-1a in hypoxia-mediated apoptosis, cell proliferation and tumour angiogenesis. Nature 1998; 394:485–490.
8. Banai S, Shweiki D, Pinson A, Chandra M, Lazarovici G, Keshet E. Up-regulation of vascular endothelial growth factor expression induced by myocardial ischemia: implications for coronary angiogenesis. Cardiovasc Res 1994; 28:1176–1179.
9. Tuder RM, Flook BE, Voelkel NF. Increased gene expression for VEGF and the VEGF receptors KDR/Flk and Flt in lungs exposed to acute or to chronic hypoxia: modulation of gene expression by nitric oxide. J Clin Invest 1995; 95:1798–1809.
10. McNeil PL, Muthukrishnan L, Warder E, D'Amore PA. Growth factors are released by mechanically wounded endothelial cells. J Cell Biol 1989; 109:811–822.
11. Gajdusek CM, Carbon S. Injury-induced release of basic fibroblast growth factor from bovine aortic endothelium. J Cell Physiol 1989; 139:570–579.
12. Haralabopoulos G, Grant DS, Kleinman HK, Maragoudakis ME. Thrombin promotes endothelial cell alignment in Matrigel in vitro and angiogenesis in vivo. Am J Physiol 1997; 273: C239–C245.
13. Maragoudakis ME, Tsopanoglou NE, Sakkoula E, Pipili-Synetos E. On the mechanism of promotion of angiogenesis by thrombin. FASEB J 1995; 9:A587.
14. Risau W, Drexler H, Mironov V, Smits A, Siegbahn A, Funa K, Heldin CH. Platelet-derived growth factor is angiogenic in vivo. Growth Factors 1992; 7:261–266.
15. Nicosia RF, Nicosia SV, Smith M. Vascular endothelial growth factor, platelet-derived growth factor and insulin-like growth factor-1 promote rat aortic angiogenesis in vitro. Am J Patholo 1994; 145:1023–1029.
16. Hom DB, Maisel RH. Angiogenic growth factors: their effects and potential in soft tissue wound healing. Ann Otol Rhinol Laryngol 1992; 101:349–354.
17. Knighton DR, Hunt TK, Scheuenstuhl N, Halliday BT, Werb Z, Bunda MJ. Oxygen tension regulates expression of angiogenesis factor by macrophages. Science 1983; 221:1283–1285.
18. Koch AE, Polverini PJ, Kunkel SL, Harlow LA, DiPietro LA, Elner VM, Elner SG, Strieter RM. Interleukin-8 (IL-9) is a potent macrophage-derived mediator of angiogenesis that is blocked by IL-8 antibody and antisense oligonucleotides. Science 1992; 258:1799–1801.
19. Polverini PJ. Macrophage-induced angiogenesis: a review. In: Sorg C, ed. Macrophage-Derived Regulatory Factors. Basel: Karger, 1977, pp. 54–73.
20. Gaudry M, Bregerie O, Andrieu V, El Benna J, Pocidalo MA, Hakim J. Intracellular pool of vascular endothelial growth factor in human neutrophils. Blood 1997; 90:4153–4161.
21. Li WW, Li VW, Casey R, Tsakayannis D, Kruger EA, Lee A, Sun Y-L, Bonar CA, Cornelius

S. Clinical trials of angiogenesis-based therapies: overview and new guiding principles. In: Maragoudakis M, ed. Angiogenesis: Models, Modulators and Clinical Applications. New York: Plenum Press, 1998, pp. 475–492.
22. Shing Y, Klagsbrun M, Folkman J. A new method for purifying heparin-binding growth factors. Ann NY Acad Sci 1989; 556:166–172.
23. Folkman J, Klagsbrun M. Angiogenic factors. Science 1987; 35:442–447.
24. Ziche M, Jones J, Gullino P. Role of prostaglandin E1 and copper in angiogenesis. J Natl Cancer Inst 1982; 69:475–482.
25. Falanga V, Margolis D, Alvarez O, Auletta M, Maggiacomo F, Altman M, Jensen J, Sabolinski M, Hardin-Young J, and the Human Skin Equivalents Investigators Group. Rapid healing of venous ulcers and lack of clinical rejection with an allogenic cultured human skin equivalent. Arch Dermatol 1998; 134:293–300.
26. Mirhoseini M, Fisher JC, Cayton MM. Myocardial revascularization by laser. Lasers Surg Med 1983; 3:241–245.
27. Horvath KA, Manning F, Cummings N. Transmyocardial laser revascularization: operative techniques and clinical results at two years. J *Thorac* Cardiovasc Surg 1996; 111:1047–1053.
28. Cianci P, Hunt TK. Adjunctive hyperbaric oxygen therapy in the treatment of diabetic wounds of the foot. In: Levin ME, O'Neal LW, Bowker JH, eds. The Diabetic Foot, 5th ed. St. Louis: Mosby–Year Book, 1993, pp. 305–319.
29. Zhao LL, Davidson JD, Wee SC, Roth SI, Mustoe TA. Effect of hyperbaric oxygen and growth factors on rabbit ear ischemic ulcers. Arch Surg 1994; 129:1043–1049.
30. Marx RE, Ehler WJ, Tayapongsak P, Pierce LW. Relationship of oxygen dose to angiogenesis induction in irradiated tissue. Am J Surg 1990; 160:519–524.
31. Tompach PC, Lew D, Stoll JL. Cell response to hyperbaric oxygen treatment. Int J Oral Maxillofac Surg 1997; 26:82–86.
32. Harada K, Friedman M, Lopex JJ, Wang SY, Li J, Pottumarthi VP, Pearlman JD, Edelman ER, Sellke FW, Simons M. Vascular endothelial growth factor administration in chronic myocardial ischemia. Am J Physiol 1996; 270:H1791–H1802.
33. Harada K, Grossman W, Friedman M, Edelman ER, Prasad P., Manning WJ, Sellke FW, Simons M. Basic fibroblast growth factor improves myocardial function in chronically ischemic porcine hearts. J Clin Invest 1994; 94:623–630.
34. Yanagisawa-Miwa A, Uchida Y, Nakamura F, Tomaru T, Kido H, Kamijo T, Sugimoto T, Kaji K, Utsuyama, Kurashima C, Ito H. Salvage of infarct myocardium by angiogenic action of basic fibroblast growth factor. Science 1992; 257:1401–1403.
35. Ono K, Matsumori A, Shoi T, Furukawa Y, Sasayama S. Enhanced expression of hepatocyte growth factor/c-Met by myocardial ischemia and reperfusion in a rat model. Circulation 1997; 95:2552–2558.
36. Schumacher B, Pecher P, von Specht BU, Stegmann Th. Induction of neoangiogenesis in ischemic myocardium by human growth factors: first clinical results of a new treatment of coronary heart disease. Circulation 1998; 97:645–650.
37. Lopez JJ, Edelman ER, Stamler A, Morgan JP, Sellke FW, Simons M. Local perivascular administration of basic fibroblast growth factor: drug delivery and toxicological evaluation. Drug Metab Dispos 1996; 24:922–924.
38. Isner JM, Pieczek A, Schainfeld R, Blair R, Haley L, Asahara T, Rosenfield K, Razvi S, Walsh K, Symes JF. Clinical evidence of angiogenesis after arterial gene transfer of phVEGF165 in patient with ischaemic limb. Lancet 1996; 348:370–374.
39. Pu L-Q, Sniderman AD, Brassard R, Lachapelle KJ, Graham AM, Lisbona R, Symes JF. Enhanced revascularization of the ischemic limb by angiogenic therapy. Circulation 1993; 88: 208–215.
40. Tsurumi Y, Kearney M, Chen D, Silver M, Takeshita S, Yang J, Symes JF, Isner JM. Treatment of acute limb ischemia by intramuscular injection of vascular endothelial growth factor gene. Circulation 1997; 96(suppl II):II-382–II-388.

41. Baffour R, Berman J, Garb JL, Rhee SW, Kaufman J, Friedmann P. Enhanced angiogenesis and growth of collaterals by in vivo administration of recombinant basic fibroblast growth factor in a rabbit model of acute lower limb ischemia: dose-response effect of basic fibroblast growth factor. J Vasc Surg 1992; 16:181–191.
42. Brown GL, Curtzinger L, Jurkiewicz MJ, Nahai F, Schulz G. Stimulation of healing of chronic wounds by epidermal growth factor. Plast Reconstr Surg 1991; 88:189–196.
43. Mustoe TA, Ahn ST, Tarpley JE, Pierce GF. Role of hypoxia in growth factor responses: differential effects of basic fibroblast growth factor and platelet-derived growth factor in an ischemic wound model. Wound Repair Regen 1994; 2:277–283.
44. Hom DB, Maisel RH. Angiogenic growth factors: their effects and potential in soft tissue wound healing. Ann Otol Rhinol Laryngol 1992; 101:349–354.
45. Grotendorst GR, Martin DR, Pencev D. Stimulation of granulation tissue formation by platelet-derived growth factor in normal and diabetic rats. J Clin Invest 1985; 76:2323–2329.
46. Steed DL and the Diabetic Ulcer Study Group. Clinical evaluation of recombinant human platelet-derived growth factor for the treatment of lower extremity diabetic ulcers. J Vasc Surg 1995; 21:71–81.
47. Krupinski J, Issa R, Bujny T, Slevin M, Kumar P, Kumar S, Kaluza J. A putative role for platelet-derived growth factor in angiogenesis and neuroprotection after ischemic stroke in humans. Stroke 1997; 28:564–573.
48. Intravenous basic fibroblast growth factor decreases brain injury resulting from focal ischemia in cats. Stroke 1997; 28:609–616.
49. Kawamata T, Alexis NE, Dietrich WD, Finkelstein SP. Intracisternal basic fibroblast growth factor (bFGF) enhances behavioral recovery following focal cerebral infarction in the rat. J Cereb Blood Flow Metab 1996; 16:542–547.
50. Kawamata T, Dietrich WD, Schaller T, Gotts JE, Cocke RR, Benowitz LI, Finkelstein SP. Intracisternal basic fibroblast growth factor enhances functional recovery and up-regulates the expression of a molecular marker of neuronal sprouting following focal cerebral infarction. Proc Natl Acad Sci USA 1997; 94:8179–8184.
51. Speliotes EK, Caday CG, Do T, Weise J, Kowall NW, Finkelstein SP. Increased expression of basic fibroblast growth factor (bFGF) following focal cerebral infarction in the rat. Mol Brain Res 1996; 39:31–42.
52. Walicke PA. Basic and acidic fibroblast growth factors have trophic effects on neurons from multiple CNS regions. J Neurosci 1988; 8:2618–2627.
53. Hayashi T, Abe K, Suzuki H, Itoyama Y. Rapid induction of vascular endothelial growth factor gene expression after transient middle cerebral artery occlusion in rats. Stroke 1997; 28:2039–2044.
54. Lin DA, Finkelstein SP. Basic fibroblast growth factor: a treatment for stroke? Neuroscientist 1997; 3:247–250.
55. Welch M. Scios ends U.S. FIBLAST trial but European study continues. Bioworld Today 1998; 140:1–2.
56. Schumacher B, Pecher P, von Specht BU, Stegmann Th. Induction of neoangiogenesis in ischemic myocardium by human growth factors: first clinical result of a new treatment of coronary heart disease. Circulation 1998; 97:645–650.
57. FGF used to get to the heart of cardiac problems. Medical Industry Today, Feb. 26, 1998.
58. Henry TD, Rocha-Singh K, Isner JM. Results of intracoronary recombinant human vascular endothelial growth factor (rhVEGF) administration trial. J Am Coll Cardiol 1998; 31:65A.
59. Baumgartner I, Pieczek A, Manor O, Blair R, Kearney M, Walsh K, Isner JM. Constitutive expression of phVEGF165 after intramuscular gene transfer promotes collateral vessel development in patients with critical limb ischemia. Circulation 1998; 97:1114–1123.
60. Robson MC. Exogenous growth factor application effect on human wound healing. Prog Dermat 1996; 30:1–7.

61. Robson MC, Phillips LG, Lawrence WT, Bishop JB, Youngerman JS, Hayward PG, Broemeling LD, Heggers JP. The safety and effect of topically applied recombinant basic fibroblast growth factor on healing of chronic pressure sores. Ann Surg 1992; 216:401–408.
62. Robson MC, Phillips LG, Cooper DM, Lyle WG, Robson LE, Odom L, Hill DP, Hanham AF, Ksander GA. The safety and effect of transforming growth factor-B2 for the treatment of venous stasis ulcers. Wound Repair Regen 1995; 3:157–167.
63. Steed DL, Donohoe D, Webster MW, Lindsley L, and the Diabetic Ulcer Study Group. Effect of extensive debridement and treatment on the healing of diabetic foot ulcers. J Am Coll Surg 1996; 183:61–64.
64. Folkman J. Angiogenic therapy of the human heart. Circulation 1998; 97:628–629.
65. Wieman TJ, Smiell JM, Su Y. Efficacy and safety of a topical gel formulation of recombinant human platelet-derived growth factor-BB (becaplermin) in patients with chronic neuropathic diabetic ulcers: a phase III randomized placebo-controlled double blind study. Diabetes Care 1998; 21:822–827.
66. Kuwabara K, Ogawa S, Matsumoto M, Koga S, Clauss M, Pinsky DJ, Lyn P, Leavy J, Witte L, Joseph-Silverstein J. Hypoxia-mediated induction of acidic/basic fibroblast growth factor and platelet-derived growth factor in mononuclear phagocytes stimulates growth of hypoxic endothelial cells. Proc Natl Acad Sci USA 1995; 92:4606–4610.
67. Pepper MS, Ferrara N, Orci L, Montesano R. Potent synergism between vascular endothelial growth factor and basic fibroblast growth factor in the induction of angiogenesis in vitro. Biochem Biophys Res Commun 1992; 3:211–220.
68. Goto F, Goto K, Weindel K, Folkman J. Synergistic effects of vascular endothelial growth factor and basic fibroblast growth factor on the proliferation and cord formation of bovine capillary endothelial cells within collagen gels. Lab Invest 1993; 69:508–517.
69. Mandriota SJ, Pepper MS. Vascular endothelial growth factor-induced in vitro angiogenesis and plasminogen activator expression are dependent on endogenous basic fibroblast growth factor. J Cell Sci 1997; 110:2293–2302.
70. Va Belle E, Witzenbichler B, Chen D, Silver M, Change L, Schwall R, Isner JM. Potentiated angiogenic effect of scatter factor/hepatocyte growth factor via induction of vascular endothelial growth factor. Circulation 1998; 97:381–390.
71. Punglia RS, Lu M, Hsu J, Kuroki M, Tolentino MJ, Keough K, Levy AP, Levy NS, Goldber MA, D'Amato RJ, Adamis AP. Regulation of vascular endothelial growth factor expression by insulin-like growth factor-1. Diabetes 1997; 46:1619–1626.
72. Edelman ER, Nugent MA, Smith LT, Karnovsky MJ. Basic fibroblast growth factor enhances coupling of intimal hyperplasia and proliferation of vasa vasorum in injured rat arteries. J Clin Invest 1992; 89:465–473.
73. Pierce GF, Mustoe TA, Altrock BW. Role of platelet-derived growth factor in wound healing. J Cell Biochem 1991; 45:319–326.
74. Yuyi I, Yoshitoshi U, Tsutsui H, Kinugawa S, Sugimachi M, Takahashi M, Yamamoto S, Tagawa H, Sunagawa K, Tskeshita A. Negative ionotropic effect of basic fibroblast growth factor on adult rat myocyte. Circulation 1997; 96:2501–2504.
75. Cuevas P, Carceler F, Ortega S, Zazo M, Nieto I, Giminez-Gallego G. Hypotensive activity of growth factor. Science 1991; 254:1208–1210.
76. Hariawala MD, Horoitz JR, Esakof D, Sheriff DD, Walter DH, Keyt B, Isner JM, Symes JF. VEGF improves myocardial blood flow but produces EDRF-mediated hypotension in porcine hearts. J Surg Res 1996; 63:77–82.
77. Lopez JJ, Laham RJ, Carrozza JP, Tofukuji M, Sellke FW, Bunting S, Simons M. Hemodynamics effects of intracoronary VEGF delivery: evidence of tachyphylaxis and NO dependence of response. Am J Physiol 1997; 273:H1317–H1323.
78. Edelman ER, Nugent MA, Karnovsky MJ. Perivascular and intravenous administration of basic fibroblast growth factor: vascular and solid organ deposition. Proc Natl Acad Sci USA 1993; 90:1513–1517.

79. Edelman ER, Mathiowitz E, Langer R, and Klagsbrun M. Controlled and modulated release of basic fibroblast growth factor. Biomaterials 1991; 12:619–626.
80. Proprietary technology: US Patent No. 5,792,453, Gene Transfer-Mediated Angiogenesis Therapy. Genrx, Collateral Therapeutics, Inc., San Diego, CA.
81. Horvath KA, Mannting F, Cummings N. Transmyocardial laser revascularization: operative techniques and clinical results at two years. J Thorac Cardiovasc Surg 1996; 111:1047–1053.
82. Riessen R, Rahimizadeh H, Blessing E, Takeshita S, Barry JJ, Isner JM. Arterial gene transfer using pure DNA applied directly to a hydrogel-coated angioplasty balloon. Hum Gene Ther 1993; 4:749–758.
83. Takeshita S, Isshiki T, Tanaka E, Eto K, Miyazawa Y, Tanaka A, Shinozaki Y, Hyodo K, Ando M, Kubota M, Tanioka K, Umetani K, Ochiai M, Sato T, Mori H, Miyashita H. Use of synchrotron radiation microangiography to assess development of small collateral arteries in a rat model of hindlimb ischemia. Circulation 1997; 95:805–808.
84. Folkman J. Angiogenesis in cancer, vascular, rheumatoid and other disease. Nature Med 1995; 1:27–31.
85. Henry TD, Rocha-Singh K, Isner JM. Results of intracoronary recombinant human vascular endothelial growth factor (rhVEGF) administration trial. J Am Coll Cardiol 1998; 31:65A.
86. Barron HV, Sciammarella M, Lenihan K, Dae MW, Botvinick EH. Repeated administration of adenosine and heparin reduces myocardial ischemia in patients with chronic stable angina. Circulation 1997; 96(suppl):I-342.
87. Hughes SE. Localization and differential expression of the fibroblast growth factor receptor (FHFR) multigene family in normal and atherosclerotic human arteries. Cardiovasc Res 1996; 32:557–569.
88. Schwartz SM, Reidy MA, O'Brien ER. Assessment of factors important in atherosclerotic occlusion and restenosis. Thromb Haemostas 1995; 74:541–551.
89. Hirschi KK, D'Amore PA. Pericytes in the microvasculature. Cardiovasc Res 1996; 32:687–698.
90. Pollock BE, Flickinger JC, Lunsford LD, Bissonette DJ, Kondziolka D. Factors that predict bleeding risk of cerebral arteriovenous malformations. Stroke 1996; 27:1–6.
91. Boulton M, Foreman D, Williams G, McLeod D. VEGF localization in diabetic retinopathy. Br J Ophthalmol 1998; 82:561–568.
92. Neely KA, Quillen DA, Schachat AP, Gardener TW, Blankenship GW. Diabetic retinopathy. Med Clin North Am. 1998; 82:847–876.
93. Hasuo K, Tamura S, Kudo S, Uchino A, Carlos R, Matsushima T, Kurokawa T, Kitamura K, Matsuura K. Moya moya disease: use of digital subtraction angiography in its diagnosis. Radiology 1985; 157:107–111.
94. Henry TD. Can we really grow new blood vessels? Lancet 1998; 351:1826–1827.
95. Rak J, Filmus J, Kerbel RS. Reciprocal paracrine interactions between tumour cells and endothelial cells: the ''angiogenesis progression'' hypothesis. Eur J Cancer 1996; 32A:2438–2450.

Index